More information about this series at http://www.springer.com/series/7412

Lecture Notes in Computer Science   11901

Yao Zhao · Nick Barnes ·
Baoquan Chen · Rüdiger Westermann ·
Xiangwei Kong · Chunyu Lin (Eds.)

Image
and Graphics

10th International Conference, ICIG 2019
Beijing, China, August 23–25, 2019
Proceedings, Part I

 Springer

Editors
Yao Zhao
Beijing Jiaotong University
Beijing, China

Nick Barnes
The Australian National University
Canberra, Australia

Baoquan Chen
Peking University
Beijing, China

Rüdiger Westermann
The Technical University of Munich
Munich, Bayern, Germany

Xiangwei Kong (iD)
Zhejiang University
Hangzhou, China

Chunyu Lin (iD)
Beijing Jiaotong University
Beijing, China

ISSN 0302-9743   ISSN 1611-3349 (electronic)
Lecture Notes in Computer Science
ISBN 978-3-030-34119-0   ISBN 978-3-030-34120-6 (eBook)
https://doi.org/10.1007/978-3-030-34120-6

LNCS Sublibrary: SL6 – Image Processing, Computer Vision, Pattern Recognition, and Graphics

This Springer imprint is published by the registered company Springer Nature Switzerland AG
The registered company address is: Gewerbestrasse 11, 6330 Cham, Switzerland

Preface

We would like to present the proceedings of the 10th International Conference on Image and Graphics (ICIG 2019), held in Beijing, China, during August 23–25, 2019.

The China Society of Image and Graphics (CSIG) has hosted this series of ICIG conferences since 2000. ICIG is a biennial conference organized by the CSIG, focusing on innovative technologies of image, video, and graphics in processing and fostering innovation, entrepreneurship, and networking. This time, the conference was organized by Tsinghua University, Peking University, and Institute of Automation, CAS. Details about the past nine conferences, as well as the current one, are as follows:

Conference	Place	Date	Submitted	Proceeding
First (ICIG 2000)	Tianjin, China	August 16–18	220	156
Second (ICIG 2002)	Hefei, China	August 15–18	280	166
Third (ICIG 2004)	Hong Kong, China	December 17–19	460	140
4th (ICIG 2007)	Chengdu, China	August 22–24	525	184
5th (ICIG 2009)	Xi'an, China	September 20–23	362	179
6th (ICIG 2011)	Hefei, China	August 12–15	329	183
7th (ICIG 2013)	Qingdao, China	July 26–28	346	181
8th (ICIG 2015)	Tianjin, China	August 13–16	345	170
9th (ICIG 2017)	Shanghai, China	September 13–15	370	172
10th (ICIG 2019)	Beijing, China	August 23–25	384	183

This time, the proceedings are published by Springer in the LNCS series. At ICIG 2019, 384 submissions were received, and 183 papers were accepted. To ease in the search of a required paper in these proceedings, the 161 regular papers have been arranged into different sections. Another 22 papers forming a special topic are included at the end.

Our sincere thanks to all the contributors, who came from around the world to present their advanced work at this event. Special thanks go to the members of the Technical Program Committee, who carefully reviewed every single submission and made their valuable comments for improving the accepted papers. The proceedings could not have been produced without the invaluable efforts of the publication chairs, the web chairs, and a number of active members of CSIG.

September 2019

Yao Zhao
Nick Barnes
Baoquan Chen
Rüdiger Westermann
Xiangwei Kong
Chunyu Lin

Organization

Organizing Committee

General Chairs

Tieniu Tan	Institute of Automation, CAS, China
Oliver Deussen	University of Konstanz, Germany
Rama Chellappa	University of Maryland, USA

Technical Program Chairs

Yao Zhao	Beijing Jiaotong University, China
Nick Barnes	ANU, Australia
Baoquan Chen	Peking University, China
Ruediger Westermann	TUM, Germany

Organizing Committee Chairs

Huimin Ma	Tsinghua University, China
Yuxin Peng	Peking University, China
Zhaoxiang Zhang	Institute of Automation, CAS, China
Ruigang Yang	Baidu, China

Sponsorship Chairs

Yue Liu	Beijing Institute of Technology, China
Qi Tian	University of Texas at San Antonio, USA

Finance Chairs

Zhenwei Shi	Beihang University, China
Jing Dong	Institute of Automation, CAS, China

Special Session Chairs

Jian Cheng	Institute of Automation, CAS, China
Gene Cheung	York University, Canada

Award Chairs

Yirong Wu	Institute of Electrics, CAS, China
Zixiang Xiong	Texas A&M University, USA
Yuxin Peng	Peking University, China

Publicity Chairs

Moncef Gabbouj   TUT, Finland
Mingming Cheng   Nankai University, China

Exhibits Chairs

Rui Li   Google, China
Jiang Liu   Meituan, China

Publication Chairs

Xiangwei Kong   Zhejiang University, China
Chunyu Lin   Beijing Jiaotong University, China

Oversea Liaison

Yo-Sung Ho   GIST, South Korea
Alan Hanjalic   Delft University of Technology, The Netherlands

Local Chairs

Xucheng Yin   USTB, China
Kun Xu   Tsinghua University, China

Tutorial Chairs

Weishi Zheng   Sun Yat-sen University, China
Chen Change Loy   NTU, Singapore

Workshop Chairs

Jiashi Feng   National University of Singapore, Singapore
Si Liu   Beihang University, China

Symposium Chair

Jinfeng Yang   Civil Aviation University of China, China

Website Chair

Bo Yan   Fudan University, China

Contents – Part I

Contents – Part II

Computer Graphics and Visualization

Computational Imaging

Color and Multispectral Processing

Artificial Intelligence

Contents – Part III

Security

Surveillance and Remote Sensing

Virtual Reality

Feature Learning for Cross-Domain Problems

Advanced Signal Processing Methods in Spectral Imaging

Computer Vision and Pattern Recognition

Superpixel-Based Saliency Guided Intersecting Cortical Model for Unsupervised Object Segmentation

Chen Wang$^{(\boxtimes)}$, Linyuan He, Shiping Ma, and Shan Gao

Air Force Engineering University (AFEU), Xi'an, China
wwangchen77@163.com, hal1983@163.com,
mashiping@126.com, 17709223919@163.com

Abstract. Unsupervised object segmentation aims to assign same label to pixels of object region with feature homogeneity, which can be applied to object detection and recognition. Intersecting cortical model (ICM) can simulate human visual system (HVS) to process image for many applications, and at the same time, saliency detection can also simulate HVS to locate the most important object in a scene. Based on saliency detection, a novel approach for unsupervised object segmentation, termed as saliency guided intersecting cortical model (SG-ICM), is proposed in this paper. Instead of using gray-scale and spatial information to motivate ICM neurons traditionally, it is better to exploit saliency characteristic to guide ICM. In this paper, we plan to do saliency detection exploiting an improved dynamic guided filtering to analyze significance of different regions in same scene. The proposed saliency feature lies on: (1) the proposed saliency detection is based on region instead of pixel; (2) the dynamic guided filter is designed to accelerate the filtering; (3) in order to improve SG-ICM for object segmentation, at the each iteration, we use adaptive and simple threshold, which can raise the speed of this model. We check the proposed algorithm on common database of DOTI, color image from public database of MSRA with ground truth annotation. Experimental results show that the proposed method is superior to the others in terms of robustness of object segmentation, furthermore, it does not need any training. In addition, this method is effective for aerial image, the detection results reveal that this model has great potential in aerial reconnaissance application.

Keywords: Unsupervised object segmentation · Intersecting cortical model (ICM) · Saliency guided intersecting cortical model (SG-ICM) · Dynamic guided filtering

1 Introduction

Image segmentation is an image preprocessing technology, which can separate an image to several parts [1] and it is the important and fundamental procedure in image analysis. In result of image segmentation, internal characteristics of each part are similar and different parts are not. Existing image segmentation methods can be divided into two categories, one is supervised image segmentation and the other is

© Springer Nature Switzerland AG 2019
Y. Zhao et al. (Eds.): ICIG 2019, LNCS 11901, pp. 3–17, 2019.
https://doi.org/10.1007/978-3-030-34120-6_1

unsupervised. In recent years, supervised image segmentation methods using deep learning such as Convolution Neural Network (CNN) [2] and Recurrent Neural Network (RNN) [3] have achieved remarkable results. At the same time, unsupervised image segmentation is also significant important in plenty of image processing applications such as disaster relief and aerial reconnaissance, which is too difficult to obtain accurate label samples for training sets beforehand.

ICM is simplified Pulse Coupled Neural Network (PCNN) model, which was introduced by Kinser [4] for the first time. This model is proposed for image processing, especially image segmentation, and it is computationally faster than PCNN model. Because of retaining the characteristics of pulse coupling, variable threshold and synchronous pulse distribution in PCNN model, ICM can effectively compensate for the discontinuity of data and help to preserve the regional information of the image. More importantly, ICM requires no training compared with tradition neural networks and novel deep learning such as BP, Hopfield and CNN. This model is based on study of the physiological and visual characteristics of mammals, many people are interested in it and many ICM-related approaches have been proposed, which have achieved competitive results. Most of existing ICM-related methods are used for gray image as external stimulus. In this paper, we also study ICM algorithm in order to propose a better object segmentation method for gray and color image segmentation.

Instead of considering the whole image, saliency detection can extract the most interest and important regions in a scene, according to people's visual habits. More accurately, saliency-based image segmentation should be called object segmentation in fact. It aims to automatically assigning labels to the main object from original image. Saliency detection technique is the study result of visual attention of HVS, which aims to capture the most visually remarkable regions in an image. Visual saliency detection is an interdisciplinary subject which integrates cognitive psychology, neurology, mathematics, statistics, and so on. The result of saliency detection is called saliency map. For decades, many saliency methods have been proposed and applied to vision problems, including object segmentation [5], image retrieval [6], and image matching [7], to name a few. In these proposed saliency detection algorithms, CNN-based saliency methods [8, 9] have been prevailed, which can get high quality saliency detection result. But this kind of method has some disadvantages of complicated network structure and more runtime.

In this work, we use a novel saliency detection method to improve ICM, which can achieve better simulation of HVS to label the important object in an image. The goal of this work is to establish a saliency guided ICM for object segmentation without any training, we named it SG-ICM, in which saliency detection result become the external stimulus of ICM neurons rather than original input image. As an iterative model, ICM in this work use a novel strategy to speed up its convergence. The proposed method is compared with other competitive approaches both on subjective evaluation and objective evaluation. Experiments demonstrate its excellent performance of the proposed method in object segmentation.

This paper is structured as follows: Sect. 2 introduces the background knowledge include the standard ICM and saliency theory. Section 3, we briefly describe the proposed SG-ICM. The result analysis of the proposed method will be done in Sect. 4. At last, we conclude the whole work in Sect. 5.

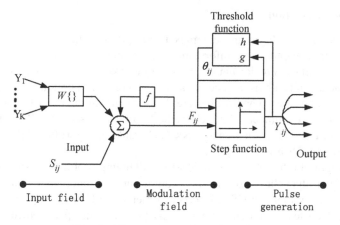

Fig. 1. The structure of standard ICM

2 Background Knowledge

2.1 The Standard ICM

Traditional ICM was presented in [4], in which each neuron consists of three parts: input field, modulation field, pulse generation. ICM is a simplified PCNN model for image processing, and each ICM neuron corresponds to an image pixel. Figure 1 shows the standard ICM structure which is widely used in image processing. The neurons with similar stimulation can pulse synchronously to form a segment of the input image. As shown in Fig. 1, each neuron consists of three parts: feeding and linking field, modulating field, and pulse generation. Neuron communicates with neighbor neurons through weight W. Its mathematic model can be described as below:

$$
\begin{aligned}
F_{ij}[n+1] &= f\,F_{ij}[n] + S_{ij} + W_{ij}\{Y\} \\
Y_{ij}[n+1] &= \begin{cases} 1 & F_{ij}[n+1] > T_{ij}[n] \\ 0 & else \end{cases} \\
T_{ij}[n+1] &= gT_{ij}[n] + hY_{ij}[n+1]
\end{aligned}
\tag{1}
$$

Where S_{ij} is the ij-th pixel value of input image. $F_{ij}[n]$ is the status function of neurons, which can remain the neuron status. That is to say, the status function $F_{ij}[n]$ for each iterative is related to last one $F_{ij}[n-1]$, which memory attenuates over time and the attenuation rate is influenced by attenuation factor f ($f < 1$). F, g, h are all scalar coefficients, and $g < f < 1$ will guarantee dynamic threshold will be less than status value of neuron at last. h is very large, so it can guarantee advance threshold after neuron firing, which not be inspired at the next iteration. $W_{ij}\{\}$ is connecting weight between neurons. T_{ij} is dynamic threshold.

We can see standard ICM is a two dimensional structure, and it is only used for gray image. Study on some algorithms based ICM, it is clear that standard ICM cannot carry out multi-region image segmentation, and cannot deal with color image directly. In order to extend standard ICM applications and improve its efficiency, many scholars have proposed ICM-related methods [10]. In this work, we improve ICM based on a novel view, which is saliency-based ICM.

2.2 Saliency Detection

Saliency detection is a basic and complex technique in computer vision, which can guide computer to capture key information in an image according to human visual habit. When a pixel (or a region) in image is uniqueness, rarity, it may be salient. State-of-the-art saliency algorithms in general can be categorized as bottom-up and top-down approaches. The bottom-up approaches are data-driven and the other type methods are task-driven. A large number of novel methods are proposed based on the existing algorithms, such as Bayesian frameworks [11], ranking algorithms [12], differential equation [13], deep learning [14], etc.

How to highlight the salient object is still challenging problem, as shown in Fig. 2. The comparison algorithms used in the example are the latest or classical methods, including frequency method: FT [15], contrast method: RC [16], graph based method: GBVS [17], deep feature based method: ELD [18] and unsupervised method: UHM [19]. We can observe that pixels will get a higher gray value in the saliency maps when they are salient. The deep feature based method such as ELD can get the best accuracy and the highest recall rate, but this type of deep learning based methods needs a complex training process, which can affect the efficiency of algorithm. On the other hand, it is difficult to get training set for unknown environments detection. So we propose a novel unsupervised saliency method based on dynamic guided filtering in this paper.

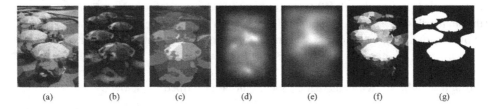

(a)   (b)   (c)   (d)   (e)   (f)   (g)

Fig. 2. Saliency detection results (a) Original image (b) FT (c) RC (d) GBVS (e) UHM (f) ELD (g) GT (Ground Truth)

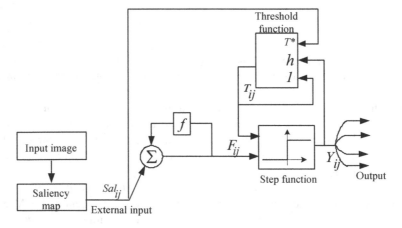

Fig. 3. The structure of SG-ICM

3 Methodology

3.1 SG-ICM

ICM is the result of research on the phenomenon of pulse-synchronous oscillation of mammalian visual cortical neurons, it comes from several visual cortex model, especially the Eckhorn model. At the same time, it absorbs advantages of other visual modes and is the product of cross-synthesis of various cerebral cortex models. It is developed based on the study of PCNN, but it is simpler than PCNN and is more suitable for object segmentation. More importantly, this model doesn't require any training.

Clearly, although ICM can effectively realize the unsupervised image segmentation, it suffers from lots of iterations and parameter setting. Visual saliency method can make the prominent of every region or pixel in an image, and it has been used in many fields computer vision. Based on traditional ICM and visual saliency detection, we proposed SG-ICM, which is improvement of existing image segmentation methods. The proposed model is designed as follows:

$$
\begin{aligned}
F_{ij}[n+1] &= fF_{ij}[n] + Sal_{ij} \\
Y_{ij}[n+1] &= \begin{cases} 1 & F_{ij}[n+1] > T_{ij}[n], n > n_{fire} \\ 0 & else, \quad n < n_{fire} \end{cases} \\
T_{ij}[n+1] &= T_{ij}[n] + hY_{ij}[n] - T^* \\
T^* &= (\max Sal(i,j) - \min Sal(i,j))/n
\end{aligned}
\tag{2}
$$

Where Sal_{ij} is the ij-th pixel's saliency, which is instead of the original image as external neuron input in the image. The saliency estimation in this work based on a pixel different level from its surroundings. Compared with the traditional ICM, the proposed model in this paper omit the natural linking term between neurons. In proposed model, all the parameters have the same definition as the traditional ICM. The novel model structure is presented in Fig. 3.

It can be noticed that proposed SG-ICM has simple structure and less parameters. Similar to the traditional ICM, we see each pixel in the saliency map as a neuron. In this novel model for object segmentation, only two parameters are preserved, namely, f and h, which maintains the main characteristics of traditional ICM and makes facilitate applications of the model. Compared with traditional ICM, the proposed SG-ICM updates the external neuron input, and can achieve better object segmentation results. At the same time, the output is accumulated, that is, the over-ignited neurons remain on fire all the time.

n is the iterative number. The traditional threshold function in ICM adopts the exponential attenuation mechanism which is more suitable for human vision. But the segmentation in this work only distinguishes objects from background, so when the pixel similarity between object and background or between different objects is poor. It will bring difficulties for segmentation, which is not conducive to implementation of segmentation algorithm. In this paper, we choose the linear decline to adjust the threshold:

$$T^* = (\max Sal(i,j) - \min Sal(i,j))/n \tag{3}$$

In addition, in order to make the saliency map play a positive role in image segmentation, it will be: 1. Salient region with larger value in saliency map should cover the most of the real object; 2. Salient region in saliency map should contain the most salient object, perhaps some background. 3. Salient object in saliency map should be the biggest in all saliency regions. Thus, our defined threshold would be closer to the average feature of the real object.

To analyze single neuron's status, the math model denotes that is no linking matric between different neurons, when neurons ignite for the first time, $n = 1$ and internal activity can be expressed as:

$$F_{ij}[n+1] = fF_{ij}[n] + Sal_{ij} = (F_{ij}[n] - \frac{Sal_{ij}}{1-f})f^n + \frac{Sal_{ij}}{1-f} \tag{4}$$

In the proposed model, we add up the output results in each iteration, that is, when the neuron is fired, it will remain on firing during subsequent iterations.

3.2   Saliency Stimulus

In our work, we regard guided filtering analysis result as external neuron input, which is great improvement of traditional unsupervised saliency detection methods. In the following, we give the definition of saliency map.

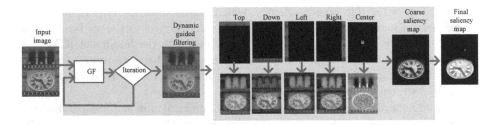

Fig. 4. Diagram of the proposed saliency detection method

As shown in Fig. 4, we model saliency detection as a dynamic guided filtering problem and proposed a two stage scheme for saliency detection. Figure 4 shows the main steps of the proposed saliency detection method. In the first stage, we exploit the iterative dynamic guided filtering to get the mainly structure of input image and remove irrelevance texture. Therefore, in the second stage, we exploit the boundaries of image and center regions prior to compute the coarse saliency of each pixel and highlight it to get the final saliency map [20, 21].

(a)	(b)	(c)	(d)	(e)	(f)

Fig. 5. The dynamic filtering with various iterative steps (a) Original image (b)–(f) Iteration outputs of guided filtering

Saliency detection aims to extract the most important regions of image according to human visual habit. The guidance image needs to get salient regions of image for us. As shows in Fig. 5, it is the dynamic filtering with various iterative steps based on guided filter. We can see guided filtering can remove the irrelevance texture, which may locate at the intra-object or intra-background. With the iteration increases, the detail in the image are gradually smoothed, at the same time, the main structure of image is preserved. Therefore, we use the dynamic guidance image to modulate input image for getting its main structure. This method is iterative and dynamic guidance image is updated at every step. The novel dynamic guided filtering kernel is given by:

$$W_{ij}(p,q^t) = \frac{1}{|\omega|^2} \sum_{k:(i,j)\in\omega_k} \left(1 + \frac{(p_i - \mu_k)(q_j^t - \mu_k)}{\sigma_k^2 + \varepsilon} \right) \tag{5}$$

where q^t is the t-*th* filtering output based on dynamic filter. Such as q^1 is the 1-st filtering output based on dynamic filter. μ_k and σ_k are the variance and the mean of dynamic guidance image q^t in window ω_k, respectively.

The novel filtering kernel uses the joint structure information of input image and dynamic guidance image, the filtering output can preserve the main structure of input image and remove the noisy texture efficiently. The filtering output is given by:

$$q_i^{t+1} = \sum_j W_{ij}(p,q^t)p_j \tag{6}$$

This novel iterative filter can also be the edge-preserving smoothing property like traditional guided filter. The guided filtering results are applied to find important pixels of image. If the pixels in image are salient, their saliency value should be high. Psychophysical study [21] shows that human attention favors center region. So pixels close to a natural image center could be salient in many cases. On the other hand, a natural image boundary could be background in many cases. With the constraints on boundary and center of image, we could get five coarse saliency results respectively.

$$S_t = \sum_{i\in\{L,a,b\}} \|c_i - m_i\|_2$$

$$m = \frac{1}{|t|} \sum_{x\in\{t\}} x \tag{7}$$

Where S_t is coarse saliency map using top boundary of image. c_i is the color vector of pixel c. m is average color of top boundary of image. Parameter t is the pixels set of top boundary. $|t|$ is the pixels number of top boundary set.

Similarly, we compute the other four maps S_b, S_l, S_r, S_c, using the bottom, left, right image boundaries and center region of image. The five saliency maps are integrated by the following process:

$$S = S_t \times S_b \times S_l \times S_r \times S_c \tag{8}$$

While most regions of the salient objects are detection in the above process, we get the coarse saliency maps, which may not be adequately highlighted. To improve the coarse results, we need to highlight the coarse results to make the salient object with higher saliency value.

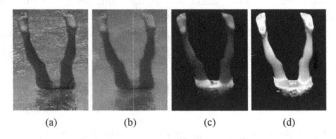

(a)   (b)   (c)   (d)

Fig. 6. Saliency map after highlighting (a) Input image (b) Filtering result (c) Integrating (d) highlighting

Coarse saliency map which we have gotten is from five different saliency maps using Eq. (8), the integration process could weaken the background pixels. Because the five saliency maps are all normalized to the range [0,1], the integration will decrease the value of salient pixels. The integration process cannot highlight object region well. Figure 6(c) shows the saliency value of salient object in coarse saliency map is not adequately high. In this work, we need to highlight the salient pixels in coarse saliency map.

We define the final saliency map using the following function:

$$SS = S' \cdot \exp(\alpha S' - 1) \tag{9}$$

where S' is the normalized integration result. SS is the final saliency map. In practice, we found the coarse saliency map is not highlighting the whole salient region. Therefore we use an exponential function in order to emphasize salient pixels. In all experiments, we use $\alpha = 2$ as the weight factor for the function. If $S' > 0.5$, then the term $\exp(\alpha S' - 1) > 1$. In other words, we magnify the pixel value of coarse saliency map when it is greater than 0.5, otherwise, reduction.

4   Experiments and Analysis

In our paper, experiments are consisted of four parts: the first part is comparing the proposed saliency method with several prior ones; The second part is doing segmentation experiment for color image; the third part is doing object segmentation for aerial image; the fourth part is doing the quantitative evaluation of segmentation accuracy.

4.1   Saliency Detection

We compare proposed saliency method with several prior ones, including contrast method: FT [15], SR [22], LC [23], HC [16], RC [16], GBVS [17], deep feature based method: ELD [18] and unsupervised method: UHM [19]. Figure 7 is saliency maps of above methods for four images from MSRA database. We can see our saliency method is superior in these comparison saliency detection algorithms. Figure 8 is the quantitative evaluation based on MSRA database. The PR-curve and F-measure express the same result that our method is the better saliency method than others.

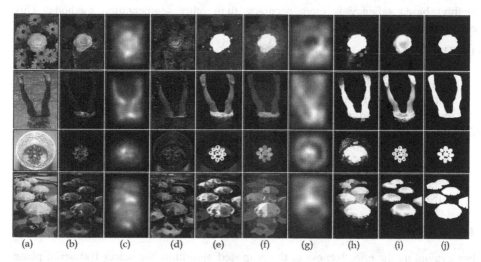

(a)   (b)   (c)   (d)   (e)   (f)   (g)   (h)   (i)   (j)

Fig. 7. Saliency detection results of different methods (a) Input (b) FT (c) GBVS (d) LC (e) HC (f) RC (g) UHM (h) ELD (i) Our (j) GT

4.2   Object Segmentation for Color Images

The proposed SG-ICM model introduces image saliency as stimulus for image segmentation. But the standard ICM is only used for gray image, and its model is a single layer two dimensional neural network. We calculate the image saliency based on

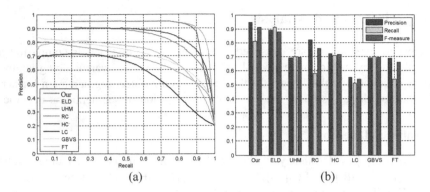

Fig. 8. Precision-recall curve and F-measure on MSRA-1000 database (a) PR curve (b) F-measure

superpixel and make it as the external neuron input in improved ICM. The color images are all selected randomly from MSRA database. The results show that the proposed method based saliency is more accuracy than other segmentation methods. This experiment is consisted of two parts:

(1) The segmentation based on different saliency detection methods
     In this part, we use manifold ranking (MR) [24] based saliency detection as the compare method. The experiment is order to test affection of the proposed saliency method in object segmentation. Figure 9 shows the different object segmentation results based on different saliency detection methods in same segmentation model.
(2) The segmentation based on different segmentation methods
     Object segmentation method have been proposed, where Kmeans [25] is a popular segmentation method in image segmentation. In this part, we use the Kmeans as compare method, and make it based on same saliency detection method. So this experiment test the ICM model is more effective in image segmentation.

4.3 Object Segmentation for Aerial Images

For evaluating the effectiveness of the proposed algorithm, we select 100 aerial plane images for this experiment. Because of uneven illumination, low contrast or fast texture change, aerial images have brought many difficulties to image processing (Fig. 10). We use several different image segmentation algorithms to segment the same set of aerial plane images, include ICM, PCNN [26], SCM [27], OSTU [28], Maximum entropy [29] and two-dimension chi-square divergence [30] in Fig. 11. Compared with the neural network-based methods (PCNN and SCM), the input term of the proposed model is improved by saliency map in this paper, which makes the neurons in the salient region more likely to be motivated, so that the segmentation results can be obtained accurately.

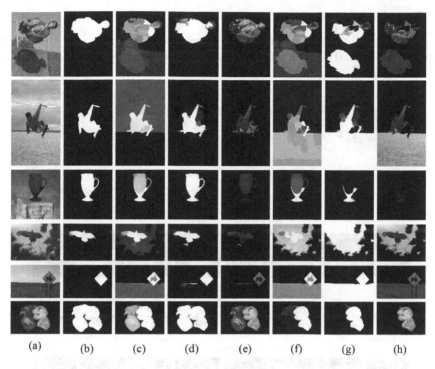

(a)   (b)   (c)   (d)   (e)   (f)   (g)   (h)

Fig. 9. Object segmentation result based on different saliency methods (a) Input (b) Saliency map based on the proposed saliency algorithm (c) Segmentation result using saliency map (e) Object segmentation result (f) Saliency map using MR (g) Segmentation result using saliency map (h) Object segmentation result

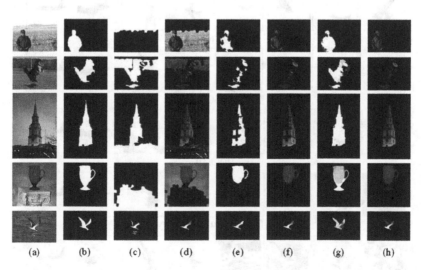

(a)   (b)   (c)   (d)   (e)   (f)   (g)   (h)

Fig. 10. Comparison with different segmentation methods (a) Input (b) GT (c) Segmentation result using Kmeans (d) Object segmentation result (e) Segmentation result using Kmeans and saliency (g) Segmentation result using the proposed algorithm (h) Object segmentation result

4.4   Quantitative Evaluation of Segmentation Accuracy

In the experiment, three kinds of indexes were used to quantitatively evaluate for different algorithms. They are Misclassification Error (EM), Mean Misclassification Error (MME) and running time.

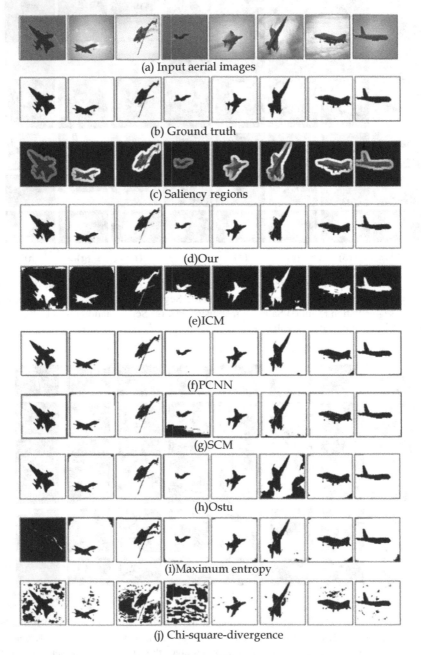

(a) Input aerial images

(b) Ground truth

(c) Saliency regions

(d)Our

(e)ICM

(f)PCNN

(g)SCM

(h)Ostu

(i)Maximum entropy

(j) Chi-square-divergence

Fig. 11. Segmentation results of different segmentation methods

Firstly, we define ME as follow:

$$ME = 1 - \frac{|B_O \cap B_T| + |F_O \cap F_T|}{|B_O| + |F_O|} \tag{10}$$

Where B_o, F_o express the real background and foreground of image, respectively. is the background and foreground of segmentation image. Table 1 shows that the ME of each image from Fig. 11, and we can get the analysis result similar to segmentation result based on the data. For testing the effectiveness of the proposed algorithm, we are calculating MME based on test image database in Table 2, and MME of the proposed algorithm is the minimum value. Table 3 show the running time based on image database.

Based on Tables 2 and 3, we can see that the MME of the proposed algorithm is close to that of PCNN, but the computational complexity of the PCNN model is too high. The running time of chi-square divergence has the most advantage in comparison algorithm, unfortunately, its MME is too high to have good segmentation performance.

Table 1. ME for different images

Image	Chi-square diverage	Maximum entropy	OSTU	SCM	PCNN	ICM	Our
1	0.2023	0.8285	0.0124	0.0469	0.0152	0.0805	0.0110
2	0.0245	0.0327	0.0132	0.0279	0.0221	0.0205	0.0036
3	0.2907	0.0389	0.0146	0.0113	0.0113	0.0050	0.0131
4	0.4121	0.0154	0.0078	0.2037	0.0207	0.3694	0.0075
5	0.0991	0.0184	0.0050	0.0320	0.0244	0.0056	0.0043
6	0.0179	0.0155	0.2075	0.0264	0.0215	0.0322	0.0055
7	0.0254	0.0202	0.0169	0.0162	0.0228	0.0031	0.0039
8	0.0183	0.0257	0.0095	0.0252	0.0073	0.0102	0.0062

Table 2. MME for test image database

	Chi-square diverage	Maximum entropy	OSTU	SCM	PCNN	ICM	Our
MME	0.2672	0.2850	0.1555	0.0587	0.0439	0.0897	0.0381

Table 3. Mean running time for test image database

	Chi-square diverage	Maximum entropy	OSTU	SCM	PCNN	ICM	Our
Running time (s)	0.0311	0.6058	0.9022	2.1106	1.0296	0.1872	0.2356

5   Conclusion

In this paper, Superpixel-based saliency guided intersecting cortical model for unsupervised object segmentation algorithm is proposed. The saliency map enables us not to calculate the global threshold of image segmentation, which facilitates the removal of unnecessary background interference in low-contrast aerial images and achieves more accurate object segmentation. It can provide reliable data quickly for object follow-up location and tracking, and can also realize automatic pre-analysis of data.

Acknowledgment. This work is supported by The National Science Foundation of China (No. 61701524).

References

1. Nie, S.P., Wang, M.: Image segmentation algorithm study for low contrast image. Chinese J. Lasers **31**(1) (2016)
2. Mortazi, A., Bagci, U.: Automatically designing CNN architectures for medical image segmentation. In: Shi, Y., Suk, H.-I., Liu, M. (eds.) MLMI 2018. LNCS, vol. 11046, pp. 98–106. Springer, Cham (2018). https://doi.org/10.1007/978-3-030-00919-9_12
3. Zheng, S., Jayasumana, S., Romera-Paredes, B., et al.: Conditional random fields as recurrent neural networks. In: ICCV, Santiago, 2015 (2015)
4. Eckhorn, R., Reitboeck, H.J., Arndt, M., et al.: Feature linking via synchronization among distributed assemblies: simulations of results from cat visual cortex. Neural Comput. **1990**(2), 293–307 (1990)
5. Zhi, X.H., Shen, H.B.: Saliency driven region-edge-based top down level set evolution reveals the asynchronous focus in image segmentation. Pattern Recogn. **80**, 241–255 (2018)
6. Papushoy, A., Bors, A.G.: Image retrieval based on query by saliency content. Digit. Sig. Process. **36**(1), 156–173 (2017)
7. Huang, S., Wang, W.Q.: Saliency-guided pairwise matching. Pattern Recogn. Lett. **97**, 37–43 (2017)
8. Li, H.Y., Cheng, J., Lu, H.C., et al.: CNN for saliency detection with low-level feature integration. Neurocomputing **2017**(226), 212–220 (2017)
9. Lee, G.Y., Tai, Y.W.: Deep saliency with encoded low level distance map and high level features. In: The IEEE Computer Vision and Pattern Recognition (CVPR), Las Vegas, USA, pp. 660–668 (2016)
10. Weng, C.T., Isa, N.A.M.: Single sperm tracking using Intersect Cortical Model-Mean Shift Method. In: International Conference on Signals & Systems (2017)
11. Xiao, L., Yanling, W., Hengliang, Z., et al.: Saliency detection based on the Bayesian model of improved convex hull. J. Comput.-Aided Des. Comput. Graph. **29**(2), 221–228 (2017)
12. Qi, W., Cheng, M.M., Borji, A., et al.: Saliency-Rank: two-stage manifold ranking for salient object detection. Comput. Vis. Media **1**(4), 309–320 (2015)
13. Liu, R.S., Cao, J.J., Lin, Z.C., et al.: Adaptive differential equation learning for visual saliency detection. In: Computer Vision and Pattern Recognition, Columbus, USA, pp. 3862–3869 (2014)
14. Li, G.B., Yu, Y.Z.: Visual saliency based on mulitscale deep feature. In: Proceedings of the 2015 IEEE Conference on Computer Vision and Pattern Recognition, Boston, USA, pp. 478–487 (2015)

15. Achanta, R., Hemanmi, S., Estrada, F.J., et al.: Frequency-tuned salient detection. In: CVPR, pp. 1597–1604. IEEE Press (2009)
16. Cheng, M.M., Zhang, G.X., Mitra, N.J., et al.: Global contrast based salient region detection. In: CVPR, pp. 409–416. IEEE Press (2011)
17. Scholdopf, B., Platt, J., Hofmann, T.: Graph-based visual saliency. In: NIPS, British Columbia, Canada, vol. 19, pp. 545–552 (2016)
18. Lee, G.Y., Tai, Y.W.: Deep saliency with encoded low level distance map and high level features. In: CVPR, Las Vegas, USA, pp. 660–668 (2016)
19. Tavakoli, H.R., Laaksonen, J.: Bottom-up fixation prediction using unsupervised hierarchical models. In: ACCV, TaiPei, Taiwan, pp. 287–302 (2016)
20. Harmann, G.: Gestalt Psychology: A Survey of Facts and Principles. Kessinger Publishing, Whitefish (2006)
21. Barris, M.C.: Vision and Art: The Biology of Seeing. Harry N. Abrams, Inc., New York (2008)
22. Hou, X.D., Zhang, L.Q.: Saliency detection: a spectral residual approach. In: Proceeding of the 2007 IEEE Conference on Computer Vision and Pattern Recognition, Minneapolis, USA, pp. 1–8 (2007)
23. Zhai, Y., Shah, M.: Visual attention detection in video sequences using spatiotemporal cues. In: the 14th Annual ACM International Conference on Multimedia, Santa Barbara, USA, pp. 815–824 (2006)
24. Yang, C., Zhang, L.H., Lu, H.C., et al.: Saliency detection via graph-based manifold ranking. In: Computer Vision and Pattern Recognition, Portland, USA, pp. 3166–3173 (2013)
25. Zhu, Y., Zhang, K.: Text segmentation using superpixel clustering. Let Image Process. **11** (7), 455–464 (2017)
26. Wang, Z., Sun, X.G., Zhang, Y.N., et al.: Leaf recognition based on PCNN. Neural Comput. Appl. **27**(4), 899–908 (2016)
27. Zhan, K., Shi, J.H., Li, Q.Q., et al.: Image segmentation using fast linking SCM. In: Proceedings of the International Joint Conference on Neural Networks (IJCNN), Killarney, Ireland, pp. 1–8 (2015)
28. Lai, Y.K., Rosin, P.L.: Efficient circular thresholding. IEEE Trans. Image Process. **23**(3), 992–1001 (2014)
29. Xiao, Y.H., Cao, Y.F., Yu, W.Y., et al.: Multi-level threshold based on artificial bee colony algorithm and maximum entropy for image segmentation. Comput. Appl. Technol. **4**(43), 343–350 (2012)
30. Chen, W., Yangyu, F., Lei, X.: Improved image segmentation based on 2-D chi-square-divergence. Comput. Eng. Appl. **50**(18), 8–13 (2014)

Object Detection for Chinese Traditional Costume Images Based GRP-DSOD++ Network

Haiying Zhao[1,2(✉)], Ting Yang[1,2], Xiaogang Hou[2,3], Hui Zhu[2,3], and Zhuoyu Yang[2,3]

[1] School of Computer Science, Beijing University
of Posts and Telecommunications, Beijing 100876, China
zhaohaiying@bupt.edu.cn
[2] Mobile Media and Cultural Calculation Key Laboratory of Beijing,
Century College, Beijing University of Posts and Telecommunications,
Beijing 102101, China
[3] Institute of Network Technology, Beijing University
of Posts and Telecommunications, Beijing 100876, China

Abstract. The image object detection methods based on deep learning have achieved remarkable results in recent years. However, as object sizes of Chinese Traditional Costume Images (CTCI-4) data set are smaller than that of natural images, and there are not enough training samples, the previous excellent object detection methods cannot achieve good detection result. To tackle this issue, mainly inspired by GRP-DSOD, we propose an effective network, namely GRP-DSOD++ network, to detect objects in the CTCI-4 data set. In order to collect multi-scale context information and capture a wider range of features, we introduce Dilated-Inception module (DI module) and applied it to object detection framework that is learned from scratch. We also applied other advanced components of several excellent object detectors to the proposed network architecture. The proposed detector in the CTCI-4 data set achieves 77.08% mAP, higher than the GRP-DSOD detector (75.33% mAP). And the detector (learning on VOC "07+12" trainval) also can achieve good performance on PASCAL VOC2007.

Keywords: GRP-DSOD++ network · Object detection · Dilated-Inception module · Chinese traditional costume images

1 Introduction

Current the advanced deep convolutional neural network has scored outstanding results in computer vision tasks such as image classification [1–4], object detection [5–13] and image segmentation [14–19]. The CNN-based object detection methods can be divided into two-stage approach and one-stage approach.

Two-stage approach is a two-step process: first generate region proposals and then classify these candidate regions. The typical representations of such method are the series of R-CNN algorithms, such as R-CNN [5], Fast R-CNN [6], Faster R-CNN [7]

and R-FCN [8]. One-stage approach does not require region proposal stage, which can directly generate the category probability and position coordinates of objects. Examples of such approach are YOLO [9], SSD [10], DSOD [11], GRP-DSOD [12] and RetinaNet [13].

In order to achieve good performance, most of the advanced object detection frameworks fine-tune models pre-trained on ImageNet. There are many deep models publicly available. Fine-tuning from the public pre-trained models requires less training data than learning object detectors from scratch. Besides, it can obtain the final model with less training time. But object detection systems that directly adopt the pre-trained networks have little flexibility to adjust the network structures. Some object detectors such as DSOD [11], GRP-DSOD [12] and RetinaNet [13] can train models from scratch through which the network structure of those can be flexibly designed according to application requirements or computing platforms. Learning object detection networks directly without pre-trained models can eliminate the learning bias due to the difference on both the loss functions and the category distributions between classification and detection tasks. Moreover, training object detectors from scratch can eliminate the mismatch between source dataset and target dataset.

Although DSOD is developed by following the single-shot detection (SSD) framework, it is the first framework that can train object detectors from scratch with outstanding performance, even with limited training data. GRP-DSOD is an extension of DSOD network, GRP-DSOD explores a new network architecture that can adaptively recalibrate the supervision intensities of layers for prediction based on input object sizes. Inspired by DSOD and GRP-DSOD, Gated Recurrent Feature Pyramid DSOD++ (GRP-DSOD++) framework is proposed in this paper.

We propose Dilated-Inception module (DI module) that combines the merits of dilated convolution and Inception structure. DI module can extract multiple range of information for prediction layers, we will describe the details in Sect. 3.2. SE block [4] is also applied in the early layer of our network, as illustrated in Fig. 5. Last but not the least, we introduce balanced confidence loss in the objective loss function which is proposed in SSD network to keep the balance between positive samples and negative samples. Details are in Sect. 3.3. We incorporate DI module, SE block and new loss function into GRP-DSOD, so the proposed network is called GRP-DSOD++, and it has achieved new state-of-the-art results on our dataset of Chinese traditional costume images.

To summarize, our main contributions are as follows:

(1) We proposed Dilated-Inception module that can aggregate multiple long-range information.
(2) In GRP-DSOD network framework, we add a Dilated-Inception module, introduce a SE block, and change standard convolution to dilation convolution/atrous convolution in partial convolution layer. Besides, we introduce a weight factor into the confidence loss function to balance the positive and negative samples.
(3) The GRP-DSOD++ network is applied to the object detection task of Chinese traditional costume images, which has higher accuracy compared with other excellent object detection networks.

2 Related Work

Classic Object Detectors: Classic object detection methods extract the features of artificial design such as LBP [20, 21], Haar [22], HOG [23], HSC [24] and SIFT [25] in each sliding window. Subsequently, SVM [26], Cascade Classifier [27] or other classifier was utilized for recognition. HOG [23] is a landmark milestone in pedestrian detection, later HOG-based deformable part model (DPM) proposed by Felzenszwalb et al. [28] had best detection performance on PASCAL [29] for several years. The sliding-window approach was the leading detection paradigm in classic computer vision. Since the resurgence of deep learning, two-stage detectors have held a dominant position in modern object detection.

Two-Stage Detectors: Two-stage detectors first generate a sparse set of candidate proposals, then classify the proposals with high-quality classifier. Selective Search [30] is a class-independent, data-driven strategy that combines the best of the intuitions of segmentation and exhaustive search and it can generate a small set of high-quality object locations for object recognition. R-CNN [5] extracts convolutional features on each selective search region proposal, and then classifies each proposal using class-specific linear SVMs. Fast R-CNN [6] network takes as input a full-image and a set of regions of interest (RoIs). The network computes the convolutional features of whole image, and gets features of each RoI by RoI projection. Thus the objector is faster than RCNN on both training and testing. Faster R-CNN [7] explores a Region Proposal Network (RPN) that shares whole-image convolutional features with the detection network, thus proposals computational cost is negligible. RPNs can generate high quality RoIs that are used by Fast RCNN for detection. R-FCN [8] develops position-sensitive score maps which achieve higher positioning accuracy than previous region-based detectors such as Fast/Faster R-CNN. Moreover, R-FCN adopts fully convolutional network, so it is more accurate and efficient.

One-Stage Detectors: YOLO [9] is a single neural network that can predict object positions and classes directly from full images in one evaluation. YOLO has extremely fast speed but with poor accuracy. SSD [10] combines the virtues of YOLO and Faster R-CNN, ensuring both speed and accuracy. The definition of default boxes in SSD is similar to the anchors of RPN network. At test time, SSD generates the per-class scores for each default box and adjusts the coordinates of those boxes to better encompass the objects.

The previous one-stage detectors have obvious advantage in speed, but the precision is inferior to the two-stage detectors. It is not until the appearance of DSOD [11] that one-stage detector can achieve the accuracy comparable to many two-stage detector with fast real-time detection speed. GRP-DSOD [12] is an extended work of the DSOD, its framework uses the gating mechanism of SENets [4] for reference to achieve adaptive feature recalibration. The performance of GRP-DSOD system outperforms a lot of advanced two-stage detection methods. RetinaNet [13] address class imbalance by using a novel, simple and highly effective focal loss. DSOD, GRP-DSOD and RetinaNet can be learned from scratch on a small dataset and achieve high precision. Our proposed method is a further improvement and development of

GRP-DSOD architecture and incorporates several excellent components of other modern detectors.

Dilated Convolution: Fisher Yu et al. [14] proposed dilated convolution module that can be applied in semantic segmentation to keep more details. The module expands the receptive field exponentially without losing resolution and does not add extra parameter, and it is also effective to increases the accuracy of image classification task. DeepLab [15–17] also utilized dilated convolution in the network structure and achieved good segmentation results. We apply dilated convolution to our detecting framework, which proves that dilated convolution is suitable for object detection task.

3 Method

In this section, we will first review the recently proposed Gated Recurrent Feature Pyramid DSOD (GRP-DSOD). Then, we will present the Dilated-Inception module (DI module), a module that combines the merits of dilation convolution and Inception, and briefly introduce the SE block of literature [4]. Following that, we will describe a balanced confidence loss. Finally, we will describe the structure of GRP-DSOD++ in detail.

3.1 GRP-DSOD

The GRP-DSOD object detection framework is an extension of DSOD, which can be trained directly without pre-training model, and the detection accuracy of small objects is improved. The Recurrent Feature Pyramid structure in the GRP-DSOD compresses rich spatial and semantic features into a single prediction layer, further reducing the number of parameters to be learned. The DSOD model needs to learn half of new features for each scale in prediction layers, while the GRP-DSOD model only needs to learn one-third of new features, the convergence speed of the GRP-DSOD model is therefore faster. In addition, the GRP-DSOD model introduces a new gating prediction strategy, which can enhance or weaken the supervision adaptively on feature maps of different scales according to the size of the input object.

The GRP-DSOD++ proposed in this paper is an improvement on the GRP-DSOD. The GRP-DSOD++ detector is described in detail below.

3.2 DI Module

DI module adopts the advantages of dilated convolution and Inception structure. The DI module takes into account the multiple long-range information with a small amount of computation added.

The Inception module in literature [2] uses different convolution kernel sizes (e.g. 3×3, 5×5) to extract the multi-scale information, the Inception module structure is shown in Fig. 1. The parallel superposition of different convolution kernel sizes in the Inception module increases the width of network and improves the adaptability of network to scale. The convolution kernel of 1×1 plays a role in reducing the thickness of feature maps, which appropriately reduces the computation of convolution.

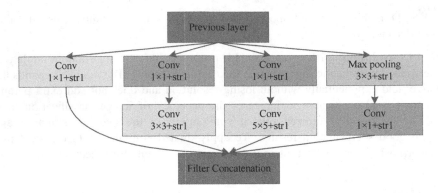

Fig. 1. An Inception module structure.

Dilated convolution with a sampling rate of r refers to the convolution of input with upsampled filters which is produced by inserting $r - 1$ zero weights between two adjacent filter weights in the vertical and horizontal directions. Standard convolution is equivalent to dilated convolution with a sampling rate of 1. The receptive field of filter can be adjusted by changing the sampling rate of dilated convolution. As shown in Fig. 2, the larger the sampling rate, the larger the corresponding receptive field of filter on the input.

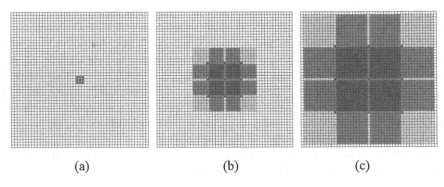

(a)   (b)   (c)

Fig. 2. Dilated convolution with kernel size 3×3 and different sampling rates. (a) dilated convolution with a sampling rate of 1 (standard convolution). (b) The sampling rate is 6, and the perceptive field of convolution kernel is 23×23. (c) The sampling rate is 12, and the perceptive field of convolution kernel is 47×47.

DI module proposed in this paper can be considered as Inception module combined with dilated convolution. The structure of DI module is shown in Fig. 3, where the convolution kernel size is 3×3 and the sampling rate is $r = \{1, 6, 12\}$. Compared with Inception module, as DI module obtains multi-scale context information through dilated convolution, it can capture longer range information and requires fewer parameters.

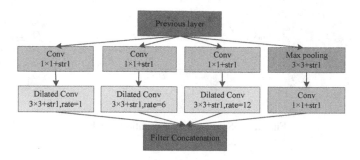

Fig. 3. A DI module structure.

3.3 SE Block

SENets [4] won the first place in ILSVRC 2017 classification competition. Inspired by SENets, we use SE block in the early layer of our network to improve the expression ability of lower level features. SE block can enhance the channel features that are useful for the later layers and suppress the channel features that are less useful.

The basic structure of a SE block is illustrated in Fig. 4. $\mathbf{U} = [u_1, u_2, \cdots, u_3] \in R^{w \times h \times c}$ is a set of feature maps. The output of SE block is $\tilde{\mathbf{U}} \in R^{w \times h \times c}$, therefore a SE block can be formulated as:

$$\tilde{\mathbf{U}} = F_{scale}(\mathbf{U}, F_{ex}(F_{sq}(\mathbf{U}))) \tag{1}$$

Squeeze operation can be thought of as a global average pooling of feature map channel by channel. $s \in R^c$ is the channel-wise descriptor obtained by squeezing U. The c-th element in s is obtained by the following calculation formula:

$$s_c = F_{sq}(u_c) = \frac{1}{w \times h} \sum_{i=1}^{w} \sum_{j=1}^{h} u_c(i,j) \tag{2}$$

The activation is composed of two fully connected layers and a sigmoid activation:

$$e = F_{ex}(s) = \sigma(W_2 \delta(W_1 s)) \tag{3}$$

Where, δ is ReLU [31] function and σ is sigmoid function. $W_1 \in R^{\frac{c}{r} \times c}$ and $W_2 \in R^{c \times \frac{c}{r}}$ are the weights of the two fully connected layers respectively (set $r = 16$ in this paper). The final output of SE block is:

$$\tilde{\mathbf{U}} = F_{scale}(\mathbf{U}, e) = e \otimes \mathbf{U} \tag{4}$$

Where, $\otimes$ stands for channel-wise multiplication.

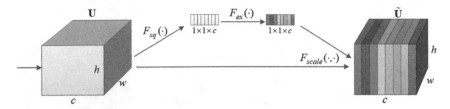

Fig. 4. A Squeeze-and-Excitation block.

3.4   Balanced Confidence Loss

We improve the objective loss function in SSD detection system [10] by introducing a weighting factor $\beta \in [0, 1]$ into confidence loss to control the balance of loss of positive and negative samples. In practice, β can be set by inverse class frequency treated as a hyperparameter set by cross validation. We write the balanced confidence loss (b-conf) of multiple classes confidences (c) as:

$$L_{b\text{-}conf}(x, c) = -\sum_{i \in Pos}^{N} \beta H_{ij}^{k} \log(\hat{c}_i^k) - \sum_{i \in Neg} (1 - \beta) \log(\hat{c}_i^0) \qquad (5)$$

where N is the number of matched default boxes, $\hat{c}_i^k = \frac{\exp(c_i^k)}{\sum_k \exp(c_i^k)}$, $H_{ij}^k = \{1, 0\}$ is an indicator that is if the i-th default box is matched to the j-th ground truth box of category k, its value is 1 otherwise is 0.

3.5   GRP-DSOD++ Architecture

DI modules integrate different range of feature information, and DI modules can be flexibly applied at any depth in an architecture. Due to the GPU memory constraints, we applied only one DI module to the proposed network, as shown in Fig. 5. The DI module is placed between the first and second prediction layer so that the features extracted by the DI module can be used by all the prediction layers except the first prediction layer.

SE block allows a network to perform feature recalibration, through which it can learn to use global information to selectively emphasize informative features and suppress less useful ones. In this paper, SE block is placed in the early layer of the proposed model to enhance the expression ability of lower level features.

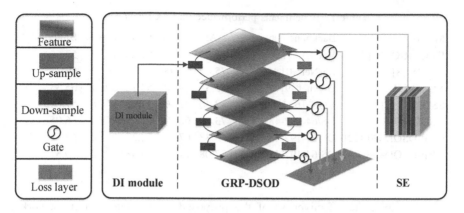

Fig. 5. GRP-DSOD++ network with all the bells and whistles.

The score of the object category in candidate boundary box that may contain an object is affected by contextual information. For example, if there is grassland in the image, the probability that the objects in the image are flowers or animals is relatively high, while the probability of office supplies is extremely low. Therefore, in this paper, the standard convolution in the last two Dense blocks of the network structure is changed to the dilated convolution to capture a wider range of feature information for the second to sixth prediction layers. A schematic view of the resulting network is depicted in Fig. 5.

4 Experiments

We conducted experiments on the CTCI data set and also on PASCAL VOC2007. Object detection performance is measured by mean Average Precision (mAP). All experiments are trained from scratch without the ImageNet [32] pre-trained models. Our code is implemented on Caffe platform. The whole network is optimized via the SGD optimizer on one GTX 1080GPU. We set the momentum to 0.9 and weight decay to 0.0005. We use the "xavier" method [33] to randomly initialize the parameters in the proposed model.

4.1 Experiments Performance and Results on CTCI Data Set

Chinese traditional costume images (CTCI) is an open and shared resource created by Beijing University of Posts and Telecommunications and Minzu University of China. By the image culture gene semantic tagging system construction, realize the traditional dress image semantic segmentation and cultural gene multi-label labeling learning. We used 1450 images from CTCI data set for this experiment, which included 4 categories, including 763 images of peony, 433 images of bird, 707 images of person and 363 images of butterfly. In the 1450 images, there are 2,462 objects for peony, 882 objects for bird, 1,881 objects for person and 748 objects for butterfly. In the experiment, we used 1020 images for training and 439 images for testing.

Table 1. Experiments performance on CTCI data set.

Method	Backbone network	mAP	Peony	Bird	Person	Butterfly
Faster R-CNN [7]	VGG-16 [1]	67.82	66.53	67.28	79.83	57.65
R-FCN [8]	ResNet-101 [3]	75.41	73.93	71.71	86.15	69.85
YOLO [9]	Darknet [9]	52.35	19.90	53.72	30.41	65.35
SSD300 [10]	VGG-16 [1]	72.40	72.00	74.38	85.58	57.64
DSOD300 [11]	DS/64-192-48-1 [11]	73.09	68.79	71.78	85.74	66.05
GRP-DSOD320 [12]	DS/64-192-48-1 [11]	75.33	67.63	78.37	87.80	67.51
GRP-DSOD++320	DS/64-192-48-1 [11]	77.08	74.55	78.48	89.12	70.37

In this paper, the effectiveness of the proposed object detection framework is illustrated by experiments on the CTCI-4 data set. Our implementation setting being the same as that of GRP-DSOD network ensure a fair comparison. We use batch size of 4 and accum_batch_size of 64 to train our models on the CTCI-4 data set. The initial learning rate is set to 0.1, then update the learning rate by multistep policy during the iteration. Different object detection frameworks are used for object detection on CTCI-4 data set, and the experimental results are shown in Table 1. It can be seen that the accuracy of GRP-DSOD (75.33% mAP) is slightly lower than that of R-FCN (75.41% mAP). Our GRP-DSOD++320 has the highest accuracy (77.08% mAP), which is better than baseline detector GRP-DSOD320 (75.33% mAP) in the same parameter settings, also higher than R-FCN. Some detection examples on the CTCI-4 data set using GRP-DSOD++320 detector are shown in Fig. 6.

4.2   Performance Analysis

We studied the effectiveness of each component in the GRP-DSOD++ model in this section. In order to prove that each component can improve detection accuracy, the control experiment is carried out on the CTCI-4 dataset, including: (1) DI module; (2) SE block; (3) Dilated convolution; (4) Balanced confidence loss.

(1) Effectiveness of DI module
DI module, SE block and dilated convolution in GRP-DSOD++ model are retained, and the improved objective loss function is used to train the model (where the value of β is fixed at 0.999). Different sampling rates are used for experiments in DI module. The experimental results are shown in Table 2. The results show that the detection performance is best when the sampling rate of DI module is set to (1, 6, 12).

The experimental results of removing SE block and dilated convolution in GRP-DSOD++ model, retaining DI module, where the sampling rate of DI module is (1, 6, 12), and training the model with the improved objective loss function are shown in Table 3. When using DI module and balanced confidence loss, the mAP is 77.05% (line 1, Table 3). Compared with using the improved loss function only (line 4, Table 3), the accuracy is improved, which proves that DI module can improve the accuracy of object detection.

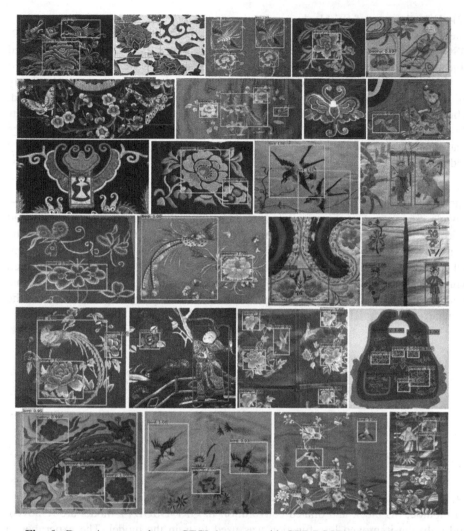

Fig. 6. Detection examples on CTCI-4 test set with GRP-DSOD++ (77.08% mAP).

Table 2. Varying sampling rates for DI module.

DI module	mAP
(1, 2, 8)	76.62
(1, 3, 6)	76.47
(1, 6, 12)	77.08

(2) **Effectiveness of SE block**
The experimental results of removing DI module and dilated convolution in GRP-DSOD++ model, retaining SE block, and training the model with the improved objective loss function are shown in Table 3. When using SE block and balanced confidence loss, the result is 76.51% (line 2, Table 3). Compared with the result of only using the improved loss function (line 4, Table 3), the accuracy is slightly improved, which can prove the effectiveness of SE block in our network. The experimental results show that these blocks can be stacked together to form GRP-DSOD++ architectures that generalise extremely effectively across on CTCI-4 that images categories vary widely.

(3) **Effectiveness of dilated convolution**
The experimental results of removing DI module and SE block in GRP-DSOD++ model, retaining dilated convolution, and training the model with the improved objective loss function (where the sampling rate of DI module is (1, 6, 12) and the value of β is 0.999) are shown in Table 3. The result of using dilated convolution and balanced confidence loss is 76.97% (line 3, Table 3), which is improved compared with the result of only using the improved loss function (line 4, Table 3). It proves that the application of dilated convolution in Dense blocks of network can improve the accuracy of object detection to some extent.

Table 3. The control experiment is carried out on the CTCI-4 data set.

DI module	SE block	Dilated convolutions	Balanced confidence loss	mAP
✓			✓	77.05
	✓		✓	76.51
		✓	✓	76.97
			✓	76.50

(4) **Effectiveness of balanced confidence loss**
DI module, SE block and dilated convolution in GRP-DSOD++ model are removed, and only the improved loss function was used to train the model. The experimental results are shown in Table 4 $mAP_{removed}$. When only balanced confidence loss is used, set different β all improves the detection accuracy, and the $mAP_{removed}$ is highest when β is 0.999.

The experimental results of retained all components in GRP-DSOD++ model, where the sampling rate of DI module is (1, 6, 12), and training the model with the improved objective loss function (using different β) are shown in Table 4 $mAP_{retained}$. The accuracy is highest when all components are applied to the network and the β value in balanced confidence loss is 0.999.

Table 4. Removed the other components and varying for balanced confidence loss.

β	mAP$_{removed}$	mAP$_{retained}$
0.999	76.50	77.08
0.99	75.88	76.45
0.90	76.22	76.28
0.75	76.30	76.50
0.5	76.32	76.11

The experimental results show that the large class imbalance encountered during training of dense detectors overwhelms the cross entropy loss. Easily classified negatives comprise the majority of the loss and dominate the gradient. While β balances the importance of positive/negative examples, it does not differentiate between easy/hard examples. Instead, we propose to reshape the loss function to down-weight easy examples and thus focus training on hard negatives.

4.3 Results on PASCAL VOC

We use batch size of 4 and accum_batch_size of 128 to train GRP-DSOD model and GRP-DSOD++ model on the PASCAL VOC 2007 train/val, and test on the PASCAL VOC 2007 test set respectively. The mAP of GRP-DSOD reached 76.51%, and the mAP of GRP-DSOD++ reached 77.06%. While we use batch size of 18 and accum_batch_size of 128 to train our model on two K80 GPUs, the mAP of GRP-DSOD++ reached 78.80%. That is because the precision of the object detector is greatly affected by the batch size, and within a certain range, the larger the batch size, the higher the precision. Due to the limitations of hardware, our batch size (18) is much smaller than the batch size (48) in GRP-DSOD. This explains why our accuracy (78.8%) is slightly lower than GRP-DSOD (79.0%) on the PASCAL VOC dataset. Our result of GRP-DSOD++320 is shown in Table 5.

Table 5. Detection results on PASCAL VOC 2007 data set.

Method	Pre-train	backbone network	mAP
Faster R-CNN [7]	✓	VGG-16 [1]	73.2
R-FCN [8]	✓	ResNet-50 [3]	77.0
R-FCN [8]	✓	ResNet-101 [3]	79.5
YOLO [9]	✓	Darknet [9]	63.4
SSD300 [10]	✓	VGG-16 [1]	74.3
DSOD300 [11]	✗	DS/64-192-48-1 [11]	77.7
GRP-DSOD320 [12]	✗	DS/64-192-48-1 [11]	79.0
GRP-DSOD++320	✗	DS/64-192-48-1 [11]	78.8

The backbone network of Faster RCNN and SSD300 is VGG-16. The backbone network of YOLO is Darknet. R-FCN adopts the ResNet-50/101 backbone network,

DSOD300, GRP-DSOD320 and GRP-DSOD++320 use DS/64-192-48-1 backbone network.

5 Conclusion

Dilated-Inception module proposed in this paper combines Inception module and dilated convolution, and DI module can be applied to one-stage object detector to improve the detection performance. Our improved GRP-DSOD++ detection framework integrates DI module, SE block, dilated convolution and class-balanced loss. A great deal of control experiments were carried out on our CTCI-4 data set to clarify the role of the individual components. GRP-DSOD++ achieves 77.08% mAP on CTCI-4 data set, that is 1.75% higher than GRP-DSOD (75.33% mAP) and 1.67% higher than R-FCN (75.41% mAP). The experiment shows that one-stage object detector learned directly without the ImageNet pre-trained models can achieve better performance than two-stage object detector. This work suggests that one-stage object detection approaches have a lot of improvement room in performance.

Acknowledgements. This work was supported by the key project of the national social science fund of China (18VDL001).

References

1. Simonyan, K., Zisserman, A.: Very deep convolutional networks for large-scale image recognition. In: ICLR International Conference on Learning Representation, pp. 1–14. ICLR, San Diego (2015)
2. Szegedy, C., Liu, W., Jia, Y., et al.: Going deeper with convolutions, pp. 1–9. IEEE, Boston (2015)
3. He, K., Zhang, X., Ren, S., et al.: Deep residual learning for image recognition. In: IEEE Conference on Computer Vision and Pattern Recognition, pp. 770–778. IEEE, Vegas (2016)
4. Hu, J., Shen, L., Sun, G.: Squeeze-and-excitation networks. In: IEEE Conference on Computer Vision and Pattern Recognition, pp. 7132–7141. IEEE, Salt Lake City (2018)
5. Girshick, R., Donahue, J., Darrell, T., et al.: Rich feature hierarchies for accurate object detection and semantic segmentation. In: IEEE Conference on Computer Vision and Pattern Recognition, pp. 580–587. IEEE, Columbus (2014)
6. Girshick, R.: Fast R-CNN. In: IEEE Conference on Computer Vision and Pattern Recognition, pp. 1440–1448. IEEE, Santiago (2015)
7. Ren, S., He, K., Girshick, R., et al.: Faster R-CNN: towards real-time object detection with region proposal networks. IEEE Trans. Pattern Anal. Mach. Intell. **39**(6), 1137–1149 (2017)
8. Dai, J., Li, Y., He, K., et al.: R-FCN: object detection via region-based fully convolutional networks. In: Advances in Neural Information Processing Systems, pp. 379–387 (2016)
9. Redmon, J., Divvala, S., Girshick, R., et al.: You only look once: unified, real-time object detection. In: IEEE Conference on Computer Vision and Pattern Recognition, pp. 779–788. IEEE, Las Vegas (2016)
10. Liu, W., Anguelov, D., Erhan, D., et al.: SSD: single shot multibox detector. In: European Conference on Computer Vision, pp. 21–37. IEEE, Cham (2016)

11. Shen, Z., Liu, Z., Li, J., et al.: DSOD: learning deeply supervised object detectors from scratch. In: IEEE International Conference on Computer Vision, pp. 1937–1945. IEEE, Venice (2017)
12. Shen, Z., Shi, H., Feris, R., et al.: Learning object detectors from scratch with gated recurrent feature pyramids, Venice (2017)
13. Lin, T.Y., Goyal P., Girshick, R., et al.: Focal loss for dense object detection. In: IEEE International Conference on Computer Vision, pp. 2999–3007. IEEE, Venice (2017)
14. Yu, F., Koltun, V.: Multi-scale context aggregation by dilated convolutions, San Juan (2016)
15. Chen, L.C., Papandreou, G., Kokkinos, I., et al.: Semantic image segmentation with deep convolutional nets and fully connected CRFs, San Diego (2015)
16. Chen, L.C., Papandreou, G., Kokkinos, I., et al.: DeepLab: semantic image segmentation with deep convolutional nets, atrous convolution, and fully connected CRFs. IEEE Trans. Pattern Anal. Mach. Intell. 40(4), 834–848 (2018)
17. Chen, L.C., Papandreou, G., Schroff, F., et al.: Rethinking atrous convolution for semantic image segmentation, Hawaii (2017)
18. Long, J., Shelhamer, E., Darrell, T.: Fully convolutional networks for semantic segmentation. IEEE Trans. Comput. Vis. Pattern Recogn. 39(4), 640–651 (2017)
19. He, K., Gkioxari, G., Dollár, P., et al.: Mask R-CNN. In: IEEE International Conference on Computer Vision, pp. 2980–2988. IEEE, Venice (2017)
20. Ojala, T., Pietikainen, M., Harwood, D.: Performance evaluation of texture measures with classification based on Kullback discrimination of distributions. In: IEEE International Conference on Pattern Recognition, pp. 582–585. IEEE, Jerusalem (1994)
21. Ojala, T., Pietikinen, M., Harwood, D.: A comparative study of texture measures with classification based on featured distributions. Pattern Recogn. 29(1), 51–59 (1996)
22. Papageorgiou, C.P, Oren, M., Poggio, T.: A general framework for object detection. In: IEEE International Conference on Computer Vision, pp. 555–562. IEEE, Bombay (2002)
23. Dalal, N., Triggs, B.: Histograms of oriented gradients for human detection. In: 2005 IEEE Computer Vision and Pattern Recognition, pp. 886–893. IEEE, San Diego (2005)
24. Ren, X., Ramanan, D.: Histograms of sparse codes for object detection. In: 2013 IEEE Conference on Computer Vision and Pattern Recognition, pp. 3246–3253. IEEE, Portland (2013)
25. Lowe, D.G.: Distinctive image features from scale-invariant keypoints. Int. J. Comput. Vis. 60(2), 91–110 (2014)
26. Chen, P.H., Lin, C.J., Schlkopf, B.: A tutorial on support vector machines. Appl. Stochast. Models Bus. Ind. 21(2), 111–136
27. Viola, P., Jones, M.J.: Robust real-time face detection. IEEE Int. J. Comput. Vis. 57, 137–154 (2004)
28. Felzenszwalb, P.F., Girshick, R.B., McAllester, D., et al.: Object detection with discriminatively trained part-based models. IEEE Trans. Pattern Anal. Mach. Intell. 32(9), 1627–1645 (2010)
29. Everingham, M., Van, G.L., Williams, C.K.I., et al.: The PASCAL visual object classes (VOC) challenge. Int. J. Comput. Vis. 88(2), 303–338 (2010)
30. Uijlings, J.R.R., Van De Sande, K.E.A., Gevers, T., et al.: Selective search for object recognition. Int. J. Comput. Vis. 104(2), 154–171 (2012)
31. Nair, V., Hinton, G.E.: Rectified linear units improve restricted Boltzmann machines. In: ICML International Conference on Machine Learning, pp. 807–814. ICML, Haifa (2010)
32. Deng, J., Dong, W., Socher, R., et al.: ImageNet: a large-scale hierarchical image database. In: IEEE Conference on Computer Vision and Pattern Recognition, pp. 248–255. IEEE, Miami (2009)

Combining Cross Entropy Loss with Manually Defined Hard Example for Semantic Image Segmentation

Zelu Deng[1,2], Jianbin Gao[1,2(✉)], Tao Huang[1,2], and James C. Gee[2]

[1] School of Resources and Environment,
University of Electronic Science and Technology of China, Chengdu, China
{dengzelu,huangtao}@std.uestc.edu.cn
[2] Center for Digital Health,
University of Electronic Science and Technology of China, Chengdu, China
{gaojb,gee}@uestc.edu.cn

Abstract. Semantic image segmentation has been one of the fundamental tasks in computer vision, which aims to assign a label to each pixel in an image. Nowadays, approaches based on fully convolutional network (FCN) have shown state-of-the-art performance in this task. However, most of them adopt cross entropy as the loss function, which will lead to poor performance in regions near object boundary. In this paper, we introduce two region-based metrics to quantitatively evaluate the performance of segmentation detail, which provides insights about the bottleneck of model. Based on this analysis, by use of a modified multi-task learning scheme, we combine cross entropy loss with manually defined hard example to propose a simple yet effective loss function named $\mathcal{L}_{\text{cehe}}$, which helps model focus on the learning of segmentation detail. Experiments show that model using $\mathcal{L}_{\text{cehe}}$ can better utilize spatial information comparing with the conventional cross entropy loss $\mathcal{L}_{\text{ce}}$. Statistically, metrics indicate that the proposed method outperforms the widely used $\mathcal{L}_{\text{ce}}$ by 1.12% in terms of MIoU on Cityscapes validation set, and by 4.15% in terms of the region-based metric MIoUiER proposed in this paper, proving that $\mathcal{L}_{\text{cehe}}$ performs better in segmentation detail.

Keywords: Semantic image segmentation · Hard example · Segmentation detail

1 Introduction

Over the past years, the performance of semantic image segmentation, a per-pixel classification problem, has been dramatically advanced by fully convolutional network (FCN) based approaches [1]. Generally, FCN can be converted from a classification model [2–4] pre-trained on ImageNet [5] by replacing fully connected layers with corresponding convolution ones. However, due to the strided layers existing in FCN, result usually performs poorly where small objects or

© Springer Nature Switzerland AG 2019
Y. Zhao et al. (Eds.): ICIG 2019, LNCS 11901, pp. 32–43, 2019.
https://doi.org/10.1007/978-3-030-34120-6_3

object boundary exists, namely segmentation detail. In order to solve this problem, atrous convolution [6] and many other novel modules [7,8] are introduced, attempting to preserve or recover spatial information when doing strided operations. However, most of the work seems to give model the capability to handle spatial information, instead of learning it.

The current mainstream loss function for semantic image segmentation is cross entropy, which treats all pixels equally. However, in the context of semantic segmentation, there are some pixels for which model is more difficult to make the right prediction. This is also in line with intuition, for human-beings, we can easily roughly circle the object in an image, but precisely segment it requiring careful consideration. This motivates us to develop a method that pays more attention to these pixels.

In this paper, we introduce two region-based metrics to analyze the performance bottleneck of model and based on this analysis, we propose a simple yet effective loss function $\mathcal{L}_{\text{cehe}}$ by combining cross entropy with hard example [9], which can alleviate the problem discovered by region-based metrics. The proposed $\mathcal{L}_{\text{cehe}}$ can be implemented as cross entropy loss $\mathcal{L}_{\text{ce}}$ with pixel-wise weight, which can replace $\mathcal{L}_{\text{ce}}$ without damage to training speed. Experiments show that model using $\mathcal{L}_{\text{cehe}}$ outperforms its counterpart $\mathcal{L}_{\text{ce}}$ by 1.12% in terms of MIoU on Cityscapes validation set, and by 4.15% in terms of the region-based metric MIoUiER proposed in this paper, indicating that our proposed method performs better in segmentation detail.

In summary, our contributions are:

- We propose two region-based metrics which can quantitatively evaluate the performance of segmentation detail.
- By analyzing model using region-based metrics, we find the key factor that limits model's performance, which can provide insights for future research.
- We propose a simple yet effective loss function $\mathcal{L}_{\text{cehe}}$, which outperforms the widely used cross entropy $\mathcal{L}_{\text{ce}}$.

2   Related Work

Approaches based on FCN have made remarkable progress in the field of semantic image segmentation. However, some properties, such as spatial in-variance, which make deep convolution networks successful in image classification, are precisely the factors that lead model failing to produce fine-grained segmentation. Quite a part of research focuses on preserving or recovering spatial information [6,8,10,11]. Besides, compared with classification, this dense prediction task has it own properties which we need take into consideration. Current methods to solve problems existing in semantic image segmentation can be divided into three categories: approaches solving intra-class inconsistency or inter-class indistinction problem, or simultaneously both [12]. In this paper, we focus on the inter-class indistinction problem.

Atrous convolution [6,13] is a solution for preserving spatial information and keeping receptive size at the same time. What's more, it will not introduce extra computation by sparsely sampling the input feature map. Currently, nearly all of the segmentation models replace the conventional convolution by the atrous one in the deep layer considering the trade-off between performance and memory usage. However, simply stacking atrous convolutions may cause the gridding issue described in [14] and they propose Hybrid Dilated Convolution (HDC) to alleviate this problem.

As for recovering spatial information, there is no general solution. The most straightforward method is to combine the low-level spatial information and the high-level semantic one by simply adding or stacking them together. This idea produces an encoder-decoder series like UNet [15], which shows good performance in the field of medical image segmentation. However, in semantic segmentation task, due to the complex content of the input image, the above method seems to cause chaos when fusing feature of different levels. Thus, various methods are proposed to alleviate this problem. For example, SegNet [7] is an instance of encoder-decoder series, which memorizes the indices of response when doing max-pooling and then uses them in decoder stage. RefineNet [8] presents a multi-path refinement network that explicitly exploits all the information available along the down-sampling process to enable high-resolution prediction using long-range residual connections.

The concept of hard example [9], proposed in the field of object detection, can be generalized to semantic segmentation task by regarding each pixel as an example. Based on this, [16] only back propagates the gradients of hard example determined by the predicted probability and a threshold, which is a fairly direct expansion of the idea stated in [9]. [17] divides a deep model into several cascade sub-models and each sub-model only operates on the hard example decided by the previous one, here hard example is also defined by a pre-defined threshold and the predicted probability. These work defines the term *hard* exactly as in [9], different from them, we define hard example based on our analysis of model, and then integrate this information into the loss function by multi-task learning scheme.

3   Method

3.1   Region Partition

Given an image, we divide it into two parts: edge region and object region. Formally, assuming (x, y) and set $I = \{(x, y)\}$ represent the coordinate of a pixel and all pixels in an image respectively, we define object region $I_{\text{object}} = I - I_{\text{edge}}$, where I_{edge} denotes edge region. From the definition of these two regions, we can see that the key step of region partition is to obtain I_{edge}. The following will introduce how to get the edge map of an image and how to efficiently and quantitatively obtain I_{edge}.

In the dataset of semantic image segmentation task, there are usually two types of images: original image and ground truth. The value of pixel in ground truth represents the target class which the corresponding pixel in the original image belongs to. Because of this unique property of ground truth, we can utilize

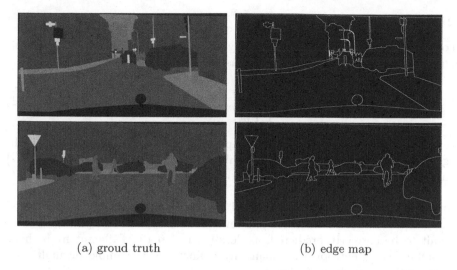

(a) groud truth   (b) edge map

Fig. 1. Using Canny algorithm to extract edge map from ground truth in Cityscapes. Best viewed in color.

Canny [18] algorithm to extract edge map of an image by setting the threshold t_1 and t_2 of Canny to 0. Some examples are illustrated with Fig. 1.

In order to efficiently and quantitatively obtain I_{edge}, we use chessboard distance as the distance between pixels in an image. Assuming that two pixels $q_i, q_j \in I$ and the distance between them is d_{ij}, the edge region I_{edge} is quantitatively defined based on d_{ij}. We let I_{canny} denote the set of edge pixels obtained by Canny algorithm described above, as shown in Fig. 1(b), then we define $I_{edge}^{(r)} = \{q \mid q \in I \text{ and } d(q, q_{canny}) < r, \exists q_{canny} \in I_{canny}\}$, where r is called the radius of edge region. By using chessboard distance, the process of computing $I_{edge}^{(r)}$ can be efficiently implemented by convolution operation in modern deep learning framework.

Assuming $E_{H \times W}$ represents the edge map obtained by Canny algorithm, where the value of pixel is 1 if it's an edge pixel, otherwise 0, as shown in Fig. 1(b). The method for efficiently computing $I_{edge}^{(r)}$ is summarized in Algorithm 1.

Algorithm 1: Efficiently compute $I_{edge}^{(r)}$

Input: edge map $E_{H \times W}$, radius r
Output: edge region $I_{edge}^{(r)}$

1 pad $E_{H \times W}$ with zero according to the radius r;
2 generate a $r \times r$ convolution kernel;
3 set all the weights of the kernel to 1;
4 let $O_{H \times W} = \text{conv}(E_{H \times W}, \text{kernel})$;
5 $I_{edge}^{(r)} = \{O_{H \times W} \neq 0\}$ (numpy style);
6 **return** $I_{edge}^{(r)}$;

3.2 Loss Function

Cross Entropy. Cross entropy is widely used as the loss function for semantic segmentation, which can be formulized as Eq. 1.

$$\mathcal{L}_{ce} = -\frac{1}{N} \sum_{i=1}^{N} \sum_{j=1}^{K} \mathcal{I}\{y_i = j\} \log p_{ij} \tag{1}$$

where N and K represent the number of pixels and classes, respectively. y_i is the target class of pixel i, and p_{ij} is the probability of pixel i assigned to class j. $\mathcal{I}\{\cdot\}$ is indicator function whose value is set to 1 if condition is satisfied, otherwise 0.

Combine Cross Entropy with Hard Example. $\mathcal{L}_{ce}$ implies that all pixels equally contribute to the total loss, however, it seems that some pixels are more difficult to be correctly predicted, as detailed in [16,19]. Different from them, we combine cross entropy with manually defined hard example by multi-task learning scheme, as shown in Eq. 2.

$$\mathcal{L}_{cehe} = \mathcal{L}_{ce} + \lambda \mathcal{L}_{he} \tag{2}$$

where $\mathcal{L}_{he}$ is the loss function for hard example and λ is a weight factor for these two losses.

We manually define pixels in edge region are hard example for semantic segmentation task, and the radius r of edge region is a hyper-parameter. The reason for this definition will be discussed in the experiment part. Function $m(i)$ indicates that whether pixel i is hard example, and it is defined as below:

$$m(i) = \begin{cases} 1, & \text{pixel } i \in I_{\text{edge}}^{(r)} \\ 0, & \text{otherwise} \end{cases} \tag{3}$$

Then we can formulate $\mathcal{L}_{he}$ as:

$$\mathcal{L}_{he} = -\frac{1}{N} \sum_{i=1}^{N} m(i) \sum_{j=1}^{K} \mathcal{I}\{y_i = j\} \log p_{ij} \tag{4}$$

Different from the conventional multi-task learning, here we compute $\mathcal{L}_{ce}$ and $\mathcal{L}_{he}$ on the same logits outputted by model, so them can be merged into a single loss function, shown as below:

$$\mathcal{L}_{cehe} = -\frac{1}{N} \sum_{i=1}^{N} \sum_{j=1}^{K} \mathcal{I}\{y_i = j\}(1 + \lambda m(i)) \log p_{ij} \tag{5}$$

3.3 Hierarchical Edge Region

In our experiments, the performance of $\mathcal{L}_{cehe}$ largely depends on the choice of λ and r. In order to alleviate this problem, we further divide edge region into

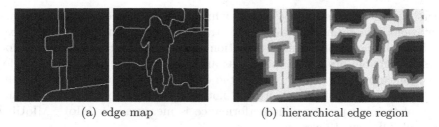

(a) edge map   (b) hierarchical edge region

Fig. 2. Visualization of hierarchical edge region of 3 levels. Colored part represents edge region and different color means different level. Best viewed in color.

different levels by the shortest distance between pixel in edge region and edge pixels obtained by Canny algorithm. Formally, $I_{\text{edge}}^{(r_1, r_2, \cdots, r_n)}$ represents an edge region of n levels, and ith region equals the set $I_{\text{edge}}^{(r_i)} - I_{\text{edge}}^{(r_{i-1})}$ when $i > 1$ or $I_{\text{edge}}^{(r_1)}$ when $i = 1$. Figure 2 shows some examples.

We re-definite $m(i)$ according to the number of levels, shown as below:

$$m(i) = \begin{cases} n - l(i) + 1, & \text{pixel } i \in I_{\text{edge}}^{(r_1, \cdots, r_n)} \\ 0, & \text{otherwise} \end{cases} \tag{6}$$

where $l(i) \in \{1, 2, \cdots, n\}$ is level index of pixel i.

4   Experiment

The purpose of this paper is to improve the performance of segmentation detail, rather than push the state-of-the-art. All experiments are conducted on a TITAN X (Pascal) GPU with 12 GB RAM, and the training parameters are detailed in the following part so that the results are easy to reproduce.

4.1   Region-Based Metric

Conventional metric like MIoU cannot quantitatively evaluate the performance of segmentation detail. In order to solve this problem, we introduce two region-based metrics: MIoUiER and MIoUiOR, which are defined on edge and object region, respectively. The calculation method of them is the same as MIoU except that MIoUiER only considers pixels belonging to set I_{edge} and MIoUiOR set I_{object}. The former can quantitatively evaluate the performance of segmentation detail. It should be noted that both of them are the function of radius r.

4.2   Dataset

We adopt Cityscapes [20] as the evaluation dataset. This dataset involves 19 semantic labels for segmentation task, which belong to 7 groups: flat, human, vehicle, construction, object, nature and sky. The dataset focuses on semantic understanding of urban street scenes, which has 5,000 fine and 20,000 coarse annotations. The former contains 2,975 (train), 500 (val) and 1,525 (test) pixel-level labeled images for training, validation and test, respectively. Previous work shows that model pre-trained on coarse annotations will have superior performance. Since our purpose is to study the effect of the proposed method rather than push the state-of-the-art, we will not use coarse annotations for the simplicity of training process. The performance is measured by MIoU, MIoUiER and MIoUiOR over 19 classes.

4.3   Implementation Details

Model. We use PyTorch framework for implementation, and we adopt DeepLab V3 Plus [11] with ResNet-50 [3] as the backbone. The output stride is set to 16. It should be noted that the proposed method can be applied to any model which uses cross entropy as the loss function.

Data Preprocess. Data augmentation is a powerful way to expand dataset and it makes the learned model robust to input varieties. Similar to previous work [11], we first scale the image by a factor randomly chosen from a pre-defined array $(0.5, 0.75, 1, 1.25, 1.5, 1.75, 2, 0)$, then randomly horizontally flip it and crop it to the size 513×513 for training. In order to make full use of the memory, the batch size is set to 12.

Learning Rate Policy. We adopt poly learning rate policy with initial learning rate 0.007 and power 0.9, the same as [11]. We train the model for 30,000 steps, considering the trade-off between accuracy and training time.

Inference. During inference stage, we do not use any data augmentation techniques. All metrics are obtained by a single-scale test on Cityscapes validation set.

Loss Function. As shown in Eq. 5, the proposed loss function $\mathcal{L}_{\text{cehe}}$ has some hyper-parameters to be specified. In all our experiments, the number of levels n is set to 3 and correspondingly, the radii for different levels are $r_1 = 7$, $r_2 = 9$ and $r_3 = 11$, the factor λ is set to 2. All these parameters are *not* carefully chosen, and performance gains may be obtained by grid search of them.

4.4   Evaluation

Metric Analysis. The proposed method helps to improve the performance of segmentation detail, thus leading to an improvement in terms of the overall performance. To evaluate $\mathcal{L}_{\text{cehe}}$, we use different loss functions to train the DeepLab model with other parameter settings and training pipeline unchanged. As listed in Table 1, our proposed $\mathcal{L}_{\text{cehe}}$ yields 72.94% in terms of MIoU, outperforming the loss function $\mathcal{L}_{\text{ce}}$ by 1.12%, which proves the effectiveness of the proposed method.

Table 1. Performance (MIoU) on Cityscapes validation set.

Loss function	MIoU
$\mathcal{L}_{ce}$	71.82
$\mathcal{L}_{cehe}$	**72.94**

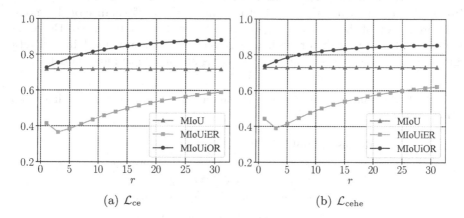

(a) $\mathcal{L}_{ce}$ (b) $\mathcal{L}_{cehe}$

Fig. 3. Performance of model trained with different loss functions. All metrics are obtained on Cityscapes validation set. Best viewed in color.

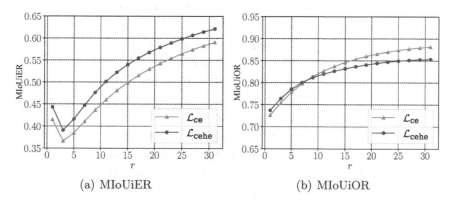

(a) MIoUiER   (b) MIoUiOR

Fig. 4. Performance (region-based metrics) of model trained with different loss functions. All metrics are obtained on Cityscapes validation set. Best viewed in color.

Region-Based Metric Analysis. The analysis based on MIoU only gives us an overall evaluation of the segmentation results, and it does not seem to be able to verify the purpose of the proposed method $\mathcal{L}_{cehe}$: improve segmentation detail. The following part will utilize two region-based metrics to (1) analyze the performance bottleneck of the model and (2) prove that the loss function $\mathcal{L}_{cehe}$ can enhance segmentation detail.

As shown in Fig. 3, generally, the curves of three metrics of these two loss functions have the same trend. Both of them perform well (MIoUiOR is larger than 80%) in object region even if radius r is just larger than 10. However, both have inferior performance in edge region comparing with their MIoUiOR. For example, MIoUiER is 45.99% and 50.13% for $\mathcal{L}_{ce}$ and $\mathcal{L}_{cehe}$ respectively when $r = 11$, which are 36.68% and 31.84% less than the value of corresponding MIoUiOR. This indicates that the factor limiting the performance of the model is the prediction of the pixels in edge region. Since the architecture of DeepLab, FCN-based feature extractor with a decoder, is commonly used in semantic segmentation, we argue that this conclusion is applicable to most models.

Table 2. Performance on the Cityscapes validation set obtained under different r settings.

Radius r	MIoUiER/MIoUiOR		
	$\mathcal{L}_{ce}$	$\mathcal{L}_{cehe}$	Improvement
1	41.53/72.62	44.34/73.69	2.81/1.07
3	36.62/75.39	39.09/76.38	2.47/0.99
5	38.49/77.81	41.59/78.49	3.10/0.68
7	41.08/79.79	44.73/80.03	3.65/0.24
9	43.64/81.39	47.64/81.14	4.00/−0.25
11	45.99/82.67	50.13/81.98	4.14/−0.69
13	48.07/83.74	52.22/82.64	**4.15**/−1.10
15	49.91/84.62	53.98/83.20	4.07/−1.42
17	51.53/85.37	55.47/83.67	3.94/−1.70
19	52.97/86.00	56.77/84.07	3.80/−1.93
21	54.26/86.53	57.90/84.42	3.64/−2.11
23	55.42/86.99	58.92/84.71	3.50/−2.28
25	56.48/87.38	59.84/84.95	3.36/−2.43
27	57.43/87.69	60.68/85.13	3.25/−2.56
29	58.30/87.94	61.44/85.25	3.14/−2.69
31	59.09/88.13	62.13/85.32	3.04/−2.81

Figure 4 illustrates the metric curves of two loss functions, and detail statistics can be found in Table 2. In terms of MIoUiER, our proposed $\mathcal{L}_{cehe}$ outperforms the widely used $\mathcal{L}_{ce}$ by a large margin (between 3% and 4%) under almost all r settings, indicating that $\mathcal{L}_{cehe}$ performs better in edge region, namely segmentation detail. Some visualized examples are shown in Fig. 5. In terms of MIoUiOR, $\mathcal{L}_{cehe}$ has inferior performance compared with $\mathcal{L}_{ce}$, which further confirms that $\mathcal{L}_{cehe}$ can improve the performance of segmentation detail since $\mathcal{L}_{cehe}$ is superior than $\mathcal{L}_{ce}$, considering the MIoU listed in Table 1. We conjecture the reason may be that the gradients of hard example are emphasized by the

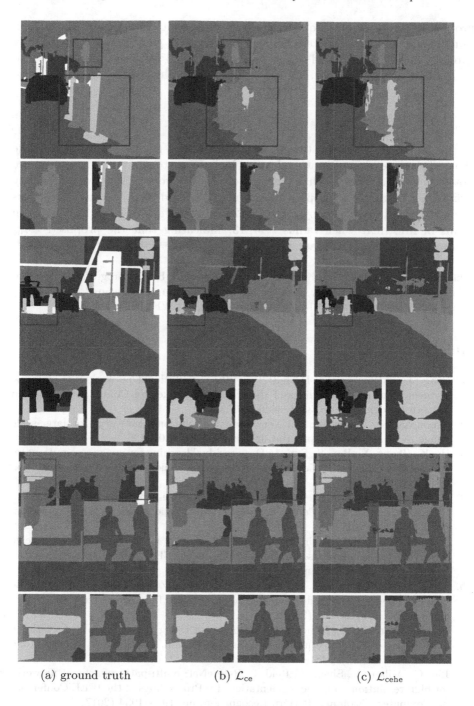

(a) ground truth   (b) $\mathcal{L}_{ce}$ (c) $\mathcal{L}_{cehe}$

Fig. 5. Visualization of segmentation results obtained by different loss functions on Cityscapes validation set. Results in red box indicate that the proposed $\mathcal{L}_{cehe}$ has better performance in segmentation detail. Best viewed in color.

product of $m(i)$ and λ in Eq. 5, so they play a leading role in the update direction of some model parameters. Further performance gains may be obtained by carefully choosing $m(i)$ and λ, and can be obtained by fusing models trained with different loss functions. We leave this for future research.

5   Conclusion

In this paper, we introduce two region-based metrics to quantitatively evaluate the performance of segmentation detail, which we use to analyze the performance bottleneck of the model. What's more, we combine cross entropy with manually defined hard example to propose a loss function named $\mathcal{L}_{\text{cehe}}$, which outperforms the widely used cross entropy $\mathcal{L}_{\text{ce}}$ by 1.12% in terms of MIoU, and by 4.15% in terms of MIoUiER when radius $r = 13$, indicating that the proposed $\mathcal{L}_{\text{cehe}}$ performs better in segmentation detail.

Acknowledgement. This work was supported in part by the programs of International Science and Technology Cooperation and Exchange of Sichuan Province under Grant 2017HH0028, Grant 2018HH0102 and Grant 2019YFH0014.

References

1. Long, J., Shelhamer, E., Darrell, T.: Fully convolutional networks for semantic segmentation. In: Proceedings of the IEEE Conference on Computer Vision and Pattern Recognition, pp. 3431–3440 (2015)
2. Simonyan, K., Zisserman, A.: Very deep convolutional networks for large-scale image recognition. arXiv preprint arXiv:1409.1556 (2014)
3. He, K., Zhang, X., Ren, S., Sun, J.: Deep residual learning for image recognition. In: Proceedings of the IEEE Conference on Computer Vision and Pattern Recognition, pp. 770–778 (2016)
4. Chollet, F.: Xception: deep learning with depthwise separable convolutions. In: Proceedings of the IEEE Conference on Computer Vision and Pattern Recognition, pp. 1251–1258 (2017)
5. Deng, J., Dong, W., Socher, R., Li, L., Li, K., Li, F.: ImageNet: a large-scale hierarchical image database. In: Proceedings of the IEEE Conference on Computer Vision and Pattern Recognition, pp. 248–255 (2009)
6. Chen, L.-C., Papandreou, G., Kokkinos, I., Murphy, K., Yuille, A.L.: Semantic image segmentation with deep convolutional nets and fully connected CRFs. arXiv preprint arXiv:1412.7062 (2014)
7. Badrinarayanan, V., Kendall, A., Cipolla, R.: SegNet: a deep convolutional encoder-decoder architecture for image segmentation. IEEE Trans. Pattern Anal. Mach. Intell. **39**(12), 2481–2495 (2017)
8. Lin, G., Milan, A., Shen, C., Reid, I.: RefineNet: multi-path refinement networks for high-resolution semantic segmentation. In: Proceedings of the IEEE Conference on Computer Vision and Pattern Recognition, pp. 1925–1934 (2017)
9. Shrivastava, A., Gupta, A., Girshick, R.: Training region-based object detectors with online hard example mining. In: Proceedings of the IEEE Conference on Computer Vision and Pattern Recognition, pp. 761–769 (2016)

10. Chen, L.-C., Papandreou, G., Kokkinos, I., Murphy, K., Yuille, A.L.: DeepLab: semantic image segmentation with deep convolutional nets, atrous convolution, and fully connected CRFs. IEEE Trans. Pattern Anal. Mach. Intell. **40**(4), 834–848 (2018)
11. Chen, L.-C., Zhu, Y., Papandreou, G., Schroff, F., Adam, H.: Encoder-decoder with atrous separable convolution for semantic image segmentation. In: Ferrari, V., Hebert, M., Sminchisescu, C., Weiss, Y. (eds.) ECCV 2018. LNCS, vol. 11211, pp. 833–851. Springer, Cham (2018). https://doi.org/10.1007/978-3-030-01234-2_49
12. Yu, C., Wang, J., Peng, C., Gao, C., Yu, G., Sang, N.: Learning a discriminative feature network for semantic segmentation. In: Proceedings of the IEEE Conference on Computer Vision and Pattern Recognition, pp. 1857–1866 (2018)
13. Chen, L.-C., Papandreou, G., Schroff, F., Adam, H.: Rethinking atrous convolution for semantic image segmentation. arXiv preprint arXiv:1706.05587 (2017)
14. Wang, P., et al.: Understanding convolution for semantic segmentation. In: IEEE Winter Conference on Applications of Computer Vision, pp. 1451–1460 (2018)
15. Ronneberger, O., Fischer, P., Brox, T.: U-Net: convolutional networks for biomedical image segmentation. In: Navab, N., Hornegger, J., Wells, W.M., Frangi, A.F. (eds.) MICCAI 2015. LNCS, vol. 9351, pp. 234–241. Springer, Cham (2015). https://doi.org/10.1007/978-3-319-24574-4_28
16. Zhuang, Y., et al.: Dense relation network: learning consistent and context-aware representation for semantic image segmentation. In: IEEE International Conference on Image Processing, pp. 3698–3702 (2018)
17. Li, X., Liu, Z., Luo, P., Change Loy, C., Tang, X.: Not all pixels are equal: difficulty-aware semantic segmentation via deep layer cascade. In: Proceedings of the IEEE Conference on Computer Vision and Pattern Recognition, pp. 3193–3202 (2017)
18. Canny, J.: A computational approach to edge detection. IEEE Trans. Pattern Anal. Mach. Intell. **8**(6), 679–698 (1986)
19. Wu, Z., Shen, C., Hengel, A.: High-performance semantic segmentation using very deep fully convolutional networks. arXiv preprint arXiv:1604.04339 (2016)
20. Cordts, M., et al.: The cityscapes dataset for semantic urban scene understanding. In: Proceedings of the IEEE Conference on Computer Vision and Pattern Recognition, pp. 3213–3223 (2016)

Attribute-Aware Pedestrian Image Editing

Xiaoyi Yin[1,2], Xinqian Gu[1,2], Hong Chang[1,2(✉)], Bingpeng Ma[2],
and Xilin Chen[1,2]

[1] Key Laboratory of Intelligent Information Processing of Chinese Academy
of Sciences (CAS), Institute of Computing Technology, CAS, Beijing 100190, China
{xiaoyi.yin,xinqian.gu}@vipl.ict.ac.cn, {changhong,xlchen}@ict.ac.cn
[2] University of Chinese Academy of Sciences, Beijing 100049, China
bpma@ucas.ac.cn

Abstract. Pedestrian image generation is a very challenging task. Existing generation methods have drawbacks including body distortion, inadequate visual details and large vague areas. In this paper, we propose Attribute-aware Pedestrian Image Editing (APIE) to address these problems based on given visual attributes. Our model denominated as APIE-Net, has three mechanisms including an attribute-aware segmentation network, a multi-scale discriminator and a latent-variable discriminator. Experiments on Market-1501 and DukeMTMC-reID datasets show that APIE-Net can generate satisfying pedestrian images with given attributes. Moreover, the generated images can augment the original datasets thus improve the performance in pedestrian-related tasks such as person re-identification (re-ID) and attribute prediction. Especially in person re-ID tasks our method outperforms state-of-the-art methods by a large margin.

Keywords: Attribute-aware · Pedestrian image editing ·
Data augmentation

1 Introduction

In recent years a surge of researches on image generation have found their applications in various real-world computer vision and multimedia tasks, e.g. facial attribute editing and animation [2,13,19], image super-resolution [4], object detection [3], and image-to-image translation [9,10,12]. Among them, some recent works focus on generating pedestrian images given pose information [16,21,26,27] or just from scratch [23]. The generated pedestrian images can be used to boost the performance of related learning tasks, such as person attribute prediction and person re-ID, through data augmentation of pedestrian

This work is partially supported by National Key R&D Program of China (No.2017YFA0700800), Natural Science Foundation of China (NSFC): 61876171 and 61572465, and Beijing Municipal Science and Technology Program: Z181100003918012.

Y. Zhao et al. (Eds.): ICIG 2019, LNCS 11901, pp. 44–56, 2019.
https://doi.org/10.1007/978-3-030-34120-6_4

image datasets. However, compared with general image generation, pedestrian image generation is more challenging, due to complex body configurations and poses, abundant details, variant lighting and backgrounds, etc.

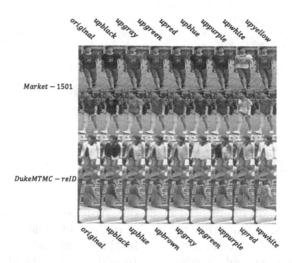

Fig. 1. The image generation examples of the APIE-Net

The main drawbacks of the aforementioned pedestrian image generation approaches are three-folds: body distortion, inadequate visual details and large vague areas. Firstly, body distortion occurs due to ambiguous locations of different body parts. For example, the generation areas might wrongly deviate to other body parts or even the background. Secondly, while obtaining global structures of pedestrians is relatively easy for most generation models, generating rich local details is problematic due to deficient given information. Though [16, 21, 26, 27] take advantage of pose annotations in person generation, the appearance details is still inadequate. Thirdly, large vague areas appear in the generated person images because of the conflict between disparate appearances from unsettled viewpoints and (or) complex background scenarios. Neither Gaussian noise nor pose labels could provide enough information to address this problem well.

To overcome the above drawbacks, we propose to introduce visual attributes in pedestrian generation. Visual attributes contain characteristics which can greatly benefit image generation: visual attributes usually describe specific regions, e.g. "upper-body black" refers to the pixels of the upper-torso; visual attributes can incorporate affluent details, e.g. "lower-body in white skirt"; visual attributes are relatively consistent, and invariant to viewpoints and backgrounds. The pose information adopted in [16, 21, 26, 27] can be considered as a kind of visual attribute. (For concise presentation we use the term "attribute" for "visual attribute" hereafter.) In this paper, we focus on more general attribute-based pedestrian generation. With a certain specified attribute, we aim to re-generate (or edit) the image of a given pedestrian with the attribute. We thus call our method **Attribute-aware Pedestrian Image Editing** (APIE). Although attributes have been used in face editing, it is the first time to study pedestrian image editing based on attributes.

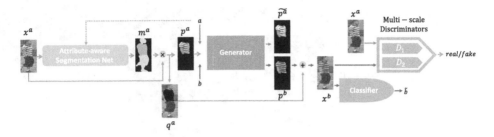

Fig. 2. The main architecture of the APIE-Net.

Though some efforts have been made on attribute-based image generation [8–13, 29], the task is more difficult in the context of pedestrian editing due to the complexity of pedestrian images. In this paper, our proposed **APIE-Net** consists of three specific mechanisms to address the difficulties. Firstly, to avoid dramatic distortions of pedestrian appearance, we use a segmentation net before image generation to locate the region of interest, so as to centralize the generation on the target body part and preclude the effects to the other regions including background. Secondly, to generate more visual details, a series of multi-scale adversarial discriminators are adopted to capture local visual information of different granularity. Thirdly, to disentangle the attributes which blend with each other or with other image appearance, we exploit adversarial training of the latent variables like the Fader-Net [14] to remove side-effects from irrelevant attributes or image regions, e.g. for white-colored and striped cloths we want to retain the strips while switching the color.

To verify the effectiveness of our method, we utilize the attribute labels provided in [28] and generate additional pedestrian images for two datasets, Market-1501 [30] and DukeMTMC-reID [22]. Some exemplars are depicted in Fig. 1. Comparisons on the generated pedestrian images with other generation methods [7,14,19] show that our method brings both quantitative and qualitative improvements. On the other hand, the generated images are beneficial to other pedestrian-related visual tasks through data augmentation. Most pedestrian datasets are "attribute-imbalanced" in which some attributes have few positive exemplars while others have many. The generated pedestrian images can augment the original datasets and make them more balanced, leading to state-of-the-art performance in both person re-ID and pedestrian attribute learning tasks. This result proves the effectiveness and practicability of our method from another view.

2   Methodology

The main architecture of the APIE-Net consisting of four parts is shown in Fig. 2. At the input-end an attribute-aware segmentation network is used to segment the input pedestrian image and extract the region corresponding to the given attributes. The generator consists of an encoder-decoder network with skip connections and a latent-variable discriminator used to erase contradictory attributes. The multi-scale discriminator guarantees the realness of the generated images. The attribute classifier ensures the generated images to posses the assigned attributes.

Let $\mathbf{x^a}$ denote an input pedestrian image $\mathbf{x} \in \mathbb{R}^{W \times H \times C}$ (W, H, C denote the width, height and the number of channels respectively) with attributes $\mathbf{a} \in \{0, 1\}^l$ (l denotes the number of attribute categories). The tth element of $\mathbf{a}$ is 1 if $\mathbf{x^a}$ has the tth attribute (e.g., white cloths) and 0 otherwise. The overall target of APIE-Net is to generate a new pedestrian image $\mathbf{x^b}$ given new attributes $\mathbf{b}$ (e.g., blue cloths).

2.1   Attribute-Aware Segmentation Network

In order to decide the location of the body part in which the attributes are to be changed, we resort to attribute-aware pedestrian segmentation. More specifically, we adopt a well-trained segmentation model [15,17]. Based on the fine-grained segmentation results and user specified target, we can obtain mask $\mathbf{m^a} \in \{0, 1\}^{W \times H}$, where the pixels with value 1 denote the foreground area and 0 the background area. Note that pose information may also be utilized like [16,21] to obtain the mask if the key points of pedestrian bodies are available. For simple and illustrative purposes, the learned mask mainly corresponds to three body parts: head, upper body, and lower body. Then, the image region to be edit can be expressed as $\mathbf{p^a} = \mathbf{x^a} \odot \mathbf{m^a}$, e.g. the upper body of a pedestrian. The other part $\mathbf{q^a} = \mathbf{x^a} \odot (1 - \mathbf{m^a})$ is considered as the background. $\odot$ denotes element-wise multiplication across channels.

We apply the above process to a set of pedestrian images $\mathbf{X^A}$ and obtain the editable and background regions represented as $\mathbf{P^A} = \mathbf{X^A} \odot \mathbf{M^A}$ and $\mathbf{Q^A} = \mathbf{X^A} \odot (1 - \mathbf{M^A})$.

2.2   The Generator

Encoder-Decoder Network with Skip Connection. The generator consists of an encoder and a decoder, denoting as $\mathbf{G}_{enc}$ and $\mathbf{G}_{dec}$ respectively, as shown in Fig. 3. The encoder $\mathbf{G}_{enc}$ projects the selected image region $\mathbf{p^a}$ to the latent variable $\mathbf{z^a}$ by use of several convolution layers. Then we randomly flip one bit of the attribute vector $\mathbf{a}$ to get the new attribute vector $\mathbf{b}$, and check the conflicts to avoid irrational attribute combination (e.g. one cannot wear in white while in red). $\mathbf{b}$ will be given at test time. The concatenation of $\mathbf{b}$ and the latent variable $\mathbf{z^a}$ passes the deconvolution layers of decoder $\mathbf{G}_{dec}$, from which a new image region $\mathbf{p^b}$ is generated. Similarly, decoding the concatenation of $\mathbf{a}$ and $\mathbf{z^a}$ outputs the reconstructed original image $\tilde{\mathbf{p}}^{\mathbf{a}}$, where "~" differentiates the estimation from the ground-truth. Skip connections are adopted between corresponding convolution and deconvolution layers, making the encoder-decoder network a U-net. The above process can be expressed more formally as following: $\mathbf{Z^A} = \mathbf{G}_{enc}(\mathbf{P^A})$, $\tilde{\mathbf{P}}^{\mathbf{A}} = \mathbf{G}_{dec}(\mathbf{Z^A}, \mathbf{A})$, $\mathbf{P^B} = \mathbf{G}_{dec}(\mathbf{Z^A}, \mathbf{B})$. The reconstructed image regions should be as close to the original ones as possible, thus the loss function for this encoder-decoder network is defined as:

$$\mathcal{L}_{rec} = \left\| \mathbf{P^A} - \tilde{\mathbf{P}}^{\mathbf{A}} \right\|^2. \tag{1}$$

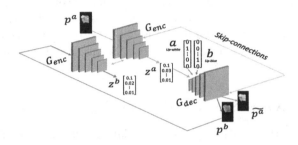

Fig. 3. The generator with skip-connections.

Discrimination Between Latent Variables. Some attributes of pedestrian images are entangled with each other, while some are mutually exclusive. For instance, the sex of a person is highly related to the styles of hair and clothes, while the concolorous shirt cannot be "white" and "blue" simultaneously. Therefore, when we edit an image according to the attribute $\mathbf{b}$, we have to pay attention to the attribute cooccurrence and contradiction issues. Instead of shielding the attributes $\mathbf{b}$ manually to avoid the conflict with the latent variable, a more ideal method is to automatically erase attributes from the latent variable in the encoding process. To this end, we resort to adversarial learning between latent variables. More specifically, the generated $\mathbf{p^b}$ is re-input to the encoder to get latent variable $\mathbf{z^b}$. Then, a discriminator $\mathbf{D}_Z$ is trained to capture the discrimination between latent variables $\mathbf{z^a}$ and $\mathbf{z^b}$, which is formulated as follows:

$$\mathcal{L}_{adv_{D_z}} = \mathbb{E}_{\mathbf{z^a}}[\mathbf{D}_Z(\mathbf{Z^A})] - \mathbb{E}_{\mathbf{z^b}}[\mathbf{D}_Z(\mathbf{Z^B})]. \tag{2}$$

The adversarial min-max process is

$$\min_{\mathbf{G}_{enc}} \max_{\|\mathbf{D}_Z\| \leq 1} \mathcal{L}_{adv_{D_z}}. \tag{3}$$

The min-max game between $\mathbf{G}_{enc}$ and $\mathbf{D}_Z$ learns latent variables invariant to attribute vectors. When the invariance is met, the decoder must use the attribute vector to generate image. In this way, attribute information is implicitly erased from the latent variables.

2.3   The Multi-scale Discriminators

We propose another adversarial learning process to guarantee the realness of the generated image $\mathbf{X^B} = \mathbf{P^B} + \mathbf{Q^A}$, which has the same size with the original image $\mathbf{X^A}$. To capture abundant local visual details, we use two discriminators in different resolutions as in [13]. As shown in Fig. 4, one discriminator, denoted as $\mathbf{D}_1$, takes in the generated images as ever. The other discriminator, denoted as $\mathbf{D}_2$, receives the generated images down sampled to the half resolution. Because of the unstable training process of the original GAN model, here we use the WGAN-GP model [18]. The formulations for the discriminators are

$$\mathcal{L}_{adv_1} = \mathbb{E}_{\mathbf{x^a}}[\mathbf{D}_1(\mathbf{X^A})] - \mathbb{E}_{\mathbf{x^b}}[\mathbf{D}_1(\mathbf{X^B})] + \lambda_{GP}\mathbb{E}_{\hat{\mathbf{x}}}[\mathbf{D}_1(\hat{\mathbf{X}})], \qquad (4)$$

$$\mathcal{L}_{adv_2} = \mathbb{E}_{\mathbf{x^a}}[\mathbf{D}_2(\mathbf{X^A})] - \mathbb{E}_{\mathbf{x^b}}[\mathbf{D}_2(\mathbf{X^B})] + \lambda_{GP}\mathbb{E}_{\hat{\mathbf{x}}}[\mathbf{D}_1(\hat{\mathbf{X}})], \qquad (5)$$

The overall loss of multi-scale discriminators is:

$$\mathcal{L}_{adv} = \mathcal{L}_{adv_1} + \mathcal{L}_{adv_2} \qquad (6)$$

The adversarial optimization process is:

$$\min_{\mathbf{G}} \max_{\|\mathbf{D}_*\| \leq 1} \mathcal{L}_{adv}, \qquad (7)$$

where $\mathbf{D}_*(\mathbf{X^B}) = \mathbf{D}_*(\mathbf{G}_{dec}(\mathbf{G}_{enc}(\mathbf{X^A} \odot \mathbf{M^A}), \mathbf{B}) + \mathbf{X^A} \odot (1 - \mathbf{M^A}))$, and $\mathbf{G}$ denotes the generator parameters in both of the encoder and the decoder. $\|\mathbf{D}_*\| \leq 1$ is the 1-Lipschitz constraint implemented by gradient penalty.

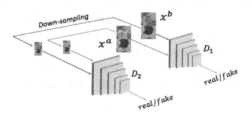

Fig. 4. The multi-scale discriminators.

2.4   The Attribute Classifier

The APIE-Net edits pedestrian images by transferring attributes, i.e., altering one attribute and keeping the others unchanged. One way to evaluate the model capability is to investigate the attributes of the generated images. We construct a classifier for this purpose, as in [14,19]. The classifier takes the generated image $\mathbf{x^b}$ as input and output the estimated attributes $\tilde{\mathbf{b}}$. Then, the cross-entropy loss between $\tilde{\mathbf{B}}$ and the ground-truth $\mathbf{B}$ is:

$$\mathcal{L}_{cls} = \sum_{\tilde{\mathbf{b}} \in \tilde{\mathbf{B}}, \mathbf{b} \in \mathbf{B}} -\mathbf{b} \log \tilde{\mathbf{b}} - (1 - \mathbf{b}) \log(1 - \tilde{\mathbf{b}}). \qquad (8)$$

It is noteworthy that the classifier share model parameters with the discriminator $\mathbf{D}_1$. The effect of attribute classification and image discrimination is reciprocal. Experiments will show the improvements on generating image details.

2.5   The Overall Loss

The generator, multi-scale discriminator and attribute classifier are trained simultaneously by optimizing the overall loss function.

$$
\begin{aligned}
&\mathcal{L}(\mathbf{C}, \mathbf{G}_{enc}, \mathbf{G}_{dec}, \mathbf{D}_1, \mathbf{D}_2, \mathbf{D}_Z) \\
&\quad = \mathcal{L}_{adv} + \lambda_C \mathcal{L}_{cls} + \lambda_R \mathcal{L}_{rec} + \lambda_{D_Z} \mathcal{L}_{adv_{D_Z}},
\end{aligned} \tag{9}
$$

where λ_* denote hyper-parameters to balance the losses. Following the adversarial training process, the parameters can be obtained by the following min-max game:

$$
\arg \min_{\mathbf{C}, \mathbf{G}_{enc}, \mathbf{G}_{dec}} \max_{\mathbf{D}_1, \mathbf{D}_2, \mathbf{D}_Z} \mathcal{L}(\mathbf{C}, \mathbf{G}_{enc}, \mathbf{G}_{dec}, \mathbf{D}_1, \mathbf{D}_2, \mathbf{D}_Z). \tag{10}
$$

3   Experiments

3.1   Datasets and Implementation

To verify the effectiveness of the proposed method, three experiments are conducted, namely pedestrian images generation, person re-ID and attribute learning. All experiments are performed on the datasets **Market-1501** [30] and **DukeMTMC-reID** [22], with labels provided in [28]. These two datasets are collected for person reID task, with 12,936/16,622 images belonging to 751/702 identities for training. The images for training have size of $128 \times 64 \times 3$, and so are the generated images. The datasets are labeled with 27/23 attributes, with 8/8 attributes for upper-body clothes and 9/7 attributes for the lower-body.

Table 1. The IS of attribute transformation on Market-1501

Models	Up-green	Up-yellow	Up-purple	Mean
CycleGAN	3.954	4.141	3.291	3.795
StarGAN	3.203	3.474	3.108	3.262
FaderNet	3.202	3.482	3.249	3.311
AttGAN	4.204	4.247	4.189	4.213
APIE(Ours)	**4.253**	**4.330**	**4.245**	**4.276**

The encoder and decoder of the generator network consist of 5 (de)convolution layers, so do the classifier and the multi-scale discriminator. The latent discriminator in the generator is a light-weighted net with only two

Table 2. The IS of attribute transformation on DukeMTMC-reID.

Models	Up-green	Up-yellow	Up-purple	Mean
CycleGAN	3.351	4.285	1.380	3.005
StarGAN	2.246	2.551	1.600	2.132
FaderNet	3.608	3.361	2.609	3.193
AttGAN	4.218	4.014	4.500	4.244
APIE(Ours)	**4.524**	**4.222**	**4.562**	**4.436**

fully connected layers. The initial learning rate is 0.0002 and it declines to one-tenth after every 10,000 iterations. We set $\lambda_R = 100, \lambda_C = 10, \lambda_{D_z} = 0.1$. We use Adam optimizer as in WGAN-GP model. The batch size is set as 50, and the number of training epochs is 400.

For the person re-ID experiments, we construct a baseline model by using the ResNet50 [25] as our backbone and tuning the network based on [20]. The input images are resized to 256×128, randomly cropped and horizontally flipped before training, and no dropout is adopted. The batch size is set to 32. The learning rate is initialized as 0.0003, and decreased to one-tenth after every 20 epochs. During training we augment the original datasets with the same amount of images generated by APIE-Net, which are supposed to have distinct identities. We use label smoothing regularization (LSR) as in [5] to balance the weights of the generated images. Mean average precision (mAP) and Cumulative Matching Characteristics (CMC) are evaluation metrics. As for attribute prediction, we use similar baseline model except two points: batch normalization is not adopted and multi-sigmoid loss is used instead of softmax loss. mAP is used for evaluation.

3.2 Pedestrian Image Editing

Pedestrian image editing is to generate new pedestrian images according to specified attributes. For illustration convenience, we select two identities from each dataset and change the color attributes of the clothes.

Some generated pedestrian examples by APIE-Net have been illustrated in Fig. 1. For comparison, we select four commonly used researches on image editing, namely CycleGAN [7], Att-GAN [19], Fader-Net [14] and StarGAN [11]. The model parameters are set as in the paper accordingly. In the Fig. 5 we demonstrate the qualitative comparison. Though these works perform well in facial attribute editing, they do poorly in the pedestrian editing scenario. The APIE-Net outperforms all of the other methods, due to its merits in image generation. First, while some attributes are changed, most details can be maintained. Second, the transformed body part can be precisely located. Third, the edited pedestrian images look natural with very few artifacts.

The quantitative comparison is listed in Tables 1 and 2. Inception score (IS) is used as the evaluation metric, which measures the diversity of the generated images. The higher value means the better result. Our method achieves the highest IS in both datasets.

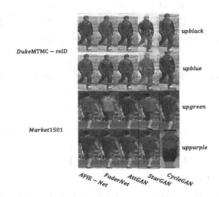

Fig. 5. The image generation examples of the **APIE**-Net.

3.3   Data Augmentation for Person Re-ID and Attribute Prediction

We further apply APIE-Net to augment the datasets in person re-ID and attribute prediction tasks. Each augmented dataset is of double size of the original one. We construct a strong "baseline" model for each task as explained above.

We compare the performance of APIE-Net with the methods that augment the datasets for person re-ID in Table 3. The results in the upper part are the highest records reported in the corresponding papers. In the lower part, the cautiously tuned baseline model performs better than the above state-of-the-art methods. The other three methods (including ours) are built on the baseline with data augmentation. However, the trivial pure-color-filling method and starGAN decrease the performance, indicating that inferior augmented images have negative effects on the task. On the other hand, our method generates images with accurate location and enough details, thus achieve the highest performance.

The comparative study on attribute learning is reported in Table 4. We compare our method with [28,31–33] with respect to "C.up" (*the color of the upper-body*), "C.low" (*the color of the lower-body*) and the mean accuracy, as we augment these attributes in the experiments. An average of 0.3% performance gain is achieved by APIE-Net over the baseline, and much larger gain over other methods on the Market1501 dataset. APIE performs lower than [32,33] on DukeMTMC for the lower baseline we used. The high performance attributed to the well generated images with accurate attribute information, which provides another verification to our method.

Table 3. Comparison on person re-ID with data augmentation.

Models	Market-1501		DukeMTMC-reID	
	rank-1	mAP	rank-1	mAP
Basel(R)+LSRO [23]	78.06	56.23	67.68	47.13
Pose-transfer [26]	87.65	68.92	78.52	56.91
IDE+CamStyle+RE [20]	89.49	71.55	78.32	57.61
baseline	92.46	78.66	83.42	68.03
pure-color filling	91.27	76.53	83.08	66.94
StarGAN	91.39	76.34	80.88	62.59
APIE(Ours)	**93.47**	**80.22**	**85.23**	**69.67**

Table 4. The attribute learning experiment using the augmented data on Market-1501 and DukeMTMC-reID.

Models	Market-1501			DukeMTMC-reID		
	C.up	C.low	Mean	C.up	C.low	Mean
SVM [31]	–	–	–	70.90	68.50	69.70
APR [28]	73.40	69.91	71.55	72.29	41.48	57.91
Sun et al. [32]	87.50	87.20	86.35	**93.90**	91.80	**92.85**
JCM [33]	92.90	93.50	93.20	92.90	**92.10**	92.50
Baseline	95.30	94.18	94.70	88.79	82.55	85.88
Ours	**95.46**	**94.59**	**95.00**	88.95	83.19	86.26

3.4 Ablation Study

Four parts are used in the APIE-Net which are alternative but useful, namely the attributes-aware segmentation net, the multi-scale discriminator ("M.D."), the latent variable discriminator ("L.D."), and the discriminator-classifier share-weight mechanism ("S.W."). The attribute-aware segmentation network and the generator-discriminator together construct the baseline set-up. "w.o. mask" means the baseline without the attribute-aware segmentation net. The combination of "M.D.", "L.D." and "S.W." is the APIE-Net. To verify the effectiveness of each model part to the final performance we make the following ablation study.

As shown in Table 5, the IS values roughly increase whenever a new model part is involved. The final combination outperforms each single part. Then we make visualization of the image generation results as shown in Fig. 6. The result of taking the attribute-aware segmentation net out of the baseline method is depicted in the second row, which is least satisfying for doing the least transformation.

Table 5. The IS comparison between different schemes

	Market-1501	DukeMTMC-reID	Mean
baseline	4.248	4.483	4.366
w.o. mask	4.228	4.238	4.233
+M.D.	4.242	4.495	4.369
+L.D.	4.246	**4.513**	4.380
+S.W.	4.233	4.486	4.360
+M.D.+S.W.+L.D.	**4.345**	4.446	**4.396**

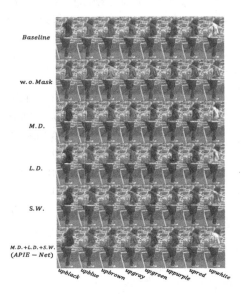

Fig. 6. Image generation results with different parts of APIE-Net.

4 Conclusion and Future Works

In this paper, we propose the problem of attribute-aware pedestrian image editing and a new model, APIE-Net, as its solution. With the aid of specified attributes, APIE-Net can generate high-quality pedestrian images, which benefit for real-world applications including person re-ID and attribute prediction. Comprehensive experiments demonstrate the effectiveness of the proposed method. In the future we will experiment on more complex attributes in pedestrian image editing.

References

1. Goodfellow, I., et al.: Generative adversarial nets. In: Advances in Neural Information Processing Systems, pp. 2672–2680 (2014)

2. Pumarola, A., Agudo, A., Martinez, A.M., Sanfeliu, A., Moreno-Noguer, F.: GAN-imation: anatomically-aware facial animation from a single image. In: Ferrari, V., Hebert, M., Sminchisescu, C., Weiss, Y. (eds.) ECCV 2018. LNCS, vol. 11214, pp. 835–851. Springer, Cham (2018). https://doi.org/10.1007/978-3-030-01249-6_50
3. Wang, X., Shrivastava, A., Gupta, A.: A-fast-RCNN: hard positive generation via adversary for object detection. In: Proceedings of the IEEE Conference on Computer Vision and Pattern Recognition, pp. 2606–2615 (2017)
4. Ledig, C., et al.: Photo-realistic single image super-resolution using a generative adversarial network. In: Proceedings of the IEEE Conference on Computer Vision and Pattern Recognition, pp. 4681–4690 (2017)
5. Szegedy, C., Vanhoucke, V., Ioffe, S., Shlens, J., Wojna, Z.: Rethinking the inception architecture for computer vision. In: Proceedings of the IEEE Conference on Computer Vision and Pattern Recognition, pp. 2818–2826 (2016)
6. Kingma, D.P., Welling, M.: Auto-encoding variational Bayes. In: Proceedings of the International Conference on Learning Representations (2014)
7. Zhu, J.Y., Park, T., Isola, P., Efros, A.A.: Unpaired image-to-image translation using cycle-consistent adversarial networks. In: Proceedings of the IEEE International Conference on Computer Vision, pp. 2223–2232 (2017)
8. Liu, M.Y., Tuzel, O.: Coupled generative adversarial networks. In: Advances in Neural Information Processing Systems, pp. 469–477 (2016)
9. Liu, M.Y., Breuel, T., Kautz, J.: Unsupervised image-to-image translation Networks. In: Advances in Neural Information Processing Systems, pp. 700–708 (2017)
10. Huang, X., Liu, M.-Y., Belongie, S., Kautz, J.: Multimodal unsupervised image-to-image translation. In: Ferrari, V., Hebert, M., Sminchisescu, C., Weiss, Y. (eds.) ECCV 2018. LNCS, vol. 11207, pp. 179–196. Springer, Cham (2018). https://doi.org/10.1007/978-3-030-01219-9_11
11. Choi, Y., Choi, M., Kim, M., Ha, J.W., Kim, S., Choo, J.: StarGAN: unified generative adversarial networks for multi-domain image-to-image translation. In: Proceedings of the IEEE Conference on Computer Vision and Pattern Recognition, pp. 8789–8797 (2018)
12. Lee, H.-Y., Tseng, H.-Y., Huang, J.-B., Singh, M., Yang, M.-H.: Diverse image-to-image translation via disentangled representations. In: Ferrari, V., Hebert, M., Sminchisescu, C., Weiss, Y. (eds.) ECCV 2018. LNCS, vol. 11205, pp. 36–52. Springer, Cham (2018). https://doi.org/10.1007/978-3-030-01246-5_3
13. Xiao, T., Hong, J., Ma, J.: ELEGANT: exchanging latent encodings with GAN for transferring multiple face attributes. In: Ferrari, V., Hebert, M., Sminchisescu, C., Weiss, Y. (eds.) ECCV 2018. LNCS, vol. 11214, pp. 172–187. Springer, Cham (2018). https://doi.org/10.1007/978-3-030-01249-6_11
14. Lample, G., Zeghidour, N., Usunier, N., Bordes, A., Denoyer, L.: Fader networks: manipulating images by sliding attributes. In: Advances in Neural Information Processing Systems, pp. 5967–5976 (2017)
15. Liang, X., Gong, K., Shen, X., Lin, L.: Look into person: joint body parsing & pose estimation network and a new benchmark. IEEE Trans. Pattern Anal. Mach. Intell. **41**, 871–885 (2019)
16. Ma, L., Jia, X., Sun, Q., Schiele, B., Tuytelaars, T., Van Gool, L.: Pose guided person image generation. In: Advances in Neural Information Processing Systems, pp. 406–416 (2017)
17. Gong, K., Liang, X., Zhang, D., Shen, X., Lin, L.: Look into person: self-supervised structure-sensitive learning and a new benchmark for human parsing. In: Proceedings of the IEEE Conference on Computer Vision and Pattern Recognition, pp. 932–940 (2017)

18. Gulrajani, I., Ahmed, F., Arjovsky, M., Dumoulin, V., Courville, A.C.: Improved training of Wasserstein GANs. In: Advances in Neural Information Processing Systems, pp. 5767–5777 (2017)
19. He, Z., Zuo, W., Kan, M., Shan, S., Chen, X.: Arbitrary facial attribute editing: only change what you want. arXiv preprint arXiv:1711.10678 (2017)
20. Zhong, Z., Zheng, L., Zheng, Z., Li, S., Yang, Y.: Camera style adaptation for person re-identification. In: Proceedings of the IEEE Conference on Computer Vision and Pattern Recognition, pp. 5157–5166 (2018)
21. Ma, L., Sun, Q., Georgoulis, S., Van Gool, L., Schiele, B., Fritz, M.: Disentangled person image generation. In: Proceedings of the IEEE Conference on Computer Vision and Pattern Recognition, pp. 99–108 (2018)
22. Ristani, E., Solera, F., Zou, R., Cucchiara, R., Tomasi, C.: Performance measures and a data set for multi-target, multi-camera tracking. In: Hua, G., Jégou, H. (eds.) ECCV 2016. LNCS, vol. 9914, pp. 17–35. Springer, Cham (2016). https://doi.org/10.1007/978-3-319-48881-3_2
23. Zheng, Z., Zheng, L., Yang, Y.: Unlabeled samples generated by GAN improve the person re-identification baseline in vitro. In: Proceedings of the IEEE International Conference on Computer Vision, pp. 3754–3762 (2017)
24. Deng, W., Zheng, L., Ye, Q., Kang, G., Yang, Y., Jiao, J.: Image-image domain adaptation with preserved self-similarity and domain-dissimilarity for person re-identification. In: Proceedings of the IEEE Conference on Computer Vision and Pattern Recognition, pp. 994–1003 (2018)
25. He, K., Zhang, X., Ren, S., Sun, J.: Deep residual learning for image recognition. In: Proceedings of the IEEE Conference on Computer Vision and Pattern Recognition, pp. 770–778 (2016)
26. Liu, J., Ni, B., Yan, Y., Zhou, P., Cheng, S., Hu, J.: Pose transferrable person re-identification. In: Proceedings of the IEEE Conference on Computer Vision and Pattern Recognition, pp. 4099–4108 (2018)
27. Qian, X., et al.: Pose-normalized image generation for person re-identification. In: Ferrari, V., Hebert, M., Sminchisescu, C., Weiss, Y. (eds.) ECCV 2018. LNCS, vol. 11213, pp. 661–678. Springer, Cham (2018). https://doi.org/10.1007/978-3-030-01240-3_40
28. Lin, Y., Zheng, L., Zheng, Z., Wu, Y., Yang, Y.: Improving person re-identification by attribute and identity learning. arXiv preprint arXiv:1703.07220 (2017)
29. Chen, X., Xu, C., Yang, X., Tao, D.: Attention-GAN for object transfiguration in wild images. In: Ferrari, V., Hebert, M., Sminchisescu, C., Weiss, Y. (eds.) ECCV 2018. LNCS, vol. 11206, pp. 167–184. Springer, Cham (2018). https://doi.org/10.1007/978-3-030-01216-8_11
30. Zheng, L., Shen, L., Tian, L., Wang, S., Wang, J., Tian, Q.: Scalable person re-identification: a benchmark. In: Proceedings of the IEEE International Conference on Computer Vision, pp. 1116–1124 (2015)
31. Kurnianggoro, L., Jo, K. H.: Identification of pedestrian attributes using deep network. In: IECON 2017–43rd Annual Conference of the IEEE Industrial Electronics Society, pp. 8503–8507 (2017)
32. Sun, C., Jiang, N., Zhang, L., Wang, Y., Wu, W., Zhou, Z.: Unified framework for joint attribute classification and person re-identification. In: Kůrková, V., Manolopoulos, Y., Hammer, B., Iliadis, L., Maglogiannis, I. (eds.) ICANN 2018. LNCS, vol. 11139, pp. 637–647. Springer, Cham (2018). https://doi.org/10.1007/978-3-030-01418-6_63
33. Liu, H., Wu, J., Jiang, J., Qi, M., Bo, R.: Sequence-based person attribute recognition with joint CTC-attention model. arXiv preprint arXiv:1811.08115 (2018)

Learning Spatial-Aware Cross-View Embeddings for Ground-to-Aerial Geolocalization

Rui Cao[1], Jiasong Zhu[3], Qing Li[4], Qian Zhang[1], Qingquan Li[3], Bozhi Liu[2], and Guoping Qiu[2,4(✉)]

[1] International Doctoral Innovation Centre and School of Computer Science, University of Nottingham, Ningbo, China
[2] College of Information Engineering and Guangdong Key Laboratory of Intelligent Information Processing, Shenzhen University, Shenzhen, China
guoping.qiu@nottingham.ac.uk
[3] Shenzhen Key Laboratory of Spatial Smart Sensing and Services, Shenzhen University, Shenzhen, China
[4] School of Computer Science, University of Nottingham, Nottingham, UK

Abstract. Image-based geolocalization is an important alternative to GPS-based localization in GPS-denied situations. Among them, ground-to-aerial geolocalization is particularly promising but also difficult due to drastic viewpoint and appearance differences between ground and aerial images. In this paper, we propose a novel spatial-aware Siamese-like network to address the issue by exploiting the spatial transformer layer to effectively alleviate the large view variation and learn location discriminative embeddings from the cross-view images. Furthermore, we propose to combine the triplet ranking loss with a simple and effective location identity loss to further enhance the performances. We test our method on a publicly available dataset and the results show that the proposed method outperforms state-of-the-art by a large margin.

Keywords: Image-based localization · Cross-view geolocalization · Image retrieval · Deep metric learning

1 Introduction

Localization is an essential component for many location-based services (LBS). Traditional outdoor localization methods rely on global positioning system (GPS). However, they do not function properly in urban areas with high-rise buildings. Image-based localization methods are regarded as promising alternatives in GPS-denied situations. They are direct and compatible with human understanding. Besides, they can also be used for place recognition when we simply want to find out where a photo is taken.

The image-based geolocalization is normally treated as an image retrieval problem. The predicted location of a query image is set as the geographical coor-

© Springer Nature Switzerland AG 2019
Y. Zhao et al. (Eds.): ICIG 2019, LNCS 11901, pp. 57–67, 2019.
https://doi.org/10.1007/978-3-030-34120-6_5

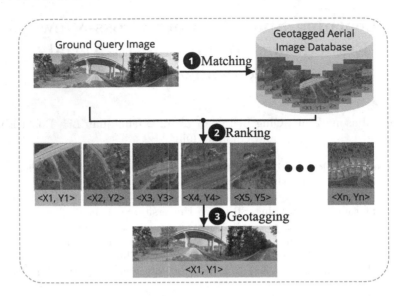

Fig. 1. Illustration of image retrieval-based ground-to-aerial geolocalization task. The goal is to find where a ground query image is taken. Normally, there are three steps: (1) matching the query image with geotagged aerial images, (2) ranking retrieved results by similarity, (3) geotagging the query image with the location of the most similar aerial image.

dinate of the most similar image from a geotagged image database. The image-based geolocalization methods can be categorized into ground-to-ground geolocalization [1,5,8,18] and ground-to-aerial geolocalization [2,3,7,10,12,13,15–17]. For ground-to-ground geolocalization, the reference image database is composed of ground-level images. This method requires a large number of accurately geotagged ground images to cover the earth surface which are difficult to acquire. While for ground-to-aerial geolocalization, the image database is made up of overhead images, as illustrated in Fig. 1. This relieves the difficulty of building a large geotagged image database because aerial images can cover the whole areas of the earth surface and are usually ready with precise geographical coordinates.

However, ground-to-aerial geolocalization is extremely difficult since the ground-level images (horizontal view) are taken in a very different perspective compared with overhead images (nadir view). The drastic viewpoint variation results in small overlap areas between the two types of images, and also leads to problems like dramatic appearance differences, occlusion, and illumination variation.

The existing cross-view image geolocalization works tackle the issues through matching building facades [2], line segments [12], and handcrafted features [3]. Some works exploit extra information such as land cover maps [13]. With the development of deep learning, the powerful deep features are also utilized [15,17].

Recently, deep metric learning has also been used to address the problem and shown to be an effective paradigm for cross-view image geolocalization [7,10,16]. It exploits the discriminative power of deep neural networks to embed cross-view images into a joint embedding metric space in which simple metrics like the Euclidean distance can be directly used to measure the semantic similarity between them.

In this paper, following the deep metric learning paradigm, we propose a novel spatial-aware Siamese-like network to address the ground-to-aerial geolocalization problem. Compared with previous methods, we exploit the spatial transformer layer (STL) [9] to tackle the large view variation problem, which can help to learn location discriminative embeddings for the challenging task. Besides, we design a loss that combines the triplet ranking loss with a simple and effective location identity loss to train the proposed network, which further enhances the geolocalization performances. We have conducted extensive experiments on a publicly available dataset of cross-view image pairs to test our method, and the results show that the proposed method has significantly outperformed the state-of-the-art.

The remainder of the paper is organized as follows. We firstly formulate the ground-to-aerial geolocalization problem in Sect. 2. In Sect. 3, we describe the proposed spatial-aware Siamese-like network as well as the loss function we use to train the network. In Sect. 4, we elaborate the experiments and analyze the results. Finally, we conclude in Sect. 5.

2   Problem Statement

The goal of ground-to-aerial geolocalization is to find the location ℓ_g^i where a ground query image I_g^i is taken, given a geotagged overhead image database $\mathcal{I}_r = \{\langle I_r^k, \ell_r^k \rangle\}$ $(k = 1, 2, ..., N)$ as reference:

$$\ell_g^i = h(I_g^i, \mathcal{I}_r). \tag{1}$$

As illustrated in Fig. 1, the task can be formulated as an image retrieval problem, i.e. finding an aerial image I_r^* from the reference image database $\mathcal{I}_r$, which is the most similar to the query image I_g^i. Then the center location ℓ_r^* of I_r^* would be regarded as the estimated location $\hat{\ell}_g^i$ of I_g^i:

$$\hat{\ell}_g^i = \ell_r^*, \text{ where } I_r^* = \arg \min_k d(f_g(I_g^i), f_r(I_r^k)), \tag{2}$$

where f_g and f_r are functions that map the ground and overhead images into a comparable embedding space $\mathbb{R}^F$ respectively, and $d(\cdot, \cdot)$ is a metric distance measuring the dissimilarity of two embedding vectors in the space. Therefore, the key to the problem is matching the ground image to the most similar aerial image.

3   Methodology

In this section, we describe our proposed network that can effectively learn spatial-aware cross-view embedding features for ground-to-aerial image matching, and we also elaborate the loss functions we use to train the network.

3.1   Spatial-Aware Siamese-Like Network for Cross-View Image Matching

Considering that the ground and aerial images are captured at totally different views and their visual contents are of large difference, we propose a network to learn spatial-aware cross-view features that can match them effectively. The architecture of the proposed network is shown in Fig. 2. It is a Siamese-like network, consisting of two sub-networks of the same structure but different parameters, whereas traditional Siamese network has two identical sub-networks of the same structure and weights. The goal of the proposed network is to learn two embedding functions $f(x; \theta_g), f(x; \theta_r) : \mathbb{R}^I \to \mathbb{R}^F$ that map the input ground and overhead images to a joint feature space so that semantically similar ground-aerial image pairs in $\mathbb{R}^I$ are metrically close in $\mathbb{R}^F$. The two functions parameterized by θ_g and θ_r represent the two sub-networks respectively.

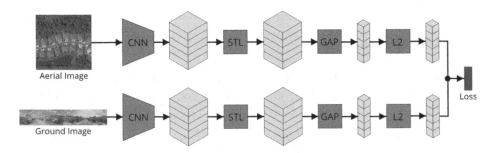

Fig. 2. Overview of the proposed network. (STL: spatial transformer layer, GAP: global average pooling, L2: L_2-normalization)

Each sub-network is fully convolutional, employing the convolutional parts of the AlexNet [11] or VGG16 [14] as the basic networks for feature extraction. The spatial transformer layer (STL) [9] is appended to the last layer of each sub-network to enable them the capacity of learning spatial transformations automatically which allow the network learns the best representation for cross-view matching. The output feature maps are then vectorized by global average pooling (GAP) to obtain fixed-length feature vectors, which are L_2-normalized to compute the final loss. In the testing phase, the L_2-normalized embedding vector can be exploited as the representative feature for cross-view image matching.

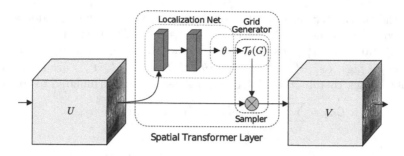

Fig. 3. Overview of the spatial transformer layer (STL).

Spatial Transformer Layer (STL). STL [9] can warp the input feature map via specified transformation. In this paper, affine transformation is exploited, which can alleviate the large view variation between cross-view images through learning translation, rotation, scale, and skew transformations, as well as cropping. STL is a learnable differentiable module that learns a spatial transformation during training and can be applied to an input feature map in a single forward manner. The architecture of it is shown in Fig. 3. It is composed of three components, i.e. a localization net, a grid generator, and a sampler. The localization net f_{loc} learns the parameters θ of the spatial transformation $\mathcal{T}_\theta$, $\theta = f_{loc}(U)$. The grid generator is then used to generate sampling points $\mathcal{T}_\theta(G)$ from the input feature map U, given the regular grid $G = \{G_i\}$ of the output feature map V:

$$\begin{pmatrix} x_i^s \\ y_i^s \end{pmatrix} = \mathcal{T}_\theta(G_i) = M_\theta \begin{pmatrix} x_i^t \\ y_i^t \\ 1 \end{pmatrix} \tag{3}$$

where (x_i^s, y_i^s) is the source point in the input feature map U, while (x_i^t, y_i^t) is the target point in the output feature map V. M_θ is a 2×3 affine transformation matrix with 6 parameters. Two fully connected layers with 32 neurons each are used for the localization net to regress the 6 parameters. The sampler generates the final output feature map V by sampling from the input feature map U according to the generated grid $\mathcal{T}_\theta(G)$ from the grid generator.

3.2 Loss Function

The overall loss function we use to train our network consists of two components, i.e. the *triplet ranking loss* $\mathcal{L}_{tri}$ and the *location identity loss* $\mathcal{L}_{id}$:

$$\mathcal{L} = \mathcal{L}_{tri} + \lambda \mathcal{L}_{id}, \tag{4}$$

where λ controls the relative importance of the two losses.

Triplet Ranking Loss. The triplet loss characterizes a relative similarity ranking order between image triplets. It has been demonstrated to be effective for cross-view image matching [7, 16]. The goal of the loss is to make an image closer to its paired cross-view image than any other cross-view images.

Let the metric that measures the similarity of images in the embedding space $\mathbb{R}^F$ be squared Euclidean distance $d(x, y) = \|x - y\|_2^2$. Then, for a triplet of images I_i^a (anchor), I_j^p (positive), and I_j^n (negative), l_a, l_p, l_n are their corresponding geotags, $\langle x_i^a, x_j^p, x_j^n \rangle = \langle f(I_i^a; \theta_i), f(I_j^p; \theta_j), f(I_j^n; \theta_j) \rangle$ are corresponding embeddings, thereby the triplet ranking loss can be formulated as follows:

$$\mathcal{L}_{tri} = \sum_{i,j} \sum_{\substack{a,p,n \\ l_a=l_p \neq l_n}} [d(x_i^a, x_j^p) - d(x_i^a, x_j^n) + \alpha]_+, \tag{5}$$

where $[x]_+$ represents $max(x, 0)$, α denotes the margin. $d(x_i^a, x_j^p)$ and $d(x_i^a, x_j^n)$ are the distances between the anchor-positive and anchor-negative pairs respectively. i and j are indicators for different types of images, $i \neq j$ and $i, j \in \{g, r\}$, with g for ground image and r for overhead reference image.

For cross-view geolocalization, there is only one paired cross-view image as the positive sample for each anchor image. In terms of the anchor image type, image triplets can be categorized into ground-to-aerial type $\langle g, r, r \rangle$ and aerial-to-ground type $\langle r, g, g \rangle$. We exhaust all the valid triplets within a mini-batch to compute the loss during training following previous works [7,16]. There would be $2m(m-1)$ valid triplets within each mini-batch of m cross-view image pairs, with $m(m-1)$ for $\langle g, r, r \rangle$ and $\langle r, g, g \rangle$ triplets each.

Location Identity Loss. For every cross-view image pair, the ground and overhead images represent the same location. However, they present very different visual contents since they are captured in totally different views. Inspired by the idea of deep feature consistency in facial attribute manipulation [6] and considering the uniqueness of the scene of every spatial location, we introduce a new location identity loss to enforce the feature consistency of cross-view image pairs, which can help preserve the unique identity of each place and learn location discriminative features. It tries to minimize the distance between the embedding features of two paired cross-view images captured at the same location. The formulation of the location identity loss is shown as follows:

$$\mathcal{L}_{id} = \sum_{k} \left\| f(I_g^k; \theta_g) - f(I_r^k; \theta_r) \right\|_2^2, \tag{6}$$

where I_g^k and I_r^k are the k-th paired ground and overhead images respectively. Intuitively, the learned embedding functions $f(x; \theta_g)$ and $f(x; \theta_r)$ should make the cross-view image pairs as close to each other as possible in the embedding space $\mathbb{R}^F$.

4 Experiments

4.1 Dataset

The CVUSA dataset [19] includes image pairs of panoramic street view images and overhead aerial imagery collected across the US. There are 35,532 image pairs for training, and 8,884 pairs for testing. The size of the ground panoramas is 224×1232, while the aerial image size is 750×750.

4.2   Experiment Setup

The street view images are resized to 112×616 in both the training and testing phase. While the aerial images are firstly resized to 300×300, then randomly cropped to 256×256 and rotated by $90n$ ($n = 0, 1, 2, 3$) degrees in training phase, and they are directly resized to 256×256 in testing phase.

The networks are implemented based on the PyTorch framework. Adam optimizer is used to train the networks, with a learning rate of 0.00001 and a batch size of 20. The parameters of convolutional layers of the networks are initialized by corresponding base network weights pretrained on ImageNet [4], while the weights of STLs are initialized by identity transformation. The maximum training iteration is set to 20 epochs. The weight λ is empirically set to 0.005. The margin α of the triplet loss is empirically set to 0.2.

Evaluation Metric. We adopt the recall accuracy at top 1% as our evaluation metric, the same as previous works [7,16,17]. A query is regarded as correct if the corresponding aerial image of the given ground query image is within the top 1% retrieval results.

4.3   Results and Analysis

To evaluate the effectiveness of the proposed network, we compare it with baselines and previous methods. There are two major types of baselines, one is Siamese network which has two identical sub-networks with shared weights, the other is Siamese-like network with two sub-networks of the same structure but different weights. Both the Siamese and Siamese-like networks have the same structure as the proposed network, except the spatial transformer layers. AlexNet and VGG16 are used as backbone networks. In addition, the proposed network and baselines are all trained with the proposed loss which combines the triplet ranking loss and location identity loss.

The top 1% recall results of the proposed network and the baselines, and the previous state-of-the-art results (reported in [7]) are presented in Table 1. As it can be seen, the VGG16-based Siamese-like network outperforms the shared-weight Siamese network (VGG16-based) by 11.1%, and the proposed network further increases the accuracy by 1.2%, reaching 95.8%, with 4.4% higher than previous state-of-the-art result of 91.4% from CVM-Net-I [7] which also use VGG16 as backbone network. The results demonstrate the efficacy of our proposed method in improving the cross-view image matching accuracy.

Ablation Study. To further validate the effectiveness of our proposed network and the location identity loss, we conduct ablation study by comparing the top 1% recall results of different networks training with different losses. There are three network architectures, i.e. Siamese network, Siamese-like network, and our proposed spatial-aware Siamese-like network. Compared to the proposed

Table 1. Top 1% recall accuracy of the proposed network, baselines, and previous state-of-the-art methods.

	Recall@1%
Workman et al. [17]	34.3%
Zhai et al. [19]	43.2%
Vo and Hays [16]	63.7%
CVM-Net-II [7]	87.2%
CVM-Net-I [7]	91.4%
Siamese (AlexNet)	58.5%
Siamese (VGG16)	83.5%
Siamese-like (VGG16)	94.6%
Ours (VGG16)	**95.8%**

network, the baseline Siamese-like networks remove the spatial transformer layers, and the Siamese networks further share weights for the two sub-networks. They are trained under the triplet ranking loss $\mathcal{L}_{tri}$ only, or the proposed loss which combines the triplet ranking loss $\mathcal{L}_{tri}$ and location identity loss $\mathcal{L}_{id}$. We also compare the results of using different backbone networks, i.e. AlexNet and VGG16.

Table 2. Top 1% recall accuracy of different network architectures training with different losses. ($\lambda = 0.005$)

		Siamese	Siamese-like	Ours
AlexNet	$\mathcal{L}_{tri}$	57.9%	65.0%	71.1%
	$\mathcal{L}_{tri} + \lambda\mathcal{L}_{id}$	58.5%	67.7%	73.6%
	Increase (Δ)	+0.6%	+2.7%	+2.5%
VGG16	$\mathcal{L}_{tri}$	78.1%	92.9%	94.2%
	$\mathcal{L}_{tri} + \lambda\mathcal{L}_{id}$	83.5%	94.6%	95.8%
	Increase (Δ)	+5.4%	+1.7%	+1.6%

The top 1% recall accuracy of the three different networks training under different losses are shown in Table 2. It can be seen that the deeper the network, the better the results, since the VGG16-based networks perform significantly better than the AlexNet-based counterparts. Siamese-like networks outperform Siamese networks dramatically, which owes to the removal of shared-weight constraint and thus increasing the model capacity to effectively learn view-specific features.

The proposed networks with STLs further improve the results noticeably under both base networks, from 65.0% to 71.1% (6.1% increase) for AlexNet and from 92.9% to 94.2% (1.3% increase) for VGG16, showing the efficacy of

STLs in alleviating large view variation by explicitly learning spatial transformations. It is also interesting that the improvement is more significant when the backbone network is not deep enough (with 6.1% increase for AlexNet and 1.3% for VGG16). Furthermore, the proposed location identity loss can improve the performances on all the networks (as shown by the increase Δ in Table 2), which demonstrates its effectiveness in regularizing the networks learning location-discriminative deep features for cross-view image matching.

Qualitative Results. Some retrieval examples of the proposed network with the best performance (95.8%) are presented in Fig. 4. There are four typical scenes, i.e. *medium residential area*, *sparse residential area*, *open cropland*, and *forest road*. For each case, the ground query image is on the leftmost column, and the top five retrieval results are listed on the right, with the ground truth enveloped by orange box.

Fig. 4. Retrieval examples of four typical scenes (from top to bottom: *medium residential area*, *sparse residential area*, *open cropland*, and *forest road*). For each ground query image, the top 5 retrieved aerial images are presented, with the ground truth enveloped by orange box. (Color figure online)

We can see that, for all the four cases, the top-5 retrieved aerial images on the right all present very similar patterns and appearances, and are clearly drawn from the same scene categories respectively. It is even hard for human eyes to find the correct matches for the ground images. This further demonstrates the effectiveness of our proposed network and the corresponding training loss in learning visually discriminative features for cross-view image matching.

However, it should also be noted that it is almost impossible to distinguish the correct paired aerial image for the *forest road* case, since the retrieved results are all well matched with the ground query image visually, and there seems to be no clues to find the real match. This indicates the limitation of simple image retrieval-based ground-to-aerial geolocalization approach as it is incapable of distinguishing visually similar image scenes. The method can be used as a coarse localization approach. To achieve accurate localization in real world applications, extra supplementary sources of data are needed or query image sequence can be exploited to reduce the ambiguity of similar scenes.

5 Concluding Remarks

Image-based ground-to-aerial geolocalization is a promising approach for localization in GPS-denied situations. However, it is very challenging due to drastic viewpoint difference between the cross-view images. In this paper, we propose a novel spatial-aware Siamese-like network to address the problem, which exploits the spatial transformer layer to explicitly learn spatial transformations between the ground and overhead images to tackle the large view variation. Moreover, we propose to combine the triplet ranking loss and the simple and effective location identity loss to train the proposed network to further enhance the performances. We evaluate our method on a publicly available dataset of cross-view image pairs, and the results demonstrate that the proposed method has achieved state-of-the-art performances. In the future, we plan to utilize extra supplementary data sources and employ image sequences, in conjunction with cross-view image matching, to meet accurate geolocalization need in real world applications.

Acknowledgments. The authors acknowledge the financial support from the International Doctoral Innovation Centre, Ningbo Education Bureau, Ningbo Science and Technology Bureau, and the University of Nottingham. This work was supported in part by the UK Engineering and Physical Sciences Research Council [grant number EP/L015463/1], the National Natural Science Foundation of China (No. 41871329), the Shenzhen Future Industry Development Funding Program (No. 201607281039561400), the Shenzhen Scientific Research and Development Funding Program (No. JCYJ20170818092931604).

References

1. Arandjelovic, R., Gronát, P., Torii, A., Pajdla, T., Sivic, J.: NetVLAD: CNN architecture for weakly supervised place recognition. In: 2016 IEEE Conference on Computer Vision and Pattern Recognition, CVPR 2016, Las Vegas, NV, USA, 27–30 June 2016, pp. 5297–5307 (2016)
2. Bansal, M., Sawhney, H.S., Cheng, H., Daniilidis, K.: Geo-localization of street views with aerial image databases. In: Proceedings of the 19th ACM International Conference on Multimedia, pp. 1125–1128. ACM, New York (2011)
3. Chu, H., Mei, H., Bansal, M., Walter, M.R.: Accurate Vision-based Vehicle Localization using Satellite Imagery. arXiv:1510.09171 [cs] (2015)
4. Deng, J., Dong, W., Socher, R., Li, L.J., Li, K., Li, F.F.: ImageNet: a large-scale hierarchical image database. In: 2009 IEEE Computer Society Conference on Computer Vision and Pattern Recognition (CVPR 2009), 20–25 June 2009, Miami, Florida, USA, pp. 248–255 (2009)
5. Hays, J., Efros, A.A.: IM2GPS: estimating geographic information from a single image. In: 2008 IEEE Computer Society Conference on Computer Vision and Pattern Recognition (CVPR 2008), 24–26 June 2008, Anchorage, Alaska, USA (2008)

6. Hou, X., Shen, L., Sun, K., Qiu, G.: Deep feature consistent variational autoencoder. In: 2017 IEEE Winter Conference on Applications of Computer Vision (WACV), pp. 1133–1141 (2017)
7. Hu, S., Feng, M., Nguyen, R.M.H., Lee, G.H.: CVM-Net: cross-view matching network for image-based ground-to-aerial geo-localization. In: 2018 IEEE Conference on Computer Vision and Pattern Recognition, CVPR 2018, Salt Lake City, UT, USA, 18–22 June 2018, pp. 7258–7267 (2018)
8. Iscen, A., Tolias, G., Avrithis, Y.S., Furon, T., Chum, O.: Panorama to Panorama matching for location recognition. In: Proceedings of the 2017 ACM on International Conference on Multimedia Retrieval, ICMR 2017, 6–9 June 2017, Bucharest, Romania, pp. 392–396 (2017)
9. Jaderberg, M., Simonyan, K., Zisserman, A., Kavukcuoglu, K.: Spatial transformer networks. In: Advances in Neural Information Processing Systems 28: Annual Conference on Neural Information Processing Systems 2015, 7–12 December 2015, Montreal, Quebec, Canada, pp. 2017–2025 (2015)
10. Kim, D.K., Walter, M.R.: Satellite image-based localization via learned embeddings. In: 2017 IEEE International Conference on Robotics and Automation, ICRA 2017, Singapore, Singapore, 29 May–3 June 2017, pp. 2073–2080 (2017)
11. Krizhevsky, A., Sutskever, I., Hinton, G.E.: ImageNet classification with deep convolutional neural networks. In: Advances in Neural Information Processing Systems 25: 26th Annual Conference on Neural Information Processing Systems 2012, 3–6 December 2012, Lake Tahoe, Nevada, US, pp. 1106–1114 (2012)
12. Li, A., Morariu, V.I., Davis, L.S.: Planar structure matching under projective uncertainty for geolocation. In: Fleet, D., Pajdla, T., Schiele, B., Tuytelaars, T. (eds.) ECCV 2014. LNCS, vol. 8695, pp. 265–280. Springer, Cham (2014). https://doi.org/10.1007/978-3-319-10584-0_18
13. Lin, T.Y., Belongie, S.J., Hays, J.: Cross-view image geolocalization. In: 2013 IEEE Conference on Computer Vision and Pattern Recognition, Portland, OR, USA, 23–28 June 2013, pp. 891–898 (2013)
14. Simonyan, K., Zisserman, A.: Very Deep Convolutional Networks for Large-Scale Image Recognition. arXiv:1409.1556 [cs] (2014)
15. Tian, Y., Chen, C., Shah, M.: Cross-view image matching for geo-localization in urban environments. In: 2017 IEEE Conference on Computer Vision and Pattern Recognition, CVPR 2017, 21–26 July 2017, Honolulu, HI, USA, pp. 1998–2006 (2017)
16. Vo, N.N., Hays, J.: Localizing and orienting street views using overhead imagery. In: Leibe, B., Matas, J., Sebe, N., Welling, M. (eds.) ECCV 2016. LNCS, vol. 9905, pp. 494–509. Springer, Cham (2016). https://doi.org/10.1007/978-3-319-46448-0_30
17. Workman, S., Souvenir, R., Jacobs, N.: Wide-area image geolocalization with aerial reference imagery. In: 2015 IEEE International Conference on Computer Vision, ICCV 2015, 7–13 December 2015, Santiago, Chile, pp. 3961–3969 (2015)
18. Zamir, A.R., Shah, M.: Accurate image localization based on Google maps street view. In: Daniilidis, K., Maragos, P., Paragios, N. (eds.) ECCV 2010. LNCS, vol. 6314, pp. 255–268. Springer, Heidelberg (2010). https://doi.org/10.1007/978-3-642-15561-1_19
19. Zhai, M., Bessinger, Z., Workman, S., Jacobs, N.: Predicting ground-level scene layout from aerial imagery. In: 2017 IEEE Conference on Computer Vision and Pattern Recognition, CVPR 2017, 21–26 July 2017, Honolulu, HI, USA, pp. 4132–4140 (2017)

Real-Time Interpretation Method
for Shooting-Range Image Based on Position
Prediction

Lijun Zhong[1,2(✉)], Qifeng Yu[1,2], Jiexin Zhou[1,2], Xiaohu Zhang[3],
and Yani Lu[4]

[1] College of Aerospace Science and Engineering,
National University of Defense Technology, Changsha, Hunan, China
trand1986@hotmail.com
[2] Hunan Provincial Key Laboratory of Image Measurement and Vision
Navigation, Changsha, China
[3] Sun Yat-Sen University, Guangzhou, China
[4] No. 560, 14 Substation, Jiuquan, Gansu, China

Abstract. When the theodolite in the shooting-range tracks the target in real
time, the theodolite may randomly jitter, causing the target a large displacement
on the next frame. While dealing with a large displacement, the tracking
methods based on the window search are easy to lose the target, and the tracking
methods based on the full-image search are time-consuming. In this paper, an
improved tracking-learning-detection (TLD) framework combining Kernelized
Correlation Filters (KCF) and target position prediction is proposed to cope with
the large displacement. First an orthogonal polynomials optimal linear filter is
used to predict the position of the target on the next frame according to the rule
of the theodolite's angle, and then the KCF is used to track the target fast in this
prediction area, which can improve the success rate and speed of tracking. If the
tracking fails and the prediction area is in the image, the detector will detect the
target in the full image. Simulation experiments have demonstrated that the
position prediction algorithm based on the optimal linear filter can accurately
predict the target position and provide KCF with a more accurate search position. The algorithm consumes only 0.6 ms per frame, and the tracking accuracy
is better than TLD and KCF. The actual task verification of the shooting-range
has proved that our method can improve the automatic interpretation of the
shooting-range and reduce manual intervention.

Keywords: Real-time interpretation · Improved TLD · Position prediction ·
Optimal linear filter

1 Introduction

At present, targets' flight parameter measurement such as missiles in the shooting-
range mainly based on optical theodolites. The theodolites can acquire and store images
in real time and can also output images and video in real time too [1]. However, the
images acquired by theodolite are mainly used for scene monitoring and interpretation

© Springer Nature Switzerland AG 2019
Y. Zhao et al. (Eds.): ICIG 2019, LNCS 11901, pp. 68–80, 2019.
https://doi.org/10.1007/978-3-030-34120-6_6

after the test. We fail to make full use of the rich information of the images. Obviously, it is of great significance to implement real-time image interpretation and measurement for it can meet the needs of real-time control and can give full play to the optical measurement's characteristics of non-contact and abundant target information. It can make optical measurement play a more important role in shooting-range measurement and control, so it will certainly become the development direction of optical measurement of the shooting-range. Compared with post-processing, real-time processing has the following advantages [2]: (1) the commander can adjust the test scheme in real time according to the measured results, making the test more flexible; (2) the data acquired in real time can provide early warning information for the security control system to facilitate timely handling of emergencies; (3) it can greatly improve the work efficiency, for the test and analysis are carried out simultaneously, without centralized data processing afterward.

Images in shooting-range generally have the following characteristics:

1. Most parts of the images are sky which lack of texture information and have big noise;
2. Target moves fast;
3. Due to the motion of the camera and random jitter, the target irregularly has a large displacement on the image, even jump out of the image;
4. Since the distance is too long, the contour of the target in the image is not clear, and the target is relatively small in the image too;
5. Images have a low image contrast and a low signal-to-noise ratio.

Due to the long distance from the theodolite to the target, the optical measurement of the shooting-range generally gives the target position through the intersection by two or more theodolites. In order to ensure the accuracy of the intersection, it is necessary to accurately give the position of the homonymy point of the target. Therefore, the interpretation in the shooting-range is different from the so-called target tracking. It is not required to accurately locate the rectangle that contains the target. The general process in shooting-range is that the interpretation point is manually selected by the user at the first frame, and then the point is accurately located by the tracking algorithm on the subsequent frame, the point selected by the user is generally the aircraft head, the missile tail or another homonymy point. There are some main difficulties in the tracking problem: First, the target situation is complex and changeable, such as with a flame or not for a missile. Second, its posture may change anytime too and the observation angle changes with the target move forward which make the different parts can be observed by the theodolite. Third, motion blur caused by the high speed of the target and occlusion also makes it hard to locate the point too. In addition, there will be random jitter when the theodolite tracks the target, which will lead to a random large displacement or even jumping out of the image on the next frame. So it is very difficult to achieve real-time high-precision automatic tracking throughout the entire process.

Because of the random jitter, the tracking algorithms based on window search can easily lead to failure, and the tracking methods based on the full-image search are time-consuming which cannot meet the requirement of real-time interpretation. So it is of great significance to study how to cope with the tracking difficulties caused by the theodolite's jitter.

In this paper a new tracking method combining position prediction, KCF and improved TLD framework is proposed. The initial tracking area is provided by the position prediction algorithm while helps KCF to overcome the problem that it cannot adapt to large displacement. When the tracker fails, the detector is used for detection. The learner completes the target feature learning.

2 Relative Works

For object tracking is a basic classic problem in computer vision, there are many researchers have proposed many excellent algorithms one after another in recent years, and many excellent algorithms are proposed one after another, such as Struck, TLD, DSST, C-COT, ECO, CREST [4–9] and so on. The performances of these above methods' are all evaluated by the error of rectangle region compared to the truth value given by people, so it is important to adapt to the target's scaling and rotation changes for these algorithms. Of course, it is decided by the nature of object tracking. So it is inevitable for these methods that they cannot accurately locate the point specified by the user, which make these methods unsuitable for real-time interpretation in the shooting-range.

KCF [10] is a discriminative tracking method that uses a circular matrix around the target to acquire positive and negative samples and uses a ridge regression to train the target detector. KCF has a fast tracking speed and can meet real-time requirements. However, when the target position of the next frame cannot be predicted, the algorithm can only search the target on the current position, and it will fail if there is a large displacement on two continues frames.

The optical flow based method can location the homonymy point by optical flow with a large amount of computation. It has achieved good results on the tracking problem while there are rich textures on the image. But most parts of the image in the shooting range are sky. As is known to all, the sky lacks texture, so the performance of the tracking methods based on optical flow did not achieve good results in the shooting-range, and they are also hard to cope with the situation that the target lost in the image.

The methods based on template matching are commonly used methods for point tracking. First, a template centered at the feature point is selected; then pixel-by-pixel matching is performed on subsequent images, the region with the greatest similarity is selected as the tracking result. There is a contradiction between the search region and the time performance for these methods, that is, the larger the search region is, the more time-consuming it is, but with more possibility to adapt to the large displacement of the target; the smaller the search region is, the less time-consuming it is, but with less possibility to adapt to the large displacement of the target.

3   Real Time Interpretation Method in Shooting-Range

3.1   TLD Framework

Origin TLD Framework

Tracking-Learning-Detection [4] decomposes the tracking problem into four steps: tracking, detection, integration, and learning. TLD first constructs the scale space according to the scale changes that may occur. According to the captured target, the initial random fern classifier and nearest neighbor classifier are trained in the semi-supervised learning framework, and the threshold parameters are adjusted by the cross-validation method. In the process, the tracker and the detector run concurrently, and the integrator gives the final result according to the success signs and trust levels of the tracker and detector; if the target meets the motion continuity constraint in the time series, the learner will learn the target feature. It uses the structural constraint-based learning algorithm (PN-Learning) to learn the new appearance of the target online. thanks to the detector, it can recapture the target when the target reappears. It has achieved good results on target tracking.

Problems with TLD in Real-Time Interpretation

Although TLD has achieved good results in post-tracking, it has the following problems when used for real-time tracking:

1. The pyramid optical flow method is relatively slow for it needs to calculate the gradient pixel-by-pixel, which cannot meet the requirements of real-time performance;
2. Since the texture of the image is relatively weak, the optical flow method works not very well, and it cannot adapt to the large displacement too.
3. The TLD algorithm performs tracking and detection at the same time. When the reliability of the tracking result is high, the full-image detection is of no significance, which also causes a large waste of computational resource.

Improved TLD Framework

Aiming to the above problems of TLD in real-time interpretation, we make the following improvements to the TLD framework:

1. We replace the optical flow tracking method with KCF for a tracking speed improvement, and use position prediction to cope with the problem that KCF cannot adapt to the large displacement of the target;
2. The integrator in TLD is removed and the detection is performed only if necessary. If the reliability of the tracking result is high, the detection will not be performed for saving the running time of the algorithm. Only if the tracker track fails and the predicted position is within the image, detection is performed.
3. When the detection is performed, first, it is performed with an interval of multi-pixel, then performed pixel-by-pixel in the region where most possibly contain the target.

The improved TLD framework is shown in Fig. 1.

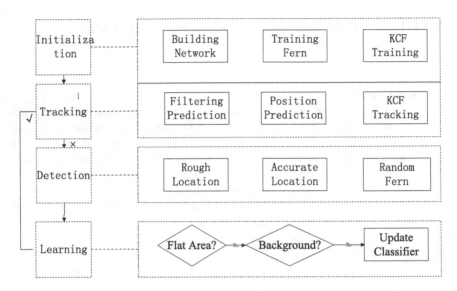

Fig. 1. Flow chart of improved online tracking.

The algorithm contains four steps: initialization, tracking, detection, and learning. The four steps contain some submodules as shown in the figure.

4   Position Prediction Based on Optimal Linear Filter

4.1   Observation Model

The theodolite records the angles of the theodolite's optical axis when it tracks the target, including the azimuth angle and elevation angle. These angles given by the theodolite are called respectively the original azimuth angle (OAA) and original elevation angle (OEA). For simplicity, they are called original angles (OAs) in this paper. The jitter mentioned in this paper will lead to the sudden change of these OAs. Based on the camera's intrinsic parameters, the target's position on the image, and the OAs, the directional angles from the theodolite's optical center to the target can be achieved, which is called the integrated azimuth angle (IAA) and the integrated elevation angle (IEA). For simplicity, they are called integrated angles (IAs) in this paper. As shown in Fig. 2, O is the origin of the coordinate system, which physically stands for the optical center of the theodolite. OA is the optical axis of the camera. OT is the line from the optical axis of the camera to target. $\angle xOC$ is the original azimuth angle (OAA). $\angle AOC$ is the original elevation angle (OEA). $\angle xOD$ is the integrated azimuth angle (IAA). $\angle TOD$ is the integrated elevation angle (IEA). Since the target's movement cannot mutate, the integrated angles do not mutate neither. By analyzing the rules of the variation of the IAs, we found that their shapes are smooth curves, so the IA of the next frame can be predicted by the previous IAs. Then the position of the target on the image can be solved on the basis of the predicted IA, the camera's intrinsic parameters and the OAs.

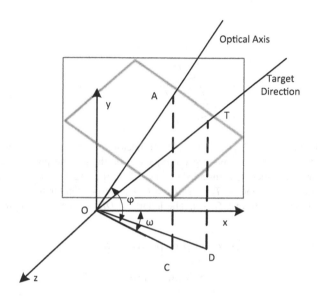

Fig. 2. Coordinate system and attitude angle definition schematic.

O is the original point of the coordinate system, which physically stands for the optical center of the theodolite. OA is the optical axis of the camera. OT is the line from the optical center of the camera to target. $\angle xOC$ is the original azimuth angle (OAA). $\angle AOC$ is the original elevation angle (OEA). $\angle xOD$ is the integrated azimuth angle (IAA). $\angle TOD$ is the integrated elevation angle (IEA).

We take the single-camera observation as an example to illustrate the application scenario of the prediction algorithm in this paper. As shown in Fig. 3, the trajectory of the target, such as a missile, can be considered to a certain curve S in a plane in a certain period of time. The straight line L is the projection of the target's trajectory on the ground plane. The curves S and line L are coplanar. The camera is installed on one side of the target's trajectory. C is the theodolite's optical center. T is the target, and E is the vertical projection of T on L. Line CD is perpendicular to line L and D is the perpendicular foot. The length of CD is d. The length of DE is l, and the length of TE is h.

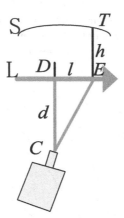

Fig. 3. Schematic of camera observation. S is the trajectory of the target, such as a missile. The straight line L is the projection of the target's trajectory on the ground plane. The curves S and line L are coplanar. The camera C is installed on one side of the target's trajectory. T is the target, and E is the vertical projection of T on L. Line CD is perpendicular to line L and D is the perpendicular foot. The length of CD is d. The length of DE is l, and the length of TE is h.

Then we define two angles α and β respectively as

$$\alpha = \tan^{-1}(l/d) \tag{1}$$

$$\beta = \tan^{-1}\left(h/\sqrt{(l^2 + d^2)}\right). \tag{2}$$

Please note that the first paragraph of a section or subsection is not indented. The first paragraphs that follows a table, figure, equation etc. does not have an indent, either.

Subsequent paragraphs, however, are indented.

4.2   Angular Rules Analysis and Prediction Model

It can be seen from the aforementioned definition that there is only one fixed difference between α and the IAA, and β equals IEA, from which we can analyze the rules of α and β through the IAs. Similarly, the rules of IA can also be analyzed by the rules of α and β, if l, d, h are known. Taking derivative from (4 1) with respect to l, we obtain

$$\Delta\alpha = \frac{\Delta l}{l^2/d + d}. \tag{3}$$

$\Delta l = v*t$ is the target's horizontal displacement between two frames. The initial value can be obtained from the speed and the photo-frequency. Assuming the speed of the target is 300 m/s and the photo-frequency is 25FPS, then Δl is 12 m. When d is 10 km, the variation of α per frame is about 1 mrad.

Derivation from (4 2) for l and h, we obtain

$$\Delta\beta = \frac{\sqrt{(l^2+d^2)}\Delta h}{l^2+d^2+h^2} + \frac{2l\Delta l}{(l^2+d^2+h^2)*\sqrt{(l^2+d^2)}} \tag{4}$$

According to the analysis above, if the target speed is constant, when the target is at the perpendicular foot D, the azimuth angle changes fastest, the farther away from the perpendicular, the slower. The change of the azimuth angle about the perpendicular symmetry, and the azimuth angle changes monotonously; The elevation angles achieve the maximum value at the perpendicular foot if the target height is consistent. In order to analyze the change rule intuitively, in the simulation, the target is set to have a uniform linear motion. In the simulation, the target speed is set to be 250 m/s; the vertical distance of the theodolite from the target trajectory is 10 km; the photo-frequency is 25FPS.

The curve of the IA and the first difference of IA curve (FDIAA and FDIEA) are shown in Fig. 4.

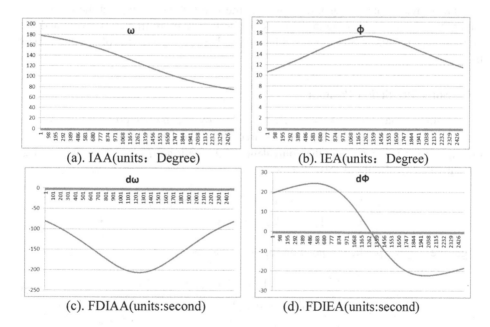

(a). IAA(units： Degree)　　　　　　　(b). IEA(units： Degree)

(c). FDIAA(units:second)　　　　　　(d). FDIEA(units:second)

Fig. 4. Diagram of integrated angles and its first difference curve. (a) is the curve of IAA, and it is a monotonic curve while target moves in one direction, (b) is the curve of IEA, and its shape is like a parabola going downwards, (c) is the curve of FDIEA, and its shape is like a parabola going upwards, (d) is the curve of FDIEA, and its shape is like a sinusoid.

It can be seen from the above simulation results that under the premise of uniform linear motion, the IA show a continuous smooth change trend, which can be fitted by a polynomial curve. In a certain time range, we can think that the target movement satisfies the uniform linear motion requirement, so the curve fitting can be used to predict the IA of the next frame.

Since l is unknown in the actual scene, and the angles given by the theodolite inevitably contain random noises, with time as the dependent variable, the orthogonal polynomial optimal linear filter [12] is used to extrapolate the integrated angular sequence to predict the IA of the next frame The prediction formula is

$$x_{N+a} = \sum_{t=1}^{N} [\frac{1}{N} + 12 * \left(t - \frac{N+1}{2}\right) * \frac{(a + \frac{N-1}{2})}{N * (N^2 - 1)}] * x_t, \tag{5}$$

where N is the filtering window. If $a < 0$, the formula is an interpolation formula. If $a > 0$, the formula is an extrapolation formula. When the amount of data is 2, we use the equal space extrapolation to predict the IA directly. When the amount of data is less than N and greater than 2, we use all current data to predict the IA. When the data amount is greater than or equal to N, we use the latest N data to predict the IA. In this paper, N is 31. So only if the initial angle of the two or more frames is given, and the prediction algorithm of this paper is able to work based on the results of the first two or more frames.

4.3   Solve Prediction Position

It is assumed that the IAA and IEA predicted by the model proposed above are $\alpha_{t+1}, \beta_{t+1}$, respectively. The actual measured OAA and OEA are $\alpha'_{t+1}, \beta'_{t+1}$, and the principal point is Cx, Cy, and the equivalent focal length are Fx, Fy. Then we can use the calculation formula of miss distance

$$\begin{aligned} \alpha_{t+1} &= \alpha'_{t+1} + (x - Cx)/(Fx * \cos(\beta'_{t+1})) \\ \beta_{t+1} &= \beta'_{t+1} + (Cy - y)/Fy \end{aligned} \tag{6}$$

to get the target's prediction position on the image as

$$\begin{aligned} x &= \left(\alpha_{t+1} - \alpha'_{t+1}\right) * \cos\left(\beta'_{t+1}\right) * Fx + Cx \\ y &= Cy - \left(\beta_{t+1} - \beta'_{t+1}\right) * Fy. \end{aligned} \tag{7}$$

The point (x, y) obtained in the Eq. (7) is the predicted position in $t + 1$ frame. If the point (x, y) is located in the image, the KCF is used to track the point as the center. If the point is outside of the image, the frame is skipped to avoid an erroneous update of the tracker.

5 Experiments Verification

The effectiveness of the position prediction algorithm and the proposed tracking algorithm is validated by simulating the tracking scene of the theodolite in the shooting-range. The simulation platform is Windows 7 and Microsoft Visual Studio 2008. The processor was an Intel(R) CoreTMi7-6820HQ 2.7 GHz.

5.1 Simulation Condition

According to the accuracy of the equipment and similar target's trajectory, a set of camera parameters and target trajectory are simulated. The theodolite coordinates are set to be (1000, 2000, 3000); the equivalent focal length is 62500, and image resolution is 720 * 576. The target's trajectory is set to be a parabola, moving from the point (−1500, 8237.5, 3500) to the point (4992, 8240, 18494), with the highest point coordinates to be (−5000, 9800, 11000). The simulation is set to be 25 frames per second, of 2500 frames, that is, the movement lasted 100 s.

In order to simulate the actual situation of the range theodolite tracking, we added a Gaussian error with a mean of 0 to the original azimuth angle and the original elevation angle, with 5 s of root mean square (RMS); In order to validate the tracking performance of the algorithm, it is necessary to make sure that the position of target each frame on the image is different. Therefore, we added a random offset to the target position on the image from the image center. This is also in accordance with the characteristics of the theodolite tracking to ensure the target is near the center of the image. For every 100 frames, one frame is selected to be added a big random displacement to emulate theodolite jitter. The angle change caused by the random displacement is then superimposed to the original angle, which simulates the random jitter of the theodolite to test the performance of the proposed position prediction algorithm.

The simulated adjacent frame target displacement is shown in Fig. 5.

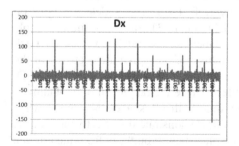 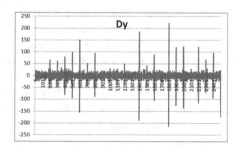

Fig. 5. Schematic diagram of target position change (unit: pixel). It can be seen from the figure that the jitter of the target between two adjacent frames is mostly within 20 pixels. The biggest jitter is 200 pixels.

It can be seen from the figure that the jitter of the target between two adjacent frames is mostly within 20 pixels, which conforms to the motion rule of the shooting range target. The occasional large jitter simulates the effect of camera jitter in the tracking process in the actual scene.

5.2   Performance Assessment

Performance of Prediction

The prediction algorithm predicts the possible position of the target based on the IA. Then we compare the predicted position with the actual position of the target and obtain the prediction bias to evaluate the performance of the prediction algorithm. The result is shown in Fig. 6.

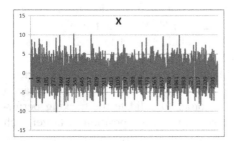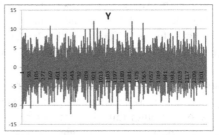

Fig. 6. The deviation of the prediction position to the truth value (unit: pixel).

It can be easily seen from the figure that the difference between the predicted position and the actual position of the target is within ±10 pixels, which proves that the prediction algorithm is effective.

Performance of Tracking

KCF can stably track the target when the target displacement is small, but the target will be lost when the target displacement is large. For TLD, track loss occasionally happens when the displacement is large, however, the target can be detected again on the next frame. In the simulation above, the TLD fails in 31 frames during the whole tracking process. The method proposed in this paper can accurately and quickly track the target. Since KCF cannot track the target stably, we only give the accuracy of the TLD algorithm and our method and remove the failure frames from the result of TLD.

The positioning accuracy is shown in Table 1. Since the target can only obtain the integer pixel precision when generating the image, there is a certain offset when compared to the ground truth. The tracking results demonstrate the effectiveness of the proposed algorithm in dealing with the large displacement brought in from the random jitter in the shooting range.

Table 1. Tracking accuracy

v	Our method	TLD
X (pixel)	0.33	7.6
Y (pixel)	0.29	5.5

The detailed results are shown in Fig. 7.

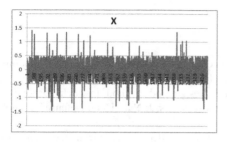 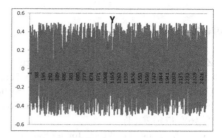

Fig. 7. Error of target tracking result (unit: pixel)

Performance of Time

The average time cost per frame was counted to evaluate the time performance of the tracking algorithm. The statistical results are shown in Table 2. Due to the tracking failure, we only counted 98 frames for KCF and all 2500 frames for TLD and our method.

Table 2. Time cost of track algorithm

Method	KCF	Ours	TLD
Time (ms)	2.15	0.60	23

It is not difficult to see from the above results that the speed of our method is faster than that of TLD and KCF, because the proposed algorithm gives the predicted position, which reduces the matching range and improves the success rate at the same time. Therefore, it takes less time than KCF. Since the tracking success rate is improved and the full-image detection after loss is avoided, our method is faster than TLD.

6 Conclusion

Aiming at the difficulties caused by the random jitter of the theodolite, a method combining position prediction, KCF and TLD is proposed. The simulation results demonstrated that the proposed method can accurately predict the position of the target based on the IA, and improved the success rate of KCF tracking, and can cope with the jitter of the theodolite effectively in the shooting-range. It only cost 0.6 ms per frame. The actual use in the shooting range demonstrated that the proposed algorithm can greatly improve the automation of the range interpretation, reduce the number of manual interventions. In addition, our method can combine with other tracking methods based on the window search effectively, our method can provide the target prediction position to the other tracking algorithms, which can effectively improve the ability to cope with the large displacement of the target of other algorithms and improve the robustness of the other tracking algorithms.

References

1. Zhang, X.: Researches on moving target detection and tracking for images of shooting ranges. National University of Defence Technology, Changsha (2006)
2. Guo, P., Ding, S., Tian, Z., et al.: System design and method research for optical measurement images real-time interpretation in test ranges. J. Natl. Univ. Defense Technol. **2**, 168–174 (2014)
3. Image Systems. TrackEye. Image Systems. TrackEye (2013). http://www.imagesystems.se/trackeye/
4. Hare, S., Saffari, A., Torr, P.H.S.: Struck: structured output tracking with kernels (2011)
5. Kalal, Z., Mikolajczyk, K., Matas, J.: Tracking-learning-detection. IEEE Trans. Softw. Eng. **34**(7), 1409–1422 (2011)
6. Danelljan, M., Häger, G., Khan, F., et al.: Accurate scale estimation for robust visual tracking. In: British Machine Vision Conference, Nottingham, 1–5 September 2014. BMVA Press (2014)
7. Danelljan, M., Robinson, A., Shahbaz Khan, F., Felsberg, M.: Beyond correlation filters: learning continuous convolution operators for visual tracking. In: Leibe, B., Matas, J., Sebe, N., Welling, M. (eds.) ECCV 2016. LNCS, vol. 9909, pp. 472–488. Springer, Cham (2016). https://doi.org/10.1007/978-3-319-46454-1_29
8. Danelljan, M., Bhat, G., Khan, F.S., et al.: ECO: efficient convolution operators for tracking. In: CVPR, vol. 1, no. 2, p. 3 (2017)
9. Song, Y., Ma, C., Gong, L., et al.: Crest: convolutional residual learning for visual tracking. In: 2017 IEEE International Conference on Computer Vision (ICCV), pp. 2574–2583. IEEE (2017)
10. Henriques, J.F., Caseiro, R., Martins, P., et al.: High-speed tracking with kernelized correlation filters. IEEE Trans. Pattern Anal. Mach. Intell. **37**(3), 583–596 (2015)
11. Tomasi, C., Kanade, T.: Detection and tracking of point features (1991)
12. Liu, L.: External Ballistic Measurement Data Processing. Beijing National Defense Industry Press (2002)

Spatial-Temporal Bottom-Up Top-Down Attention Model for Action Recognition

Jinpeng Wang and Andy J. Ma[✉]

Sun Yat-sen University, Guangzhou, China
majh8@mail.sysu.edu.cn

Abstract. Driven by the importance of capturing non-local information in video understanding, we propose Spatial-temporal Bottom-up Top-down Attention Module (STBTA). Features are processed across in multiple scales and then combined to best capture the spatial relationships associated with the region of interest and the surrounding environment in a complicated scene. Attention maps are used for adaptive feature refinement. STBTA can be plugged into any feedforward network architectures and is end-to-end trainable along with CNN. Extensive experiments on UCF101, HMDB51, Kinetics-400 datasets demonstrate that the proposed method can improve the performance for action recognition.

Keywords: Attention mechanism · Bottom-up top-down · Action recognition

1 Introduction

Non-local information is found to be of central importance for video understanding and image recognition [3,25]. By stacking a series of convolutional layers, CNN is capable of capturing non-local information [25]. However, each of the learned filters in a special layer operates in a local receptive field and consequently, each corresponding unit of the transformation output is unable to exploit global information outside of this local receptive field. This problem becomes more severe in the lower layers of the network [8].

Stacked Hourglass Networks (SHN) [14] repeats bottom-up, top-down processing with intermediate supervision to improve the performance of human pose estimation. A single pipeline with skip layers is used to preserve spatial information on each scale. Bottom-up top-down mechanism combines multi-scale information and filters operate in a non-local receptive field, can be considered as another way to capture non-local information. But videos/images own much irrelevant and background information [3]. Nevertheless, SHN considers multi-scale feature maps as the same without adaptive feature refinement.

Attention mechanism has been proven to be an efficient way to help the network see important parts and diminishes background responses [29]. On cognition theory, people focus sequentially on different parts of the scene to extract

© Springer Nature Switzerland AG 2019
Y. Zhao et al. (Eds.): ICIG 2019, LNCS 11901, pp. 81–92, 2019.
https://doi.org/10.1007/978-3-030-34120-6_7

Fig. 1. Visualization of some samples on UCF101 using Grad-CAM [16]. The ground-truth label is shown on the left of each input image. We compare the visualization results of the STBTA network(STBTA + Inception v3) with baseline(Inception v3). The Grad-CAM visualization highlight the class-specific discriminative regions, which is calculated for the last convolutional outputs. These visualizations show STBTA network focus on target objects more properly

relevant information [13]. Attention mechanism has been shown to achieve promising results of image caption generation, machine translation, image recognition [22,23,29].

Our goal is to increase representation power by using Bottom-up Top-down mechanism and attention mechanism: capturing non-local information both in space and temporal and focusing on important features. In this paper, we design two efficient module: Spatial Attention Module (SAM) and Temporal Attention Module (TAM), which is different from existing attention module. Based on these modules, we propose Spatial Bottom-up Top-down Attention Module (STBTA) as an efficient and a general component for capturing non-local spatial dependencies and to obtain more discriminative attentional maps. As shown in Fig. 1, an STBTA-integrated network focus on class-discriminative objects more properly compared with baseline. There are several advantages of using STBTA. (a) STBTA can generate temporal-wise statistics and spatial grids statistics, which increases the sensitivity to informative features and choose useful information. (b) Our method can be considered as a general module which is feedforward fashion and can be inserted into any CNNs directly. (c) STBTA can improve the visual recognition performance efficiently.

2   Related Work

Attention Mechanism. Human perception does not tend to process the whole scene at once and focus selectively on parts of the visual space to acquire information when and where it is needed [13]. Soft attention developed in recent work can be trained end-to-end for convolutional neural network [23]. CBAM [27] emphasizes meaningful features along two principal dimensions: channel and spatial axes. In our model, we first propose a Spatial Attention Module I(SAM I) based on SE Net [8], then we design a new grid-wise spatial attention module II(SAM II) with depthwise convolution. Otherwise, driven by the intuition that different frame play different role for action recognition, we design a fully new temporal attention model.

Residual Network. Deep residual learning [7] is designed to learn residual of identity mapping. This method has proved to be an efficient way to prevent overfitting and increase the depth of the feedforward neuron network. Inception-Resnet architectures [18] showed that the network can achieve competitive accuracy by embedding multi-scale processes in the deep residual network. In our work, we use the residual connection to add different scale feature maps with origin feature maps together.

Multi-scale Fusion. The work in [20] uses multiple resolution banks in parallel and capture features at a variety of scales. Based on this method, bottom-up (from high resolutions to low resolutions) and top-down (from low resolutions to high resolutions) [14] is proposed to capture information at every scale. This approach uses a single pipeline with skip layers has the capacity to capture full body information and bring it to the next layer. Residual attention network [23] uses bottom-up top-down mechanism as attention mask. Our network design partly builds off of their work, exploring how to capture information across scales and adapting their method of combing features across different resolutions. Instead, we don't use intermediate supervision process and introduce attention mechanism which is different from previous work.

To the best of our knowledge, this is the first single-pipeline end-to-end feedforward attention module that encoding non-local information with bottom-up top-down mechanism about action recognition.

3   Proposed Method

STBTA: A STBTA net based on Inception-v3 [19] and TSN [24] for action recognition is illustrated in Fig. 2. All of these submodules in STBTA are residual modules and STBTA performs like a big residual block. For each STBTA, max pooling layer with stride 2 is used to process features down to a very low resolution. We use t to denote the number of downsample and upsample times of this paper, which is 1 default.

There exists a residual submodule between any adjacent layer during downsampling and upsampling (We have not visualized the residual submodule in

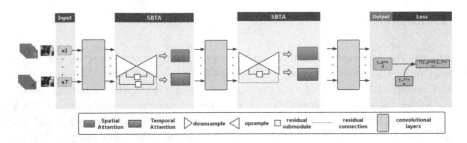

Fig. 2. A STBTA network based on Inception-v3 and TSN. The first STBTA with $t = 2$ is add after inception 3a. The second STBTA with $t = 1$ is add after inception 4e . T is the number of temporal segments, 3 in our experiment. t is the number of downsample and upsample times in STBTA

Fig. 2 for simplicity). The design of residual submodule is the same as SHN [14]. We downsample the input feature map several times in this module. After reaching the lowest resolution, the module begins the sequences of bilinear upsample and combination features across scales by a symmetrical top-down architecture. Furthermore, we add spatial and channel attention module to emphasize the features of key local regions and further improve the performance of the network. The output size is the same as the input feature map.

Global content information and temporal information are both important for action recognition. Most simple actions can be recognitioned by a few frames or a still frame. But for complicated actions, recognition highly rely on temporal information. Based on this, we design two branch which are added after upsample. The first branch is spatial attention module, which focus on spatial information and process on feature maps which combined all scale information. Spatial attention module is added after upsample to control computing cost. Only one channel attention module is added into the last part of STBTA for simplifying and process on all channels which combined all scale information.

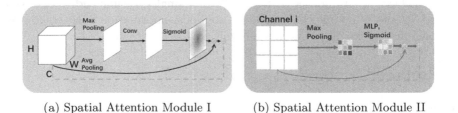

(a) Spatial Attention Module I   (b) Spatial Attention Module II

Fig. 3. The design of SAM I and SAM II.

Spatial Attention: Inspired by the design of channel attention recently [8]. For action recognition, we care about 'where' is an informative part, which is symmetric with the channel attention branch. The design of spatial attention module

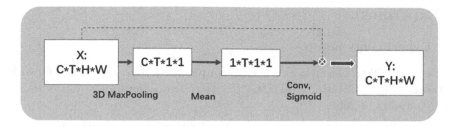

Fig. 4. Temporal attention module

has two ways. As shown in Fig. 3a. The first form is computed a 2D descriptor that encodes channel information at each pixel across the channel, which named Spatial Attention Module I (SAM-I). Formally, given an intermediate feature map $F \in \mathbb{R}^{C \times H \times W}$ as input, using channel max pooling and channel avg pooling, generate two 2D maps: $F_{avg}^s \in \mathbb{R}^{1 \times H \times W}$ and $F_{max}^s \in \mathbb{R}^{1 \times H \times W}$. Then do element-wise-addition between them and convolved by a standard convolution layer to produce 2D spatial attention map, sigmoid activate function is added in the last. Then we get spatial coefficients:

$$M_s(F) = \sigma(f^{conv}(F_{max}^s)) \tag{1}$$

where f^{conv} represents a convolution operation and σ denotes the sigmoid function. Then $M_s(F)$ is multiplied with each channel and add with origin feature map to get the output.

Fusion channel may weaken distinguish information, so we design spatial attention module in a new way. In the second form, the spatial dimension is $W \times H$ for every channel. We divide every channel into $N \times N$ grids, N is chosen to be 3 in our experiments. Max pooling is performed with each grid, and then a conv layer and one softmax activation function are used to produce coefficients for these grids. We use depthwise separable convolution here to not change channel dependence. Which named Spatial Attention Module II(SAM-II). The details of SAM-II are in Fig. 3b.

Temporal Attention: Intuitively, every temporal information play different role for action, some temporal information may be key frame which has high distinction. Inspired by this intuitively, Temporal attention module(TAM) mainly consider relations along temporal dimensions. First, we reshape the input feature map as $B \times C \times T \times H \times W$. As shown in Fig. 4(batch size B is not shown for conveniently). Notice 2D CNNs without temporal sampling is a special situation, when $T = 1$. Firstly use 3D max pooling to get max response, then we calculate mean along channel dim, following with conv layer and sigmoid activation function too. Then we use the output to re-weight the input feature map. The benefit of this design is the computation overhead is negligible and strengthen the key information along temporal dim.

4 Experiments

4.1 Experiments Setup

We use the PyTorch framework for CNN implementation and all the networks are trained on 4 NVIDIA 1080Ti GPUs. Here, we describe the datasets and implementation details.

Datasets. Three well-known benchmarks, UCF101 [17], HMDB-51 [11] and Kinetics-400 [10] are used in the evaluations of action recognition. UCF101 consists of 13,320 manually labeled videos from 101 action categories. It has three train/test splits, each split has around 9,500 videos for training and 3,700 video for testing. HDMB51 is a realistic and challenging dataset. It consists of 6,766 manually labeled clips from 51 categories. Kinetics-400 contains around 246K training videos and 20k validation videos from 400 categories.

Implement Details. For 2D networks, all of our network are based on TSN [24]. To conduct fair comparison, we keep most of the settings same as TSN. Random cropping and horizontal flipping are used for data augmentation. We train network by using the SGD optimizer with a mini-batch size of 64. The learning rate drops down by 10 every 30 epochs and we set the dropout radio at 0.7 to prevent over-fitting. We use a weight decay of 0.0005 with a momentum of 0.9 and set the initial learning to 0.001. The spatial size is 224×224 pixels. We train our module for 100 epochs. In the resting stage, 25 segments are sampled from RGB and optical flow. For 3D networks, we add our module on 3D Inception-v1 [1] and 3D ResNext-101 [6]. For 3D Inception-v1, we follow the design in [1]. What's different is in our practice we sample 10 clips randomly from a full-length video and compute the softmax scores, the final result is averaged of these scores. For 3D ResNext-101, we follow the implement details as [6] to conduct fair comparison. We choose ResNext as the back bone because the good performance. What's different is that we use fine-tune strategies which be describe in Sect. 4.3.

Table 1. Ablation study on our proposed module. We show RGB top-1 classification accuracy on split 1 of UCF-101.

Method	BNInception
baseline	84.30%
baseline + SAM-I	84.63%
baseline + SAM-II	85.02%
baseline + TAM	84.71%
basline + SAM-I + TAM	85.27%
baseline + SAM-II + TAM	85.66%

4.2   The Efficiencies of STBTA

First we add our proposed module on BNInception [9], the ablation study result is shown in Table 1. Both SAM and TAM can improve the recognize performance and combine them lead to better result.

Table 2. We compare 1, 2 STBTA be added to the BNInception(the first with $t = 2$ before inception (3c) and the second with $t = 1$ before inception (4d)), Inception-v3(the first with $t = 3$ before mixed_5b and the second with $t = 2$ before mixed_7a) and Inception-Resnet-v2(the first with $t = 3$ before mixed_5b and the second with $t = 2$ before mixed_7a). We show RGB top-1 classification accuracy on split 1 of UCF-101

Method	BNInception	Inception-v3	Inception-Resnet v2
baseline	84.30%	84.88%	86.49%
+ 1 STBTA	85.27%	85.93%	87.95%
+ 2 STBTA	85.76%	86.59%	88.44%

In order to show the efficiencies of STBTA, we use BNInception [9], Inception-v3 [19] and Inception-Resnet-v2 [18] as baseline and all pretrained on ImageNet. Table 2 shows the results of different number of STBTA be added to the baseline. A network with STBTA leads to a better result in general. It is noteworthy that add one STBTA lead to 1% improvement generally. Considering calculation overhead, we add 2 STBTA to baseline in this paper as default. Furthermore, to demonstrate our module's general applicability. We use our STBTA on Kinetics-400, which is two orders of magnitude larger than HMDB51 and UCF101 and is very time-consuming to train. Limited to the hardware resources, we only one STBTA with $t = 3$ on Inception-v3(before mixed_5b). The result is shown in Table 3. In Table 4, we list some recently comparable methods. Our result is based on Inception-Resnet-v2 baseline(the first STBTA with $t = 3$ is added before mixed_5b and the second STBTA with $t = 2$ is added before mixed_7a), we call this STBTA net. Only use RGB frame as input and pretrained on ImageNet, our method outperforms MiCT-Net by 1.4% on UCF101. In addition, use SAM-II, we can obtain an extra gain about 0.4% but time-consuming. We use SAM-I in STBTA as default in rest.

Table 3. We show video top-1 classification accuracy for RGB input on Kinetics-400. Report on the val sets.

Method	Inception V3
baseline	72.5%
+ 1 STBTA	73.7%

Table 4. Performance comparison to the state-of-the-arts methods on UCF-101 over three splits for RGB as input.

Method	RGB
TSN [24]	86.01%
I3D [1]	84.5%
MiCT-Net [30]	87.3%
STBTA net (SAM-I)	88.70%
STBTA net (SAM-II)	**89.10%**

4.3   Fine-Tune Strategy

Due to the large number of 3D ConvNets's parameters, small datasets can be easily over-fitting. One would fine-tune existing networks that are trained on Kinetics or Sports1M. There are three general guidelines for fine-tuning if new dataset is similar to the original dataset. The first common practice is to truncate the last layer. The second common practice is to use a smaller learning rate to train all the network. The third method is to freeze the weights of the first few layers and train others later.

A general solution is the first few layers capture universal features like curves and edges. But ignore data imbalanced totally. In this paper, we propose a new engineering strategy to fine-tune neural networks. Give different learning rates, according to the depth of neural networks, achieve an impressive performance advancing. Which be formulated with.

$$\beta_l = \sin\left(\frac{l}{L} * \frac{\pi}{2}\right) * \alpha \tag{2}$$

L is the network's depth, l is current layer's depth. α is the learning rate now. $\beta_l, l = 1, 2...L$ is the learning rate of the l layer.

Table 5 show the results of fine-tune strategy and a single STBTA added to 3D Inception-v1. We inflated a 2D Inception-v1 follow [1] and pretrained on Kinetics-400. Fine-tune strategy can lead to 1% improvement over the baseline. And with additional 1 STBTA can further lead to 0.8% improvement.

Table 5. We show top-1 result based on 3D Inception-v1. Report on the split1 of UCF101.

Method	3D Inception-v1
baseline	92.72%
+ fine-tune strategy	93.75%
+ 1 STBTA	94.55%

In order to show our method's effectiveness, we visualize several examples for the behavior of a SBTA be added to the baseline in Figs. 5 and 1. Our module

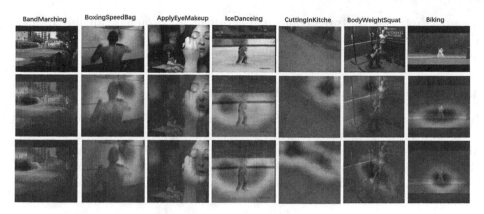

Fig. 5. We compare the visualization results of SBTA-integrated network(Inception-v3+SBTA) with baseline(Inception-v3). All the convnets are based on TSN. The grad-CAM visualization is calculated for the convolutional outputs after Mixed7D. The first row is input image, the second row is baseline's results and the third row is our SBTA-integrated network's results.

Table 6. Comparisons with state-of-the-art results on UCF101 and HMDB51 over 3 splits.

Method	UCF101	HMDB51
TSN [24]	94.0%	68.5%
ST-ResNet [4]	93.5%	66.4%
TLE [2]	**95.6%**	71.1%
Attention Cluster [12]	94.6%	69.2%
STP [26]	94.6%	68.9%
Two Stream MiCT-Net [30]	94.7%	70.5%
ActionVLAD [5]	92.7%	66.9%
CoViAR + optical flow [28]	94.9%	70.2%
ISPAN(30 frames) [3]	95.5%	70.7%
Two Stream STBTA Net	95.20%	**71.1%**

can learn to find meaningful relational clues in long distance and pas attention to more specific and accurate action regions in every frame.

4.4   Comparison with 2D State-of-the-Arts

To prove the effectiveness, we further evaluate our STBTA net on all 3 splits of UCF-101 and HMDB-51 with only use ImageNet pre-trained in Table 6. We list recent state-of-the-art and comparable methods. Two stream STBTA net obtain the improved performance 95.2%/71.1%, which is on pair with TLE. It can be noticed that our proposed STBTA's performance is better on HMDB51 (a hard dataset). Note that the two-stream architecture numbers on individual RGB and Flow streams can be interpreted as a simple baseline, which applies a ConvNet independently on 25 uniformly sampled frames then average the predictions.

4.5 Comparison with 3D State-of-the-Arts

In Table 7, we compare 3D state-of-the-arts method on UCF101 and HMDB51 with only RGB as input. ResNext-101 are pre-trained on Kinetics-400. Our 3D STBTA obtain an extra gain about 1.3% on UCF101 and about 1.2% on HMDB51. The reason why STBTA's result on HMDB51 isn't competing with UCF101 may be HMDB51's samples is too small for 3D ConvNets.

Table 7. Comparisons with state-of-the-art results on UCF101 and HMDB51 over 3 splits.

Method	UCF101	HMDB51
C3D [21]	82.3%	–
RGB-I3D(64f) [1]	95.6%	74.8%
P3D Resnet + IDT [15]	93.7%	–
ResNext-101(64f) [6]	94.5%	70.2%
ResNext-101(64f) [6] + STBTA	95.8%	71.4%
+ fine-tune strategy	96.0%	72.2%
RGB-I3D(64f) [1] + STBTA	96.1%	75.4%
+ fine-tune strategy	**96.3%**	**75.8%**

5 Conclusions

We propose a novel Spatial Bottom-up Top-down Attention Module (STBTA), which can encoding non-local information and achieve adaptive feature refinement via Bottom-up Top-down and attention mechanism. Experimental results show that the proposed module can improve the recognition performance for the task of video classification. Even a simple addition of one STBTA in a baseline CNN can achieve significant improvement over the baseline.

For the future work, we will exploit different applications of our module such as action detection and image segmentation to better explore Bottom-up Top-down mechanism and attention mechanism for different tasks.

References

1. Carreira, J., Zisserman, A.: Quo vadis, action recognition? A new model and the kinetics dataset. In: CVPR, pp. 4724–4733. IEEE (2017)

2. Diba, A., Sharma, V., Van Gool, L.: Deep temporal linear encoding networks. In: CVPR, vol. 1 (2017)
3. Du, Y., Yuan, C., Li, B., Zhao, L., Li, Y., Hu, W.: Interaction-aware spatio-temporal pyramid attention networks for action classification. arXiv preprint arXiv:1808.01106 (2018)
4. Feichtenhofer, C., Pinz, A., Wildes, R.: Spatiotemporal residual networks for video action recognition. In: Advances in Neural Information Processing Systems, pp. 3468–3476 (2016)
5. Girdhar, R., Ramanan, D., Gupta, A., Sivic, J., Russell, B.: Actionvlad: learning spatio-temporal aggregation for action classification. In: CVPR, vol. 2, p. 3 (2017)
6. Hara, K., Kataoka, H., Satoh, Y.: Can spatiotemporal 3D CNNs retrace the history of 2D CNNs and imagenet? In: Proceedings of the IEEE Conference on Computer Vision and Pattern Recognition, pp. 6546–6555 (2018)
7. He, K., Zhang, X., Ren, S., Sun, J.: Deep residual learning for image recognition. In: CVPR, pp. 770–778 (2016)
8. Hu, J., Shen, L., Sun, G.: Squeeze-and-excitation networks. arXiv preprint arXiv:1709.01507 7 (2017)
9. Ioffe, S., Szegedy, C.: Batch normalization: Accelerating deep network training by reducing internal covariate shift. arXiv preprint arXiv:1502.03167 (2015)
10. Kay, W., et al.: The kinetics human action video dataset. arXiv preprint arXiv:1705.06950 (2017)
11. Kuehne, H., Jhuang, H., Stiefelhagen, R., Serre, T.: HMDB51: a large video database for human motion recognition. In: Nagel, W., Kröner, D., Resch, M. (eds.) High Performance Computing in Science and Engineering 2012, pp. 571–582. Springer, Heidelberg (2013). https://doi.org/10.1007/978-3-642-33374-3_41
12. Long, X., Gan, C., de Melo, G., Wu, J., Liu, X., Wen, S.: Attention clusters: purely attention based local feature integration for video classification. In: CVPR, pp. 7834–7843 (2018)
13. Mnih, V., Heess, N., Graves, A., et al.: Recurrent models of visual attention. In: NIPS, pp. 2204–2212 (2014)
14. Newell, A., Yang, K., Deng, J.: Stacked hourglass networks for human pose estimation. In: Leibe, B., Matas, J., Sebe, N., Welling, M. (eds.) ECCV 2016. LNCS, vol. 9912, pp. 483–499. Springer, Cham (2016). https://doi.org/10.1007/978-3-319-46484-8_29
15. Qiu, Z., Yao, T., Mei, T.: Learning spatio-temporal representation with pseudo-3D residual networks. In: ICCV (2017)
16. Selvaraju, R.R., et al.: Grad-cam: visual explanations from deep networks via gradient-based localization. In: ICCV, pp. 618–626 (2017)
17. Soomro, K., Zamir, A.R., Shah, M.: Ucf101: a dataset of 101 human actions classes from videos in the wild. arXiv preprint arXiv:1212.0402 (2012)
18. Szegedy, C., Ioffe, S., Vanhoucke, V., Alemi, A.A.: Inception-v4, inception-ResNet and the impact of residual connections on learning. In: AAAI, vol. 4, p. 12 (2017)
19. Szegedy, C., Vanhoucke, V., Ioffe, S., Shlens, J., Wojna, Z.: Rethinking the inception architecture for computer vision. In: CVPR, pp. 2818–2826 (2016)
20. Tompson, J.J., Jain, A., LeCun, Y., Bregler, C.: Joint training of a convolutional network and a graphical model for human pose estimation. In: NIPS, pp. 1799–1807 (2014)
21. Tran, D., Bourdev, L., Fergus, R., Torresani, L., Paluri, M.: Learning spatiotemporal features with 3D convolutional networks
22. Vaswani, A., et al.: Attention is all you need. In: NIPS, pp. 5998–6008 (2017)

23. Wang, F., et al.: Residual attention network for image classification. arXiv preprint arXiv:1704.06904 (2017)
24. Wang, L., et al.: Temporal segment networks: towards good practices for deep action recognition. In: Leibe, B., Matas, J., Sebe, N., Welling, M. (eds.) ECCV 2016. LNCS, vol. 9912, pp. 20–36. Springer, Cham (2016). https://doi.org/10.1007/978-3-319-46484-8_2
25. Wang, X., Girshick, R., Gupta, A., He, K.: Non-local neural networks. In: CVPR (2018)
26. Wang, Y., Long, M., Wang, J., Philip, S.Y.: Spatiotemporal pyramid network for video action recognition. In: CVPR, vol. 6, p. 7 (2017)
27. Woo, S., Park, J., Lee, J.-Y., Kweon, I.S.: CBAM: convolutional block attention module. In: Ferrari, V., Hebert, M., Sminchisescu, C., Weiss, Y. (eds.) ECCV 2018. LNCS, vol. 11211, pp. 3–19. Springer, Cham (2018). https://doi.org/10.1007/978-3-030-01234-2_1
28. Wu, C.Y., Zaheer, M., Hu, H., Manmatha, R., Smola, A.J., Krähenbühl, P.: Compressed video action recognition. In: CVPR, pp. 6026–6035 (2018)
29. Xu, K., et al.: Show, attend and tell: neural image caption generation with visual attention. In: ICML, pp. 2048–2057 (2015)
30. Zhou, Y., Sun, X., Zha, Z.J., Zeng, W.: MiCT: mixed 3D/2D convolutional tube for human action recognition. In: CVPR, pp. 449–458 (2018)

Hierarchical Graph Convolutional Network for Skeleton-Based Action Recognition

Linjiang Huang[1,3], Yan Huang[1,3], Wanli Ouyang[5], and Liang Wang[1,2,3,4(⊠)]

[1] National Laboratory of Pattern Recognition (NLPR),
Center for Research on Intelligent Perception and Computing (CRIPAC),
Beijing, China
wangliang@nlpr.ia.ac.cn
[2] Center for Excellence in Brain Science and Intelligence Technology (CEBSIT),
Institute of Automation, Chinese Academy of Sciences (CASIA),
Beijing, China
[3] University of Chinese Academy of Sciences (UCAS), Beijing, China
[4] Chinese Academy of Sciences Artificial Intelligence Research (CAS-AIR),
Beijing, China
[5] University of Sydney, Sydney, Australia

Abstract. Skeleton-based action recognition has drawn much attention recently. Previous methods mainly focus on using RNNs or CNNs to process skeletons. But they ignore the topological structure of the skeleton which is very important for action recognition. Recently, Graph Convolutional Networks (GCNs) achieve remarkable performance in modeling non-Euclidean structures. However, current graph convolutional networks lack the capacity of modeling hierarchical information, which may be sub-optimal for classifying actions which are performed in a hierarchical way. In this work, a novel Hierarchical Graph Convolutional Network (HiGCN) is proposed to deal with these problems. The proposed model includes several Hierarchical Graph Convolutional Layers (HiGCLs). Each layer consists of an attention block and a hierarchical graph convolutional block, which are used for salient feature enhancement and hierarchical representation learning, respectively. To represent hierarchical information of human actions, we propose a graph pooling method, which is differentiable and can be plugged into GCN in an end-to-end manner. Extensive experiments on two benchmark datasets show the state-of-the-art performance of our method.

Keywords: Action recognition · Hierarchical graph convolutional network · Skeleton

1 Introduction

Action recognition is a fundamental task in computer vision with many applications such as robotics, video surveillance, etc. [11]. Due to the development of

© Springer Nature Switzerland AG 2019
Y. Zhao et al. (Eds.): ICIG 2019, LNCS 11901, pp. 93–102, 2019.
https://doi.org/10.1007/978-3-030-34120-6_8

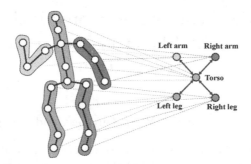

Fig. 1. Illustration of our main idea. We aggregate the nodes of body joint into nodes of body part. The physical relation between body parts are used for constructing the adjacency matrix for graph convolution.

depth sensors and pose estimation methods [13], skeleton-based action recognition has drawn much attention recently. Different from other modalities, human skeletons only focus on spatial configurations and temporal evolution of human poses, which are robust to variations of viewpoints, body scales and motion speeds [20].

The main challenge of skeleton-based action recognition is how to model the spatial-temporal patterns of skeletons. Recent methods mainly rely on deep models, *e.g.*, Recurrent Neural Networks (RNNs) [3,17] and Convolutional Neural Networks (CNNs) [2,5,7], which are suitable for regular representations, *e.g.*, sequential data and images. However, if we view spatial-temporal connections between body joints as a graph, the RNNs and CNNs may be not enough to handle the graph-shaped topology of skeleton.

Recently, Graph Convolutional Networks (GCNs) are applied to various applications [1,6] with graph-shaped data, and obtains impressive performances. Two recent works in [22] and [9] first propose to employ the graph convolutional networks to automatically learn the spatial-temporal patterns of human skeletons. They construct a spatial graph based on the physical structure of human body and add temporal connections between corresponding joints in adjacent frames. Nevertheless, these methods only focus on the spatial-temporal patterns of body joints, ignoring the hierarchical information, *i.e.*, the movement of human body parts in action recognition. Moreover, graph convolutional networks inherently lack the capacity of modeling hierarchical structure [23], which restricts the ability of predicting action labels for entire graph.

To address these limitations, we propose a novel Hierarchical Graph Convolutional Network (HiGCN). The proposed model is the stack of several Hierarchical Graph Convolutional Layers (HiGCLs), each of which consists of an attention block and a hierarchical graph convolutional block. The attention block is added first to emphasize the salient spatial-temporal nodes of skeletons. The hierarchical graph convolutional block is designed to model hierarchical information of human actions, the main idea is shown in Fig. 1. Specifically, it has two branches. One branch is a regular graph convolutional layer for modeling spatial-temporal patterns of body joints, and another branch consists of two graph pooling layers and a graph convolutional layer for hierarchical representation learning and reasoning.

Our contributions are summarized below:

- We present a novel hierarchical graph convolutional network, *i.e.*, HiGCN, which can model the spatial-temporal patterns of skeletons in a hierarchical way.
- We propose a graph pooling method which can be elegantly plugged into GCN in an end-to-end manner.
- Our method obtains the state-of-the-art results on two widely used benchmarks.

2 Method

In this section, the overall framework of Hierarchical Graph Convolutional Network (HiGCN) is introduced. In Sect. 2.1, we briefly introduce the original spatial-temporal graph convolutional network. In Sect. 2.2, we describe the proposed hierarchical graph convolution networks in detail.

2.1 Preliminaries

We consider an undirected graph $\mathcal{G} = \{\mathcal{V}, \mathcal{E}\}$, where $\mathcal{V}$ is the set of vertices and $\mathcal{E}$ is the set of edges. Let $\boldsymbol{A}$ denotes the adjacency matrix, whose element a_{ij} is the weight assigned to the edge (i, j). We set $a_{ij} = 1$ if vertices i and j are connected and $a_{ij} = 0$ otherwise. For the skeleton sequence, a spatial-temporal graph is constructed based on the physical structure of human body and chronological order.

Here, we adopt the similar implementation of graph convolution as in [6]. For spatial dimension, the graph convolution operation at layer l can be formulated as:

$$H^{(l+1)} = \widetilde{D}^{-\frac{1}{2}} \widetilde{A} \widetilde{D}^{-\frac{1}{2}} H^{(l)} W \qquad (1)$$

where $\boldsymbol{H}^{(l)} \in \mathbb{R}^{T \times V \times C_{in}}$ and $\boldsymbol{H}^{(l+1)} \in \mathbb{R}^{T \times V \times C_{out}}$ are the input feature and the output feature, respectively. C_* denotes the number of channels, T denotes the sequence length and V denotes the number of joints. $\widetilde{\boldsymbol{A}} = \boldsymbol{A} + \boldsymbol{I}_V$ and $\widetilde{\boldsymbol{D}} \in \mathbb{R}^{V \times V}$ is the degree matrix, whose element $\widetilde{d}_{ii} = \sum_j \widetilde{a}_{ij}$. We also adopt the partition strategy, which is similar to the sampling function in CNN. However, different from [22], we employ the distance partitioning, which can be applied to any graph rather than skeletons. For temporal dimension, it is straightforward to perform graph convolution similar to the regular convolution. Specifically, we utilize a $K_t \times 1$ convolution to simulate the temporal graph convolution operation. For more details, please refer to [22].

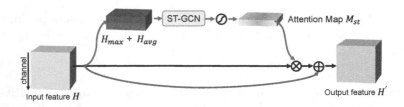

Fig. 2. The overview of the attention block. As illustrated, the attention block utilizes the outputs of average pooling and max pooling and feed them to a spatial-temporal graph convolutional layer to get the attention map.

2.2 Hierarchical Graph Convolutional Network

The proposed hierarchical graph convolutional network mainly consists of several hierarchical graph convolutional layers (HiGCLs). The framework of a HiGCL is shown in Fig. 3, and the details are described as follows.

Input Features: Inspired by [15], we try to utilize more powerful features as input. We concatenate the original coordinates, relative coordinates as well as the temporal displacement into a new feature, we denote it as *Hybrid Features*. The effect of *Hybrid Features* will be evaluated in Sect. 3.

Graph Attention Block: Before feeding the feature for hierarchical representation learning, an additional attention block is employed to highlight the discriminative nodes. The overview of graph attention block is shown in Fig. 2. We first aggregate channel information of the input feature $H \in \mathbb{R}^{T \times V \times C}$ by employing the average pooling and max pooling operation. The generated features H_{avg} and H_{max} are concatenated to form a spatial-temporal descriptor $H_{att} \in \mathbb{R}^{T \times V \times 2}$. The descriptor is forwarded to a spatial-temporal graph convolutional block. We utilize the sigmoid function to make the values of the attention map to be between 0 and 1. During multiplication, the spatial-temporal attention map $M_{st} \in \mathbb{R}^{T \times V \times 1}$ is broadcasted along channel dimension. In addition, we employ the skip connection to preserve information of the input feature. The whole attention process is represented as:

$$H^{'} = H \otimes M_{st} + H \tag{2}$$

where $\otimes$ is the Hadamard product, and $H^{'} \in \mathbb{R}^{T \times V \times C}$ is the refined feature.

Hierarchical Graph Convolutional Block: The recent GCN-based approaches [9,15,22] focus on modeling spatial-temporal patterns of body joints. However, human actions are always performed in a hierarchical way. For example, we can easily distinguish *wave* from *kick ball* only by the movement of body parts, but for some fine-grained classes, *e.g.*, reading and writing, we need more discriminative information such as the movement of body joints. Nevertheless, current graph convolutional networks lack such capacity of modeling hierarchical information.

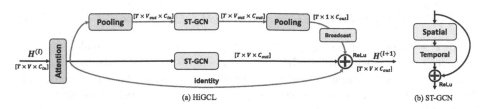

Fig. 3. (a) The overview of Hierarchical Graph Convolutional Layer (HiGCL). The HiGCL mainly consists of an attention block and a hierarchical graph convolutional block. (b) The spatial-temporal graph convolutional network. Here we adopt the distance partitioning in [22].

To solve the above problem, we propose a hierarchical graph convolutional block which aims at modeling spatial-temporal evolutions of body joints and body parts in a two branch fashion, as illustrated in Fig. 3(a). The first branch is a regular graph convolutional network, which decouples the spatial-temporal graph convolution into two successive convolutional components. The second branch aims to explore the spatial-temporal relationship in a hierarchical way. Thus we need a pooling operation over graph to aggregate the nodes into super nodes.

Inspired by [23], we develop a graph pooling operation which can learn hierarchical representations of graph in an end-to-end fashion. Let $H \in \mathbb{R}^{V_{in} \times C}$ [1] represents the input feature of graph pooling layer. C denotes the number of channels, and V_{in} denotes the number of nodes. The graph pooling operation aims to aggregate feature H into compact feature $H_p \in \mathbb{R}^{V_{out} \times C}$, where V_{out} denotes the number of pooled nodes. In general, we need $V_{out} < V_{in}$. We propose to generate the pooling matrix as:

$$P = Relu(\mathcal{G}(H, A)) \tag{3}$$

where $P \in \mathbb{R}^{V_{in} \times V_{out}}$ is the learned pooling matrix, $\mathcal{G}(\cdot, \cdot)$ is graph convolution operation and A is the adjacency matrix. $Relu(\cdot)$ is utilized for meeting the definition of pooling operation. However, different from other types of data, graph data has to take into account the intrinsic geometrical structure. Especially, it is essential to learn a new adjacency matrix after pooling for down-stream graph convolutional layers. For human body, it is easy to find a hierarchical structure, i.e., body joints and body parts. A body part can be viewed as an abstraction of several body joints, which means we can aggregate the nodes of body joints into the nodes of body parts. Moreover, the physical relation between body parts are obvious, so we can easily construct a new adjacency matrix for further reasoning the relation between human body parts.

Specifically, we manually define a mask $M \in \mathbb{R}^{V_{in} \times V_{out}}$, whose element $m_{ij} \in \{0, 1\}$. The rows of mask M are one-hot vectors, and each column of M represents one body part. $m_{ij} = 1$ means that the i-th joint belongs to the j-th body part, not vice versa. According to the physical structure of human body, we define several major body parts, i.e., torso, two arms and two legs. The pooling operation is formulated as:

---

[1] We omit temporal dimension for simplicity.

$$H_p = \bar{P}^T H, \quad \bar{P} = Softmax(P + (1 - M) \times n) \tag{4}$$

where P is the same as in Eq. (3), $\mathbf{1} \in \mathbb{R}^{V_{in} \times V_{out}}$ is a matrix whose elements are all 1. n is a large negative number, we set it as -9×10^5 in experiments. The softmax operation is implemented in a column-wise fashion. After graph pooling, the features of nodes are projected into a part-level space, and the pooled nodes have explicit semantic information.

We introduce the pooling method into the second branch for modeling hierarchical information. Specifically, the second branch includes two graph pooling layers and a spatial-temporal graph convolutional block, as shown in Fig. 3(a). The first pooling layer aggregates body joints into body parts for each frame, and the graph convolutional block is used for reasoning spatial-temporal relation of body parts. The new adjacency matrix are constructed based on the physical relation between body parts. The second pooling layer outputs a body-level feature, which is global-aware and discriminative. The outputs of two branches and an identity mapping of the input are summed up to form the output of the HiGCL.

3 Experiment

In this section, we first introduce the implementation details. Then, we compare our method with several state-of-the-art methods on two benchmark datasets. Next, we comprehensively investigate some ablation studies. Finally, we visualize the learned attention maps.

3.1 Datasets

NTU RGB+D [12]: This dataset consists of 56880 actions with 60 classes. The benchmark evaluations include Cross-Subject (CS) and Cross-View (CV). In the CS evaluation, training samples come from one subset of actors and networks are evaluated on samples from remaining actors. In the CV evaluation, samples captured from cameras 2 and 3 are utilized for training, while samples from camera 1 are employed for testing.

Northwestern-UCLA dataset (N-UCLA) [19]: This dataset contains 1494 videos of 10 actions. These actions are performed by 10 subjects, repeated 1 to 6 times. Each subject has 20 joints. There are three views in this dataset. Usually, two of the views are used for training and the other one is used for testing.

3.2 Implementation Details

The proposed hierarchical graph convolutional network is the stack of nine hierarchical graph convolutional layers. Before the first HiGCL, an embedding layer is employed to project the dimension of the input feature to 64. The number of output channels for each layer are 64, 64, 64, 128, 128, 128, 256, 256 and 256, respectively. After that, a global average pooling layer is performed and the final output is feeded to a fully connection layer and a softmax layer to get the prediction. The batch size is set to 64 for NTU RGB+D and 16 for N-UCLA. The learning rates for both datasets are 0.1 initially, reduced by 0.1 after 20 epochs and 50 epochs. The training procedure stops at 80 epochs.

3.3 Experimental Results

Comparison with the State-of-the-Art Methods. The experiments of our method on two widely used benchmark datasets (NTU RGB+D [12] and N-UCLA [19]) are shown in Tables 1 and 2, respectively. We first compare our method [16] with traditional method based on hand-crafted features. As we can see, our method significantly outperforms these approaches, which shows the superiority of deep learning methods over hand-crafted approaches. Then our method is compared with recent deep learning methods. We can see that our method outperforms the state-of-the-arts on both datasets. Specifically, our method achieves the highest accuracy of 87.9% and 93.8% using CS and CV pro-

Table 1. Comparison on NTU RGB+D.(%)

Methods	CS	CV	Year
Lie Group [16]	50.1	82.8	2014
HBRNN [4]	59.1	64.0	2015
Part-aware LSTM [12]	62.9	70.3	2016
Geometric Features [24]	70.3	82.4	2017
Two-Stream CNN [7]	83.2	89.3	2017
Deep STGC$_K$ [9]	74.9	86.3	2018
ST-GCN [22]	81.5	88.3	2018
SR-TSL [14]	84.8	92.4	2018
HCN [8]	86.5	91.1	2018
PB-GCN [15]	87.5	93.2	2018
HiGCN	**87.9**	**93.8**	

Table 2. Comparison on N-UCLA. (%)

Methods	V3	V2	V1	Average	Year
HOJ3D [21]	54.5	–	–	–	2015
AE [18]	76.0	–	–	–	2015
LARP [16]	74.2	–	–	–	2015
HBRNN-L [3]	78.5	83.5	**79.3**	80.5	2016
ESV [10]	86.1	–	–	–	2017
ESV (Synthesized+Pre-trained) [10]	**92.6**	–	–	–	2017
HiGCN	88.9	**85.4**	77.6	**83.9**	

tocols respectively on NTU RGB+D, and obtains the best performance 85.4% for V2 setting and 83.9% for Average setting on N-UCLA. Note that, the performance of ESV (Synthesized+Pre-trained) [10] is higher than ours on the N-UCLA dataset. However, they synthesize more data for training and benefit from the pre-trained model on large scale image datasets. By contrast, our method is trained from scratch and we can achieve better performance compared with ESV which is trained from scratch only using the original data.

Evaluation of Components of HiGCN. We evaluate several components in our network to show their effectiveness on skeleton-based action recognition. We give the results on the NTU RGB+D dateset as shown in Table 3. As we can see, our method significantly improves the performances by 6.4% in cross-subject and 5.5% in cross-view over the baseline model, *i.e.*, ST-GCN [22]. Even without the *Hybrid Features*, our method still outperforms the baseline model. Moreover, after the removal of the attention block and the hierarchical block, the performances drop significantly, indicating the two proposed blocks are very useful for action recognition.

Table 3. Ablation study on NTU RGB+D.(%)

Methods	CS	CV
ST-GCN [22]	81.5	88.3
ST-GCN [22] + Hybrid Features	84.0	89.8
HiGCN w/o Hybrid Features	82.0	88.5
HiGCN w/o Attention Block	87.0	93.2
HiGCN w/o Hierarchical Block	85.9	90.7
HiGCN	**87.9**	**93.8**

Visualization of Attention Maps. Visualization of attention maps of the first three HiGCLs is shown in Fig. 4. We find that the attention maps gradually

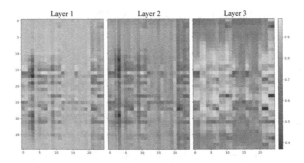

Fig. 4. Visualization of learned attention maps of the first three HiGCLs.

fucus on the salient spatial-temporal patterns. This demonstrates that the attention maps at the later layers can effectively capture important spatial-temporal information of the skeleton sequence.

4   Conclusion

In this paper, we have proposed a novel Hierarchical Graph Convolutional Network (HiGCN) for skeleton-based action recognition. The construction of HiGCN is mainly based on the Hierarchical Graph Convolutional Layers (HiGCLs). The HiGCL is comprised of an attention block and a hierarchical graph convolutional block. We have evaluated our HiGCN on two publicly available datasets, *i.e.*, NTU RGB+D and Northwestern-UCLA, and achieved the state-of-the-art performance on both datasets. We also show the effectiveness of different components of our method based on the experimental analysis. In the future we plan to focus more on modeling temporal dynamics of actions.

Acknowledgements. This work is jointly supported by National Key Research and Development Program of China (2016YFB1001000), National Natural Science Foundation of China (61525306, 61633021, 61721004, 61420106015, 61806194), Capital Science and Technology Leading Talent Training Project (Z181100006318030), Beijing Science and Technology Project (Z181100008918010), and CAS-AIR.

References

1. Defferrard, M., Bresson, X., Vandergheynst, P.: Convolutional neural networks on graphs with fast localized spectral filtering. In: Advances in Neural Information Processing Systems, pp. 3844–3852 (2016)
2. Du, Y., Fu, Y., Wang, L.: Skeleton based action recognition with convolutional neural network. In: Asian Conference on Pattern Recognition, pp. 579–583. IEEE (2015)
3. Du, Y., Fu, Y., Wang, L.: Representation learning of temporal dynamics for skeleton-based action recognition. IEEE Trans. Image Process. **25**, 3010–3022 (2016)
4. Du, Y., Wang, W., Wang, L.: Hierarchical recurrent neural network for skeleton based action recognition. In: IEEE Conference on Computer Vision and Pattern Recognition, pp. 1110–1118. IEEE (2015)
5. Kim, T.S., Reiter, A.: Interpretable 3d human action analysis with temporal convolutional networks. In: IEEE Conference on Computer Vision and Pattern Recognition Workshops, pp. 1623–1631. IEEE (2017)
6. Kipf, T.N., Welling, M.: Semi-supervised classification with graph convolutional networks. arXiv preprint arXiv:1609.02907 (2016)
7. Li, C., Zhong, Q., Xie, D., Pu, S.: Skeleton-based action recognition with convolutional neural networks. In: IEEE International Conference on Multimedia & Expo Workshops, pp. 597–600. IEEE (2017)
8. Li, C., Zhong, Q., Xie, D., Pu, S.: Co-occurrence feature learning from skeleton data for action recognition and detection with hierarchical aggregation. In: International Joint Conference on Artificial Intelligence, pp. 786–792 (2018)

9. Li, C., Cui, Z., Zheng, W., Xu, C., Yang, J.: Spatio-temporal graph convolution for skeleton based action recognition. arXiv preprint arXiv:1802.09834 (2018)
10. Liu, M., Liu, H., Chen, C.: Enhanced skeleton visualization for view invariant human action recognition. Pattern Recogn. **68**, 346–362 (2017)
11. Poppe, R.: A survey on vision-based human action recognition. Image Vis. Comput. **28**, 976–990 (2010)
12. Shahroudy, A., Liu, J., Ng, T.T., Wang, G.: Ntu rgb+d: A large scale dataset for 3d human activity analysis. In: IEEE Conference on Computer Vision and Pattern Recognition, pp. 1010–1019. IEEE (2016)
13. Shotton, J., et al.: Real-time human pose recognition in parts from single depth images. In: IEEE Conference on Computer Vision and Pattern Recognition, pp. 1297–1304. IEEE (2011)
14. Si, C., Jing, Y., Wang, W., Wang, L., Tan, T.: Skeleton-based action recognition with spatial reasoning and temporal stack learning. arXiv preprint arXiv:1805.02335 (2018)
15. Thakkar, K., Narayanan, P.: Part-based graph convolutional network for action recognition. arXiv preprint arXiv:1809.04983 (2018)
16. Vemulapalli, R., Arrate, F., Chellappa, R.: Human action recognition by representing 3d skeletons as points in a lie group. In: IEEE Conference on Computer Vision and Pattern Recognition, pp. 588–595. IEEE (2014)
17. Wang, H., Wang, L.: Modeling temporal dynamics and spatial configurations of actions using two-stream recurrent neural networks. In: IEEE Conference on Computer Vision and Pattern Recognition, pp. 499–508. IEEE (2017)
18. Wang, J., Liu, Z., Wu, Y., Yuan, J.: Learning actionlet ensemble for 3d human action recognition. IEEE Trans. Pattern Anal. Mach. Intell. **36**, 914–927 (2014)
19. Wang, J., Nie, X., Xia, Y., Wu, Y., Zhu, S.C.: Cross-view action modeling, learning and recognition. In: IEEE Conference on Computer Vision and Pattern Recognition, pp. 2649–2656. IEEE (2014)
20. Wang, P., Li, W., Ogunbona, P., Wan, J., Escalera, S.: Rgb-d-based human motion recognition with deep learning: a survey. Comput. Vis. Image Understand. **171**, 118–139 (2017)
21. Xia, L., Chen, C.C., Aggarwal, J.K.: View invariant human action recognition using histograms of 3d joints. In: IEEE Conference on Computer Vision and Pattern Recognition Workshops, pp. 20–27. IEEE (2012)
22. Yan, S., Xiong, Y., Lin, D.: Spatial temporal graph convolutional networks for skeleton-based action recognition. arXiv preprint arXiv:1801.07455 (2018)
23. Ying, Z., You, J., Morris, C., Ren, X., Hamilton, W., Leskovec, J.: Hierarchical graph representation learning with differentiable pooling. In: Advances in Neural Information Processing Systems, pp. 4801–4811 (2018)
24. Zhang, S., Liu, X., Xiao, J.: On geometric features for skeleton-based action recognition using multilayer lstm networks. In: IEEE Winter Conference on Applications of Computer Vision, pp. 148–157. IEEE (2017)

Constrained Dual Graph Regularized NMF for Image Clustering

Shaodi Ge, Hongjun Li, and Liuhong Luo$^{(\boxtimes)}$

College of Science, Beijing Forestry University, Beijing 100083, China
{shaodi,lihongjun69,llh7667}@bjfu.edu.cn

Abstract. Non-negative matrix factorization (NMF) becomes an important dimension reduction and feature extraction tool in the fields of scientific computing and computer vision. In this paper, for using the known label information in the original data, we put forward a semi-supervised NMF algorithm called constrained dual graph regularized non-negative matrix factorization (CDNMF). The new algorithm employs hard constraints to retain the priori label information of samples, constructs two association graphs to encode the geometric structures of the data manifold and the feature manifold, and incorporates the additional bi-orthogonal constraints to improve the identification ability of data in the new representation space. We have also developed an iterative optimization strategy for CDNMF and proved its convergence. Finally the clustering experiments on five standard image data sets show the effectiveness of the proposed algorithm.

Keywords: NMF · Dual graph regularized · Semi-supervised learning · Image clustering

1 Introduction

Dimension reduction is a classical data processing technology in statistical analysis and intelligence computation. In addition to many classical dimension reduction methods, such as LPP [1], PCA [2] and LDA [3], non-negative matrix factorization (NMF) [4] has been becoming a popular dimension reduction and feature extraction method in recent years [5–9] and widely used in image clustering. NMF decomposes a non-negative data matrix into two non-negative factors and approximately linearizes the original matrix in terms of a product of factor matrices. Therefore, the sparse data representation can be obtained and the potential structure of the data can be effectively mined.

Research and development over the years have resulted in a variety of improved NMF methods, based on whether the priori label information of the original data samples is used or not, these improved algorithms can be roughly divided into two categories: unsupervised and semi-supervised. Cai et al. [10] proposed an unsupervised learning algorithm called graph regularized NMF (GNMF). GNMF creates the nearest neighbor graph from the original data

© Springer Nature Switzerland AG 2019
Y. Zhao et al. (Eds.): ICIG 2019, LNCS 11901, pp. 103–117, 2019.
https://doi.org/10.1007/978-3-030-34120-6_9

to encode the geometric information of the data space and integrates graph structure into the objective function, so that the data maintains the neighborhood information of the high-dimensional space in the low-dimensional space, which improves the learning performance significantly and obtains good results in image and document clustering. The graph Laplacian regularizer has been widely used in various NMF frameworks since its application in GNMF. In addition, some of the more classic unsupervised algorithms, including Topological structure regularized NMF (TNMF) [11], Penalized NMTF (PNMT) [12], graph dual regularization NMF (DNMF) [13], and graph-preserving sparse NMF (GSNMF) [14], achieve better performance than classical NMF, but ignore the class information carried by label samples in practical applications.

Semi-supervised algorithms that use a small amount of label information can enhance learning accuracy [15–17] and have become a hot topic in recent years. Liu et al. [18] introduced a constrained NMF (CNMF) algorithm, in which existing label information could serve as an additional hard constraint and data points from the same class should be merged into the new presentation space. Since then, the hard constraint based on priori label information has been widely used in various semi-supervised NMF. Sun et al. [19] presented a graph regularized and sparse NMF with hard constraints (GSNMFC), which consolidates graph regularization, hard constraints based on label information and sparse constraint, demonstrates excellent performance in clustering accuracy and mutual information. Sun and Wang et al. [15] combined regular constraints and sparse constraints of dual graphs as the additional constraints for revealing the geometric and discriminating structures in data space and feature space, and based on these they put forward sparse dual graph-regularized NMF (SDGNMF), which reduces the dimension of image data and greatly improves the clustering performance on image clustering as well.

Hence, based on the concepts of manifold learning and semi-supervised learning, we propose a novel constrained dual graph regularized NMF (CDNMF) algorithm. CDNMF takes the known label information of data samples as an additional hard constraint, and constructing two associated graphs to encode the geometric information of the data space and the feature space. By combining with the independent bi-orthogonal constraints as penalty terms, CDNMF eliminate the correlation among the basis vectors and the similarity of the data in the new representation space. We design a new NMF objective function incorporating the three aspects above and discuss the corresponding optimization problem. We have also proved that the propose objective function do not increase under the corresponding multiplicative update rules. And finally we perform a lot of clustering experiments on five image benchmark data sets to verify the effectiveness of the proposed algorithm.

The remainder of this paper is organized as follows: Sect. 2 explicitly discuss the present CDNMF algorithm, including the multiplicative update rules and convergence analysis. Section 3 includes the image clustering experiments and the corresponding result analysis. Finally, the conclusion is made in Sect. 4.

2   CDNMF

Our algorithm uses the geometric information of both the data manifold and the feature manifold, combining with the priori label information of data points and independent bi-orthogonal penalty terms. And we have deduced the multiplicative iterative rules of the algorithm and proved the convergence.

2.1   Problem Formulation

Given a non-negative data matrix $X = [x_1, x_2, \ldots, x_n] \in R_+^{m \times n}$, in which $x_i = X_{:,i} \in R_+^{m \times 1}, i = 1, 2 \ldots, n$ are m-dimensional non-negative column vectors (the ith data point). We first use the 0–1 weighting scheme to construct a k-nearest neighbor data graph, and its vertices correspond to $\{x_1, x_2, \cdots, x_n\}$. As shown in [10], the data weight matrix is defined as:

$$W_{ij}^V = \begin{cases} 1, & \text{if } x_i \in N_k(x_j) \text{ or } x_j \in N_k(x_i) \\ 0, & \text{otherwise} \end{cases} \tag{1}$$

where $N_k(x_j)$ represents the set of k nearest neighbor data points of the data point x_j.

If data points x_i and x_j are close to each other, their corresponding representation $f(x_i) = v_i$ and $f(x_j) = v_j$ in the low dimensional space should also be close:

$$\frac{1}{2} \sum_{i,j=1}^n (f(x_i) - f(x_j))^2 W_{ij}^V = \sum_{i,j=1}^n f(x_i)^2 D_{ii}^V - \sum_{i,j=1}^n f(x_i) f(x_j) W_{ij}^V$$

$$= \sum_{i,j=1}^n v_i^T v_i D_{ii}^V - \sum_{i,j=1}^n v_i^T v_j W_{ij}^V = Tr(V^T D^V V) - Tr(V^T W^V V)$$

$$= Tr(V^T L_V V) \tag{2}$$

where D^V is a diagonal matrix defined as $D_{ii}^V = \sum_j W_{ij}^V$, and the $L_V = D^V - W^V$ [20] is the Laplacian matrix of data graph. Then let $V = AZ$ where A is the hard constraint matrix based on priori label information of the original data [18], so the regularizer of the constraint data graph $Tr(Z^T A^T L_V AZ)$ is obtained.

Like the construction of the data graph, a k-nearest neighbor feature graph is also constructed by using the 0–1 weighting scheme. Its vertices correspond to $\{y_1, y_2, \ldots, y_m\}$, where $y_i = X_{i,:}^T \in R_+^{n \times 1}, i = 1, 2 \ldots, m$ denote the transposition of the ith row of the data matrix X. We can get $Tr(U^T L_U U)$ as the regularizer term of the feature graph, where $D_{ii}^U = \sum_j W_{ij}^U$, $L_U = D^U - W^U$.

Combined with independent bi-orthogonal constraints, we obtain the objective function of CDNMF as follows:

$$D_{CDNMF} = \| X - UZ^T A^T \|_F^2 + \alpha Tr(Z^T A^T L_V AZ) + \beta Tr(U^T L_U U)$$
$$+ \lambda (Tr(AZEZ^T A^T) + Tr(UEU^T)) \qquad s.t \quad U \geq 0, Z \geq 0. \tag{3}$$

where α, β and λ are non-negative regularization parameters that balance the contribution of regularizers in the objective function Eq. (3), $Tr(UEU^T) = \sum_{i \neq j} u_i^T u_j$ and $Tr(VEV^T) = \sum_{i \neq j} v_i^T v_j (V = AZ)$ are penalty items to ensure the orthogonality of U and V (or near orthogonal). When $\alpha = \beta = \lambda = 0$, the objective function degenerates to CNMF.

2.2 Optimization and Update Rules

The objective function of CDNMF in Eq. (3) is not convex for both U and Z together. Therefore, it is unrealistic to find a global minimum. So, we propose an iterative updating rules to look for a local minimum of Eq. (3).

Equation (3) can be rewritten by applying the matrix properties $Tr(AB) = Tr(BA)$ and $Tr(A) = Tr(A^T)$ as

$$
\begin{aligned}
D_{CDNMF} &= Tr((X - UZ^T A^T)(X - UZ^T A^T)^T) + \alpha Tr(Z^T A^T L_V AZ) \\
&+ \beta Tr(U^T L_U U) + \lambda(Tr(AZEZ^T A^T) + Tr(UEU^T)) \\
&= Tr(XX^T) - 2Tr(XAZU^T) + Tr(UZ^T A^T AZU^T) + \alpha Tr(Z^T A^T L_V AZ) \\
&+ \beta Tr(U^T L_U U) + \lambda(Tr(AZEZ^T A^T) + Tr(UEU^T))
\end{aligned}
\tag{4}
$$

Let $\Psi = [\psi_{ij}]$ and $\Phi = [\phi_{ij}]$ be Lagrange multipliers of constraints $U_{ij} \geq 0$ and $Z_{ij} \geq 0$ respectively, then the Lagrangian function ζ is

$$
\begin{aligned}
\zeta &= Tr(XX^T) - 2Tr(XAZU^T) + Tr(UZ^T A^T AZU^T) + \alpha Tr(Z^T A^T L_V AZ) + \beta Tr(U^T L_U U) \\
&+ \lambda(Tr(AZEZ^T A^T) + Tr(UEU^T)) + Tr(\Psi U^T) + Tr(\Phi Z^T)
\end{aligned}
\tag{5}
$$

The partial derivatives of ζ with respect to U and Z are:

$$
\frac{\partial \zeta}{\partial U} = -2XAZ + 2UZ^T A^T AZ + 2\beta L_U U + 2\lambda UE + \Psi
$$

$$
\frac{\partial \zeta}{\partial Z} = -2A^T X^T U + 2A^T AZU^T U + 2\alpha A^T L_V AZ + 2\lambda A^T AZE + \Phi
\tag{6}
$$

Using the KKT conditions, $\psi_{ij} u_{ij} = 0$ and $\phi_{ij} z_{ij} = 0$. By substituting $L_V = D^V - W^V$ and $L_U = D^U - W^U$, the following equations for u_{ij} and z_{ij} can be obtained:

$$
(XAZ + \beta W^U U - UZ^T A^T AZ - \beta D^U U - \lambda UE)_{ij} u_{ij} = 0
$$
$$
(A^T X^T U + \alpha A^T W^V AZ - A^T AZU^T U - \alpha A^T D^V AZ - \lambda A^T AZE)_{ij} z_{ij} = 0
\tag{7}
$$

The following update rules can be further acquired:

$$
U \leftarrow U \cdot \frac{XAZ + \beta W^U U}{UZ^T A^T AZ + \beta D^U U + \lambda UE}
\tag{8}
$$

$$
Z \leftarrow Z \cdot \frac{A^T X^T U + \alpha A^T W^V AZ}{A^T AZU^T U + \alpha A^T D^V AZ + \lambda A^T AZE}
\tag{9}
$$

As a summary, the iterative update rules of the optimization problem Eq. (3) is shown in Algorithm 1.

Algorithm 1. CDNMF

Input: data matrix X, constraint matrix A, the clustering number c,
    the neighbor number k, parameters α, β and λ, iterations t_{max}.
Output: U and Z.
1: Randomly initialize two non-negative matrices U, Z.
2: Use the 0-1weighting scheme for constructing the k-nearest neighbor data graph W^U
    and feature graph W^V
3: Update matrix $D_{ii}^U = \sum_j W_{ij}^U, D_{ii}^V = \sum_j W_{ij}^V, E = \bar{1} - I.$
4: for $t = 1, 2, ..., t_{max}$, Do
    a):update U as
$$U \leftarrow U \frac{XAZ + \beta W^U U}{UZ^T A^T AZ + \beta D^U U + \lambda UE}$$
    b):update Z as
$$Z \leftarrow Z \cdot \frac{A^T X^T U + \alpha A^T W^V AZ}{A^T AZU^T U + \alpha A^T D^V AZ + \lambda A^T AZE}$$
    end
5: End CDNMF.

2.3 Proof of Convergence

Theorem 1. *The objective function Eq. (3) does not increase under the update rules in Eqs. (8) and (9).*

In order to prove Theorem 1, the following definition and three Lemmas are introduced first.

Definition 1. [21] $G(x, x')$ *is an auxiliary function for $F(x)$ if the conditions*

$$G(x, x') \geq F(x), \qquad G(x, x) = F(x). \tag{10}$$

are satisfied.

Lemma 1. [21] *If $G(x, x')$ is an auxiliary function of $F(x)$, then $F(x)$ is non-increasing under the update*

$$x^{t+1} = arg\ min_x G(x, x^t) \tag{11}$$

Proof.

$$F(x^{t+1}) \leq G(x^{t+1}, x^t) \leq G(x^t, x^t) = F(x^t) \tag{12}$$

For any element z_{ij} in Z, let $F_{ij}(z_{ij})$ be the part about Z in the objective function Eq. (3).

Then, we can get:

$$F_{ij}'(z_{ij}) = (-2A^T X^T U + 2A^T AZU^T U + 2\alpha A^T L_V AZ + 2\lambda A^T AZE)_{ij}$$
$$F_{ij}''(z_{ij}) = 2(A^T A)_{ii}(U^T U)_{jj} + 2\alpha(A^T L_V A)_{ii} + 2\lambda(A^T A)_{ii} E_{jj}.$$

Lemma 2. *The function*

$$G(z_{ij}, z_{ij}^t) = F_{ij}(z_{ij}^t) + F_{ij}'(z_{ij} - z_{ij}^t)$$
$$+ \frac{(A^T AZU^T U + \alpha A^T D^V AZ + \lambda A^T AZE)_{ij}}{z_{ij}^t}(z_{ij} - z_{ij}^t)^2 \tag{13}$$

is an auxiliary function for $F_{ij}(z_{ij})$.

Proof. The Taylor series expansion of $F_{ij}(z_{ij})$ is defined as below:

$$
\begin{aligned}
F_{ij}(z_{ij}) = F_{ij}(z_{ij}^t) &+ F_{ij}'(z_{ij} - z_{ij}^t) \\
&+ [(A^T A)_{ii}(U^T U)_{jj} + \alpha(A^T L_V A)_{ii} + \lambda(A^T A)_{ii}E_{jj}](z_{ij} - z_{ij}^t)^2 \quad (14)
\end{aligned}
$$

Compare Eqs. (13) with (14), it is easy to find that $G(z_{ij}, z_{ij}^t) \geq F_{ij}(z_{ij})$ is equivalent to

$$
\frac{(A^T AZU^T U + \alpha A^T D^V AZ + \lambda A^T AZE)_{ij}}{z_{ij}^t} \geq (A^T A)_{ii}(U^T U)_{jj} + \alpha(A^T L_V A)_{ii} + \lambda(A^T A)_{ii}E_{jj}
$$

since

$$
\begin{aligned}
(A^T AZU^T U + \lambda A^T AZE)_{ij} &= \sum_{l=1}^{r}(AA^T Z)_{il}(U^T U)_{lj} + \lambda \sum_{l=1}^{r}(AA^T Z)_{il}E_{lj} \\
&\geq (AA^T Z)_{ij}(U^T U)_{jj} + \lambda(AA^T Z)_{ij}E_{jj} \\
&\geq \sum_{l=1}^{r}(A^T A)_{il}z_{ij}^t(U^T U)_{lj} + \lambda \sum_{l=1}^{r}(A^T A)_{il}z_{ij}^t E_{lj} \\
&\geq z_{ij}^t(A^T A)_{ii}(U^T U)_{jj} + \lambda z_{ij}^t(A^T A)_{ii}E_{jj} \\
\alpha(A^T D^V AZ)_{ij} &= \alpha \sum_{q=1}^{r}(A^T D^V A)_{iq}z_{qj}^t \geq \alpha(A^T D^V A)_{ii}z_{ij}^t \\
&\geq \alpha(A^T(D^V - W^V)A)_{ii}z_{ij}^t = \alpha(A^T L_V A)_{ii}z_{ij}^t
\end{aligned}
$$

Thus, $G(z_{ij}, z_{ij}^t) \geq F_{ij}(z_{ij})$ and obviously $G(z_{ij}, z_{ij}) = F_{ij}(z_{ij})$. According to the definition of auxiliary function, Lemma 2 is proved.

For any element u_{ij} in U, the similar results can be obtained as z_{ij} in Z. They are

$$
\begin{aligned}
F_{ij}'(u_{ij}) &= (-2XAZ + 2UZ^T A^T AZ + 2\beta L_U U + 2\lambda U E)_{ij} \\
F_{ij}''(u_{ij}) &= 2(Z^T A^T AZ)_{jj} + 2\beta(L_U)_{ii} + 2\lambda E_{jj}.
\end{aligned}
$$

Lemma 3. *The function*

$$
\begin{aligned}
G(u_{ij}, u_{ij}^t) = F_{ij}(u_{ij}^t) &+ F_{ij}'(u_{ij} - u_{ij}^t) \\
&+ \frac{(UZ^T A^T AZ + \beta D^U U + \lambda U E)_{ij}}{u_{ij}^t}(u_{ij} - u_{ij}^t)^2 \quad (15)
\end{aligned}
$$

is an auxiliary function for $F_{ij}(u_{ij})$.

The proof is similar to that of Lemma 2. With these 3 Lemmas, the proof of Theorem 1 can be done as following.

Proof of Theorem 1. In order to get the minimum value of $G(z, z')$ (see Eq. (13)), we let $\frac{\partial G(u_{ij}, u_{ij}^t)}{\partial u_{ij}} = 0$ and $\frac{\partial G(z_{ij}, z_{ij}^t)}{\partial z_{ij}} = 0$. Then, we can obtained:

$$u_{ij}^{t+1} \leftarrow u_{ij}^t \cdot \frac{(XAZ + \beta W^U U)_{ij}}{(UZ^T A^T AZ + \beta D^U U + \lambda UE)_{ij}} \tag{16}$$

$$z_{ij}^{t+1} \leftarrow z_{ij}^t \cdot \frac{(A^T X^T U + \alpha A^T W^V AZ)_{ij}}{(A^T AZU^T U + \alpha A^T D^V AZ + \lambda A^T AZE)_{ij}} \tag{17}$$

Equations (17) and (16) have the same form as Eqs. (8) and (9) after rewriting while $G(z_{ij}, z_{ij}^t)$ and $G(u_{ij}, u_{ij}^t)$ are the auxiliary functions of $F_{ij}(z_{ij})$ and $F_{ij}(u_{ij})$. It is easy to deduce from Lemmas 2 and 3 that $F_{ij}(z_{ij})$ and $F_{ij}(u_{ij})$ are not increased under the update rules of Eqs. (8) and (9), so Theorem 1 ensures that our CDNMF algorithm is convergent.

3   Experiments and Result Analysis

3.1   Compared Algorithms and Data Sets

We choose seven representative algorithms, k-means, NMF [21], GNMF [10], DNMF [13], CNMF [18], GSNMFC [19] and SDGNMF [15], to compare with our proposed CDNMF, for the data representation obtained by NMF-based algorithms, we use K-means clustering to get the clustering results.

In the experiments, five well-known baseline image data sets of different features and sizes are used, including UMIST[1], JAFFE [22], COIL20 [23], Optdigits[2] and Pointing04[3]. The important statistical summary of these data sets is shown in Table 1. All experiments were run 20 times independently and then averaged.

Table 1. Description of the five image data sets

Data sets	No. of features (m)	No. of samples (n)	No. of classes (c)	Data types
UMIST	10304	575	20	Face image
JAFFE	4096	213	10	Facial expression
COIL20	1024	1440	20	Object image
Optdigit	64	3823	10	Handwritten digit
Pointing04	1728	1395	15	Head pose

In order to measure the clustering performance, two common evaluation indexes namely clustering accuracy (AC) [24] and normalized mutual information (NMI) are utilized [24].

---

[1] http://images.ee.umist.ac.uk/danny/database.html.
[2] http://archive.ics.uci.edu/ml/machine-learning-datasets/optdigits/.
[3] http://www-prima.inrialpes.fr/Pointing04/data-face.html/.

3.2 Sensitivity to Parameters

There are four main parameters for our proposed CDNMF, which are the two dual-graph regularization parameters (α and β), the bi-orthogonal parameter λ and a nearest neighbor number k. This part discusses the influence of different parameter values on the performance of CDNMF in five data sets. We select all samples from each data set for clustering, and 20% of each class of samples is randomly selected as labeled samples for the construction of a constraint matrix A. The experiments comprise of three major parts, and Fig. 1 shows the influence of different parameters on the performance of CDNMF.

– Set λ ranging in [0.0001, 0.001, 0.01, 0.1, 0.5, 1, 10, 50, 100, 500, 1000]. We can see that CDNMF are sensitive to parameter λ, whereby proper selection of the parameter λ can improve the performance of the proposed algorithm. The bi-orthogonal parameter λ may not be sufficient to affect factor matrices U and AZ when it is too small, but performance degradation may be due to the distortion of U and AZ when λ is too large.
– Set α ranging in [0.1, 1, 10, 100, 500, 1000], and set β to the same value for simplicity. It is worth noticing that when α is small, the clustering performance of CDNMF is not ideal because the small α plausibly neglects the geometric structure of data manifold and feature manifold. The clustering performance of CDNMF has been improving with the increase of α. When α is greater than 10, the clustering performance of the algorithm tends to be stable in most case, indicating the robustness of the proposed algorithm relative to the change of regularization parameters (α and β).
– CDNMF captures the hidden geometric structures of data manifold and feature manifold through data graph and feature graph. The increase of k does not improve the clustering performance of CDNMF because the graphs cannot reflect the hidden structure of data manifold and feature manifold as the neighborhood range increases.

3.3 Analysis of Convergence

The updating rules used to minimize the CDNMF objective function is essentially an iterative process. It is proved theoretically that these renewal rules are convergent in Theorem 1. In this part, the convergence rate of these updating rules is studied experimentally.

Figure 2 displays the convergence curves of CDNMF on five data sets. In the experiments, all samples of each data sets are factorized. Although the sizes of the data sets are different, CDNMF typically converge in 5 iterations, it is shown that CDNMF has a relatively stable and fast convergence rate.

3.4 Analysis of Clustering Results

In order to show the performance of clustering, we compare our algorithm with other related methods in this part.

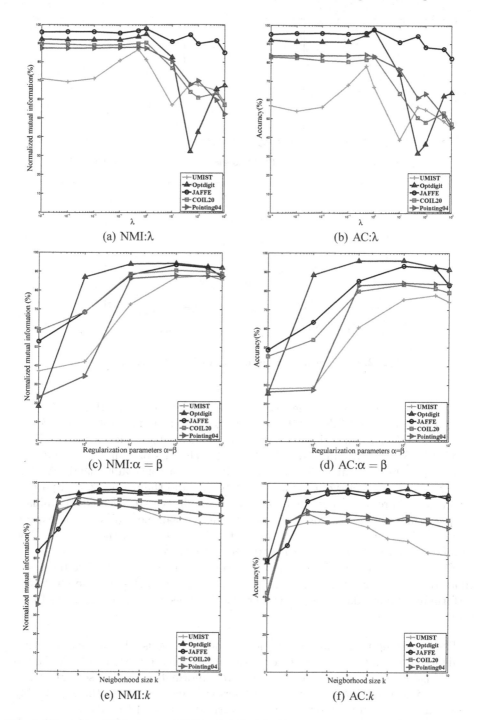

(a) NMI:λ

(b) AC:λ

(c) NMI:α = β

(d) AC:α = β

(e) NMI:*k*

(f) AC:*k*

Fig. 1. Clustering performance NMI and AC of CDNMF versus the value of parameters

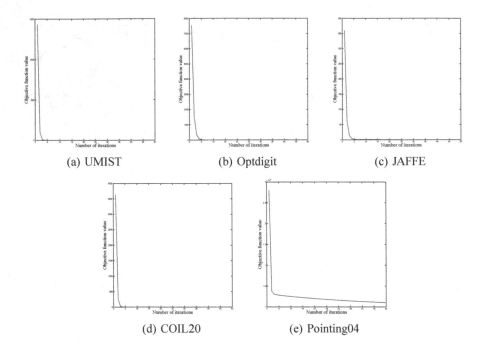

(a) UMIST   (b) Optdigit   (c) JAFFE

(d) COIL20   (e) Pointing04

Fig. 2. Convergence curves of CDNMF on five data sets

All selected graph-based algorithms (GNMF, DNMF, GSNMFC, SDGNMF and our proposed CDNMF) are 0–1 weighted, while the number of nearest neighbors is set to 5 as recommended in [10]. For semi-supervised algorithms (CNMF, GSNMFC, SDGNMF and CDNMF), 20% of each class of samples is randomly selected as labeled samples for the construction of a constraint matrix A. The bi-orthogonal parameter of CDNMF is set to 0.5, and the regularization parameters α and β are set to 100. Different cluster numbers c are selected to evaluate the performance of the algorithms under different sample scales in the experiments, and we choose a subset of the first c categories from the data set. The parameter values of each algorithm are set as suggested.

Figure 3 show the performance of different algorithms on the five data sets under different clustering numbers c, and Table 2 shows the average clustering performance on five data sets. Several crucial findings could be observed as listed below:

1: The clustering performance of k-means and NMF is usually lower than other algorithms and CNMF has better performance than k-means and NMF algorithms in most cases as it utilizes the label information of data set. But these three algorithms often show poorer performance than the others probably because they ignore the inherent geometric structure of the data.

2: Compared to k-means and NMF, the algorithms (GNMF, GSNMFC, DNMF, SDGNMF, and CDNMF) demonstrate better performance in most cases due

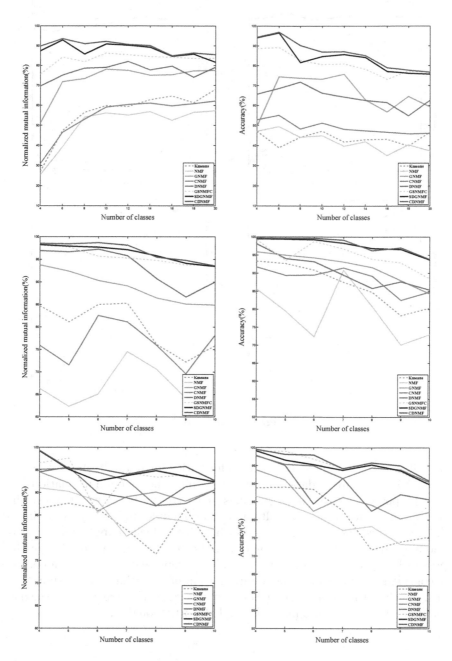

Fig. 3. Clustering performance NMI and AC on the five data sets (from top to bottom: UMIST, Optdigit, JAFFE, COIL20 and Pointing04).

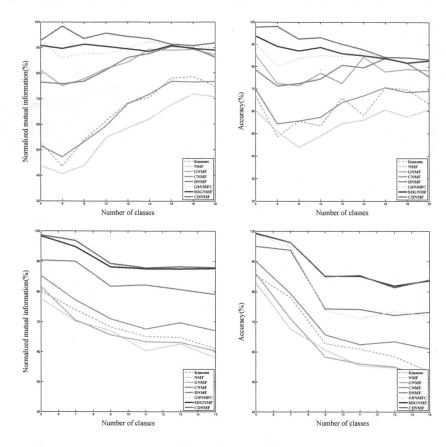

Fig. 3. (*continued*)

to utilizing the manifold structure information of data. It shows that the geometric structure of data is useful in learning the hidden information.

3: The proposed CDNMF algorithm has achieved the best average clustering results for all five data sets, although they include high complexity due to different types and sizes of images such as faces, facial expressions, objects, and handwritten numbers. This finding indicates that our proposed algorithm possess better learning ability for parts-based representations of data because it considers all kinds of valuable information of the original data simultaneously, such as the prior-label information, the geometry structures of the data manifold and the feature manifold, and the orthogonality.

Table 2. Comparison of average clustering performance on five data sets

Data sets	Kmeans	NMF	GNMF	CNMF	DNMF	GSNMFC	SDGNMF	CDNMF
(a) NMI								
Normalized Mutual Information (%)								
UMIST	56.31	50.34	73.05	54.85	83.28	77.24	87.61	**89.31**
Optdigit	80.05	67.11	88.87	76.39	95.59	93.43	96.33	**96.79**
JAFFE	83.19	85.71	90.08	91.85	93.34	91.36	94.51	**95.38**
COIL20	64.61	57.04	83.87	64.53	89.67	83.42	89.96	**93.59**
Pointing04	68.92	66.04	67.45	72.98	91.27	84.03	90.68	**91.48**
(b) AC								
Accuracy (%)								
UMIST	43.51	42.21	65.53	49.02	80.81	64.17	83.92	**85.96**
Optdigit	86.82	78.87	91.72	88.36	94.77	90.51	97.63	**97.86**
JAFFE	81.42	79.15	85.74	94.05	95.35	89.14	94.78	**95.93**
COIL20	61.36	54.83	77.57	64.03	84.21	77.54	86.41	**90.07**
Pointing04	66.52	61.92	62.91	70.36	88.74	80.43	88.52	**89.12**

4   Conclusions

In this work, a novel NMF method called CDNMF is proposed by combining manifold learning and semi-supervised learning. The algorithm extracts the hidden geometric structure information of the data and the feature spaces by constructing a data graph and a feature graph. Moreover, CDNMF also combines the priori label information and independent bi-orthogonal constraints to enhance the discriminating ability for data in the new representation space. Herein, the multiplicative update rules of CDNMF is proposed, and the convergence of update rules is proved. The experimental results on the five image benchmark data sets verify the effectiveness of the proposed algorithm. Compared with several existing graph embedding and semi-supervised algorithms, CDNMF demonstrate better clustering performance on both AC and NMI.

Acknowledgments. This work is supported by the Fundamental Research Funds for the Central Universities No. 2015ZCQ-LY-01, and the National Natural Science Foundation of China under Grant No. 61571046.

References

1. He, X., Yan, S., Yuxiao, H., Niyogi, P., Zhang, H.: Face recognition using Laplacianfaces. IEEE Trans. Pattern Anal. Mach. Intell. **27**, 328–340 (2005)
2. Wold, S., Esbensen, K., Geladi, P.: Principal component analysis. Chemometr. Intell. Lab. Syst. **2**(1), 37–52 (1987)
3. Belhumeur, P.N., Hespanha, J.P., Kriegman, D.J.: Eigenfaces vs. Fisherfaces: recognition using class specific linear projection. In: Buxton, B., Cipolla, R. (eds.) ECCV 1996. LNCS, vol. 1064, pp. 43–58. Springer, Heidelberg (1996). https://doi.org/10.1007/BFb0015522

4. Lee, D.D., Seung, H.S.: Learning the parts of objects by non-negative matrix factorization. Nature **401**(6755), 788–791 (1999)
5. Wang, C., Song, X., Zhang, J.: Graph regularized nonnegative matrix factorization with sample diversity for image representation. Eng. Appl. Artif. Intell. **68**, 32–39 (2018)
6. Meng, Y., Shang, R., Jiao, L., Zhang, W., Yuan, Y., Yang, S.: Feature selection based dual-graph sparse non-negative matrix factorization for local discriminative clustering. Neurocomputing **290**, 87–99 (2018)
7. Yang, S., Zhang, L., He, X., Yi, Z.: Learning manifold structures with subspace segmentations. IEEE Trans. Cybern. **PP**(99), 1–12 (2019). https://ieeexplore.ieee.org/document/8645761
8. Song, M., Peng, Y., Jiang, T., Li, J., Zhang, S.: Accelerated image factorization based on improved NMF algorithm. J. Real-Time Image Proc. **15**(1), 93–105 (2018)
9. Nikunen, J., Diment, A., Virtanen, T.: Separation of moving sound sources using multichannel NMF and acoustic tracking. IEEE/ACM Trans. Audio Speech Lang. Process. **26**(2), 281–295 (2017)
10. Cai, D., He, X., Han, J., Huang, T.S.: Graph regularized nonnegative matrix factorization for data representation. IEEE Trans. Pattern Anal. Mach. Intell. **33**(8), 1548–1560 (2011)
11. Zhu, W., Yan, Y., Peng, Y.: Topological structure regularized nonnegative matrix factorization for image clustering. Neural Comput. Appl., 1–19 (2018). https://link.springer.com/article/10.1007/s00521-018-3572-4
12. Wang, S., Huang, A.: Penalized nonnegative matrix tri-factorization for co-clustering. Expert Syst. Appl. **78**, 64–73 (2017)
13. Shang, F., Jiao, L.C., Fei, W.: Graph dual regularization non-negative matrix factorization for co-clustering. Pattern Recogn. **45**(6), 2237–2250 (2012)
14. Ruicong, Z., Markus, F., Qiuqi, R., Bastiaan, K.W.: Graph-preserving sparse non-negative matrix factorization with application to facial expression recognition. IEEE Trans. Syst. Man Cybern. B Cybern. **41**(1), 38–52 (2011)
15. Sun, J., Wang, Z., Sun, F., Li, H.: Sparse dual graph-regularized NMF for image co-clustering. Neurocomputing **316**, 156–165 (2018)
16. Zhu, W., Yan, Y.: Label and orthogonality regularized non-negative matrix factorization for image classification. Sig. Process. Image Commun. **62**, 139–148 (2018)
17. Meng, Y., Shang, R., Jiao, L., Zhang, W., Yang, S.: Dual-graph regularized nonnegative matrix factorization with sparse and orthogonal constraints. Eng. Appl. Artif. Intell. **69**, 24–35 (2018)
18. Liu Haifeng, W., Zhaohui, L.X., Deng, C., Huang, T.S.: Constrained nonnegative matrix factorization for image representation. IEEE Trans. Pattern Anal. Mach. Intell. **34**(7), 1299–1311 (2012)
19. Sun, F., Meixiang, X., Xuekao, H., Jiang, X.: Graph regularized and sparse non-negative matrix factorization with hard constraints for data representation. Neurocomputing **173**(P2), 233–244 (2016)
20. Logothetis, N., Sheinberg, D.: Visual object recognition. Annu. Rev. Neurosci. **19**, 577–621 (1996)
21. Lee, D.D., Sebastian Seung, H.: Algorithms for non-negative matrix factorization. In: Leen, T.K., Dietterich, T.G., Tresp, V. (eds.) Advances in Neural Information Processing Systems 13, pp. 556–562. MIT Press, Cambridge (2001)
22. Lyons, M.J., Budynek, J., Akamatsu, S.: Automatic classification of single facial images. IEEE Trans. Pattern Anal. Mach. Intell. **21**(12), 1357–1362 (1999)

23. Nene, S.A., Nayar, S.K., Murase, H.: Columbia object image library (COIL-20). Technical report CUCS-005-96, Department of Computer Science, Columbia University, February 1996
24. Salehani, Y.E., Arabnejad, E., Cheriet, M.: Graph and sparse-based robust nonnegative block value decomposition for clustering. IEEE J. Sel. Top. Signal Process. **12**(6), 1561–1574 (2018)

A Spiking Neural Network Architecture for Object Tracking

Yihao Luo[1], Quanzheng Yi[1], Tianjiang Wang[1(✉)], Ling Lin[1], Yan Xu[1], Jing Zhou[1,2], Caihong Yuan[1,3], Jingjuan Guo[1,4], Ping Feng[1,5], and Qi Feng[1]

[1] School of Computer Science and Technology,
Huazhong University of Science and Technology, Wuhan 430074, China
tjwang@hust.edu.cn
[2] School of Mathematics and Computer Science, Jianghan University,
Wuhan 430056, Hubei, China
[3] School of Computer and Information Engineering, Henan University,
Kaifeng 475004, China
[4] School of Information Science and Technology, Jiujiang University,
Jiujiang 332005, China
[5] School of Information, Guizhou University of Finance and Economics,
Guiyang 550025, Guizhou, China

Abstract. Spiking neural network (SNN) has the advantages of high computational efficiency, low energy consumption, low memory resource consumption, and easy hardware implementation. But its training algorithm is immature and inefficiency which limits the applications of SNN. In this paper, we propose a SNN architecture named SiamSNN for object tracking to avoid the training problems. Specifically, we propose a more comprehensive parameter conversion scheme with the processes of standardization, retraining, parameter transfer, and weight normalization, in order to convert a trained CNN to a similar SNN. Then we propose an encoder named Attention with Average Rate Over Time (AAR) in order to encoding images to spiking sequences. By using IF model, the accuracy decreases by only 0.007 on MNIST compared to the original method. Our approach applies SNN to object tracking and achieves certain effects, which is a reference for SNN applications in other computer vision areas in the future.

Keywords: Spiking neural network · Object tracking · Conversion · Encoder

1 Introduction

Spiking Neural Network (SNN) is known as the "third-generation neural network", which simulates the information processing mechanism of biological neurons and has a high degree of bionics. It has become the focus of research in pattern recognition such as image classification. It belongs to the frontier technology research topic in the field of artificial intelligence, and has the advantages

© Springer Nature Switzerland AG 2019
Y. Zhao et al. (Eds.): ICIG 2019, LNCS 11901, pp. 118–132, 2019.
https://doi.org/10.1007/978-3-030-34120-6_10

of high computational efficiency, low energy consumption, low resource consumption, easy hardware implementation, etc. It is an ideal choice for researching brain-like calculation and coding strategies. Through the theoretical and applied research on SNN, it is of great significance to promote the development of artificial neural networks. It can also promote the research of edge devices such as new artificial intelligence chips that are not Von Neumann architecture.

At present, there are some preliminary results for SNN research, but its application is still in its infancy, mainly used for handwritten digit recognition, image segmentation, etc., and it is difficult to be applied to complex visual scenes. The key to this problem is that the neuron function in SNN cannot be differentiated and its hard to be trained using traditional backpropagation. However, the training algorithm with too low efficiency cannot overcome the training problem of complex SNN model, which brings a bottleneck to the popularization and application of SNN.

On the other hand, object tracking is an important research in the field of computer vision. It has specific applications in many fields such as autonomous driving, security, behavior recognition and human-computer interaction. In recent years, deep learning models based on Convolutional Neural Network (CNN) [1] and automatic encoder (AE) [2] have made a lot of progress in tracking technology. This is because that the depth model has significant feature extraction capabilities. However, due to its large amount of computation, large resources, and the need to rely on top-level GPU acceleration, these models cannot be applied to edge devices. However, if it can take advantages of computationally efficient and easy hardware implementation in SNN, it is possible to apply the target tracking algorithm to the edge devices. However, SNN has not been applied to object tracking.

Therefore, this paper combines deep neural network with spiking neural network, and constructs an object tracking model based on SNN. We avoid the difficulties of SNN training through a conversion scheme, which can advance the progress of SNN related theoretical research and provide reference for SNN to apply to more computer vision problems in the future. On the other hand, the tracking model based on SNN can reduce the computational resource occupation, reduce the power consumption generated by the calculation and the degree of dependence on hardware when a certain tracking effects is achieved. It provides a new method for the application of complex deep learning techniques to edge devices such as tracking.

2   Related Work

Although deep neural networks are historically brain-inspired, there are fundamental differences in their structure, neural computations, and learning rule compared to the brain [3]. SNN has good biomimetic properties, delivering and processing information with precise spiking sequences. And it has theoretically been shown to have Turing-equivalent computing power [4]. The combination of SNN and current deep learning models includes how to build more complex deep

SNNs, overcome training problems and convert pre-trained model parameters to SNN.

On conversion research Neil et al. [5] studied to convert the model parameters of the trained fully connected neural network to the SNN model weights with similar structure through mapping, and achieved the similar accuracy on MNIST while reducing the power consumption and delay of the model. With the development of CNN, the effect on computer visual tasks is better than fully connected networks, which promotes the research of the combination of CNN and SNN. Xu et al. [6] proposed CSNN structure, it inputs the features extracted by CNN to the classification layer of SNN, achieving an accuracy of 88% on MINST, which verifies the possibility of end-to-end training CNN and SNN. Furthermore, some researchers have tried to convert the weighted CNN structure into a SNN structure, it can be used with less operation and consume less energy [7], which can apply CNN to edge devices. Diehl et al. [8] used weight normalization to improve the architecture for reducing performance loss. Rueck-auer et al. [9] proposed several conversion criteria to support the biasing of the original CNN and maximal pooling layer converted to SNN, and they tried to identify targets that are more difficult than MNIST (e.g. CIFAR-10 [10] and ImageNet [11]). However, this conversion method is complicated and limited by the specific neuron model, and it is difficult to apply it in SNN of a different neuron model. [12,13] studies how to convert deeper depth CNN structures such as ResNet [14] to SNN.

On object tracking algorithms based on deep feature similarity have recently made significant progress. They can use a large amount of training data for offline learning, and these models can achieve high accuracy [15]. SiamFC [16] uses full convolution and similarity learning to solve the tracking problem, and finally determines the location of the object through the response score map. In this paper we propose a tracking model with SNN based on SiamFC. SiamRPN [17] introduced region proposal network (RPN) [18] in the field of object detection, avoiding multi-scale testing through network regression, it improves the speed while directly obtaining a more accurate target position through the regression of RPN. DaSiamRPN [19] proposed a distractor-aware feature learning scheme based on SiamRPN which significantly improves the discriminative power of the network, it obtains state-of-the-art accuracy and speed.

Some experimental results show that SNN methods have similar effects as the traditional deep learning methods, but SNN usually requires less operation [9]. In addition, SNN is a structure that mimics the human brain and has a potential to perform better than traditional neural network in the future, it has important research value.

3   Conversion Scheme

3.1   Background

The spiking neuron model used for our work is the integrate-and-fire (IF) model. The membrane potential $V(t)$ of a spiking neuron in the SNN architecture is updated at each time step by the following equation:

$$V(t) = V(t-1) + L + \sum_{n=1}^{N} w_i p_i(t)$$

$$\text{If } V(t) \geq \theta, \text{ spike and reset } V(t) = 0$$

$$\text{If } V(t) < Vmin, \text{ reset } V(t) = Vmin$$

(1)

We save the last membrane potential $V(t-1)$, and then calculate the current $V(t)$ corresponding to the voltage by the current spike. $p_i(t)$ represents the upper layer of neurons output spike which is present at the current time, the value is 1 when it produces a spike, and the value is 0 when there is no spike. Where L is the constant parameter, $\sum_{n=1}^{N} w_i p_i(t)$ is the summed input at time t from all synapses connected into the neuron. Whenever $V(t)$ exceeds the voltage threshold θ, the neuron fires and produces a spike (output 1), and its membrane potential $V(t)$ is reset to zero. The membrane potential V is not allowed to go below its resting state $Vmin$ which is usually set to 0, but it can be also changed to allow V to go negative. The parameters used in our simulation can be found in Sect. 5.

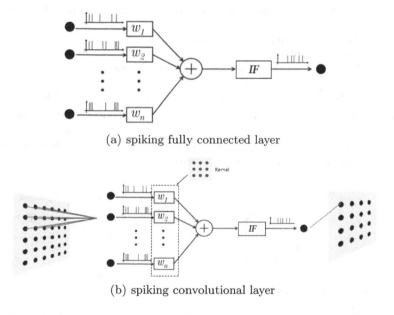

(a) spiking fully connected layer

(b) spiking convolutional layer

Fig. 1. How the spiking fully connected layer and spiking convolutional layer work.

As shown in Fig. 1, spiking fully connected layer has a very similar form to the fully connected layer in the traditional neural network. The latter directly inputs the result of the operation into the activation function to generate the output value of the neuron, and the former needs to accumulate the IF neuron membrane voltage continuously during the simulation time. The IF neurons are equivalent to the activation function in the traditional neural network, which increase the ability of neurons by nonlinear processing.

In the convolution calculation, the same weight matrix is used for the different receptive field regions in an image. Spiking convolution layer also uses this strategy to achieve spiking calculation. On each input image, the local spiking feature map of each receptive field region are calculated convolution operator. Therefore, the calculation method is equivalent to fully connected layer. Membrane voltage of the IF neuron is cumulatively calculated, and then it obtains a spiking feature map.

3.2   Challenges in Conversion

In the specific calculation, the SNN is determined to be different because the input is a spiking sequence. But the parameters can be the same as CNN. In theory, the similarity between structure and parameters can support the conversion from CNN to SNN. But directly using parameter transfer have many problems and challenges, which will cause excessive loss of accuracy [7]:

(1) Negative output values:
   a. The activation function tanh() has output values between −1.0 and 1.0
   b. Weights and biases can be negative may causing the output value to be negative
   c. preprocessing may produce negative values.
(2) Representation problems:
   a. The biases in each convolution layer can be positive or negative, which cannot be represented easily in SNN.
   b. Max-pooling requires two layers of spiking networks. This approach requires more neurons and can cause accuracy loss due to the added complexity.
   c. Softmax layer, batch normalization layer (BN), local response normalization layer (LRN) cannot be represented directly.

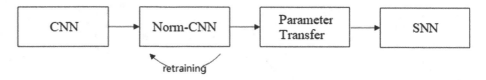

Fig. 2. Flow-diagram of converting a CNN into SNN architecture.

To solve these problems from converting CNN to SNN, we need two steps as Fig. 2:

(1) Model normalization: Modify the structure of the CNN to become the specific structure of the Norm-CNN, and then retrain it.
(2) Parameter Transfer: Transfer the parameters of Norm-CNN to SNN with weight normalization.

3.3   Model Normalization

We need process CNN into a similar form (Norm-CNN) by some operations for converting to SNN easily. These operations include:

(1) Remove biases from all convolution and the fully connected layers, the kernel size and initialization settings are unchanged.
(2) Where the original activation function is used, it is replaced with the ReLU function, in order to avoiding the negative numbers and reducing the loss of precision after conversion. If the original structure is not activated after the convolutional layer or fully connected layer, the ReLU layer needs to be added later.
(3) If the network uses a single-spike output neuron, the pooling layer maintains the original Max-Pooling layer or the Average-Pooling layer. If a multi-spike output neuron is used, the pooling layer needs to modify the Max-Pooling layer to the Average-Pooling layer.
(4) Expect the output layer, we use L2 regularization during training in order to accelerating the convergence of weights to a smaller range and avoiding model overfitting.
(5) Remove LRN, BN, etc. layer that cannot be directly represented in SNN. Meanwhile, to avoid the model doesnt converge during training, the input image needs to be normalized in a positive range.
(6) Model compression can be performed by converting weights to 16-bit floats.

Studies have shown that in a convolutional neural network, the use of a 16-bit floating-point type can achieve the same effect as a 32-bit floating-point type [20]. The simulation of the network computing process on the GPU, with the underlying support of CUDA, is twice as fast as the 32-bit floating point type. In addition, the model is quantized and compressed to reduce the bit width, which is more conducive to hardware implementation, breaking the limitation of hardware on type accuracy and size.

Because we modified the network structure, we should retrain the model to get the parameters.

3.4   Parameter Transfer

Because of the similarity between the Norm-CNN and SNN, the activation function in the original model is replaced with the IF neurons. Other parts of the SNN model can transfer the trained parameters.

After that, the SNN model can perform feature extraction or classification, but it may not achieve the desired result. This is because the original weight is matched to the discrete eigenvalues instead of the spiking value (e.g. 0 or 1) of the previous layer. In the network with voltage threshold obtained by converted, when the forward process calculation is performed according to Eq. 1, it is likely that the result of the accumulated membrane voltage will far exceed the threshold value, or cannot be reached at all, resulting in high spiking rate or very low spiking rate [8].

In order to perform spiking activation more effectively, we fixedly set the membrane voltage threshold of each neuron in the network to 1, and the resting potential to 0. The weights are normalized so that the membrane voltage threshold and weight can be adapted for producing spiking properly [7]. This process requires the participation of some training samples, but does not require labels.

If the weight of the lth layer in the network is W^l, we can get all the output values greater than 0 in this layer by inputting the samples into Norm-CNN, and sort them from small to large. Then we select the Kth (often set to 99.9%) one as the scaling factor λ^l of this layer [9], and the new weight is calculated according to Eq. 2. This completes the normalized calculation process and sets the normalized parameters as final parameters.

$$W^l \rightarrow W^l \cdot \frac{\lambda^{l-1}}{\lambda^l} \tag{2}$$

Weight normalization is an optional operation, and there are different ways of dealing with complex networks or different network model. In addition, K can also be adjusted according to the specific application effect.

Before applying the converted SNN model to a specific problem such as classification, the following operations are required:

(1) Encoding: Input image should be encoded to become a spiking sequence before forward calculation.
(2) Output: The output spiking sequence should be processed. Since the Softmax layer is removed, it is necessary to count the output result obtained at each moment of the last layer, calculate the total number of spiking in each neuron. Then output the classification category according to the category of the neuron which produces the maximum spiking numbers.

Now the entire CNN to SNN conversion process is complete. SNN model can be used to process visual tasks through the conversion method which indirectly overcome the problem of end-to-end training inefficiently. Its possible to promote SNN to more complex fields.

4   SiamSNN: A SNN Architecture for Tracking

4.1   AAR: A Spiking Encoder

SNN is different from CNN, the image needs to be encoded into a spiking sequence before input. There are two main types of spiking coding methods in terms of temporal coding and rate coding [21]:

(1) Temporal coding
    The pixel value is encoded with the precise firing time of the information. The specific time at which pixel should be spiking is determined according to the size of the pixel value and the total encoding time. The encoding process in this way is relatively fast, but only produces a single spiking sequence.

The spiking sequence is too sparse, so that membrane voltage may not be accumulated to reach the threshold when calculating, which has a great influence on the result. However, the sparse spiking sequence can generate facilitates the training algorithm to determine the order of the precise firing timing of the spiking, which makes weight adjustment easier. Therefore, the temporal coding mechanism is suitable for the case of end-to-end training SNN.

(2) Rate coding

The pixel value is encoded with the average firing rate of the neurons. This scheme presupposes that the information content is hidden at the rate of spiking. It is necessary to count the number of spiking generated in a certain period to determine the rate of the spiking sequence for obtaining the result. The coding efficiency of this method is relatively low, and each pixel point is spiking at the first moment by default, so the firing time is not accurate. However, the densely distributed spiking sequence is suitable for the converted SNN, which is beneficial to accumulating membrane voltage to reach the threshold in time and realizing the forward transmission of information.

Because we use the converted SNN, this paper proposes a new method based on the rate coding scheme named Attention with Average Rate Over Time (AAR) in order to improve the effect of the converted SNN.

For the input image, the average rate coding scheme is calculated according to Eqs. 3 and 4. $P_{i,j}$ is the value of each pixel in the image, the maximum pixel value in the whole image is $Pmax$, the minimum pixel value is $Pmin$, the total spiking time is T. And the maximum number of spiking is S, which is produced by the pixel with $Pmax$ (The maximum spiking rate equals S/T). The number of spiking for each pixel is $s_{i,j}$, and the corresponding rate is $f_{i,j}$. The result spiking sequence for each pixel is averaged over the total spiking time T by rate $f_{i,j}$.

$$s_{i,j} = 1 + (S - 1) \cdot \frac{P_{i,j} - Pmin}{Pmax - Pmin} \tag{3}$$

$$f_{i,j} = \frac{T}{s_{i,j}} \tag{4}$$

The ability of SNN to extract features is weak relatively, and the original image may have some noise, which affects the effect of the model in applications. Therefore, we convolve the original image to extract the edge features, which can extract better spiking convolution features and suppress the noise. Meanwhile, since the maximum value of the pixel is too large, the proportion of pixels that can reach the maximum number of spiking is small, resulting in information is easily lost. So AAR does the following operations:

(1) Pre-convolution: Use $w = \begin{bmatrix} -2 & 1 & -2 \\ 1 & 5 & 1 \\ -2 & 1 & -2 \end{bmatrix}$ 3×3 receptive field filter to convolute
the original image to obtain a feature map. In a specific application, the size
and value of the filter can be adjusted according to the effect.

(2) Attention processing: Set the maximum eigenvalues for the top 20% of the
eigenvalues in all the eigenvalues of the feature map. This will ensure enough
maximum spiking rate.

(3) The encoding of each pixel is performed according to Eqs. 3 and 4, then the
final spiking map is obtained (Fig. 3).

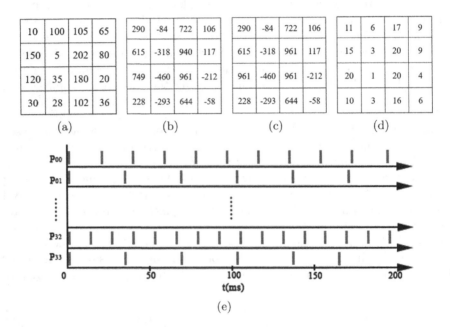

Fig. 3. AAR Calculation schematic diagram. (a) is original image, assume that the
maximum number of spiking $S = 20$, the total spiking time $T = 200$. After the first
AAR convolution, the result is shown in (b). After the attention processing, the result
is shown in (c). The number of spiking for each pixel is shown as (d). The final spiking
coding result is shown in (e).

4.2   SiamSNN Construction

According to the above research, we propose an object tracking model SiamSNN.
It is based on SiamFC [16].

As shown in Fig. 4 the architecture is fully-convolutional with respect to the
input image. The output is a scalar-valued score map whose dimension depends

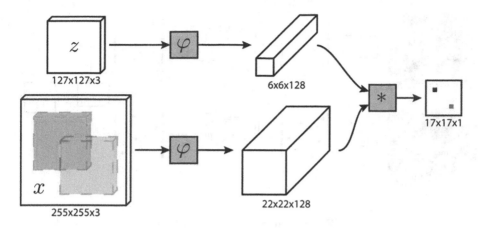

Fig. 4. SiamSNN architecture.

on the size of the search image. It uses similarity learning to solve the tracking problem. The model producing a higher score if the objects are similar, and the target position is predicted by the position of the maximum value in the score graph. During training, SiamFC preprocesses the training set, the images are resized to 255×255 and the center 127×127 area is the ground truth. In the same video, two images in a certain interval are selected to input the network for training. The positive and negative samples are calibrated by the distance from the center point in the response map. In the prediction, the template frame is processed in the same way, and the template frame branch is calculated only once. Then select three region proposals of different scales in the position of the previous frame to obtain three response maps, and select the maximum response value to get the final result.

We construct SiamSNN by the following steps:

(1) Convert CNN part of SiamFC into Norm-CNN, and the modified structure is shown in Table 1. The type of all parameters are float16, and we use ReLU activation function after each conv layer.
(2) Re-train Norm-CNN by using the original training method and data set in SiamFC, and perform weight normalization. Add IF neurons on the Norm-CNN to construct the SNN and transfer the trained parameters.
(3) Add AAR encoder after input layer. We use the same convolution operation to evaluate the similarity of two spiking features.

The input and output sizes of the SiamFC and SiamSNN models have no difference, so we use the same strategy in preprocessing, scale selection, and response map processing. In order to guarantee the calculation speed, we continue to use the convolution operation to measure the similarity between the pulse characteristics. But it is not accurate enough, which is one of the main reasons for the decline in accuracy. We will next study the similarity evaluation method for spiking features to replace the convolution operation (Fig. 5).

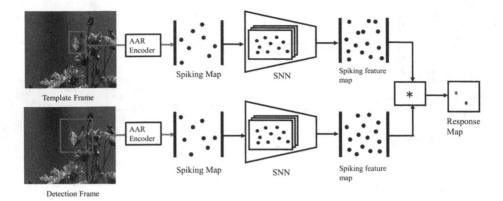

Fig. 5. SiamSNN architecture.

Table 1. Architecture of convolutional embedding function in Norm-CNN. In-channels are the kernel channels matched to input, Out-channels are the numbers of kernels.

Layer	Kernel-size	In-channels	Out-channels	Strides	Detection-size	Template-size
Input	–	3	–	–	127×127	255×255
Conv1	11×11	3	96	2	59×59	123×123
Avg-pooling1	3×3	–	–	2	29×29	61×61
Conv2	5×5	96	256	1	25×25	57×57
Avg-pooling2	3×3	–	–	2	12×12	28×28
Conv3	3×3	256	384	1	10×10	26×26
Conv4	3×3	384	384	1	8×8	24×24
Conv5	3×3	384	256	1	6×6	22×22

Our SiamSNN architecture theoretically takes advantage of high computational efficiency, low energy consumption, low resource consumption, easy hardware implementation in SNN, which makes it possible to apply the model to edge devices.

5   Experimental Results

5.1   Converted SNN on MNIST

We modify LeNet [22] to Norm-LeNet based on the method in Sect. 3.

In the process of modifying CNN to Norm-CNN, although the biases were removed and the Max-pooling layer was changed, the accuracy was only reduced by 0.003 after retraining in Table 2. In this adjustment process, the loss is not a lot. And the parameters are all the float16 type, which is significantly lower than the original model's occupation of space resources.

Then we verify the effects of different total spiking time, normalization, and the maximum spiking rate. The voltage threshold was chosen to be 1 in the experiment. The first number in the model name indicates whether the SNN

Table 2. Comparative experiment on MNIST

Model	Accuracy	Space usage
LeNet	0.983	843 KB
Norm-LeNet	0.980	420 KB

model uses weight normalization, 1 indicates use, and 0 indicates no use. The second number represents the total spiking time. The maximum spiking rate S/T is set to 0.3, 0.4, 0.5, 0.6, 0.7, 0.8, 0.9.

Table 3. Converted SNN on MNIST with different parameters

Model	The maximum spiking rate & accuracy						
	0.3	0.4	0.5	0.6	0.7	0.8	0.9
SNN-0-200	0.562	0.628	0.584	0.677	0.631	0.606	0.591
SNN-0-300	0.595	0.634	0.577	0.679	0.632	0.606	0.595
SNN-1-200	0.972	0.972	0.972	0.975	0.974	0.972	0.972
SNN-1-300	0.971	0.971	0.972	0.974	0.973	0.973	0.971

As shown in Table 3, when weight normalization is performed, the total spiking time is selected to be 200 ms and the maximum spiking rate is 0.6, which can achieve good conversion effect. Finally, we verify AAR encoder. It can be seen from the Table 4 that the AAR coding scheme can achieve better accuracy than the original rate coding scheme, and the accuracy decreases by only 0.007 on MNIST compared to the original method.

5.2 SiamSNN for Tracking

In SiamSNN, we set the total spiking time to be 200 ms, the maximum spiking rate is 0.6, use weight normalization, the constant parameter L in IF model is 0, and the voltage threshold is 1.

As shown in the Fig. 6, the red boxes are the results predicted by SiamSNN, the blue boxes are the ground truth. It can be seen from the image sequence of the first video that the SiamSNN model can achieve the desired tracking effect when the target has no background interference and there is no such problem as blur or severe deformation. In the second video, the model can basically track correctly, but the prediction result and the ground truth are too different. Starting from frame 268, the prediction result is larger than the ground truth. Although the tracking can be completed, it is not very accurate. In the third video, after 1713 frames, the tracking object was lost because of similar interference, and there was no retargeting in the subsequent sequences.

Table 4. Comparative experiment between AAR and original rate coding scheme

Model	Accuracy
LeNet	**0.983**
Rate coding SNN	0.975
AAR-SNN	0.976

Fig. 6. The first row is a video for tracking a walking man. The second row is a video for tracking a surfer's head. The third row is a video for tracking a car. (Color figure online)

It is found that SiamSNN can successfully track the target in most of the previous frames. However, in the case where the disturbance is large and the similarity is too large, the effect of the tracking cannot be ensured. How to improve the robustness of the SNN model after conversion to improve the accuracy of tracking, is the place to be studied in the follow-up research.

6   Conclusion

In this paper, we propose a SNN architecture named SiamSNN for object tracking to avoid the training problems. Specifically, we propose a more comprehensive parameter conversion scheme with the processes of standardization, retraining, parameter transfer, and weight normalization, in order to convert a trained CNN to a similar SNN. Then we propose an encoder named Attention with Average Rate Over Time (AAR) in order to encoding images to spiking sequences. We verify the effect of AAR encoder by experiments, and the converted SNN can reduce the resource consumption and reduce the complexity of the model in the case of classification or tracking.

Meanwhile, there is still a defect that the accuracy is reduced in the converted SNN model. This is mainly caused by the modification of pooling layer and the removal of the original BN layer. And the current similarity matching algorithm of spiking features is not accurate enough. In addition, the robustness of SNN is not as good as CNN. Later, we will work on these places to make SNN achieve better results in the applications.

References

1. Krizhevsky, A., Sutskever, I., Hinton, G.E.: Imagenet classification with deep convolutional neural networks. In: Advances in Neural Information Processing Systems (NIPS), pp. 1097–1105. MIT Press, Lake Tahoe (2012)
2. Zhuang, B., Wang, L., Lu, H.: Visual tracking via shallow and deep collaborative model. Neurocomputing **218**(61), 71 (2016)
3. Tavanaei, A., Ghodrati, M., Kheradpisheh, S.R., et al.: Deep learning in spiking neural networks. Neural Netw. **111**(47), 63 (2019)
4. Maass, W.: Lower bounds for the computational power of networks of spiking neurons. Neural Comput. **8**(1), 1–40 (1996)
5. Neil, D., Pfeiffer, M., Liu, S.: Learning to be efficient: algorithms for training low-latency, low-compute deep spiking neural networks. In: ACM, pp. 293–298 (2016)
6. Xu, Q., Qi, Y., Yu, H., et al.: CSNN: an augmented spiking based framework with perceptron-inception. In: International Joint Conference on Artificial Intelligence (IJCAI), pp. 1646–1652. Morgan Kaufmann, Stockholm (2018)
7. Cao, Y., Chen, Y., Khosla, D.: Spiking deep convolutional neural networks for energy-efficient object recognition. Int. J. Comput. Vis. (IJCV) **113**(1), 54–66 (2015)
8. Diehl, P.U., Neil, D., Binas, J., et al.: Fast-classifying, high-accuracy spiking deep networks through weight and threshold balancing. In: International Joint Conference on Neural Networks (IJCNN), pp. 1–8. IEEE, Killarney (2015)
9. Rueckauer, B., Lungu, I.A., Hu, Y., et al.: Conversion of continuous-valued deep networks to efficient event-driven networks for image classification. Front. Neurosci. **11**, 682 (2017)
10. Krizhevsky, A., Hinton, G.: Learning multiple layers of features from tiny images. Technical report, University of Toronto (2009)
11. Deng, J., Dong, W., Socher, R., et al.: Imagenet: a large-scale hierarchical image database. In: Computer Vision and Pattern Recognition (CVPR), pp. 248–255. IEEE, Miami (2009)
12. Hu, Y., Tang, H., Wang, Y., et al.: Spiking deep residual network. arXiv preprint arXiv:1805.01352 (2018)
13. Sengupta, A., Ye, Y., Wang, R., et al.: Going deeper in spiking neural networks: VGG and residual architectures. Front. Neurosci. **13**, 95 (2019)
14. He, K., Zhang, X., Ren, S., et al.: Deep residual learning for image recognition. In: Computer Vision and Pattern Recognition (CVPR), pp. 770–778. IEEE, Las Vegas (2016)

15. Danelljan, M., Hager, G., Shahbaz, K.F., et al.: Convolutional features for correlation filter based visual tracking. In: Proceedings of the IEEE International Conference on Computer Vision Workshops (ICCV), pp. 58–66. IEEE, Santiago (2015)

16. Bertinetto, L., Valmadre, J., Henriques, J.F., Vedaldi, A., Torr, P.H.S.: Fully-convolutional siamese networks for object tracking. In: Hua, G., Jégou, H. (eds.) ECCV 2016. LNCS, vol. 9914, pp. 850–865. Springer, Cham (2016). https://doi.org/10.1007/978-3-319-48881-3_56

17. Li, B., Yan, J., Wu, W., et al.: High performance visual tracking with siamese region proposal network. In: Computer Vision and Pattern Recognition (CVPR), pp. 8971–8980. IEEE, Salt Lake City (2018)

18. Ma, J., Shao, W., Ye, H., et al.: Arbitrary-oriented scene text detection via rotation proposals. IEEE Trans. Multimed. **20**(11), 3111–3122 (2018)

19. Zhu, Z., Wang, Q., Li, B., Wu, W., Yan, J., Hu, W.: Distractor-aware siamese networks for visual object tracking. In: Ferrari, V., Hebert, M., Sminchisescu, C., Weiss, Y. (eds.) ECCV 2018. LNCS, vol. 11213, pp. 103–119. Springer, Cham (2018). https://doi.org/10.1007/978-3-030-01240-3_7

20. Farabet, C.: Towards real-time image understanding with convolutional networks. Master Degree thesis. University Paris-Est, Paris (2013)

21. Krause, T.U., Wrtz, P.D.D.R.: Rate coding and temporal coding in a neural network. Master Degree thesis. University of Bochum, Germany (2014)

22. LeCun, Y., Bottou, L., Bengio, Y., et al.: Gradient-based learning applied to document recognition. Proc. IEEE **86**(11), 2278–2324 (1998)

Image Dehazing Framework Using Brightness-Area Suppression Mechanism

Shengkui Dai[✉], Xiangcheng Chen, and Ziyu Wang

College of Information Science and Engineering, Huaqiao University, Xiamen, China
D.S.K@126.com, kris9575@sina.com, 904504002@qq.com

Abstract. Since more and more outdoor images are often degraded by haze and suffer from bad visibility, haze removal has become an important task of image restoration in recent decades. A systematic dehazing framework based on Koschmieder model is proposed in this paper, which adopts a novel brightness-area suppression mechanism. Firstly, global brightness-area suppression blending the large-scale atmospheric veil with the result of edge-preserving filtering, could protect the white objects not becoming darker. Then, the local brightness-area suppression based on sky detection could prevent the sky region from over saturation. In addition, post-processing procedures are designed in this dehazing system in order to generate haze-free image with better visual perception. This framework is on-limits and extensible, in that, it can accept other better dehazing technique as one of the core steps inside. Experiments show that the performance of this framework outperforms multiple state-of-the-art dehazing algorithms.

Keywords: Brightness suppression · Transmission · Dehazing · Haze removal · Image restoration

1 Introduction

Haze can seriously accelerate the degeneration of the optical image or video. Therefore, image dehazing research has been one of the hot-spots in both academic and industrial societies [1, 2].

Currently, the mainstream dehazing methods are almost based on atmospheric scattering model [3], or called Koschmieder model. In this model, the key process is the computation of the transmission of the air-light, and the core procedure of it is Edge Preserving Smoothing Filter/Filtering (EPF), which is the research emphasis in recent years.

He [4] proposed dark channel prior and guided image filter to compute the air-light transmission. Essentially, dark channel prior could suppress the pixel value and increase the transmission value around dark pixel. Tarel [5] used median filter as smoothing filter, but median filter was not strict edge-preserving, which resulted in serious halo effect. Berman [6] presented a haze-line dehazing algorithm, which used the WLS-filter [7] as EPF; Cai [8] built the Spatio-temporal Markov Random Field to estimate the transmission. Gibson [9] adopted a two-step method to calculate Wiener

© Springer Nature Switzerland AG 2019
Y. Zhao et al. (Eds.): ICIG 2019, LNCS 11901, pp. 133–145, 2019.
https://doi.org/10.1007/978-3-030-34120-6_11

filter as EPF. Meng [10] used geometric boundary constraint to estimate the accurate transmission with regularization of optimization. Kim [11] proposed an adaptive contrast enhancement algorithm, which computed the transmission map by iterating the optimization function of contrast and information loss. Park [12] raised the optimization process of information entropy and image fidelity to obtain the transmission map, which is also performed in iterative manner.

In recent years, many novel dehazing algorithms have been proposed instead of tradition atmospheric scattering model. MSCNN [13] used a multi-scale Convolutional Neural Network (CNN) to train the dataset that remove the haze on the target fog images. Double-DIP [14] proposed a unified framework for unsupervised layer decomposition of a single image, which based on coupled Deep-Image-Prior (DIP) networks. These novel algorithms had excellent dehazing effects for most haze images, but some images are still slightly deficient.

The above-mentioned methods based on Koschmieder model focused on EPF rather than the systematic framework. They're not entire solutions of haze removal, Therefore, these methods usually exist some defects, e.g. the whole result becomes dark, the sky area is over-saturated, and the details are not sufficient, and so on.

Therefore, in this paper, we attempt to develop a systematic framework for Brightness-Area Suppression Mechanism Dehazing (BASMDE), which includes new brightness-area suppression mechanism and necessary post-processing operations, what's more, it can integrate all other relevant techniques in this solution.

2   Methodology

This section describes an entire dehazing framework based on Koschmieder model. The whole flowchart is shown in Fig. 1, and the model is expressed as follows:

$$J(x) = I(x)t(x) + \mathbf{A}(1 - t(x)) \tag{1}$$

where $J(x)$ is hazy image, $I(x)$ is haze-free image desired, other parameters are need to be evaluated: $\mathbf{A}$ is the intensity of atmospheric light, $t(x)$ is transmission describing the atmospheric light how reach the camera.

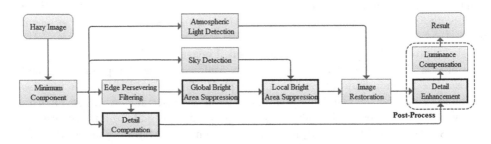

Fig. 1. The flowchart of the proposed dehazing framework

2.1 Sky Detection

The sky area is usually abnormity in color and occurrence of colour spots causes over-saturated after using above algorithms. Dai [15] proposed a method to detect sky region. The parameters k was obtained through the sky analysis function as:

$$k = a_0(\text{B} \times \beta \times \text{Mean}) + 1 \tag{2}$$

where a_0 is a constant and parameter B is a Boolean value to indicate whether there exists sky in hazy image, and parameter β is the confidence coefficient of parameter B. Parameter $Mean$ represent the mean value of the sky area in the hazy image $J(x)$. The parameter k is a scalar of local brightness suppression that will be adopted in Sect. 2.4 to suppress the local brightness-area.

2.2 Atmospheric Light Detection

In order to faster estimate the air light, this paper proposes a statistical method based on the gray intervals of histogram. $J_{min}(x)$ is defined as the minimum component of RGB image, which can be obtained as below:

$$J_{min}(x) = min_{c \in \{r,g,b\}}(J^c(x)) \tag{3}$$

where, $J^c(x)$ is the input hazy image, c is the index of RGB channel. Then, the histogram of J_{min} was computed. When the accumulative value of this histogram accounted for 98% and 99.5% of the total pixel number, the corresponding gray values are A_0 and A_1. Moreover, the atmospheric light values of the RGB channels are as follows:

$$A^c = \frac{\sum J^c(\text{x})}{\text{m}}, \text{s.t.} \{x | A_0 \leq J_{min}(x) \leq A_1\} \tag{4}$$

$$A = \frac{A^r + A^g + A^b}{3} \tag{5}$$

where m is the total number of pixels in the interval of $[A_0, A_1]$ in $J_{min}(x)$.

2.3 Edge-Preserving Filtering

The core step of the dehazing algorithm based on the model (1) is edge-preserving filtering. In the systematic framework of this paper, all kinds of EPFs could suit and be adopted in our EPF module, which is presented as a fixed module. Let the image J_{EPF} be the result of EPF on J_{min}. As shown in Fig. 2(b). the detail layer J_{detail} is obtained by the basic layer J_{EPF} after using the formula (7). The detail layer is as shown in Fig. 2 (c).

$$J_{EPF}(x) = \text{EPF}(J_{min}(x)) \tag{6}$$

$$J_{detail}(x) = J_{min}(x) - a_1 \times J_{EPF}(x) \tag{7}$$

where, a_1 is the detail control parameter of the detail image, and its value is within the interval [0.8, 1.0].

(a) $\mathbf{J}_{min}$ (b) $\mathbf{J}_{EPF}$ (c) $\mathbf{J}_{detail}$

(d) $\mathbf{J}_{veil}$ (e) $\mathbf{J}_{global}$ (f) $\mathbf{J}_{local}$

Fig. 2. Intermediate results of brightness-area suppression

2.4 Brightness-Area Suppression

There are sky areas and white objects in the hazy images, the white areas of hazy images will darken after dehazing, such as: sky, white cars, the zebra of road signs, etc.

In theory, for the part of the brighter, the J_{EPF} value is larger generally, which will lead to a smaller transmission and the luminance of these brightness areas will be suppressed in the restored result. In order to solve this defects, a brightness-area suppression method is proposed to decrease the brightness of J_{EPF}.

A large-scale mean filtering was performed on J_{min} to obtain the atmospheric veil J_{veil}, as shown in Fig. 2(d). Then, by fusing J_{EPF} and J_{veil} with formula (8), the global brightness-area suppression is achieved.

$$J_{global}(x) = \min(a_2 \times J_{EPF}(x), \ J_{veil}(x)) \tag{8}$$

where a_2 is the fusion coefficient, J_{global} is the image after global brightness-area suppression, as shown in Fig. 2(e). By comparing between Fig. 2(b) and (e), it can be seen that the brightness of the sky areas, roads and building are suppressed as a whole.

After performing the above operation, if J_{global} value of the sky-area is still larger, the transmission will get smaller. It would lead to the color of the sky-area be abnormal and over-saturated. Therefore, a local brightness-area suppression method is proposed to overcome this disadvantages as formula (9).

$$J_{local}(x) = \min\left(\frac{|gray(x) - A|}{k}, 1\right) \times J_{global}(x) \tag{9}$$

where the gray is the grayscale of the input hazy image, k is the local brightness suppression parameter proposed in Sect. 2.1, and J_{local} is the result of local brightness-area suppression as shown in Fig. 2(f). By comparison of Fig. 2(e) and (f), the brightness of the sky region of Fig. 2(f) is significantly suppressed, which will increase the value of the sky region of the transmission map, thus obtaining a without anomalous dehazing image.

Consequently, transmission map could be deduced as formula (10).

$$t(x) = 1 - \frac{J_{local}(x)}{A} \tag{10}$$

By further deformation [16] of t via formula (11),

$$t(x) = 1 + t_0 - \omega \frac{J_{local}(x)}{A} \tag{11}$$

where ω is a parameter to control the degree of haze removal, and its value is in the interval [0, 1]. t_0 is a control parameter to increase the overall brightness of the restored image, and its value interval is [0.05, 0.15].

2.5   Image Restoration

According to the physical model (1), the restored result J_{dehaze} of the hazy image is listed as:

$$J_{dehaze}^c(x) = \frac{J^c(x) - \mathbf{A}^c}{t(x)} + \mathbf{A}^c \qquad (12)$$

where $\mathbf{A}$ is the atmospheric light values calculated by the Eq. (5), $t(x)$ is the transmission map after the brightness-area suppression, $J(x)$ is the hazy image. The result image J_{dehaze} is dehazing image after restore, as shown in Fig. 4(a).

2.6   Post-processing Procedures

Comparing with the Fig. 3(b), it is obvious that, in Fig. 4(a), after brightness-area suppression, the sky area is no longer distorted and the road looks more natural, however, the result image is still depressing with worse perceptive effect. In order to further improve the visual perception, a post-processing procedure is performed on the restored result J_{dehaze}.

(a) t from $\mathbf{J}_{\text{EPF}}$ (b) result of (a)   (c) t from $\mathbf{J}_{\text{local}}$

Fig. 3. Comparisons of brightness-area suppression

The first step shown in the formula (13) is to enhance the details for J_{dehaze}.

$$J_{sum}^c(x) = J_{dehaze}^c(x) + J_{detail}(x) \qquad (13)$$

The next step is to improve the luminance of J_{sum}, which adopts the fitting function [17] of weber's law curve.

$$J_{final}^c(x) = \frac{J_{sum}^c(x) \times (255 + a_3)}{J_{sum}^c(x) + a_3} \qquad (14)$$

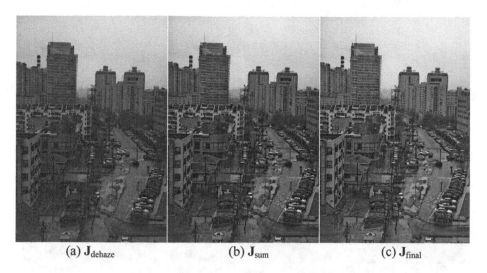

(a) J_{dehaze} (b) J_{sum} (c) J_{final}

Fig. 4. Demonstration of post-processing

where a_3 is the controlling parameter, and the smaller the value a_3 is, the greater the brightness adjustment degree will be.

The results of the post-processing operations are shown in Fig. 4(b) and (c). It can be seen that the final output of the proposed framework has better degree of naturalness, without obvious defects on the other side.

3 Experimental Results

In this section, the results of dehazing image of three comparison algorithms for Meng [10], Berman [6] and Cai [8] are listed in Fig. 5(b)–(d). Several classic EPFs such as Bilateral Filter (BF) [18], WLS Filter [7], RTV Filter [19], and Recursive Filtering (RF) [20] are selected as the optional EPF in the EPF module in framework. Dehazing images using different EPFs in BASMDE framework of this paper are shown in Fig. 5 (e)–(h). Furthermore, Fig. 6 is a comparison of the other dehazing results of BASMDE framework with MSCNN [13] and Double-DIP [14].

To evaluate the superiority of the proposed method, it is compared with those of Meng [10], Berman [6], Cai [8], MSCNN [13], Double-DIP [14] and the EPF module in this framework using different EPFs dehazing result. Ten classic test images were listed for subjective and objective evaluation.

In the following experiments, for all competitor algorithms and BASMDE, parameter $\omega = 0.9$, $a_0 = 1.1$, $a_1 = 0.95$, $a_2 = 0.95$, and $a_3 = 255 + 4 * \text{mean}(J_{sum})$.

3.1 Subjective Evaluation

As can be seen from Fig. 5(a)–(d), all the haze-free images resulted from the Meng, Berman and Cai have a certain degree of distortion in the sky area. The detail of Cai's

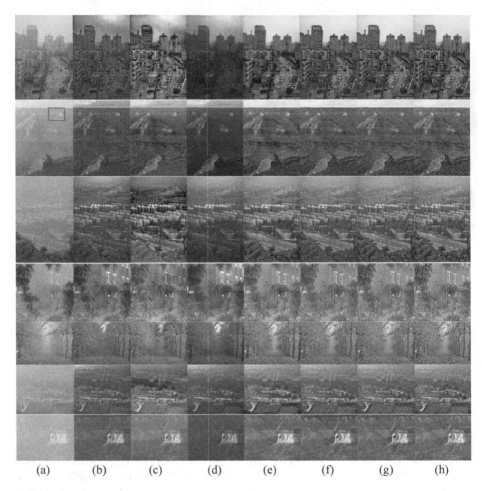

Fig. 5. Result comparison of four dehazing methods and four EPFs in framework (a) Original, (b) Meng [10], (c) Berman [6], (d) Cai [8], (e) BASMDE(BF), (f) BASMDE (WLS), (g) BASMDE (RF), (h) BASMDE (RTV). (The last row is the detail enlarged of images at the 2nd row)

result is very obscure in the first and second images. For Meng's and Cai's results, the road in the first and third images are lackluster and not natural. Berman's method is over-saturated for half of test images. As one can see from Fig. 5(e)–(h), the dehazing images by using four different EPFs in BASMDE framework without obvious disadvantages, and all the results have steady and excellent effects.

In Fig. 6, one can find that the two algorithms are darkened in the bottom half of the image Aerial. In the image Mountain, the color of the MSCNN is a slight imbalance, and the Double-DIP has a good result, whereas the detail is not as good as us. In the image of Cones and Train, the dehazing of their algorithm result insufficient degree, and Double-DIP has color anomaly in the Cones. In terms of detail enhancement and

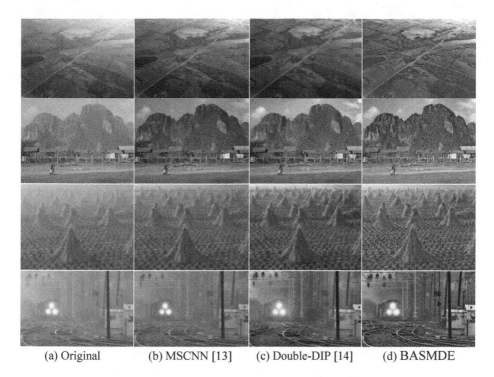

(a) Original   (b) MSCNN [13]   (c) Double-DIP [14]   (d) BASMDE

Fig. 6. Result comparison with MSCNN and Double-DIP

brightness preservation, our results are more moderated and comfortable. In short, the proposed method is able to achieve better visual perception than other dehazing algorithms under comparison.

3.2   Objective Evaluation

There are many typical image quality objective evaluation indicators for dehazing algorithm evaluation, as follows: Natural image quality evaluator(NIQE) [21], Image Visibility Measurement (IVM) [22], Visual Contrast Measure (VCM) [23], Structural Similarity index (SSIM) [24] and Universal Quality Index (UQI) [25], etc. Through experimental comparison and analysis, NIQE and SSIM are adopted as evaluation indicator in this paper as no-reference and full-reference, respectively.

NIQE is a no-reference image quality evaluation algorithm proposed by Mittal [21], in good consistency with the visual perception of human vision system. The smaller the value is, the better the image quality is. The corresponding results in terms of NIQE are shown in Tables 1 and 2.

SSIM is an index to measure the similarity between two images. The structural similarity index defines structural information from the perspective of image composition as independent of brightness and contrast, reflecting the properties of the object structure in the scene. Using the mean as the estimate of the brightness, the standard

Table 1. Indicator NIQE of the images in Fig. 5

NIQE	b	c	d	e	f	g	h
1st row	**2.36**	3.73	3.36	**2.85**	3.34	3.25	3.59
2nd row	2.68	2.66	3.05	2.32	**2.29**	**2.20**	2.37
3rd row	**2.13**	2.33	2.63	**2.17**	2.29	2.28	2.21
4th row	**2.95**	**3.14**	3.83	3.22	3.35	3.68	3.60
5th row	3.14	3.23	3.51	**2.93**	3.00	3.20	**2.82**
6th row	2.63	2.61	2.82	2.59	**2.57**	**2.53**	2.60
Average	**2.65**	2.95	3.20	**2.68**	2.80	2.86	2.86

Table 2. Indicator NIQE of the images in Fig. 6

NIQE	a	b	c	d
1st row	3.82	3.51	**3.35**	3.43
2nd row	2.59	2.34	**2.15**	2.69
3rd row	4.02	**3.43**	3.58	4.10
4th row	3.50	3.42	2.83	**2.54**
Average	3.48	3.17	**2.98**	3.19

deviation as the estimate of the contrast, and the covariance as a measure of the degree of structural similarity. The mathematical definition of the SSIM index is as follows:

$$SSIM(x, y) = \frac{(2\mu_x\mu_y + c_1)(2\sigma_{xy} + c_2)}{(\mu_x^2 + \mu_y^2 + c_1)(\sigma_x^2 + \sigma_y^2 + c_2)} \quad (15)$$

$$c_1 = (k_1 L)^2, \quad c_2 = (k_2 L)^2 \quad (16)$$

Where x is the dehazing image, y is the ground truth. μ_x and μ_y are the average values of x and y. σ_x and σ_y are the variances of x and y, respectively. σ_{xy} is the covariance of x and y, and L is the gray dynamic range. $k_1 = 0.01$, $k_2 = 0.03$.

In Tables 1 and 2, The result of BASMDE framework have smaller NIQE values, indicating that the method has excellent dehazing effect. Although Meng's and Double-DIP's NIQE value is the smallest, combined with subjective perception, their results are not the best. From the above statement, our algorithm is more moderated and natural than others.

500 hazy images were randomly selected from the RESIDE [26] as the SSIM evaluation dataset of this paper. In the dataset, the SSIM values were calculated from the dehazing image of the comparison algorithm and ground truth. Then calculate the SSIM average of 500 images, as shown in Table 3. It can be seen from Table 3 that the SSIM value of our method is the largest. It shows that the BASMDE framework results have higher structural similarity with ground truth than other algorithm results. It also shows that the dehazing framework of this paper can remove the hazy well.

Table 3. Indicator SSIM of the images dataset [26]

Meng	Berman	Cai	BASMDE (BF)	BASMDE (WLS)	BASMDE (RF)	BASMDE (RTV)
0.810	0.790	0.746	**0.834**	0.806	0.807	0.801

3.3 Computational Complexity

This sections provides a comparison of the computational complexity of this dehazing framework and other classic dehazing algorithms. All these methods are tested using their MATLAB R2016a implementations and run on the same machine with an Intel Core i5-4590 3.30 GHz CPU and 8 GB RAM. The average running times (average of 10 runs per image) of the classic methods for the six test images in Fig. 5 are documented in Table 4.

Table 4. Average running time (In seconds) of different algorithms for enhancing the 6 images

Meng	Berman	Cai	BASMDE (BF)	BASMDE (WLS)	BASMDE (RF)	BASMDE (RTV)
2.406	1.444	**0.052**	**0.379**	1.254	0.397	0.774

It can be seen that the dehazing framework proposed in this paper is faster than Meng [10] and Berman [6] in computationally, except the Cai [8]. A closer examination shown that using different EPFs in the framework can make the different calculation times for the algorithm. Therefore, a fast and accurate EPF is especially important for BASMDE framework.

4 Conclusion

An entire image dehazing framework is proposed in this paper with brightness suppression. Global brightness-area suppression could fuse the edge-preserving filtering with the atmospheric veil to protect white objects. Local brightness-area suppression could prevent the sky area or highlight region from over saturation. In addition, a simple but effective procedure of detail and contrast enhancement is designed in this framework. In theory, this framework can integrate all of other dehazing methods. Experiments show this dehazing framework works well with better visual perception.

Acknowledgments. This work is supported by Science and Technology Planned Project of Quanzhou (No. 2018C016). We thank the referees for their comments and suggestions which make the paper much improved.

References

1. Gibson, K., Vo, D., Nguyen, T.: An investigation of Dehazing effects on image and video coding. IEEE Trans. Image Process. **21**(2), 662–673 (2012)
2. Xu, Y., Wen, J., Fei, L., et al.: Review of video and image defogging algorithms and related studies on image restoration and enhancement. IEEE Access **4**, 165–188 (2016)
3. Middleton, W.: Vision Through the Atmosphere, p. 250. University of Toronto Press, Toronto (1952)
4. He, K., Sun, J., Tang, X.: Single image haze removal using dark channel prior. IEEE Trans. Pattern Anal. Mach. Intell. **33**(12), 2341–2353 (2011)
5. Tarel, J., Hautiere, N.: Fast visibility restoration from a single color or gray level image. In: IEEE International Conference on Computer Vision (ICCV), pp. 2201–2208 (2009)
6. Berman, D., Treibitz, T., Avidan, S.: Non-local image dehazing. In: IEEE Computer Vision and Pattern Recognition (CVPR), pp. 1674–1682 (2016)
7. Farbman, Z., Fattal, R., Lischinski, D., Szeliski, R.: Edge-preserving decompositions for multi-scale tone and detail manipulation. ACM Trans. Graph. (TOG) **27**(3) (2008)
8. Cai, B., Xu, X., Tao, D.: Real-time video dehazing based on spatio-temporal MRF. In: Chen, E., Gong, Y., Tie, Y. (eds.) PCM 2016. LNCS, vol. 9917, pp. 315–325. Springer, Cham (2016). https://doi.org/10.1007/978-3-319-48896-7_31
9. Gibson, K., Nguyen, T.: Fast single image fog removal using the adaptive wiener filter. In: IEEE International Conference on Image Processing, pp. 714–718 (2013)
10. Meng, G., Wang, Y., Duan, J., Xiang, S., Pan, C.: Efficient image dehazing with boundary constraint and contextual regularization. In: IEEE International Conference on Computer Vision (ICCV), pp. 617–624, December 2013
11. Kim, J., Jang, W., Sim, J., et al.: Optimized contrast enhancement for real-time image and video dehazing. J. Vis. Commun. Image Represent. **24**(3), 410–425 (2013)
12. Park, D., Park, H., Han, D., Ko, H.: Single image dehazing with image entropy and information fidelity. In: IEEE International Conference on Image Processing (ICIP), pp. 4037–4041, October 2014
13. Ren, W., Liu, S., Zhang, H., Pan, J., Cao, X., Yang, M.-H.: Single image dehazing via multi-scale convolutional neural networks. In: Leibe, B., Matas, J., Sebe, N., Welling, M. (eds.) ECCV 2016. LNCS, vol. 9906, pp. 154–169. Springer, Cham (2016). https://doi.org/10.1007/978-3-319-46475-6_10
14. Gandelsman, Y., Shocher, A., Irani, M.: 'Double-DIP': unsupervised image decomposition via coupled deep-image-priors (2018)
15. Dai, S., Tarel, J.: Adaptive sky detection and preservation in dehazing algorithm. In: International Symposium on Intelligent Signal Processing and Communication Systems (ISPACS), pp. 634–639 (2015)
16. Xu, H., Guo, J., Liu, Q., Ye, L.: Fast image dehazing using improved dark channel prior. In: IEEE International Conference on Information Science and Technology, pp. 663–667, March 2012
17. Wang, W., Dai, S.: Fast haze removal method based on image fusion and segmentation. J. Image Graph. **19**(8), 1155–1161 (2014). in Chinese
18. Tomasi, C., Manduchi, R.: Bilateral filtering for gray and color images. In: IEEE International Conference on Computer Vision (ICCV), pp. 839–846 (1998)
19. Xu, L., Yan, Q., Xia, Y., Jia, J.: Structure extraction from texture via relative total variation. ACM Trans. Graph. (TOG) **31**(6) (2012)
20. Gastal, E., Oliveira, M.: Domain transform for edge-aware image and video processing. ACM Trans. Graph. (TOG) **30**(4) (2011)

21. Mittal, A., Soundararajan, R., Bovik, A.: Making a 'completely blind' image quality analyzer. IEEE Sig. Process. Lett. **20**(3), 209–212 (2013)

22. Hautiere, N., Tarel, J.-P., Aubert, D., Dumont, E.: Blind contrast enhancement assessment by gradient ratioing at visible edges. Image Anal. Stereol. **27**(6), 87–95 (2008)

23. Jobson, D.J., Rahman, Z.-U., Woodell, G.A., Hines, G.D.: A comparison of visual statistics for the image enhancement of FORESITE aerial images with those of major image classes. In: Proceedings of SPIE the International Society for Optical Engineering, Visual Information Processing XV, SPIE 2006, pp. 624601-1–624601-8

24. Wang, Z., Bovik, A.C., Sheikh, H.R., Simoncelli, E.P.: Image quality assessment: from error visibility to structural similarity. IEEE Trans. Image Process. **13**(4), 600–612 (2004)

25. Wang, Z., Bovik, A.C.: A universal image quality index. IEEE Sig. Process. Lett. **9**(3), 81–84 (2002)

26. Li, B., Ren, W., Fu, D., et al.: Benchmarking single-image dehazing and beyond. IEEE Trans. Image Process. **28**(1), 492–505 (2019)

Parallel-Structure-based Transfer Learning for Deep NIR-to-VIS Face Recognition

Yufei Wang, Yali Li, and Shengjin Wang$^{(\boxtimes)}$

Beijing National Research Center for Information Science and Technology,
Department of Electronic Engineering, Tsinghua University, Beijing 100086, China
wgsgj@tsinghua.edu.cn

Abstract. This paper considers a heterogeneous face recognition problem, *i.e.*, matching near-infrared (NIR) to visible (VIS) face images. The significant domain gap between the NIR and VIS modalities poses great challenges to accurate face recognition. To overcome the domain gap problem, previous works usually adopted a series structure to transfer high-level features. This paper proposes a **P**arallel-**S**tructure-based **T**ransfer learning method (PST), which fully utilizes multi-scale feature map information. Specifically, PST consists of two parallel streams of network, *i.e.,* a source stream (S-stream) and a transfer stream (T-stream). S-stream is pre-trained on a large-scale VIS database, and its parameters are fixed. It preserves the discriminative ability learned from the large-scale source dataset. T-stream absorbs multi-scale feature maps from S-stream and transfers the NIR and VIS face embeddings to a unique feature space, which is agnostic to the input image modality. The proposed PST method achieves state-of-the-art performance on CASIA NIR-VIS 2.0 Database, the largest near-infrared face database.

Keywords: Heterogeneous face recognition · Near-infrared · Transfer learning · Multi-scale

1 Introduction

In recent years, the performance of ordinary face recognition algorithms has almost reached saturation. However, extreme lighting scenes, such as a dark environment, severely limit the application of face recognition in daily life. Near-infrared face images maintain high image quality regardless of lighting conditions, making face recognition possible in dark environments. Matching near-infrared face images to visible light face images, *i.e.*, heterogeneous face recognition, is important in realistic face recognition systems [7,16]. Compared with the canonical face recognition, it is faced with the following challenges:

This work was supported by the state key development program in 13th Five-Year under Grant No. 2016YFB0801301 and the National Natural Science Foundation of China under Grant Nos. 61701277, 61771288.

© Springer Nature Switzerland AG 2019
Y. Zhao et al. (Eds.): ICIG 2019, LNCS 11901, pp. 146–156, 2019.
https://doi.org/10.1007/978-3-030-34120-6_12

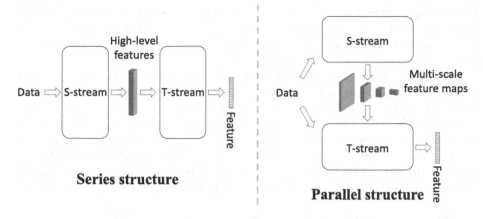

Fig. 1. Comparison of series structure and parallel structure. Classic **series** structure means that the lower-level layer of the network, S-stream, is pre-trained and its parameters are fixed. The high-level layer of the network, T-stream, is trainable and uses only the high-level features of the S-stream as input. Our **parallel** structure means that one branch of the network, S-stream, is pre-trained and its parameters are fixed. Another branch of the network, T-stream, absorbs multi-scale feature maps extracted by S-stream and transfer face embedded features.

Insufficient Data. The matching task of near-infrared and visible light faces requires each subject to have both VIS images and NIR images. The collection of such data is difficult, and the existing NIR and VIS face databases are small in scale [12,14] compared with ordinary face databases which are composed of visible light facial images collected by web crawlers.

Cross-domain recognition is the fundamental difficulty of heterogeneous face recognition [12,14]. Images from different domains tend to vary greatly in appearance, attributes, and distribution. To match pictures from diverse domains, the algorithm needs to ignore the attributes of the domain and extract the domain-independent features.

Considering the above challenges, we argue that transfer learning and cross-domain are two key issues. First, we believe that parallel structure using multi-scale feature maps for transfer learning is an optimal choice. Since deep learning is a data starvation algorithm, training a deep model with insufficient samples is prone to overfitting. Therefore, the common practice is to pre-train the model on a large-scale VIS database (source domain) and then fine-tune the last few layers on a small-scale NIR database (target domain) [9,10]. This is a series structure that only transfers high-level features. To facilitate comparison with our parallel structure, we refer to the untunable low-level layers in the series structure as S-stream, and the tunable high-level layers as T-stream. In the series structures, T-stream only uses the high-level features generated by S-stream to transfer the face embedded features, which loses a considerable amount of diagnostic information and is not competent enough. In addition, it is difficult to determine the number of layers to be fine-tuned. For the sake of compromise, we propose a parallel two-stream architecture that pre-trains S-stream on a large-scale VIS database to generate multi-scale feature maps and transfers them with T-stream which contains fewer parameters to avoid overfitting. The comparison of series and parallel structure is illustrated in Fig. 1.

Second, we believe that both NIR and VIS features should be transferred to a new unique modality agnostic feature space. For general transfer learning tasks, we tend to focus on model performance only in the target domain, ignoring the source domain. For NIR-to-VIS cross-domain face recognition, we must focus on both domains and transfer discriminative features of different domains to a unique space to eliminate the effects of modality. Using mixed NIR-VIS images, i.e., each subject has both NIR and VIS images, the classification loss automatically optimizes the features of different domains to the modality agnostic feature space.

The main contributions of our work are summarized as follows:

- We propose a parallel architecture consisting of two streams of network to achieve cross-domain identification. With multi-scale feature maps from S-stream, T-stream transfers face embeddings of different domains to a modality agnostic feature space.
- We optimize loss selection in heterogeneous face recognition and use the margin-based loss to help the model learn more expressive representations.
- The CASIA NIR-VIS 2.0 Face database [8] and Megaface [6] are combined to form a more challenging NIR-to-VIS MegaFace scenario.

2   Related Work

Face Recognition. FaceNet [15] proposes triplet loss, minimums the distance between an anchor and a positive, and maximizes the distance between the anchor and a negative, which is an effective metric learning method. Sphereface [11], NormFace [19], CosFace [18,20] and ArcFace [2] improve the softmax loss used for classification to accommodate metric learning, reaching state-of-the-art of face recognition at present. Our approach also benefits from this improvement.

Heterogeneous Face Recognition. IDR [4] divided high-level layer into two orthogonal subspaces that contain modality-invariant identity information and modality-variant spectrum information, respectively. W-CNN [5] aims to achieve the minimization of the Wasserstein distance between the NIR distribution and VIS distribution for invariant deep feature representation of heterogeneous face images. DVR [21] makes use of the Disentangled Variational Representation (DVR) for cross-modal matching. They both pre-train the model on big databases and keep the parameters of low-level layers fixed, which are the series-structure-based transfer learning methods.

PTU [24] argues that parameter transferability differs with domains and networks and should not be simply divided into random, fine-tune, and frozen. They propose a parameter transfer unit to tackle the above limitations. In our method, feature fusion units connect the S-stream used to preserve the discriminative

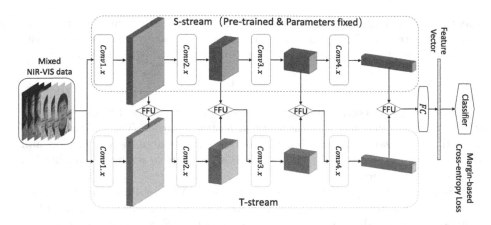

Fig. 2. An overview of the proposed PST architecture. The proposed architecture consists of two parallel streams of network, *i.e.,* a source stream (S-stream) and a transfer stream (T-stream). During training, we first pre-train S-stream on large-scale VIS data. Accordingly, parameters of S-stream are fixed, and we train T-stream with mixed NIR and VIS data. Intermediate feature maps of S-stream, in multiple resolutions, are absorbed by feature fusion units (FFU) and fused with the corresponding feature maps of T-stream. Finally, margin-based classification loss is applied on top of the fused features.

ability learned on large databases and the T-stream used for transfer, which is similar to PTU. Moreover, feature fusion units use the multi-scale feature maps from S-stream to help T-stream learn expressive representations.

3   Method

In this section, we first present the details of the proposed architecture (Sect. 3.1) and explain how we transfer both NIR and VIS features to a unique space (Sect. 3.2). Subsequently, we briefly introduce the margin-based loss and explain its importance for representation learning in heterogeneous face recognition (Sect. 3.3). Finally, we introduce the more challenging NIR-to-VIS MegaFace scenario (Sect. 3.4).

3.1   Architecture

The overall structure of PST is illustrated in Fig. 2. It is a parallel two-stream architecture consisting of a source stream (S-stream), a transfer stream (T-stream), and feature fusion units (FFUs). The S-stream adopts a Resnet-like network structure; the T-stream is a lite version of the S-stream. $Convn.x$ (where n = 1, 2, 3, 4) in Fig. 2 denotes units that contain several convolution layers, activation layers and residual units. FFUs absorb multi-scale intermediate feature maps of S-stream and fuse them with corresponding feature maps of T-stream. Given the structure of Resnet [3], which consists of 4 scale feature maps, we used

4 feature fusion units to connect the S-stream and the T-stream so that feature maps of different resolutions could be transferred. A feature fusion unit consists of several convolution layers and activation layers, which is formulated by:

$$f = FFU(f_s, f_t) = H(Concat(f_s, f_t)) \tag{1}$$

where f_s and f_t are corresponding intermediate feature maps of S-stream and T-stream, respectively, and f is the result of feature transfer. These three tensors have the exact same shape. $Concat(\cdot)$ denotes a concatenation operation and $H(\cdot)$ denotes a residual unit composed of two depth-wise separable convolution and activation layers.

After S-stream is pre-trained with a large-scale VIS database and its parameters are fixed, we use the original NIR or VIS image and the corresponding multi-scale intermediate feature maps extracted by S-stream as input to the T-stream and FFUs. While T-stream generates feature maps of an NIR or VIS image, it absorbs the corresponding feature maps generated by S-stream through FFUs to form transferred feature maps. The top-level post-transfer features form face embeddings, which are optimized by margin-based classification loss.

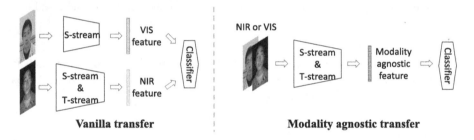

Fig. 3. Description of the transferring target. Left. Vanilla transfer. The model only transfers NIR features while retaining VIS features extracted by S-stream. Right. Modality agnostic transfer. The model transfers both the VIS and NIR features to a new modality agnostic feature space.

3.2   Modality Agnostic Transfer

Since the S-stream in our model is pre-trained on large-scale VIS data and its parameters remain fixed, the model can extract discriminative features of the VIS domain only by S-stream. The vanilla transfer approach (left of Fig. 3) keeps the VIS feature space unchanged (using the features extracted only by S-stream) and transfers NIR features to the VIS feature space. However, forcing only NIR features to be transferred into the VIS feature space would cause the model to lose focus on the VIS features. The key point of cross-domain identification is that we must focus on both the source and target domains, and it is not appropriate to place emphasis on either side. We argue that the model should perform "modality agnostic transfer" on both NIR and VIS data (right of Fig. 3). This key phrase means that we indiscriminately transfer the features of both VIS and NIR data to a new modality agnostic unique feature space. Modality agnostic transfer ensures that the model balances both the source and target domains. The features of these two domains are simultaneously transferred to a new unique feature space, where features from different domains of the same subject are optimized more closely. Our empirical study further proves this.

3.3 Margin-Based Loss

Recently, many works [2,11,18–20] have improved the cross entropy loss function to accommodate metric learning, which significantly enhances the performance of face recognition. Cross entropy loss is formulated as follows:

$$L = -log\frac{e^{s \cdot score(\theta_y)}}{\sum_{j=1}^{n} e^{s \cdot score(\theta_j)}} \tag{2}$$

where $score(\theta_y)$ denotes the category score belonging to the y-th class and n is the class number of the dataset. The details of improvement are mainly reflected in the increase of margin(m) into the category score(θ_y) of loss, which is formulated by the following three main types:

$$score(\theta_y) = cosine(m \cdot \theta_y), m > 1 \tag{3}$$
$$score(\theta_y) = cosine(\theta_y) - m, m > 0 \tag{4}$$
$$score(\theta_y) = cosine(\theta_y + m), m > 0 \tag{5}$$

They are proposed in Sphereface [11], Cosface [20] and Arcface [2], respectively.

Improvements in loss design have been fully validated on large databases, but there are only poor works applying them to heterogeneous face recognition. The sample categories of the large-scale face databases are very rich, while the CASIA NIR-VIS 2.0 database [8] has less than 400 subjects as the training set. The margin-based loss design described above can effectively ensure the distribution of features remains more compact. Small databases with poor categories often require larger margins than large databases (Subsect. 4.2).

3.4 NIR-to-VIS MegaFace Scenario

Both CASIA NIR-VIS 2.0 database and Megaface's evaluation protocol are for face identification [6,8]. CASIA evaluates the performance of algorithms matching near-infrared and visible images. The probe set contains of near-infrared images, and the gallery set contains visible images. The probe set of Megaface evaluation is even smaller than CASIA, but there are as many as a million distractors in the gallery set, making the task much more challenging. We add the Megaface's distractors into the CASIA's gallery set to enhance the challenge of the NIR-to-VIS task. Specific evaluation details will be given in Subsect. 4.3.

4 Experiments

Our approach adopts a two-step training strategy. We first pre-train the S-stream on a large-scale VIS database, keeping its parameters fixed. Then, we train the T-stream and FFUs with mixed NIR-VIS images and the corresponding multi-scale feature maps extracted by S-stream. All experiments are based on Pytorch [13].

4.1 Pre-train S-stream with Large-Scale VIS Data

As mentioned in Sect. 1, pre-training model is a common and essential process. Our implementation is similar to Sphereface [11], Cosface [20] and Arcface [2]. We select a Resnet50-like backbone as in Arcface. The output of the network is embedded features with a length of 512. WebFace [23] and VggFace2 [1] are used to pre-train the S-stream.

4.2 Train T-stream and FFUs with Mixed NIR-VIS Data

To the best of our knowledge, the CASIA NIR-VIS 2.0 Face Database [8] is the largest public near-infrared face database that is the only accessible dataset designed for face recognition. All of our experiments are conducted on this database. The database contains a total of 725 subjects, each subject with 1–22 visible light images and 5–50 near-infrared images. We follow the 10-fold protocol in View 2 of the database. There are approximately 8,600 NIR or VIS images from around 360 subjects as training fold. Each testing fold consists of a probe set and a gallery set. The probe set contains over 6,100 NIR images of the remaining 358 subjects. The gallery set contains only one visible light image of these subjects. In the experiment, we compare the proposed method with a strong baseline and the state-of-the-art to prove the effectiveness of our method.

Fine-Tune Baseline. As mentioned in Subsect. 3.2, we argue that fine-tuning with only the last fully connected layer achieves a more convincing and the best performance based on the fine-tune method. We can be clear from Fig. 4 that when the margin is small, the model performance can even be degraded (compared with the results of the pre-training). Compared to large databases, we need to set a larger margin.

Modality Agnostic Transfer. The architecture of the network is illustrated in Fig. 2. The model performs modality agnostic feature transfer, *i.e.*, VIS features transferred by T-stream and FFUs. The model performs vanilla transfer, *i.e.*, the VIS features are extracted only by S-stream. It can be seen from Table 1 that performing modality agnostic transfer, *i.e.*, indiscriminately transferring NIR and VIS to a unique space, can significantly improve the cross-domain recognition performance.

Table 1. Vanilla transfer vs. modality agnostic transfer.

Method	Rank-1
Vanilla transfer	99.02%
Modality agnostic transfer	99.93%

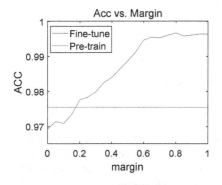

Fig. 4. Fine-tuning rank-1 accuracy under various margins.

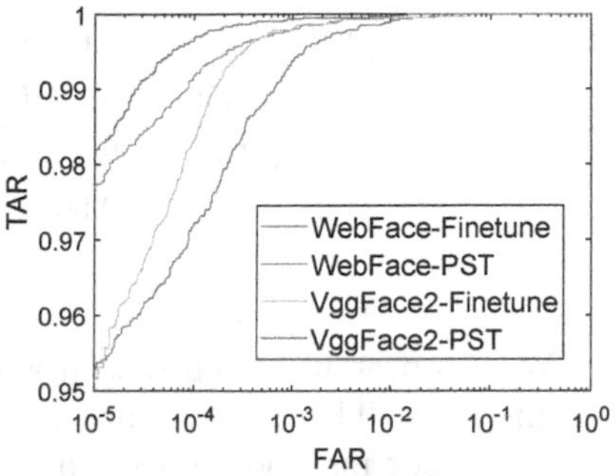

Fig. 5. ROC curves on CASIA. The legend is described in the following format—{pre-train dataset}-{method}. (Color figure online)

Table 2. Performance on CASIA. Pre-trained with WebFace (top) & Pre-trained with VggFace2 (bottom).

Method	Rank-1	TAR@FAR $= 1e-3$	TAR@FAR $= 1e-4$
Pre-train	97.55%	–	–
Fine-tune baseline	99.657%	99.379%	97.191%
PST	99.934%	99.804%	99.183%
Pre-train	98.63%	–	–
Fine-tune baseline	99.902%	99.804%	98.301%
PST	99.967%	99.935%	99.657%

4.3   Evaluation

CASIA Protocol. We present the ROC curve (Fig. 5) and rank-1 identification performance (Table 2) of the fine-tune baseline and PST with different pre-training databases. We can see from Fig. 5 that the TAR of PST is far superior to the fine-tune baseline under any FAR (yellow vs. blue & purple vs. red). The rank-1 error rates of our algorithm are reduced from 3.43% to 0.66% and from 0.98% to 0.33%. Compared with the fine-tune baseline, PST reduces the rank-1 error rate by 0.2–0.3.

Table 3. Comparisons with recent methods. Ours mean S-stream of PST is pre-trained with VggFace2.

Method	Rank-1(%)	TAR@FAR $= 1e-3$(%)
TRIVET [12]	95.7 ± 0.5	91.0 ± 1.3
IDR [4]	97.3 ± 0.4	95.7 ± 0.7
ADFL [17]	98.2 ± 0.3	97.2 ± 0.3
CDL [22]	98.6 ± 0.2	98.3 ± 0.1
W-CNN [5]	98.7 ± 0.3	98.4 ± 0.4
DVR [21]	99.7 ± 0.1	99.6 ± 0.3
Ours	99.967 ± 0.016	99.92 ± 0.01

We also present a comparison with some recent heterogeneous face recognition methods. These results are from the DVR [21]. Compared with the state-of-the-art, our method reduces the rank-1 error rate by 11%, dropping from 3% to 0.33%. As shown in Table 3, we achieve the best performance.

NIR-to-VIS MegaFace Scenario. As shown in Tables 2 and 3, the improvement of rank-1 accuracy seems to be limited, which is to a great extent because the scale of the gallery set under the original CASIA evaluation is too small, and the performance is approaching saturation. Therefore, we add Megaface's distractors into the CASIA's gallery set to enhance the challenge of the NIR-to-VIS face recognition task as described in Subsect. 3.4.

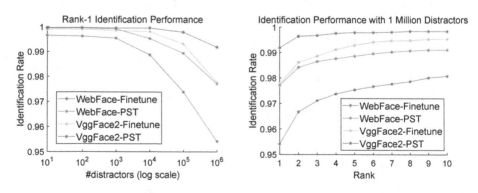

Fig. 6. Identification performance on NIR-to-VIS MegaFace. The legend is described in the following format—{pre-train dataset}-{method}

We use MTCNN to detect and align face images in the MegaFace distractors. We follow the MegaFace evaluation to measure the results of rank-n under different distractor terms (Fig. 6). Particularly, a model with a similar performance

Table 4. Identification performance under different protocols. The Model is described in the following format—{pre-train dataset}-{method}

Model	CASIA protocol	NIR-to-VIS MegaFace
WebFace-Finetune	99.657%	95.411%
WebFace-PST	99.934%	97.714%
VggFace2-Finetune	99.902%	97.763%
VggFace2-PST	99.967%	99.183%

in the original CASIA protocol shows a more significant performance gap on MegaFace scenario (Table 4). The rank-1 accuracy rates are slightly improved (from 99.902% to 99.967%) under the original CASIA protocol while greatly improved (from 97.763% to 99.183%) in the Megaface scenario. The Megaface scenario better reflects the model performance gaps and confirms the superiority of our algorithm.

5 Conclusion

In this paper, we propose a NIR-to-VIS face recognition architecture consisting of two parallel streams of network, *i.e.*, a source stream (S-stream) and a transfer stream (T-stream). S-stream is pre-trained on a large-scale VIS database, and its parameters are fixed. Utilizing multi-scale intermediate feature maps extracted by S-stream, T-stream transfers both the NIR and VIS features to a new unique modality agnostic space. Only transferring NIR features to the VIS feature space is not competent and modality agnostic transferring is essential. Finally, we validate the effectiveness of our approach in the NIR-to-VIS MegaFace scenario.

References

1. Cao, Q., Shen, L., Xie, W., Parkhi, O.M., Zisserman, A.: VGGFace2: a dataset for recognising faces across pose and age. In: 2018 13th IEEE International Conference on Automatic Face and Gesture Recognition (FG 2018), pp. 67–74. IEEE (2018)
2. Deng, J., Guo, J., Zafeiriou, S.: ArcFace: additive angular margin loss for deep face recognition. arXiv preprint arXiv:1801.07698 (2018)
3. He, K., Zhang, X., Ren, S., Sun, J.: Deep residual learning for image recognition. In: Proceedings of the IEEE Conference on Computer Vision and Pattern Recognition, pp. 770–778 (2016)
4. He, R., Wu, X., Sun, Z., Tan, T.: Learning invariant deep representation for NIR-VIS face recognition (2017)
5. He, R., Wu, X., Sun, Z., Tan, T.: Wasserstein CNN: learning invariant features for NIR-VIS face recognition. IEEE Trans. Pattern Anal. Mach. Intell. **41**, 1761–1773 (2018)
6. Kemelmacher-Shlizerman, I., Seitz, S.M., Miller, D., Brossard, E.: The MegaFace benchmark: 1 million faces for recognition at scale. In: Proceedings of the IEEE Conference on Computer Vision and Pattern Recognition, pp. 4873–4882 (2016)

7. Klare, B.F., Jain, A.K.: Heterogeneous face recognition using kernel prototype similarities. IEEE Trans. Pattern Anal. Mach. Intell. **35**(6), 1410–1422 (2013)
8. Li, S., Yi, D., Lei, Z., Liao, S.: The CASIA NIR-VIS 2.0 face database. In: Proceedings of the IEEE Conference on Computer Vision and Pattern Recognition Workshops, pp. 348–353 (2013)
9. Li, Y., Wang, S., Tian, Q., Ding, X.: Feature representation for statistical-learning-based object detection: a review. Pattern Recogn. **48**(11), 3542–3559 (2015)
10. Li, Y., Wang, S., Tian, Q., Ding, X.: A survey of recent advances in visual feature detection. Neurocomputing **149**, 736–751 (2015)
11. Liu, W., Wen, Y., Yu, Z., Li, M., Raj, B., Song, L.: SphereFace: deep hypersphere embedding for face recognition
12. Liu, X., Song, L., Wu, X., Tan, T.: Transferring deep representation for NIR-VIS heterogeneous face recognition. In: 2016 International Conference on Biometrics (ICB), pp. 1–8. IEEE (2016)
13. Paszke, A., et al.: Automatic differentiation in pytorch (2017)
14. Reale, C., Nasrabadi, N.M., Kwon, H., Chellappa, R.: Seeing the forest from the trees: a holistic approach to near-infrared heterogeneous face recognition. In: Proceedings of the IEEE Conference on Computer Vision and Pattern Recognition Workshops, pp. 54–62 (2016)
15. Schroff, F., Kalenichenko, D., Philbin, J.: FaceNet: a unified embedding for face recognition and clustering. In: Proceedings of the IEEE Conference on Computer Vision and Pattern Recognition, pp. 815–823 (2015)
16. Socolinsky, D.A., Selinger, A., Neuheisel, J.D.: Face recognition with visible and thermal infrared imagery. Comput. Vis. Image Underst. **91**(1–2), 72–114 (2003)
17. Song, L., Zhang, M., Wu, X., He, R.: Adversarial discriminative heterogeneous face recognition. arXiv preprint arXiv:1709.03675 (2017)
18. Wang, F., Cheng, J., Liu, W., Liu, H.: Additive margin softmax for face verification. IEEE Signal Process. Lett. **25**(7), 926–930 (2018)
19. Wang, F., Xiang, X., Cheng, J., Yuille, A.L.: NormFace: L2 hypersphere embedding for face verification. In: Proceedings of the 2017 ACM on Multimedia Conference, pp. 1041–1049. ACM (2017)
20. Wang, H., Wang, Y., Zhou, Z., Ji, X., Liu, W.: CosFace: large margin cosine loss for deep face recognition
21. Wu, X., Huang, H., Patel, V.M., He, R., Sun, Z.: Disentangled variational representation for heterogeneous face recognition. arXiv preprint arXiv:1809.01936 (2018)
22. Wu, X., Song, L., He, R., Tan, T.: Coupled deep learning for heterogeneous face recognition. arXiv preprint arXiv:1704.02450 (2017)
23. Yi, D., Lei, Z., Liao, S., Li, S.Z.: Learning face representation from scratch. arXiv preprint arXiv:1411.7923 (2014)
24. Zhang, Y., Zhang, Y., Yang, Q.: Parameter transfer unit for deep neural networks. arXiv preprint arXiv:1804.08613 (2018)

MF-SORT: Simple Online and Realtime Tracking with Motion Features

Heng Fu[1], Lifang Wu[1], Meng Jian[1(✉)], Yuchen Yang[1], and Xiangdong Wang[2]

[1] Beijing University of Technology, Beijing 100124, China
mjian@bjut.edu.cn
[2] Sports Science Research Institute of the State Sports General Administration, Beijing, China

Abstract. Multiple object tracking (MOT) plays a key role in video analysis. On MOT, DeepSORT (Simple Online and Realtime Tracking with a deep association metric) performs effectively by combining features of appearance and motion for estimating data association. However, computing with multiple features are time consuming. In certain applications, cameras are static, such as pedestrian surveillance, sports video analysis and so on. Here, without camera movement the motion trajectories of objects are generally possible to estimate. The introduction of more features cannot improve the performance of object tracking discriminatively. Furthermore, the time cost rises evidently. To address this problem, we propose a novel Simple Online and Realtime Tracking with motion features (MF-SORT). By focusing on the motion features of the objects during data association, the proposed scheme is able to take a trade-off between performance and efficiency. The experimental results on the MOT Challenge benchmark and MOT-SOCCER (newly established in this work) demonstrate that the proposed method is much faster than DeepSORT with the comparable accuracy.

Keywords: Multiple Object Tracking · Online tracking · Static camera video · Benchmark · Video analysis

1 Introduction

Multiple object tracking (MOT) is an essential task in video analysis, such as video pedestrian surveillance [1, 2], sport players analysis [3, 4], autopilot [5], etc. Currently, the state-of-the-art methods of MOT are primarily based on a tracking-by-detection paradigm [6–11], taking advantage of progress in object detection. The key challenge in this framework is data association, which aims to accurately associate existing object trajectories, according to the detection results in each frame.

The existed MOT schemes can be categorized into three classes: online tracking [12–15], near-online tracking [15] and offline tracking [16, 17]. DeepSORT [9] is one of representative online tracking algorithms with high tracking accuracy but slow processing speed, due to introducing the objects' appearance features.

In real application scenarios such as sports video analysis, pedestrian surveillance and so on, the videos are captured in view of the static cameras. The object trajectory is generally predictable and appearance features are not necessary.

Y. Zhao et al. (Eds.): ICIG 2019, LNCS 11901, pp. 157–168, 2019.
https://doi.org/10.1007/978-3-030-34120-6_13

Motivated by the above, we propose a scheme of Simple Online and Realtime Tracking with motion features (MF-SORT). The framework of the proposed scheme is as illustrated in Fig. 1. First, the location of tracking boxes is estimated based on Kalman filter. Then, the data from the object detections (measurements) and the predicted estimations (tracking boxes) are matched based on motion features. Finally, according to the matching results, initialization, update and deletion modules are determined and implemented to produce tracking results. The experimental results demonstrate that the proposed scheme is more adaptable to the static camera video scene.

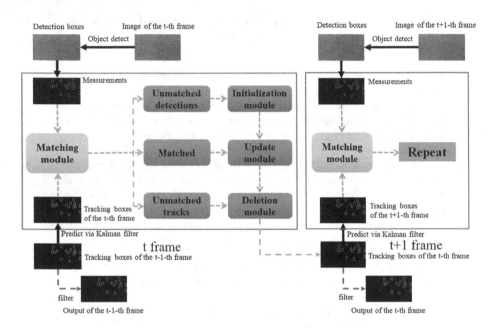

Fig. 1. The framework of the proposed MF-SORT.

The popular benchmark database for evaluating MOT algorithms is MOT Challenge. It focuses on video surveillance and provides numerous false positive (false detection) and false negative (missed detection) detection results. It is one of the bottlenecks that influences the effectiveness of the MOT algorithms. In addition, in this paper, we establish a supplementary database referred as MOT-SOCCER. It consists of 10 clips of static camera sports videos with annotations. This benchmark provides high-quality public detection whose F1-score is over 90%. An exemplary frame from MOT Challenge and MOT-SOCCER are shown in Fig. 2.

The main contributions of this work are as follows:

1. We propose a novel simple online and realtime object tracking algorithm MF-SORT. Simply with motion features in data association, it is able to track the objects in the static cameras effectively and efficiently. The comparative experimental results demonstrate that the proposed scheme can achieve competitive results with less computation complexity in MOT Challenge and MOT-SOCCER benchmark.

(a) From MOT Challenge   (b) From MOT-SOCCER

Fig. 2. Example frames in the MOT Challenge and MOT-SOCCER.

2. We establish a benchmark MOT-SOCCER which provides a high-quality detection. The benchmark consists of 10 clips of sports videos with static camera. It helps to enrich the performance assessments of MOT researches.

2 The Proposed Scheme

2.1 The Framework of the Proposed Scheme

The scheme is proposed by modifying DeepSORT in the initialization and matching stages. The framework is shown in Fig. 1. Assume that there are M detection boxes in the (t)-th frame. And there are N tracking boxes from the Kalman filter based on the results in the $(t - 1)$-th frame. The model of Kalman filter is defined on the eight-dimensional state space $(u; v; a; h; \dot{u}; \dot{v}; \dot{a}; \dot{h})$, which contains the center of the bounding box $(u; v)$, the aspect ratio a and height h of the bounding box. It is intuitive to employ the output of the Kalman filter as the tracking boxes. The M detection boxes and N tracking boxes are fed into the matching modules for association matching. The similarity between detection boxes and the tracking boxes are computed in matching module, based on their motion features.

There are three possible cases in matching results: (1) Matched: It means that some detection boxes and tracking box are successfully matched. Suppose that $M1$ boxes are matched. (2) Unmatched detections: It means that some detection boxes have not been matched to the tracking boxes. These boxes possibly are the new objects in the (t)-th frame. The number should be M-$M1$. (3) Unmatched tracks: It means that some tracking boxes have not been matched with the detection boxes. The number of boxes should be N-$M1$. Following each case, the corresponding operation is then elaborately designed. For case "matched", the bounding boxes of the objects are updated from the tracking box to the corresponding detection boxes. For case "unmatched detections", these detection boxes are initialized as the bounding boxes of the new objects. For case "unmatched tracks", the objects of these tracking boxes may not stay in this frame, they are deleted. The remaining of this section would introduce the corresponding details of matching, initialization, update and deletion module respectively.

2.2 Matching Module

In order to improve the matching efficiency, the priority of all the tracking boxes are estimated based on the *time_since_update*. Sequentially, cascade matching [9] is implemented based on the priorities. For the tracking boxes which have not been matched in the cascade stage, global matching is further employed, in which the similarity between all the unmatched tracking boxes and unmatched the detection boxes are computed by appropriate metrics.

Because the videos are collected with static cameras, the trajectory of objects is predictable and motion features are robust and sufficient for data association. Mahalanobis distance has the characteristic of scale independence. Therefore, we introduce the squared Mahalanobis distance of motion features instead of the cosine distance of appearance features in DeepSORT to measure the similarity between the tracking box and detection box:

$$d\,(i, j) = (x_j - y_i)^T C_i^{-1} (x_j - y_i) \tag{1}$$

where the projection distribution of the *(i)-th* tracking box is represented as (y_i, C_i), which can be obtained from the Kalman filter directly. And the *(j)-th* detection bounding box is represented as x_j. The metric computation is faster than appearance feature based in DeepSORT, and it is more reliable than the IoU (Intersection-over-Union) metric in SORT [8]. The detailed algorithm is summarized in Algorithm1.

Algorithm 1 Matching Module

Input: Trackers $T = \{t_1, t_2, \dots t_n\}$, Detections $D = \{d_1, d_2, \dots d_m\}$, Maximum age A_{max}

Output: Matched M, Unmatched detections U_d, Unmatched tracks U_t

---

Select confirmed tracks T_c and tentative tracks T_t from T
Initialize set of unmatched detections $U_d \leftarrow D$
Initialize set of matched $M \leftarrow \Phi$
for $l \in \{1, 2 \dots \dots A_{max}\}$ **do**
  Select tracks by age $T_l \leftarrow \{t_i \in T_c \,|t_i'age = l\}$
  $C \leftarrow$ compute_cost_matrix(T_l, U_d) using eq.1
  $M_l \leftarrow$ hungarian_assignment (C)
  for $(x_i, y_j) \in \overline{M}_l$ **do**
    if $C[(x_i, y_j] > th_{ca}$ (9.488) **then** $M_l \leftarrow$ remove(x_i, y_j) from M_l **end**
    else $U_d \leftarrow$ remove d_{y_j} from U_d **end**
  end for
  $M \leftarrow M \cup M_l$
end for
Initialize set of unmatched tracks $U_t \leftarrow \Phi$
for $t_i \in T$ **do**
  if $i \notin M$ **then** $U_t \leftarrow U_t \cup t_i$ **end**
end for
$U_t \leftarrow U_t \cup T_t$
$C \leftarrow$ compute_cost_matrix(U_t, U_d) using eq.1
$M_g \leftarrow$ hungarian_assignment (C)
for $(x_i, y_j) \in \overline{M}_g$ **do**
    if $C[(x_i, y_j] > th_{go}$ (13.277) **then** $M_g \leftarrow$ remove(x_i, y_j) from M_g **end**
    else $U_d \leftarrow$ remove d_{y_j} from U_d $U_t \leftarrow$ remove t_{x_j} from U_t **end**
end for
$M \leftarrow M \cup M_g$
return M, U_d, U_t

Further, it is necessary to delete the impossible associations by setting a threshold of the Mahalanobis distance. In cascade matching, the threshold th_{ca} for Mahalanobis distance is set as 9.488 (this threshold corresponds to a confidence value 0.95 in four-dimensional chi-square distribution). While in global matching stage, the threshold th_{go} is set as 13.277 (this threshold corresponds to a confidence value 0.99 in four-dimensional chi-square distribution), to obtain broader range of matching result.

2.3   Initialization, Update and Deletion Module

As shown in Sect. 2.1, there are three cases for matching results: matched, unmatched detections and unmatched tracks. For each case, one of the corresponding operations (initialization, update and deletion) are then conducted respectively.

The update and deletion module in DeepSORT [9] are remained in the proposed MF-SORT method. When the defined Kalman filter estimates the tracking boxes in each frame [21], the time interval (*time_since_update*) will be increased by 1. This value is reset to 0 in the update module after each successful match. When a tracker has not been successfully matched for a long time, this variable will be accumulated with each frame of Kalman filter estimation until it exceeds the maximum age we set (*max_age* = 5), and then the tracker will be deleted. More details in the update module and the deletion module are preserved for tentative tracker. In the update module, trackers with more than 3 successful matches hits (*hits* = 3) can be set to a confirmed state. In the deletion module, the tentative tracker will be deleted immediately when it does not successfully match in matching module.

In the initialization module, an additional gating method is introduced into the initialization module. The aim is to reduce the false trackers initialized by erroneous detection and avoid subsequent adverse impacts on tracking. In this work, IoU between each unmatched detection box and all tracking boxes are evaluated. In case that the IoU is higher than the given threshold (th_{gating} = 0.7), it means that the detection box is a false positive detection. It is initialized as the bounding box of a new object. The detailed initialization algorithm is shown in Algorithm 2.

Algorithm 2 Initialization Module

Input: Trackers $T = \{t_1, t_2, \ldots t_n\}$, Unmatched Detections $U_d = \{d_{y_1}, d_{y_2}, \ldots d_{y_j}\}$
Output: Trackers T

---

for $d_{y_j} \in U_d$ **do**
  if IoU $(d_{y_j}, T) > th_{gating}$ (0.7) **then** $U_d \leftarrow$ remove d_{y_j} from U_d **end**
end for
for $d_{y_j} \in U_d$ **do**
  Initialize Kalman filter tracker t_{new} with d_{y_j}
  Set t_{new}'s state = tentative
  Set t_{new}'s time_since_update = 0
  Set t_{new}'s hits = 0
  Set t_{new}'s id = the_next_id
  $T \leftarrow T \cup t_{new}$
end for
return T

---

3 Benchmark[1]

3.1 Overview

In most tracking-by-detection algorithms, the results are influenced greatly by the performance of object detection. In other words, the quality of detection boxes seriously impacts the performance of these methods. The MOT Challenge benchmark [18] are usually used for evaluating MOT algorithms, while the quality of public detection in MOT16 or MOT17 is not proper due to its complicated background. This directly results in that some of the estimated detection boxes are false. To alleviate the problem, MOT-SOCCER benchmark is established.

The dataset consists of 10 clips of amateur soccer videos that are collected with a static camera installed in a straight view of high position. It provides the detection boxes with F1-score over 90%. Some example frames in MOT-SOCCER are shown in Fig. 3.

Different from other tracking tasks, the objects in MOT-SOCCER display smaller scale changes as well as relatively similar appearance features. Although MOT-SOCCER is collected from soccer matches, it includes many specific cases in MOT Challenge such as inter-target occlusion, target disappearing and complex movement. Therefore, the MOT-SOCCER can also make sense of realistic MOT task.

We have compiled total 10 clips, half of which are applied to training and the rest to testing. An overview of this benchmark is shown in Table 1.

Table 1. Overview of the sequences currently included in the MOT-Soccer benchmark

Name	FPS	Resolution	Length	Tracks	Boxes	Density
MOT-S-01	30	2704 × 1520	920 (00:30)	20	11733	12.7
MOT-S-03	25	4096 × 2160	644 (00:25)	23	14398	22.3
MOT-S-05	30	2720 × 1530	900 (00:30)	27	16136	17.9
MOT-S-07	30	4096 × 2160	899 (00:30)	22	19800	22
MOT-S-09	30	4096 × 2160	735 (00:24)	22	16192	22
Total training			4098 (02:19)	114	78259	19.1
MOT-S-02	30	2704 × 1520	867 (00:28)	21	11733	16.1
MOT-S-04	25	4096 × 2160	644 (00:25)	23	14398	21.8
MOT-S-06	30	2720 × 1530	900 (00:30)	27	16136	19.2
MOT-S-08	30	4096 × 2160	899 (00:30)	22	19800	22
MOT-S-10	30	4096 × 2160	659 (00:21)	22	16192	22
Total testing			3969 (02:14)	114	79679	20.1

---

[1] MOT-SOCCER benchmark can be downloaded at https://github.com/jozeeandfish/motsoccer.

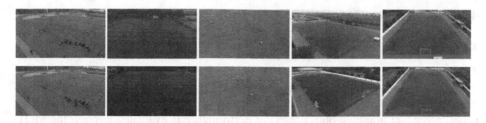

Fig. 3. An overview of the MOT-Soccer dataset. Top: training sequences; bottom: test sequences.

3.2 Detection

In order to support multiple object tracking methods, we provide a high-quality public detection results on MOT-SOCCER database, which is generated by LFFD object detection [20]. Its F1-score reaches 93.62%. It is much higher than that in MOT Challenge benchmark. The detailed performance is shown in Table 2.

Table 2. Public detection performance provided in each benchmark. The IoU threshold used in the evaluation is set to 0.5.

	Detection method	Precision	Recall	F1-score
MOT-SOCCER	LFFD [20]	**94.42**	**92.84**	**93.62**
MOT Challenge	MOT16-DPM [18]	60.34	42.89	50.14
	MOT17-FRCNN [19]	68.82	45.18	54.55
	MOT17-SDP [19]	**72.82**	**53.47**	**61.66**
	POI [7]	69.83	51.20	59.08

3.3 Data Format

The data format in MOT-SOCCER are definitely consistent with the MOT Challenge benchmark [18]. All images are converted into JPEG format and named sequentially to a 5-digit file name (e.g. 00001.jpg). Detection and annotation files are comma-separated text files. Each line represents one object instance. It contains 9 properties including frame number, tracking id, coordinates of the bounding box (x, y, w, h), confidence score, and category. In case of any property absent, 1 or −1 is used to fill this vacancy according to the criterion in MOT Challenge [18].

4 Experiments

4.1 Implement Details

The parameters of the proposed method referred in Sect. 2 have been determined on training sequences, which are provided by MOT-SOCCER. In the reproduced source code, we conduct experiments with the default parameters set in the corresponding paper. Moreover, multiple object tracking performance is evaluated through the MOT Challenge Development Kit [19] provided by A. Milan. The computing device hardware for the experimental application is i7 7700HQ (2.80 GHz), Nvidia GTX 1060.

4.2 Evaluation on MOT Benchmarks

Many existing methods used POI [7] public detection as inputs in their work, they did not try the SDP public detection or others updated in MOT17 [19] to evaluate tracking performance. Therefore, the best-performance public detection in the benchmark (See Table 2) MOT17-SDP is applied as inputs, and the annotation of MOT17 acts as a ground truth. In this case, the performance of the proposed MF-SORT scheme is compared to that of DeepSORT. The results are shown in Table 3. In addition, we also compared the performance and efficiency of MF-SORT with several state-of-the-art methods as shown in Fig. 4.

Table 3. Tracking results on the MOT Challenge training sequences with SDP detection input.

Sequence	Camera	Method	FP↓	FN↓	IDS↓	FM↓	MOTA↑	Hz↑
MOT16-02	**Static**	MF-SORT	1015	**9109**	215	**271**	**44.4**	**46**
		DeepSORT	**342**	9987	**128**	298	43.7	16
MOT16-04	**Static**	MF-SORT	226	**10890**	143	**175**	**76.3**	**34**
		DeepSORT	**56**	12626	**131**	352	73.1	12
MOT16-05	Moving	MF-SORT	674	**2341**	94	**107**	55.1	**173**
		DeepSORT	**83**	2733	**56**	112	**58.5**	42
MOT16-09	**Static**	MF-SORT	71	**1702**	48	**46**	**65.8**	**129**
		DeepSORT	**9**	1904	**29**	55	63.5	20
MOT16-10	Moving	MF-SORT	1036	**3241**	169	**243**	65.4	**49**
		DeepSORT	**379**	3715	**92**	244	**67.4**	8
MOT16-11	**Moving**	MF-SORT	517	**2308**	65	**80**	**69.4**	**123**
		DeepSORT	**174**	2710	**46**	100	68.9	44
MOT16-13	Moving	MF-SORT	1114	**4908**	74	120	47.6	**50**
		DeepSORT	**390**	5528	**31**	**89**	**48.9**	23
Total		MF-SORT	4653	**34499**	808	**1042**	**64.4**	**60**
		DeepSORT	**1433**	39203	**513**	1250	63.4	18

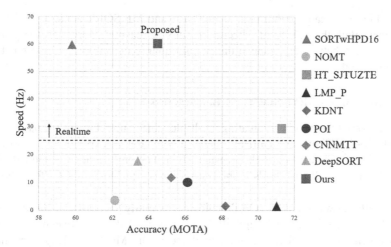

Fig. 4. Benchmark performance of the proposed scheme (MF-SORT) in relation to several state-of-the-art trackers.

The results show that the proposed MF-SORT has obtained higher MOTA (multiple object tracking accuracy) scores than that of DeepSORT in the MOT Challenge training sequences. It is shown that MF-SORT achieves the best performance in videos from static cameras (MOT 16-02, MOT 16-04 and MOT 16-09). Most importantly, the improved scheme is capable to produce a satisfying trade-off between tracking performance and efficiency. The results in Fig. 4 demonstrate that the proposed MF-SORT achieves competitive results with less computational complexity than existing SOTA methods.

4.3   Comparison of Tracking Performance with Different Detections

In order to investigate how the quality of detection boxes influences the tracking performance of our proposed scheme, we utilize the detection boxes from POI and MOT17-SDP (The detection performance is shown in Table 2.) and the ground truth (GTP) as inputs respectively. In the videos from static cameras (MOT 16-02, MOT 16-04 and MOT 16-09), the tracking performance of the proposed MF-SORT is compared with that of DeepSORT. The results are shown in Table 4.

Table 4. Tracking results in the videos from static camera with different detection quality.

Detection input	Method	FP↓	FN↓	IDS↓	FM↓	MOTA↑
POI	MF-SORT	6558	**23298**	354	441	**57.7**
	DeepSORT	**1651**	30033	**122**	**395**	55.5
SDP	MF-SORT	1312	**21701**	406	**492**	**67.2**
	DeepSORT	407	24517	**288**	705	64.7
GTP	MF-SORT	378	**189**	**0**	**0**	99.2
	DeepSORT	19	351	25	13	**99.4**

From Fig. 4 we can see that both DeepSORT and MF-SORT achieve performance improvement with the quality of detection results increasing. Moreover, the proposed scheme achieves better performance under high-quality detection and also has higher processing speed.

4.4   Evaluation on MOT-SOCCER Benchmarks

Aiming at comprehensively evaluating multiple object tracking performance of the proposed MF-SORT in static camera videos, a comparative experiment is carried out on the MOT-SOCCER benchmark we established. The performance of the MF-SORT compared to DeepSORT methods in the test sequences of MOT-SOCCER is shown in Table 5.

The result shows that MF-SORT achieves a slightly increasing MOTA score in MOT-SOCCER compared to DeepSORT, and made a balance between performance and processing speed, which is similar to those in the MOT Challenge benchmark. Since the detection quality in MOT-SOCCER is better than that in the MOT Challenge, we could conclude that the proposed scheme is more effective and efficient than DeepSORT in the condition of good detection quality.

Table 5. Tracking results on the test sequences of MOT-SOCCER benchmark.

Sequence	Method	FP↓	FN↓	IDS↓	FM↓	MOTA↑	Hz↑
MOT-S-02	MF-SORT	192	**647**	28	**50**	**93.8**	**18**
	DeepSORT	**163**	760	**15**	62	93.3	8
MOT-S-04	MF-SORT	145	**221**	19	**39**	**97.3**	**32**
	DeepSORT	**102**	277	**6**	44	**97.3**	14
MOT-S-06	MF-SORT	234	**248**	18	**46**	**97.1**	**95**
	DeepSORT	**210**	311	24	48	96.8	27
MOT-S-08	MF-SORT	46	**417**	28	**49**	**97.5**	**44**
	DeepSORT	**33**	516	**27**	63	97.1	8
MOT16-10	MF-SORT	30	**51**	3	**19**	**99.4**	**28**
	DeepSORT	**23**	95	**2**	19	99.2	6
Total	MF-SORT	647	**1584**	96	**203**	**97.1**	**44**
	DeepSORT	**531**	1959	**74**	236	96.8	12

5   Conclusion

In this paper, we propose a novel simple online and realtime tracking with motion features (MF-SORT). It utilizes the motion features instead of appearance features in data association in the tracking-by-detection paradigm, which helps improve efficiency of data association. The experimental results demonstrate that the proposed MF-SORT achieves competitive results with less computational costs compared with state-of-the-art methods. It produces a satisfactory trade-off between performance and efficiency, which is more competent for realtime application scenarios. We also establish an open-download MOT benchmark MOT-SOCCER, which provides a high-quality detection. It comes to enrich the assessments of MOT methods.

Acknowledgement. This work was supported by the National Natural Science Foundation of China under Grant 61702022 and 61802011, in part by the Beijing Municipal Education Committee Science Foundation under Grant KM201910005024, in part by "Ri Xin" Training Programme Foundation for the Talents by Beijing University of Technology.

References

1. Yang, B., Huang, C., Nevatia, R.: Learning affinities and dependencies for multi-target tracking using a CRF model. In: CVPR 2011, pp. 1233–1240. IEEE (2011)
2. Pellegrini, S., et al.: You'll never walk alone: modeling social behavior for multi-target tracking. In: ICCV 2009, pp. 261–268. IEEE (2009)
3. Lu, W., et al.: Learning to track and identify players from broadcast sports videos. IEEE Trans. Pattern Anal. Mach. Intell. **35**(7), 1704–1716 (2013)
4. Xing, J., Ai, H., Liu, L., et al.: Multiple player tracking in sports video: a dual-mode two-way bayesian inference approach with progressive observation modeling. IEEE Trans. Image Process. **20**(6), 1652–1667 (2011)
5. Koller, D., Weber, J., Malik, J.: Robust multiple car tracking with occlusion reasoning. In: Eklundh, J.-O. (ed.) ECCV 1994. LNCS, vol. 800, pp. 189–196. Springer, Heidelberg (1994). https://doi.org/10.1007/3-540-57956-7_22
6. Feng, W., et al.: Multiple object tracking with multiple cues and switcher-aware classification. arXiv preprint arXiv:1901.06129 (2019)
7. Yu, F., Li, W., Li, Q., Liu, Y., Shi, X., Yan, J.: POI: multiple object tracking with high performance detection and appearance feature. In: Hua, G., Jégou, H. (eds.) ECCV 2016. LNCS, vol. 9914, pp. 36–42. Springer, Cham (2016). https://doi.org/10.1007/978-3-319-48881-3_3
8. Bewley, A., Ge, Z., Ott, L., Ramos, F., Upcroft, B.: Simple online and realtime tracking. In: 2016 IEEE International Conference on Image Processing (ICIP), pp. 3464–3468. IEEE (2016)
9. Wojke, N., Bewley, A., Paulus, D.: Simple online and realtime tracking with a deep association metric. In: 2017 IEEE International Conference on Image Processing (ICIP), pp. 3645–3649. IEEE (2017)
10. Long, C., Haizhou, A., Zijie, Z., Chong, S.: Real-time multiple people tracking with deeply learned candidate selection and person re-identification. In: 2018 IEEE International Conference on Multimedia and Expo (ICME), San Diego, pp. 1–6 (2018)
11. Yoon, Y., et al.: Online multiple object tracking with historical appearance matching and scene adaptive detection filtering. In: 2018 15th IEEE International Conference on Advanced Video and Signal Based Surveillance (AVSS). IEEE (2018)
12. Milan, A., Rezatofighi, S.H., Dick, A.R., Reid, I.D., Schindler, K.: Online multi-target tracking using recurrent neural networks. In: AAAI, vol. 2, p. 4 (2017)
13. Zhu, J., Yang, H., Liu, N., Kim, M., Zhang, W., Yang, M.-H.: Online multi-object tracking with dual matching attention networks. In: Ferrari, V., Hebert, M., Sminchisescu, C., Weiss, Y. (eds.) ECCV 2018. LNCS, vol. 11209, pp. 379–396. Springer, Cham (2018). https://doi.org/10.1007/978-3-030-01228-1_23

14. Fang, K., Xiang, Y., Li, X., et al.: Recurrent autoregressive networks for online multiple object tracking. In: 2018 IEEE Winter Conference on Applications of Computer Vision (WACV), pp. 466–475. IEEE (2018)
15. Choi, W.: Near-online multi-target tracking with aggregated local flow descriptor. In: Proceedings of the IEEE International Conference on Computer Vision. IEEE (2015)
16. Henschel, R., Leal-Taixe, L., Cremers, D., Rosenhahn, B.: Fusion of head and full-body detectors for multiple object tracking. In: Computer Vision and Pattern Recognition Workshops (CVPRW) (2018)
17. Tang, S., Andriluka, M., Andres, B., Schiele, B.: Multiple people tracking by lifted multicut and person reidentification. In: Proceedings of the IEEE Conference on Computer Vision and Pattern Recognition, pp. 3539–3548. IEEE (2017)
18. Milan, A., et al.: MOT16: a benchmark for multiple object tracking. arXiv preprint arXiv: 1603.00831 (2016)
19. Multiple object tracking benchmark. https://motchallenge.net. Accessed 26 Apr 2019
20. Xu, D., et al.: LFFD: a light and fast face detector for edge devices. arXiv preprint arXiv: 1904.10633 (2019)
21. Kalman, R.: A new approach to linear filtering and prediction problems. J. Basic Eng. 82(Series D), 35–45 (1960)

Semantic Segmentation of Street Scenes Using Disparity Information

Hanwen Hu and Xu Zhao[✉]

Department of Automation, Shanghai Jiao Tong University, Shanghai, China
zhaoxu@sjtu.edu.cn

Abstract. In this work, we address the task of semantic segmentation in street scenes. Recent approaches based on convolutional neural networks have shown excellent results on several semantic segmentation benchmarks. Most of them, however, only exploit RGB information. Due to the development of stereo matching algorithms, disparity maps can be more easily acquired. Structural information encoded in disparity can be treated as supplementary information of RGB images, which is expected to boost performance. Therefore, in this work we propose to fuse disparity information in street scene understanding task. And we design four methods to incorporate disparity information into semantic segmentation framework. They are summation, multiplication, concatenation and channel concatenation. Besides, disparity map can be utilized as ground truth of a regression task, guiding the learning of semantic segmentation as a loss term. Comprehensive experiments on KITTI and Cityscapes datasets show that each method can achieve performance improvement. The experimental results validate the effectiveness of disparity information to street scene semantic segmentation tasks.

Keywords: Semantic segmentation · Disparity

1 Introduction

Semantic segmentation is one of the fundamental topics and challenging tasks in computer vision. The goal is to make pixel-wise prediction for a given image.

Since the introduction of the fully convolutional networks (FCN) [16], FCN based approaches [7,15,23] have achieved great success in semantic segmentation tasks. However, most of them only exploit RGB information. In addition to using the appearance information provided by RGB images, semantic segmentation tasks can benefit from structural information of the scene, *e.g.* depth information.

In this work, we utilize both RGB images and disparity maps to address street scene semantic segmentation tasks. The difficulty of street scene segmentation mainly lies in the complex scenes, occlusion and illumination variation, which make it still an open problem at present. Only using RGB appearance information may lead to incorrect predictions in two situations: (1) adjacent pixels that share the same semantic categories but different appearances; (2)

© Springer Nature Switzerland AG 2019
Y. Zhao et al. (Eds.): ICIG 2019, LNCS 11901, pp. 169–181, 2019.
https://doi.org/10.1007/978-3-030-34120-6_14

Fig. 1. Training examples from KITTI [2] dataset. From top to bottom: original RGB images, disparity maps generated by PSMNet [3] and ground truths. Disparity maps can be used as supplementary information to RGB images.

adjacent pixels that have different semantic labels but with similar appearances. Disparity maps provide structural information as well as hierarchical relationships between objects of the scenes, which can be exploited as supplementary information to RGB images.

Previous approaches utilizing depth information mainly focus on indoor scene understanding tasks. Several approaches [11,21] treat depth map as an additional input channel, utilizing FCN based models to segment RGB-D images. Gupta et al. [12] transform depth maps to HHA image[1] and [13] employs two branches of CNNs to extract RGB and HHA image features respectively. In our work, we use disparity maps as depth information to address the task of semantic segmentation in street scenes. Disparity maps are acquired by applying PSMNet [3], which is a state-of-the-art stereo matching algorithm. Training examples are illustrated in Fig. 1.

In this work, we employ MobileNetV2 [19] and Xception [8] to extract RGB image features. VGG [20] like fully convolutional network is utilized as disparity feature extractor. We propose four fusion methods, *i.e. summation, multiplication, concatenation* and *channel concatenation*, to incorporate disparity information into semantic segmentation framework without introducing much computation complexity. Besides, we treat disparity map as ground truth of a regression task, imposing extra constraints to guide training process along with semantic segmentation task. Each method is evaluated on two street scene datasets: KITTI [2] and Cityscapes [9]. The experimental results indicate that each fusion method can improve the performance of semantic segmentation and validate the effectiveness of disparity information to street scene understanding tasks.

---

[1] HHA image consists of three channels: horizontal disparity, height of the pixels and norm angle.

2   Related Work

2.1   Semantic Segmentation

Long et al. [16] transforms classification-purposed CNN models into Fully Convolutional Networks (FCN) by replacing fully connected layers with convolutional layers. Recent methods [7,15,23] mainly focus on: (1) encoding context information; (2) decoding semantic features and (3) providing structured outputs.

Encoding Context Information. It is of great importance to utilize context information of an image in semantic segmentation tasks. Dilated convolution is utilized in [22] to control the resolution of extracted features and aggregate multi-scale contextual information. Besides, PSPNet [23] and Deeplab [4–7] exploiting spatial pyramid pooling modules to encode context information. PSPNet implements spatial pyramid pooling at several grid scales. While Deeplabv2 [5] introduces *Atrous Spatial Pyramid Pooling* (ASPP) module that applies several parallel dilated convolutions with different dilation rates.

Decoding Semantic Features. Because of pooling layers and strided convolutional layers, encoder module gradually reduces the resolution of feature maps and captures high-level semantic information. Several approaches have been proposed to recover resolution from semantic features. Bilinear interpolation [6] and transposed convolution [17] are commonly used operations to enlarge resolution. Skip connections [15,18] between encoder parts and decoder parts have been adopted to acquire accurate information.

Providing Structured Outputs. Conditional Random Fields (CRF) is often utilized to generate sharper results. Deeplabv1 [4] applies DenseCRF [14] as a post-processing method to refine predictions along object boundaries. CRF-as-RNN [24] interprets mean-field of CRFs as a recurrent neural network, allowing for training CNN and CRF parameters in an end-to-end differentiable network.

In this work, we utilize dilated convolution to enlarge receptive field without reducing the resolution of feature maps. Moreover, we adopt ASPP module to aggregate multi-scale and context information. Bilinear interpolation is used in decoding stage to recover the resolution.

2.2   RGB-D Semantic Segmentation

The availability of low-cost range sensors advances the progress of RGB-D semantic segmentation. However, RGB-D semantic segmentation mainly focuses on indoor scene understanding tasks. Several approaches [11,21] treat depth map as an additional input channel, utilizing FCN based models to segment RGB-D images. Gupta et al. [12] transform depth to HHA image, which consists of three channels: horizontal disparity, height of the pixels and norm angle. [13] employs two branches of CNNs to extract RGB and HHA image features respectively, achieving promising results on several indoor scene semantic segmentation benchmarks.

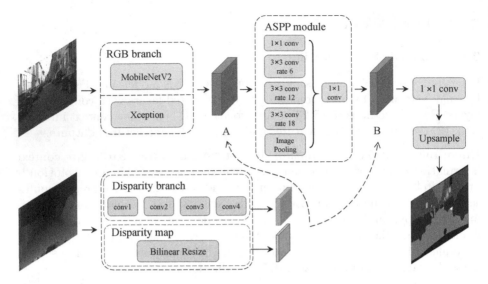

Fig. 2. Network structure. Two branches of CNNs are employed to extract RGB and disparity features respectively. MobileNetV2 and Xception are employed as RGB feature extractor, followed by ASPP module. The disparity features or resized disparity maps are fused with RGB features. Two possible fusion options are marked with A and B. Position A stands for *early fusion* while position B strands for *late fusion*.

In this work, we focus on street scene semantic segmentation instead of indoor scenes. Although both disparity maps and depth maps encode 3D information, disparity maps can be effectively computed using stereo matching algorithms with no need of other sensors. Hence, we exploit disparity map as an additional information to perform street scene semantic segmentation.

3   Methods

3.1   Network Architecture

The architecture of our network is illustrated in Fig. 2. We have two branches of networks to extract features from RGB images and disparity maps, respectively.

In RGB branch, we employ two CNNs as backbone networks: MobileNetV2 [19] and Xception [8]. Both of them leverage depthwise separable convolution as basic building block, greatly reducing the number of parameters. And residual connections are essential in helping with convergence. Besides, we enlarge the resolution of the final feature maps using dilated convolution. We set the stride of last pooling or strided convolutional layer to 1 and replace all subsequent convolutional layers with dilated convolutional layers. This allows us to extract denser feature maps without sacrificing receptive field. Although convolutional features have shown remarkable ability to implicitly represent scale, explicitly accounting for object scale can further improve the ability to handle

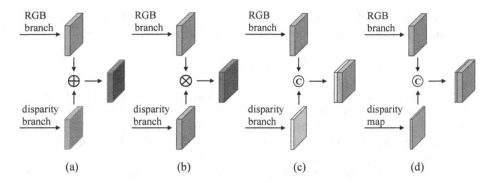

Fig. 3. Illustration of fusion methods. (a) Element-wise summation. (b) Element-wise multiplication. (c) Concatenation. (d) Channel concatenation. $\oplus$, $\otimes$, $\copyright$ represent summation, multiplication and concatenation respectively.

objects with various size. Thus, we adopt *atrous spatial pyramid pooling* (ASPP) module, which introduces dilated convolution into spatial pyramid pooling, to aggregate multi-scale and context information. We stack ASPP module on the top of backbone networks following the spirit of [7].

In disparity branch, we apply a light-weight fully convolutional network to extract disparity features. The network follows the VGG [20] style and has fewer channels in convolutional layers. It consists of alternating convolutional blocks and max pooling layers. Disparity features are then fused with RGB features.

3.2   Fusion Strategies

As illustrated in Fig. 2, we have two options to fuse disparity information:

Early Fusion. Since ASPP module is used to extract multi-scale features, fusion of disparity and RGB features before ASPP module is expected to simultaneously capture context information of RGB features and disparity information.

Late Fusion. The second possible fusion position is after ASPP module. Fused features are only passed through one 1×1 convolutional layer to obtain final logits, which are the feature maps before softmax activation. The motivation is to let disparity information directly influence prediction process.

Four fusion methods are proposed to incorporate disparity information into semantic segmentation framework without introducing much computation complexity. Let f_{RGB}, f_{Disp} and F denote RGB features, disparity features and fused features respectively. Proposed methods are illustrated in Fig. 3 and described as follows.

Summation. Inspired by [13], we apply element-wise summation to fuse features. Both disparity branch and ASPP module generate 256-channel features. When *late fusion* is adopted, element-wise summation is performed directly upon disparity features and RGB features. However, when *early fusion* is applied, the output of RGB branch is first reduced by a 1×1 convolutional layer, generating 256-channel feature maps. We formulate this fusion method as follows:

$$F = f_{RGB} \oplus f_{Disp}. \tag{1}$$

Multiplication. In this method, we regard disparity feature maps as masks, multiplying RGB features element-wisely. Because of ReLU non-linearity, the value of each element in feature maps is greater or equal to 0. Hence, it is expected that disparity features would provide additional information to corresponding areas in RGB features. Opreation of *early fusion* follows the same operation in *summation* method. This method can be written as:

$$F = f_{RGB} \otimes f_{Disp}. \tag{2}$$

Concatenation. Another intuitive method is to concatenate RGB feature maps and disparity feature maps together, automatically fusing them by convolutional layers. Therefore, the most useful disparity features would be assigned higher weights in convolution. As additional information, the proportion of disparity features should not exceed the proportion of RGB features. Thus, we modify the disparity branch to generate 128-channel features, which are then concatenated to 256-channel RGB features. While in *early fusion* the disparity features are directly concatenated to the output of RGB branch. The fused feature is as below:

$$F = [f_{RGB}, f_{Disp}]. \tag{3}$$

Channel Concatenation. This method treats disparity map as an addition channel to RGB features. Disparity maps are bilinearly resized to the resolution of RGB feature maps, *i.e.* $1/16$ of input image size, and then directly concatenated to them. This method utilizes original disparity maps and introduces minimal computation complexity. It is noteworthy that the only operation we perform on disparity maps is normalization, so that the range of disparity maps is $[-1, 1]$. This method can be represented by the following equation with $Disp$ standing for bilinearly resized disparity maps:

$$F = [f_{RGB}, Disp]. \tag{4}$$

All the four methods are applied at each fusion position. Fused feature F is utilized to compute final output. Thus we incorporate disparity information into semantic segmentation. In conclusion, we propose four methods and two options to fuse disparity information with RGB features. We also combine RGB images and disparity maps into four-channel RGB-Disparity images, which are directly fed into the network. The network has no disparity branch in this case.

3.3   Disparity Loss Regularization

The disparity information can also guide learning of semantic segmentation as a loss term. We treat disparity map as ground truth of a regression task. Based on the final RGB feature map, we employ a convolutional layer to predict disparity of the image. The total loss is expressed as

$$L = \frac{1}{N} \sum_i L_{cls}(y_{label}^i, y_{pred}^i) + \lambda \cdot L_{reg}(d_{label}^i - d_{pred}^i), \tag{5}$$

where L_{cls} is cross-entropy loss and L_{reg} is disparity regression loss. y_{label}^i and y_{pred}^i are ground truth class and predicted class of semantic segmentation task respectively. d_{label}^i and d_{pred}^i are ground truth disparity and predicted disparity of disparity regression task respectively. We adopt Huber loss as the disparity regression loss, which is expressed as below:

$$L_{reg}(x) = \begin{cases} \frac{1}{2}x^2 & \text{if } |x| < 1 \\ |x| - \frac{1}{2} & \text{oterwise.} \end{cases} \tag{6}$$

The hyper-parameter λ controls the balance between the two task losses and we set $\lambda = 10.0$ empirically.

When training the network, the disparity regression loss L_{reg} is propagated back to the RGB feature extractor. Along with the basic cross-entropy loss L_{cls}, disparity regression loss L_{reg} imposes extra constraints to guide training process. We note that the network has no disparity branch in this case.

4   Experiments

4.1   Datasets and Implementation Details

We evaluate our approach on two challenging semantic segmentation datasets: KITTI [2] and Cityscapes [9]. Both of them focus on street scenes segmentation and provide rectified stereo image pairs. We apply PSMNet [3] to calculate dense disparity maps for both datasets.

KITTI. KITTI [2] semantic segmentation benchmark consists of 200 semantically annotated train images as well as 200 test images. The images were recorded while driving around a mid-size city, in rural areas and on highways. And the resolution of images is mainly 375×1242. We further split the whole training set into a training set (160 images) and a validation set (40 images).

Cityscapes. Cityscapes [9] dataset is a large-scale dataset that focuses on semantic understanding in urban street scenes. It consists of 5000 images (2975, 500, 1525 for the training, validation, and test sets respectively) with fine annotations and another 20000 images with coarse annotations. In this work, experiments are only conducted on the fine-annotated dataset.

Implementation Details. We employ MobileNetV2 and Xception, which have been pretrained on ImageNet dataset [10], as network backbone to extract features of RGB images. The parameters in disparity branch are initialized randomly. We follow the training protocol as in [6]. Input images are scaled randomly from 0.5 to 2.0 and are flipped randomly left-right during training stage.

Table 1. Results on KITTI validation set. The name of the network indicates network backbone, fusion position and fusion method.

Network	mIoU	Network	mIoU
MobileNetV2_baseline	49.40	Xception_baseline	52.76
MobileNetV2_RGB-D	50.27	Xception_RGB-D	54.54
MobileNetV2_earlyfusion_sum	51.88	Xception_earlyfusion_sum	56.66
MobileNetV2_earlyfusion_mul	51.62	Xception_earlyfusion_mul	**58.96**
MobileNetV2_earlyfusion_concat	51.77	Xception_earlyfusion_concat	57.62
MobileNetV2_earlyfusion_channel	**55.74**	Xception_earlyfusion_channel	58.75
MobileNetV2_latefusion_sum	50.62	Xception_latefusion_sum	57.37
MobileNetV2_latefusion_mul	52.63	Xception_latefusion_mul	56.90
MobileNetV2_latefusion_concat	**54.15**	Xception_latefusion_concat	**59.71**
MobileNetV2_latefusion_channel	50.47	Xception_latefusion_channel	59.04

The final logits output by the network are upsampled to the size of ground truths to preserve annotation details. SGD optimizer with momentum 0.9 is used to train the network. And the initial learning rate is multiplied by $(1 - \frac{iter}{max\_iter})^{0.9}$ to decrease learning rate. The performance is measured in terms of mIoU, which is the mean value of classwise intersection-over-union. All experiments are built with TensorFlow [1] framework on a single NVIDIA Titan X (Pascal) GPU.

4.2 Experimental Results

KITTI. Baseline network only adopts RGB branch and is trained with the training protocols described above. In addition, we also conduct experiment with RGB-Disparity input where disparity map is treated as an additional input channel. We note that the network has no disparity branch in this case and parameters of the first convolutional layer are randomly initialized.

The results of MobileNetV2 and Xception based networks are summarized in Table 1. Furthermore, mIoU curves on KITTI validation set during training are provided in Fig. 4. The curves indicate that networks have converged at the end of training. We observe that each fusion method, no matter *early fusion* or *late fusion*, outperforms the performance of baseline. The four-channel RGB-D input method achieves the smallest performance improvement compared with baseline. Best performing *MobileNetV2_earlyfusion_channel* and *Xception_latefusion_concat* increase performance by 6.34% and 6.95%, respectively. Interestingly, performance gain can be achieved as well by the simplest *channel concatenation* fusion method. Thus, experimental results reveal that disparity information can be exploited as supplementary information to street scene understanding tasks.

Comparing the results of MobileNetV2 and Xception based networks, performance gain can be observed on both of them and Xception based networks

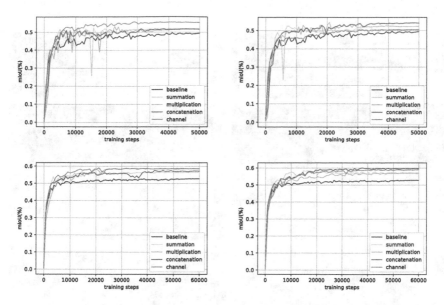

Fig. 4. mIoU curves on KITTI validation set. The first row stands for MobileNetV2 based networks and the second row stands for Xception based networks. The first column represents *early fusion* and the second column represents *late fusion*.

perform better because of more representative features. The results are accord with our expectation and also validate that disparity information is effective to different RGB convolutional features. Although each fusion method can improve performance, it is hard to tell which fusion method is the best. However, *channel concatenation* can be used to boost performance in resource constrained environments. At last, qualitative results on KITTI validation set are illustrated in Fig. 5. It shows that cars and poles are segmented better with the help of disparity maps.

Cityscapes. Experimental results are summarized in Table 2. Compared with the baseline, the performance of RGB-Disparity input method decreases slightly by 0.52%. We speculate that randomly initialized parameters of the first convolutional layer make the network difficult to learn representative features. Except RGB-Disparity method, each fusion method obtains performance gain on Cityscapes dataset, which is consistent with the results on KITTI dataset. *Xception_latefusion_sum* achieves best result and improves the performance by 4.08% over baseline. Besides, the results of *summation, concatenation*, and *channel concatenation* fusion methods are close to each other, surpassing the performance of baseline by about 3–4%. Hence, it can be concluded that disparity maps provide useful information to street scene understanding tasks.

We note that the disparity maps of Cityscapes dataset are calculated by PSMNet, which is trained on KITTI 2015 stereo dataset. Better results can be expected if we obtain more accurate disparity maps of Cityscapes.

Fig. 5. Visualization results on KITTI validation set. From top to bottom: RGB images, disparity maps, ground truths, results of *Xception_baseline*, results of *Xception_latefusion_concat*.

Table 2. Results on Cityscapes validation set. The name of the network indicates network backbone, fusion position and fusion method.

Network	mIoU
Xception_baseline	69.34
Xception_RGB-Disparity	68.82
Xception_earlyfusion_sum	72.27
Xception_earlyfusion_mul	71.56
Xception_earlyfusion_concat	72.92
Xception_earlyfusion_channel	**73.23**
Xception_latefusion_sum	**73.42**
Xception_latefusion_mul	71.86
Xception_latefusion_concat	73.31
Xception_latefusion_channel	72.27

Disparity Loss Regularization. When we treat disparity map as ground truth of a regression task, a 3×3 convolutional layer is employed on the final RGB feature map of Xception to predict disparity of the image. In this case the network has only RGB branch. Table 3 shows experimental results on KITTI and Cityscapes datasets. Compared with baseline, the performance of disparity regularized methods increase 1.62% and 0.89% on KITTI and Cityscapes respectively. Experimental results manifest that disparity information can guide the learning of semantic segmentation in street scenes.

Table 3. mIoU results of disparity regularized method on KITTI and Cityscapes dataset.

Network	KITTI	Cityscapes
Xception_baseline	52.76	69.34
Xception_regression	54.38	70.23

5   Conclusions

We leverage disparity maps as supplementary information to address semantic segmentation in street scenes. Two branches of CNNs are employed to extract RGB and disparity features respectively. We propose four methods and two options to incorporate disparity information into semantic segmentation framework. Experimental results on KITTI and Cityscapes validate the effectiveness of disparity information to street scene understanding tasks. Moreover, each fusion method can improve the performance of semantic segmentation over baseline without introducing much computation complexity. We note that the proposed fusion methods are easy to implement and can be applied to fuse other information. Besides, we treat disparity map as ground truth of a disparity regression task and employ a convolutional layer on the final RGB feature map to predict disparity of the scene. Experimental results validate that disparity information can guide learning of street scene semantic segmentation.

Acknowledgments. This work is supported by: National Natural Science Foundation of China (U1764264, 61673269, 61273285).

References

1. Abadi, M., et al.: Tensorflow: a system for large-scale machine learning. In: OSDI 2016, pp. 265–283 (2016)
2. Alhaija, H.A., Mustikovela, S.K., Mescheder, L., Geiger, A., Rother, C.: Augmented reality meets deep learning for car instance segmentation in urban scenes. In: British Machine Vision Conference, vol. 1, p. 2 (2017)

3. Chang, J.R., Chen, Y.S.: Pyramid stereo matching network. In: Proceedings of the IEEE Conference on Computer Vision and Pattern Recognition, pp. 5410–5418 (2018)
4. Chen, L.C., Papandreou, G., Kokkinos, I., Murphy, K., Yuille, A.: Semantic image segmentation with deep convolutional nets and fully connected CRFs. In: ICLR, San Diego, United States, May 2015. https://hal.inria.fr/hal-01263610
5. Chen, L.C., Papandreou, G., Kokkinos, I., Murphy, K., Yuille, A.L.: Deeplab: Semantic image segmentation with deep convolutional nets, atrous convolution, and fully connected CRFs. IEEE Trans. Pattern Anal. Mach. Intell. **40**(4), 834–848 (2018)
6. Chen, L.C., Papandreou, G., Schroff, F., Adam, H.: Rethinking atrous convolution for semantic image segmentation. arXiv preprint arXiv:1706.05587 (2017)
7. Chen, L.C., Zhu, Y., Papandreou, G., Schroff, F., Adam, H.: Encoder-decoder with atrous separable convolution for semantic image segmentation. arXiv preprint arXiv:1802.02611 (2018)
8. Chollet, F.: Xception: deep learning with depthwise separable convolutions. arXiv preprint arXiv:1610.02357 (2017)
9. Cordts, M., et al.: The cityscapes dataset for semantic urban scene understanding. In: CVPR, pp. 3213–3223 (2016)
10. Deng, J., Dong, W., Socher, R., Li, L.J., Li, K., Fei-Fei, L.: Imagenet: a large-scale hierarchical image database. In: 2009 IEEE Conference on Computer Vision and Pattern Recognition, CVPR 2009, pp. 248–255. IEEE (2009)
11. Eigen, D., Fergus, R.: Predicting depth, surface normals and semantic labels with a common multi-scale convolutional architecture. In: Proceedings of the IEEE International Conference on Computer Vision, pp. 2650–2658 (2015)
12. Gupta, S., Girshick, R., Arbeláez, P., Malik, J.: Learning rich features from RGB-D images for object detection and segmentation. In: Fleet, D., Pajdla, T., Schiele, B., Tuytelaars, T. (eds.) ECCV 2014. LNCS, vol. 8695, pp. 345–360. Springer, Cham (2014). https://doi.org/10.1007/978-3-319-10584-0_23
13. Hazirbas, C., Ma, L., Domokos, C., Cremers, D.: FuseNet: incorporating depth into semantic segmentation via fusion-based CNN architecture. In: Lai, S.-H., Lepetit, V., Nishino, K., Sato, Y. (eds.) ACCV 2016. LNCS, vol. 10111, pp. 213–228. Springer, Cham (2017). https://doi.org/10.1007/978-3-319-54181-5_14
14. Krähenbühl, P., Koltun, V.: Efficient inference in fully connected CRFs with Gaussian edge potentials. In: Advances in Neural Information Processing Systems, pp. 109–117 (2011)
15. Lin, G., Milan, A., Shen, C., Reid, I.: Refinenet: multi-path refinement networks for high-resolution semantic segmentation. In: CVPR (2017)
16. Long, J., Shelhamer, E., Darrell, T.: Fully convolutional networks for semantic segmentation. In: Proceedings of the IEEE Conference on Computer Vision and Pattern Recognition, pp. 3431–3440 (2015)
17. Noh, H., Hong, S., Han, B.: Learning deconvolution network for semantic segmentation. In: ICCV, pp. 1520–1528 (2015)
18. Ronneberger, O., Fischer, P., Brox, T.: U-Net: convolutional networks for biomedical image segmentation. In: Navab, N., Hornegger, J., Wells, W.M., Frangi, A.F. (eds.) MICCAI 2015. LNCS, vol. 9351, pp. 234–241. Springer, Cham (2015). https://doi.org/10.1007/978-3-319-24574-4_28
19. Sandler, M., Howard, A., Zhu, M., Zhmoginov, A., Chen, L.C.: MobileNetV2: inverted residuals and linear bottlenecks. In: Proceedings of the IEEE Conference on Computer Vision and Pattern Recognition, pp. 4510–4520 (2018)

20. Simonyan, K., Zisserman, A.: Very deep convolutional networks for large-scale image recognition. arXiv preprint arXiv:1409.1556 (2014)
21. Wang, J., Wang, Z., Tao, D., See, S., Wang, G.: Learning common and specific features for RGB-D semantic segmentation with deconvolutional networks. In: Leibe, B., Matas, J., Sebe, N., Welling, M. (eds.) ECCV 2016. LNCS, vol. 9909, pp. 664–679. Springer, Cham (2016). https://doi.org/10.1007/978-3-319-46454-1_40
22. Yu, F., Koltun, V.: Multi-scale context aggregation by dilated convolutions. arXiv preprint arXiv:1511.07122 (2015)
23. Zhao, H., Shi, J., Qi, X., Wang, X., Jia, J.: Pyramid scene parsing network. In: IEEE Conference on Computer Vision and Pattern Recognition (CVPR), pp. 2881–2890 (2017)
24. Zheng, S., et al.: Conditional random fields as recurrent neural networks. In: ICCV, pp. 1529–1537 (2015)

Pulmonary DR Image Anomaly Detection Based on Deep Learning

Zhendong Song, Lei Fan, Dong Huang, and Xiaoyi Feng[✉]

Northwestern Polytechnical University, Shaanxi, China
secszd@163.com, {dear,huangdong137}@mail.nwpu.edu.cn,
fengxiao@nwpu.edu.cn

Abstract. The morbidity and mortality in lung cancer is increasing which makes the diagnosis of abnormal lungs particularly important. Because of the advantages in DR image, this paper aimed at two problems in current medical image research: first, it is difficult to completely segment the lung of DR image only used traditional image segmentation methods. This paper replaces the padding in the U-net network model with zero padding to maintain the image size and apply it to the lung DR image segmentation, and finally uses the lung DR image dataset to fine-tuning. Secondly, the results of anomaly detection experiments show that the algorithm would get more complete segmentation of lung DR images. Secondly, because of the insufficient of training set, the idea of multi-classifier fusion is used. Combining Gabor-based SVM classification, 3D convolutional neural network, and transfer learning to achieve a more complete description of features and make full use of the classification advantages of multi-classifiers. The experimental results show that the classification accuracy of this algorithm is 6% higher than that of the Transfer-ImageNet algorithm, 5% higher than SVM, 15% higher than 3D convolutional neural network, and improved 2.5% compared with FT-Transfer-DenseNet3D algorithm.

Keywords: DR image · SVM · Transfer learning · 3D convolutional neural network · Lung field segmentation

1 Introduction

1.1 Research Background and Significance

With the changes in the global environment, the air quality is getting worse. At the same time, the improvement of living standards and the faster pace of work make life more convenient. The resulting irregularities in life is increasingly detrimental to people's health. According to the latest data, 546,259 cases of tracheal, bronchial and lung cancer deaths occurred in China in 2013, accounting for one-third of the number of deaths of this type of cancer worldwide. According to the World Health Organization's research, between 2008 and 2018, the incidence of cancer in the world is rising linearly. At the same time, the probability of patients surviving within 5 years is only

© Springer Nature Switzerland AG 2019
Y. Zhao et al. (Eds.): ICIG 2019, LNCS 11901, pp. 182–198, 2019.
https://doi.org/10.1007/978-3-030-34120-6_15

6–13%. The reason to this phenomenon is that 65% of patients are already in precancerous stage. If the symptoms of early stage lung cancer can be found through related techniques, there will be a great survival rate within 5 years. Therefore, the detection rate of early lung cancer can greatly improve the treatment of thousands of patients, and can also reduce the medical burden of patients with advanced cancer.

At present, the lung image can be said to be an extremely important and common way in the medical diagnosis methods. The doctor can find the abnormal part to carry out further diagnosis and treatment. DR imaging and CT tomography are the two main methods of pulmonary diagnosis. Among them, DR images have less radiation and higher image quality than other X-ray images [7], and the price is cheaper. Therefore, automatic reading through DR images will have a larger application space.

In terms of edge detection and contour extraction, Yao et al. proposed a method for segmentation of lung parenchyma from coarse to improve the segmentation accuracy [3]. In this method, 246 high-quality labeled DR images are used as data sets. Wang et al. proposed a DR chest slice contour extraction method based on fully convolution network [14]. Chernuhi et al. used edge detection and target extraction for DR images, proposed a target detection and edge extraction method for X-ray medical images, and represented the object boundaries with vectors [1]. Candemir et al. implemented an automatic lung field segmentation algorithm that has a good segmentation effect on non-rigid registration [6]. The overall accuracy of the JSRT dataset is 95.4%. Luo et al. considered that X chest radiographs have different regional features and edge features in pneumoconiosis, which makes the segmentation results not accurate enough [14]. An algorithm combining wavelet transform and Snake model algorithm is proposed. The experimental results show that the proposed method has a high recognition rate and stable recognition effect.

In terms of image enhancement and noise suppression, Feng et al. described an image enhancement technique CWGCE based on neighborhood contrast and wavelet transform coefficients, which can adaptively enhance edge details and effectively suppress noise [2]. Du et al. proposed a cell membrane optimization algorithm to calculate the spatial position of the optimal pixel in the image for the disadvantages of sharp noise, improper exposure, thick tissue and uneven distribution in the DR image [13]. Xu Yanli et al. constructed a Gaussian pyramid and a Laplacian pyramid, used a specific function to adjust its coefficients, and repeatedly extended the image. Finally, the results were added to reconstruct the original image to enhance the detail of image [12].

Hong et al. divided 100 test samples into 50 normal and 50 abnormal, and used SVM classifier to train the training set of chest DR images to obtain decision function [4]. Khatami et al. proposed a three-step framework for multi-category X-ray image classification. The denoising techniques based on wavelet transform (WT) and Kolmogorov Smirnov (KS) statistics are first used to remove noise and insignificant features in the image; then unsupervised deep belief network (DBN) is used to learn unlabeled features; The more descriptive features serve as input to the classifier [5]. Torrents-Barrena et al. used a method to automatically analyze X-ray images based on tissue density to determine the presence of normal and suspicious regions in the breast [14]. Luo Haifeng et al. proposed a method for feature extraction and classification by combining gray level co-occurrence matrix with BP neural network [10]. The average accuracy of the network can reach 68.3%. Li Bo et al. designed a relatively independent classification framework by combining the grayscale features, texture features, shape features and features extracted from the frequency domain with medical image segmentation methods. The accuracy of traditional medical image classification methods has been significantly improved [11]. Song Yuqing et al. proposed a classification method based on feature-level data fusion and decision-level data fusion based on the shortcomings of any feature that could not correctly express medical images [2].

In summary, medical image still has great prospects in the field of artificial intelligence. However, the main problems at present are: first, the DR medical images in the lungs are difficult to obtain, and there is no system for automatic labeling. Second, in the difficulty of image segmentation algorithms, the morphology of the lungs is different under different disease conditions. The manual extraction method such as template matching cannot find the edge lines in every cases. Thirdly, in the anomaly detection algorithm, manual extraction of features and the construction of classifiers through machine learning algorithms usually can be used because of small sample of the dataset. On the other side, the deep techniques have been used in many fields successfully and achieved promising results [15, 16]. Therefore, this paper studies the application of automatic labeling algorithm and deep learning used in lung field segmentation and anomaly detection.

2   Lung DR Image Segmentation

Because it is impossible to obtain clear and complete segmentation results while retaining the lung texture. Considering the good segmentation effect of U-net network for segmentation of medical cells under a magnifying glass, U-net network is used in this paper. First, gradation and histogram equalization for the low contrast of the lung DR image in some cases; second, minimize the lung image in order to preserve the lung information as much as possible.

ChinaSet_AllFiles dataset construct by the National Medical Library of Malan State and the Third People's Hospital of Shenzhen Medical College. A chest X-ray database created in Montgomery County, the NLM-MontgomeryCXRSet dataset created by the National Library of America and the Montgomery County Department of Health and Human Services in Maryland. DR_date data set collected by one hospital in Xi'an. CT images dataset which is published in Kaggle competition.

2.1 Data Preprocessing

Three data sets used in this paper including lung DR images, CT images, and ImageNet data sets. The overall DR image data included 700 cases of normal cases and 700 cases of tuberculosis, some of them are shown in Fig. 1.

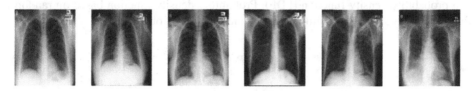

Fig. 1. Normal dataset and abnormal dataset sample

Data Enhancement

On the one hand, some of the annotated data is translated, rotated, and cut, to increase the quantity and diversity of data, as shown in Fig. 2. On the other hand, semi-supervising the unlabeled samples. After data enhancement processing, the sample size of the data has increased by more than 200.

Histogram Equalization

For each image, different gray levels have different total pixels. After histogram equalization, the input pixel distribution of the original input image can be changed, so that the pixel numbers of each gray level are different, so that the gray histogram of the image is smoothed.

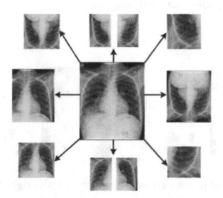

Fig. 2. Translated, rotated, and cut some of the annotated data

Suppose an image has a gray level of g, the range is [(0, 1)], and passes through the mapping function *s*. For g in any range, according to the form of the transformation:

$$s = T(g), \quad 0 \le g \le 1 \tag{1}$$

The gray level of the lung DR image is set to a random number of [0, 1], expressed as a probability density function (PDF). Probability density function based on random variables r and s, the probability density function can be obtained:

$$P_s(s) = P_r(r)\left|\frac{dr}{ds}\right| \tag{2}$$

The histogram of the image is obtained:

$$P_s(s)ds = P_r(r)dr \tag{3}$$

If $P_s(s) = s$, c is a constant and the histogram equalization formula is obtained:

$$\int_0^s sds = \int_0^r P_r(r)dr \Rightarrow s = \frac{1}{c}\int_0^r P_r(x)dx \tag{4}$$

As shown in Fig. 3, the second one is the histogram of the original image. It can be seen that the gray scale distribution of the whole image is concentrated at about 200, and the gray portion of the bright portion is less distributed; the third one shows the lung after the histogram equalization of the DR image, it can be seen from the histogram that the gray level of the entire image is still concentrated at 200, but the number of gray values of the bright portion is increased, and the overall gray scale distribution is enhanced, so that the image is enhanced. At this point, the image has a higher contrast, so the details can be displayed.

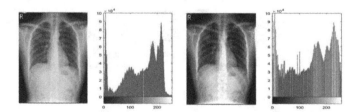

Fig. 3. The first and second pictures are the histograms and it's original image, and the last two are the histograms after the histogram equalization and it's image.

2.2   Lung Field Segmentation Based on U-Net Network

In this paper, the lung field segmentation of lung DR images based on U-net network is mainly divided into three processes. First, grayscale the image. Then perform histogram equalization on grayscale images to denoise and enhance image contrast. Finally, the segmentation result of the DR image is obtained by training the U-net network.

As shown in Fig. 4, the U-net network set a 2×2 pooling at layers with even sizes, which can keep the same size [9]. This architecture, the so-called "full convolutional network", can get more accurate segmentation result with fewer training images after modified and extended. The main idea is to supplement the usual shrinking network by successive layers, where the pooling operator is replaced by the upsampling operator. Therefore, these layers increase the output resolution, combining high resolution features from the contracted path with the upsampled output. Finally, the convolutional layer can learn more accurate output based on this network learning. An important advantage in this architecture is that in the upsampling section, there are a large number of feature channels that allow the network to propagate context information to higher resolution layers. So the expanded portion and the contracted portion have symmetrical characteristics, forming a U-shaped structure. Since the full connection layer is not used in the entire network architecture, only the effective part of the convolution is used, so the result of the segmentation can only obtain the image pixel portion, it has the connections in the upper and lower pixel. By overlapping block strategies, it can seamlessly segmenting arbitrarily large images. In order to predict pixels in the boundary region of the image, the missing pixel portion is extrapolated by mirroring the input image. This tiling strategy can make the network used in large images, so that resolution is no longer limited by GPU memory.

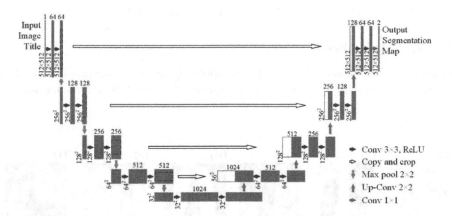

Fig. 4. U-net networks: Each blue box corresponds to a multi-channel feature map. The number of channels is indicated at the top of the box. The x-y size is located on the lower left edge of the box. The white box indicates the copied feature map. The arrows indicate different operations [8]. (Color figure online)

Fig. 5. The image of the lung DR image and the label is divided according to the Grab cut edge detection, then combined U-net network segmentation with the gradient information

The basic U-net architecture is shown in Fig. 5. The network has 23 convolution layers. It consists of a shrink path and an extension path. The entire network architecture is similar to a convolutional neural network architecture. Two such convolutional blocks are constructed from a 3×3 convolutional layer, a linear rectifying unit and a 2×2 pooling layer as a convolutional block, then a downsampling of one step. In each downsampling step, the number of feature channels is doubled. Each step in the extended path upsamples the feature map and then performs a 2×2 up-convolution, halving the number of feature channels in tandem with the corresponding cropped feature map from the shrink path. Finally, a 1×1 convolution is applied to the 64-component feature vector to obtain the corresponding class.

The original image and its segmentation map are used as labels for the training network, and implement TensorFlow's stochastic gradient descent. There are two improvements compared with original network. On the one hand, changing the no-fill in the original convolution network to zero padding keeps the same image boundary width, and the step size is changed to 1. On the other hand, the original network training set is the cell medical image of the human body under the microscope. Since the morphological characteristics of the cells are different from the DR images, the parameters after training in the cell medical image are retained.

2.3 Segmentation Result

In view of the small amount of training data, applying elastic deformation to make data enhancement. Allowing the network to learn the invariance of this deformation without

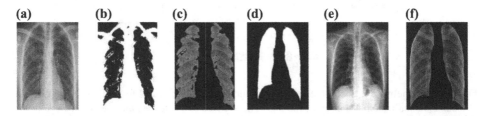

Fig. 6. (a) original image, (b) result based on Otsu; (c) result based on Watershed algorithm; (d) the result based on CNN; (e) the result based on Crab cut; (f) the result based on U-net

having to see these transformations in the annotated image corpus, which is especially important in biomedical segmentation because deformation is the most common change in tissue and can effectively simulate real Deformation.

In order to increase the CPU usage, large input slices are used in the case of large quantities, thereby reducing the batch size to a single image. Therefore, the use of high momentum (0.99) allows a large number of training samples used in the previous iteration to participate in the update in the current optimization step.

As shown in Fig. 6, the original lung texture is preserved and a considerable part of the edge features are also lost in detail; Automatic segmentation of the lung field based on convolutional neural networks shows that the contours of the lungs can be displayed evenly, but the texture of the lungs is not preserved. and small lung nodules will be lost after segmentation, and it is not easy to detect early abnormal state; Grab cut segmentation result shows that the contour of the lungs is very good, and the overlap rate reached 95.9%. The segmentation result of the Grab cut is added to the network training as a training set, and the segmentation is performed in combination with the gradient information. The results show that the contrast of the lungs is deeper, the contours of the lungs are more pronounced, and the lungs can be completely segmented without losing edge information.

3 Anomaly Detection Algorithm for Lung DR Images

This chapter is aimed at the shortcomings of the training set samples in the DR image anomaly detection and the low classification accuracy. Considering the advantage that the model fusion can combine the different classifiers to improve the classification accuracy, based on the improved bagging idea, combined SVM, 3D CNN, and transfer learning. First, establish the Gabor feature matrix, select the radial basis kernel function as the kernel function, train the SVM classifier. Second, use the Inception-v3 model as the transfer object on ImageNet dataset, and fine tune on the DR image data; thirdly, constructing a 3D CNN training on the preprocessed CT data set, and testing on the DR image dataset. Finally, combined these algorithms.

3.1 Model Fusion

Using the idea of bagging to construct three classification models and carry out a certain degree of fusion, and improve on the basis of bagging, first bagging is to form multiple data sets by sampling the same sample set multiple times. The set is set to three relatively large data sets, including the CT dataset, the DR dataset, and the JPG format DR data, and the data of three different data formats are combined to construct the classifier. Selecting a combination strategy of relative majority voting to combine the advantages of three different classifiers to extract different features; secondly, the relative majority vote in bagging randomly selects a result when the results of multiple

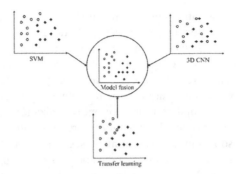

Fig. 7. Model fusion. The decision of the classifier will be combined according to the respective results of these component classifiers using a relative majority voting method.

classifiers are different, so it is proposed in this case Samples that did not receive a majority of the voting results were retrained and predicted again. Get the final abnormal diagnosis result.

Since there are 3 independent data sets, DR data D1 in JPG format, DR data D2 in DIM format, and CT data set D3 in lung. The classifier C1 established by the machine learning SVM is constructed with the three different kinds of data, the classifier C2 established by the transfer learning algorithm and the classifier C3 established by the 3D CNN are generated, finally, The decision of the classifier will be combined according to the respective decision results of these component classifiers using a relative majority voting method. The different training models have different weights under the same classification model. Therefore, it is possible to integrate the connection between different medical data and models to a certain extent to achieve abnormal diagnosis, as shown in Fig. 7.

3.2   SVM-Based Pulmonary Abnormality Detection

The detection algorithm is based on SVM. First, input the original image, normalize the image to a 512 × 512 matrix size as input. Second, through the threshold segmentation, corrosion, connectivity processing, to obtain the pre-processed image. Third, establish a total of 24° Gabor feature matrices in the four directions of 0°, 45°, 90°, 135° and six scales 7, 9, 11, 13, 15, 17. The frequency determines the wavelength of the Gabor filter, and the direction is determined. The size of the Gaussian window. Fourth, the feature matrix established by the training set sample is taken as the input of the radial basis kernel function map. The penalty parameter of the error term is used as the trade-off between the smoothing of the control decision boundary and the correct classification training point. Set it to 0.01 for SVM classifier training. Finally, according to the test set picture, the Gabor feature matrix with corresponding dimensionality reduction is established as the classifier input for class prediction, and finally the prediction result is output.

3.3   Pulmonary Abnormality Detection Based on Transfer Learning

Network Construction
Transfer learning can extract and leverage relevant features from existing data to complete new learning tasks. It is possible to adapt a trained model to a new problem with a simple fine-tuning process.

The transfer learning algorithm in this paper uses the trained Inception-v3 model on the ImageNet dataset to solve the problem of abnormal classification of lung DR images. The entire network is centered on the Inception-v3 model. By retaining all the convolutional layer (bottleneck layer) parameters in the trained Inception-v3 model, the DR image is fine-tuned on the latter m layer, replacing only the last layer of the full connection layer, set the number of categories to 2.

Transfer Learning Model
This transfer learning uses Inception-v3 for model migration. The Inception-v3 model is pre-trained on the ImageNet dataset to obtain the pre-training parameters of the network. The parameters in the pre-n-layer region of the pre-training model are transferred to another model. The same position, while the convolutional neural network m layer to fine-tune the DR data set, extract the deep features of the network and then perform anomaly diagnosis, as shown in Fig. 8.

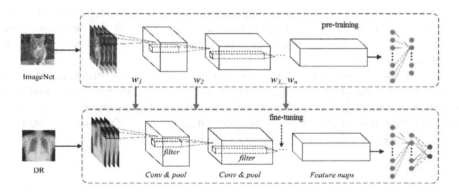

Fig. 8. Transfer learning.

3.4   Pulmonary Abnormality Detection Based on 3D Convolutional Neural Network

The 3D convolutional neural network is built on the CT dataset and anomaly detection is performed. The experimental results show that the feature extraction by convolutional neural network can extract the deep features of the original medical image and assist the anomaly detection. Anomaly detection can be realized by combining deep feature association between CT image data and DR image data.

Model Building

3D convolutional neural network architecture for lung abnormality detection using CT image data sets is described in this section. The 3D convolutional neural network consists of a hardwire layer, 3 convolutional layers, 2 downsampling layers, and a fully connected layer. A 7D image consists of a 3D convolution kernel. As shown in Fig. 49, seven 20×20 frames centered on the current frame are used as inputs to the 3D convolutional neural network model. First, a set of hardwire layers is used to generate multiple channels from the input frame, and 33 feature maps are obtained in the second layer of five different channels. A convolution with a kernel size of $3 \times 3 \times 3$ (space dimension of 3×3, time dimension of 3) is applied to each channel. Two different convolution operations are used at each location to increase the number of feature maps, and two sets of feature maps are generated in the C2 layer, each group consisting of 23 feature maps. Thereafter, a 3D pooling layer with a kernel size of $2 \times 2 \times 2$ is applied to each feature map in the C2 layer, and spatial resolution is reduced by the same number of feature maps. Next, applying a 3D convolution with a kernel size of $3 \times 3 \times 3$ on each channel of S3, 13 feature maps of size 22×22 can be obtained. Similarly, the 3D pooling layer, with a kernel size of $2 \times 2 \times 4$ applied to each feature map in the C4 layer, can obtain the same number of feature maps with a size of 11×11 as the C4 layer. Finally, a dropout layer with a ratio of 0.8 is added after the normalized S5 layer is 28×28 size. After the completion of the neural network training, a SoftMax classification layer with a learning rate of 0.001 was added for classification (Fig. 9).

After the 3D convolutional neural network is built, the lung CT image is used as the training set, and the preprocessing is performed by unified. After the lung field segmentation, the training is performed as the input of the 3D convolutional neural network, and the training parameters are obtained. Finally, the lung DR image is obtained. The data is used as a test set for abnormal diagnosis.

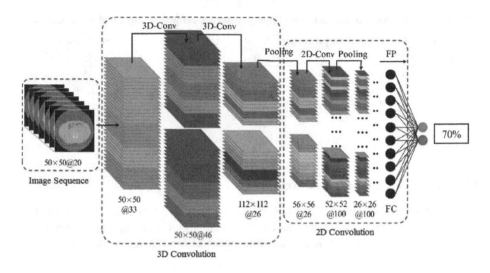

Fig. 9. The structure of 3D CNN

4   Experimental Result

4.1   SVM

The training data is divided into two types of samples, that is, the image in which no lesion occurs in the lung is a positive sample, the label is 1; the abnormal lesion has a negative sample, and the label is −1. Firstly, the lung image is preprocessed, the lung field is segmented, and the Gabor feature is extracted as the feature matrix. After Gabor feature extraction for each image, PCA is performed on 24 different feature matrices of 6 angles and 6 directions in the Gabor feature. Dimensionality reduction, and finally a matrix vector representation containing the largest information entropy is obtained, as shown in Fig. 43. After training the SVM classifier with the training set feature matrix vector as input, the feature matrix vector constructed by the test set is classified, and the radial basis kernel function is trained by 10-fold cross-validation to predict the result.

Table 1 shows the results of the existing research. In the case where the sample size is also small, the feature extraction and LS-SVM are combined to identify the pulmonary nodules. The final average accuracy of the experiment is about 60%. Figure 44 the results of the SVM experiment are the results of the algorithm used in this paper, and the graph (a) is the curve of the test accuracy as the number of iterations increases. It can be seen from the figure that the accuracy rate starts to decrease after the 9th iteration. After 10 iterations, the accuracy basically converges, the overall average accuracy rate reaches 80%, and the highest accuracy rate can reach 85%. From the ROC curve, you can see Out, with the increase of the number of iterations, when the false positive is 1, the true positive can reach 89%, and the diagnosis result is ideal. It can make the true positive reach the ideal value when the false positive is 1.

Table 1. Pulmonary nodule detection based on feature extraction and SVM classification.

Number	Kernel function parameter	Penalty factor	Number of samples	Accuracy
1	0.1	10	20	55%
2	0.1	100	20	55%
3	0.1	1000	20	60%
4	0.2	10	20	60%
5	0.2	100	20	65%
6	0.2	1000	20	65%

4.2   3D Convolutional Neural Network

In the experiment, CT images and DR images formed by 1500 patients were used as training data sets and test data sets, and a file containing the data tags was generated to act on the 3D convolutional neural network model of lung image recognition.

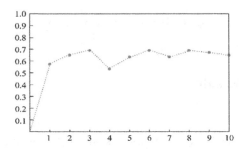

Fig. 10. The result trained on 3D CNN

First, define the weights and offsets of the convolutional neural network. The first convolution layer is set to 3 × 3 × 3 patches, 1 channel, 32 feature maps; the second convolution layer is set to 3 × 3 × 3 patches, 32 channels, 64 features. The full connection layer has a weight of 1024 and an output weight of 1024. The offset of the first convolutional layer is 32, corresponding to 32 feature maps, the offset of the second convolutional layer is 64, the offset of the fully connected layer is 1024, and the offset of the output is 2. Then define the fit function, the activation function and the loss formula, and optimize them with the Adam Optimizer optimizer. The CT scan data is trained as a training set, and the DR image data set is tested to obtain an accurate result curve, as shown in figure. The results show that after 10 iterations, the 3D convolutional neural network model test on the DR dataset achieve 70% accuracy (Fig. 10).

4.3   Transfer Learning

Firstly, the ImageNet data set is pre-trained on the Inception-v3 model, the trained parameter model is saved, and the parameters of the first 90 convolutional layers are frozen, and the next six convolutional layers are The DR image data set is fine-tuned, and the number of output nodes (number of categories) of the last layer is set to 2, and the final classification result is obtained.

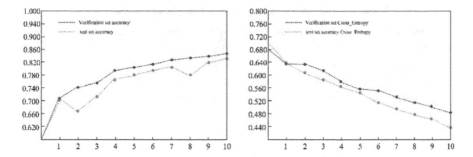

Fig. 11. The result of transfer learning. The x-axis represents the batch_size in units of 100, that is, the iteration number is 10 times, the y-axis represents the cross entropy loss value and the accuracy of the test set, and the red curve and the blue curve represent the result distribution curves of the verification set and the test set (Color figure online)

This paper first proposes the application of migration learning to DR medical image anomaly detection. Figure 11 shows the loss function and accuracy curve of the migration learning network after training. It can be seen the accuracy tends to converge after 10 iterations, and the cross entropy loss function is decreasing, eventually reaching 45%, and the test accuracy of the test sample can reach about 80%.

4.4 Model Fusion

The model fusion results are obtained by model fusion of the above three classifier models, as shown in Table 2. The results show that, first, after the model fusion, the accuracy of the test set of the whole anomaly detection is 5% higher than that of the base classifier, and the final accuracy can reach 85%. The classification accuracy rate is 5% higher than SVM, 15% higher than 3D convolutional neural network, and 6% higher than Transfer-ImageNet; AUC value is 4% higher than SVM. The 3D convolutional neural network increased by 7% compared to Transfer-ImageNet by 9%. Second, the classification accuracy of the SVM algorithm in the experimental results is higher than that of the 3D convolutional neural network, and it is not much different from the Transfer-ImageNet. This is mainly because the classifier has only 700 samples of positive and negative samples on the small sample data. When classifying, the SVM can still map the positive and negative samples according to the distribution of the sample to the corresponding high-dimensional space to find the optimal classification plane. However, in the deep convolution network, the size of the data is proportional to the classification effect. The larger the amount, the higher the classification accuracy. Thirdly, the model migration based on Inception-v3 can achieve fine classification results with SVM after the parameter is fine-tuned, indicating that migration learning

Table 2. Comparison of experimental results in this paper

Algorithm	Accuracy on test set	AUC	Accuracy on verification set
SVM	0.80	0.82	0.88
3D CNN	0.70	0.81	0.85
Transfer-ImageNet	0.79	0.80	0.83
Model fusion	0.85	0.86	0.92

Table 3. Comparison of existing experimental results

Algorithm	Accuracy
SVM	0.800
3D CNN	0.700
Transfer-ResNet3D31	0.782
FT-Transfer-ResNet3D31	0.821
Transfer-DenseNet3D31	0.797
FT-Transfer-DenseNet3D31	0.825
Transfer-Image	0.790
Model fusion	0.850

can achieve better classification effect when dealing with small sample data. And have better generalization ability for lung DR images in different situations. Fourth, the model fusion improves the final classification accuracy by 5% on the basis of the base classifier by combining the feature advantages of different classifiers.

Table 3 shows a comparison of the existing related methods with the results of this paper. The data shows that the existing migration learning algorithm migrates the model parameters on the ResNet3D and DenseNet3D networks by pre-training the relevant data on the 2D convolutional neural network, and the accuracy rate can reach 82.5%. The migration learning was based on the Inception-v3 model, pre-training on the ImageNet dataset, fine-tuning through the lung DR image, and model fusion with the SVM, 3D convolutional neural network. The experimental results show that the accuracy of the model fusion is 6% higher than that of the Transfer-ImageNet algorithm, 5% higher than SVM, and 15% higher than 3D convolutional neural network, and FT-Transfer-DenseNet3D. The algorithm is up 2.5%. This is an increase of 2.9% compared to the FT-Transfer-ResNet3D algorithm.

5   Summary

Based on the relevant theories of deep learning and the starting point, this paper combines traditional machine learning and deep learning to construct the network model. The final anomaly classification accuracy is 6% higher than that of Transfer-

ImageNet. FT-Transfer-DenseNet3D increased by 2.5%, SVM increased by 5%, and 3D convolutional neural network increased by 15%.

There is still a great distance from the research stage to practice. For the future development direction of this research, we can improve from the improvement of DR image dataset, lung field segmentation of lung DR image and lung abnormality detection. Firstly, combined with the doctor's accurate judgment to obtain high-quality annotated data, it is expected to establish a sufficiently complete database; secondly, use a deeper network on larger data; finally, classify and normalize the classification types of lung abnormalities, and convert the two classification networks into multi-class networks, which can detect abnormal types.

References

1. Chernuhin, N.A.: On an approach to object recognition in X-ray medical images and interactive diagnostics process. In: Computer Science and Information Technologies (CSIT), pp. 1–6 (2013)
2. Feng, J., Xiong, N., Shuoben, B.: X-ray image enhancement based on wavelet transform. In: Asia-Pacific Services Computing Conference (APSCC 2008), pp. 1568–1573 (2008)
3. Yao, F., Jun-Feng, W., Lin, G., et al.: Lung field segmentation based on ASM and GrabCut on DR image. Mod. Comput. 4, 25–30 (2015)
4. Hong, S., Tian-yu, N., Yan, K., et al.: Chest DR image classification based on support vector machine. In: Education Technology and Computer Science (ETCS), vol. 1, pp. 170–173 (2010)
5. Khatami, A., Khosravi, A., Nguyen, T., et al.: Medical image analysis using wavelet transform and deep belief networks. Expert Syst. Appl. 86, 190–198 (2017)
6. Pan, S.J., Yang, Q.: A survey on transfer learning. IEEE Trans. Knowl. Data Eng. 22(10), 1345–1359 (2010)
7. Zhang, L., Lu, L., Nogues, I., et al.: DeepPap: deep convolutional networks for cervical cell classification. IEEE J. Biomed. Health Inform. 21(6), 1633–1643 (2017)
8. DeSantis, C.E., Ma, J., Goding Sauer, A., et al.: Breast cancer statistics, 2017, racial disparity in mortality by state. CA: Cancer J. Clin. 67(6), 439–448 (2017)
9. Szegedy, C., Vanhoucke, V., Ioffe, S., et al.: Rethinking the inception architecture for computer vision. In: Proceedings of the IEEE Conference on Computer Vision and Pattern Recognition, pp. 2818–2826 (2016)
10. Wu, J., Li, H., Xia, Z., et al.: Screen content image quality assessment based on the most preferred structure feature. J. Electron. Imaging 27(3), 033025 (2018)
11. Fang, M., Yin, J., Zhu, X.: Transfer learning across networks for collective classification. In: International Conference on Data Mining (ICDM), pp. 161–170 (2013)
12. Pan, W.: Transfer learning in collaborative filtering (2012)
13. Saxena, A.: Convolutional neural networks: an illustration in TensorFlow. Crossroads 22(4), 56–58 (2016)

14. Diba, A., Fayyaz, M., Sharma, V., et al.: Temporal 3D ConvNets: new architecture and transfer learning for video classification. In: Computer Vision and Pattern Recognition, vol. 9, no. 11, p. 76 (2017)
15. Xia, Z., Lin, J., Feng, X.: Trademark image retrieval via transformation-invariant deep hashing. J. Vis. Commun. Image Represent. **59**, 108–116 (2019)
16. Xia, Z., Hong, X., Gao, X., Feng, X., Zhao, G.: Spatiotemporal recurrent convolutional networks for recognizing spontaneous micro-expressions. IEEE Trans. Multimed. (2019, online)

A Stackable Attention-Guided Multi-scale CNN for Number Plate Detection

Yixuan Wang[1], Shangdong Zheng[1], Wei Xu[2], Yang Xu[1], Tianming Zhan[1], Peng Zheng[1], Zhihui Wei[1], and Zebin Wu[1(⊠)]

[1] School of Computer Science and Engineering, Nanjing University of Science and Technology, Nanjing 210094, China
wuzb@njust.edu.cn
[2] China Railway Shanghai Group Co., Ltd., Nanjing Power Supply Section, NangJing 210011, China

Abstract. High-speed railway transportation faces various risks due to its increasing laying area. A key to find the abnormal conditions is to locate the position of a pillar nearby. The complex background and different size of number plate (NP) makes the detection of number plate in overhead catenary system (OCS) a hard work. In this paper, we propose a novel framework and solution with two main advances: (1) a stackable attention model, which can improve the robustness of the system in complex scenarios; (2) multi-scale feature fusion stage, improve detection accuracy of distance pillar number plate. Both of them can be integrated into any CNN architectures seamlessly with negligible overheads and are end-to-end trainable along with base CNNs. We demonstrate the effectiveness of our method with experiments on different high-speed train lines and one benchmark dataset – Pascal VOC [4].

Keywords: Number plate detection · Attention model · Multi-scale

1 Introduction

Overhead catenary system (OCS) is an indispensable part of the high-speed railway network which is the transmission line set up over the railway line to supply power to electric locomotives. Number plates are installed in OCS and each of these represents a fixed position in railway lines. Once any fault occurs on OCS, the power supply to the train will have an impact. The first step to address these problems in OCS is to locate the position of abnormal situation precisely.

This work was supported in part by the National Natural Science Foundation of China under Grant 61772274, Grant 61701238, Grant 61471199, Grant 11431015, and Grant 61671243, in part by the Jiangsu Provincial Natural Science Foundation of China under Grant BK20180018 and Grant BK20170858, in part by the Fundamental Research Funds for the Central Universities under Grant 30919011103, Grant 30917015104, and Grant 30919011234, and in part by the China Postdoctoral Science Foundation under Grant 2017M611814.

© Springer Nature Switzerland AG 2019
Y. Zhao et al. (Eds.): ICIG 2019, LNCS 11901, pp. 199–209, 2019.
https://doi.org/10.1007/978-3-030-34120-6_16

Therefore, accurate and robust automatic number plate detection (ANPD) is an important issue for locating the abnormal conditions in OCS.

Vehicle license plate (LP) detection task, which has same properties as our problem, is a hot research topic in computer vision due to its important practical significance. Traditional methods [3,6] mostly depend on manual-selecting feature, lead to slow speed and low precision. The advances in deep Convolution Neural Networks (CNNs) have demonstrated high capability in many computer vision tasks including LP detection. Most of the recent approaches can be divided into 2 branches: anchor-based methods and region proposal methods [1,7,9,17] utilize CNNs to learn robust feature representations from images, and have demonstrated impressive detection accuracy, but usually at the expense of very complicated network architecture and much time. The success of Yolo networks [10–12] and SSD [8] inspired many recent works, targeting real-time performance for LP detection [5,14,17].

Fig. 1. Examples of OCS pillar number plate in our dataset, including different place, character length and direction.

Unfortunately, we notice that OCS images exhibit great variations in lighting conditions, occlusion, background and shooting angle [16], which makes automatic detection of number plates a challenging task. As shown in Fig. 1, first, railway network covers all parts of China so that the pictures might be captured from cities, mountains, forests, deserts and other scenes. These constraints result in extremely complex background in these images. Second, affected by a long period of running and large-amplitude sloshing, OCS images exhibit great variations in lighting changes, illumination conditions and distortion. Third, the number plates in OCS could be divided into different types according to the direction and length. Even some number plates fall off or break down due to long-term corrosion. For reasons already enumerated, existing methods are not completely suitable for automatic number plate detection in OCS. As far as we know, this paper is the first to utilize neural network to complete the automatic number plate detection in OCS.

In conclusion, the main contributions of this paper are:

- We proposed a Stackable Attention model which picks out the most important features of current stage and enhance them to improve the accuracy of detention phase effectively.
- A multi-scale strategy was designed to obtain more precise detection results from different size of NPs.

The rest parts of this paper are as follows: In Sect. 2, we briefly review the researches related to our work. Then, details of the proposed method are given in Sect. 3. In Sect. 4, we will introduce our datasets and evaluation protocols. Finally, Sect. 5 summarizes our methods and gives perspectives for some future work.

2 Related Work

In this work, ANPD is the task of finding the number plates in OCS images which is similar to vehicle license plate (LP) detection. Before the deep learning era, traditional methods for LP detection mostly depend on manual-selected feature. Kaushik et al. [2] put forward a HSI color model which is used for detecting candidate regions and vehicle license plate regions are verified and detected by using position histogram. Mahmood et al. [6] proposed a filtering method called region-based in order to smooth the uniform and background areas of an image. The author also applied the Sobel operator and morphological filtering to extract the candidate regions and vertical edges respectively. However, these approaches are mostly depending on manual-selecting feature, leading to slow speed and low precision.

With the rapid development of deep learning, more and more scholars employ neural network due to its high accuracy and strong robustness for vehicle license plate detection. Most of the recent approaches can be divided into 2 branches: region proposal methods and regression-based methods. Orhan et al. [1] put forward 2-stage plate localization and failure identification using CNN features. However, they tend to be computationally inefficient as a result of not sharing calculations. Li et al. [7] utilized a Region Proposal Network to create candidate license plate regions and feature maps which are cropped by RoI Pooling layer.

Then, the next sub-network will estimate these candidates regions containing a license plate or not. However, this method does not work well in some complex scenarios. One-stage methods, as typified by Yolo [11], frames object detection as a regression problem to spatially separate bounding boxes and associated class probabilities [10]. Yolo network divides an image into $S * S$ grid cells, which are responsible for prediction an object if the center of the object falls into the grid. In order to make the network more stable, the author designed loss function to maintain a good balance among three aspects: coordinate(x, y, w, h), confidence and classification. On this foundation, Josephed al. [11] also put forward many improvements. First, the distribution of input in each layer has been changing at each batch, which will make the training process more difficult to convergence. This paper addressed this problem by adding batch normalization after each convolution layer to regularize the model. Second, Yolo, previously, used the feature maps of the full connection layer to complete the prediction of the border, resulting in the spatial information loss and inaccurate positioning. In this version, the author draw lessons from [13], using anchor idea to receive more richer spatial information leads to a higher recall. With the success of Yolo, many researches tend to modified version of the Yolov1 and Yolov2 networks to detect the LP. Silva et al. [14] proposed an end-to-end DL-ALPR system for Brazilian license plate based on Yolov2. By prior knowledge of the number of characters, their method achieved state-of-the-art performance at that time. However, the paper focuses on the seven-characters license plate and does not work well in other type of LP datasets. Hsu et al. [5], where the authors improve the number of detection by enlarging the output granularity and they also set the probabilities for two classes. Despite advancing, this method does not evaluate over scenarios where the car is far from the camera.

Fig. 2. The same pillar number plate captured at different shooting distances.

3   The Proposed Method

In this paper, we focus on the detection of number plates in high-speed OCS. Our framework is based on fine-tuning Yolov2, thus we adopt the same loss function and prediction method with Yolov2.

3.1   Stackable Attention-Guided NP Detection in Complex Scene

Additionally, we find that the number plates in OCS images have more complex deformations caused by oblique camera views. And all pictures also exhibit great variations in lighting changes, illumination conditions, occlusion and shooting angle. These constraints make automatic detection of number plates in OCS a challenging task. Thus, we design the guiding attention model to focus on the distorted number plate in complex scene better. Figure 4 illustrates the architecture of our stackable attention-guided number plate detection network.

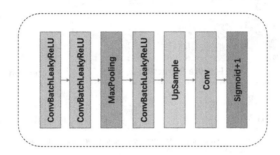

Fig. 3. The structure of the proposed attetion model.

The Stackable Attention Model can pick out the most important features of current stage and enhance them to improve the accuracy of detention phase effectively. It is particularly meaningful for complex and changeable scenes detection for us. As shown in Fig. 3, our attention model contains four convolution layers, one down-sample layer, one up-sample layer and an active layer. The attention module will be used repeatedly in our framework.

Inspired from [15], at the end of the attention model, a sigmoid layer normalizes the output range to $[0, 1]$. The attention map is given by

$$F_{att}(X) = Sigmoid\big(F_{ai}(X; \theta)\big) \oplus 1 \tag{1}$$

Where θ is the parameter to be learned for the attention information extraction network (F_{ai}) and X represents the input of F_{ai}. Symbol $\oplus$ denotes element-wise adding. This algorithm makes sure that even if our attention model does not work at all, the performances should be no worse than its counterpart without attention model [15].

Let a feature map as input to our attention model, we can obtain an attention map – $F_att(X)$. Every element in this matrix is ranging from $[1, 2]$, representing the importance of the feature at that position. Then, our method generates the output x_{t+1} on the $t-th$ step. Details of the attention model are given in Fig. 3

$$X_{t+1} = X_t \otimes F_{att}(X_t) \qquad (2)$$

Where $\otimes$ represents element-wise product. In order to judge the importance of the feature map, the surrounding features should be taken into consideration. Thus, our attention model downsamples the feature map to expand the receptive field to obtain more richer contextual information. Then, we restore the down-sampled features to their original size by deconvolution. This pair of up-down sampling not only expands the receptive field to improve the accuracy but also takes less time.

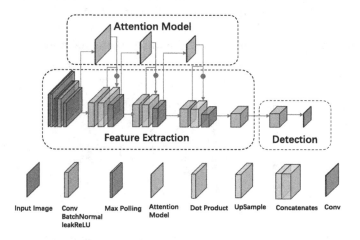

Fig. 4. The structure of stackable attention-guided NP detection network.

However, the experiments indicate that the guiding attention model is hard to focus on all the NP where the number plate is far from the camera, as indicated in Fig. 2. Due to this limit, we put forward an adaptive NP detection method based on multi-scale which will be introduced in Sect. 3.2.

3.2 Multi-scale Based Adaptive NP Detection

The results of region detection mostly rely on either feature maps from the last layer or last few layers, which poses challenges to accurate region detection.

As is indicated in Fig. 2, running of the train on a railway track, the number plate in OCS is becoming smaller and harder to detect.

Due to this limit, our method, inspired by [12], taking out the features before every down-sampling layer as supplementary information. Figures 5 and 6 illustrate that all the candidate boxes will be obtained on the 1/32, 1/16 and 1/8 size of the original image size which greatly alleviates the small size problem.

$$p_i = Det_i(d_i), \qquad i = 1, 2, 3 \qquad (3)$$

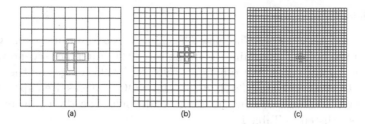

Fig. 5. A schematic drawing the multi-scale feature fusion detection. Receptive fields, focused by the same size of kernels in different resolution of feature maps are shown in (a), (b) and (c).

To obtain more precise detection results, our framework makes predictions on three scales. All the candidate bounding boxes will be performed non-maximum suppression to retain the most appropriate results.

$$d_i = \begin{cases} Conv(f_i) & i = 1, \\ Conv\big(F_{fusion}(d_{i-1}, f_i)\big) & i = 2, 3. \end{cases} \tag{4}$$

$$F_{fusion}(d_{i-1}, f_i) = Conv\big(Upsample(d_{i-1} \oplus f_i)\big) \tag{5}$$

This equation illustrates the fusion of d_{i-1} and f_i where $\oplus$ represents concatenation.

4   Experiments

We conduct experiments on our datasets, including different types of images taken by Catenary-Checking Video Monitor System. We will also compare the performance of our method with other classical or state-of-the-art models on one benchmark dataset – Pascal VOC.

Table 1. Data information of OCS images.

Train lines	Number of images
Hefei-Fuzhou Railway	3019
Beijing-Shanghai Railway	2532
Nanjing-Qidong Railway	5427
NanJing-Anqing Railway	1447
Total	**12425**

4.1   Dataset

Our dataset are taken by the Catenary-Checking Video Monitor System which is installed on the head of the train, the frame rate is 20 frame per second with the spatial size of 1620 * 1234. We randomly selected 12425 images as the dataset, dated from October 2017 to March 2019. These OCS images are captured from different train lines and various imaging conditions. The data information of OCS images is listed in Table 1.

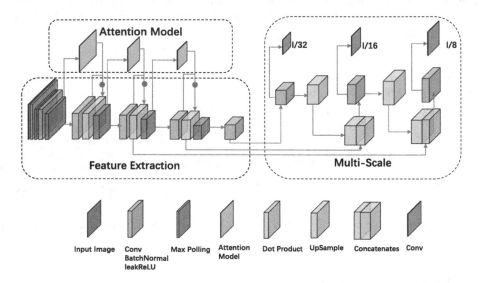

Fig. 6. The network architecture. The architecture consists of three parts: 1. feature extraction model, which extracts a feature sequence from the input image; 2. attention model, which picks out the most important features of current stage; 3. multi-scale feature fusion detection model, which fuses multi-scale features to detect.

4.2   Attention Model

We use three attention models in different stages of feature extraction, because different attention models of depths capture different types of feature. As shown in Fig. 7, in a shallow layer, network focus more on corner Feature; at a deeper layer, network pay more attention on more complex features; at the third attention model, the network can find target approximately.

4.3   Performance Evaluation

Our work uses batch-GD with a mini-batch size of 64. We use a weight decay of 0.0005 with a momentum of 0.9 and set the initial learning rate to 0.0001. The

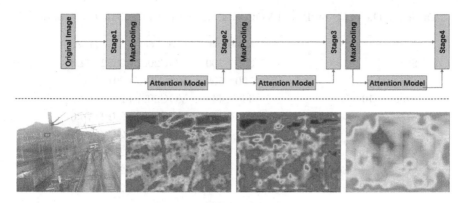

Fig. 7. At different stage, attention modules capture different types of feature.

learning rate decayed at 400, 700, 1000, 40k and 60k iterations. The overall the hyper parameters setting is described in Table 2.

After trying several basic detection networks, we choose Darknet-19 as our feature extraction network finally. Therefore, we mainly compare our approach with the original Yolov2.

Table 2. The hyper parameters be used in our network.

Parameter	Iteration	Value
Batch	Always	64
Momentum	Always	0.9
Decay	Always	0.0005
Learning rate	1–1000	0.0001
	1001–40000	0.001
	40001–60000	0.0001
	>60000	0.00001

We compare our framework with some previous methods on Pascal VOC and summarize the results in Table 3. Our approach improves the performance by up to 81.05%, measured by mean average precision (mAP), while the original Yolov2 achieves 78.6%. Even if we only merge attention model or multi-scale strategy into our framework the accuracy is still higher.

As is shown in Table 3, combing three attention models and multi-scale feature fusion with the prediction results of original image improves significantly to 96.35% measure by mAP while the fine-tuning baseline Yolov2 achieves 95.15%. If we only integrate attention model into our framework, the mAP is as high as 95.96%. The mAP of adoption multi-scacle strategy into our framework is 96.07%.

Table 3. The mAP on Pascal VOC and our dataset of different network.

Network	Resolution	DataSet	mAP
SSD [8]	300 * 300	VOC2007+2012	74.3
SSD [8]	500 * 500	VOC2007+2012	76.8
Yolov1 [10]	448 * 448	VOC2007+2012	63.4
Yolov2 [11]	544 * 544	VOC2007+2012	78.6
Yolov3 [12]	544 * 544	VOC2007+2012	79.6
Ours (Attention Model only)	544 * 544	VOC2007+2012	80.01
Ours (Multi-Scale only)	544 * 544	VOC2007+2012	80.91
Ours	544 * 544	VOC2007+2012	**81.05**
Yolov2	544 * 544	Our DataSet	95.15
Ours (Attention Model only)	544 * 544	Our DataSet	95.96
Ours (Multi-Scale only)	544 * 544	Our DataSet	96.07
Ours	544 * 544	Our DataSet	**96.35**

5   Conclusion

In this work, we propose a new effective method for high-speed railway pillar number plate region detection. The benefits of our network are in two folds. The first benefit lies in that different attention modules capture different types of feature to improve network learning. The second benefit comes from multi-scale feature fusion detection, can improve detection accuracy of small object. Lots of experiments prove that our framework achieves good performance in various complex railway scenarios. In the future, we will conduct the research on designing an end-to-end framework to complete pillar number plate detection and recognition at the same time.

References

1. Bulan, O., Kozitsky, V., Ramesh, P., Shreve, M.: Segmentation-and annotation-free license plate recognition with deep localization and failure identification. IEEE Trans. Intell. Transp. Syst. **18**(9), 2351–2363 (2017)
2. Deb, K., Hossen, M.K., Khan, M.I., Alam, M.R.: Bangladeshi vehicle license plate detection method based on HSI color model and geometrical properties. In: 2012 7th International Forum on Strategic Technology (IFOST), pp. 1–5. IEEE (2012)
3. Deb, K., Jo, K.H.: HSI color based vehicle license plate detection. In: 2008 International Conference on Control, Automation and Systems, pp. 687–691. IEEE (2008)
4. Everingham, M., Van Gool, L., Williams, C.K., Winn, J., Zisserman, A.: The Pascal visual object classes (VOC) challenge. Int. J. Comput. Vision **88**(2), 303–338 (2010)
5. Hsu, G.S., Ambikapathi, A., Chung, S.L., Su, C.P.: Robust license plate detection in the wild. In: 2017 14th IEEE International Conference on Advanced Video and Signal Based Surveillance (AVSS), pp. 1–6. IEEE (2017)
6. Lalimi, M.A., Ghofrani, S., McLernon, D.: A vehicle license plate detection method using region and edge based methods. Comput. Electr. Eng. **39**(3), 834–845 (2013)

7. Li, H., Wang, P., Shen, C.: Towards end-to-end car license plates detection and recognition with deep neural networks. Corr abs/1709.08828 (2017)

8. Liu, W., et al.: SSD: single shot multibox detector. In: Leibe, B., Matas, J., Sebe, N., Welling, M. (eds.) ECCV 2016. LNCS, vol. 9905, pp. 21–37. Springer, Cham (2016). https://doi.org/10.1007/978-3-319-46448-0_2

9. Long, S., He, X., Ya, C.: Scene text detection and recognition: the deep learning era. arXiv preprint arXiv:1811.04256 (2018)

10. Redmon, J., Divvala, S., Girshick, R., Farhadi, A.: You only look once: unified, real-time object detection. In: Proceedings of the IEEE Conference on Computer Vision and Pattern Recognition, pp. 779–788 (2016)

11. Redmon, J., Farhadi, A.: YOLO9000: better, faster, stronger. In: Proceedings of the IEEE Conference on Computer Vision and Pattern Recognition, pp. 7263–7271 (2017)

12. Redmon, J., Farhadi, A.: Yolov3: an incremental improvement. arXiv preprint arXiv:1804.02767 (2018)

13. Ren, S., He, K., Girshick, R., Sun, J.: Faster R-CNN: towards real-time object detection with region proposal networks. In: Advances in Neural Information Processing Systems, pp. 91–99 (2015)

14. Silva, S.M., Jung, C.R.: Real-time Brazilian license plate detection and recognition using deep convolutional neural networks. In: 2017 30th SIBGRAPI Conference on Graphics, Patterns and Images (SIBGRAPI), pp. 55–62. IEEE (2017)

15. Wang, F., et al.: Residual attention network for image classification. In: Proceedings of the IEEE Conference on Computer Vision and Pattern Recognition, pp. 3156–3164 (2017)

16. Wu, X., Yuan, P., Peng, Q., Ngo, C.W., He, J.Y.: Detection of bird nests in overhead catenary system images for high-speed rail. Pattern Recogn. **51**, 242–254 (2016)

17. Xie, L., Ahmad, T., Jin, L., Liu, Y., Zhang, S.: A new CNN-based method for multidirectional car license plate detection. IEEE Trans. Intell. Transp. Syst. **19**(2), 507–517 (2018)

Residual Joint Attention Network with Graph Structure Inference for Object Detection

Chuansheng Xu[1,2], Gaoyun An[1,2(✉)], and Qiuqi Ruan[1,2]

[1] Institute of Information Science, Beijing Jiaotong University, Beijing 100044, China
{chshxu,gyan,qqruan}@bjtu.edu.cn
[2] Beijing Key Laboratory of Advanced Information Science and Network Technology, Beijing 100044, China

Abstract. Most object detectors include three main parts, CNN feature extraction, proposal classification, and duplicate detection removal. In this work, focusing on the improvement of the feature extraction, we propose Residual Joint Attention Network, a convolutional neural network using a residual joint attention module which is composed of a spatial attention branch, a channel attention branch, and a residual learning branch within an advanced object detector with graph structure inference. An attention map generated by the joint attention mechanism is used to weight the original features extracted from a specific layer of VGG16 aiming at performing feature recalibration. Besides, the residual learning mechanism is complementary to the joint attention mechanism and keeps good attributes of the original features. Experimental results show that different branches of our residual joint attention module do not contradict each other. By combining them together, the proposed network obtains higher mAP than many advanced detectors including the baseline on VOC dataset.

Keywords: Joint attention · Residual learning ·
Graph structure inference · Object detection

1 Introduction

In recent years, thanks to the advances of deep convolutional neural networks, a large number of computer vision tasks have enjoyed significant progress, including segmentation [7,8], image classification [1–3], object detection [4–6]. Among them, object detection is one of the fundamental problems that has been widely studied. Currently, there are two mainstream frameworks to solve the problem of object detection: the one-stage frameworks such as SSD [9] and YOLO [10], which directly transform the problem of object border positioning into a regression problem without extracting proposals; and the two-stage frameworks such as Fast R-CNN [5] and Faster R-CNN [6] which generate proposals by RPN layers [6] and then apply classification and regression to each proposal.

© Springer Nature Switzerland AG 2019
Y. Zhao et al. (Eds.): ICIG 2019, LNCS 11901, pp. 210–221, 2019.
https://doi.org/10.1007/978-3-030-34120-6_17

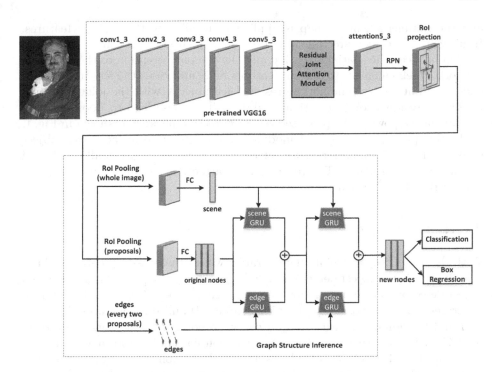

Fig. 1. The framework of our method. We feed an image into VGG16 pre-trained on ImageNet dataset to get a feature map named conv5_3. Then the residual joint attention module recalibrates conv5_3 feature map. Next, the feature map named attention5_3 is passed to an RPN layer followed by the graph structure inference part which involves two contextual information into the inference of node state. Eventually, the final state of each node is used to predict the category and refine the location of the corresponding RoI.

Most object detectors include three main parts, CNN feature extraction, proposal classification, and duplicate detection removal. For these three parts, improving the quality of the features of ConvNet backbones is a straightforward idea through which a lot of algorithms have made major breakthroughs [12–15]. Most of them use effective methods to increase the receptive field or semantic information of the feature maps extracted from ConvNet backbones. However, all of them do not consider utilizing the spatial and channel information of the feature maps when improving the detection accuracy.

Motivated by the success of the attention modules in image classification field [16]. We consider the combination of the spatial attention and the channel attention. As indicated in SENet [16], the channel-wise features can be adaptively recalibrated by effectively modeling the interdependencies between the channels of the feature map extracted from a ConvNet backbone. Similar to SENet [16], we also model the interdependencies between the spatial features. Spontaneously, the joint and multiplicative result of the spatial attention map and the channel attention map is applied to recalibrate the original features. Intuitively, We conjecture this adds the complementary and compatible information between the spatial attention and the channel attention to the proposed network which enhances the useful features and suppresses the less informative ones. In addition, we combine the residual learning [3] with the joint attention to form a residual joint attention module. All of these lead to boost model's discriminative power. The proposed object detection network is shown in Fig. 1.

In this paper, the proposed module incorporates into an advanced object detector [11] with graph structure inference only increasing a small number of parameters. In principle, The residual joint attention module is universal and not restricted to object detection.

2   Related Work

With the rise of deep convolutional neural networks, the two-stage detectors have rapidly dominated object detection over the past few years [4–6]. These advanced object detectors prevailingly follow the pioneering work R-CNN [4]. R-CNN first generates object proposals by Selective Search [21] and then operates classification and bounding box regression on every proposal. But the biggest problem of R-CNN is repetitive convolutional operation consuming too much time. To speed up, Fast R-CNN [5] introduces a novel RoI pooling layer to extract features for each proposal from the shared ConvNet feature map of the whole image. Whereas proposal generators are still not trained together with Fast R-CNN. To solve this problem, Faster R-CNN [6] develops RPN which can generate precise proposals and be trained together with detection subnetwork. Different from the two-stage detectors, the one-stage detectors remove proposal generators and directly operate classification and regression on a series of pre-computed anchors for real-time detection. Anyway, these state-of-the-art methods only consider the appearance features of the objects without considering the connections between the context and the objects in an image. Consequently, it is natural to utilize contextual information to improve object detection.

Many papers have proposed that scene information or relations between objects help object detection [17–19]. However, After the rise of deep learning, There haven't been significant breakthroughs in using contextual information to explore object detection until the emergence of [11,20]. In SIN [11], two kinds of contextual information are introduced: one is scene-level context, the other is instance-level relationships. These two complementary contextual information are combined through GRU [22] to help detection. Hu *et al.* [20] proposes an object relation module for object detection. By modeling the interdependencies between object appearance features and object geometry features, the object relation module can be used for instance recognition.

Most object detectors include CNN feature extraction, proposal classification, and duplicate detection removal. Actually, Using contextual information is working in the proposal classification part. Another way to improve object detection is promoting the quality of the features of ConvNet backbones. At present, many works are focusing on increasing the receptive field and semantic information of the features extracted from ConvNet backbones [12–15]. To involve multi-scale features, FPN [12] utilizes the hierarchical feature maps from different depths of CNN. DES [13] augments the low-level feature maps of VGG16 with strong semantic information which is trained by week bounding-box level segmentation ground-truth. In order to make the feature maps own higher resolution and larger receptive field at the same time, DetNet [14] designs a new backbone. RFB [15] adds dilated convolution layers on the basis of SSD [9] to effectively increase the receptive field of the feature maps.

Attention can be seen as a way of allocating limited computational resources to the most useful parts of an image. Therefore, attention can be used to improve the quality of the features of ConvNet backbones by selectively emphasizing the informative features and suppressing noises. However, as far as we know, there is only one work [13] that applies attention mechanism to ConvNet backbones in object detection.

3 Method

In this section, we present the details of the proposed network. Firstly, we describe the graph structure inference part, next elaborate the residual joint attention module.

3.1 Graph Structure Inference

Contextual information plays an important role in accurate object detection. Therefore advanced detectors not only consider object visual appearance, but also take advantage of two kinds of structured contextual information: scene-level information and object relationship information. SIN [11] is one of them which considers object detection as the problem of graph structure inference. Given an image, the objects will be treated as graph nodes while the relationships between the objects will be regarded as graph edges jointly under the supervision of the scene context formed by the whole image. More specifically, an object will receive information passed from other objects and scene which is closely related to it. By this way, the object state is finally confirmed by both its appearance features and the contextual information. For encoding different information into objects, SIN chooses Gated Recurrent Units (GRU) [22] as the tool of graph structure inference. The graph structure inference part is shown in Fig. 1. The specific operation steps are described as follows.

Initially, an image is passed through pre-trained VGG16 and the residual joint attention module. The features map named attention5_3 is extracted and then sent to the graph structure inference part. After RPN, a fixed number of RoIs (Region of Interest) are obtained. To get the descriptors about the graph nodes of 4096 dimension, operation of RoI pooling followed by an FC layer is performed on per-RoIs. The conv5_3 feature map is extracted as the scene by the same layer as the graph nodes. As for the descriptors of the graph edges of 4096 dimension, object-object relationships are modeled by both the spatial features and the visual features of the objects. Eventually, GRU whose input and initial state are respectively the 4096-dimension scene or the edge vectors and the 4096-dimension object vectors iteratively updates two steps to determine the node final state.

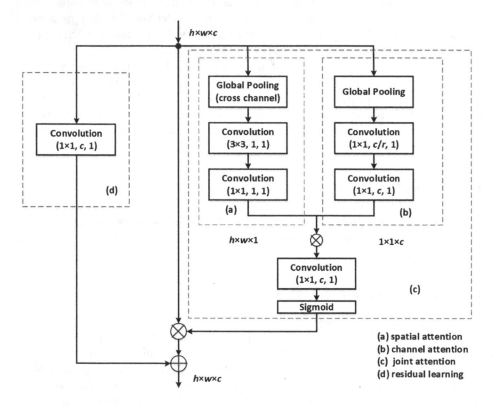

Fig. 2. The structure of residual joint attention module. The three items in the block of a convolution layer are filter shape, filter number, and stride.

3.2   Residual Joint Attention Module

It can be seen from Fig. 2 that our residual joint attention module is the union of the spatial attention, the channel attention, and the residual learning. The spatial attention aims at choosing spatially important features (not related to channels), while the channel attention is dedicated to seeking vital channels for our task. The ultimate goal of them is promoting the quality of the features of ConvNet backbones by performing feature recalibration. Intuitively, if they can be compatible and complementary to each other in functionality, the combination of the spatial attention and the channel attention should apply attention mechanism to every pixel of a specific feature map leading to better performance than any single attention. At the same time, the residual learning is proposed to keep good attributes of the original features. To the end, we package the spatial attention, the channel attention, and the residual learning into a module which can conveniently be embedded everywhere in a CNN with only a small number of additional parameters.

Mathematically, let $X \in R^{h \times w \times c}$ be the input to a residual joint attention module where h, w, c respectively denotes dimension in height, width, channel of the input feature map. Then X goes through three branches: the channel attention branch which produces a weight map $C \in R^{1 \times 1 \times c}$, the spatial attention branch which produces a weight map $S \in R^{h \times w \times 1}$, and the residual learning branch. Eventually, we select a natural way to combine two weight maps together to get the final weight map $A \in R^{h \times w \times c}$:

$$A = C \times S \tag{1}$$

Next, we describe the designs of three branches in details.

Channel Attention Branch. Our channel attention branch is derived from SENet [16] aiming at promoting the quality of features of its convolutional neural networks by effectively modeling the interactions between the channels from a specific layer of a network. There are two main steps in it. First is squeeze operation (0 parameters) which produces a channel descriptor by squeezing global spatial information named GAP. Squeeze stage will produce $Z \in R^{1 \times 1 \times c}$:

$$Z_i = \frac{1}{h \times w} \sum_{h,w} X_{hwi} \tag{2}$$

Next is excitation operation ($\frac{2c^2}{r}$ parameters) which aims to capture the interactions between the channels. We use two convolutions to generate $C \in R^{1 \times 1 \times c}$:

$$C = \text{RELU}(W_2 \times \text{RELU}(W_1 Z)) \tag{3}$$

where $W_1 \in R^{1 \times 1 \times \frac{c}{r}}$, $W_2 \in R^{1 \times 1 \times c}$ (to simplify the notation, bias terms are omitted). In our work, we keep $r = 4$ to reduce parameters.

Spatial Attention Branch. To explicitly modeling the interactions between the spatial features of a convolution layer, we imitate SENet [16] to build a spatial attention branch. First step is to eliminate channel information by a global cross-channel averaging pooling operation (0 parameters) producing $M \in R^{h \times w \times 1}$. The operation is defined as follows:

$$M_{i,j} = \frac{1}{c} \sum_c X_{ijc} \tag{4}$$

A convolution operation with the kernel size of 3×3 (9 parameters) is applied to M next. This filter purposes to model the interactions between the spatial features. Lastly, we use a harmonic convolution operation of a 1×1 filter (1 parameter) to obtain $S \in \mathbf{R}^{h \times w \times 1}$.

After getting S and C two attention maps, we adopt tensor multiplication to combine them. But the union of the spatial attention and the channel attention is not inherent. So it is necessary to further use a $1 \times 1 \times c$ convolution (c^2 parameters) to make the combination more harmonious. A sigmoid activation function maps the combination into the range between 0.5 and 1.

Residual Learning Branch. In the experiment, we notice that the dot production of the original features and the joint attention map who ranges from 0.5 to 1 will degrade the values of the original features. In fact, the values of the useless features decrease more significantly than the useful features. But to ease the situation, we apply residual learning to the joint attention mechanism. According to ResNet [3], if the attention module can be built as identical mapping, the performance should be no worse without attention. The new output is expressed as:

$$X^{'} = X + A \times X \qquad (5)$$

To harmoniously fusing the original features with the weighted features, we deploy a convolutional operation with a $1 \times 1 \times c$ filter (c^2 parameters) for the original features before fusing.

4 Experiments

In our experiments, we evaluate our model on VOC dataset [23]. At the same time, several ablation studies are conducted on our various branches to verify the effectiveness of our method. All experiments are evaluated by using VOC metric with $IOU = 0.5$.

4.1 Experimental Settings

During training and testing, the proposal number is set to 128 because too many proposals lead to out of memory when inferencing graph structure. Specifically, we follow the popular split which takes the combination of VOC2007 trainval and VOC2012 trainval as the train data, and takes VOC2007 test as the test data. The training steps are set to 130000. In the previous 80,000 iterations, we use a learning rate of 0.0005 while the learning rate is reduced by 10 times for the next 50000 iterations. We use momentum gradient descent with momentum 0.9 and batch size of 1 to train the parameters of our network.

4.2 Overall Performance

The results are shown in Table 1. To illustrate the superiority of our method, the ConvNet backbone for all methods in the Table 1 is VGG16. Comparing the results of the baseline, our mAP is higher than SIN [11], which proves that the residual joint attention module really helps our detector to achieve better

Table 1. Overall performance on VOC2007 test.

Method	Faster R-CNN [6]	ION [24]	SIN [11]	Shrivastava et al. [25]	Ours
Backbone	VGG16	VGG16	VGG16	VGG16	VGG16
mAP	73.2	75.6	76.0	76.4	76.7
aero	76.5	79.2	77.5	79.3	79.5
bike	79.0	83.1	80.1	80.5	80.4
bird	70.9	77.6	75.0	76.8	76.4
boat	65.5	65.6	67.1	72.0	68.4
bottle	52.1	54.9	62.2	58.2	63.4
bus	83.1	85.4	83.2	85.1	86.0
car	84.7	85.1	86.9	86.5	86.9
cat	86.4	87.0	88.6	89.3	88.3
chair	52.0	54.4	57.7	60.6	59.8
cow	81.9	80.6	84.5	82.2	85.5
table	65.7	73.8	70.5	69.2	71.4
dog	84.8	85.3	86.6	87.0	86.1
horse	84.6	82.2	85.6	87.2	86.5
mbike	77.5	82.2	77.7	81.6	77.1
perpon	76.7	74.4	78.3	78.2	78.6
plant	38.3	47.1	46.6	44.6	50.2
sheep	73.6	75.8	77.6	77.9	77.3
sofa	73.9	72.7	74.7	76.7	74.1
train	83.0	84.2	82.3	82.4	82.8
tv	72.6	80.4	77.1	71.9	75.1

detection accuracy. Interestingly, on some specific classes, it is found that our model performs very well including *aero, bird, bus, chair, plant* and so on. Our method is also better than ION [24] which is a network with explicitly modeling of contextual information using RNN, and Shrivastava et al. [25] which exploits segmentation information in the framework of Faster R-CNN. We show some detection examples in Fig. 3. The top column is the results of the original SIN, and the bottom column is the results of our network. From these examples, it can see that our method is good at detecting objects in complex situations like a dog only with a head, a blurry ship, obscured cows and obscured sheep. These also directly indicate that our residual joint attention module makes the original inapparent features more powerful and differentiated through feature recalibration.

Table 2. Ablation studies on VOC2007 test.

Method	Baseline	Baseline+SA	Baseline+CA	Baseline+SA+CA	Ours
mAP	76.0	76.2	76.3	76.5	76.7
aero	77.5	78.8	78.0	78.7	79.5
bike	80.1	79.9	80.2	79.5	80.4
bird	75.0	76.3	76.2	75.6	76.4
boat	67.1	67.5	65.7	69.5	68.4
bottle	62.2	61.9	61.2	61.5	63.4
bus	83.2	85.9	86.4	85.6	86.0
car	86.9	87.0	86.9	86.7	86.9
cat	88.6	89.3	87.8	89.2	88.3
chair	57.7	60.2	61.1	59.7	59.8
cow	84.5	83.1	84.2	84.5	85.5
table	70.5	70.9	71.0	70.0	71.4
dog	86.6	84.2	86.3	86.8	86.1
horse	85.6	87.6	87.1	86.0	86.5
mbike	77.7	77.4	77.6	78.8	77.1
perpon	78.3	78.1	78.3	78.4	78.6
plant	46.6	51.0	47.9	47.6	50.2
sheep	77.6	78.6	77.7	76.3	77.3
sofa	74.7	72.2	73.4	74.2	74.1
train	82.3	78.8	83.1	83.4	82.8
tv	77.1	75.8	76.4	77.0	75.1

4.3   Ablation Studies

In order to verify the effectiveness of each branch in our proposed method, we
conduct several ablation studies which still use the same dataset settings as
above. Table 2 shows the results of different branches, where Baseline stands for
SIN, SA stands for the spatial attention branch, CA stands for the channel attention branch. Comparing with the baseline, there is a slight increase by adding
any kind of attention mechanism to the baseline. This shows that the feature
recalibration through attention mechanisms is effective. What's more, The combination of the spatial attention and the channel attention improves more than
any single attention which proves our preliminary conjecture that the spatial
attention and the channel attention are complementary and compatible. Similarly, the residual learning continues to optimize our model that demonstrates
the validity of the residual learning.

Fig. 3. Examples of detection results. Top: SIN. Bottom: ours.

5   Conclusion

In this paper, we proposed a residual joint attention module embedded in an advanced network with graph structure inference. The graph structure inference part is used for the detection subnetwork of the detector and the residual joint attention module composed of the spatial attention, the channel attention and the residual learning follows VGG16. Due to the complementarity and compatibility of the spatial attention and the channel attention, the joint attention mechanism more significantly improves the representational power of a network by performing feature recalibration than any single attention. Moreover, the residual learning keeps good attributes of the original features. Quantitative evaluations show that our residual joint attention module boosts model's discriminative power. We hope that this paper can provide reference for researchers to use attention.

Acknowledgment. This work was supported partly by the National Natural Science Foundation of China (61772067, 61472030, 61471032).

References

1. Krizhevsky, A., Sutskever, I., Hinton, G.E.: ImageNet classification with deep convolutional neural networks. In: International Conference on Neural Information Processing Systems, NIPS 2012, pp. 1097–1105. Curran Associates Inc. (2012)
2. Simonyan, K., Zisserman, A.: Very deep convolutional networks for large-scale image recognition. In: ICLR (2015)
3. He, K., Zhang, X., Ren, S., Sun, J.: Deep residual learning for image recognition. In: Proceedings of the IEEE Conference on Computer Vision and Pattern Recognition, CVPR 2016 (2016)

4. Girshick, R., Donahue, J., Darrell, T., Malik, J.: Region-based convolutional networks for accurate object detection and segmentation. IEEE Trans. Pattern Anal. Mach. Intell. **38**(1), 142–158 (2016)
5. Girshick, R.: Fast R-CNN. In: Proceedings of the IEEE International Conference On Computer Vision, ICCV 2015, pp. 1440–1448 (2015)
6. Ren, S., He, K., Girshick, R., Sun, J.: Faster R-CNN: towards real-time object detection with region proposal networks. In: Advances in Neural Information Processing Systems, NIPS 2015, pp. 91–99 (2015)
7. He, K., Gkioxari, G., Dollár, P., Girshick, R.: Mask R-CNN. In: Proceedings of the IEEE International Conference on Computer Vision, ICCV 2017 (2017)
8. Long, J., Shelhamer, E., Darrell, T.: Fully convolutional networks for semantic segmentation. In: Proceedings of the IEEE Conference on Computer Vision and Pattern Recognition, CVPR 2015, pp. 3431–3440 (2015)
9. Liu, W., et al.: SSD: single shot multibox detector. In: Leibe, B., Matas, J., Sebe, N., Welling, M. (eds.) ECCV 2016. LNCS, vol. 9905, pp. 21–37. Springer, Cham (2016). https://doi.org/10.1007/978-3-319-46448-0_2
10. Redmon, J., Divvala, S., Girshick, R.: You only look once: unified, real-time object detection. In: Proceedings of the IEEE Conference on Computer Vision and Pattern Recognition, CVPR 2015, pp. 779–788 (2015)
11. Liu, Y., Wang, R., Shan, S., Chen, X.: Structure inference net: object detection using scene-level context and instance-level relationships. In: IEEE Conference on Computer Vision and Pattern Recognition, CVPR 2018 (2018)
12. Lin, T.-Y., Dollár, P., Girshick, R., He, K., Hariharan, B., Belongie, S.: Feature pyramid networks for object detection. In: IEEE Conference on Computer Vision and Pattern Recognition, CVPR 2017 (2017)
13. Zhang, Z., Qiao, S., Xie, C., Wei, S.: Single-shot object detection with enriched semantics. In: IEEE Conference on Computer Vision and Pattern Recognition, CVPR 2018 (2018)
14. Li, Z., Chao, P., Gang, Y., Zhang, X., Jian, S.: DetNet: a backbone network for object detection. In: IEEE Conference on Computer Vision and Pattern Recognition, CVPR 2018 (2018)
15. Liu, S., Huang, D., Wang, Y.: Receptive field block net for accurate and fast object detection. In: IEEE Conference on Computer Vision and Pattern Recognition, CVPR 2018 (2018)
16. Hu, J., Shen, L., Sun, G.: Squeeze-and-excitation networks. arXiv preprint arxiv:1709.01507 (2017)
17. Divvala, S.K., Hoiem, D., Hays, J.H., Efros, A.A., Hebert, M.: An empirical study of context in object detection. In: IEEE Conference on Computer Vision and Pattern Recognition, CVPR 2009 (2009)
18. Galleguillos, C., Rabinovich, A., Belongie, S.: Object categorization using co-occurrence, location and appearance. In: IEEE Conference on Computer Vision and Pattern Recognition, CVPR 2008 (2008)
19. Torralba, A., Murphy, K.P., Freeman, W.T., Rubin, M.A.: Context-based vision system for place and object recognition. In: IEEE Conference on Computer Vision and Pattern Recognition, CVPR 2003 (2003)
20. Hu, H., Gu, J., Zhang, Z., Dai, J., Wei, Y.: Relation networks for object detection. In: IEEE Conference on Computer Vision and Pattern Recognition, CVPR 2018 (2018)
21. Uijlings, J.R., Van, K.E., Gevers, T., Smeulders, A.W.: Selective search for object recognition. Int. J. Comput. Vis. **104**(2), 154–171 (2013)

22. Cho, K., Merrienboer, B.V., Bahdanau, D., Bengio, Y.: On the properties of neural machine translation: encoder-decoder approaches. In: SSST-8 (2014)
23. Everingham, M., Gool, L.V., Williams, C.K., Winn, J., Zisserman, A.: The pascal visual object classes (VOC) challenge. Int. J. Comput. Vis. **88**(2), 303–338 (2010)
24. Bell, S., Zitnick, C.L., Bala, K., Girshick, R.: Inside-outside net: detecting objects in context with skip pooling and recurrent neural networks. In: IEEE Conference on Computer Vision and Pattern Recognition, CVPR 2016 (2016)
25. Shrivastava, A., Gupta, A.: Contextual priming and feedback for faster R-CNN. In: Leibe, B., Matas, J., Sebe, N., Welling, M. (eds.) ECCV 2016. LNCS, vol. 9905, pp. 330–348. Springer, Cham (2016). https://doi.org/10.1007/978-3-319-46448-0_20

Saliency Detection Based on Foreground and Background Propagation

Qing Xing[✉], Suoping Zhang, Mingbing Li, Chaoqun Dang, and Zhanhui Qi

National Ocean Technology Center, Tianjin 300112, China
1636858663@qq.com

Abstract. In recent years, image saliency detection has become a research hotspot in the field of computer vision. Although significant progress has been witnessed in visual saliency detection, several existing saliency detection methods still cannot highlight the complete salient object when under complex background. For the purpose of improving the robustness of saliency detection, we propose a novel salient detection method via foreground and background propagation. In order to take both foreground and background information into consideration, we obtain a background-prior map by computing the dissimilarity between superpixels and background labels. A foreground-prior map is obtained by calculating the difference of superpixels between the inner and outer of a convex hull. Then we use label propagation algorithm to propagate saliency information based on foreground and background prior maps. Finally, the two saliency maps are integrated to generate an accurate saliency map. The experimental results on two public available data sets MSRA and ECSSD demonstrate that the proposed method performs well against the state-of-the-art methods.

Keywords: Saliency detection · Foreground prior · Background prior · Propagation

1 Introduction

Image saliency detection has become a research hotspot in the field of computer vision. It filters out redundant visual information of an image by imitating the human visual attention mechanism, and selects objects of interest to human eyes, which are called salient objects. Efficient saliency detection models have been applied to numerous computer vision scenarios, such as image classification [1], object detection [2], image retrieval [3], and so forth.

Corresponding to the visual attention mechanism, existing saliency detection models can be broadly categorized as bottom-up [4–6], or top-down approaches [7, 8]. Since saliency detection based on bottom-up models which mainly focus on color, intensity, contrast and other low-level visual features of the image, the detection speed is fast. The top-down approaches add a higher level of prior knowledge, which is more complex. In this paper, our work focus on bottom-up salient detection models. Many salient detection models use low-level features such as color and contrast [9, 10] of the image. They perform well in many situations, but still struggle in complex images.

© Springer Nature Switzerland AG 2019
Y. Zhao et al. (Eds.): ICIG 2019, LNCS 11901, pp. 222–231, 2019.
https://doi.org/10.1007/978-3-030-34120-6_18

Based on the assumption that saliency objects are mostly located in the center of the image, some researchers regard the image boundary region as the background and propose background-prior saliency detection models [11, 12]. They perform well in many cases, except for salient objects that appearing at the edge of the image. Since the background template of the image contains the foreground noise, the detection result has a poor performance.

In this paper, we propose a novel saliency detection method based on foreground and background propagation. First, we obtain a background template by selecting the image border superpixels, which have been proved to be good indicators for background-prior in saliency detection [13, 14]. Second, we calculate the color and spatial distances between each superpixel and the background labels to obtain the background-prior map. Third, we calculate the convex hull of the image, and obtain the foreground-prior map by computing the difference of superpixels between the inner and outer of the convex hulls. Finally, we use label propagation algorithm to propagate saliency information based on foreground-prior and background-prior maps. The saliency result is obtained by integrating two propagated maps.

2 Related Work

Significant progress has been made in visual saliency detection in recent years. A quantitative analysis of different saliency methods can be found in [15]. Most bottom-up saliency detection methods based on contrast and background. It is has been verified that contrast is an effective cue for satisfying results. Furthermore, contrast-based methods can be roughly divided into local and global contrast-based techniques. The model of Itti et al. [4] is one of the typical local contrast-based methods. They define saliency by calculating center-surround differences in color, density, and orientation from images. Harel et al. [16] improved the Itti's method and added Markov chains to the calculation of saliency maps. Achanta et al. [17] proposed a multi-scale contrast saliency detection method by calculating contrasts of feature vectors between the inner and outer regions of a sliding window. Hou et al. [18] proposed a frequency-based method, which uses the spectral residual in the frequency domain to extract saliency regions of the image. Global contrast-based approaches use contrast relationships of the whole image to calculate the saliency of single pixels or image regions. Goferman et al. [6] proposed a context-aware saliency detection method that computes the average of salient values at multiple scales to obtain saliency regions. Achanta et al. [19] propose a frequency-tuned method that directly defines saliency as the differences of image color distances. Cheng et al. [5] proposed a saliency detection method based on color contrast and spatial position features, using color contrast weighted by spatial distance to define the salient regions.

Background-based methods typically regard image border pixels as background labels, which calculate saliency by background labels query or propagation mechanism. Yang et al. [20] proposed a saliency detection algorithm based on manifold ranking, which ranks the similarity of the image's superpixels with the background labels on a graph to obtain saliency maps. Li et al. [21] regard boundary superpixels as background templates, and obtain salient results by constructing sparse and dense errors. Wang et al.

[22] proposed a method based on foreground and background seed selection, which uses image border superpixels to generate foreground and background saliency maps. Zhu et al. [23] proposed a saliency detection method based on boundary connectivity, which characterizes the spatial layout of image regions. Zhang et al. [24] proposed an approach based on local structure propagation, which updates saliency values under the guidance of local propagation.

3   Proposed Method

In this section, we present an efficient and effective saliency detection method that integrates foreground and background information of the image, as shown in Fig. 1. We first abstract the image into superpixels using simple linear iterative clustering (SLIC for short). Then, we compute background-prior map by selecting boundary labels, and use convex hulls to generate foreground-prior map. Finally, we use label propagation saliency (LPS for short) respectively on foreground-prior and background-prior map, and integrate the two propagated saliency maps to generate a pixel-wise saliency map.

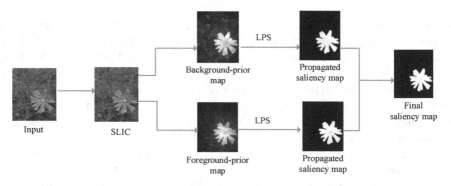

Fig. 1. Flowchart of the proposed method. It contains three stages: abstracting the image; computing background-prior map and foreground-prior map; generating a pixel-wise saliency map that integrates two propagated saliency maps.

3.1   Background Labels Selection

The strategy of selecting background labels affects the precision of the salient map. Based on the observation of saliency models that the object is likely to appear at the center of an image, the border near the image center is selected as the background labels. However, this method may contain foreground noise, which affects the results of saliency detection. Therefore, we propose a mechanism based on image boundary information to remove the foreground noises and select background labels from the border superpixels.

We first use SLIC algorithm to abstract the image into N uniform and compact regions. Then we select the superpixels whose centroids locate within a certain number of pixels as image background template. Since the most obvious boundary of the image is likely to be the contour between the object and background, we can roughly remove the image superpixels with strong boundaries, which are regarded as the foreground noises.

We adopt the probability of boundary (PB for short) [25] to detect image boundary. The boundary feature of the i-th superpixel is calculated from the average PB value of the pixel along the edge contour of the superpixel i, as follows:

$$PB_i = \frac{1}{|B_i|} \sum_{I \in B_i} I^{pb} \tag{1}$$

Where B_i is the edge pixel set of superpixel i and $|B_i|$ is the number of the template. I^{pb} is the PB value of pixel I. The larger the PB value, the more obvious the boundary feature of the superpixel. Since superpixels with obvious boundary feature is more likely to be the object, we remove the superpixels whose boundary features are larger than the adaptive gray threshold derived by Otsu [26]. Then the remaining superpixels in the background template are selected as background labels. As shown in Fig. 2, the selected background labels have less foreground noise than the background template.

Fig. 2. Illustration of the main phases of our method. (a) Input image. (b) Superpixels generated by SLIC. (c) The convex hull. (d) The prior map based on background template. (e) The prior map based on selected background labels. (f) The foreground-prior map.

3.2 Background-Prior Saliency

If a superpixel has larger color differences to the background labels, it is more likely to be a salient superpixel. In addition, the background labels contribute more for the salient value of the closer superpixels while less to a farther one. Therefore, we use spatial weight based color contrast to define a salient superpixel p_i, as follow:

$$S_i = \sum_{i=1, i \neq j}^{N} d_c(p_i, p_j) \times \exp(1 - d_s(p_i, p_j)) \tag{2}$$

Where $d_c(p_i, p_j)$ and $d_s(p_i, p_j)$ are respectively the Euclidean color and spatial distances between the i-th superpixel and the j-the superpixel which belongs to the background template. Both distances are normalized to [0, 1].

3.3   Foreground-Prior Saliency

We use image corner points to select foreground seeds. Traditional Harris point detection algorithm only considers the gray information of the image, which leads to more invalid corner or contour points being detected in complex cases. Since the color boosted Harris detection algorithm combines brightness information and color information to make most points locate around salient objects, the detection result is more stable. Therefore, in this paper we use Harris point detection algorithm [27] to detect corners or contour points of salient objects in the image.

Since salient points are usually not located in image boundary, we eliminate those near the image boundary, and enclose all the remaining salient points to compute a convex hull. Then we select the foreground seeds based on the difference of the superpixels between the inner and outer of the convex hull. We define the superpixels set in the convex hull as I, and the superpixels set outside the convex hull as O. The difference of superpixels between the inner and outer of the convex hull is defined as follow:

$$w(p_i, p_j) = \exp\left(-\frac{1}{2\sigma^2}||p_i - p_j||^2\right), i \in I \tag{3}$$

$$d_i = \sum_{i=1} w(p_i, p_j) \cdot ||c_i - c_j||, i \in I, j \in O \tag{4}$$

Where $||c_i - c_j||$ and $w(p_i, p_j)$ are respectively the Euclidean color and spatial distances between the j-th superpixel and the i-th superpixel which belongs to the convex hull. Both distances are normalized to [0, 1]. We remove the superpixels in the convex hull that are less than the average difference and regard the remaining superpixels as foreground seeds. The average difference is defined as $\alpha = \frac{1}{I}\sum_{j=1} d_j, j \in I$, where $|I|$ is the number of superpixels in a convex hull. Figure 2 shows the foreground-prior map of our method.

3.4   Graph Construction

We create a graph $G = (V, E)$ with N nodes $\{n_1, n_2, \cdots, n_N\}$ and edges E. Node n_i corresponds to the i-th image superpixel or patch and edge e_{ij} link nodes n_i and n_j to each other. The similarity of two nodes is measured by a defined distance of the mean color features in each region. We define w_{ij} as the weight of the edge between node n_i and node n_j. The affinity matrix $W = [w_{ij}]_{N \times N}$ indicate the similarity between superpixels:

$$w_{ij} = \begin{cases} \exp\left(-\frac{||c_i - c_j||_2}{\sigma^2}\right) & , j \in N(i) \\ 0, \; otherwise \end{cases} \tag{5}$$

Where $N(i)$ indicates the set of the neighboring nodes of superpixel i, $||c_i - c_j||_2$ represents the average color distance of two superpixels on the CIE LAB color space. σ^2 is a tuning parameter, which controls strength of the similarity. The degree matrix of graph G. $D = \text{diag}\{d_1, d_2, \cdots, d_N\}$, where $d_i = \sum_j w_{ij}$ is the degree of node i, and a row-normalized affinity matrix:

$$A = D^{-1} \times W \qquad (6)$$

3.5 Label Propagation Saliency

In [28], a label propagation that get information about unlabeled nodes based on the provided labelled nodes is proposed. Given a data set $X = \{x_1, \cdots, x_i, x_{i+1}, \cdots, x_n\} \in R^{m \times n}$, the former i. ta points are labelled and the rest need to be propagated according to their relevance to the labelled points. We seek out a function $V = [V_{m_1}, V_{m_2}, \cdots, V_{m_N}]^T$ such that $V : R \to [0, 1] \in R^{N \times 1}$ indicates the possibility of how similar each data point is to the labels. The similarity measure $V(r_i)$ satisfies:

$$V_{t+1}(r_i) = \sum_{j=1}^{N} w_{ij} V_t(r_i) \qquad (7)$$

Where, w_{ij} is the affinity entry defined in Eq. 5 and t is the recursion step. The similarity measure of the labeled tag in the recursive process is fixed to 1, and the initial measure of the unlabeled objects is set to 0. The final similarity of the region to the label is influenced by the features of the surroundings. In other words, the similarity $V(r_i)$ is iteratively learned by the propagation of the similarity measure of its neighbor $V(r_j)$.

Then, we integrate the two saliency maps: the foreground-propagated based one which highlight the whole object and the one based on background propagation which reduce the background noses, as follow:

$$S = S_f + S_b \qquad (8)$$

Where S_f is a foreground-propagated saliency map, S_b is a background-propagated saliency map.

4 Experimental Results

In this section, we evaluate our method on two public available data sets. One is MSRA [19] data set that contains 1000 images, which equipped with pixel-wise ground truth. The other is ECSSD [29] data set which contains 1000 natural images under complex background.

We compare the performance of the proposed method with other 7 state-of-the-art methods: IT [4], FT [19], BSCA [12], GBMR [20], BFS [22], DSR [21], and LPS [28]. The evaluation standards are precision and recall (PR) curve and F-measure. We segment the saliency map using a threshold between 0 and 255 for each method, then compare the binary map obtained by different thresholds with ground truth to obtain the precision and recall curve. F-measure is used to comprehensively evaluate the performance of recall and precision, we set $\beta^2 = 0.3$ [5] in order to emphasize precision:

$$F - Measure = \frac{\left(1 + \beta^2\right) \cdot Precision \cdot Recall}{\beta^2 \cdot Precision + Recall} \tag{9}$$

Figure 3 shows the precision and recall curve (PR curve for short) and F-measure comparison results of our method with other methods on MSRA-1000 data set and ECSSD data set. For the MSRA-1000 data set, the background of images are relatively simple. The precision and recall curve obtained by our method is obviously superior to other saliency detection algorithms. Although LPS and DSR methods also achieved better results in precision and recall curve and F-measure, the value of F-measure in our proposed method are 1.7% and 2.1% higher than LPS and DSR methods respectively. For the ECSSD data set, the background of images are more complicated. Although our method and LPS have achieved good results, the precision and recall of our method are 77% and 68% respectively, higher than the LPS algorithm with 71% precision and 64% recall.

Figure 4 shows several images for visual comparison to previously published methods. From these samples, we can see that our method achieves the best performance on these images. Most saliency methods can detect the complete saliency objects when the background is relatively simple. When the background of the image is

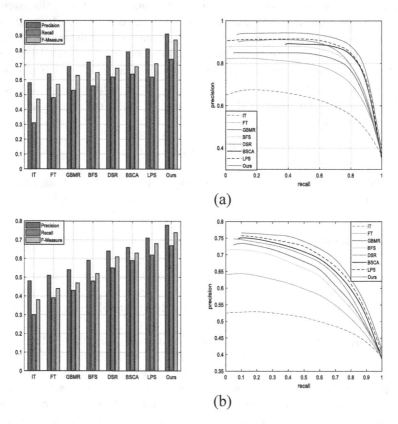

Fig. 3. Quantitative PR-curve and F-measure evaluation of different methods on two data sets: (a) MSRA, (b) ECSSD

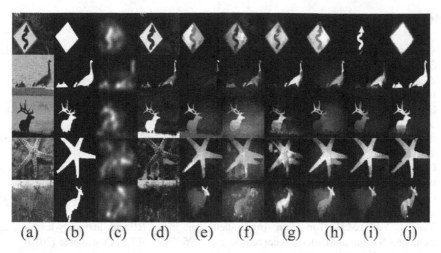

Fig. 4. Visual comparison of our method with seven state-of-the-art methods. From left to right: (a) Input image, (b) Ground Truth, (c) IT [4], (d) FT [19], (e) GBMR [20], (f) BFS [22], (g) DSR [21], (h) BSCA [12], (i) LPS [28], (j) Ours.

more complicated, the results of other saliency methods either contain background noise or are incomplete. Since we consider both foreground and background information, our method can effectively suppress background noises. With the help of the label propagation, our method can assign high salient values to candidate objects based on the differences between the labels. Furthermore, our experimental results are closer to Ground Truth, and conserve a more complete boundary of the salient objects.

5   Conclusion

In this paper, we propose a novel saliency detection algorithm based on foreground and background propagation. First, we select the image border superpixels to obtain background information and calculate a background-prior map. Second, we use salient points detected by the color boosted Harris algorithm to obtain a convex hull and compute a foreground-prior map. Third, we use label propagation algorithm to propagate saliency on the two prior saliency maps respectively. The final saliency map is obtained by integrating the foreground and background propagated saliency map. Results on two benchmark data sets show that our methods achieve superior performance compared with the state-of-the-art methods.

Acknowledgement. Thanks for the support of National Key R&D Program Key Projects NQI (No.2017YFF0206400), the Natural Science Foundation of Tianjin (No.17JCYBJC16300), and Innovation Fund of National Ocean Technology Center (No.K51700404).

References

1. Schmid, C., Jurie, F., Sharma, G.: Discriminative spatial saliency for image classification. In: Computer Vision and Pattern Recognition, pp. 3506–3513. IEEE, Providence (2012)
2. Zhao, R., Ouyang, W., Wang, X.: Unsupervised salience learning for person re-identification. In: Computer Vision and Pattern Recognition, pp. 3586–3593. IEEE, Portland (2013)
3. Cheng, M.M., Mitra, N.J., Huang, X., et al.: Salient shape: group saliency in image collections. Vis. Comput. **30**(4), 443–453 (2014)
4. Itti, L., Koch, C., Niebur, E.: A model of saliency-based visual attention for rapid scene analysis. IEEE Trans. Pattern Anal. Mach. Intell. **20**(11), 1254–1259 (1998)
5. Cheng, M.M., Zhang, G.X., Niloy, J., et al.: Global contrast based salient region detection. In: Computer Vision and Pattern Recognition, pp. 409–416. IEEE, Colorado Springs (2011)
6. Goferman, S., Zelnik-Manor, L., Tal, A.: Context-aware saliency detection. Pattern Anal. Mach. Intell **34**(10), 1915–1926 (2012)
7. Yang, J., Yang, M.H.: Top-down visual saliency via joint CRF and dictionary learning. In: Computer Vision and Pattern Recognition, pp. 2296–2303. IEEE, Providence (2012)
8. Kanan, C., Tong, M.H., Zhang, L., et al.: SUN: top-down saliency using natural statistics. Vis. Cognit. **17**(6–7), 979–1003 (2009)
9. Huo, L., Jiao, L., Wang, S., Yang, S.: Object-level saliency detection with color attributes. Pattern Recogn. **49**, 162–173 (2016)
10. Zhou, L., Yang, Z., Yuan, Q., et al.: Salient region detection via integrating diffusion-based compactness and local contrast. IEEE Trans. Image Process. **24**(11), 3308–3320 (2015)
11. Wei, Y., Wen, F., Zhu, W., Sun, J.: Geodesic saliency using background priors. In: Fitzgibbon, A., Lazebnik, S., Perona, P., Sato, Y., Schmid, C. (eds.) ECCV 2012. LNCS, vol. 7574, pp. 29–42. Springer, Heidelberg (2012). https://doi.org/10.1007/978-3-642-33712-3_3
12. Qin, Y., Lu, H., Xu, Y., Wang, H.: Saliency detection via cellular automata. In: Computer Vision and Pattern Recognition, pp. 110–119. IEEE, Boston (2015)
13. Li, X., Lu, H., Zhang, L., et al.: Saliency detection via dense and sparse reconstruction. In: IEEE International Conference on Computer Vision, pp. 2976–2883. IEEE, Sydney (2013)
14. Wang, Q., Zheng, W., Piramuthu, R.: GraB: visual saliency via novel graph model and background priors. In: IEEE Conference on Computer Vision and Pattern Recognition. IEEE, Nevada (2016)
15. Borji, A., Cheng, M.M., Jiang, H., et al.: Salient object detection: a benchmark. IEEE Trans. Image Process. **24**(12), 5706–5722 (2015)
16. Harel, J., Koch, C., Perona, P.: Graph-based visual saliency. Adv. Neural Inf. Process. Syst. **19**, 545–552 (2007)
17. Achanta, R., Estrada, F., Wils, P., Süsstrunk, S.: Salient region detection and segmentation. In: Gasteratos, A., Vincze, M., Tsotsos, John K. (eds.) ICVS 2008. LNCS, vol. 5008, pp. 66–75. Springer, Heidelberg (2008). https://doi.org/10.1007/978-3-540-79547-6_7
18. Hou, X., Zhang, L.: Saliency detection: a spectral residual approach. In: IEEE Conference on Computer Vision and Pattern Recognition, pp. 1–8. IEEE, Hawaii (2007)
19. Achanta, R., Hemami, S., Estrada, F., et al.: Frequency-tuned salient region detection. In: IEEE International Conference on Computer Vision and Pattern Recognition, pp. 1597–1604. IEEE, Florida (2009)
20. Yang, C., Zhang, L., Lu, H., et al.: Saliency detection via graph-based manifold ranking. In: IEEE International Conference on Computer Vision and Pattern Recognition, pp. 1665–1672. IEEE, Portland (2013)

21. Li, X., Lu, H., Zhang, L., et al.: Saliency detection via dense and sparse reconstruction. In: IEEE International Conference on Computer Vision, pp. 2976–2883. IEEE, Portland (2013)
22. Wang, J., Lu, H., Li, X., et al.: Saliency detection via background and foreground seed selection. Neurocomputing **152**(25), 359–368 (2015)
23. Zhu, W.J., Liang, S., Wei, Y.C., Sun, J.: Saliency optimization from robust background detection. In: Proceedings of the 2014 IEEE Conference on Computer Vision and Pattern Recognition, pp. 2814–2821. IEEE, Columbus (2014)
24. Zhang, M., Pang, Y., Wu, Y., et al.: Saliency detection via local structure propagation. J. Vis. Commun. Image Represent. **52**, 131–142 (2018). S1047320318300129
25. Martin, D.R., Fowlkes, C.C., Malik, J.: Learning to detect natural image boundaries using local brightness, color, and texture cues. Pattern Anal. Mach. Intell. **26**(5), 530–549 (2004)
26. Otsu, N.: A threshold selection method from gray-level histograms. IEEE Trans. Syst. Man. Cybern. **9**(1), 62–66 (1979)
27. Van de, W.J., Gevers, T., Bagdanov, A.D.: Boosting color saliency in image feature detection. IEEE Trans. Pattern Anal. Mach. Intell. **28**(1), 150–156 (2006)
28. Li, H., Lu, H., Lin, Z., et al.: Inner and inter label propagation: salient object detection in the wild. IEEE Trans. Image Process. **24**(10), 3176–3186 (2015)
29. Yan, Q., Xu, L., Shi, J, Jia, J.: Hierarchical saliency detection. In: Proceedings of the 2013 IEEE Conference on Computer Vision and Pattern Recognition, pp. 1155–1162. IEEE, Portland (2013)

Online Handwritten Diagram Recognition with Graph Attention Networks

Xiao-Long Yun[1,2], Yan-Ming Zhang[1], Jun-Yu Ye[1,2], and Cheng-Lin Liu[1,2(✉)]

[1] National Laboratory of Pattern Recognition (NLPR),
Institute of Automation, Chinese Academy of Sciences,
95 Zhongguancun East Road, Beijing 100190, People's Republic of China
{xiaolong.yun,ymzhang,junyu.ye,liucl}@nlpr.ia.ac.cn
[2] University of Chinese Academy of Sciences, Beijing, People's Republic of China

Abstract. Handwritten text recognition has been extensively researched over decades and achieved extraordinary success in recent years. However, handwritten diagram recognition is still a challenging task because of the complex 2D structure and writing style variation. This paper presents a general framework for online handwritten diagram recognition based on graph attention networks (GAT). We model each diagram as a graph in which nodes represent strokes and edges represent the relationships between strokes. Then, we learn GAT models to classify graph nodes taking both stroke features and the relationships between strokes into consideration. To better exploit the spatial and temporal relationships, we enhance the original GAT model with a novel attention mechanism. Experiments on two online handwritten flowchart datasets and a finite automata dataset show that our method consistently outperforms previous methods and achieves the state-of-the-art performance.

Keywords: Flowchart recognition · Diagram recognition · Stroke classification · Graph attention networks

1 Introduction

Handwriting is one of the most natural and efficient ways for human to record information. As the widespread usage of smartphone, tablet computer and electrical whiteboard, recording information in intelligence devices has become a major choice for its convenience. As a result, handwritten text recognition has been intensively studied over the last decades and widely applied in many fields. However, the recognition and analysis of 2D diagrams, such as flowchart, circuit and music score, are still challenging because of the complex 2D structures and great writing style variation.

Existing methods for online handwritten diagram recognition and interpretation can be roughly divided into two categories: bottom-up [4–6,15,27] and top-down ones [1,11,14,21,22]. Bottom-up approaches sequentially perform a symbol segmentation step and a recognition step. However, due to the error accumulation, these methods often lead to low recognition accuracy. On the other

© Springer Nature Switzerland AG 2019
Y. Zhao et al. (Eds.): ICIG 2019, LNCS 11901, pp. 232–244, 2019.
https://doi.org/10.1007/978-3-030-34120-6_19

hand, top-down approaches integrate the two steps in one framework, such as probabilistic graphical models (PGM), and perform segmentation and recognition simultaneously. Typically, top-down methods can achieve higher accuracy results but suffer from high computational cost because of the complicated learning and inference algorithms. We review these methods in more details in the next section.

In this work, we propose an efficient and high-accuracy method for online handwritten diagram recognition. In particular, we treat diagram stroke classification as a graph node classification problem and solve it with attention-based graph neural networks (GNN) [13,20]. Compared with PGM, such as conditional random fields (CRF) and Markov random fields (MRF), GNN is more powerful and flexible in learning the stroke representation and exploiting the contextual information. Unlike PGM, the learning and inference algorithms of GNN are very simple and efficient, which makes it very suitable for large-scale applications.

We highlight the main contributions of this work as follows. First, we propose a general online handwritten diagram recognition method based on GNN. Second, to better exploit the relationships between strokes, we enhance the original GAT [20] by introducing a novel attention mechanism. Third, on three popular benchmark datasets, our method consistently outperforms the existing methods and achieves the state-of-the-art results.

In the rest of this paper, we first provide a general review of existing online handwritten diagram recognition works and a brief review of GNN in Sect. 2. In Sect. 3, we give a detailed introduction to the proposed method. The experimental setting and comparison results are described in Sect. 4. Finally, Sect. 5 draws our concluding remarks.

2 Related Works

2.1 Diagram Recognition

Since it is a difficult task to classify text and non-text strokes or segment symbols precisely in early stage for diagram recognition, some works only considered graphic symbols, and others imposed some constraints on users. Qi et al. [16] presented a recognition system for flowchart recognition using Bayesian conditional random fields, but their dataset only included very simple graphic symbols rather than texts. Yuan et al. [27] proposed a hybrid model combing support vector machine (SVM) and hidden Markov models (HMM) for programming teaching. Miyao et al. [15] presented a flowchart recognition and editing system that segmented the symbols based on loop structure and recognized them using SVMs. Although [15,27] allowed the flowcharts to contain both symbols and texts, there were many constraints on users. [27] required users to draw each symbol with only one stroke, and [15] required users to differentiate texts from graphic symbols.

Awal et al. [1] proposed two methods—bottom-up and top-down approaches for flowchart recognition from different viewpoints. For the former, texts and graphic symbols were classified based on the entropy of strokes, then time delayed neural network (TDNN) or SVM was applied for graphical symbols recognition. Moreover, they introduced a global recognition architecture based on the TDNN and dynamic programming (DP) algorithm.

Flowchart diagrams are document with complex 2D structures, thus previous statistical approaches [1,15,27] have reported limited performance because of the ignorance of structure information. Lemaitre et al. [14] proposed a method that tried to handle the segmentation and recognition simultaneously. Their model integrated structural and syntactic prior of flowchart with Enhanced Position Formalism (EPF) language [8], then they used Description and MOdification of the Segmentation (DMOS) method [8] to segment and recognize the flowchart in one step. Their method achieved great progress in stroke labeling and symbol recognition compared with [1], but it is too restricted and rigid to adapt to other domains and it is impossible to describe every symbol with variation.

For exploring structure information, Carton et al. [7] presented a human-like perceptive mechanism approach that incorporated both structural and statistical information of a flowchart. Same as [14], the work made use of DMOS to express circular symbols and quadrilateral symbols, then proposed a deformation measure to quantify what was a good quadrilateral.

In handwritten diagrams, arrows are variable in appearance and are difficult to recognize compared to other symbols using identical classifier. Bresler et al. [4–6] proposed a new framework that strokes were firstly classified as text or non-text, then non-text strokes were clustered and uniform symbols were classified with SVM, lastly the arrows were detected. For structure analysis, they modeled whole flowchart excluding texts as a max-sum problem and applied integer linear programming to solve it [3]. This approach achieved the state-of-the-art results in three handwritten datasets. However, the recognition system has some severe flaws, such as each arrow must consist of a shaft and a head, which may lead to recognition failures if one of them is absent.

Wang et al. [21,22] proposed a general model, max-margin MRF, which combines MRF and structural SVM to perform stroke segmentation and recognition simultaneously. By exploiting temporal and spatial relationship between strokes, their model greatly improved the stroke labeling accuracy. To lower the complexity in evaluating stroke grouping candidates in a diagram, Julca-Aguilar et al. [11] applied the Faster R-CNN model [17] to the detection of online handwritten graphics through converting the original online data to offline images. Despite the overall high performance of flowchart symbol detection, the arrow detection accuracy is not satisfactory, and the conversion into images causes loss of temporal information of strokes.

2.2 Graph Neural Networks

In recently years, graph neural networks (GNN) have received extensive attention and become one of the most popular research highlights in deep learning field. With its capability of capturing the dependency between objects and operating on non-Euclidean domain [28], GNN have obtained great success in many tasks, such as relational reasoning [2] and text classification [24]. Kipf et al. [13] proposed a simple and efficient layer-wise propagation rule for graph convolutional networks (GCN) based on spectral graph convolution and their model achieved significant raise in several graph-structured datasets. Veličković et al. [20] put forward a novel GNN architecture—graph attention networks (GAT), which introduced masked self-attention mechanism to tackle some key challenges of GNN. Recently, Ye et al. [26] proposed a new GAT framework for stroke classification, which demonstrated the great potential for online handwritten document recognition with GNN.

3   Method

We are given N labeled online handwritten diagrams $D = \{(X^i, Y^i) | i \in [1, N]\}$, where each diagram X^i is composed of a sequence of strokes $X^i = \{X^i_s | s \in [1, M_i]\}$ (M_i is the number of strokes in X^i) and Y^i_s is the label of X^i_s which takes discrete semantic annotation, such as process, decision and arrow. Our target is to learn a model from the training set D that can predict the labels of strokes in testing diagrams as accurate as possible.

Roughly speaking, our method models each diagram with a graph in which nodes represent strokes and edges represent the relationships between strokes. Then, we treat diagram stroke classification as a graph node classification problem and solve it with attention-based GNN. The proposed method is composed of three modules, including the construction of the diagram graph, extraction of node and edge features from raw signals and graph attention networks, which will be introduced separately as follows.

3.1   Graph Building

Here we introduce a new approach to abstract the structure information in the diagram that each handwritten diagram is formulated as a space-time relationship graph (STRG). Every stroke s_i is represented as a vertex $v_i \in V$ and the relevance in space and time between strokes s_i and s_j is noted as edge $e_{ij} \in E$ in graph $G(V, E)$, where V is the vertex set and E is the edge set in G.

Specifically, from the time perspective, we build the edges $E_T = \{(t, t+1) | t \in [1, n-1]\}$ between every temporal adjacent strokes in the diagram, where n is the number of strokes.

In view of spatial relationship, for the stroke s_s, the edge set $e_{s,N(s)}$ is added to $E(G)$, where $N(s)$ are all space neighbors of stroke s_s. If any stroke pairs' minimal Euclidean distance is less than the spatial neighbor threshold (SNT), they are regarded as neighbors each other. The hyperparameter SNT is elaborately tuned on validation set. We also try to build more complex STRG of a document, but it has little effect to the experimental result. Figure 1 shows an example of flowchart rendering from original data and its corresponding STRG.

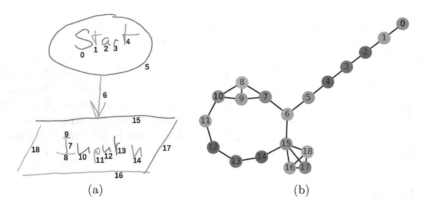

Fig. 1. An example of handwritten diagram and its corresponding space-time relationship graph (STRG). The numbers in the figure indicate the temporal order of strokes. (a) An example handwriting diagram and (b) STRG.

3.2 Feature Extraction

For each stroke in an online document, 10 local features and 13 context features [25] are extracted as node features in STRG. These features have been proven to be very effective in previous works [19,25]. In addition, 19 edge features [25] are extracted from stroke pairs for modeling the relations between strokes. In feature pre-processing, we conduct power transformation with the coefficient 0.5 and normalization with mean μ and standard deviation σ. Therefore, the original feature h become:

$$h' = \text{sign}(h)\sqrt{|h|} \tag{1}$$
$$h'' = (h' - \mu)/\sigma \tag{2}$$

where $\text{sign}(\cdot)$ is the sign function.

3.3 Graph Attention Networks

In this section, we introduce the enhanced GAT model, which is constructed by stacking multiple graph attention layers.

The input to each graph attention layer are a set of node features, $\mathbf{H} = \left\{ \overrightarrow{h}_1, \overrightarrow{h}_2, \ldots, \overrightarrow{h}_{|V|} \right\}$, $\overrightarrow{h}_i \in \mathbb{R}^C$, and a set of edge features $\mathbf{F} = \left\{ \overrightarrow{f}_{ij} | (i,j) \in E \right\}$, $\overrightarrow{f}_{ij} \in \mathbb{R}^D$, where $|V|$ is the number of nodes, and C, D are the dimensionality of node features and edge features, respectively. The layer generates a new set of node features, $\mathbf{H}' = \left\{ \overrightarrow{h}'_1, \overrightarrow{h}'_2, \ldots, \overrightarrow{h}'_{|V|} \right\}$, $\overrightarrow{h}'_i \in \mathbb{R}^{C'}$, where C' is the dimension of output features.

In each layer, the first step is applying a shared linear transformation to every node, then a shared attention mechanism is performed to compute attention coefficients utilizing self-attention on the nodes:

$$c_{ij} = a\left(\mathbf{W}_h\,\vec{h}_i, \mathbf{W}_h\,\vec{h}_j\right) \tag{3}$$

where $\mathbf{W}_h$ is a shared learnable weight matrix for the node-wise feature transformation. The node attention mechanism $a : \mathbb{R}^{C'} \times \mathbb{R}^{C'} \to \mathbb{R}$ used in this work is the additive attention parameterized by a learnable weight $\vec{a}_h \in \mathbb{R}^{C'}$ with an activation function σ, which is formulated as:

$$c_{ij} = \sigma\left(\vec{a}_h^T\left(\mathbf{W}_h\,\vec{h}_i + \mathbf{W}_h\,\vec{h}_j\right)\right). \tag{4}$$

In addition to computing attention coefficients by self-attention mechanisms, we also incorporate edge features to measure the importance of edges by applying an one-layer feedforward neural network:

$$c'_{ij} = \sigma\left(\vec{a}_f^T\sigma\left(\mathbf{W}_f\,\vec{f}_{ij} + \vec{b}_f\right)\right) \tag{5}$$

where $\mathbf{W}_f \in \mathbb{R}^{C' \times D}, \vec{b}_f \in \mathbb{R}^{C'}, \vec{a}_f \in \mathbb{R}^{C'}$ are all learnable parameters. In this work, we use Leaky ReLU as the activation function.

It should be noted that, the coefficients mentioned above are not comparable across different nodes. Consequently, they are normalized across all neighbors using the softmax function:

$$\alpha_{ij} = \text{softmax}_j\left(c_{ij} + c'_{ij}\right) = \frac{\exp\left(c_{ij} + c'_{ij}\right)}{\sum_{k \in N(i)} \exp\left(c_{ik} + c'_{ik}\right)} \tag{6}$$

where $N(i)$ is the neighborhood of node i.

The final output features for every node are computed by aggregating weighted node features of neighbors with attention coefficients:

$$\vec{h}'_i = \sigma\left(\sum_{j \in N(i)} \alpha_{ij}\mathbf{W}_h\,\vec{h}_j\right). \tag{7}$$

Following Veličković et al. [20], we also adopt multi-head attention in our model. Specifically, K independent attention mechanisms execute the transformation of Eq. 7, and then their features are concatenated:

$$\vec{h}'_i = \|_{k=1}^{K}\sigma\left(\sum_{j \in N(i)} \alpha_{ij}^k\mathbf{W}_h^k\,\vec{h}_j\right) \tag{8}$$

in which $\|$ denotes concatenation operation, and α_{ij}^k are normalized attention coefficients calculated by the k-th attention mechanism a^k.

In the final layer, we perform average operation instead of concatenation, and then apply the softmax function to output predicted values:

$$\overrightarrow{h}'_i = \sigma\left(\frac{1}{K}\sum_{k=1}^{K}\sum_{j\in N(i)}\alpha_{ij}^k \mathbf{W}_h^k \overrightarrow{h}_j\right) \tag{9}$$

$$\overrightarrow{p}_i = \text{softmax}\left(\mathbf{W}_o \overrightarrow{h}'_i\right) \tag{10}$$

where $\mathbf{W}_o \in \mathbb{R}^{C'\times L}$ is a learnable weight matrix that transforms features to outputs.

The standard cross-entropy loss on the training set is used to train the GAT model, which is formulated as:

$$L(\mathbf{W}) = -\sum_{i=1}^{N}\sum_{s=1}^{M_i}\log\overrightarrow{p}_s\left(Y_s^i\right), \tag{11}$$

in which $\mathbf{W}$ encompasses all learnable parameters. $\mathbf{W}$ is initialized with Glorot initialization [9] and learned using mini-batch gradient descent. In practice, parameter optimization is performed with the Adam SGD optimizer [12].

4   Experiments

4.1   Dataset

In this work, we evaluate our method on three publicly accessible online handwritten diagram databases: FC_A [1], FC_B [5] and FA [6]. FC_A and FC_B are two flowchart databases which include text and six graphical symbols: terminator, connection, decision, data, process and arrow. FA is a finite automata database which encompasses state (circle), final state (pairwise concentric circles), arrow and label. Table 1 shows the details of the three databases.

Table 1. Online handwritten diagram datasets overview.

Dataset	Partition	#Writers	#Templates	#Diagrams	#Strokes	#Symbols
FC_A	Train	31	14	248	23355	5540
	Test	15	14 others	171	15696	3791
FC_B	Train	10	28	280	30443	6195
	Validation	7	28	196	20102	4342
	Test	7	28	196	20139	4343
FA	Train	11	12	132	6792	3631
	Validation	7	12	84	4059	2307
	Test	7	12	84	4125	2323

4.2   Experiments Setup

For all experiments, we adopt a seven-layer GAT model with 32 neurons for each hidden layer employing residual connections [10]. Following [20], we employ the Leaky ReLU activation function with the negative slope of 0.2 as the attention functions in Eqs. 4 and 5. We also introduce dropout rules [18] with the dropout probability of 0.1 for all layers. In addition, every hidden layer consists of 8 attention heads for flowchart datasets (FC_A and FC_B) and 6 attention heads for FA. We use 2 output attention heads for all networks.

We train all models for 200 epochs by minimizing the standard cross-entropy loss on the training set with the early stopping strategy. The optimization is performed using the Adam optimizer [12] with an initial learning rate of 0.005 for flowchart datasets and 0.003 for FA. The decay rate r is set to 0.1 and the number of patience round $r = 15$ for flowchart datasets and $r = 17$ for FA. To train networks more efficiently, we adopt the mini-batch trick with 8 graphs for flowchart datasets and 6 graphs for FA.

The feature extraction module is implemented by C++ and both training and inference algorithms are implemented with Pytorch and Deep Graph Library (DGL)[1]. Training of GAT is conducted on a server with a NVIDIA Geforce GTX 980 GPU, while testing is performed on a PC with four Intel (R) Core (TM) i5-7400 CPU @ 3.00 GHz. Unless otherwise specified, we repeat each experiment for 10 times using the same configurations and report the average results.

4.3   Results and Discussion

In this work, we use the stroke classification accuracy to evaluate our method. Each stroke in a diagram is assigned a predefined symbol-level label by the model, which is then compared against the ground truth. Stroke classification accuracy on FC_A, FC_B and FA are reported in Tables 2, 3 and 4, respectively. Numbers in boldface show the best results. Results of comparison methods are directly cited from the original papers. GAT denotes the method that uses the original GAT model, while GAT with EFA denotes the method proposed in this work which enhances GAT by introducing edge feature attention.

As we can see, on all datasets, GAT with EFA outperformes previous methods and achieved the best overall accuracy results. One notable phenomenon is that the performance of previous methods vary dramatically across different labels, while GAT with EFA consistently delivers accurate prediction for all labels. In addition, GAT with EFA improves GAT with a large margin for all experiments, which demonstrates that the proposed edge feature attention plays an important role in capturing the complicated temporal and spatial relationships between strokes. Furthermore, our method is very efficient in both training and testing. For example, on FC_A, it takes about 38 min to train our model and 70 ms to classify all strokes of one flowchart under the settings described in Sect. 4.2.

---

[1] https://github.com/dmlc/dgl.

Table 2. Stroke classification accuracy on FC_A (%).

Class	Lemaitre [14]	Bresler [4]	Carton [7]	Wang [22]	Bresler [5]	Wu [23]	Wang [21]	GAT	GAT with EFA
Arrow	79.6	88.7	83.8	92.1	87.5	87.4	**98.18**	84.75 ± 1.63	97.28 ± 0.36
Connection	80.0	94.1	80.3	79.3	94.1	73.7	83.22	80.81 ± 4.02	**94.96 ± 1.39**
Data	84.7	**96.4**	84.3	87.6	95.3	87.6	93.28	84.70 ± 2.45	93.81 ± 1.13
Decision	84	90.9	90.9	89.7	88.2	89.7	**95.69**	82.83 ± 2.65	93.61 ± 1.03
Process	85.7	95.2	90.4	93.5	**96.3**	91.8	93.13	85.68 ± 1.01	93.71 ± 1.12
Terminator	79.6	90.2	69.8	78.9	90.7	91.6	89.29	85.16 ± 2.37	**93.88 ± 1.78**
Text	97.8	**99.3**	97.2	99.0	99.2	98.8	96.81	98.53 ± 0.20	98.48 ± 0.23
Overall	91.1	96.5	92.4	95.8	96.3	94.9	96.19	93.16 ± 0.36	**97.27 ± 0.20**

Table 3. Stroke classification accuracy on FC_B (%).

Class	Bresler [5]	GAT	GAT with EFA
Arrow	**98.3**	93.39 ± 0.88	97.16 ± 0.52
Connection	88.4	84.55 ± 4.63	**93.39 ± 2.35**
Data	96.1	93.89 ± 1.41	**96.30 ± 1.85**
Decision	90.3	92.01 ± 2.32	**95.94 ± 0.82**
Process	**98.4**	93.16 ± 1.85	96.90 ± 1.32
Terminator	**99.7**	94.90 ± 1.12	97.89 ± 0.29
Text	99.6	99.44 ± 0.10	**99.86 ± 0.06**
Overall	98.4	97.78 ± 0.10	**99.04 ± 0.10**

Table 4. Stroke classification accuracy on FA (%).

Class	Bresler [6]	Bresler [4]	Wang [22]	Bresler [5]	GAT	GAT with EFA
Label	99.1	**99.8**	99.0	99.7	98.87 ± 0.20	**99.83 ± 0.13**
Arrow	89.3	94.9	97.7	98.0	96.62 ± 0.66	**99.10 ± 0.30**
Initial arrow	78.5	85.0		98.6		
State	95.2	96.9	91.6	**98.3**	91.71 ± 3.12	97.38 ± 0.50
Final state	96.1	**99.2**	96.5	**99.2**	95.50 ± 1.06	97.12 ± 0.27
Overall	94.5	97.4	97.8	99.0	97.34 ± 0.33	**99.22 ± 0.13**

4.4 Error Analysis

In Fig. 2, we show three examples of recognized flowchart from FC_A with typical errors. If a stroke is far away from the symbol which it should belong to, but close to the neighboring one, it is more likely to be misclassified, as Fig. 2(a) and (b). For isolated *text* strokes next to the *arrow*, it is more possible to be predicted as *arrow* by mistake. Some recognition errors could be eliminated through postprocess, such as a *process* stroke surrounded by *texts*, as shown in (c). The confusion matrix for strokes classification result on test set of FC_A is

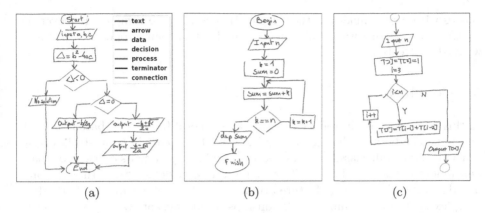

(a)	(b)	(c)

Fig. 2. Examples of misrecognized flowchart from FC_A dataset. Every recognized stroke is colored and explained in the legend. In (a) a *data* stroke is misclassified as *arrow* and an *arrow* stroke is misclassified *decision*, besides, a *text* stroke is misclassified as *data*. (b) two strokes in the symbol *decision* are misclassified as *data* and *process*, respectively. (c) a *text* stroke enclosed by a *process* is misclassified as *process*, and another *text* beside on *arrow* is misclassified as *arrow* (Color figure online).

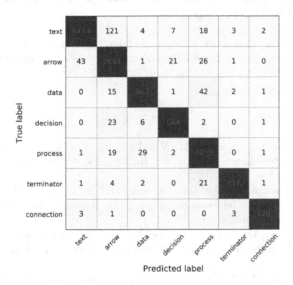

Fig. 3. Confusion matrix for stroke classification result on test set of FC_A (overall precision is 97.27%). Each row in the matrix encompasses all strokes of the ground-truth class, and each column encompasses all strokes of the predicted class.

presented in Fig. 3. Since some symbols are ambiguous in appearance, such as *process* and *data*, they are likely to be misclassified in highly confidence. Another very important factor for misclassification is that the number of training samples of different classes are imbalanced severely in nature, which has a serious side effect on recognition performance: the classifier is more likely to predict stroke classes with less samples as other classes that have more samples. However, in contrast to previous work, this effect is moderate in our proposed framework with edge attention mechanism.

5   Conclusions

In this work, we have introduced a novel and general framework based on GAT for online handwritten diagram recognition. We formulate diagram stroke classification as the node classification task in a graph. Experiments on two flowchart benchmark datasets and one finite automata dataset demonstrate that the proposed framework with edge feature attention mechanism is capable of encoding complex spatial and temporal relationships in an efficient way for stroke classification. Our method outperforms several recently proposed approaches by a prominent margin. Our model is computationally efficient, which is suitable for large-scale applications in mobile devices. Moreover, the classification performances have a great potential to be improved from our analysis of the failure cases. In the future work, we will investigate how to extend our framework to perform stroke grouping and symbol recognition of handwritten diagrams, as well as structure analysis of diagrams.

Acknowledgements. This work is supported by the National Key Research and Development Program Grant 2018YFB1005000, the National Natural Science Foundation of China (NSFC) Grants 61773376, 61721004.

References

1. Awal, A.M., Feng, G., Mouchere, H., Viard-Gaudin, C.: First experiments on a new online handwritten flowchart database. In: Document Recognition and Retrieval XVIII (2011)
2. Battaglia, P., Pascanu, R., Lai, M., Rezende, D.J., et al.: Interaction networks for learning about objects, relations and physics. In: Advances in Neural Information Processing Systems (2016)
3. Bresler, M., Průša, D., Hlaváč, V.: Modeling flowchart structure recognition as a max-sum problem. In: International Conference on Document Analysis and Recognition (2013)
4. Bresler, M., Průša, D., Hlaváč, V.: Detection of arrows in on-line sketched diagrams using relative stroke positioning. In: IEEE Winter Conference on Applications of Computer Vision (2015)
5. Bresler, M., Průša, D., Hlaváč, V.: Online recognition of sketched arrow-connected diagrams. Int. J. Doc. Anal. Recogn. **19**(3), 253–267 (2016)
6. Bresler, M., Van Phan, T., Průša, D., Nakagawa, M., Hlaváč, V.: Recognition system for on-line sketched diagrams. In: International Conference on Frontiers in Handwriting Recognition (2014)
7. Carton, C., Lemaitre, A., Coüasnon, B.: Fusion of statistical and structural information for flowchart recognition. In: International Conference on Document Analysis and Recognition (2013)

8. Coüasnon, B.: DMOS, a generic document recognition method: application to table structure analysis in a general and in a specific way. Int. J. Doc. Anal. Recogn. 8(2–3), 111–122 (2006)

9. Glorot, X., Bengio, Y.: Understanding the difficulty of training deep feedforward neural networks. In: International Conference on Artificial Intelligence and Statistics (2010)

10. He, K., Zhang, X., Ren, S., Sun, J.: Deep residual learning for image recognition. In: IEEE Conference on Computer Vision and Pattern Recognition (2016)

11. Julca-Aguilar, F.D., Hirata, N.S.: Symbol detection in online handwritten graphics using Faster R-CNN. In: International Workshop on Document Analysis Systems (2018)

12. Kingma, D.P., Ba, J.: Adam: a method for stochastic optimization. In: International Conference on Learning Representations (2015)

13. Kipf, T.N., Welling, M.: Semi-supervised classification with graph convolutional networks. In: International Conference on Learning Representations (2017)

14. Lemaitre, A., Mouchère, H., Camillerapp, J., Coüasnon, B.: Interest of syntactic knowledge for on-line flowchart recognition. In: Kwon, Y.-B., Ogier, J.-M. (eds.) GREC 2011. LNCS, vol. 7423, pp. 89–98. Springer, Heidelberg (2013). https://doi.org/10.1007/978-3-642-36824-0_9

15. Miyao, H., Maruyama, R.: On-line handwritten flowchart recognition, beautification and editing system. In: International Conference on Frontiers in Handwriting Recognition (2012)

16. Qi, Y., Szummer, M., Minka, T.P.: Diagram structure recognition by Bayesian conditional random fields. In: IEEE Conference on Computer Vision and Pattern Recognition (2005)

17. Ren, S., He, K., Girshick, R., Sun, J.: Faster R-CNN: towards real-time object detection with region proposal networks. In: Advances in Neural Information Processing Systems (2015)

18. Srivastava, N., Hinton, G., Krizhevsky, A., Sutskever, I., Salakhutdinov, R.: Dropout: a simple way to prevent neural networks from overfitting. J. Mach. Learn. Res. 15(1), 1929–1958 (2014)

19. Van Phan, T., Nakagawa, M.: Combination of global and local contexts for text/non-text classification in heterogeneous online handwritten documents. Pattern Recogn. 51, 112–124 (2016)

20. Veličković, P., Cucurull, G., Casanova, A., Romero, A., Lio, P., Bengio, Y.: Graph attention networks. In: International Conference on Learning Representations (2018)

21. Wang, C., Mouchère, H., Lemaitre, A., Viard-Gaudin, C.: Online flowchart understanding by combining max-margin Markov random field with grammatical analysis. Int. J. Doc. Anal. Recogn. 20(2), 123–136 (2017)

22. Wang, C., Mouchere, H., Viard-Gaudin, C., Jin, L.: Combined segmentation and recognition of online handwritten diagrams with high order Markov random field. In: International Conference on Frontiers in Handwriting Recognition (2016)

23. Wu, J., Wang, C., Zhang, L., Rui, Y.: Offline sketch parsing via shapeness estimation. In: International Joint Conference on Artificial Intelligence (2015)

24. Yao, L., Mao, C., Luo, Y.: Graph convolutional networks for text classification. arXiv preprint arXiv:1809.05679 (2018)

25. Ye, J.Y., Zhang, Y.M., Liu, C.L.: Joint training of conditional random fields and neural networks for stroke classification in online handwritten documents. In: International Conference on Pattern Recognition (2016)

26. Ye, J.Y., Zhang, Y.M., Yang, Q., Liu, C.L.: Contextual stroke classification in online handwritten documents with graph attention networks. In: International Conference on Document Analysis and Recognition (2019)
27. Yuan, Z., Pan, H., Zhang, L.: A novel pen-based flowchart recognition system for programming teaching. In: Leung, E.W.C., Wang, F.L., Miao, L., Zhao, J., He, J. (eds.) WBL 2008. LNCS, vol. 5328, pp. 55–64. Springer, Heidelberg (2008). https://doi.org/10.1007/978-3-540-89962-4_6
28. Zhou, J., Cui, G., Zhang, Z., Yang, C., Liu, Z., Sun, M.: Graph neural networks: a review of methods and applications. arXiv preprint arXiv:1812.08434 (2018)

CNN-Based Erratic Cigarette Code Recognition

Zhi-Feng Xie[1,2]([✉]), Shu-Han Zhang[1]([✉]), and Peng Wu[1]([✉])

[1] Department of Film and Television Engineering, Shanghai University,
Shanghai 200072, China
zhifeng_xie@shu.edu.cn, 1993jerryzhang@163.com, 1097829202@qq.com
[2] Shanghai Engineering Research Center of Motion Picture Special Effects,
Shanghai 200072, China

Abstract. Cigarette code is a string printed on the wrapper of cigarette packet as a basis of distinguishing illegal sales for tobacco administration. In general, the code is excerpted and entered to administration system manually during on-site inspection, which is quite time-consuming and laborious. In this paper, we propose a new solution based on convolutional neural network for intelligent transcription. Our recognition method is composed of four components: detection, identification, alignment, and regularization. First of all, the detection component fine-tunes an end-to-end detection network to obtain the bounding box region of cigarette code. Then the identification component constructs an optimized CNN architecture to recognize each character in the region of cigarette code. Meanwhile the alignment component trains a CPM-based network to estimate the positions of all characters including some missing characters. Finally, the regularization component develops a matching algorithm to produce a regularized result with all characters. The experimental results demonstrate that our proposed method can perform a better, faster and more labor-saving cigarette code transcription process.

Keywords: Cigarette code · Optical Character Recognition · Convolutional neural network

1 Introduction

Cigarette code is a string with 32 characters printed on the wrapper of cigarette packet, which can be used to distinguish illegal cigarette sales in China, and example is shown in Fig. 1. Presently, the code is excerpted and entered to administration system manually during on-site inspection. This manual recording method is quite time-consuming and laborious. Thus it is urgent that an intelligent method can be proposed to simplify manual operations and improve inspection efficiency.

OCR (Optical Character Recognition) is an ordinary method to recognize text from an image. However, our task to recognize cigarette code faces several

© Springer Nature Switzerland AG 2019
Y. Zhao et al. (Eds.): ICIG 2019, LNCS 11901, pp. 245–255, 2019.
https://doi.org/10.1007/978-3-030-34120-6_20

difficulties with classical OCR methods [9,19,29] or modern CNN-based OCR methods [1,2,5,7,8,13,14,22,24] : (a) Erratic layouts. The cigarette codes are printed by different administrations, their layouts, font types, and font sizes are miscellaneous. (b) Complicated backgrounds. Normally, cigarette code is printed on the its wrapper randomly, which may cause cigarette code contaminated by its background. (c) Geometric deformation. Due to imperfect printing technique, characters are often printed with distortion. (d) Man-made sabotage. In order to evade punishment, some retail stores with illegal sale would make sabotage to cigarette code that results in some indiscernible characters. (e) Semantic demands. Even if some characters are unable to recognize, the tobacco monopoly administration demands these characters to be regularized by '*' character.

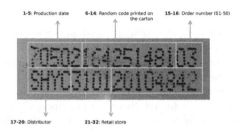

Fig. 1. Information of cigarette code. This carton of cigarettes is manufactured in May 2, 2017, it is the third one in a large box, and distributed from Shanghai Tobacco Monopoly Administration to a retail store with '310120104842' identifier.

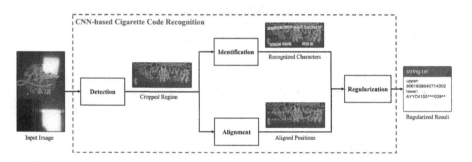

Fig. 2. Overview. Our CNN-based recognition method consists of four main components: detection, identification, alignment, and regularization.

To deal with these issues, we proposed a new CNN-based solution to achieve the efficient recording of erratic cigarette code in a single image. As shown in Fig. 2, our pipeline proceeds in four components: detection, identification, alignment, and regularization. Given an image with erratic cigarette code, the detection component first employs transfer training technique to fine-tune an end-to-end detection network [23], which can classify the categories of cigarette code (black and white) and obtain the bounding box region of cigarette code. Then the identification component constructs an optimized CNN architecture by strengthening feature extraction and defining multi-parallel region proposal networks, which can recognize and locate each character in the cropped region of cigarette code. At the same time, the alignment component trains a CPM-based (Convolutional Pose Machine) network [31] to estimate the positions of all 32 characters in the cropped region, including some missing characters. Finally, the regularization component develops a matching algorithm to set up the mapping relationship between the identification and alignment results, and fill some $'*'$ characters to produce a regularized result with all 32 characters. The experimental results show that our proposed method can yield the detection accuracy of over 98%, and the recognition accuracy (all of 32 characters are correct in an image) of over 90% in the testing dataset.

2   Related Work

Text recognition is always a key task to recognize text content accurately in the process of OCR. Recently, a lot of state-of-the-art methods have been proposed to achieve high-accuracy text recognition. Wang et al. [29] proposed a system rooted in generic object recognition to achieve superior performance of text recognition. Neumann et al. [19] further proposed an ERs-based (Extremal Regions) real-time scene text recognition. Jaderberg et al. [9] present the text recognition problem as multi-class classification task with large number of labels. Since then, a lot of CNN-based methods are introduced into the term, which can be mainly divided into three categories: LSTM (Long Short-Term Memory) + CTC (Connectionist Temporal Classification) [2,7,13,16], LSTM + Attention [1,3–6,8,14,15,17,24] and object detection [21,22,30].

The perfect recognition of erratic cigarette code not only depends on identification accuracy for each character, but also need to fill some missing characters and produce a regularized recognition result with all 32 characters by estimating all positions of missing characters. Thus a excellent localization technique is very important for the regularization of erratic cigarette code. At present, many alignment algorithms have been applied successfully to locate key points, especially in human face and human pose. For face alignment, a number of CNN-based methods [25], such as TCDCN [35], TCNN [32], MTCNN [34], DAN [11], and so on, can produce the key points of face with partial occlusion efficiently and accurately. For pose estimation, many state-of-the-art methods [10,26,27,31] can also construct CNN-based models to yield a great performance even under body occlusion.

3   CNN-Based Recognition for Erratic Cigarette Code

Since erratic cigarette code is more complex and distinctive than traditional text, a number of state-of-the-art OCR techniques fail to produce satisfying recognition results. Here, we propose a new CNN-based solution to recognize erratic cigarette code effectively. As shown in Fig. 2, its pipeline consists of four key components: detection, identification, alignment, and regularization.

3.1 Detection

In the detection component, we employ a end-to-end convolutional neural network to obtain a bounding box region of cigarette code and point out its category (black code or white code). Our end-to-end network concatenates three sub-networks: feature extraction, region proposal, regression and classification.

Our detection component refers to the concept of inductive transfer learning [20] for network training. We first collect tens of thousands of images with cigarette code, and their categories and bounding box coordinates are manually annotated. Then we construct the end-to-end detection network and achieve the fine-tuned training based on the VGG-16, RPN, and RCNN architectures. Finally, we apply the trained model and the NMS (Non-Maximum Suppression) algorithm to predict the bounding box and category of cigarette code accurately.

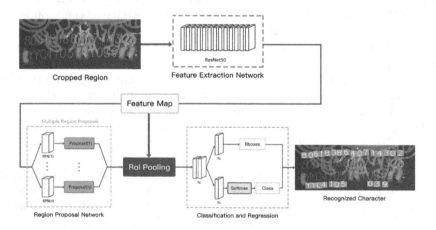

Fig. 3. Network of identification component. The identification network consists of three sub-networks: ResNet-based feature extraction, multi-parallel region proposal, classification and regression.

3.2 Identification

In the identification component, we construct a new convolutional neural network to identify and locate each character in the cropped region of cigarette code. As shown in Fig. 3, the identification network consists of three sub-networks: ResNet-based feature extraction, multi-parallel region proposal, classification and regression.

Inspired by the concept of CNN ensemble [33], we introduce extra anchors, with $0.25, 0.333, 0.5, 1, 2$ in ratio and $4, 6, 8, 12, 16$ in scale, yielding $5 \times 5 = 25$ anchors. The optimized architecture defines more small and tall anchors for the character shapes of cigarette code.

As shown in Fig. 4, the characters in two cropped cigarette codes and their bounding boxes are correctly marked by our identification component. However, since some characters in the white example are destructed deliberately, their positions cannot be specified and the identification result is also incomplete. Thus we must further estimate the positions of missing characters, introduce a special character '*' to fill them and produce the recognized result with all 32 characters.

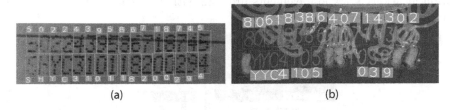

(a)   (b)

Fig. 4. Two examples of cigarette code identification. (a) Recognized characters of black cigarette code. (b) Recognized characters of white cigarette code.

3.3   Alignment

In the alignment component, we integrate the popular DeepPose's [28] concept to localize all characters of erratic cigarette code, especially including some missing characters. We can apply the fine-tuned training to optimize the network stage by stage, and then yield a predicted model to produce a final alignment result.

First of all, we annotate the bounding boxes of 32 characters in the cigarette code region, $(x_i^1, y_i^1, x_i^2, y_i^2), i \in [1, 32]$, where (x^1, y^1) and (x^2, y^2) are the top left and bottom right points of each bounding box. Then, we compute their center points as the ground-truth positions $Z = (z_1, \ldots, z_{32})$, where $z_i = \{(x_i^1 + x_i^2)/2, (y_i^1 + y_i^2)/2\}, i \in [1, 32]$, and create the ideal belief map $b_*^p(z)$ for each position z by putting Gaussian peaks at ground truth locations of the p-th position. Next we construct the CPM-based alignment network and define the cost function of each stage $t \in [1, 6]$:

$$f_t = \sum_{p=1}^{32} \sum_{z \in Z} \|b_t^p(z) - b_*^p(z)\|_2^2 \tag{1}$$

where b_t^p denotes the belief map of the p-th estimated position by stage t, b_*^p denotes the ideal belief map of the p-th ground-truth position. Finally, we add the losses at each stage $F = \sum f_t$, and use standard stochastic gradient descend to jointly train all stages in the network. As shown in Fig. 5, the alignment network can effectively estimate character positions even with characters indiscernible.

3.4   Regularization

In the regularization component, we further propose a matching algorithm to set up a corresponding relationship between the identification and alignment results, and then employ a special character $'*'$ to fill some missing characters and produce a regularized result with all 32 characters.

First of all, we obtain the locations of identification by computing the central points of bounding boxes Y^{rb} in the identification result. Then we denote the locations of identification as Y^r and the positions of alignment as Y^a. The mathematical model for our matching task can be defined as a typical assignment problem [18]. Based on this, we introduce Hungarian algorithm [12] to minimize Φ and calculate the mapping matrix $\mathbf{X}$ in order to match the identification locations Y^r whose elements are ≤ 32 with the estimated 32 alignment positions Y^a.

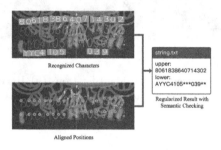

Fig. 5. Regularization results.

With the output of mapping matrix $\mathbf{X}$, we can assign the identification result Y^r_i into the j-th position of the output string for each $(x_{ij} \neq 0) \in \mathbf{X}$. Since the length of output string is fixed to 32, we need to fill the rest positions in output string with $'*'$ if Y^r has less than 32 characters. If there exists a unique matching between the output string and the dictionary element, we can replace these corresponding $'*'$'s with certain characters from the semantic dictionary. As shown in Fig. 5, by our matching algorithm, we can yield the output string with 32 characters "8061838640714302" and "*YYC4105***039**".

4   Data Preparation and Network Training

To train our networks, we need to prepare a lot of images with annotations. We first collect 21, 861 images including the black and white cigarette codes for the detection network. The annotation of each input image contains the bounding box of cigarette code and its category $C_r = \{black, white\}$. Then we train the detection network to predict the bounding box and category of cigarette code, which can be used to produce the cropped region image of cigarette code. We collect these cropped images to construct the identification dataset, including

21, 740 images of the black cigarette code and 17, 044 images of the white code. The annotation of each cropped region image image contains the bounding box of each character and its corresponding category $C_c = \{0–9, A–Z, *\}$, where $'*'$ denotes the indiscernible character. We randomly pick 20, 000, 20, 000, 16, 000 samples for training and make the rest 1, 861, 1, 740, 1, 044 samples for testing.

With all the data preparation done, we can start our network training process. For the detection network, we set some necessary training parameters, including the total iterations of 40, 000 with step learning rate at $\lambda = 0.001$, gamma $= 0.1$, stepsize $= 50,000$, momentum set at 0.9, weight decay at $5e − 4$. The training parameters of the identification network is similar with the detection network except a total iterations of 100, 000. For the alignment network, we integrate all identification datasets as the alignment dataset, and compute the central points of their bounding boxes as the ground-truth positions. We set its necessary training parameters, including the total iterations of 62, 500 with step learning rate at $\lambda = 4e − 6$, gamma $= 0.333$, stepsize $= 13,275$, momentum set at 0.9, weight decay at $5e − 4$.

5 Experimental Results

In this section, we perform a lot of experiments to evaluate the detection, iden- tification, and alignment components one by one. On the other hand, we also demonstrate the excellent end-to-end performance of our proposed method.

5.1 Evaluation of Detection Component

We evaluate the detection component with accuracy and region correctness. The accuracy of category classification is near 99.6%, and 8 images with cigarette code aren't classified correctly.

For region detection, we expand the bounding box of detected region by 6.25% horizontally and 3.125% vertically, and define its correctness if the anno- tated bounding box B_a is fully included by the expanded bounding box region B_r, that is $B_a \subset B_r$, it is different with traditional definition. The detection com- ponent achieves the 98.6% accuracy of region detection in total testing dataset, including 98.8% and 98.3% in two testing datasets of black and white codes respectively.

5.2 Evaluation of Identification Component

We employ the detection component to produce the cropped region of cigarette code as identification dataset.

To perform a fair comparison, all the state-of-the-art methods are trained and tested in our same identification dataset, and the definition of identification correctness is that all characters excluding $'*'$ in cigarette code region must be correctly recognized. As shown in Table 1, we observe that most of state-of- the-art methods can achieve the good identification of simple black code but

Table 1. Accuracy of identification component.

	Correct results (black)	Accuracy (black)	Correct result (white)	Accuracy (white)
Deep text spotter [2]	1,489/1,740	85.6%	792/1,044	75.9%
Attention OCR [5]	1,522/1,740	87.5%	851/1,044	81.5%
SEE [1]	1,445/1,740	83.0%	718/1,044	68.8%
TextSpotter [8]	1,501/1,740	86.3%	830/1,044	79.5%
Aster [24]	1,529/1,740	87.9%	857/1,044	82.1%
Our identification component	1,578/1,740	**90.7%**	901/1,044	**86.3%**

be difficult to handle the complex white code. In contrast, our identification component can reach the higher identification accuracy on both two testing datasets.

5.3 Evaluation of Alignment Component

The definition of alignment correctness is that the estimated character position locates inside our beforehand artificially annotated bounding box with a 10% expansion both horizontally and vertically. Our alignment component achieves 92.7% accuracy in the testing dataset, including 94.7% accuracy on black code and 89.4% accuracy on white code respectively.

5.4 End-to-End Performance

To evaluate the overall performance of our proposed method, we execute an end-to-end verification in the detection testing dataset. We define the principle of correctness transcription as follow: all of the characters are recognized correctly, with $'*'$ character labeling unrecognized characters, and all the characters must be in right order. Our proposed method achieves 92.2% accuracy in total, 95.8% on black code and 87.2% on white one respectively.

During on-site cigarette inspection, we randomly pick 500 cartons of cigarettes and make comparison with artificial transcription of cigarette code. Our transcription system achieves 90.8% accuracy of recognition, which is slightly lower than 95.2% accuracy of artificial transcription. But our system only takes 43 min to finish the whole process, which is higher-efficiency and more labour-saving than 382 min of artificial transcription. The comparison result also demonstrates that our transcription system can further simplify manual operations and improve inspection efficiency.

6   Conclusion

In this paper, we mainly propose a new solution to detect and recognize erratic cigarette code accurately and rapidly. Although some existing techniques are applied into our solution, we still put forward some new ideas and make some important contributions. First of all, we collect more than 40 thousands images of cigarette code and annotate their bounding boxes and character information. Compared with a number of traditional OCR datasets, our cigarette code dataset is more complex and distinctive, such as erratic layouts, various fonts, complicated backgrounds, geometric deformation, man-made sabotage, and so on. It is a new challenge to solve these issues by existing state-of-the-art OCR techniques. In the future, we will share our cigarette code dataset with all researchers. Secondly, our new solution not only integrates the existing models but also further optimizes them to improve the recognition accuracy of erratic cigarette code. On one hand, we construct multi-parallel RPN units to strengthen the effect of region proposal and avoid missing some characters with different shapes. On the other hand, we propose a novel regularization method by training CPM-based network and developing an optimal string matching algorithm. The experimental results have demonstrated the effectiveness of our new improvement. Finally, with a view to the practical application, we employ our new solution to implement an intelligent transcription system of cigarette code.

Although our proposed method can achieve a higher-efficiency recognition for erratic cigarette code, its recognition accuracy is still slightly lower than artificial transcription. Therefore, we must first extend the training dataset of cigarette code, and further optimize the network architecture of our model to solve some bottleneck problems, such as rotation, various character shapes, alignment accuracy with many missing characters, and so on.

References

1. Bartz, C., Yang, H., Meinel, C.: SEE: towards semi-supervised end-to-end scene text recognition. arXiv preprint arXiv:1712.05404 (2017)
2. Busta, M., Neumann, L., Matas, J.: Deep TextSpotter: an end-to-end trainable scene text localization and recognition framework. In: IEEE International Conference on Computer Vision, pp. 2223–2231 (2017)
3. Cheng, Z., Xu, Y., Bai, F., Niu, Y., Pu, S., Zhou, S.: AON: towards arbitrarily-oriented text recognition. In: 2018 IEEE Conference on Computer Vision and Pattern Recognition, CVPR 2018, Salt Lake City, UT, USA, 18–22 June 2018, pp. 5571–5579 (2018). https://doi.org/10.1109/CVPR.2018.00584
4. Deng, Y., Kanervisto, A., Ling, J., Rush, A.M.: Image-to-markup generation with coarse-to-fine attention. arXiv preprint arXiv:1609.04938 (2016)
5. Deng, Y., Kanervisto, A., Rush, A.M.: What you get is what you see: a visual markup decompiler (2016)
6. Ghosh, S.K., Valveny, E., Bagdanov, A.D.: Visual attention models for scene text recognition. arXiv preprint arXiv:1706.01487 (2017)
7. He, P., Huang, W., Qiao, Y., Chen, C.L., Tang, X.: Reading scene text in deep convolutional sequences, vol. 116, no. 1, pp. 3501–3508 (2015)
8. He, T., Tian, Z., Huang, W., Shen, C., Qiao, Y., Sun, C.: An end-to-end textspotter with explicit alignment and attention. In: 2018 IEEE Conference on Computer Vision and Pattern Recognition, CVPR 2018, Salt Lake City, UT, USA, 18–22 June 2018, pp. 5020–5029 (2018). https://doi.org/10.1109/CVPR.2018.00527

9. Jaderberg, M., Simonyan, K., Vedaldi, A., Zisserman, A.: Synthetic data and artificial neural networks for natural scene text recognition. Eprint Arxiv (2014)
10. Jain, A., Tompson, J., LeCun, Y., Bregler, C.: MoDeep: a deep learning framework using motion features for human pose estimation. In: Cremers, D., Reid, I., Saito, H., Yang, M.-H. (eds.) ACCV 2014. LNCS, vol. 9004, pp. 302–315. Springer, Cham (2015). https://doi.org/10.1007/978-3-319-16808-1_21
11. Kowalski, M., Naruniec, J., Trzcinski, T.: Deep alignment network: a convolutional neural network for robust face alignment. In: IEEE Conference on Computer Vision and Pattern Recognition Workshops, pp. 2034–2043 (2017)
12. Kuhn, H.W.: The hungarian method for the assignment problem. Nav. Res. Logist. Q. **2**(1–2), 83–97 (1955)
13. Li, W., Cao, L., Zhao, D., Cui, X.: CRNN: integrating classification rules into neural network. In: International Joint Conference on Neural Networks, pp. 1–8 (2013)
14. Liu, W., Chen, C., Wong, K., Su, Z., Han, J.: STAR-NET: a spatial attention residue network for scene text recognition (2016)
15. Liu, W., Chen, C., Wong, K.K.: Char-Net: a character-aware neural network for distorted scene text recognition. In: Proceedings of the Thirty-Second AAAI Conference on Artificial Intelligence, (AAAI 2018), The 30th Innovative Applications of Artificial Intelligence (IAAI 2018), and The 8th AAAI Symposium on Educational Advances in Artificial Intelligence (EAAI 2018), New Orleans, Louisiana, USA, 2–7 February 2018, pp. 7154–7161 (2018)
16. Liu, Y., Wang, Z., Jin, H., Wassell, I.: Synthetically supervised feature learning for scene text recognition. In: Ferrari, V., Hebert, M., Sminchisescu, C., Weiss, Y. (eds.) ECCV 2018. LNCS, vol. 11209, pp. 449–465. Springer, Cham (2018). https://doi.org/10.1007/978-3-030-01228-1_27
17. Liu, Z., Li, Y., Ren, F., Goh, W.L., Yu, H.: SqueezedText: a real-time scene text recognition by binary convolutional encoder-decoder network. In: Proceedings of the Thirty-Second AAAI Conference on Artificial Intelligence, AAAI 2018, The 30th Innovative Applications of Artificial Intelligence (IAAI 2018), and The 8th AAAI Symposium on Educational Advances in Artificial Intelligence (EAAI 2018), New Orleans, Louisiana, USA, 2–7 February 2018, pp. 7194–7201 (2018)
18. Mulmuley, K., Vazirani, U.V., Vazirani, V.V.: Matching is as easy as matrix inversion. Combinatorica **7**(1), 105–113 (1987). https://doi.org/10.1007/BF02579206
19. Neumann, L., Matas, J.: Real-time scene text localization and recognition. In: 2012 IEEE Conference on Computer Vision and Pattern Recognition (CVPR), pp. 3538–3545. IEEE (2012)
20. Pan, S.J., Yang, Q., et al.: A survey on transfer learning. IEEE Trans. Knowl. Data Eng. **22**(10), 1345–1359 (2010)
21. Prasad, S., Kong, A.W.K.: Using object information for spotting text. In: Ferrari, V., Hebert, M., Sminchisescu, C., Weiss, Y. (eds.) ECCV 2018. LNCS, vol. 11220, pp. 559–576. Springer, Cham (2018). https://doi.org/10.1007/978-3-030-01270-0_33
22. Redmon, J., Farhadi, A.: YOLOv3: an incremental improvement. CoRR arXiv:abs/1804.02767 (2018)
23. Ren, S., He, K., Girshick, R., Sun, J.: Faster R-CNN: towards real-time object detection with region proposal networks. In: Advances in Neural Information Processing Systems, pp. 91–99 (2015)
24. Shi, B., Yang, M., Wang, X., Lyu, P., Yao, C., Bai, X.: ASTER: an attentional scene text recognizer with flexible rectification. IEEE Trans. Pattern Anal. Mach. Intell. 1 (2018). https://doi.org/10.1109/TPAMI.2018.2848939

25. Sun, Y., Wang, X., Tang, X.: Deep convolutional network cascade for facial point detection. In: IEEE Conference on Computer Vision and Pattern Recognition, pp. 3476–3483 (2013)
26. Tompson, J., Goroshin, R., Jain, A., Lecun, Y., Bregler, C.: Efficient object localization using convolutional networks, pp. 648–656 (2014)
27. Tompson, J., Jain, A., Lecun, Y., Bregler, C.: Joint training of a convolutional network and a graphical model for human pose estimation. Eprint Arxiv, pp. 1799–1807 (2014)
28. Toshev, A., Szegedy, C.: DeepPose: human pose estimation via deep neural networks. In: Proceedings of the IEEE Conference on Computer Vision and Pattern Recognition, pp. 1653–1660 (2014)
29. Wang, K., Babenko, B., Belongie, S.: End-to-end scene text recognition. In: 2011 IEEE International Conference on Computer Vision (ICCV), pp. 1457–1464. IEEE (2011)
30. Wang, T., Wu, D.J., Coates, A., Ng, A.Y.: End-to-end text recognition with convolutional neural networks. In: 2012 21st International Conference on Pattern Recognition (ICPR), pp. 3304–3308. IEEE (2012)
31. Wei, S.E., Ramakrishna, V., Kanade, T., Sheikh, Y.: Convolutional pose machines. In: The IEEE Conference on Computer Vision and Pattern Recognition (CVPR), June 2016
32. Wu, Y., Hassner, T., Kim, K., Medioni, G., Natarajan, P.: Facial landmark detection with tweaked convolutional neural networks. IEEE Trans. Pattern Anal. Mach. Intell. $PP(99)$, 1 (2015)
33. Xu, Y., et al.: End-to-end subtitle detection and recognition for videos in East Asian languages via CNN ensemble. Sig. Process. Image Commun. 60, 131–143 (2018)
34. Zhang, K., Zhang, Z., Li, Z., Qiao, Y.: Joint face detection and alignment using multitask cascaded convolutional networks. IEEE Sig. Process. Lett. 23(10), 1499–1503 (2016)
35. Zhang, Z., Luo, P., Loy, C.C., Tang, X.: Facial landmark detection by deep multitask learning. In: Fleet, D., Pajdla, T., Schiele, B., Tuytelaars, T. (eds.) ECCV 2014. LNCS, vol. 8694, pp. 94–108. Springer, Cham (2014). https://doi.org/10.1007/978-3-319-10599-4_7

Modified Capsule Network for Object Classification

Sheng Yi, Huimin Ma$^{(\boxtimes)}$, and Xi Li

Department of Electrical Engineering, Tsinghua University, Beijing, China
{yis18,lixi16}@mails.tsinghua.edu.cn, mhmpub@tsinghua.edu.cn

Abstract. The recognition of images in complex scenes is essential to intelligent unmanned systems. The CapsNet performs well on MNIST datasets with overlapping numbers, but it has too many parameters on real scene datasets. In this paper, we proposes three methods to reduce its excessive parameters: (1) proposing the CapsNetPr network, in which the shallow feature extraction network is introduced, to reduce the data dimension of the input capsule layer. (2) utilizing the method of decomposing the transformation matrix to reduce space consumption and time consumption. (3) sharing the transformation matrix on the same location to reduce the number of matrices in the low-level capsule layers. The study successfully reduces the number of parameters of the capsule network and accelerates training and testing at the same time, which is of great value to the promotion and use of the capsule network.

Keywords: Capsule · Parameter · CapsNetPr · Channel Sharing · Matrix Decomposition

1 Introduction

Neural networks play more and more critical roles in CV tasks, such as image classification, object detection, object localization, VQA, and object tracking. These methods performs much better than traditional methods.

CNN is one kind of main neural networks. It uses convolution layers to extract image features and solve the translation invariance problem. Therefore, CNN achieves high accuracy in most computer vision tasks, and the most efficient network architectures in the latest work are mainly based on CNN, such as ResNet [4], DenseNet [6], DeepLab [1], and SSD [12].

However, CNN has its inherent fault. Firstly, it is insensitive to change of object location in images, which is an advantage for image classification but are not suitable for other tasks such as semantic segmentation and object localization. Secondly, it is difficult for CNN to deal with complex conditions. For example, if the object changes its angle in the image in object detection tasks,

This work was supported by National Key Basic Research Program of China (No. 2016YFB0100900) and National Natural Science Foundation of China (No. 61773231).

from top or bottom, network performance will decrease. These two shortcomings reduce output accuracy and restrict the application of CNN.

Some studies have been conducted to solve these problems. To locate objects in images, R-CNN [3] takes advantage of proposal method. It adopts SelectiveSerach [17] method to propose candidate boxes. Hinton's CapsNet [15] is another attempt to address CNN's shortcomings. CapsNet uses a set of neurons rather than a single neuron as the basic unit of the layer. This basic unit, named capsule, contains a vector with its length representing the probability of the feature in the capsule. CapsNet produces more abundant output information and can solve the problem of translation invariance, because the direction of the vector in capsules can encode rich characteristic information, instead of having only location information as do traditional networks.

However, the above works are imperfect. The proposal method in R-CNN doesn't solve the problem but bypasses it. The problem of the CapsNet is that the network is slower and more cumbersome than the other networks. Using vectors instead of real numbers as the basic unit means that, as the length of the vector grows, the amount of computation and memory used will increase dramatically. Because the length of the capsule cannot be too short and will grow as the network depth increases to ensure the capacity of the capsule layer. For example, on VOC dataset or COCO dataset, the CapsNet requires dozens or even hundreds of GPU, which is intolerable. If we downsample the input image, it will cause a severe drop in network performance.

We propose a new network based on CapsNet, adopting two network parameter reduction methods, and appropriately introducing the convolution layer to achieve the goal of applying the capsule network to more complex object classification datasets. Our contributions are as follows. First, we test a variety of structures of the capsule network, comparing the depth and performance of feature extraction. Then we design a suitable network named CapsNetPr, which means the capsule network designed for parameters reduction. Second, we analyze the topic source of the number and the primary constraint conditions of parameters in capsule network and adopt the method of transformation matrix decomposition to save space and time cost with a finite precision loss. Third, based the place-coded features, we account for most of the parameters of the lower capsule layers and adopt the method of sharing transformation matrix in the same position of different channels, reducing the storage space needed for the network. We test the above methods on MNIST, CIFAR10 and Princeton CAD datasets, achieve fast and accurate results.

2   Related Work

Based on CapsNet, many works have been done. Some research focus on CapsNet on complex datasets [20], confirming that CapsNet is not performing well in real scenes images. Due to the parameter limitation, this work only tested the performance of CapsNet on the CIFAR10 dataset. [10] reduces the number of connections between capsules by changing the rules of capsule routing, and introduces a feedback connection, which has achieved good results on complex datasets. In contrast, our work changes the transformation matrix and reduces the number of parameters required without changing the routing rules. [13] adopt a complex feature extraction network to outperform the CNN network in specific indicators, while our CapsNetPr uses a simple feature extraction network to have better performance than CNN, with the same complexity as CNN. [5] optimizes the format of the capsule and the routing algorithm to improve network performance, while not addressing the problem of excessive parameters. Unlike that, our work optimizes the transformation matrix and network structure to reduce the number of parameters. [13,21] use a weight sharing method similar to our work, but the method we use is more productive and more efficient.

Other works, [2,7–9,11,14,18,19,21] apply capsule network to various practical problems, replacing the traditional convolutional network with the capsule network, effectively improving the performance on these specific tasks, but there is little or no optimization of the capsule layer itself, which is different from our work.

3   Method

We figure out why the parameter size of the capsule network is much larger than that of the CNN network. By comparing the two network, we find that the parameters of the capsule network are mainly concentrated on the transformation matrix between capsule layers. Since the unit of the capsule network change from scalars to vector, the transformation matrix dimension between the two capsule layers needs to rise by one dimension. Therefore, the total amount of data of the transformation matrix will increase by a square of the length of the capsule. Our main work is to design new network structure and modify the capsule layer for this part of the parameters so that the parameter amount can be significantly reduced under the condition that the performance of the capsule network is nearly not reduced.

3.1   CapsNetPr

CapsNet [15] is a three-layer simple network, with one convolution layer and two capsule layers. This network is not suitable for VOC, COCO and other complex datasets, because the parameters are too large. We need to design a new capsule network model with acceptable parameters, and the new network model performs better on complex datasets than the CNN with the same complexity.

We test the most straightforward way: design a classification network, extract features by convolution and pooling operations, and then connect to the capsule layers to process the extracted features to predict the classification results. Experiments show that the modified network has neither increased nor decreased in performance compared to the original network. Because the convolution and pooling operations lose low-level semantic information while extracting high-level semantic information, the former is precisely what the capsule layer needs.

Experimental details: we use the resnet20 model and replaced its fully connected layers with the capsule layers. Experiments are implemented on the cifar10 dataset, and the results are in Table 1. After the above experiments and discussions, we find that to achieve the goal; we need to modify both the network structure and the capsule layer.

Table 1. The test result of the feature extraction module and the capsule layer. ResNet (we use) is the resnet20 network we implemented. ResNet+capsule is the result of resnet20 removing the last fully connected layers and replacing with the capsule layers. ResNet (in paper [4]) is the result of ResNet's original paper.

Network	ResNet (we use)	ResNet+capsule	ResNet (in paper [4])
Accuracy (%)	90.02	90.03	91.25

The network structure we designed is shown in Fig. 1. First, there are two convolutional layers, and a pooling layer follows each one. Behind these are three capsule layers. Each of the capsule layers has been modified by Matrix Decomposition and Channel Sharing methods in this paper, and the parameter amount is significantly reduced.

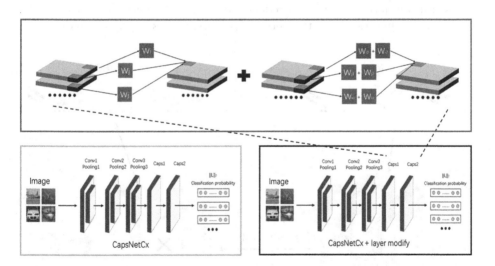

Fig. 1. The above is our work: the CapsNetPr network and modified capsule layer. In the yellow box is the CapsNetPr network structure we designed, using three layers of convolution layers, and then two capsule layers. In the blue box is the optimization methods of capsule layer, and details are given in Sects. 3.2 and 3.3. The CapsNetPr and modified capsule layer are combined in black box. (Color figure online)

3.2   Matrix Decomposition

The decomposition transformation matrix method reduces the number of parameters by breaking a large matrix into two small matrices. An N by M matrix can be broken down into a N by 1 matrix and a 1 by M matrix. According to this method, the size of a transformation matrix can be significantly reduced from $N * M$ to $N + M$. Moreover, in the operation of matrix multiplication, this method also reduces the amount of computation of the same ratio. However, this method has a severe prerequisite: if the matrix split is entirely equivalent, the column rank of the split matrix must equal to 1. This prerequisite is difficult to meet for the transformation matrix.

A sparse matrix can also carry out the matrix splitting, and difference before and after decomposition are acceptable. So, we train a capsule network on the MNIST dataset and check the transformation matrix to see if it satisfies sparsity. We directly observe the characteristics of the matrix and find that it is approximately sparse. The performance of the network after using the matrix decomposition method also proves that it is feasible to adopt this method for the transformation matrix.

Implementation details: as Fig. 2, we perform matrix decomposition on a matrix that composed of a specific two-dimensional structure in the transformation matrix, and the capsules of the two layers before and after being passed through the two-dimensional matrix for feature transfer. Matrix decomposition of this transformation matrix can reduce the number of parameters, reduce the amount of computation, and the network performance is nearly not reduced.

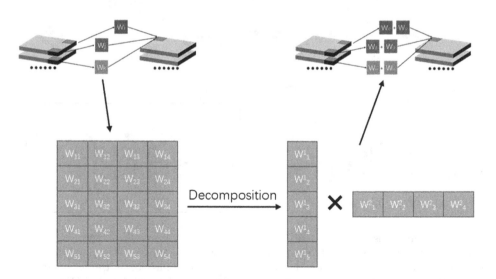

Fig. 2. Example of Matrix Decomposition. Decompose the transformation matrix between two capsules into the product of two vectors.

3.3   Channel Sharing

Channel sharing method reduces the number of parameters required by using multiple channels to reuse the same transformation matrix. In the capsule layer, the place-coded feature is that the position of the lower capsule in the overall capsule matrix can represent some feature of the input image, which may be the location, color, texture, and so on. Because this feature is related to the position of capsule, it is called place-coded. Given that place-code is encoded by location, it is assumed that the capsule at the same location in different channels in the lower layers of the network should have similar encoding characteristics. The transition matrix between the capsule with a similar encoding feature and the same capsule below should also be similar, and the source of this similarity has considerable credibility. Since these transformation matrices are similar, can similar transformation matrices be combined and replaced by the same transformation matrix?

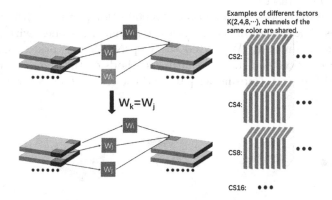

Fig. 3. Example of Channel Sharing. The rectangles of different colors represent different transformation matrices. In the original capsule network, any two transformation matrices are different, as shown in the upper left of the figure. The channel sharing method is to share the transformation matrix between the capsule at the same position of different channels in the same layer and a capsule of the next layer, as shown in the bottom left of the figure. All channels in a layer can be shared, or sharing parameters of adjacent K channels, and K must be a factor of the number of channels, as shown on the right: each row is an example of a different parameter sharing factor(CS2, CS4,...). (Color figure online)

As shown in the Fig. 3, the transformation matrix between the original capsule in the same position of multiple channels and the same capsule in the next layer is different, but we can replace these transformation matrixes with the same one. The parameters of new capsule layers will be less than that of original capsule layers, reduced by the same number of channels sharing the same transformation matrix. The optimal layer to use channel sharing method is the underlying capsule layer, which is easy to understand: the lower the capsule layer, the more obvious its place-coded feature. Meanwhile, the underlying capsule layer has the most parameters, and modifying them is most significant for network optimization.

4   Experiments

4.1   Datasets

We evaluate the proposed method on three different datasets. The datasets include: MNIST dataset, CIFAR10 dataset and Princeton CAD dataset.

The MNIST dataset contains 70,000 handwritten digital pictures, 10 categories, and every picture is grayscale with size 28 * 28. Since the images in the MNIST dataset are small, have few categories, and are pure in content, high accuracy can be achieved without using the CapsNetPr network, so we only tested the effects of transformation matrix decomposition and channel sharing methods on this dataset.

The CIFAR10 dataset is made up of 60,000 color pictures with size 32 * 32, ten categories. The images in the dataset are all natural scenes, with animals or vehicles as the main body. We tested the CapsNetPr network, transformation matrix decomposition, and channel sharing methods on this dataset. In contrast, we also tested CapsNet, CapsuleNet [20], and the effect of a CNN with the same number of layers.

Fig. 4. Examples of CAD dataset, projected from CAD modeling of 20 common objects. Each type of object has multiple CAD models, and each model projects pictures with a size of 224 * 224 in seven directions.

The Princeton CAD dataset is from a public CAD library of Princeton University. We project each CAD model in 7 directions, and the resulting image is saved in 224 * 224 size, 20 categories, totaling 25,000 pictures. Some examples of the Princeton CAD dataset are in Fig. 4. We tested the CapsNetPr network, transformation matrix decomposition, and channel sharing methods on this dataset. For comparison, we examined the performance of VGG16 and CNN on this dataset. CapsNet cannot be used in this dataset because of parameter limitation.

4.2 Results

The MNIST dataset is a common dataset. The results are in Table 2. The results show that the MD method is better than the CS method with the same parameter reduction ratio. The parameter size of the MD method on CapsNet is reduced to 0.1875 times, and CS2 is 0.5 times, while the accuracy of the two is similar.

CIFAR10 dataset has ten classes, with 32 * 32 natural color pictures. Compared to MNIST dataset, CIFAR10 are more complex. In this dataset, we examine the results of a CNN and our CapsNetPr. Our network has the highest accuracy. As with the results on the MNIST dataset, the MD method has less loss of accuracy than the CS method. There is also a surprising discovery that the results of CS4 are higher than CS2.

The Princeton CAD dataset has 20 classes, with 224 * 224 grey images. The images in this dataset are too large for the CapsNet, we only test CNN, VGG 16 and CapsNetPr on this dataset. CaspNetCx has higher results than the unpretrained VGG16 network, which proves that CaspNetCx does have excellent performance. The Performance of the CS method and the MD method are similar to those on other datasets (Tables 3 and 4).

Table 2. MNIST dataset results. Classification accuracy (%). Trained 30 epochs. CS means Channel Sharing, MD means Matrix Decomposition.

Network	Original	CS2	CS4	CS8	CS16	CS32	MD	MD + CS32
CapsNet [15]	99.23	99.13	98.76	–	–	–	99.12	98.67

Table 3. CIFAR10 dataset results. Classification accuracy (%). Trained 30 epochs. CS means Channel Sharing, MD means Matrix Decomposition.

Network	Original	CS2	CS4	CS8	CS16	CS32	MD	MD + CS32
CapsuleNet [20]	71.50	–	–	–	–	–	–	–
CapsNet [15]	68.15	–	–	–	–	–	–	–
CNN (3Conv+2FC)	75.09	–	–	–	–	–	–	–
CapsNetPr (ours)	**78.78**	**78.21**	**78.65**	**78.13**	**77.44**	**77.34**	**78.68**	**76.99**

Table 4. Princeton CAD dataset results. VGG16 (pretrained) has been pretrained on IMAGENET. Classification accuracy (%). Trained 30 epochs. CS means Channel Sharing, MD means Matrix Decomposition.

Network	Original	CS2	CS4	CS8	CS16	CS32	MD	MD + CS32
CNN (3Conv+2FC)	64.35	–	–	–	–	–	–	–
CapsNetPr (ours)	**89.69**	**89.13**	**89.92**	**89.72**	**89.04**	**87.33**	**89.19**	**85.21**
VGG16 [16]	87.21	–	–	–	–	–	–	–
VGG16 (pretrained)	92.04	–	–	–	–	–	–	–

5   Discussion

Our CapsNetPr network model, MD and CS methods have achieved significant results across different datasets. Especially on the Princeton CAD dataset, our network performs better than VGG16 network, which is much more complicated than CapsNetPr. The reason why CapsNetPr has a good performance on the Princeton CAD dataset is related to the characteristics of the dataset and the capsule layer. This dataset is composed of images projected from objects at different angles and contains a large number of unfamiliar perspectives. Other datasets come from real photos, following a few fixed perspectives. Therefore, in the Princeton CAD dataset, the high-level information obtained by feature extraction layers may be entirely different for images of the same category. The capsule network can keep the underlying information in the capsule, with better rotation invariance and excellent performance on this dataset. This result proves that our optimized capsule network can perform excellently under complex conditions.

Observing the relationship between the CS method's reduced parameter ratio and the final accuracy, we can find two characteristics. First, as the CS method's parameter reduction ratio increases, the network results have different changes in two stages. In the first stage, each time the CS ratio is doubled (for example, from CS2 to CS4, CS4 to CS8), the accuracy only changes little, less than 0.3%. In the second stage, after crossing a certain CS ratio (CS8 for CIFAR10, CS16 for CAD), the accuracy of the network drops sharply. From this feature, we can infer the relationship between the number of network parameters and the accuracy of the network: The network parameter quantity has a threshold. When the parameter is lower than the threshold, increasing the network parameter quantity will significantly improve the result, but the effect is not significant when the parameter quantity is above the threshold. This threshold is the appropriate target for our task. Second, when the CS ratio is 4, the network result is better than when CS is 2. With fewer parameters, the network results have improved, which proves that sharing some of the parameters helps the network find more general feature links on the picture.

References

1. Chen, L.C., Papandreou, G., Kokkinos, I., Murphy, K., Yuille, A.L.: DeepLab: semantic image segmentation with deep convolutional nets, atrous convolution, and fully connected CRFs. IEEE Trans. Pattern Anal. Mach. Intell. **40**(4), 834–848 (2018)
2. Deng, F., Pu, S., Chen, X., Shi, Y., Yuan, T., Pu, S.: Hyperspectral image classification with capsule network using limited training samples. Sensors **18**(9), 3153 (2018)
3. Girshick, R., Donahue, J., Darrell, T., Malik, J.: Rich feature hierarchies for accurate object detection and semantic segmentation. In: Proceedings of the IEEE Conference on Computer Vision and Pattern Recognition, pp. 580–587 (2014)
4. He, K., Zhang, X., Ren, S., Sun, J.: Deep residual learning for image recognition. In: Proceedings of the IEEE Conference on Computer Vision and Pattern Recognition, pp. 770–778 (2016)
5. Hinton, G.E., Sabour, S., Frosst, N.: Matrix capsules with EM routing (2018)
6. Huang, G., Liu, Z., Van Der Maaten, L., Weinberger, K.Q.: Densely connected convolutional networks. In: 2017 IEEE Conference on Computer Vision and Pattern Recognition (CVPR), pp. 2261–2269. IEEE (2017)
7. Islam, K.A., Pérez, D., Hill, V., Schaeffer, B., Zimmerman, R., Li, J.: Seagrass detection in coastal water through deep capsule networks. In: Lai, J.-H., et al. (eds.) PRCV 2018. LNCS, vol. 11257, pp. 320–331. Springer, Cham (2018). https://doi.org/10.1007/978-3-030-03335-4_28
8. Jaiswal, A., AbdAlmageed, W., Wu, Y., Natarajan, P.: CapsuleGAN: generative adversarial capsule network. In: Leal-Taixé, L., Roth, S. (eds.) ECCV 2018. LNCS, vol. 11131, pp. 526–535. Springer, Cham (2019). https://doi.org/10.1007/978-3-030-11015-4_38
9. Kruthika, K., Maheshappa, H., Alzheimer's Disease Neuroimaging Initiative, et al.: CBIR system using capsule networks and 3D CNN for Alzheimer's disease diagnosis. Inform. Med. Unlocked **14**, 59–68 (2019)
10. Li, H., Guo, X., Dai, B., Ouyang, W., Wang, X.: Neural network encapsulation. In: Ferrari, V., Hebert, M., Sminchisescu, C., Weiss, Y. (eds.) ECCV 2018. LNCS, vol. 11215, pp. 266–282. Springer, Cham (2018). https://doi.org/10.1007/978-3-030-01252-6_16
11. Li, Y., et al.: The recognition of rice images by UAV based on capsule network. Cluster Comput. 1–10 (2018)
12. Liu, W., et al.: SSD: single shot multibox detector. In: Leibe, B., Matas, J., Sebe, N., Welling, M. (eds.) ECCV 2016. LNCS, vol. 9905, pp. 21–37. Springer, Cham (2016). https://doi.org/10.1007/978-3-319-46448-0_2
13. Nair, P., Doshi, R., Keselj, S.: Pushing the limits of capsule networks. Technical note (2018)
14. Ramasinghe, S., Athuraliya, C.D., Khan, S.H.: A context-aware capsule network for multi-label classification. In: Leal-Taixé, L., Roth, S. (eds.) ECCV 2018. LNCS, vol. 11131, pp. 546–554. Springer, Cham (2019). https://doi.org/10.1007/978-3-030-11015-4_40
15. Sabour, S., Frosst, N., Hinton, G.E.: Dynamic routing between capsules. In: Guyon, I., et al. (eds.) Advances in Neural Information Processing Systems, vol. 30, pp. 3856–3866. Curran Associates, Inc. (2017). http://papers.nips.cc/paper/6975-dynamic-routing-between-capsules.pdf

16. Simonyan, K., Zisserman, A.: Very deep convolutional networks for large-scale image recognition. arXiv preprint arXiv:1409.1556 (2014)
17. Uijlings, J.R., Van De Sande, K.E., Gevers, T., Smeulders, A.W.: Selective search for object recognition. Int. J. Comput. Vis. **104**(2), 154–171 (2013)
18. Wang, Q., Qiu, J., Zhou, Y., Ruan, T., Gao, D., Gao, J.: Automatic severity classification of coronary artery disease via recurrent capsule network. In: 2018 IEEE International Conference on Bioinformatics and Biomedicine (BIBM), pp. 1587–1594. IEEE (2018)
19. Wang, S., Liu, G., Li, Z., Xuan, S., Yan, C., Jiang, C.: Credit card fraud detection using capsule network. In: 2018 IEEE International Conference on Systems, Man, and Cybernetics (SMC), pp. 3679–3684. IEEE (2018)
20. Xi, E., Bing, S., Jin, Y.: Capsule network performance on complex data. arXiv preprint arXiv:1712.03480 (2017)
21. Zhu, K., Chen, Y., Ghamisi, P., Jia, X., Benediktsson, J.A.: Deep convolutional capsule network for hyperspectral image spectral and spectral-spatial classification. Remote Sens. **11**(3), 223 (2019)

Insulator Segmentation Based on Community Detection and Hybrid Feature

Yuanpeng Tan[1], Chunyu Deng[1], Aixue Jiang[2],
and Zhenbing Zhao[2(✉)]

[1] Artificial Intelligence Application Department, China Electric Power Research Institute, Beijing 100761, China
[2] North China Electric Power University, Baoding 071003, China
zhaozhenbing@ncepu.edu.cn

Abstract. Image segmentation is an important prerequisite for automatic detection of insulator's surface defects. It is difficult to remove the interference by using traditional methods to segment insulator from aerial images with wires adhering to insulators, pole tower in a large proportion and complex background. To solve these problems, this paper proposes a segmentation method with complex network community detection and hybrid feature for aerial insulator image segmentation. We implement community segmentation for achieving a higher accuracy by using the similarity between pixels. In this method, the image is segmented into super-pixels, then the features of color and texture are calculated to get hybrid feature. Next, we set up a super-pixel network by calculating Gauss similarity of hybrid feature information. Finally, we use the complex network community detection method to extract the insulator. The experiment results demonstrate the presented method is robust, efficient and accurate.

Keywords: Insulator · Community detection · Hybrid feature · Super-pixel · Complex network · Aerial image

1 Introduction

Long-term variable weather and mechanical fatigue affect the state of insulators on the transmission lines. As a result, insulators would be cracked or defected on their surface, which affects the safety of transmission line [1]. Manual detection not only wastes time,

This paper is supported by the project "Research on Deep Detection and Defect Recognition Technology for Electric Power Equipment Facing Transmission Line Patrol Inspection of Unmanned Aerial Vehicle" under grant number AI83-19-006 in China Electric Power Research Institute and "Research and Application of Operation Robot System in Electric Power Industry" under grant number 2018YFB1307400 in National Key R&D Program of China and the National Natural Science Foundation of China (NSFC) under grant number 61871182, 61401154, 61773160, 61302163, by Beijing Natural Science Foundation under grant number 4192055, by the Natural Science Foundation of Hebei Province of China under grant number F2016502101, F2017502016, F2015502062, by the Fundamental Research Funds for the Central Universities under grant number 2018MS095, 2018MS094, and by the Open Project Program of the National Laboratory of Pattern Recognition (NLPR) under grant number 201900051.

Y. Zhao et al. (Eds.): ICIG 2019, LNCS 11901, pp. 267–283, 2019.
https://doi.org/10.1007/978-3-030-34120-6_22

but also is dangerous. Currently, aerial patrol, as an efficient way of transmission lines monitoring, has been put into a practical application. Extracting the insulator targets accurately from images is the essential premise for detection of surface defects by analyzing large size aerial video images.

Threshold segmentation methods can extract the target based on its gray component [2]. However, inordinate amount of edge details is lost. For images with complex background, they can hardly do anything to segment target. Ma and An [3] adopt region location method to extract insulator targets from images. Combining OTSU method, it locates blue insulators by using images' chrominance and saturation. It is the problem that the method just considers gray feature and ignores spatial features of pixels. It's hard to get ideal segmentation results of aerial images, which are complex, noisy and low resolution. Iruansi [4] and Wu [5] all use active contour model for insulator segmentation. Although the results of their methods are better than others, manual selection of initial contour would increase operations of user interaction. Their method is only suitable for images with simple background and large strings of insulators. Moreover, their time complexity is high due to the multiple iterations. Literature [6] puts forward a kind of composite insulator images segmentation method based on improved color differences. An image segmentation method for insulator based on ant colony classification and weighted variable fuzzy c-means (FCM) is proposed in this study to overcome the limitations of the traditional method for insulator in transmission line [7].

The above methods get a better result for images with simple background and grayscale difference between insulators and other targets. An additional worry is that minor local changes in the complex image target (such as, partial occlusion) maybe lead to image segmentation or edge extraction failure. The shortcomings contribute to low efficiency and weak robustness of segmentation. In addition, their level of automation needs to be improved. Complex network community detection is based on similarity of every node and link strength of every edge to partition image network. In the paper of Girvan and Newman [8], they proposed an algorithm about complex network community structure detection, namely famous GN algorithm.

Literature [9] combines community detection with minimum spanning trees to implement image segmentation processing. However the method only segments some simple background images and can't segment images with target adhesion. Amiri and Abin [10] select single feature as the input of community detection. However the image can only be segmented into some monochrome targets by using the method. It can't extract target as an entirety and different parts of the target are different in color.

Community detection method is sensitive to the entirety of target. It uses community to express target. So it should to be possible to extract insulator as an entire community from aerial image with target adhesion for the uniqueness and integrity of insulators. To segment the images with complex background, we use hybrid feature as the input of community detection, and it combines the color feature and texture feature by setting an appropriate weight. The hybrid feature makes community detection appropriate for most aerial images.

In order to solve the problems about extracting insulator target from aerial images with wires adhering to insulators, a large proportion of pole tower and complex background, we introduce the complex network community detection into the insulator image segmentation and propose a segmentation method for aerial insulator image based on community detection and hybrid feature. The method not only removes false targets that are similar to insulator effectively, but also improves segmentation efficiency.

The rest of the paper is organized as follows. Section 2 describes the specific implementation processes and procedures of the proposed method; Sect. 3 mainly shows results of our experiments and their analysis and comparison; Sect. 4 is conclusion of the proposed method.

2   Segmentation Based on Complex Networks Community Detection

In this paper, we propose an insulator segmentation method based on community detection and hybrid feature. By observing a large number of aerial images, there are many significant characteristics, such as complex background, similar feature, more false targets and so on. Overview of this method is shown in Fig. 1.

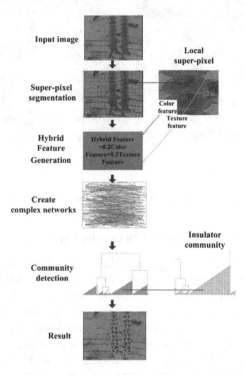

Fig. 1. Overview of the aerial insulator segmentation method based on complex network community and hybrid feature.

2.1 Super-Pixel Segmentation

Complex network community detection has two problems, which are low accuracy of community segmentation and time-consuming. If we consider pixels as nodes to detect communities, the accuracy of the results will be affected, and the amount of calculation will be greatly increased. At present, super-pixel segmentation [11] is a method of image pre-processing, which can keep more information of image and reduce the computational burden of image post-processing. There are many adjacent pixels with similar features of targets and background in aerial insulator image. It clusters similar pixels together to form some large pixels that will be regarded as nodes of image network, which improves accuracy of community detection segmentation and reduces time complexity.

The super-pixel method used in this paper is Turbopixel (TP) [12]. It can directly control the number of super-pixel and has a fast computation speed. Segmentation result of super-pixel is shown in Fig. 2.

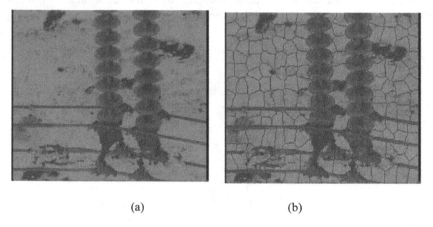

(a)   (b)

Fig. 2. Super-pixel segmentation result. (a) The aerial image; (b) Result of super-pixel for (a).

2.2 Hybrid Feature Generation and Networks Creation

At present, most aerial insulator images are still processed with single feature. If using separate feature entirely, we can't remove wires, pole towers (that have similar texture features to insulator), which affect segmentation accuracy greatly.

Insulator images are complex and have many false targets. As shown in the right picture of Fig. 3, every super-pixel has regular shape and contains many characteristics, such as color, texture and so on. When the image is segmented only by using color feature or texture feature, we can't get ideal results that can be accepted. To solve the problem, we select hybrid feature combining color feature and texture feature with a certain weight value ratio. Segmenting image by using the hybrid feature can improve the quality of results. However, different images have different weight value to get ideal segmentation results. Through making a lot of experiments, we come to a conclusion that texture feature is more important than color feature in aerial insulator images. And we use different weight values to set experiments. In the aerial insulator experiments, using 2:5 weight value ratios is better than others between color feature and texture feature to get hybrid feature.

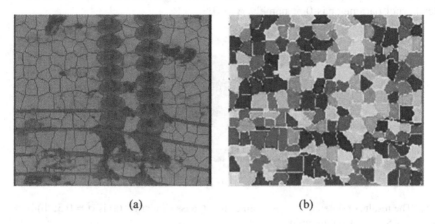

(a)   (b)

Fig. 3. Distribution of super-pixel segmentation. (a) Result of super-pixel; (b) Image patch distribution. (Color figure online)

According to the hybrid feature, we consider every super-pixel as node to calculate the Gauss similarity by Eq. (1).

$$S_{ij} = \begin{cases} \exp\left(-\frac{1}{2\sigma^2}\|x_i - x_j\|^2\right), & i \neq j \\ 0, & i = j \end{cases} \tag{1}$$

Where, S_{ij} is an element of the network' adjacency matrix that means the similarity between node i and node j.

The calculated results are be defined a path between nodes. So we use the paths and nodes to create the image network (Fig. 4). Every part of the complex network has its

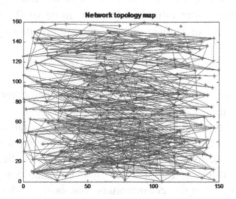

Fig. 4. Image network map.

own characteristics. For example, the links of target region are concentrated in the network. We can extract it separately by the characteristics of insulator.

As shown in Fig. 5, the surrounding part of the green line is the extracted insulator part. We make an experiment to set different kernel parameter of Gauss similarity. The better kernel parameter is 0.5, namely σ = 0.5.

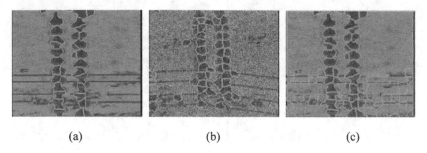

| (a) | (b) | (c) |

Fig. 5. The results of different kernel parameter of Gauss similarity. (a) is σ = 0.5; (b) is σ = 1; (c) is σ = 1.5. (Color figure online)

2.3   Complex Networks Community Detection

Complex networks community detection is based on modularity [13]. The method uses Eq. (2) to calculate modularity Q between features of each node in the network.

$$Q = \sum_i \left(e_{ii} - a_i^2 \right) \tag{2}$$

The value of Q is the standard to judge the quality of community. Greater value of Q can get better community. In Eq. (2), e_{ij} is the value of edge between community i and community j, a_i is the value of edge between nodes in one community i.

The value of Q only judge quality of separate community. According to the fast algorithm proposed by Jordi and Alex [14], we use Eq. (3) to calculate the value of relative modularity ΔQ and cluster communities with the maximum ΔQ to get a new community. Because the greater value of ΔQ is, the more similar the two communities will be. Repeat the process until there is only one big community finally.

$$\Delta Q = e_{ij} + e_{ji} - 2a_i a_j = 2 \left(e_{ij} - a_i a_j \right) \tag{3}$$

Where, e_{ij} is the value of edge between community i and community j, and a_i is the value of edge between nodes in one community.

As shown in the right picture of Fig. 6, after n times' iteration, the image network will be represented as an entirety. Each iteration will generate a new community, and the new community will be used for the next iteration. Because the similarities of every two communities are different, the sizes of final communities are different. In the picture, we can find 7 bigger communities obviously, which include insulator community.

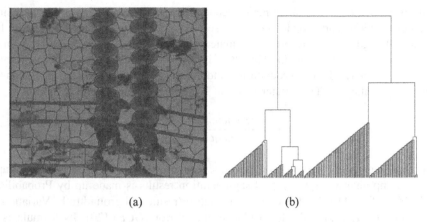

(a)   (b)

Fig. 6. Super-pixel segmentation. (a) Super-pixel segmentation result of 7 times iteration; (b) n times' iteration of network.

In every community, internal nodes are linked closely in each separated small community, but external nodes are linked sparsely.

As shown in Fig. 7, all of the insulators' super-pixels are expressed in an entire separated community, which don't contain the wires that adhering to insulator. We still need to do more experiments to find ideal parameters to remove the small inference in the aerial image.

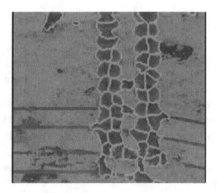

Fig. 7. The result of insulator segmentation.

3 Experimental Results Analysis

In order to verify the efficiency and robustness of the method, we design three groups of experiment images, which are with simple background (Sect. 3.1), complex background (Sect. 3.2) and with different noise (Sect. 3.3). The images of experiment 1 are downloaded from Internet, the images of experiment 2 are aerial insulator images and the images of experiment 3 are collected from the aerial video of transmission line. The aerial insulator images contain complex scenes such as pole tower, insulator adhesion, a number of wires, and ground scene whose color feature is similar to insulator. The accuracy of segmentation results is evaluated by the value of intersection-over-union (IOU) and comprehensive quality [15]. IOU is defined as the ratio between intersection of segmentation result and groundtruth and their union. (Groundtruth is gotten by manual segmentation.) The greater value of IOU is, the better results would be.

$$IOU = \frac{segment \cap groudtruth}{segment \cup groudtruth} \tag{4}$$

Where, *segment* is the result of experiment and *groundtruth* is the standard graph.

The comprehensive quality Y of segmentation results is made up by Probabilistic Rand Index (*PRI*) [16–18] between segmentation result and groundtruth, Variation of Information (*VoI*) [19] and Global Consistency Error (*GCE*) [20]. Its formula is as following:

$$Y = \frac{1}{3} * \left(PRI + \frac{1}{VoI} + GCE \right) \tag{5}$$

As shown in Eq. (5), smaller *VoI* means better results and greater Y can get better performance.

The parameters of the experiment: the number of super-pixel is 200, the kernel parameter of Gauss similarity is 0.5 and the hybrid feature weight ratios is 2:5.

The presented method is compared with the 5 methods:

1. OTSU;
2. Segmentation based on color feature;
3. Segmentation based on texture feature;
4. Method about extracting insulator with active contour model [7];
5. Segmentation based on community detection and minimum spanning trees [9];

Method 4 uses a new active contour model to extract in homogeneous insulators from aerial images for overcoming the difficulties caused by texture inhomogeneity. Method 5 is community detection method by using two rounds of minimum spanning trees.

According to the thought of the literatures [7, 9] respectively, we implement method 4 and method 5.

3.1 Segmentation Experiments About Insulators with Different Material and Shape

As shown in Fig. 8(a), experimental samples are insulators with different material and shape. The segmentation results of the 6 methods are shown in Fig. 8(b) to (g).

As shown in Fig. 8, aiming at insulator images with simple background, method 1 can't remove wires and other false targets; method 2 can only remove the false targets with different color to insulator, and it is not applicable to other false targets; method 3 just removes the targets with different texture feature to insulator; method 4 can get

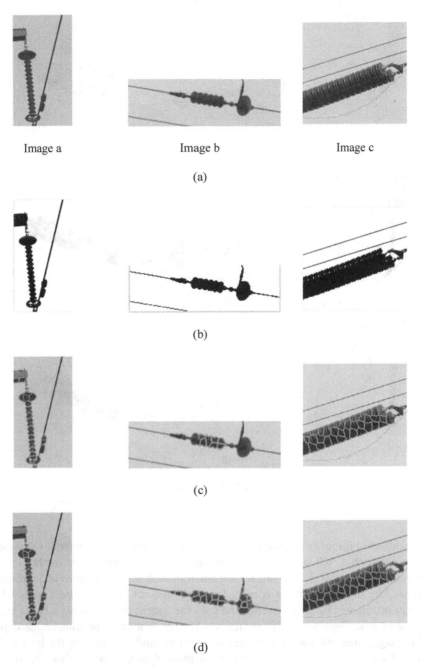

Image a   Image b   Image c

(a)

(b)

(c)

(d)

Fig. 8. The results of 6 segmentation methods. (a) Original image; (b) The results of method 1; (c) The results of method 2; (d) The results of method 3; (e) The results of method 4; (f) The results of method 5; (g) The results of the presented method. (Color figure online)

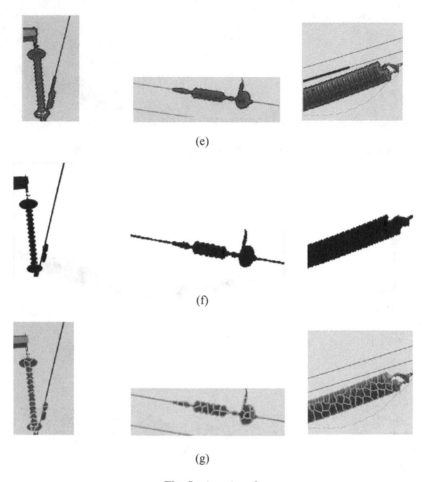

(e)

(f)

(g)

Fig. 8. (*continued*)

good results, but it mostly relies on the amount of iteration and the manual selection of initial contour that will undoubtedly increase the user's interaction; method 5 gets better results than other 4 methods, it uses undecimated wavelet feature and watershed algorithm, the method' runtime is more long and can not remove false targets which adhere to insulator. The presented method uses the hybrid feature to get a better result.

As shown in Figs. 9 and 10, insulator's texture feature is prominent for experimental images, method 3 is close to the presented method. But putting the information in Figs. 9 and 10 all together, the presented method is still better than the other 5 methods.

The presented method can get the best results. It generates hybrid feature by combining color feature with texture feature and uses relative modularity based detect communities to increase segmentation accuracy. Most importantly, it can remove false targets and background efficiently.

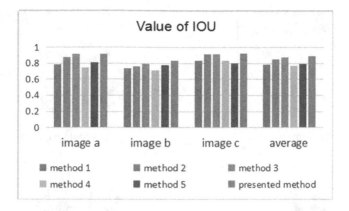

Fig. 9. The comparison of the 6 segmentation methods' IOU values.

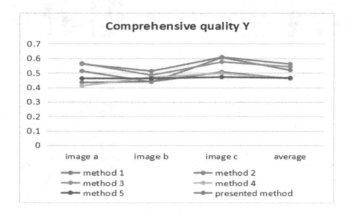

Fig. 10. The comparison of 6 segmentation methods' Y values.

3.2 Segmentation Experiments About Aerial Insulator Images

We use the 6 methods to segment aerial insulator images with different background. The results of experiments about aerial insulator images are shown in Fig. 11.

As shown in Fig. 11, the presented method can get the best segmentation results in these 6 methods, especially in aerial image (b).

In these methods, method 2 and method 3 only use single feature (color feature in method 2 and texture feature in method 3) to implement segmentation. But the super-pixels contain many characteristics, we can't get ideal results when we just use single feature. Because method 2 and method 3 all use super-pixel method, their results are better than method 1, method 4 and method 5. Method 1 just removes small part of background, because the method can only get better segmentation results for the images with single peak between-class variance. For the images, method 4 and method 5 can remove most false targets. Method 4 can't remove these wires due to the manual selection of initial contour, and it is a difficult job to select a contour that doesn't

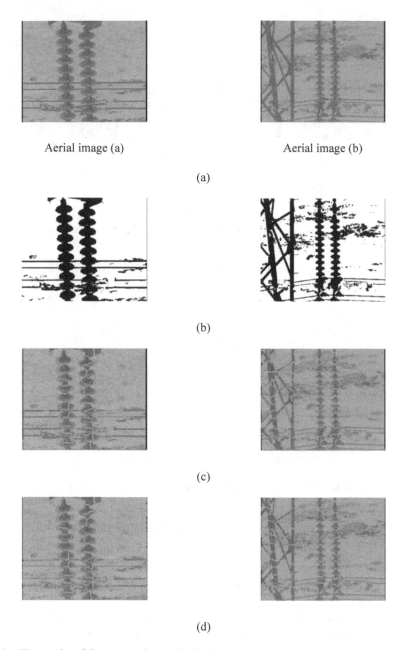

Aerial image (a)   Aerial image (b)

(a)

(b)

(c)

(d)

Fig. 11. The results of 6 segmentation methods. (a) Original image; (b) The results of method 1; (c) The results of method 2; (d) The results of method 3; (e) The results of method 4; (f) The results of method 5; (g) The results of the presented method. (Color figure online)

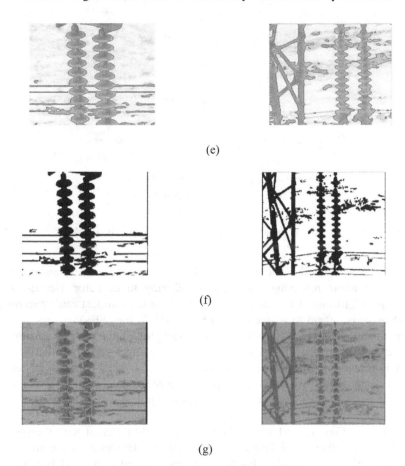

(e)

(f)

(g)

Fig. 11. (*continued*)

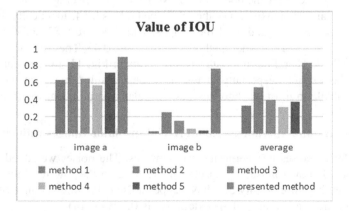

Fig. 12. The comparison of 6 segmentation methods IOU values.

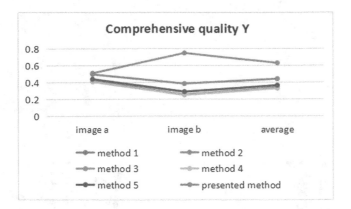

Fig. 13. The comparison of 6 segmentation methods Y values.

contain wires. For method 5, the undecimated wavelet feature and watershed method lead to failure about removing false targets adhering to insulator. Because of the disadvantages of method 4 and method 5, the two methods almost can't remove pole tower when we use them to segment aerial image (b) of Fig. 11(a).

Even if the processed images have complex background, the presented method can get the best results. It uses hybrid feature to calculate super-pixels' similarity, which can find a little different between super-pixels. Then it uses community detection method to cluster super-pixels. Finally, super-pixels of insulators can be contained in a separate community. Most importantly, it can remove pole tower and wires adhering to insulators.

As shown in Figs. 12 and 13, method 2 has a little different with the presented method for segmenting aerial image (a) of Fig. 11(a). However, when we segment aerial image (b) of Fig. 11(a), advantages of the presented method become more obvious. The values of IOU and Y are all better than other 5 methods. Most importantly, the presented method can remove pole tower entirely and reserve all insulators.

As shown in Figs. 9 and 10, the average IOU of the proposed method is 0.89 and the maximum of the other methods is 0.87. Moreover, the average Y of the proposed method is 0.56 and the maximum of the other methods is 0.54. Identically, in Figs. 11 and 12, the average IOU and Y of the proposed method are 0.83 and 0.62, but the maximum IOU and Y of the other methods are 0.55 and 0.43. They are far less than the proposed method. Therefore, no matter in Fig. 8, or in Fig. 11, the presented method can get the best results, especially in Fig. 11. These results show that the presented method is suitable to most insulator images, especially aerial images with pole tower.

3.3 Segmentation Experiments About Aerial Insulator Images with Noise

We add different noises into aerial insulator images. The noises we added are salt & pepper noise, Gaussian noise and speckle noise. Due to the low resolution of aerial images, we set the noise density of salt & pepper noise is 0.1, the variance of Gaussian noise is 0.01 and the variance of speckle noise is 0.1 (Fig. 14).

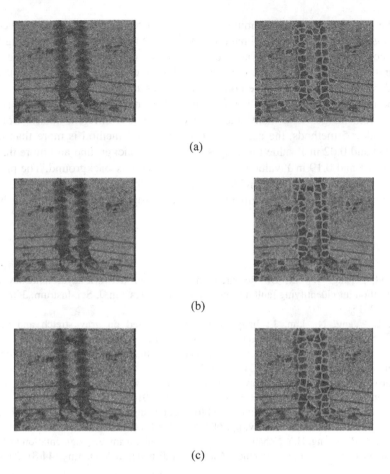

(a)

(b)

(c)

Fig. 14. The segmentation results of different noise images processed by the proposed method. (a) Image with salt & pepper noise and Segmentation result; (b) Image with Gaussian noise and Segmentation result; (c) Image with speckle noise and Segmentation result.

The segmentation results of different noise images processed by the proposed method. (a) Image with salt & pepper noise and Segmentation result; (b) Image with Gaussian noise and Segmentation result; (c) Image with speckle noise and Segmentation result.

4   Conclusion

In this paper, we introduce the complex network community detection method into the insulator image segmentation. Based on complex network community and hybrid feature, we propose a segmentation method for aerial insulator image. It allows us to extract insulator as an entirety community from images with complex background. On the one hand, the TP super-pixel reduces the amount of the calculation. On the other hand, the complex network community method reduces the interference between different targets. And the insulator image is set up an entire community by using hybrid feature combined color feature and texture feature of each pixel. As long as the community is found, we can get all insulator information. Equally, other targets can be got, and we only need to find the corresponding communities.

Experimental results show that the presented method can efficiently segment insulator from images with many interference targets and different noises. Compared with the other 5 methods, the accuracy of the proposed method is more than 0.02 in IOU value and 0.02 in Y value for images with simple background and more than 0.28 in IOU value and 0.19 in Y value for images with complex background. The presented method is robust and has considerable accuracy and efficiency advantages. We achieve the automatic aerial insulator segmentation by the proposed segmentation method.

References

1. Gao, Q., Yang, W., Li, Q.: Research on deep belief network layer tendency and its application into identifying fault images of aerial images. Chin. J. Sci. Instrum. **36**(6), 1267–1274 (2015)
2. Liu, L., Yang, N., Lan, J.: Image segmentation based on gray stretch and threshold algorithm. Opt.-Int. J. Light. Electron Opt. **126**(6), 626–629 (2015)
3. Ma, S.Y., An, J.B., Chen, F.M.: Segmentation of the blue insulator images based on region location. Electr. Power Constr. **31**(3), 14–17 (2010)
4. Iruansi, U., Tapamo, J.R., Davidson, I.E.: An active contour approach to insulator segmentation. In: AFRICON IEEE, pp. 1–5. IEEE (2015)
5. Zhang, X.Y., An, J.B., Wu, Q.G.: Method for recognizing insulator from airborne image. In: ICICTA-IEEE Computer Society, pp. 604–607. IEEE (2012)
6. Huang, X.B., Zhang, H.Y., Zhang, Y.: Composite insulator images segmentation technology based on improved color difference. Gaodianya Jishu/High Volt. Eng. **44**(8), 2493–2500 (2018)
7. Ke, H.C., Wang, H., Li, B.: Image segmentation method of insulator in transmission line based on weighted variable fuzzy c-means. J. Eng. Sci. Technol. Rev. **10**(3), 115–123 (2017)
8. Girvan, M., Newman, M.E.J.: Community structure in social and biological networks. arXiv preprint, pp. 7821–7826 (2001)
9. Wu, J.S., Li, X.X.: Minimum spanning trees for community detection. Phys. A: Stat. Mech. Its Appl. **392**(9), 2265–2277 (2013)
10. Amiri, S.H., Abin, A.A., Jamzad, M.: CDSEG: community detection for extracting dominant segments in color images. In: International Symposium on Image and Signal Processing and Analysis, pp. 177–182. IEEE (2015)
11. Ren, X.F., Malik, J.: Learning a classification model for segmentation. In: Proceedings of the 9th IEEE International Conference on Computer Vision, pp. 10–17. IEEE Comput Society, Washington DC (2013)
12. Levinshtein, A., Stere, A., Kutulakos, K.N.: TurboPixels: fast superpixels using geometric flows. IEEE Trans. Pattern Anal. Mach. Intell. **31**(12), 2290–2297 (2009)
13. Newman, M.E.J., Girvan, M., Kutulakos, K.N.: Finding and evaluating community structure in networks. Phys. Rev. E: Stat. Nonlinear Soft Matter Phys. **69**(2), 026113 (2004)
14. Duch, J., Arenas, A.: Community detection in complex networks using extremal optimization. Phys. Rev. E: Stat. Nonlinear Soft Matter Phys. **72**(2), 027104 (2005)

15. Mignotte, M.: A label field fusion model with a variation of information estimator for image segmentation. Inf. Fusion **20**(2), 7–20 (2014)
16. Hoffman, M., Steinley, D., Brusco, M.J.: A note on using the adjusted rand index for link prediction in networks. Soc. Netw. **42**(3), 72–79 (2015)
17. Zhang, S.H., Wong, H.S., Shen, Y.: Generalized adjusted rand indices for cluster ensembles. Pattern Recognit. **45**(6), 2214–2226 (2012)
18. Unnikrishnan, R., Pantofaru, C., Hebert, M.: A measure for objective evaluation of image segmentation algorithms. In: IEEE Computer Society Conference on Computer Vision and Pattern Recognition, pp. 34–41. IEEE (2005)
19. Marin, M.: Comparing clusterings: an axiomatic view. In: Proceedings of the 22nd International Conference on Machine Learning, Bonn, Germany (2005)
20. Martin, D., Fowlkes, C., Tal, D.: A database of human segmented natural images and its application to evaluating segmentation algorithms and measuring ecological statistics. In: Proceedings Eighth IEEE International Conference on Computer Vision, pp. 416–423. IEEE (2002)

Saliency Detection Based on Manifold Ranking and Refined Seed Labels

Shan Su, Ziguan Cui[✉], Yutao Yao, Zongliang Gan, Guijin Tang,
and Feng Liu

Image Processing and Image Communication Lab,
Nanjing University of Posts and Telecommunications, Nanjing 210003, China
cuizg@njupt.edu.cn

Abstract. Graph-based manifold ranking has been exploited for saliency detection with seed labels. However, when the selected labels are not accurate, these methods can't emphasize the foreground and suppress the background effectively. In this paper, we propose a novel saliency detection approach through manifold ranking and refined seed labels. We first construct a half-two layers graph based on the nodes after superpixel segmentation, which is generated by connecting each node to neighboring nodes and the half of the most similar nodes that share common boundaries with neighboring nodes. Then we compute superpixel saliency using manifold ranking with refined labels by two-step manner. After clustering superpixel with K-means, the background-based detection is obtained by refined background labels, which are those clusters containing boundary. The foreground-based detection is acquired with the refined foreground labels which are the complete cluster after thresholding the background-based detection. The proposed method has been tested on four universal datasets: ASD, CSSD, ECSSD and SOD. Experimental results show that our method performs better than prior similar state-of-the-art methods in various assessment indexes.

Keywords: Saliency detection · Manifold ranking · K-means · Graph model

1 Introduction

Recently, salient object detection has acquired much research interest, which aims to locate interesting and important regions in an image [1]. The output of saliency can be benefit to numerous applications such as object recognition, object tracking, image segmentation, image compression, image retrieval, and image quality assessment.

Generally, based on data processing mechanisms, saliency detection can be categorized as either bottom-up [1–4] or top-down [5–7] schemes. The bottom-up model is a fast, unconscious, data-driven and open-loop visual attention mechanism which base on the characteristics of the visual scene. In contrast, top-down model is a slow, conscious, task-driven and closed-loop visual attention mechanism which relies on the observer's expectations. Saliency detection methods can also be classified as salient region detection and eye fixation prediction. In this paper, we focus on the bottom-up salient object detection task.

© Springer Nature Switzerland AG 2019
Y. Zhao et al. (Eds.): ICIG 2019, LNCS 11901, pp. 284–296, 2019.
https://doi.org/10.1007/978-3-030-34120-6_23

Most bottom-up saliency detection methods are based on low-level features, such as color contrast, Euclidean distance and orientation. Itti et al. [1] proposed a conceptual model for saliency detection by performing multi-feature extraction and multi-scale decomposition of the input image, then fused the feature map linearly. Cheng et al. [3] presented a histogram contrast-based (HC) method, which considered the regional contrast with respect to the entire image and pixel-wise color separation to produce saliency map. Zhai et al. [8] calculated the global luminance contrast (LC) of pixel over the entire image to detect saliency. Hou et al. [9] established a spectral residual (SR) model of the image to obtain the saliency map. Achanta et al. [10] computed the saliency likelihood of each pixel by a frequency-tuned method based on luminance and color. By combining color uniqueness and spatial distribution, Perazzi et al. [11] applied a high-dimensional Gaussian filter to generate pixel-map. Zhou et al. [12] generated pixel saliency map by integrating diffusional compactness and local contrast (DCLC) cues.

However, those low-level features based methods maybe ignore the intrinsic connection between pixels and regions in images. To solve this problem, the graph-based methods are put forward. Harel et al. [13] explored a graph based visual saliency algorithm, which uses certain features to form activation map and then highlights the area of interest by normalizing. Gopalakrishnan et al. [2] detected seed nodes by Markov random walk model, which is carried out with the sparse k-regular graph and the complete graph, then the estimated location of the most notable region in an image is determined by seed nodes. By graph-based manifold ranking (MR) method, Yang et al. [4] utilized the boundary regions as background labels to generate initial saliency map and extracted foreground labels from initial map to obtain the final saliency map. In [14], a co-transduction algorithm is devised to fuse both boundary and objectness labels based on inter propagation scheme (LPS). Zhang et al. [15] adopted a linear scheme to fuse texture saliency map and color saliency map (TC) by manifold ranking. Zhou et al. [16] detected salient regions via diffusion process on sparse graph (DSG), and calculated background seed vectors by a compactness measure. Yuan et al. [17] removed foreground labels from background prior by reversion correction and built the regularized random (RCRR) walk ranking model to generate pixel-wise saliency map.

Among the graph-based methods, the boundary-based model outperforms most of the state-of-the-art saliency detection methods and is more computationally efficient. However, there still are some drawbacks that prevent from optimal performance. Firstly, most constructed graphs such as proposed in [4, 17] are full connected, each node connects to those nodes neighboring it as well as sharing common boundaries with its neighboring nodes. However, if the nodes of salient objects are inhomogeneous or incoherent, the full connected graph may lead to errors and seldom detect complete foreground. Secondly, background regions usually have a wider distribution over the entire image. The four boundaries of the image are treated as background labels for background-based saliency detection in [4, 17]. It's insufficient and maybe fail due to the negative influence when foreground objects touch the boundary.

In order to overcome above-mentioned problems, we propose half-two layers graph and select accurate seed labels by clustering for saliency detection. Firstly, we construct a half-two layers graph model, which is generated by connecting each node to neighboring nodes and the half of the most similar nodes that share common boundaries with neighboring nodes. This method effectively removes redundant nodes and fully uses the local spatial information. Then we apply the K-means to cluster image superpixels and those clusters containing boundary are regarded as background. Due to foreground objects may touch the boundary, we employ reversion correction method [17] to remove foreground in these background labels. The background saliency map is obtained based on background labels by manifold ranking. Finally, we binarize the background saliency map and use those complete clusters as the foreground labels. And we use foreground labels based manifold ranking method to get the final saliency map.

The residual of this paper is organized as follows. Section 2 shows the overall flow of our algorithm, including the construction of the graph model, the selection of foreground labels and background labels. The experimental results for ASD, CSSD, ECSSD and SOD datasets are shown in Sects. 3, and 4 is conclusion.

2   The Proposed Method

The framework of our proposed algorithm is shown in Fig. 1.

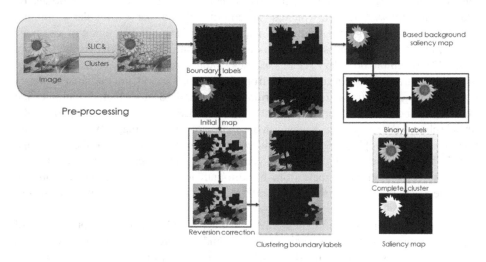

Fig. 1. Principal steps of our method.

Firstly, we perform the SLIC algorithm [18] to generate superpixels and construct a half-two layers graph. Secondly, we employ the K-means to cluster the superpixels. Thirdly, we select the background labels that those clusters contain boundary and remove the foreground labels. Finally, the complete cluster is regarded as foreground label after using an adaptive threshold, and then we apply the manifold ranking [16] to obtain the final saliency map.

2.1 Graph Construction and Clustering

In order to improve the performance of salient object detection, we use the SLIC algorithm to divide the input image into homogeneous and compact superpixels using the color means. Then we construct a graph $G = (V, E)$ depend on the superpixels of image, where each node V denotes a superpixel produced by the SLIC algorithm and edge E denote that V_i connects to V_j. The node set V consists of superpixels $X = \{x_1, \ldots, x_q, x_{q+1}, \ldots, x_n\} \in \mathbb{R}^m$. Some nodes are used as queries, and the remaining nodes need to be ranked according to their relevance to the queries. Let $f : X \rightarrow \mathbb{R}$ denote a ranking function, which assigns a ranking value f_i to each block x_i, and f can be regarded as a vector $f = [f_1, \ldots, f_n]^T$. Let $y = [y_1, \ldots, y_n]^T$ denotes an indication vector, where $y_i = 1$ if x_i is a query, and $y_i = 0$ otherwise. We use manifold ranking [4] as the ranking function, which is written as:

$$f = (D - \alpha W)^{-1} y \tag{1}$$

where α denote a constant, the affinity matrix is denoted by $W = \{w_{ij}\}_{N \times N}$, and $D = diag\{d_{11}, d_{22}, \ldots, d_{NN}\}$ is the degree matrix, where $d_{ii} = \sum_j w_{ij}$. More manifold ranking details could be found in [4, 19].

We define the weight w_{ij} between two nodes as

$$w_{ij} = e^{\frac{-\|c_i - c_j\|}{\sigma^2}} \tag{2}$$

where c_i and c_j denote the mean of color of nodes V_i and V_j in Lab color space, σ is constant factor which controls the weight.

Generally, most graph-based methods construct a full connection, each node connects to those neighboring nodes $D_1(j)$ as well as those nodes sharing common boundaries with its neighboring nodes $D_2(j)$, which may obtain erroneous local relation. Thus, in this paper, we propose a half-two layers graph for calculating saliency. As shown in Fig. 2, the half-two layers graph generated by connecting each node to its neighboring nodes and the half of the most similar nodes p that share common boundaries with neighboring nodes. It's well known that the second layer contains some local information, and some redundant information is adulterated in. To reduce redundancy and retain more local information, we retain the half of the most similar nodes, which is denoted as:

$$D(p) = \{q \in D_2(j) : w_{ij} > v\} \tag{3}$$

where v is the weight means of the second layer nodes $D_2(j)$, q is the node in $D_2(j)$, and p is the node whose weight larger than v.

Moreover, each node of the four boundaries of the image must be connected in pairs, and we describe the image as a closed-loop graph. Thus, the constructed graph model effectively removes redundant nodes and fully uses the local spatial distribution feature, which shows the obvious advantages compared with others graph models.

(a)   (b)

Fig. 2. The two-half layer graph model. (a) Input image. (b) Edge connection between nodes. A node (illustrated by a pink dot) connects to both its adjacent nodes (yellow dot) and the half of the most similar nodes (green dot) sharing common boundaries with its adjacent nodes. Each pair of boundary nodes are connected to each other (red dot and connection). (Color figure online)

We then employ K-means algorithm to cluster the N superpixels of the image into K clusters. Considering Lab color space is more related to human perception, we use three-dimensional Lab color feature to cluster.

2.2   Background-Based Saliency Estimation

Usually most of background regions are near the boundary, which are sparse and have a wider spatial distribution over the entire image compared with foreground regions. However, it's not adequate that simply utilizes the boundary labels as background labels. Therefore, we extend the background labels by clustering the image, each cluster contains one superpixel at least, and those clusters that contain boundary background are regarded as background labels. With the increase of the background labels, when calculating the background prior of the image, it's more effective to detect the foreground saliency object and uniformly highlight the entire salient region.

To select the background labels more accurately, we first calculate the initial saliency map using the boundary regions as [4] and remove the boundary-adjacent foreground regions from the boundary clusters by reverse correction method [17]. The initial map is generated via the separation and combination (SC) scheme, that is, we construct four background prior maps with boundary labels and then multiply them each other as the initial map. Then we use reverse correction method to mark the foreground regions with 1 and the background regions with 2. Specifically, for each boundary, the mean of the cluster that contains boundary background is called L_{label}. Given pre-defined threshold Th1 = 1, if Th1 smaller than L_{label}, we will repute that those clusters contain foreground regions in background regions, and then we will remove those regions and acquire exact background labels. Figure 3 shows examples of background labels, we can see that compare with general background labels (Fig. 3 (b)) and undoing reverse correction background labels (Fig. 3(c)), our background labels (Fig. 3(d)) are more precise.

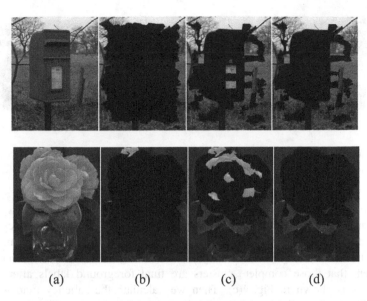

(a)   (b)   (c)   (d)

Fig. 3. Examples of background labels. From left to right: (a) Input image. (b) General background labels. (c) Not reverse correction background labels. (d) Our background labels.

After, we calculate background saliency maps by the manifold ranking. Taking top labels as an example, the queries are the exact background labels and the remaining regions are ranked. Thus, the indication vector y_i is obtained, and all the nodes are ranked based on Eq. (1) in f_b, which means each superpixel relevance to the exact background labels. The background saliency S_b based on top labels is calculated as:

$$S_b(i) = 1 - f_b(i) \qquad (4)$$

where $f_b(i)$ denotes the normalize vector, and the range of $f_b(i)$ is between 0 and 1.

We generate the other three saliency maps using the queries that selected via the similar method. And then the background-based saliency S_B is obtained by the following procedure:

$$S_B(i) = \prod_{b=1}^{k} S_b(i) \qquad (5)$$

Where k denotes the number of boundary.

2.3   Foreground-Based Saliency Estimation

Through the above steps, the most saliency regions are highlighted. However, there are some background regions which may not be inhibited. By the adaptive threshold method could diminish this problem, but the picked foreground labels may adulterate some background labels, as is shown in Fig. 4(b). To select the foreground labels more reasonable, we regard the extracted labels belonging to the complete clusters as foreground labels.

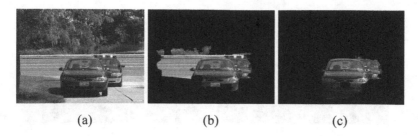

(a)	(b)	(c)

Fig. 4. Example of foreground labels. From left to right: (a) Input Image. (b) Adaptive threshold labels. (c) Adaptive threshold labels and the same cluster labels.

We separate the background saliency map by binary threshold, which exploits the adaptive threshold Th2 defined as the mean saliency over the whole saliency map. If $S_B(i) >$ Th2, the $S_B(i)$ is treated as foreground labels. The K-means algorithm divides the image into three categories: intra-object, intra-background and object-background, so we deem that those complete clusters are final foreground labels after adaptive threshold, as is shown in Fig. 4(c). Then we calculate the saliency map with final queries in each superpixel using Eq. (1). The foreground-based saliency map S_F is defined:

$$S_F(i) = \bar{f}(i) \tag{6}$$

where $\bar{f}(i)$ denote the normalized vector.

By the above method, the final saliency map will be greatly improved. As shown in Fig. 5. We notice that our method can stress the foreground evenly and suppress the background in effect.

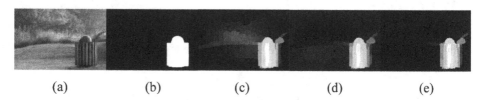

(a)	(b)	(c)	(d)	(e)

Fig. 5. An saliency example by our method. (a) Input image. (b) GT. (c) Saliency map based on half-two layers, (d) Saliency map based on background labels. (e) Saliency map based on foreground labels.

3   Experimental Results

3.1   Experimental Setup

We test the proposed method on four datasets. The ASD dataset [10] contains 1000 images. The second one is SOD dataset [20], which contains 300 images with multiple objects. The CSSD [21] is the third dataset, which contains diversified patterns in both the foreground and background. And the last one is ECSSD dataset [21], which is an extension of CSSD to express natural circumstances.

There are four parameters in the experiment which need to be set. In all experiments, we empirically set the number of superpixel nodes N = 200. σ is the edge weight, which controls the fall-off rate of the exponential function. In manifold ranking algorithm, α balances the smooth and fitting constraints. We empirically set σ = 0.1, and α = 0.99. The parameter K is the number of cluster in K-means, through experiment we set K = 70. As shown in Fig. 6, we varied it from 30 to 90 in intervals of 10 to determine an appropriate value for K with ASD dataset.

To evaluate the performance of different methods, we use the average precision-recall curve and the F-measure as evaluation criterion. We vary the threshold from 0 to 255 and compute the precision and recall at each threshold by comparing the binary mask and the ground truth to compare the accuracy of the different saliency maps. Then we apply the sequence of precision-recall pairs to plot the precision-recall curve. The F-measure is calculated using:

$$F_\beta = \frac{(1+\beta^2)Precision \times Recall}{\beta^2 Precision + Recall}$$

(7)

Following [4], we set $\beta^2 = 0.3$.

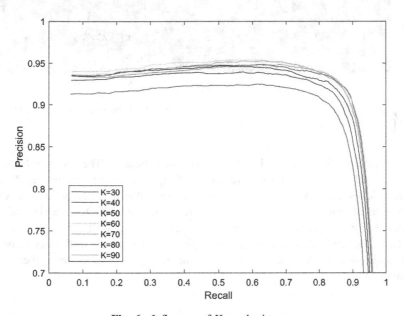

Fig. 6. Influence of K on the image.

3.2   Performance Comparison

We compare our method with 8 state-of-the-art algorithms, namely HC [3], MR [4], LC
[8], DCLC [12], LPS [14], TC [15], DSG [16], and RCRR [17]. As shown in Fig. 7,
our method acquires better subjective performance, and uniformly stress foreground
salient object and suppress background even for complex natural images.

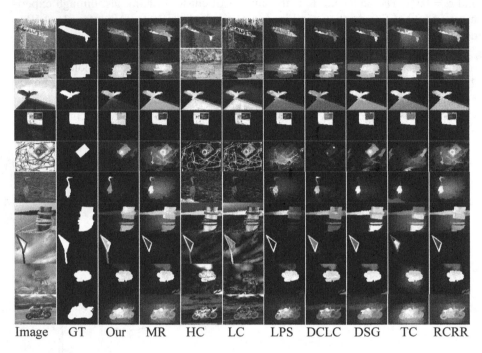

| Image | GT | Our | MR | HC | LC | LPS | DCLC | DSG | TC | RCRR |

Fig. 7. Saliency detection results of different methods. The proposed algorithm consistently
highlight foreground and suppress background.

We calculate P-R curve and F-measure on four databases. The result of F-measure
is listed in Table 1. The P-R curves are shown in Fig. 8 and the precision, recall and F-
measure indexes are shown in Fig. 9. Compared with other representative methods, the
performance of our method is better in F-measure for CSSD, ECSSD and SOD
databases. From the P-R curves, our algorithm performs also well, and it is competitive
to DCLC, MR, and RCRR. Although the performance of the P-R curve does not
surpass other algorithms by a large margin, our method obtains better subjective saliency map.

Table 1. F-measure results on ASD, CSSD, ECSSD and SOD databases.

	Our	MR	HC	LC	LPS
ASD	0.9115	0.9067	0.7264	0.5477	0.9009
CSSD	**0.8377**	0.8197	0.5196	0.4680	0.7922
ECSSD	**0.7425**	0.7355	0.4205	0.3793	0.6962
SOD	**0.6395**	0.6294	0.4157	0.4028	0.5868
	DCLC	DSG	TC	RCRR	
ASD	**0.9121**	**0.9164**	0.8600	0.9067	
CSSD	0.8275	**0.8352**	0.7183	0.8213	
ECSSD	0.7311	**0.7445**	0.6703	0.7390	
SOD	0.6169	0.6211	0.5785	**0.6311**	

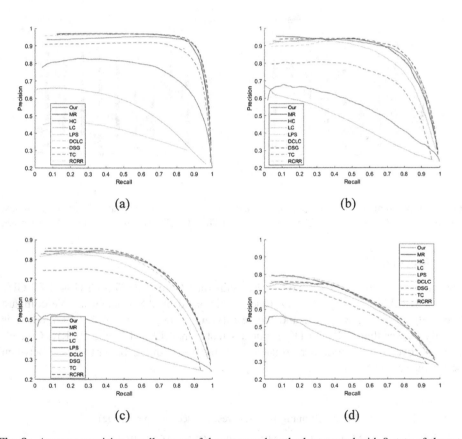

(a)

(b)

(c)

(d)

Fig. 8. Average precision-recall curves of the proposed method compared with 8 state-of-the-art methods. (a) the ASD database. (b) the CSSD database. (c) the ECSSD database. (d) the SOD database.

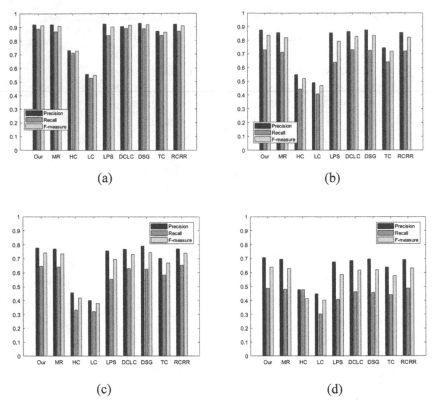

Fig. 9. F-measure of the proposed method compared with 8 state-of-the-art methods. (a) The ASD database. (b) The CSSD database. (c) The ECSSD database. (d) The SOD database.

3.3   Running Time

The running time is tested on a 64-bit PC with Intel Core i5-3337U CPU @ 1.80 GHz and 4 GB RAM. Average running time is calculated on ASD database. We compare five methods in recent years, and the results are shown in Table 2. Our method is slightly slower than MR and DSG, but it's faster than LPS, LC and RCRR. Considering the overall evaluation performances, our method acquires better trade-off between performance and complexity.

Table 2. Running time test results (seconds per image).

Method	Our	MR	LPS	DSG	TC	RCRR
Time (s)	0.834	0.667	1.287	0.630	1.664	1.531

4 Conclusion

We propose a bottom-up method to extract saliency region by calculating the relevance using manifold ranking with refined background and foreground labels. Our proposed half-two layers graph model alleviates the limitations in the prior graph models. In addition, we pick up the more precise labels using the cluster with k-means algorithm. The refined background and foreground labels can help to improve the performance of manifold ranking. By comparing with state-of-the-art saliency algorithms on four databases, it's confirmed that our method acquires better performance and can suppress background region and highlight foreground region accurately.

Acknowledgements. This work is supported by National Natural Science Foundation of China (NSFC) (61501260, 61471201, 61471203), Jiangsu Province Higher Education Institutions Natural Science Research Key Grant Project (13KJA510004), The peak of six talents in Jiangsu (RLD201402), and "1311 Talent Program" of NJUPT.

References

1. Itti, L., Koch, C., Niebur, E.: A model of saliency-based visual attention for rapid scene analysis. IEEE TPAMI **20**(11), 1254–1259 (1998)
2. Gopalakrishnan, V., Hu, Y., Rajan, D.: Random walks on graphs for salient object detection in images. IEEE TIP **19**(12), 3232–3242 (2010)
3. Cheng, M.M., Zhang, G.X., Mitra, N.J., Huang, X., Hu, S.M.: Global contrast based salient region detection. In: CVPR, pp. 409–416 (2011)
4. Yang, C., Zhang, L., Lu, H., Ruan, X., Yang, M.H.: Saliency detection via graph-based manifold ranking. In: CVPR, pp. 3166–3173 (2013)
5. Gao, D., Vasconcelos, N.: Discriminant saliency for visual recognition from cluttered scenes. In: Advances in Neural Information Processing Systems, pp. 481–488 (2004)
6. Yang, J., Yang, M.H.: Top-down visual saliency via joint CRF and dictionary learning. In: CVPR, pp. 2296–2303 (2012)
7. Itti, L., Sihite, D.N., Borji, A.: Probabilistic learning of task-specific visual attention. In: CVPR, pp. 470–477 (2012)
8. Zhai, Y., Shah, M.: Visual attention detection in video sequences using spatiotemporal cues. In: ACM Multimedia, pp. 815–824 (2006)
9. Hou, X., Zhang, L.: Saliency detection: a spectral residual approach. In: CVPR, pp. 1–8 (2007)
10. Achanta, R., Hemami, S., Estrada, F., Susstrunk, S.: Frequency-tuned salient region detection. In: CVPR, pp. 1597–1604 (2009)
11. Perazzi, F., Krahenbuhl, P., Pritch, Y., Hornung, A.: Saliency filters: contrast based filtering for salient region detection. In: CVPR, pp. 733–740 (2012)
12. Zhou, L., Yang, Z., Yuan, Q., Zhou, Z., Hu, D.: Salient region detection via integrating diffusion-based compactness and local contrast. IEEE TIP **24**(11), 3308–3320 (2015)
13. Harel, J., Koch, C., Pietro, P.: Graph-based visual saliency. In: Advances in Neural Information Processing Systems, pp. 545–552 (2006)
14. Li, H., Lu, H., Lin, Z., Shen, X., Price, B.: Inner and inter label propagation: salient object detection in the wild. IEEE TIP **24**(10), 3176–3186 (2015)

15. Zhang, Q., Lin, J., Tao, Y., Li, W., Shi, Y.: Salient object detection via color and texture cues. Neurocomputing **243**, 35–48 (2017)
16. Zhou, L., Yang, Z., Zhou, Z., Hu, D.: Salient region detection using diffusion process on a 2-layer sparse graph. IEEE TIP **26**(12), 5882–5894 (2017)
17. Yuan, Y., Li, C., Kim, J., Cai, W., Feng, D.D.: Reversion correction and regularized random walk ranking for saliency detection. IEEE TIP **27**(3), 1311–1322 (2018)
18. Achanta, R., Shaji, A., Smith, K., Lucchi, A., Fua, P., Susstrunk, S.: SLIC superpixels compared to state-of-the-art superpixel methods. IEEE TPAMI **34**(11), 2274–2282 (2012)
19. Zhou, D., Weston, J., Gretton, A., Bousquet, O., Scholkopf, B.: Ranking on data manifolds. In: Advances in Neural Information Processing Systems, pp. 169–176 (2014)
20. Movahedi, V., Elder, J.H.: Design and perceptual validation of performance measures for salient object segmentation. In: CVPRW, pp. 49–56 (2010)
21. Yan, Q., Xu, L., Shi, J., Jia, J.: Hierarchical saliency detection. In: CVPR, pp. 1155–1162 (2013)

Hierarchical Convolution Feature for Target Tracking with Kernel-Correlation Filtering

Jing Zhang[3](✉), Dong Hu[1,2,3], Biqiu Zhang[3], and Yuwei Pang[3]

[1] Education Ministry's Key Lab of Broadband Wireless Communication and Sensor Network Technology, Nanjing, China
[2] Education Ministry's Engineering Research Center of Ubiquitous Network and Heath Service, Nanjing, China
[3] Jiangsu Province's Key Lab of Image Procession and Image Communications, Nanjing University of Posts and Telecommunications, Nanjing 210003, China
{1217012312,hud}@njupt.edu.cn

Abstract. Target tracking is widely used in many fields, but tracking performance still needs to be improved due to factors such as deformation, illumination and occlusion. In this paper, we propose a scale adaptive target tracking solution based on hierarchical convolution features and establish a kernel correlation filtering target tracking framework that combines multi-layer convolution features. The improved convolutional neural network is used to extract multi-layer features, and the correlation filters of each layer are separately trained to perform weighted fusion to obtain the target position. Then, the edge box algorithm is adopted to obtain the size of the actual tracking frame to achieve exact target tracking. An extensive evaluation on OTB-2013 with public test sequences are conducted. Experimental results and analysis indicate that our method is better than other known advanced tracking algorithms even in video sequences with many uncertain factors, while the speed and accuracy of tracking can be effectively improved.

Keywords: Object tracking · Convolution neural network · Kernel correlation filter · Edge boxes

1 Introduction

Visual tracking is a fundamental computer vision task with a wide range of applications. Although much process has been made in the past decade, tremendous challenges still exist in designing a robust tracker that can well handle significant appearance changes, pose variations, severe occlusions and background clutters. In order to adapt to the actual application scenarios and ensure the accuracy and robustness of target tracking, lots of efforts have been paid in theoretic studies and applications.

Existing appearance-based tracking methods adopt either generative or discriminative models. The generative methods establish a model for the target region in the current frame, and then match the next frame, such as particle filter [1], Kalman filter [2] and so on. Discriminative methods treat the tracking problem as a two-category

© Springer Nature Switzerland AG 2019
Y. Zhao et al. (Eds.): ICIG 2019, LNCS 11901, pp. 297–306, 2019.
https://doi.org/10.1007/978-3-030-34120-6_24

problem that find the decision boundary of the target and background. For example, the Stuck algorithm [3] and the Compressed Tracking (CT) algorithm [4]. Whereas, this kind of tracking algorithm is slow. Subsequently, correlation filtering algorithms are introduced into the target tracking. For example, the Circulant Structure of Tracking-by-detection with Kernels (CSK) algorithm [5] opens the beginning of correlation filtering research; Kernel Correlation Filter (KCF) [6] uses the Histograms of Orients Gradients (HOG) feature to convert a single channel into multiple channels. Although this algorithm is faster, the disadvantage is that the target rotation and occlusion problems cannot be solved.

Driven by the emergence of large-scale visual data sets and fast development of computation power, Deep Neural Networks (DNNs), especially Convolutional Neural Networks (CNNs), with their strong capabilities of learning feature representations, such as Hierarchical Convolutional Features (HCF) [7]; Recursive neural networks are used to track targets, such as Recurrently Target-Attending Tracking (RTT) [8]. But how to design neural networks and tracking processes to achieve speed improvement, there is still a lot of research space.

Considering the robustness and real-time of tracking, in our method, the multi-layer depth feature is extracted by convolutional neural network. Based on the hierarchical convolution feature and KCF target tracking, the tracking problem is deeply studied in complex environment.

The rest of the paper is organized as follows. The Sect. 2 is the related technical analysis of convolutional neural network and kernel correlation filters. The Sect. 3 discusses the detailed algorithm, the scale adaptive target tracking method based on layered convolution characteristics. The Sect. 4 provides a discussion concerning experimental simulation. Finally, the summary is delivered in Sect. 5.

2 Related Work

In our algorithm, the features of the convolutional neural network are applied to the kernel correlation filter tracking framework. Therefore, this section first introduces the principle of convolutional neural networks, and secondly introduces the kernel correlation filter tracking algorithm.

2.1 Convolutional Neural Network

Convolutional neural network is a multi-layer neural network. The whole structure includes convolution layer, nonlinear activation function, pooling layer. The high-level information is obtained from the original data layer by layer. The main role of the convolutional layer is to use the convolution kernel for feature extraction and feature mapping. The input image is first convoluted with the convolution kernel, and the result is used as the input to the nonlinear activation function. The activation function is used to add nonlinear factors. The commonly used activation functions are Tanh function, ReLU function and so on. The pooling layer has the effect of quadratic feature extraction, which can reduce the dimension of the feature map. The typical pooling operation is the average pooling and the largest pooling. The fully connected layer is a classifier of the convolutional neural network. For classification tasks, SVM is usually used because it can be combined with CNN to solve different classification tasks.

Among commonly used convolutional neural networks, e.g. AlexNet, VggNet, and ResNet, we use VGGNet-19 [9] network, because it is easy to migrate to other image recognition projects. Besides, VGGNet trained parameters can be download for a good initialization weight operation.

2.2 Tracking by Kernel Correlation Filters

The kernel correlation filtering algorithm is to train better classifiers to find the decision boundary of the target and background. The purpose of training is to find a function $f(x) = w^T x$ that minimizes the error function. The objective function can be expressed as Eq. (1):

$$\min_{w} \sum_{i} (f(x_i) - y_i)^2 + \lambda ||w||^2 \tag{1}$$

where λ is a regularization coefficient used to control overfitting. We can get the solution in complex domain $w = (X^H X + \lambda I)^{-1} X^H y$, where the matrix X has one sample x_i per line, each element of y is a regression target y_i, X^H represents a complex conjugate transpose matrix, and I represents the identity matrix. Using the properties of the diagonalization of the circulant matrix to obtain the simplified ridge regression of the Fourier diagonalization, the following formula $\hat{w} = diag(\frac{\hat{x}^*}{\hat{x}^* \otimes \hat{x} + \lambda})\hat{y}$ is obtained, where $\hat{x}^*$ is the complex conjugate of $\hat{x}$, $\otimes$ represents the dot multiplication of the element. Since most of the cases are nonlinear, high-dimensional solutions and kernel functions [10] have been introduced. The objective function can be expressed as:

$$f(z) = w^T z = \sum_{i=1}^{n} \alpha_i \kappa(z, x_i) \tag{2}$$

After the ridge regression is nucleated, the form of the frequency domain solution is $\hat{\alpha} = \frac{\hat{y}}{\hat{k}^{xx'} + \lambda}$, where $k^{xx'}$ represents the kernel correlation of any two vectors x and x', the symbol $\wedge$ represents the DFT transform of the vector. It is easy to prove that the kernel matrix between all training samples and all candidate image blocks satisfies the condition of the cyclic matrix, so that the regression function in the frequency domain of all candidate image blocks can be obtained: $\hat{f}(z) = \hat{k}^{xz} \otimes \hat{\alpha}$. In particular, when the kernel function is a Gaussian kernel, we can get Gaussian kernel related $k^{xx'} = \exp(-\frac{1}{\sigma^2}(||x||^2 + ||x'||^2 - 2F^{-1}(\hat{x}^* \otimes \hat{x}')))$. By the maximum value of the positioning, the relative motion of the tracking target can be obtained.

3 Scale Adaptive Tracking Based on Hierarchical Convolution Features

In this section, we will give a detailed description of the algorithm we proposed. This section will be divided into two modules. The first module is used for target positioning. We extract improved multi-layer convolution features for VGG-Net and combine them with kernel correlation filtering algorithms for target localization. The second module is used for target scale estimation. We use the edge box algorithm to obtain the size of the actual tracking frame and achieve scale adaptation.

3.1 Target Position Estimation

Considering the robustness and accuracy, it is proposed to combine the multi-layer convolution feature with the kernel correlation filtering algorithm to achieve the target position estimation.

The network model used here is VGGNet-19, including 5 sets of convolutional layers, a total of 16 layers, and the last three are fully connected layers. In order to deeply understand the characterization ability of each layer feature of the convolutional neural network, the single-layer convolution feature is applied to the KCF correlation filtering tracking algorithm respectively, and the tracking results are compared and analyzed. The experiment performed on 35 color video sequences in the OTB-2013. The result is expressed in terms of precision, which is calculated as the center point of the target position estimated by the tracking algorithm and the center point of the manually labeled target. For a given threshold, the distance between the two is less than the percentage of the video frame, and the general threshold is set to 20 Pixel.

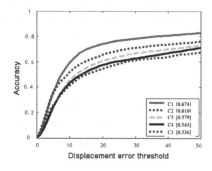

Fig. 1. OPE precision map using different convolution layers

Figure 1 is an OPE accuracy map obtained using different convolution features. It can be seen that the first layer convolution feature has high resolution and can accurately locate the target, the fourth layer and the fifth layer convolution feature contain more semantic information and can roughly locate the target. Therefore, we use Conv1-2, Conv4-4, Conv5-4 layer convolution features for correlation filtering target tracking algorithm. We take the output of each convolutional layer as a multi-channel feature

[5], and take the virtual samples obtained from all the cyclic shift of feature X as training samples. Each cyclic shift sample has a corresponding Gaussian distribution label $y_{ij} = e^{-\frac{(i-M/2)^2 + (j-N/2)^2}{2\sigma^2}}$, where σ is the kernel width. Learn about correlation filters of the same x size by addressing the following minimization issues:

$$W^* = \arg \min_w \sum_{m,n} ||W \cdot X_{m,n} - y(m,n)||^2 + \lambda||W||_2^2 \tag{3}$$

where linear product is defined as $w \cdot x_{ij} = \sum_{d=1}^{D} w_{ijd}^D x_{ijd}$. The filter learned in the frequency domain on the dth $(d \in \{1,\ldots,D\})$ channel is:

$$W^d = \frac{Y \odot \bar{X}^d}{\sum_{n=1}^{D} X^i \odot \bar{X}^i + \lambda} \tag{4}$$

where Y is the Fourier transform form of y_{ij}, the horizontal line on the letter indicates the complex conjugate. Let z be expressed as the feature vector on the lth layer and the size is M × N × D, and then the lth correlation response map can be calculated by the following formula:

$$f(z) = F^{-1}\left(\sum_{d=1}^{D} W^d \odot Z^d\right) \tag{5}$$

where operator F^{-1} represents inverse FFT transform. Let $(\hat{m}, \hat{n}) = \arg\max_{m,n} f_l(m,n)$ denote the position of the maximum value on the lth layer, then the best position of the target in the l-1th layer is expressed as:

$$\arg\max_{m,n} f_{l-1}(m,n) + \gamma f_l(m,n)$$
$$s.t.\ |m - \hat{m}| + |n - \hat{n}| \leq r \tag{6}$$

The constraint indicates that only the region centered at $(\hat{m}, \hat{n})$ and r is the radius is searched for in the in the l-1th layer correlation response graph. The response value from the latter layer is weighted as a regularization term and then propagated back to the response graph of the previous layer. In this way, the maximum value in the response graph of the last layer is the predicted position of the target.

3.2 Target Scale Adaptation

Based on target position estimation, our method proposes a scale adaptive target tracking by edge frame detection algorithm [11]. The edge frame detection algorithm traverses the entire image in a sliding window manner, and scores the bounding box of each sample, selects the top 200 candidate frames with the highest score, and performs a convolution operation on the candidate frame and the filter to obtain a response graph. The maximum response value in the candidate target can be expressed as $f_{max} = \max(f_{max,1}, f_{max,2}, \ldots, f_{max,n})$, where $f_{max,1}, f_{max,2}, \ldots, f_{max,n}$ is the maximum response value in the response graph of each candidate target, n is the number of candidate targets. If f_{max} less than f_p (f_p is the maximum response of the correlation filter by using the layered convolution feature), this means that the detection algorithm is most likely to find that the position of the target is not as accurate as the target position estimated by the convolution feature. Thus, abandoning the candidate target detected by the detection algorithm, and the target size remains unchanged. Otherwise updating the position and size using the damping factor γ. The scale update method is as follows:

$$(w_t, h_t) = \begin{cases} (w_{t-1}, h_{t-1}) + \gamma[(w_{p,t}, h_{p,t}) - (w_{t-1}, h_{t-1})], & \text{if } f_{max} > f_p \\ (w_{t-1}, h_{t-1}), & \text{if } f_{max} < f_p \end{cases} \quad (7)$$

where w_{t-1}, h_{t-1}, $w_{p,t}$, $h_{p,t}$ respectively indicate the width and height of the t-1th candidate frame with the largest response value in the target and the t-th frame, which γ is set to 0.5 as the learning rate. The target location is updated as follows:

$$I_t = \begin{cases} I_{d,t} + \gamma(I_{p,t} - I_{d,t}), & \text{if } f_{max} > f_p \\ I_{d,t}, & \text{if } f_{max} < f_p \end{cases} \quad (8)$$

where $I_{d,t}$ is the target position based on the hierarchical convolution feature, $I_{p,t}$ is the target position of the maximum response value corresponding to the t frame. Finally, the location and size of the target are estimated to achieve target tracking.

4   Experiments

In this section we evaluate our algorithm from two aspects. Firstly, a qualitative comparison is provided, we display the tracking effect for scale change and occlusion test sequence on OTB-2013 dataset. Secondly, through the quantitative analysis, the tracking effects of several excellent open source trackers in the visual tracker benchmark test were compared.

The simulation environment for this experiment is MATLAB, and the experimental environment is on an i7 machine with 8 GB of memory. To assess accuracy and success rate, we compared four advanced tracking algorithms in literature: HCF, Stuck, KCF, CT. In these experiments, three video sequences in the standard target tracking library OTB-2013 were tested.

4.1   Qualitative Experiment Verification

In this section, we select three video sequences from OTB-2013 for qualitative analysis, which are Dog1, Singer1, and CarScale video sequences.

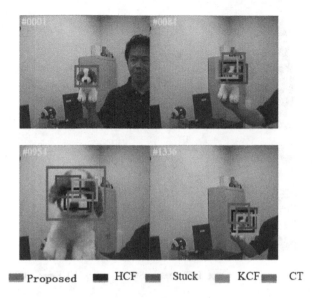

Proposed   HCF   Stuck   KCF   CT

Fig. 2. Dog1 video sequence renderings

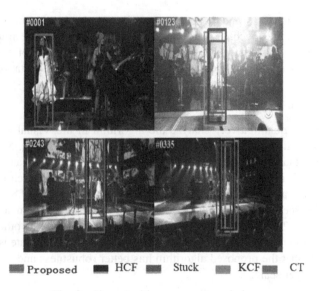

Proposed   HCF   Stuck   KCF   CT

Fig. 3. Singer1 video sequence renderings

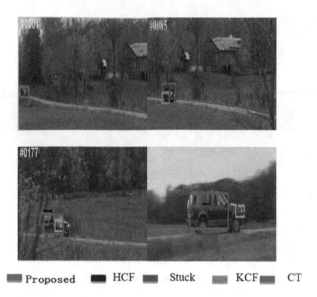

Proposed ■ HCF ■ Stuck ■ KCF■ CT

Fig. 4. CarScale video sequence renderings

As shown in Figs. 2, 3 and 4, there are obvious scale changes in the three video sequences. If the target becomes larger, the sample will lose some important information. HCF, CT, Stuck and KCF can only track a small part of the target. Our algorithm can track the target accurately and achieve the scale adaptation.

4.2 Quantitative Experiment Verification

In this section, we quantitatively analyze the algorithm in the Visual Tracker Benchmark and compare it with several popular algorithms, as shown in Fig. 5. It can be seen from (a) and (b) that compared with other algorithms, the two indicators of the algorithm have the best results, the average accuracy reaches 81.2%, which is 0.3% higher than HCF; the average success rate reached 65.8%, an increase of 9.8% compared to HCF. It can be seen from (c) and (d) that the proposed algorithm achieves better tracking results in 28 scale-changing video sequences compared to other algorithms. The average accuracy is improved by 2.6% and the average success rate is increased by 14.6%. It shows that the proposed algorithm has better robustness and can better adapt to changes in target scale.

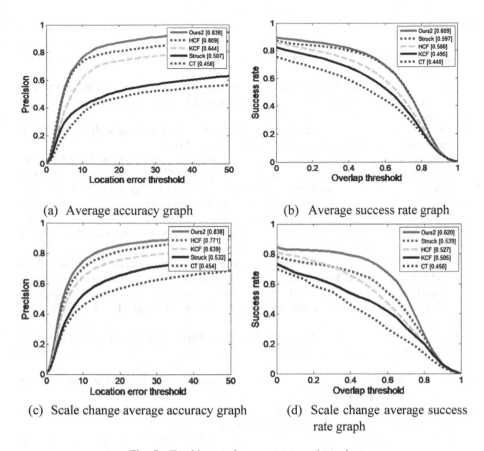

(a) Average accuracy graph

(b) Average success rate graph

(c) Scale change average accuracy graph

(d) Scale change average success rate graph

Fig. 5. Tracking performance comparison chart

5 Conclusions

In this paper, we empirically present some important properties of CNN features under the viewpoint of visual tracking. Based on these attributes, we propose a tracking algorithm for pre-training image classification tasks using complete convolution network. The improved convolutional neural network extraction feature is applied to the kernel correlation filtering tracking algorithm to achieve accurate target location. At the meanwhile, the edge frame detection algorithm is used to generate the target positional bounding box. The problem of fast scale change in target tracking is solved, and the accuracy and robustness of target tracking are improved.

References

1. Chang, C., Ansari, R.: Kernel particle filter for visual tracking. IEEE Signal Process. Lett. **12**(3), 242–245 (2005)
2. Torkaman, B., Farrokhi, M.: Real-time visual tracking of a moving object using pan and tilt platform: A Kalman filter approach. In: 20th Iranian Conference on Electrical Engineering, Iran, pp. 56–67 (2012)
3. Hare, S., Saffari, A., Torr, P.H.S.: Struck: structured output tracking with kernels. IEEE Trans. Pattern Anal. Mach. Intell. **38**(10), 2096–2109 (2015)
4. Zhang, K., Zhang, L., Yang, M.-H.: Real-Time compressive tracking. In: Fitzgibbon, A., Lazebnik, S., Perona, P., Sato, Y., Schmid, C. (eds.) ECCV 2012. LNCS, vol. 7574, pp. 864–877. Springer, Heidelberg (2012). https://doi.org/10.1007/978-3-642-33712-3_62
5. Henriques, J.F., Caseiro, R., Martins, P., Batista, J.: Exploiting the circulant structure of tracking-by-detection with kernels. In: Fitzgibbon, A., Lazebnik, S., Perona, P., Sato, Y., Schmid, C. (eds.) ECCV 2012. LNCS, vol. 7575, pp. 702–715. Springer, Heidelberg (2012). https://doi.org/10.1007/978-3-642-33765-9_50
6. Henriques, J.F., Caseiro, R., Martins, P., et al.: High-Speed tracking with kernelized correlation filters. IEEE Trans. Pattern Anal. Mach. Intell. **37**(3), 583–596 (2015)
7. Ma, C., Huang, J.B., Yang, X., et al.: Hierarchical convolutional features for visual tracking. In: IEEE International Conference on Computer Vision 2015, pp 111–121. IEEE, Chile (2015)
8. Cui, Z., Xiao, S., Feng, J., et al.: Recurrently target-attending tracking. In: IEEE Conference on Computer Vision and Pattern Recognition 2016, pp. 1449–1458. IEEE, LAS VEGAS (2016)
9. Simonyan, K., Zisserman, A.: Very Deep Convolutional Networks for Large-Scale Image Recognition. Computer Science, pp. 569–577 (2014)
10. Schölkopf, B., Smola, A.: Learning with kernels: support vector machines, regularization, optimization, and beyond. Am. Stat. Assoc. **98**(462), 489 (2002)
11. Dollár, P., Zitnick, C.L.: Structured forests for fast edge detection. In: IEEE International Conference on Computer Vision 2013, pp. 854–863. IEEE, Sydney (2013)

Non-negative Representation Based Discriminative Dictionary Learning for Face Recognition

Zhe Chen[1], Xiao-Jun Wu[1(✉)], and Josef Kittler[2]

[1] Jiangsu Provincial Engineering Laboratory of Pattern Recognition
and Computational Intelligence, School of IoT Engineering, Jiangnan University,
Wuxi 214122, China
wu_xiaojun@jiangnan.edu.cn
[2] CVSSP, University of Surrey, Guildford GU2 7XH, UK

Abstract. In this paper, we propose a non-negative representation based discriminative dictionary learning algorithm (NRDL) for multi-category face classification. In contrast to traditional dictionary learning methods, NRDL investigates the use of non-negative representation (NR), which contributes to learning discriminative dictionary atoms. In order to make the learned dictionary more suitable for classification, NRDL seamlessly incorporates non-negative representation constraint, discriminative dictionary learning and linear classifier training into a unified model. Specifically, NRDL introduces a positive constraint on representation matrix to find distinct atoms from heterogeneous training samples, which results in sparse and discriminative representation. Moreover, a discriminative dictionary encouraging function is proposed to enhance the uniqueness of class-specific sub-dictionaries. Meanwhile, an inter-class incoherence constraint and a compact graph based regularization term are constructed to respectively improve the discriminability of learned classifier. Experimental results on several benchmark face data sets verify the advantages of our NRDL algorithm over the state-of-the-art dictionary learning methods.

Keywords: Face recognition · Discriminative dictionary learning · Non-negative representation

1 Introduction

Due to the non-repeatability and uniqueness of human faces, face recognition has been the hottest topic in object classification applications [1–4]. Over the past few years, sparse representation theory has been deeply studied in the field of face recognition. The representative one is sparse representation-based classification (SRC) [5] algorithm. SRC aims at linearly representing an input data with a few atoms from given training samples. Zhang et al. [6] considered that the inter-class collaboration plays a more important role than sparsity. Therefore,

© Springer Nature Switzerland AG 2019
Y. Zhao et al. (Eds.): ICIG 2019, LNCS 11901, pp. 307–319, 2019.
https://doi.org/10.1007/978-3-030-34120-6_25

they proposed a collaborative representation-based classification (CRC) algorithm by imposing an L_2-norm constraint on representation coefficients which can achieve competitive classification accuracy in less time than L_1-norm based methods. Besides, Xu et al. [7] proposed a new discriminative sparse representation method, which suggests that the discriminative representation can be obtained by suppressing the relevance of inter-class reconstructions. More recently, Xu et al. [8] found that non-negative representation can strengthen the representation ability of homogeneous samples while weakening the negative effects caused by heterogeneous samples. Note that above representation based algorithms can only work well when the samples are collected under well-controlled conditions. However, in reality, face images often include severe illumination, pose, expression and occlusion changes that can destroy the subspace structure of data. So it is necessary to learn some distinct atoms which are beneficial for representation. Recently, as a major branch of sparse representation, dictionary learning has attracted extensive interests in face recognition field. The main idea of dictionary learning is to extract competitive and discriminative features from original training samples, while removing the useless information that is bad for reconstruction.

According to whether the label information of training samples is used, dictionary learning algorithms are usually divided into two categories: supervised or unsupervised. Xu et al. [9] proposed a sample-diversity and representation-effectiveness based robust dictionary learning algorithm (SDRERDL) by taking advantages of the mirroring samples to address the small-sample-size problem. Different from unsupervised situation, supervised dictionary learning algorithms can extract more discriminative features from data with the use of label information. Zhang et al. [10] and Jiang et al. [11] proposed a discriminative KSVD (D-KSVD) algorithm and a label-consistent K-SVD (LC-KSVD) algorithm, respectively. Yang et al. [12] proposed a famous Fisher discrimination criterion to encourage the discriminability of learned dictionary by simultaneously minimizing within-class scatter and maximizing inter-class scatter of representation. Recently, based on the observation that locality of data may be more significant than sparsity, Li et al. [13] proposed a locality-constrained and label embedding dictionary learning algorithm (LCLE), which considers the locality and label information of samples together during the dictionary learning process.

It is worth noting that aforementioned dictionary learning algorithms exist a common problem: the representation coefficients for training samples may include negative values. According to [8], since there is no non-negative constraint on representation, a query sample will be represented by both heterogeneous and homogeneous samples. As a result, the obtained coefficients have both negative and non-negative values, which brings about a difficult physical interpretation. Based on the assumption that a query sample should be approximated by homogeneous samples as much as possible, in this paper, we propose a non-negative representation based discriminative dictionary learning algorithm (NRDL) for face recognition. Specifically, the contributions of the proposed NRDL algorithm are presented as follows.

(1) By restricting the coding values to be non-negative, the NRDL model will select some useful features from heterogeneous samples to form dictionary atoms, which naturally leads to discriminability and sparsity on representation.

(2) Due to the combination of dictionary learning and multi-class classifier learning, the learned dictionary is efficient for reconstruction and classification simultaneously.

(3) NRDL minimizes the inter-class reconstruction to encourage the discriminability of learned dictionary instead of directly forcing the representation to be a block-diagonal structure which may result in overfitting problem.

(4) For better classification, NRDL uses an inter-class incoherence term and a compact graph structure to improve the robustness of learned linear classifier.

2   Non-negative Representation for Classification (NRC)

Among previously mentioned representation based classification methods [5,6], there still exists controversy between sparsity and collaborative mechanism, that is to say which mechanism is more significant when using training samples to approximate a query sample? Nevertheless, Xu et al. [8] thought it is meaningless to determine an eventual winner. Despite the successful applications of sparsity and cooperativity, they proposed a simple but efficient non-negative representation-based classification algorithm (NRC) by introducing a non-negative constraint on coding coefficients instead of using sparse constraints (e.g., l_1 or l_2 norm). Assuming that we have N training samples from C classes, denoted by $X = [X_1, X_2, \ldots, X_C] \in R^{d \times N}$, where $X_i \in R^{d \times n} (n = N/C)$ denotes the training samples of ith class. d is the sample dimensionality. Given a query sample $y \in R^d$, the model of NRC can be formulated as:

$$\min_{\theta} \|y - X\theta\|_2^2 \quad s.t. \quad \theta > 0 \tag{1}$$

where $\theta \in R^N$ is the coding vector. NRC argues that the non-negative coefficients over the homogeneous samples are crucial to classify y. By restricting the values of coding vector θ to be non-negative, the contributions of homogeneous samples can be enlarged, meanwhile, eliminating the negative effects caused by heterogeneous samples. Instead of using original training samples as the dictionary, we extend the non-negativity theory of representation to dictionary learning method, so that the learned dictionary atoms are more high-quality and homogeneous. In addition, we also investigate how to enhance the discriminability and compact of learned dictionary and classifier, respectively.

3   Non-negative Representation Based Dictionary Learning Algorithm (NRDL)

In this section, we present a novel non-negative representation-based dictionary learning algorithm (NRDL), in which dictionary learning and linear classifier learning are incorporated into a joint framework. In NRDL, we want to learn a discriminative and reconstructive dictionary by leveraging the label information of training samples under the condition that the coding values are non-negative. Because representation of samples can also be regarded as feature for classification, the classification error term is also included in our learning model to make the learned dictionary efficient for classification as well. Thus, the optimization framework of NRDL can be defined as follows

$$<D, S, W> = arg \min_{D,S,W} h(D, S) + g(H, W, S) \ s.t. \ S \geq 0 \qquad (2)$$

where $h(D)$ is the discriminative dictionary encouraging function and $g(H, W, S)$ is the robust linear classifier learning function. $H = [h_1, h_2, \ldots, h_n] \in R^{C \times N}$ are the class labels of training samples X. Specifically, $h_i = [0, 0, \ldots, 1, \ldots, 0, 0] \in R^C$ is a label vector of ith training sample x_i where the position of element '1' indicates the class of x_i. $D = [D_1, D_2, \ldots, D_C] \in R^{d \times K}$ is the learned dictionary and $S = [S_1, S_2, \ldots, S_C] \in R^{K \times N}$ are the sparse codes of training samples, where D_i and S_i $(i = 1, 2, \ldots, C)$ denote the sub-dictionary and sub-representation corresponding to class i, respectively. $S \geq 0$ represents the non-negative constraint on coding matrix S. $W \in R^{C \times K}$ is the projection matrix of linear classifier. K is the number of dictionary atoms. Now, we will detail the objective functions of $h(D, S)$ and $g(H, W, S)$.

3.1   Non-negative Discriminative Dictionary Encouraging Function

Obviously, for the lth class training samples X_l, we have $X_l = DS_l$. Supposing S_{ii} is the coding coefficients of X_i over sub-dictionary D_i, note that S_{ll} should be capable of well reconstructing X_i with D_i. So it is reasonable to assume that $X_i \approx D_i S_{ii}$. What's more, it is expected that X_i can only be well reconstructed by S_{ii}, rather than by inter-class representation $S_{ji}(j = 1, 2, \ldots, C, j \neq i)$. In order to learn a discriminative and reconstructive dictionary, Zhang et al. [23] argued the representation matrix to present an ideal '0-1' block-diagonal structure $Q = [q_1, q_2, \ldots, q_n] \in R^{K \times n}$ to capture more structured information from data. Let sample x_i belongs to class l, then the coding coefficients in q_i for D_l are all 1 s, while the remaining elements are all 0 s. However, the samples from the same class often have different coding coefficients because of the diversity of samples. To address above problems, we propose the following non-negative discriminative dictionary encoding function for our NRDL method

$$h(D, S) = \|X - DS\|_F^2 + \lambda \|D(A \odot S)\|_F^2 \ s.t. \ S \geq 0 \qquad (3)$$

where λ is a positive parameter. $A = O - Q$, where $O = 1_K 1_N^T \in R^{K \times N}$ denotes all 1 s matrix. $Q = [q_1, q_2, \ldots, q_N] \in R^{K \times N}$ denotes the '0-1' block-diagonal matrix, where q_i has the form of $[0, 1, 1, 1, \ldots]^T$. If sample x_i belongs to class L, then the coding in q_i over D_L are all 1 s, while all the others are 0 s. $\|X - DS\|_F^2$ is the reconstruction error term. We can see that $A \odot S$ is actually the off-block-diagonal components of representation matrix S. Based on the observation that the samples corresponding to a certain class should only be well represented by the sub-dictionary from the same class, hence it is natural to assume the off-block-diagonal reconstruction with dictionary D, i.e., $(D(A \odot S))$, should be minimized as much as possible, thus enhancing the uniqueness of class-specific sub-dictionaries.

3.2   Robust Linear Classifier Learning Function

Similar to [10] and [11], we consider to introduce the classification error term in our NRDL framework to learn a dictionary which is also suitable for classification. Besides, we also propose another two regularization terms to improve the discriminability of learned classifier. The proposed robust classifier learning function can be formulated as

$$
\begin{aligned}
g(H, W, S) = \alpha \|H - WS\|_F^2 \\
+ \beta \sum_{i=1}^{C} \sum_{j=1, j\neq i}^{C} \|(WS_i)^T(WS_j)\|_F^2 + \gamma \sum_{v=1}^{N} \sum_{u=1}^{N} M_{uv} \|WS_v - WS_u\|_2^2
\end{aligned} \quad (4)
$$

where α, β and γ control the weights of regularization terms. $\|H - WS\|_F^2$ represents the reconstruction error term. Given a sample $x_i \in X$, $f(x_i; W) = Wx_i$ is the linear predictive classifier. The second term is the inter-class classification projections incoherence promoting term. WS_i and WS_j denote the projection results of ith class and jth class, respectively. By minimizing the second term, the independency of inter-class classification projections will be encouraged, hence improving the discriminability of classifier. Although the classifier is discriminative, the intra-class compact is also crucial to robust classifier learning. Inspired by the idea of manifold learning, the intra-class compact graph is built in the transformed space. In our model the transformed space is generated by projecting the coding vector of samples and each graph node, i.e., WS_u, represents the classification projection of coding vector S_u which corresponding to sample x_u. Because the label vectors of the samples from the same classes are the same, we think that their projections should be maintained close together. In the third term of formula (4), M_{uv} denotes the weight between two graph nodes corresponding to two different samples x_u and x_v. Referring to LPP [22], we use the sample similarity to define the weight of compact graph. In our paper, the similarity between samples is calculated under following unsupervised form

$$
M_{uv} = \frac{1}{1 + \|x_u - x_v\|_2^2} \quad (5)
$$

From formula (5), we can find if samples x_u and x_v have the same labels, although their distance $\|x_u - x_v\|_2^2$ is intrinsically small, their similarity (graph weight) M_{uv} is big. Inversely, if x_u and x_v are from different classes, their distance is intrinsically big, but their similarity is small. Hence, minimizing the third term will simultaneously enhance the intra-class compactness and the inter-class discriminability of classifier. According to (5), we can rewrite function (4) as

$$
\begin{aligned}
g(H, W, S) = \alpha \|H - WS\|_F^2 \\
+ \beta \sum_{i=1}^{C} \sum_{j=1, j\neq i}^{C} \|(WS_i)^T(WS_j)\|_F^2 + \gamma tr(WSLS^TW^T)
\end{aligned} \quad (6)
$$

where $L = Z - M$ is the graph laplacian matrix. Z is a diagonal matrix and $Z_{ii} = \sum_{j=1}^{N} M_{ij}$. By combining the non-negative discriminative dictionary encouraging function (3) and the robust linear classifier learning function (6), the final formulation of the proposed NRDL algorithm is

$$<D, S, W> = arg \min_{D,S,W} \|X - DS\|_F^2 + \lambda\|D(A \odot S)\|_F^2$$

$$+ \alpha\|H - WS\|_F^2 + \beta \sum_{i=1}^{C} \sum_{j=1, j \neq i}^{C} \|(WS_i)^T(WS_j)\|_F^2$$

$$+ \gamma tr(WSLS^TW^T) \quad s.t. \ S \geq 0 \tag{7}$$

4   Solving the Optimization Problem of NRDL

We can find it is impossible to directly solve problem (7) because the variables, i.e., D, S, W, are interactional. In this section, we use the alternating direction method of multipliers (ADMM) [15] to update variables one by one, which means when updating one variable, all the others should be fixed. We first introduce two auxiliary variables to make the problem separable. Therefore, problem (7) can be rewritten as

$$<D, S, W, P, J> = arg \min_{D,S,W,P,J} \|X - DS\|_F^2 + \lambda\|D(A \odot S)\|_F^2$$
$$+ \alpha\|H - WS\|_F^2 + \beta \sum_{i=1}^{C} \sum_{j=1,j \neq i}^{C} \|P_i^T P_j\|_F^2 + \gamma tr(JLJ^T) \tag{8}$$
$$s.t. \ WS = P, WS = J, S \geq 0$$

Then, the augmented Lagrangian function L_μ of problem (8) is defined as

$$L_\mu(D, S, W, P, J, C_1, C_2) = \|X - DS\|_F^2 + \lambda\|D(A \odot S)\|_F^2$$
$$+ \alpha\|H - WS\|_F^2 + \beta \sum_{i=1}^{C} \sum_{j=1,j \neq i}^{C} \|P_i^T P_j\|_F^2 + \gamma tr(JLJ^T)$$
$$+ \mu\|WS - P + \tfrac{C_1}{\mu}\|_F^2 + \mu\|WS - J + \tfrac{C_2}{\mu}\|_F^2 - \tfrac{1}{\mu}(\|C_1\|_F^2 + \|C_2\|_F^2) \tag{9}$$
$$s.t. \ S \geq 0$$

where C_1 and C_2 are the Lagrangian multipliers, and $\mu > 0$ is the penalty parameter. Next starting the iterations:

Update S: By fixing variables J, W, P, D, C_1 and C_2, we update S as

$$S^{k+1} = arg \min_{S} \|X - DS\|_F^2 + \lambda\|D(A \odot S)\|_F^2$$
$$+ \alpha\|H - WS\|_F^2 + \mu\|WS - P + \tfrac{C_1}{\mu}\|_F^2 + \mu\|WS - J + \tfrac{C_2}{\mu}\|_F^2 \ s.t. \ S \geq 0 \tag{10}$$

Because $A = O - Q$, $D(A \odot S) = D[(O - Q) \odot S] = D(S - Q \odot S) = DS - D(Q \odot S)$. Let $R = D(Q \odot S^k)$, $\|D(A \odot S)\|_F^2$ becomes $\|DS - R\|_F^2$. By making the derivation of (10) with respect to S, the optimal S can be calculated as

$$S^{k+1} = [(\lambda + 1)D^k(D^k)^T + (\alpha + 2\mu)(W^k)^T W^k]^{-1}[(D^k)^T(X + \lambda R)$$
$$+ (W^k)^T(\alpha H + \mu P^k - C_1^k + \mu J^k - C_2^k)] \tag{11}$$

then in each iteration, we change the negative elements in S to be 0, thus generating non-negative representation.

Update W, P, J and D:

$$W^{k+1} = (\alpha H + \mu P_k - C_1^k + \mu J^k - C_2^k)(S^{k+1})^T[(\alpha + 2\mu)S^{k+1}(S^{k+1})^T]^{-1} \quad (12)$$

$$P_i^{k+1} = \left(\sum_{j=1,j\neq i}^{C} P_j^k (P_j^k)^T + \mu I \right)^{-1} (\mu W^{k+1} S_i^{k+1} + C_{1i}^k) \quad (13)$$

$$J^{k+1} = (\mu W^{k+1} S^{k+1} + C_2^k)(\gamma L + \mu I)^{-1} \quad (14)$$

$$D^{k+1} = X S^{k+1}[S^{k+1}(S^{k+1})^T + \lambda M M^T]^{-1} \quad (15)$$

Update Lagrangian multipliers C_1 and C_2:

$$C_1^{k+1} = C_1^k + \mu(W^{k+1} S^{k+1} - P^{k+1}) \quad (16)$$

$$C_2^{k+1} = C_2^k + \mu(W^{k+1} S^{k+1} - J^{k+1}) \quad (17)$$

Check the convergence:

$$if \max\{\|W^{k+1} S^{k+1} - P^{k+1}\|_\infty, \|W^{k+1} S^{k+1} - J^{k+1}\|_\infty\} \leq tol. \quad (18)$$

then stop the iterations.

5   Classification

After the discriminative dictionary $D = [D_1, D_2, \ldots, D_C] \in R^{d \times K}$ and robust linear classifier $W \in R^{C \times K}$ are obtained by NRDL, given a test sample $x_{new} \in R^d$, we first calculate its coding vector over the learned dictionary D. Because the dictionary and classifier are both learned with non-negative representation, here we use the NRC [8] model to solve the coding vector of x_{new}

$$\min_{\eta} \|x_{new} - D\eta\|_2^2 \quad s.t. \quad \eta > 0 \quad (19)$$

The used NR model is primarily the non-negative least squares problem, which does not have a closed-form solution. Referring to literature [8], we also utilize ADMM [15] to solve the NR model. Once we get the optimal coding vector $\hat{\eta}$, the classification of sample x_{new} is similar to algorithms [13] (See Algorithm 1 for details).

6   Experiments

In this section, some experiments were performed on five benchmark face databases: ORL [16], GT [17], CMU PIE [18], Extended Yale B [19], and Labeled Faces in the wild (LFW) [20]. To evaluate the performance of proposed NRDL

Algorithm 1. The classification approach based on NR model

Input: Learned dictionary D and classification projection matrix W, outside test sample x_{new}

1. Normalize x_{new} to have unit l_2 norm;
2. Code x_{new} over dictionary D via solving problem (19):
3. Calculate the classifier projection of coding vector $\hat{\eta}$: $f = W\hat{\eta}$;

Output: $Label(x_{new}) = arg\max_i\{f_i\}$, where f_i represents the ith entry of f.

---

algorithm, we compared it with the LLC [9], SRC [5], LRC [21], D-KSVD [10], LC-KSVD2 [11], FDDL [12], SDRERDL [9], and LCLE [13] algorithms. LLC solves the coding coefficients by utilizing the approximate LLC method. LLC, LRC, and SRC have no dictionary learning method and directly use the original training samples as their dictionary. All the compared dictionary learning algorithms used the same classification approach (linear classification). We used the LC-KSVD algorithm to initialize the dictionary. Following [11], sparsity factor $\xi = 30$ was used in the K-SVD, D-KSVD, LC-KSVD2, SDRERDL, and LCLE algorithms. For the sake of fairness, the number of local bases was identical to ξ in the LLC algorithm. Besides, in order to verify the effects of non-negative representation, we also tested the performance of our algorithm without non-negative constraint on representation (NRDL-test), then the CRC model [6] was used to solve the coding coefficients of test samples in NRDL-test. As shown in [9] and [13], the average recognition rates of almost all algorithms increased with the increase in the number of dictionary atoms. This is mainly because the reconstruction and discriminative ability of the dictionary improve with the increase of the number of atoms. Thus, in this paper, we tested all the dictionary learning algorithms with setting the number of dictionary atoms to the size of original training samples. The brief description of these datasets are shown in Table 1.

Table 1. Brief description of the used five datasets.

Dataset	Info.				
	Classes	Features	Total Num.	Train Num.	K
ORL	40	2576	400	240	240
GT	50	2000	750	250	250
CMU PIE	68	1024	11554	680	680
Extended Yale B	38	1024	2414	760	760
LFW	86	1024	1251	688	688

6.1   Experimental Results on Five Face Databases

The average recognition rates on five face databases are reported in Table 2. The sign ± represents the standard deviation of ten times results. As shown in Table 2, the proposed NRDL algorithms achieves almost the best recognition rates in comparison with all of the compared algorithms on different databases. This indicates that our NRDL algorithm is capable of effectively learning a discriminative dictionary from original samples under the condition of enforcing representation to be non-negative, and the learned dictionary is simultaneously robust for classification. Moreover, we can see that the recognition performance of NRDL is better than NRDL-test in all databases, especially in GT, CMU PIE and LFW databases. This is mainly because that the non-negative representation is beneficial for learning distinct features and eliminating useless information from heterogeneous data.

Table 2. Average recognition rates (%) of different algorithms on five databases

Algorithms	Databases				
	ORL	GT	CMU PIE	Extended Yale B	LFW
LLC	93.1	60.0	53.7 ± 0.016	88.9 ± 0.010	34.8 ± 0.011
LRC	**94.4**	59.4	61.6 ± 0.021	92.4 ± 0.008	37.1 ± 0.014
SRC	**94.4**	**63.8**	72.1 ± 0.008	95.3 ± 0.005	38.1 ± 0.011
D-KSVD	93.8	56.6	71.9 ± 0.008	83.0 ± 0.026	33.4 ± 0.016
LC-KSVD2	92.5	56.0	72.3 ± 0.009	92.7 ± 0.008	32.2 ± 0.012
FDDL	93.8	62.4	70.6 ± 0.020	93.0 ± 0.008	41.7 ± 0.016
SDRERDL	93.8	57.6	77.0 ± 0.006	96.0 ± 0.004	37.3 ± 0.013
LCLE	91.9	58.6	75.6 ± 0.009	95.8 ± 0.005	38.8 ± 0.009
NRDL-test	92.5	56.6	75.0 ± 0.009	93.2 ± 0.009	35.0 ± 0.010
NRDL (ours)	**94.4**	63.6	**81.0 ± 0.009**	**96.3 ± 0.004**	**43.1 ± 0.014**

6.2   Experimental Results with 'Salt and Pepper' Noise

To investigate the robustness of NRDL algorithm, we tested the performance of proposed NRDL algorithm and some relatively new dictionary algorithms, i.e., FDDL, SDRERDL and LCLE, by adding artificial noise in samples. In our experiments, we obtain contaminated images by using the Matlab function "imnoise" to impose 'Salt&Pepper' noise on all the original face images. In order to verify the performance with different degrees of contaminations, the density of noise (i.e., the third parameter of "imnoise" function) are set to 0.01, 0.02, and 0.03, respectively. The way of selecting samples of different databases are the same as the previous section. All experiments ran ten times and the rate was averaged. The average recognition rates of different algorithms are shown in Fig. 1. Figure 1 shows that the average recognition rates of the comparison algorithms and the proposed NRDL algorithm all decrease with the increase in the contamination degree of images. This is mainly because the 'Salt&Pepper' noise can destroy the subspace structure of data so that the learned features are not sufficient and distinct. Moreover, we can see the accuracy of our NRDL algorithm outperforms all the comparison algorithms on four databases, which demonstrates that NRDL is indeed more robust to artificial noise.

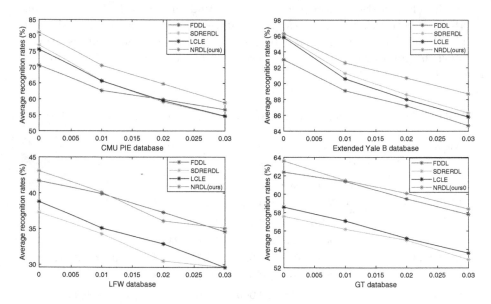

Fig. 1. Average recognition rates under different degrees of 'Salt&Pepper' noise contamination

6.3 Convergence Validation

The convergence of the proposed NRDL algorithm on four databases is illustrated in Fig. 2. As expected, the convergence of the objective function (7) is very fast.

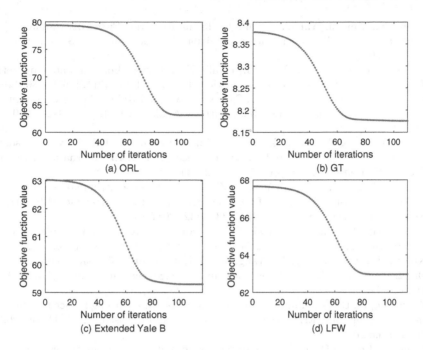

Fig. 2. Objective function value versus the number of iterations of the proposed NRDL algorithm on four databases.

7   Conclusion

In this paper, a non-negative representation based discriminative dictionary learning algorithm (NRDL) is proposed for multi-class face classification. Different from other dictionary learning methods, NRDL considers to learn a more discriminative and reconstructive dictionary with non-negative representation constraint. Specifically, NRDL is designed to incorporate non-negative representation learning, discriminative dictionary learning and robust linear classifier learning into a unified framework. As a result, the learned dictionary is effective for classification. Experimental results on five face databases indicate that the NRDL algorithm is superior to the nine state-of-the-art sparse coding and dictionary learning algorithms.

References

1. Chen, Z., Wu, X.J., Kittler, J.: A sparse regularized nuclear norm based matrix regression for face recognition with contiguous occlusion. Pattern Recogn. Lett. **125**, 494–499 (2019)
2. Wu, X.-J., Kittler, J., Yang, J.-Y., et al.: A new direct LDA (D-LDA) algorithm for feature extraction in face recognition. In: 2004 Proceedings of the 17th International Conference on Pattern Recognition, ICPR 2004, vol. 4, pp. 545–548. IEEE (2004)

3. Zheng, Y., Yang, J., Yang, J., et al.: A reformative kernel Fisher discriminant algorithm and its application to face recognition. Neurocomputing **69**(13–15), 1806–1810 (2006)
4. Zheng, Y.J., Yang, J.Y., Yang, J., et al.: Nearest neighbour line nonparametric discriminant analysis for feature extraction. Electron. Lett. **42**(12), 679–680 (2006)
5. Wright, J., Yang, A., Ganesh, A., Sastry, S., Ma, Y.: Robust face recognition via sparse representation. IEEE Trans. Pattern Anal. Mach. Intell. **31**(2), 210–227 (2009)
6. Zhang, L., Yang, M., Feng, X.: Sparse representation or collaborative representation: which helps face recognition? In: Proceedings of IEEE International Conference on Computer Vision, pp. 471–478 (2011)
7. Xu, Y., Zhong, Z., Yang, J., You, J., Zhang, D.: A new discriminative sparse representation method for robust face recognition via l2 regularization. IEEE Trans. Neural Netw. Learn. Syst. **28**(10), 2233–2242 (2017)
8. Xu, J., An, W., Zhang, L., et al.: Sparse, collaborative, or nonnegative representation: which helps pattern classification? Pattern Recogn. **88**, 679–688 (2019)
9. Xu, Y., Li, Z.M., Zhang, B., Yang, J., You, J.: Sample diversity, representation effectiveness and robust dictionary learning for face recognition. Inf. Sci. **375**, 171–182 (2017)
10. Zhang, Q., Li, B.: Discriminative K-SVD for dictionary learning in face recognition. In: Proceedings of IEEE Conference on CVPR, San Francisco, CA, USA, pp. 2691–2698, June 2010
11. Jiang, Z., Lin, Z., Davis, L.S.: Learning a discriminative dictionary for sparse coding via label consistent K-SVD. In: Proceedings of IEEE CVPR, Providence, RI, USA, pp. 1697–1704, June 2011
12. Yang, M., Zhang, L., Feng, X., Zhang, D.: Fisher discrimination dictionary learning for sparse representation. In: Proceedings of 13th ICCV, pp. 543–550, November 2011
13. Li, Z., Lai, Z., Xu, Y., et al.: A locality-constrained and label embedding dictionary learning algorithm for image classification. IEEE Trans. Neural Netw. Learn. Syst. **28**(2), 278–293 (2015)
14. Fang, X., Xu, Y., Lai, Z., Wong, W., Fang, B.: Regularized label relaxation linear regression. IEEE Trans. Neural Netw. Learn. Syst. 1–13 (2017)
15. Boyd, S., Parikh, N., Chu, E., Peleato, B., Eckstein, J.: Distributed optimization and statistical learning via the alternating direction method of multipliers. Found. Trends Mach. Learn. **3**(1), 11–22 (2011)
16. Samaria, F.S., Harter, A.C.: Parameterisation of a stochastic model for human face identification. In: Proceedings of 2nd IEEE Workshop on Applications of Computer Vision, pp. 138–142 (1994)
17. Goel, N., Bebis, G., Nefian, A.: Face recognition experiments with random projection. In: Proceedings of the SPIE, vol. 5779, pp. 426–437 (2005)
18. Gross, R., Matthews, I., Cohn, J., Kanade, T., Baker, S.: Multi-PIE. In: Proceedings of the IEEE International Conference on Automatic Face and Gesture Recognition, pp. 1–8, September 2008
19. Georghiades, A.S., Belhumeur, P.N., Kriegman, D.: From few to many: illumination cone models for face recognition under variable lighting and pose. IEEE Trans. Pattern Anal. Mach. Intell. **23**(6), 643–660 (2001)

20. Huang, G.B., Ramesh, M., Berg, T., Learned-Miller, E.: Labeled faces in the wild: a database for studying face recognition in unconstrained environments. Technical report 07–49, College of Computer Science, University of Massachusetts, Amherst, MA, USA, October 2007

21. Naseem, I., Togneri, R., Bennamoun, M.: Linear regression for face recognition. IEEE Trans. Pattern Anal. Mach. Intell. **32**(11), 2106–2112 (2010)

22. He, X., Niyogi, P.: Locality preserving projections. In: Proceedings of Neural Information Processing Systems, pp. 153–160 (2003)

Iterative Face Detection
from the Global to Local

Jingdong Ma$^{(\boxtimes)}$ and Yupin Luo

Tsinghua National Laboratory for Information Science and Technology (TNList),
Department of Automation, Tsinghua University, Beijing 100084, China
majd14@mails.tsinghua.edu.cn

Abstract. To balance between accuracy and speed is a dilemma in face detection, and scale variance problem is one of the main causes. In this paper we propose fast and accurate iterative face detection method processing from the global to local. We define a class of object called "probable regions" which contain small faces. "Probable regions" are detected iteratively in order to enlarge the small face parts. Thus small faces turn into large faces after several iterations. We design a strategy of training samples augmentation to meet the requirement for two object classes, so that extra annotation is unneeded. Our method is simple and clear to deploy. Experiments show that our method achieves competitive accuracy with real time speed. Detection time consumption will not explicitly grow when resolution of sample image increases. The speed is merely related to actual amount of faces, which adapts to real world applications.

Keywords: Face detection · Scale variance · Iterative method ·
Real-time detection

1 Introduction

The task of face detection aims at locating all the faces in a image, as to calculate their positions and scales [1]. Recently the demand for face detection in the wild arouses. In applications like online image retrieval and video surveillance, the scale of faces vary incredibly. It is challengeable to balance between detection accuracy and speed, and scale variance problem is one of the main causes, attracting many researchers' attention. With massive images gathered on the Internet and the wide deployment of GPU, deep convolutional neural networks (CNNs) play an important part in computer vision [2]. Considering scale variation problem, face detection methods based on CNNs can be divided into two types. The first one is "separate method". These methods detect faces with different scales separately. Large faces and small faces are detected in pyramidal images or different brunches of the networks. Another one is "integrate method". These methods treat large and small faces as a whole group, and train "integrate models" to detect faces of different scales simultaneously. Generally

© Springer Nature Switzerland AG 2019
Y. Zhao et al. (Eds.): ICIG 2019, LNCS 11901, pp. 320–331, 2019.
https://doi.org/10.1007/978-3-030-34120-6_26

"separate methods" are more accurate while "integrate methods" achieve faster detection speed.

However these methods are all designed to weigh small faces equally with large ones. Theoretically it takes the same long time of these methods to locate a small face and a large one. Therefore they whether process in large face region redundantly or omit details in small face region. We propose an iterative detection method that process from the global to local, treating variant scales of faces diversely. We define a class of object called "probable region", which contains small faces. As an example shown in Fig. 1, we enlarge small faces through iteratively detecting "probable region", which serves similarly as proposal stage. Eventually small faces are turned into large ones. It is simpler to locate larger faces resulting in less mistakes. The final detection of large faces can be regarded as a refinement stage. Our iterative method is adaptive according to the scale of target face. Furthermore experiments show that its time consumption is not necessarily related to resolution of input images, but rather affected by amount of faces, which is another superiority of our method.

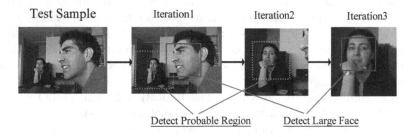

Fig. 1. An example of our iterative detection method

The following parts are as below. Section 2 is about related works, including "separate methods" and "integrate methods". In Sect. 3 we define the "probable region" and turn the task of detecting small faces into the task of detecting two classes of object: "probable regions" and large faces. In Sect. 4 we test the performance of proposed method. The conclusions are in Sect. 5.

2   Related Works

Many studies in "separate methods" are pursuing faster detection speed meanwhile retaining high accuracy. DDFD [3] generates candidate boxes with sliding windows, and afterwards classifies the boxes with a state-of-the-art networks structure in ILSVRC-2010 [4] called AlexNet [5]. However, the size of faces in the wild varies in a wide range of scales. CNNs lack the ability to deal with scale variation naturally [6]. DDFD has to generate candidate boxes on pyramidal images of scales with sliding windows. The amount of candidate boxes rises quadratically as the resolution of sample image increases [2], which leads to detection speed sharply dropping. Several studies focus on speeding up "separate" methods. CASCNN [7] and MTCNN [8] take advantage of cascaded structures. Although it is valid to reject non-faces candidate boxes in early stages by fast shallow networks, it is only a compromise on scale variation problem, since that the pyramidal images are still necessary in testing processes. RSA [6] and SAFD [9] prove the capability of CNNs to estimate the potential scales of faces, which helps cutting down number of scales in image pyramids. Innovated by SPP-Net [10], Faster-RCNN [11] introduces Region Proposal Networks with anchors to generate candidate boxes instead of sliding windows on pyramidal images. As pointed out in [12], it is difficult to classify anchors of different sizes with the same set of features. Still it is not satisfied to real-time detection demands [13]. SSH [14] tries to solve the scale variation problem by designing three brunches in networks to separately deal with large, medium and small faces. However, the features learned by brunches from different scales are not shared, leading to low efficiency.

"Integrate methods" are free of candidate boxes generating step, and implement the detection process in one shot by a single model. Therefore they are faster than "separate methods". YOLO [15] is a typical "integrate method" with high detection speed. The sample is resized to a fixed resolution as $P \times P$. Then grids with fixed $N \times N$ shape are initialized, and each grid is responsible for multiple L targets. If the center of a face lies within a grid, the grid will converge towards the coordinate of the face. Train YOLO networks with annotated samples containing faces coordinates. YOLO achieves real-time speed which is much faster than "separate" methods. The limitation is that YOLO performs well merely on samples with little variation in scales of face size. G-CNN [16] is another "integrate method". In order to cover various size scales, grids of multiple sizes are placed on the image as initial bounding boxes. Then iteratively converge the bounding boxes towards target boxes by stepwise regression model. G-CNN noticed the potential advantage of iterative detection. However its iteration number is fixed and thus unable to change due to testing sample resolution. No matter how large the resolution is, it takes same long time for detection. Other studies like S3FD [17] and FaceBoxes [18] implement scale-invariance by linking the output layer with intermediate ones in order to enhance feature extraction capability on different scales. The drawback is that small instances may confuse with large ones in such integrate models and false detections happen commonly.

3   Iterative Detection of "Probable Region"

In the wild when size scales of faces vary in wide range, YOLO performs poor on balancing between accuracies of large faces and small faces. The accuracy on small faces are limited due to previous resizing process which omit the details of image. We define a new class of object called "probable region" to handle small faces. We limit the variation range of its size scale, leading to more accurate detection. The definition of "probable region" is stated in Sect. 3.1. We propose iterative detection method in Sect. 3.2, making use of "probable region" to balance between the large and the small, meanwhile sticking on real-time speed. In

Sect. 3.3 we show how to convert small face annotation into "probable region" annotation in preparation of training samples. The conversion of annotations is the key point.

3.1 Definition of "Probable Region"

Face targets are treated as two separate classes of large ones and small ones. Suppose the area ratio of annotation bounding box in the sample image is α. Set the threshold to separate small faces from large faces as R. Label large faces with area ratio in $R \le \alpha_0 < 1.0$ as class "F". Thus the range of small faces area ratio is $r \le \alpha_0 < R$, and r is the least face area ratio which the system is capable of dealing with. The area ratio of "probable region" is similarly limited. Set the least area ratio as q. Any subregions in the sample image whose area ratio is in the range of $q \le \alpha_X < 1.0$, and moreover of which there exists at least one small face in the center, are regarded as "probable region", which is labeled as class "X". The area size of these two classes of objects varies in a confined range, leading to simplicity for detection. An example of "probable region" is illustrated in Fig. 2.

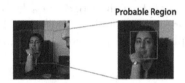

Fig. 2. An example of probable region

3.2 Iterative Detection Process

The original single-class face detection task is converted to a two-classes detection task, i.e. to detect large faces and "probable regions". The detection pipeline is shown in Fig. 3. We adopt YOLO structure as the two-classes object detection networks, setting object classes as two instead of one. Notice that "probable region" which is a subregion of the sample image, is naturally smaller than the sample image. Hence the area ratio of small face in the "probable region" is enlarged than originally in the sample image. Repeat the detection process iteratively in "probable regions" and the small faces in the sample image are gradually turned into large faces in some subregions. Eventually, large faces detected within any "probable regions" in every iterations are picked out as the final outputs.

Our method is inherently similar to cascaded structures like [7,8,19,23] to conduct from coarse to fine. In iterative detection the time consumption varies according to the size of faces. The larger ones are rapidly picked out on the global level in early iterations, while the smaller ones are on the local level in latter iterations. Our method balance accuracy and speed. A compare of detection speed will be presented in latter experiment part.

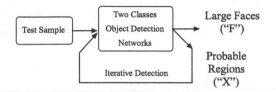

Fig. 3. The process of iterative detection from the global to local

3.3 Annotation of Train Samples

Existing datasets contain only one class of annotation as human face. Iterative detection requires annotated training samples of two classes as large faces and "probable regions". We convert the original one class annotations to the two classes annotations. The annotations of large faces are inherited. Samples with annotations of small faces are used for generating samples with annotations of "probable regions". As illustrated in Fig. 4(a), the center of "probable region" bounding box labeled as "X" (dashed line box) coincides with a small face, while the size is related to both the size of small face and the size of sample image. On one hand, select a parameter $t(t > 1.0)$ as expansion ratio of small face, ensuring in the next iteration the small face shall become a large one to be picked out. One the other hand, according to definition in Sect. 3.1 the least area ratio of "probable region" to sample image is q. Therefore, the area of "probable region" S_X is

$$S_X = max(t \cdot S_{Face}, q \cdot S_{Image}) \tag{1}$$

S_{Face} is the area of small face, and S_{Image} is the area of sample image. In such a way, the area ratio of "probable region" to sample image varies adaptively to small face size within a restricted range.

Fig. 4. An example of generating training samples (artificially annotated small face as dot line box, "X" annotation as dash line box, and "F" annotation as solid line box).

As stated in Sect. 3.2, eventually the output is merely the detected bounding boxes of large faces. In order to accurately detect small faces, we carry on the process of cropping the "probable regions" mapping back to a certain region in the high resolution testing sample. If small face within the region is not large enough, i.e. $\alpha < R$, another training sample is generated and annotated as shown in Fig. 4(b). Repeat such annotate-and-crop process until the area ratio of face $\alpha \geq R$, label the face as "F", as solid line box shown in Fig. 4(c). After such a training samples augmentation, all the training samples are labeled only with large faces and "probable regions".

4   Experiments and Analysis

The backbone networks in our iterative method are GoogLeNet [20] which is relatively high in accuracy while small in model size. Since the performance is proved to be better to finetune on ImageNet [4] pretrained models [21], we also follow the same finetuning strategy in our experiments. We randomly select 15663 images for training and 674 for validation from AFLW dataset [22]. The testing dataset contains totally 721 images with 1078 faces in various scales. In Sect. 4.1 we test the amount of "probable region" and the accuracy in each iteration. In Sect. 4.2 we analyse the impact of average times of iteration on face detection recall rate and detection speed. In Sect. 4.3 we compare our iterative method with other face detection methods on precision and recall rate. In Sect. 4.4 we evaluate efficiency of our iterative method by detection speed.

4.1   "Probable Region" in Each Iteration

In this section we test and analyse the effectiveness of "probable region". "Probable Region" is the key in our iterative detection method. The amount of "probable region" effects detection efficiency. The most ideal condition is that for every small face only one "probable region" is proposed in each iteration. More pieces of "probable region" lead to lower detection speed, and may bring about duplicated detection on the same small face.

We conduct our iterative detection on the testing dataset with the following parameters. Input data shape of object detection networks is $P \times P = 320 \times 320$. The shape of grids is $N \times N = 5 \times 5$. The number of detected objects in each grid is $L = 1$. The number of classes is $M = 2$, including class "F" for large faces and class "X" for "probable regions". The threshold to separate small faces from large ones is $R = 0.16$. The least area ratio of "probable region" to image sample is $q = 0.04$. As analysed in Sect. 3.3 we select the parameter $t = 4.0$. The threshold for filtering "probable region" is $Threshold_X = 0.05$.

Table 1. Detected probable region in each iteration

Order of iterations	λ_0	λ_1	δ
1	2.30	1.71	72.02%
2	3.52	2.70	74.11%
3	2.14	1.91	57.05%
4	2.14	1.86	44.44%

We count the total numbers of "probable regions" and small faces in each iteration, and list their ratio λ_0 in the first column of Table 1. The results show that λ_0 are greater than 1.0 in all the four iterations, revealing that two or more "probable regions" may correspond to a single small face. The cause of this is due to the fact that small faces are not always positioned completely inside a grid, instead they possibly intersect with two or four grids. In such conditions, several grids nearby all regress towards the same small face target, which results in the existance of the same small face in several "probable regions".

We reduce the amount of "probable regions" which stay very close by Non-Maximum-Suppression (NMS) algorithm. The ratio of total numbers of "probable regions" after NMS to small faces are listed in the second column of Table 1 as λ_1. The total number of "probable regions" reduces by 10%–26% after NMS.

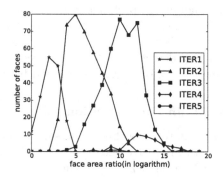

Fig. 5. The distribution of detected face sizes in each iteration (smaller to the right)

The accuracy of "probable region" also affects detection efficiency. The most ideal phenomenon is that all "probable regions" contain at least one small face. To evaluate the accuracy of "probable region", we list the proportions of "probable regions" which contain small faces for every iteration as δ in the third column of Table 1. We find that the proportions are higher in the early two iterations, but drop down in the latter two iterations. Regions in sample image are enlarged in iterations, and noises effect detection accuracy, leading to proposed "probable regions" which contain no faces in latter iterations. "Probable regions" proposed in K-th iteration are test samples for $(K + 1)$-th iteration.

Figure 5 shows the size distribution of faces picked out in each iteration. Most faces are picked out in the early three iterations. So detection efficiency is mostly affected by the "probable regions" proposed in the early two iterations. Though the dropped accuracy rate in latter iterations brings negative effects on detection efficiency, the effects are slight. Experiments in this section illustrate that it is feasible and efficient to detect small faces by proposing "probable regions". Figure 5 also shows that it requires adaptive numbers of iterations to faces with various sizes. Smaller faces require more iterations, which meets with our purpose.

4.2   Average Number of Iterations

It is possible to recall more small faces with more average number of iterations. The following experiments are on the effects of average number of iterations to detection performance. Original YOLO and Tri-Nets method is for comparison. YOLO is equivalent to conducting one iteration of detection for a testing sample (1.0*). Tri-Nets method is an instinctive method. Training samples are divided into three groups, as large, medium and small faces. Then train three independent networks with each group of training samples. In testing process each network only picks out faces within the corresponding size range. Such a method relieve the hardness of balance on recall rates of different sizes to some degree. It is equivalent to conducting three iterations of detection for a testing sample (3.0*). Tri-Nets method is not a contribution, but simply a plain idea to demonstrate the benefit of our iterative method.

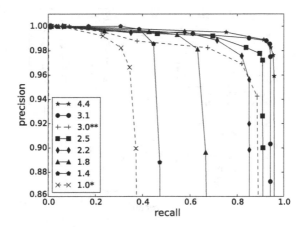

Fig. 6. The Precision-Recall-Curve under various average iteration times

We illustrate Precision-Recall-Curve of our iterative method, YOLO and Tri-Nets under a variety of average number of iterations in Fig. 6. The performance of our iterative method is notably better than YOLO. The recall rate surpass Tri-Nets with less iterations as 2.5. Locating small faces through "probable regions" is purposeful, while Tri-Nets compute in background regions three times redundantly resulting in time consumption.

With more iterations, the recall rate increases while detection speed also drops. When the average number of iterations is above 3.1, the growth of recall rate is subtle. So choosing 3.1 iterations is a reasonable balance between recall rate and detection speed.

Most false positive instances and failures of recall are attribute to low scores of small faces which are undistinguishable from regions in background. In our method for each face instance the last iteration is naturally detection of large face, which has high score distinguishable from the background, leading to less mistakes. In Fig. 6, the curve of 2.5 iterations our method is better than YOLO and Tri-Nets. Our method is suitable for application conditions that pursue low false positive rate.

4.3   Comparison of Detection Performances

We compare the performance of our method with several other methods. Precision-Recall-Curves are shown in Fig. 7. The compared methods include two types. The first type is "separate method", including RSA [6], CASCNN [7], MTCNN [8], Faster-RCNN [11] and SSH [14]. The second type is "integrate method", including YOLO [15], S3FD [17] and FaceBoxes [18]. "Separate methods" are generally better than "integrate methods" in accuracy. Our method is competitive even compared with "separate methods".

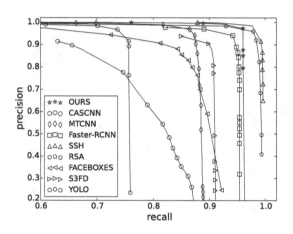

Fig. 7. Comparison of Precision-Recall-Curve with other methods

4.4   Detection Efficiency

Basically detection time increases as testing sample image becomes larger. We evaluate the effect of image resolution on several detection methods, as illustrated in Fig. 8. The horizontal axis is the size of image, presented in height × width. Images within a size range are gathered as group 0 to 6, larger to the right. The vertical axis is average detection time for a particular method on each group in logarithm. Experiments are all conducted under same condition, on a computer with Intel Core i7-3770 CPU and one NVIDIA GTX TITAN X GPU. In "separate methods" detection time sharply increases as the image becomes larger. On the other hand, in "integrate methods" detection time remains unchanged as a result of their fixed input data shape, which often leads to waste computational resource on larger ones or omission on details of smaller ones.

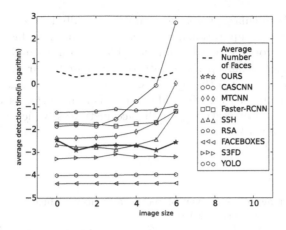

Fig. 8. Average detection time for different image sizes

It doesn't mean that there always have to be more faces in a high resolution image. We present the average numbers of faces in each group, illustrated in Fig. 8 as dashed line. We believe that actual number of faces in the image is a better factor than the size of image, which time consumption of detection should correspond to. In our method the first iteration is processed on the global of image avoiding unnecessary search on local regions where no face exists at all. The latter iterations focus purposefully on local regions where smaller faces probably exist avoiding missing recalls of the small faces. Our method's curve is highly related to the dashed line, revealing that the time consumption varies according to the number of faces in testing image.

Table 2. Detection speeds of "separate methods"

Methods	**Ours**	SSH	Faster-RCNN	MTCNN	RSA	CASCNN
FPS	**12.91**	10.26	5.29	3.92	3.11	0.30

Average detection speeds of "separate methods" are shown in Table 2. Benefited from the iterative structure, our method is the fastest among "separate methods", reaching real-time speed. Our method balances well on accuracy and speed. With accuracy level of "separate method", a faster speed is the advantage.

5 Conclusions

We propose a solution to scale variance problem as iterative detection method that detect faces from the global to local regions. For large faces it is a direct regression. While for small faces the process is iterative, through "probable region" proposals to enlarge small faces and eventually locate them. It is a reasonable allocation of computational resource to focus on small faces. Experiments show that with "probable region" it is feasible to locate small faces efficiently and accurately, as the false positive rate is suppressed. Our method adapts to application conditions when the size scale of faces in sample images, or the resolution of sample images vary in wide range. But it performs not so well on extremely intersected or scattered small face instances, which requires further study in the future works.

References

1. Zhang, C., Zhang, Z.: A survey of recent advances in face detection. Technical report, Microsoft Research (2010)
2. Chang, L., Deng, X., Zhou, M., et al.: Convolution neural network in image understanding. Autom. J. **42**(9), 1300–1312 (2016)
3. Farfade, S.S., Saberian, M.J., Li, L.J.: Multi-view face detection using deep convolutional neural networks. In: Proceedings of the 5th ACM on International Conference on Multimedia Retrieval, ICMR 2015, pp. 643–650. ACM, New York (2015)
4. Russakovsky, O., et al.: ImageNet large scale visual recognition challenge. IJCV **115**, 211–252 (2015)
5. Krizhevsky, A., Sutskever, I., Hinton, G.E.: ImageNet classification with deep convolutional neural networks. In: Neural Information Processing Systems Conference, vol. 25, no. 2, pp. 109–1105 (2012)
6. Liu, Y., Li, H., Yan, J., Wei, F., Wang, X., Tang, X.: Recurrent scale approximation for object detection in CNN. In: IEEE International Conference on Computer Vision (2017)
7. Li, H., Lin, Z., Shen, X., Brandt, J., Hua, G.: A convolutional neural network cascade for face detection. In: Proceedings of the IEEE Conference on Computer Vision and Pattern Recognition, pp. 5325–5334 (2015)
8. Zhang, K., Zhang, Z., Li, Z., Qiao, Y.: Joint face detection and alignment using multi-task cascaded convolutional networks. IEEE Sig. Process. Lett. **23**(10), 1499–1503 (2016)
9. Hao, Z., Liu, Y., Qin, H., Yan, J., Li, X., Hu, X.: Scale-aware face detection. In: The IEEE Conference on Computer Vision and Pattern Recognition (CVPR) (2017)
10. He, K., Zhang, X., Ren, S., Sun, J.: Spatial pyramid pooling in deep convolutional networks for visual recognition. In: Fleet, D., Pajdla, T., Schiele, B., Tuytelaars, T. (eds.) ECCV 2014. LNCS, vol. 8691, pp. 346–361. Springer, Cham (2014). https://doi.org/10.1007/978-3-319-10578-9_23
11. Ren, S., He, K., Girshick, R., Sun, J.: Faster R-CNN: towards real-time object detection with region proposal networks. In: Advances in Neural Information Processing Systems, pp. 91–99 (2015)
12. Li, H., Liu, Y., Ouyang, W., et al.: Zoom out-and-in network with map attention decision for region proposal and object detection. Int. J. Comput. Vis. **127**(3), 225–238 (2019)
13. Zhang, H., Wang, K., Wang, F., et al.: Advances and perspectives on applications of deep learning in visual object detection. Autom. J. **43**(8), 1289–1305 (2017)
14. Najibi, M., Samangouei, P., Chellappa, R., Davis, L.S.: SSH: single stage headless face detector. In: IEEE International Conference on Computer Vision (ICCV) (2017)

15. Redmon, J., Divvala, S., Girshick, R., Farhadi, A.: You only look once: unified, real-time object detection. In: CVPR (2016)
16. Najibi, M., Rastegari, M., Davis, L.S.: G-CNN: an iterative grid based object detector. In: IEEE Conference on Computer Vision and Pattern Recognition, pp. 2369–2377 (2016)
17. Zhang, S., Zhu, X., Lei, Z., Shi, H., Wang, X., Li, S.Z.: S3FD: single shot scale-invariant face detector. In: IEEE International Conference on Computer Vision (ICCV) (2017)
18. Zhang, S., Zhu, X., Lei, Z., Shi, H., Wang, X., Li, S.Z.: FaceBoxes: a CPU real-time face detector with high accuracy. In: IJCB (2017)
19. Viola, P., Jones, M.: Robust real-time face detection. IJCV **57**, 137–154 (2004)
20. Szegedy, C., et al.: Going deeper with convolutions. Technical report (2014)
21. Agrawal, P., Girshick, R., Malik, J.: Analyzing the performance of multilayer neural networks for object recognition. In: Fleet, D., Pajdla, T., Schiele, B., Tuytelaars, T. (eds.) ECCV 2014. LNCS, vol. 8695, pp. 329–344. Springer, Cham (2014). https://doi.org/10.1007/978-3-319-10584-0_22
22. Kostinger, M., Wohlhart, P., Roth, P.M., Bischof, H.: Annotated facial landmarks in the wild: a large-scale, real-world database for facial landmark localization. In: ICCV Workshops, pp. 2144–2151 (2011)
23. Shi, X., Shan, S., Kan, M., et al.: Real-time rotation-invariant face detection with progressive calibration networks. In: Proceedings of the IEEE Conference on Computer Vision and Pattern Recognition, pp. 2295–2303 (2018)

Challenges Driven Network
for Visual Tracking

Jiaming Wei, Huimin Ma$^{(\boxtimes)}$, and Ruiqi Lu

Department of Electronic Engineering,
Tsinghua University, Beijing 100084, China
{wjm16,lrq17}@mails.tsinghua.edu.cn,
mhmpub@tsinghua.edu.cn

Abstract. Tracking-by-detection is an effective approach for regular objects tracking. However, trackers still drift under challenges like fast motion and occlusion in only several frames. In this paper, we propose the challenges driven network (CDNet) for motion and occlusion handling. First, as for proposal, we provide a motion-aware proposal method utilizing cues of optical flow and previous target states. Second, as for classification, a challenges driven feature learning approach is constructed by combining dual multi-domain learning and category prior learning. The dual multi-domain method improves the network responses to both challenge and common targets, and the steady category prior features are utilized to detect and avoid tracking drift. Experiments on the OTB100 [19] and VOT2015 [10] datasets demonstrate that our CDNet can achieve competitive performance against state-of-art trackers.

Keywords: Visual tracking · Challenge targets · Motion change · Scale change · Occlusion

1 Introduction

Tracking-by-detection framework is widely used in current state-of-art trackers, which consists of two main stages: proposal and classification. This framework is highly effective for regular objects tracking. However, it usually fails under the challenge of dynamic motion and occlusion. In the proposal stage, the Gaussian sampling used in [8,13,14] only relies on the previous target location and can not deal with fast motion and deformation. Selective search [18] and region proposal networks [17] only consider the contrast between objects and usually ignore the small and occluded target. As for the classification stage, networks on-line updating suffers from frequent sudden changes of the target appearance features caused by motion and occlusion. Meanwhile, the challenge targets are usually small and textureless, leading to misclassification as backgrounds.

The challenges driven network (CDNet) is proposed to focus on the properties of challenge targets and handle dynamic motion and occlusion. Main contributions of our work are summarized as the following three folds:

© Springer Nature Switzerland AG 2019
Y. Zhao et al. (Eds.): ICIG 2019, LNCS 11901, pp. 332–344, 2019.
https://doi.org/10.1007/978-3-030-34120-6_27

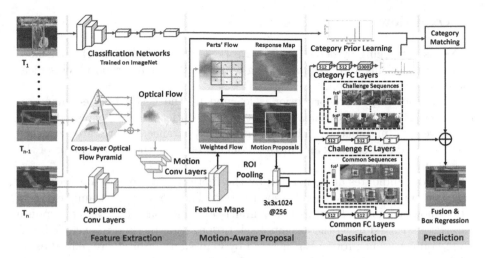

Fig. 1. Overview of our challenges driven tracking model. First, the target category prior is learned in the first frame. The motion and appearance features are extracted during tracking. Second, region proposals are generated by the motion-aware proposal method. Third, the proposals are classified by the multi-path FC layers which are trained by the challenges driven learning method. Finally, high score proposals are fused and the box regression is performed to refine the target state

(1) Motion-aware proposal: We provide the proposal method for motion and deformation handling by making use of optical flow and previous target states. It utilizes the cross-layer pyramid to obtain optical flow as motion features. The motion features are further weighted by target response map and are used to refine proposals extracted by Gaussian sampling.

(2) Challenges driven feature learning: The learning approach is proposed to adapt to target feature variation under motion and occlusion. The dual multi-domain learning method incorporates the properties of challenge and common objects respectively and improves the network response to challenge targets. The category classifier learns the target category prior to resist the influence of other objects and avoid tracking drift.

(3) Motion and occlusion handling: We handle the challenge of motion and occlusion in one framework. The motion features which are extracted by the cross-layer optical flow pyramid provide richer features for both proposal and classification stages under challenges. Furthermore, the target classifiers have been pre-trained to adapt to the motion and appearance feature changes under dynamic motion and occlusion.

Our CDNet is evaluated on two large-scale tracking benchmarks, the OTB100 [19] and VOT2015 [10] datasets. The experiments demonstrate that our tracker can effectively deal with dynamic motion and occlusion, and achieve competitive performance against state-of-art trackers.

2   Overview

CDNet is a tracking-by-detection framework as shown in Fig. 1. This tracking framework contains four parts including feature extraction, proposal, classification and target prediction.

First, the target category prior is learned by the networks pre-trained on ImageNet [11] in the first frame. During tracking, optical flow is calculated by the cross-layer pyramid method. The motion and appearance features are extracted by corresponding networks and further concatenated as the feature maps of the current image.

Second, candidates are generated by the motion-aware proposal which combines two optical flow and target response map. The cross-layer paramid extracts optical flow as motion features under multi-scale motions. The target response map is constructed using the weights back propagation method to exclude background objects. The proposals generated by Gaussian sampling are refined using weighted part-based motion features. RoI pooling is performed to generate convolutional features for each proposal.

Third, the candidates scores are calculated by multi-path FC layers trained using challenges driven feature learning method, which consists of the category prior learning and the dual multi-domain learning. The category features of each proposal are extracted by classification networks trained on ImageNet [11] dataset and are compared with the category prior of the true target. The challenge and common FC layers which are trained using dual multi-domain method have higher response to challenge or common objects respectively.

Finally, the target score of each proposal is computed by fusing scores of multi-path FC layers. High score proposals will be intergrated and further refined by a linear box regression. The refined box is predicted as the optimal target state of the current frame.

3   Motion-Aware Proposal

3.1   Cross-Layer Optical Flow Pyramid

We propose a cross-layer pyramid method to calculate multi-scale optical flow, in order to adapt to different motion and different sizes of targets. The cross-layer pyramid is shown in the feature extraction stage of Fig. 1. The previous and current frames are resized to $\frac{1}{8}$, $\frac{1}{4}$ and $\frac{1}{2}$ of original size. The $\frac{1}{8}$ size optical flow is calculated in the first place by Horn-Schunck method [9] and used as initialization of $\frac{1}{4}$ size optical flow, and so on. Small size images are sensitive to large motion and large images are sensitive to small motion. Finally, optical flows of all sizes are averaged as the motion features of the target.

3.2 Target Response Map

In order to exclude background motions in the target box, the Target response map is constructed by weighted averaging feature response maps of the current frame. The weights are contributions of feature maps to the target classification score.

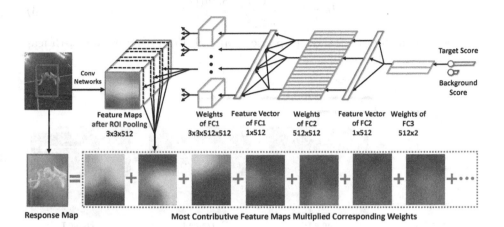

Fig. 2. A figure illustrating our one-to-three weights back propagation method. Target response map is obtained by weighted averaging the feature maps which are most contributive to the classification. We first obtain target scores of proposals through feeding forward process. Then, the multiplicative weights of feature maps are back propagated through the fully connected layers using the one-to-three method

A one-to-three weight back propagation method is proposed to calculate the target classification contributions of the feature maps as Fig. 2. First, the target scores are obtained by feeding the feature maps of proposals forward through fully connected layers. Second, the back propagation procedure is only conducted on the proposal with the highest target score, formulated as Eq. 1.

$$C(l_m, S_t) = C(l_m, l_1) \times C(l_1, l_2) \times C(l_2, S_t) \tag{1}$$

Where C denotes contribution, l_m, l_1 and l_2 denote contributive channels of feature maps, the first and the second fully connected layers respectively, S_t denotes the target score of the most positive proposal. Adjacent contributions are defined as Eq. 2.

$$C(l_1, l_2) = \max(0, F_{l_1} * W_{l_1}^{l_2}) \tag{2}$$

Where $F \in \mathbb{R}^{w \times h \times l_1}$ denotes feature maps or feature vectors, $W \in \mathbb{R}^{w \times h \times l_1 \times l_2}$ denotes weights from current layer to the next layer, l_1 is channel of current layer and l_2 is channel of the next layer, $C \in \mathbb{R}^{l_1 \times l_2}$ is the contribution matrix from current layer to the next layer, $*$ denotes the convolution operator. For each contributive column of C, the one-to-three method finds the three largest elements, and get the row indexes representing contributive channels of the previous layer.

Finally, contributions from feature maps to target score can be calculated by multiplying contributions between adjacent layers. We average the feature maps of current frame weighted by $C(l_m, S_t)$ to construct target response map as Eq. 3.

$$H = \frac{\sum_{m=1}^{n} C(l_m, S_t) h(l_m)}{\sum_{m=1}^{n} C(l_m, S_t)} \tag{3}$$

Where H denotes the target response map and $h(l_m)$ denotes the l_m-th feature map before ROI pooling.

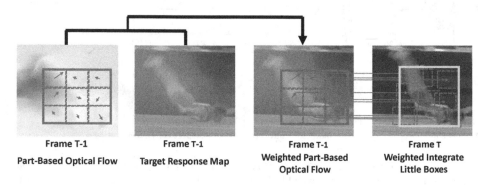

Frame T-1	Frame T-1	Frame T-1	Frame T
Part-Based Optical Flow	Target Response Map	Weighted Part-Based Optical Flow	Weighted Integrate Little Boxes

Fig. 3. An illustrative figure of our motion-aware proposal method. Part-based optical flow are used to calculate motion of each part. The motion is weighted by the target response map of the previous frame to exclude motions of background. The refined proposal box is obtained by weighted integrating relocated little boxes

3.3 Motion Refined Gaussian Sampling

We construct the motion-aware proposal by improving the Gaussian sampling and refining the candidates utilizing part-based motion features as shown in Fig. 3. In each frame, 256 candidates are generated by Gaussian sampling. We refined 128 samples using weighted part-based optical flow for fast motion handling. The other 128 samples are not refined for making use of previous target states.

First, Gaussian sampling is performed to randomly place candidate boxes around the previous target. Only one box of those candidates is shown in Fig. 3 for convenience. Each candidate box is divided into nine little boxes as parts of the target. Optical flows in the little boxes indicate the movements of parts between two frames.

Second, the target response map is obtained by weights back propagation method as introduced in the previous section.

Next, the part-based optical flows are weighted by the target response map to exclude background motions. The weights of little boxes are obtained by averaging target response in boxes.

Finally, the little boxes are relocated in the current frame according to the magnitude and direction of optical flow in the box. The relocated part boxes are further integrated into the refined proposal as Eq. 4.

$$\bar{X}_l = \frac{h_i \bar{x}_{li} + h_j \bar{x}_{lj} + h_k \bar{x}_{lk}}{h_i + h_j + h_k} \tag{4}$$

Where $\bar{X}_l$, $\bar{x}_{li}$, $\bar{x}_{lj}$ and $\bar{x}_{lk}$ denote the x-coordinate of the left border of the refined proposal and three left-most part boxes, h_i, h_j and h_k denote the mean target response in little boxes as weights to exclude backgrounds. Other borders of the refined samples can be calculated in a similar way.

4   Challenges Driven Feature Learning

4.1   Dual Multi-domain Learning

The dual multi-domain learning is a three-step training method using Stochastic Gradient Descent. Both dual paths are binary classifiers for target and background classification.

First, the challenge path is trained by the multi-domain method [14] using the challenge sequences. The dataset is divided into challenge sequences and common sequences according to tracking results of state-of-art tracker like MDNet [14]. The challenge sequences are the half sequences of the training dataset with more tracking drifts. The first two fully connected layers (fc4–5) are shared layers of the challenge path. The last layers (fc6) are domain-specific layers and have K branches corresponding to K domains in the challenge sequences. In each training iteration, only samples extracted in one sequence and corresponding fc6 layers are used. The training process is performed sequence by sequence until the network is converged. The common properties of challenge targets are modeled in the shared challenge layers after training. Second, the shared layers of common path are trained as the first step using the common sequences. The loss function is binary softmax cross-entropy loss.

Third, the challenge path and the common path are jointly trained during tracking. Only shared layers of each path are reserved and fc6 is randomly initialized. Both dual paths are initialized with samples extracted in the first frame. During tracking, the sample pools of each path are maintained independently. The positive samples of a path are chosen under two conditions. The target scores of the samples are greater than 0.5 or the path has higher target scores on the samples.

4.2   Categories Prior Learning

The updating of target and background classifier could be broken under heavy occlusion or large motion. However, the category of the target will never change

Fig. 4. The results of target and background categories classification in the first frame (Lemming, BlurFace and Tiger2). Targets are shown in the green boxes with fuzzy categories. Background objects with high scores are shown red boxes. During tracking process, we award candidates with fuzzy target categories and punish candidates with background categories (Color figure online)

during the tracking process. Hence, we propose the category path which is a 1000 categories classifier to make use of target categories prior. This path is pre-trained on the ImageNet [11] dataset and never updated during tracking. As shown in Fig. 1, the category prior of the target is calculated using the pre-trained networks in the first frame. During tracking, the category features of samples are generated by the category path. The inner products of the target and samples category features are regarded as the category scores of the samples.

The results of target and background categories classification in the first frame are demonstrated in Fig. 4. The category features of target and backgrounds are different in most cases. For some special cases as pedestrians tracking, although the category features could be deceptive, the target can still be classified using appearance and motion features.

4.3   Multi-path Fusion

The target state is predicted by averaging candidates with the highest target scores which are calculated by fusing the scores of the three paths. We first fuse the challenge path and the common path. Then, the category prior is used to award candidates with target categories. The final target score $\bar{S}^+(x_i)$ of sample x_i is calculated as Eq. 5.

$$\bar{S}^+(x_i) = \max(P_{clg}^+(x_i), P_{cmn}^+(x_i)) + S^C(x_i) \tag{5}$$

Where $P_{clg}^+(x_i)$ and $P_{cmn}^+(x_i)$ denote the target probability of the challenge and the common path, $S^C(x_i)$ denotes the category score after category matching. Finally, the optimal target state x^* is predicted by Eq. 6.

$$x^* = \arg\max_{x_i} \bar{S}^+(x_i) \tag{6}$$

4.4   Implementation Detail

Network Architecture: Our networks are modified from VGG-M [2] networks. The first three layers are convolution layers trained to extracted the appearance

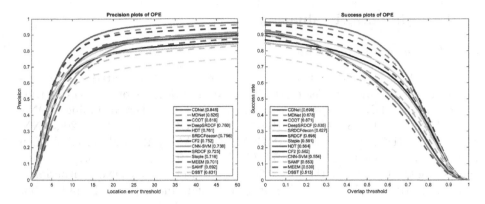

Fig. 5. Precision and success plots of OPE on OTB100 [19], comparing our tracker CDNet with state-of-art trackers

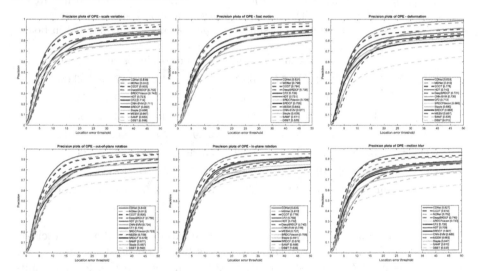

Fig. 6. Precision plots of OPE under different attributes on OTB100 [19]. The attributes include scale variation, fast motion, deformation, out-plane-rotation, in-of-plane rotation, and motion blur

and motion features. The next two layers are fully connected layers with 512 output units following by ReLUs and dropouts. The final layers have 2 or 1000 output units for target or category classification respectively. We use small networks for efficient online training and a higher response to the small target.

Training Data Sampling: The positive samples have > 0.7 IoU overlap ratios with the target and are extracted by Gaussian sampling around the target. Meanwhile, the negative samples have < 0.5 IoU overlap ratios with the target and are extracted by Uniform sampling in the whole image. Hard negative mining is further performed to find the most distracting negative samples.

5 Experimental Evaluation

5.1 Evaluation on OTB100

OTB100 [19] contains 100 fully annotated videos with different attributes including occlusion, fast motion, deformation, etc. We perform the one-pass evaluation (OPE) on our CDNet comparing with 12 state-of-art trackers including MDNet [14], CCOT [7], DeepSRDCF [4], HDT [16], SRDCFdecon [6], CF2 [20], CNN-SVM [15], SRDCF [5], staple [1], MEEM [21], SAMF [12] and DSST [3]. Our tracker is pre-trained on VOT2015 [10]. Figure 5 illustrates the precision and success plots of OPE on the OTB100 [19] dataset. Our CDNet outperforms state-of-art trackers by 1.9% on location precision and 2.1% on bounding box overlap.

Table 1. Accuracy evaluation and ablation study of CDNet compared with state-of-art trackers under challenges. CDNet-P only utilizes the motion-aware proposals and CDNet-C only uses the multi-path classifier. Top performance is highlighted in bold.

	empty	camera	illum	motion	occlu	size	mean
SRDCF [5]	58.00	53.57	69.16	48.56	42.11	47.16	53.09
C-COT [7]	57.03	53.51	65.98	46.42	44.49	49.06	52.75
DeepSRDCF [4]	62.81	56.24	66.54	49.65	45.19	53.19	55.60
MDNet [14]	64.95	60.91	68.03	55.95	54.33	56.01	60.03
CDNet-C(Ours)	**65.31**	61.93	69.20	56.26	55.10	57.32	60.85
CDNet-P(Ours)	64.10	61.86	70.33	**57.03**	54.80	**58.61**	61.12
CDNet(Ours)	65.11	**62.07**	**70.57**	56.92	**56.31**	58.39	**61.56**

We further evaluate trackers under different attributes and plot the precision curves as shown in Fig. 6. Our CDNet performs better on most attributes. We make progress by 2.7% on scale variation, 2.3% on fast motion, 2.3% on out-plane rotation, 2.8% on out-plane rotation, 2.2% on in-plane rotation and 1.7%

Table 2. Robustness evaluation and ablation study of our CDNet compared with state-of-art trackers under challenges. Top performance is highlighted in bold.

	empty	camera	illum	motion	occlu	size	mean
SRDCF [5]	16.00	43.00	8.00	36.00	22.00	21.00	24.33
C-COT [7]	11.00	24.00	2.00	20.00	14.00	13.00	14.00
DeepSRDCF [4]	9.00	25.00	**0.00**	23.00	26.00	**8.00**	15.17
MDNet [14]	6.20	20.00	1.07	15.43	13.93	11.20	11.36
CDNet-C(Ours)	**4.00**	17.00	**0.00**	13.00	13.00	9.00	9.33
CDNet-P(Ours)	7.00	18.00	**0.00**	14.00	15.00	11.00	10.83
CDNet(Ours)	5.00	**13.00**	**0.00**	**12.00**	**12.00**	**8.00**	**8.33**

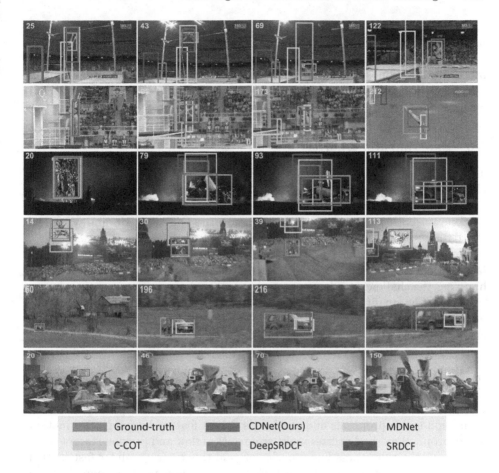

Fig. 7. Qualitative results of our CDNet comparing with other state-of-art trackers, including MDNet [14], C-COT [7], DeepSRDCF [4] and SRDCF [5]

on motion blur. The experiment results demonstrate that our CDNet can handle multiple dynamic challenges in one framework.

5.2   Evaluation on VOT2015

VOT2015 [10] contains 60 sequences providing different challenges, including motion change, occlusion, size change, camera motion and illumination change. We perform ablation study and compare our tracker CDNet with state-of-art trackers including MDNet [14], DeepSRDCF [4], CCOT [7] and SRDCF [5]. Our tracker is pre-trained using sequences in OTB100 [19] datasets. Table 1 illustrates the accuracy evaluation of trackers under challenges. Our CDNet achieves better performance on most challenges, especially on size change and occlusion. We improve the accuracy by 2.3% on size change, 2.0% on occlusion and 1.5% in the average. And the robustness evaluation is illustrated in Table 2. The robustness of our CDNet is also improved under most challenges, especially on camera motion and size change.

According to the results of ablation study shown in Tables 1 and 2, the motion-aware proposal contributes most on accuracy under fast motion and size change. The multi-path classifier mainly improves the robustness under challenges.

5.3   Experimental Analysis

The experiments mainly evaluate the accuracy and robustness of trackers under challenges. We summarize the results of experiments in three folds:

First, our CDNet achieves competitive performance against state-of-art trackers on both accuracy and robustness on the OTB100 [19] and VOT2015 [10] datasets.

Second, CDNet achieves better performance on multiple challenges, especially fast motion, deformation, and occlusion. Our CDNet is an effective model for handling the challenge of dynamic motion and occlusion.

Third, the qualitative results which are shown in Fig. 7 demonstrate that our model can generate better proposals and make better classification on the challenge targets.

6   Conclusion

The proposed CDNet tracking framework handles dynamic motion and occlusion by motion-aware proposals and challenges driven feature learning. The motion-aware proposal method utilizes multi-scale optical flow and previous target states to improve the robustness under occlusion and the accuracy under fast motion. The challenges driven learning approach improves the network response to both challenge and common targets by dual multi-domain learning. The learning approach also utilizes the category prior to detect and avoid tracking drift. We combine the motion-aware proposal and the challenges driven feature learning to improve accuracy and robustness of CDNet under challenges like motion, occlusion and scale variation, and achieve a competitive performance against state-of-art trackers.

Acknowledgements. This work is supported by National Key Basic Research Program of China (No. 2016YFB0100900) and National Natural Science Foundation of China (No. 61773231).

References

1. Bertinetto, L., Valmadre, J., Golodetz, S., Miksik, O., Torr, P.H.: Staple: complementary learners for real-time tracking. In: Proceedings of the IEEE Conference on Computer Vision and Pattern Recognition, pp. 1401–1409 (2016)

2. Chatfield, K., Simonyan, K., Vedaldi, A., Zisserman, A.: Return of the devil in the details: delving deep into convolutional nets. arXiv preprint arXiv:1405.3531 (2014)
3. Danelljan, M., Häger, G., Khan, F., Felsberg, M.: Accurate scale estimation for robust visual tracking. In: British Machine Vision Conference, Nottingham, 1–5 September 2014. BMVA Press (2014)
4. Danelljan, M., Hager, G., Shahbaz Khan, F., Felsberg, M.: Convolutional features for correlation filter based visual tracking. In: Proceedings of the IEEE International Conference on Computer Vision Workshops, pp. 58–66 (2015)
5. Danelljan, M., Hager, G., Shahbaz Khan, F., Felsberg, M.: Learning spatially regularized correlation filters for visual tracking. In: Proceedings of the IEEE International Conference on Computer Vision, pp. 4310–4318 (2015)
6. Danelljan, M., Hager, G., Shahbaz Khan, F., Felsberg, M.: Adaptive decontamination of the training set: a unified formulation for discriminative visual tracking. In: Proceedings of the IEEE Conference on Computer Vision and Pattern Recognition, pp. 1430–1438 (2016)
7. Danelljan, M., Robinson, A., Shahbaz Khan, F., Felsberg, M.: Beyond correlation filters: learning continuous convolution operators for visual tracking. In: Leibe, B., Matas, J., Sebe, N., Welling, M. (eds.) ECCV 2016. LNCS, vol. 9909, pp. 472–488. Springer, Cham (2016). https://doi.org/10.1007/978-3-319-46454-1_29
8. Han, B., Sim, J., Adam, H.: BranchOut: regularization for online ensemble tracking with convolutional neural networks. In: Proceedings of IEEE International Conference on Computer Vision, pp. 2217–2224 (2017)
9. Horn, B.K., Schunck, B.G.: Determining optical flow. Artif. Intell. 17(1–3), 185–203 (1981)
10. Kristan, M., et al.: The visual object tracking VOT2015 challenge results. In: Proceedings of the IEEE International Conference on Computer Vision Workshops, pp. 1–23 (2015)
11. Krizhevsky, A., Sutskever, I., Hinton, G.E.: ImageNet classification with deep convolutional neural networks. In: Advances in Neural Information Processing Systems, pp. 1097–1105 (2012)
12. Li, Y., Zhu, J.: A scale adaptive Kernel correlation filter tracker with feature integration. In: Agapito, L., Bronstein, M.M., Rother, C. (eds.) ECCV 2014. LNCS, vol. 8926, pp. 254–265. Springer, Cham (2015). https://doi.org/10.1007/978-3-319-16181-5_18
13. Nam, H., Baek, M., Han, B.: Modeling and propagating CNNs in a tree structure for visual tracking. arXiv preprint arXiv:1608.07242 (2016)
14. Nam, H., Han, B.: Learning multi-domain convolutional neural networks for visual tracking. In: Proceedings of the IEEE Conference on Computer Vision and Pattern Recognition, pp. 4293–4302 (2016)
15. Niu, X.X., Suen, C.Y.: A novel hybrid CNN-SVM classifier for recognizing handwritten digits. Pattern Recogn. 45(4), 1318–1325 (2012)
16. Qi, Y., et al.: Hedged deep tracking. In: Proceedings of the IEEE Conference on Computer Vision and Pattern Recognition, pp. 4303–4311 (2016)
17. Ren, S., He, K., Girshick, R., Sun, J.: Faster R-CNN: towards real-time object detection with region proposal networks. In: Advances in Neural Information Processing Systems, pp. 91–99 (2015)
18. Uijlings, J.R., Van De Sande, K.E., Gevers, T., Smeulders, A.W.: Selective search for object recognition. Int. J. Comput. Vis. 104(2), 154–171 (2013)
19. Wu, Y., Lim, J., Yang, M.H.: Object tracking benchmark. IEEE Trans. Pattern Anal. Mach. Intell. 37(9), 1834–1848 (2015)

20. Ma, C., Huang, J.B., Yang, X., Yang, M.H.: Hierarchical convolutional features for visual tracking
21. Zhang, J., Ma, S., Sclaroff, S.: MEEM: robust tracking via multiple experts using entropy minimization. In: Fleet, D., Pajdla, T., Schiele, B., Tuytelaars, T. (eds.) ECCV 2014. LNCS, vol. 8694, pp. 188–203. Springer, Cham (2014). https://doi.org/10.1007/978-3-319-10599-4_13

Towards Photo-Realistic Visible Watermark Removal with Conditional Generative Adversarial Networks

Xiang Li[1], Chan Lu[2], Danni Cheng[3], Wei-Hong Li[4], Mei Cao[1], Bo Liu[5], Jiechao Ma[1(✉)], and Wei-Shi Zheng[1]

[1] Sun Yat-sen University, Guangdong, China
lixiang651@gmail.com, majch7@mail2.sysu.edu.cn
[2] Shanghai University of Finance and Economics, Shanghai, China
[3] Shanghai Jiao Tong University, Shanghai, China
[4] The University of Edinburgh, Edinburgh, UK
[5] Zhejiang University, Zhejiang, China

Abstract. Visible watermark plays an important role in image copyright protection and the robustness of a visible watermark to an attack is shown to be essential. To evaluate and improve the effectiveness of watermark, watermark removal attracts increasing attention and becomes a hot research top. Current methods cast the watermark removal as an image-to-image translation problem where the encode-decode architectures with pixel-wise loss are adopted to transfer the transparent watermarked pixels into unmarked pixels. However, when a number of realistic images are presented, the watermarks are more likely to be unknown and diverse (i.e., the watermarks might be opaque or semi-transparent; the category and pattern of watermarks are unknown). When applying existing methods to the real-world scenarios, they mostly can not satisfactorily reconstruct the hidden information obscured under the complex and various watermarks (i.e., the residual watermark traces remain and the reconstructed images lack reality). To address this difficulty, in this paper, we present a new watermark processing framework using the conditional generative adversarial networks (cGANs) for visible watermark removal in the real-world application. The proposed method enables the watermark removal solution to be more closed to the photo-realistic reconstruction using a patch-based discriminator conditioned on the watermarked images, which is adversarially trained to differentiate the difference between the recovered images and original watermark-free images. Extensive experimental results on a large-scale visible watermark dataset demonstrate the effectiveness of the proposed method and clearly indicate that our proposed approach can produce more photo-realistic and convincing results compared with the state-of-the-art methods.

Keywords: Visible watermark · Watermark removal · Conditional generative adversarial networks

© Springer Nature Switzerland AG 2019
Y. Zhao et al. (Eds.): ICIG 2019, LNCS 11901, pp. 345–356, 2019.
https://doi.org/10.1007/978-3-030-34120-6_28

1   Introduction

Nowadays, people tend to post photos and videos on the Internet for sharing and preserving memories of events and so on. To protect the copyright of photos and videos, the visible watermark is commonly used. Typically, those watermarks are opaque or semi-transparent images containing names or logos, overlaying on the original images. Despite billions of online images have been embedded with visible watermarks for ownership declaration by watermarking techniques, they always suffer from a security flaw that watermarks may be affected and damaged by various watermark processing methods. To evaluate and improve the robustness of watermarks, a number of scientists [2,3,5,11,12,14,16] attempt to attack it by removing watermarks from images.

Due to the existence of diverse categories and patterns of visible watermarks, developing an advanced visible watermark removal method remains as a difficult task. More specifically, visible watermarks often contain complex structures (e.g., the texts, symbols, graphic, thin lines and shadows are diverse (Fig. 1(a))), leading to the challenge of removing unknown and diverse patterns of watermarks from images without user supervision or prior information in practical situation.

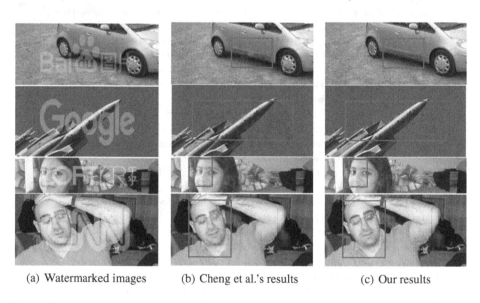

(a) Watermarked images   (b) Cheng et al.'s results   (c) Our results

Fig. 1. In real-world scenarios, visible watermarks usually contain complex structures (a). Compared with Cheng et al.'s [2] recovered results which remain a few residual watermark traces and are perceptually not photo-realistic (b), our framework is able to generate more convincing results (c).

Most of existing watermark removal methods are unable to tackle those aforementioned challenges on removing watermarks from images. Although these models are designed for estimating and wiping off the watermark regions, they either highly depend on the prior knowledge [5, 11, 14] or assume that the watermarked images have the same watermark pattern [3, 16], which are not suitable for removing watermarks in real-world scenarios where the watermarks may be unknown and the watermarks in different images are more likely to be distinct. Recently, Cheng et al. [2] cast the watermark removal as an image-to-image translation problem and used a fully convolutional architecture to transfer the watermarked pixels to the original unmarked pixels, which provided a reasonable solution for watermark removal. However, directly training a generator with pixel-wise loss to estimate pixel relation mapping is difficult. In addition, the watermark-free images recovered by this kind of approach mostly contains a few residual watermark traces and are perceptually not photo-realistic in human visual sense (Fig. 1(b)).

To make the results of watermark removal more photo-realistic and convincing (e.g., to recover the watermarked patch without any residual watermarks and make it more photo-realistic), in this work, we propose a new watermark removal framework with conditional generative adversarial networks (cGANs) [10]. Specifically, an effective cGAN model is widely adopted to form a framework for photo-realistic watermark removal. To achieve this, we introduce a new loss function, consisting of an adversarial loss and a pixel-wise content loss. In particular, the adversarial loss working with a patch-based discriminator network enables our method to reconstruct a photo-realistic watermark-free image. Here, the patch-based discriminator network is conditioned on the input watermarked images and is trained to differentiate the difference between the recovered images and original watermark-free images. Additionally, we use a content loss motivated by perceptual similarity and pixel similarity [2, 7], which consists of the L1 loss and the perceptual loss. With both the adversarial loss and the content loss, our framework is able to generate more convincing recovered results from images marked by diverse unknown watermarks (Fig. 1(c)).

In summary, our contributions are twofold. Firstly, to the best of our knowledge, this is the *first* work to exploit the concept of cGAN to design an effective framework to solve the visible watermark removal problem in a realistic setting. Our cGAN-based watermark removal framework is much more principled than existing approaches. Secondly, we introduce an effective watermark removal cGAN model with a new loss function, which is comprised of an adversarial loss and a pixel-wise content loss. This can drive the reconstruction of the watermark regions to be more photo-realistic. Moreover, extensive experiments are conducted on a large-scale visible watermark dataset for evaluation. The results demonstrate that our proposed model is capable of addressing the visible watermark removal problem confronted in real-world scenarios, achieving more convincing reconstruction than state-of-the-art methods.

2   Methodology

In this section, we present our watermark removal framework which is build based on the concept of cGANs [6, 10]. In recent, the cGANs [6] are commonly

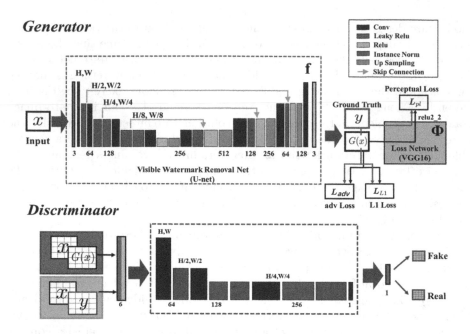

Fig. 2. The architecture of our visible watermark removal framework.

adopted to reconstruct the hidden information which is obscured in original image. In this work, as we aim at restoring the original images from the water-marked images, we adopt the idea of the cGANs and propose a cGANs-based framework for watermark removal. The architectures of our proposed watermark removal cGANs is illustrated in Fig. 2.

Our network mainly embodies a generator and a discriminator. In the generator, we leverage a U-net based architecture [13] to transform a watermarked image to a watermark-free one. In our discriminator, we use a patch-based classifier [6] conditioned on the input watermarked images to distinguish those recovered images generated by the generator from the ground-truth watermark-free images in a patch level.

More specifically, our network takes as input a watermarked image and exploits the generator **G** to generate a photo-realistic watermark-free image. To enable the image restored by the generator to be similar to the ground-truth watermark-free image as much as possible, we introduce a new objective which is the combination of the L1 loss, perceptual loss and the patch-based adversarial loss to restrain the training of the generator. In the meanwhile, an adversarially trained discriminator **D** is employed to detect the "fake" images (i.e., the images which is generated by the generator and is not distinguished as the real watermark-free images) from those ground-truth images (i.e., the real watermark-free images). We detail the adversarial network architecture and loss functions individually in Sects. 2.1 and 2.2.

2.1 Adversarial Network Architectures

We formulate our generator and discriminator architecture inspired by [6]. Figure 2 shows the details of our architectures, and the key features will be illustrated below.

Generator (U-Net). Typically, watermark removal can be cast as an image-to-image translation problem [2]. Analyzing the watermark removal task, we find that the unmarked image areas share the pixel values as the input while the watermarked pattern needs to be removed to meet the visual requirements. Unlike the general encode-decode structure [1], which directly transforms an image in source domain to a target image through a series of convolution module in the network, we adopt a U-net based architecture as our generator in our work, followed by the fully convolutional network proposed in [2]. In our system, the U-net takes the advantage of its skip connection structure, which combines the low level feature and the high level features, allowing the sharing of global information and edge details between the input and the output. Specifically, our generator comprises of six standard modules, which are down-blocks or up-blocks. In down-blocks, the channels of feature map are doubled and the its side size is reduced by half, while the up-blocks go the opposite. In addition, there are skip connections between every i^{th} layer and the $(n-i)^{th}$ layer, where n is the total number of layers. Each skip connection simply concatenates all channels of the ith layer with those of the $(n-i)^{th}$ layer as shown in Fig. 2.

Discriminator (Patch-Based). Different from common GAN discriminators, which map the input into one scalar representing the probability of the input sample attributed to "real". In our work, as shown in Fig. 2, we employ the patch-based network [6] as our discriminator. The structure of our discriminator is a full convolutional network, which maps the combination of watermarked image and watermark removed image to a feature map, representing the class probabilities i.e., "fake" or "real" of the patches of the input. Since the point in the feature map can be traced back to the receptive field in the original image, thus each value in the output matrix refers the probability that the patch in the original image is "real", and we calculate the probability of the input image is "real" as the average of the all patches are "real".

Observing the watermark images, we can find that the difference between the watermarked image and the original watermark-free one solely exists in some parts of the image. Since the watermarked area is relatively small compared with the whole image, it is critical for the discriminator to identify the most different patches of two input images and focus more on minimizing loss of these patches. Thus, introducing patch-based discriminator can make our cGANs based network be more powerful for removing visible watermark.

2.2 Objective Function

The objective functions play a very important role in training the network. In this work, we aim to learn a solution $\mathbf{G}^*$ to minimize the loss function defined as below:

$$G^* = \arg \min_G \max_D \underbrace{\boldsymbol{L}_{adv}(G, D)}_{adversarial\ loss} + \underbrace{\alpha \boldsymbol{L}_{l_1}(G) + \beta \boldsymbol{L}_{per}(G)}_{content\ loss}. \tag{1}$$

During each stage of training our cGANs-based watermark removal model the generator **G** and the discriminator **D** are trained alternately, where **G** is trained to minimize this objective against an adversarial discriminator **D**, which is trained to maximize the loss.

Specially, the generator G is trained by minimizing the loss. The task of generator is not only to fool the discriminator but also to generate a image that is closed to the ground truth watermark-free image in visual. Therefore, the objective of the generators consists of a content loss and an adversarial loss, where the perceptual loss and l_1 loss comprise the content loss. In the Eq. (1), $\boldsymbol{L}_{l_1}$, $\boldsymbol{L}_{per}$ and $\boldsymbol{L}_{adv}$ refer to the l_1 loss, perceptual loss and adversarial loss respectively, where α and β are weights to balance the l_1 loss, perceptual loss and adversarial loss. The discriminator **D** is trained alternately to avoid being fooled by the generators by distinguishing the inputs as either real or fake. Thus, the adversarial loss is defined as the opposite loss function as that in the generator. We detail the content loss and adversarial loss in the following sections.

Content Loss. At present, the most commonly used content loss in image-to-image tasks is MSE loss. It obtained the state-of-art PSNR results in many image-to-image task such as super-resolution and image style translation and etc. However, it is proved to be blur occurrence in the generated images and the output results do not satisfy human visual sense. To solve this problem, we use the l_1 distance rather than the l_2, which is defined as:

$$\boldsymbol{L}_{l_1}(G) = \|G(x) - y\|_1, \tag{2}$$

$G(x)$ denotes the output of generators and y denote the ground truth watermark-free image. Apart from the l_1 loss, it is beneficial to inject the perceptual loss for watermark removal [2]. The perceptual loss function of our network can be expressed as:

$$\boldsymbol{L}_{per}^{\Phi, j}(G) = \frac{1}{C_j H_j W_j} \|\Phi_j(G(x)) - \Phi_j(y)\|_2^2. \tag{3}$$

Here, we define Φ as the convolutional transformation for calculating the perceptual loss, which refers to the pertrained $VGG16$ [15] in our work. The feature size of the j_{th} convolutional layer of loss network is $C_j \times H_j \times W_j$. Specifically, the weight of the $VGG16$ is frozen and the outputs of the **relu2_2** are extracted as features to calculate the semantic difference between the input and output of generator as [7].

Adversarial Loss. Different from the general GAN [4], the discriminator of conditional GAN observes not only the output of G, but also the input x. Mathematically, the adversarial loss can be formulated as:

$$\boldsymbol{L}_{adv}(G, D) = \mathbb{E}_{x,y}[\log D(x, y)] + \mathbb{E}_x[\log(1 - D(x, G(x)))]. \tag{4}$$

Here, rather than training G to minimize $\log(1 - D(x, G(x)))$, we instead train to maximize $\log(D(x, G(x)))$ [4].

3   Experiments

3.1   Datasets and Settings

Dataset. To evaluate the performance of our framework on the large visible watermark image dataset, we conduct extensive experiments on the Large-scale Visible Watermark Dataset (LVW) [2], containing 60k watermarked images made of 80 watermarks, with 750 images per watermark. In this dataset, the original images in the training and testing sets are randomly chosen from the train/val and test sets in PASCAL VOC2012 dataset with replacement, respectively. The 80 categories of watermarks covering a vast quantity of patterns (e.g., the watermarks contain English and Chinese), are collected from renowned E-commercial brand, websites, organization, personal, and etc. Moreover, the size, location and transparency of each watermark in different images are distinct and set randomly. Example images of LVW dataset are shown in Fig. 3.

Training Details. PyTorch platform was applied to construct the proposed deep architecture. All experiments are conducted on a computer cluster equipped with NVIDIA Tesla K80 GPU with 12 GB memory. To optimize the generative adversarial networks, we follow the training strategy in [4] to alternate between one gradient descent step on the discriminator, then one step on generator. During training, we use mini batch SGD (the batch size is set to be 1) and apply the Adam solver [8], with the initial learning rate of 2e−4 and momentum parameters (i.e., $\beta1 = 0.5$, $\beta2 = 0.999$). And we evaluated the proposed framework with default setting ($\alpha = 10$, $\beta = 1e-4$).

Evaluation Setting and Metrics. In our experiments, the watermarks in training set are different from those in testing set. In LVW dataset, around 80% sorts of watermark are used for training, and the remaining 20% are for test (see Fig. 3), which is the same as [2] . This setting meets the requirements of unknown watermarks removal in real-world scenarios well. Both of the Peak signal to noise ratio (PSNR) and structural dissimilarity image index (DSSIM), measuring the similarity between the recovered image and the ground truth one, are adopted as evaluation metrics by previous work (e.g., [2,3]). However, both metrics fail to capture and accurately assess image quality with respect to the human visual system [9]. Therefore, in addition to two aforementioned evaluation metrics, we used the mean opinion score (MOS) testing [9] to further quantify the ability of different methods of reconstructing photo-realistic and convincing watermark-free images from watermarked images. Specifically, we asked 10 raters to assign an integral score from 1 (bad quality, i.e., watermarked images) to 5 (excellent quality, i.e., original images) to the recovered images for assessing the quality of the recovered images.

Train Set

Test Set

Fig. 3. The example images of LVW dataset.

Table 1. Evaluation of different loss functions.

Loss	PSNR	DSSIM	MOS
L1	30.42	0.045	2.85
Perceptual	29.86	0.051	3.17
L1 + Perceptual	**30.86**	**0.043**	3.23
L1 + Perceptual + GAN	30.33	0.049	3.31
L1 + Perceptual + cGAN	30.69	0.045	**4.08**

3.2   Results and Analysis

Analysis of the Objective Function. The objective function of our framework has three components terms in Eq. (1), including the l_1 loss term, the perceptual loss term and the adversarial loss term. In this section, we conduct experiments to analyze the effect of these loss terms.

In Table 1, 'L1' and 'Perceptual' indicate that the generator network only using the l_1 loss (Eq. (2)) and perceptual loss (Eq. (3)), respectively. 'L1 + Perceptual' represents the generator network using the combination loss of the l_1 loss as well as the perceptual loss. As shown in Table 1, 'L1 + Perceptual' performs clearly better than the 'L1' and 'Perceptual', demonstrating that the combination of l_1 loss and perceptual loss can incorporate the strength of both losses to reconstruct the fine details.

To evaluate the effect of adversarial loss term, we further show the performance of the combination of adversarial loss and L1 loss and the perceptual loss. Specifically, we compare the model using a discriminator conditioned on the input (adversarial loss of cGAN, Eq. (4)) with the model using an unconditional discriminator (adversarial loss of GAN). They are respectively denoted as 'L1 + Perceptual + cGAN' and 'L1 + Perceptual + GAN' in Table 1. Although combining the l_1 loss and the perceptual loss with the adversarial loss causes a slight drop in the PSNR and DSSIM values, it achieves higher MOS scores, indicating that the recovered results are more photo-realistic. The reason is that the GAN-based procedure encourages the reconstructions to move towards regions with high probability of containing photo-realistic images in searching space and thus closer to the convincing results. Moreover, the results in Table 1 also show clearly that cGAN performs much better than GAN, verifying the effectiveness of the conditional discriminator. This suggests that it is important that the loss measure the quality of the match between input (watermarked images) and output recovered images.

Evaluation of the Patch-Based Discriminator. As the patch-based discriminator is essential in our proposed framework to model the discriminative information of local image patches. To investigate the effect of patch-based discriminator, we compared our model with the model using the conventional image-based discriminator, which classifies if whole image region is real or fake (i.e., in image level). Note that, in this section all experiments are conducted with the 'L1 + perceptual + cGAN' loss (Eq. (1)).

Table 2. Evaluation of different discriminators.

Discriminator	PSNR	DSSIM	MOS
Image-based	29.72	0.052	3.46
Patch-based	**30.69**	**0.045**	**4.08**

The results are shown in Table 2. Compared with our patch-based discriminator, the image-based discriminator gets a considerably worse performance. Specifically, the image-based discriminator identify the difference between two images in a image level, which alleviate the effect of the difference in local areas and hamper the results (i.e., the generated images are not photo-realistic). In addition, the image-based discriminator has much more parameters and deeper than the patch-based discriminator. This can slow down the speed of the watermark removal model and it is harder to train the model, which makes this kind of discriminator unscalable to the real-world application. In other words, as the patch-based discriminator is a light weight model, it can run faster even on arbitrarily large image and it is shown to perform better. This strongly suggests that our proposed model is more suitable to be applied to train the cGANs-based model for visible watermarks removal in realistic data.

Comparison with State-of-the-Art. To justify the effectiveness of the proposed model, we performed experiments to compare our method with Cheng et al. [2]. As shown in Table 3, our model obtained the comparable results in PSNR and DSSIM, and the MOS results indicate that our method outperforms Cheng et al. by a large margin in human visual system.

Table 3. Comparisons with state-of-the-art method on LVW dataset.

Metrics	Input	Cheng *et al.* [2]	Ours	Ground truth
PSNR	20.65	30.86	30.69	∞
DSSIM	0.103	0.043	0.045	0
MOS	1.0	3.23	**4.08**	5.0

We also visualized the watermark removal results of test examples in LVW dataset and show them in Fig. 4. The results in the figure further demonstrate that the performance of the proposed method is noticeably convincing than the ones of existing methods, suggesting that the adversarial model is more suitable for solving the visible watermark removal problem.

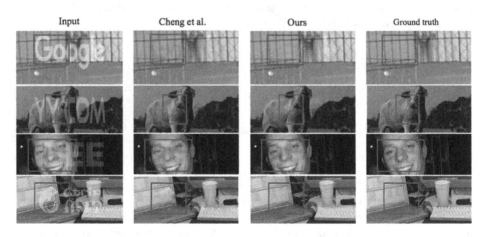

Fig. 4. Example results on watermark removal. From left to right: watermarked images, Cheng et al.'s results, our results, ground truth. This experiment demonstrates that our method is effective to generate more photo-realistic recovered images. (best view in color)

3.3 Discussion and Future Work

Our experiments show that our proposed framework can effectively remove the unknown and diverse visible watermarks, resulting more satisfactory recovered images. The focus of this work was the photo-realistic quality of reconstruction rather than obtaining better performance in standard quantitative evaluation metrics such as PSNR and SSIM, which can not accurately capture and evaluate the quality of images associated with the human visual system. The experimental results further verify that a deep convolutional architecture using the concept of cGANs to form an adversarial loss is useful for photo-realistic watermark removal in real-world scenarios. Significantly, our original intention is to increase the awareness on the copyrights of online images, reminding that visible watermarks should be designed to be more resistant against removal attacking. Developing a more robust watermarking technique for copyright protection is challenging and part of future work.

4 Conclusion

In this work, we introduced a new watermark processing framework for more photo-realistic visible watermark removal, which augments the conventional L1 and perceptual loss function with an adversarial loss by training a conditional generative adversarial network. The proposed model is able to drive the reconstruction of watermark regions towards the photo-realistic results producing perceptually more convincing solutions. Extensive experiments are conducted on a large-scale visible watermark dataset to verify the feasible of our method. Experimental results clearly demonstrated the superior performance of the proposed framework compared to existing methods.

Acknowledgment. This work was supported by NSFC(U1811461). Xiang Li and Chan Lu equally contributed to this work.

References

1. Badrinarayanan, V., Kendall, A., Cipolla, R.: SegNet: a deep convolutional encoder-decoder architecture for image segmentation. IEEE Trans. Pattern Anal. Mach. Intell. **39**(12), 2481–2495 (2017)
2. Cheng, D., et al.: Large-scale visible watermark detection and removal with deep convolutional networks. In: Lai, J.H., et al. (eds.) PRCV 2018. LNCS, vol. 11258, pp. 27–40. Springer, Cham (2018). https://doi.org/10.1007/978-3-030-03338-5_3
3. Dekel, T., Rubinstein, M., Liu, C., Freeman, W.T.: On the effectiveness of visible watermarks. In: Proceedings of the IEEE Conference on Computer Vision and Pattern Recognition, pp. 2146–2154 (2017)
4. Goodfellow, I., et al.: Generative adversarial nets. In: Advances in Neural Information Processing Systems, pp. 2672–2680 (2014)
5. Huang, C.H., Wu, J.L.: Attacking visible watermarking schemes. IEEE Trans. Multimed. **6**(1), 16–30 (2004)

6. Isola, P., Zhu, J.Y., Zhou, T., Efros, A.A.: Image-to-image translation with conditional adversarial networks. In: Proceedings of the IEEE Conference on Computer Vision and Pattern Recognition, pp. 1125–1134 (2017)

7. Johnson, J., Alahi, A., Fei-Fei, L.: Perceptual losses for real-time style transfer and super-resolution. In: Leibe, B., Matas, J., Sebe, N., Welling, M. (eds.) ECCV 2016. LNCS, vol. 9906, pp. 694–711. Springer, Cham (2016). https://doi.org/10.1007/978-3-319-46475-6_43

8. Kingma, D.P., Ba, J.: Adam: a method for stochastic optimization. arXiv preprint arXiv:1412.6980 (2014)

9. Ledig, C., et al.: Photo-realistic single image super-resolution using a generative adversarial network. In: Proceedings of the IEEE Conference on Computer Vision and Pattern Recognition, pp. 4681–4690 (2017)

10. Mirza, M., Osindero, S.: Conditional generative adversarial nets. arXiv preprint arXiv:1411.1784 (2014)

11. Pei, S.C., Zeng, Y.C.: A novel image recovery algorithm for visible watermarked images. IEEE Trans. Inf. Forensics Secur. **1**(4), 543–550 (2006)

12. Qin, C., He, Z., Yao, H., Cao, F., Gao, L.: Visible watermark removal scheme based on reversible data hiding and image inpainting. Sig. Proces.: Image Commun. **60**, 160–172 (2018)

13. Ronneberger, O., Fischer, P., Brox, T.: U-Net: convolutional networks for biomedical image segmentation. In: Navab, N., Hornegger, J., Wells, W.M., Frangi, A.F. (eds.) MICCAI 2015. LNCS, vol. 9351, pp. 234–241. Springer, Cham (2015). https://doi.org/10.1007/978-3-319-24574-4_28

14. Santoyo-Garcia, H., Fragoso-Navarro, E., Reyes-Reyes, R., Sanchez-Perez, G., Nakano-Miyatake, M., Perez-Meana, H.: An automatic visible watermark detection method using total variation. In: 2017 5th International Workshop on Biometrics and Forensics (IWBF), pp. 1–5. IEEE (2017)

15. Simonyan, K., Zisserman, A.: Very deep convolutional networks for large-scale image recognition. arXiv preprint arXiv:1409.1556 (2014)

16. Xu, C., Lu, Y., Zhou, Y.: An automatic visible watermark removal technique using image inpainting algorithms. In: 2017 4th International Conference on Systems and Informatics (ICSAI), pp. 1152–1157. IEEE (2017)

Online Detection of Welding Quality Based on ZYNQ and Data Mining

Yicheng Zhang, Jing Han, Lianfa Bai, and Zhuang Zhao[✉]

Jiangsu Key Laboratory of Spectral Imaging and Intelligent Sense,
Nanjing University of Science and Technology, Nanjing 210094, China
zhaozhuang@njust.edu.cn

Abstract. With the rapid development of manufacturing industry, traditional quality detection methods can no longer meet the demand. As an important part of intelligent manufacturing, the research of online quality monitoring technology of arc welding is imminent. The welding electrical signal reflects various changes of arc composition in the welding process and contains abundant information about the welding quality. Therefore, an online quality monitoring method based on Apriori algorithm is proposed. The ZYNQ board is used to sample welding electric signals under three shielding gas flow rates levels. Apriori algorithm is used to mine the potential distinguishing rules under three levels of shielding gas flow rates, and Verilog hardware language is used to design a suitable rule to monitor the shielding gas flow rate automatically. Finally, ZYNQ board is used to control the shielding gas flow rate of welding machine. Abundant online experiments have demonstrated that the classification results obtained by Apriori algorithm can distinguish the current signals of three shielding gas flow rates.

Keywords: Online quality detection · Data mining · Welding current

1 Preface

In the past decades, China has made great progress in machine manufacturing industry. As an important part of machine manufacturing, welding technology plays a vital role in the development of manufacturing industry. Traditional quality monitoring method cannot obtain the defect information by non-destructive method and the timeliness is insufficient, which cannot meet the needs of society. People prefer to use external sensor science to realize online quality monitoring.

CMT (Cold Metal Transition) is a new technology developed by Fronius Company for continuous alternate welding of "hot-cold-hot". Compared with traditional MIG/MAG welding technology, there are two different kinds: firstly, the wire feeding of CMT is for the reciprocating motion of feeding/retracting; secondly, the droplet transfer is performed under the condition of no current. CMT has the advantages of low heat input, stable arc, no spatter and high reliability. It effectively improves the defects of MIG/MAG and is widely used in various fields.

During the welding process, the electrical signal of the arc exhibits periodic fluctuations, and the electrical signal of the arc is closely related to the stability of the

© Springer Nature Switzerland AG 2019
Y. Zhao et al. (Eds.): ICIG 2019, LNCS 11901, pp. 357–368, 2019.
https://doi.org/10.1007/978-3-030-34120-6_29

welding. The effective information of the arc electric signal is extracted by statistical method, and the relationship model between the welding quality and electric signal is established. For GMAW, Li et al. extracted the independent component of current and voltage signal matrix in the welding process, and realized the online quality detection of GMAW process [3]. According to the assessment on stability of aluminum alloy pulsed MIG welding, Shi et al. proposed an analytical method based on the probability density of arc voltage signal. The ratio of first peak value and second peak value in voltage signal probability density distribution was taken as the criterion to assess the pulsed MIG welding of aluminum alloy [4]. For the stability analysis of CO_2 arc welding process, Gao et al. respectively adopted autocorrelation distribution, probability density distribution, time frequency distribution and other methods to quantitatively analysis the stability of welding process [5–7]. Aiming at the process stability of underwater wet electrodes arc welding, Hu et al. collected instantaneous value of arc voltage for underwater wet manual arc welding at different welding conditions, and then, the instantaneous value of voltage was processed statistically and evaluated quantitatively based on standard deviation [8]. However, these methods can not realize online detection in the welding process, lacking timeliness and cannot effectively find the causes of defects formation and adjust the parameters timely. Therefore, it is necessary to research online defect detection system to monitor welding defects and improve production efficiency.

In order to find out the hidden relationship between electrical signals of different shielding gas flow rates and realize online defect detection, this paper applies data mining algorithms to analyze electrical signal data. In the field of data mining algorithms, Apriori algorithm is a classic algorithm for mining frequent itemsets and association rules. The Apriori algorithm uses a priori knowledge or assumption to define the problem, and then mines frequent itemsets and potential rules. These rules will help us identify the cause of weld quality defects and adjust welding parameters timely.

This paper mainly research the difference of electrical signals in CMT welding process under different shielding gas flow rates. The Apriori data mining algorithm is used to mine different frequent sets of electrical signals in CMT welding process under different shielding gas flow rates. The FPGA is designed to automatically acquire frequent sets of CMT welding cycles and determine whether the confidence level is reached. Therefore the shielding gas flow rate levels are identified and use the mining result to control shielding gas flow rate.

2 Experimental Device

The ZYNQ7000 chip model is XC7Z020CLG400-2 from Xilinx. The AD conversion module is the AD7607. The CMT welding process was used in the experiment. The specific welding parameters are shown in Table 1. The physical diagram of system device is shown in Fig. 1.

Table 1. Welding parameters

Category	Parameter
Welding current	130 A
Welding speed	24 cm/min
Shielding gas	$1.5\%O_2 + 98.5\%Ar$
Power type	FRONIUS
Welding base metal	sus304
Welding wire material	ER316L

Fig. 1. Physical diagram of system device

3 System Design

The online quality detection system designed in this paper adopts the Zynq7000 board based on ARM+FPGA SOC technology, which is responsible for data sampling, data analysis and processing, and Keypad communication with the welder control system. The overall block diagram of the system solution is shown in Fig. 2.

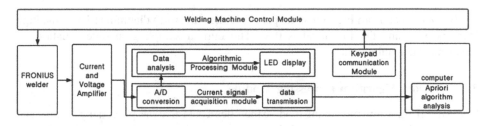

Fig. 2. System block diagram

3.1   Current Signal Sampling Module

Welding current signal sampling module includes AD conversion module and Gigabit network port transmission module. It can realize the real-time sampling and transmission of current signals.

The model of AD conversion module is AD7607, which supports the sampling rate of 8 channels at 200KSPS with 16 bits accuracy. The chip integrates input amplifier, over-voltage protection, filter and high-speed interface. It can satisfy requirement of the experiment. The converted data is stored in DDR3, and finally the Gigabit Ethernet port is controlled to upload the data in DDR3 to the PC for later analysis by the Apriori algorithm. The experimental network port transmission uses the LWIP protocol, which is a lightweight IP protocol that is applicable to both the operating system and stand-alone operation on FPGA.

Before experiment, we should verify the correctness of the current signal sampling module. Signal generator is used to provide a sinusoidal signal with a frequency of 1 kHz, a high level of 5 V, and a low level of 0 V to the CH1 channel of AD7607. The ZYNQ board is connected to the PC through the network port and opens the PC software. The waveform data displayed is shown in Fig. 3.

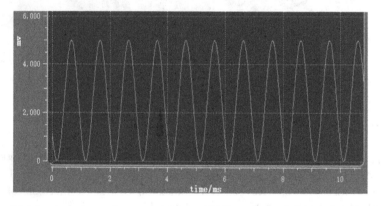

Fig. 3. Waveform simulation diagram of current signal sampling module

It can be seen from Fig. 3 that the period of the waveform diagram is 1 ms, the high level is 5 V, and the low level is 0 V. The current sampling module satisfies the experimental requirements.

3.2   Apriori Algorithm Processing Module

The Apriori algorithm processing module includes a data analysis processing module and an LED display module. The data analysis processing module first arranges the collected current data into an $m \times n$ matrix L on the PC (n represents the number of samples in a CMT cycle, and m represents the number of CMT cycles. Due to the instability of the CMT cycle, the shortest CMT duration is taken as a CMT period). Then discretize the sampled current data and obtain matrix D. The specific process is as follows:

Each column of the matrix corresponds to the same time in the CMT cycles, and each column of current data can be equally divided, and expressed as follows:

$$z_j = [MAX(L_j) - MIN(L_j)]/N \tag{1}$$

In Eq. (5), L_j is all current data in j-th column; z_j is the interval value in the j-th column current data N is equally divided. MAX (L_j) is the maximum value in j-th column current data, MIN (L_j) is the minimum value in j-th column current data.

Discretizing the current data of each column, the current data has different chronological order, that is, the same current value belongs to different items at different period. So the interval of the values after discretization of each column current data is different. The equation is expressed as follows:

$$\begin{cases} l_{ij} < MIN(L_j); c = 1 \\ MIN(L_j) + (c-2) \times z_j < l_{ij} \le MIN(L_j) + (c-1) \times z_j; c \in [2, N+1] \\ l_{ij} > MAX(L_j); c = N+2 \end{cases} \tag{2}$$

$$d_{ij} = (N+2) \times (j-1) + c \tag{3}$$

In Eq. (6), $l_{ij}(i = 1, 2\ldots m; j = 1, 2\ldots n)$ is the real current value in matrix L, and in Eq. (7), d_{ij} is the discretized value in matrix D.

For each l_{ij}, we use Eq. (6) to obtain c, and the used Eq. (7) to obtain the discretized value d_{ij}. Finally, we set appropriate support and confidence for data mining, and obtain the classification results.

When carry out online welding experiment, firstly, the interval thresholds of several correlation points in the classification result are set on the ZYNQ board, and the number of periods satisfying the correlation points is counted, and the ratio of the period of the correlation point to the total count of the CMT period is used as the judgment index. According to the obtained classification result, the corresponding LED light is lighted.

3.3 Keypad Communication Module

In this device, Keypads on ZYNQ board are used to send a signal to IO port to control shielding gas flow rate. Keyboard KEY1 sends 01 to IO port to increase the shielding gas flow rate, and key KEY2 sends 10 to IO port to reduce it. The welder can select the applicable shielding gas flow rate, according to the classified result.

4 Association Rules and Definition of Apriori Algorithm

(1) Items and ItemSets

$$I = \{i_1, i_2, \ldots \ldots i_n\} \tag{4}$$

In Eq. (1), I is a set of all items, called itemset. Among them, i_k (k = 1, 2 … , n) is called item, and the set of items containing K items is called K itemset.

(2) Transactions and transaction sets

Each transaction T consists of several items, and T belongs to a subset of itemset I. Each independent transaction T has a unique identifier TID to identify. A transaction set D is composed by all independent transactions together.

(3) Association Rules

Association rules can be expressed as an implication of X => Y ($' \Rightarrow '$ is called Association operation), where X and y are subsets of itemset I, but not empty sets, however the intersection of X and Y is empty sets. The equation is expressed as follows:

$$\begin{cases} X \subset I \\ Y \subset I \\ X \cap Y = \Phi \end{cases} \tag{5}$$

(4) Support and Confidence

The support degree sup of association rules is defined as follows:

$$\sup(X \Rightarrow Y) = P(X \cup Y) \tag{6}$$

Represents the probability of a transaction containing $X \cup Y$ in transaction set D. That is, the ratio of the transaction containing the item sets X and Y to the transaction set D, and the degree of support represents the frequency of occurrence in rules.

The confidence con of association rules is defined as follows:

$$\text{con}(X \Rightarrow Y) = \frac{SUP(X \cup Y)}{\sup(X)} \tag{7}$$

Represents the probability that in transaction set D, a transaction that contains X will contain Y. Confidence reflects the probability that another event will appear at the same time after an event occurs. The higher the confidence, the higher the credibility of the rule and the more trustworthy it is.

Apriori algorithm is a classical data mining algorithm for mining frequent itemsets and association rules. Apriori algorithm extracts frequently occurring data sets from a large data set, and then deriving strong rules based on these frequent itemsets. The process consists of connection (class matrix operation) and pruning (removal of intermediate items that do not conform to support). The goal of Apriori algorithm is to find the largest itemset. It uses an iterative method to first count the set of 1 items that satisfy the support, which is the set of frequent 1 items. Then connect the frequent 1 itemsets, remove those 2 itemsets that do not satisfy the support degree, and the frequent 2 itemsets are obtained. Iterating these processing until no maximum frequent candidate set is produced.

5   Experimental Design and Analysis

5.1   Current Signal Sampling and Analysis

First sample current data under three different shielding gas flow rates and the result is shown in Fig. 4. It can be seen from the figure that there is a significant difference among the electrical signals under different shielding gas flow rates, and the length of the CMT welding cycle is also inconsistent. The result shows one CMT phase is larger than 16 ms and the corresponding sample point is 800.

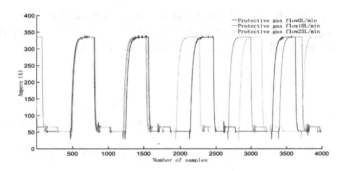

Fig. 4. Current signal under different shielding gas flow rates

In order to use Apriori algorithm to mine the relationship between current data under different shielding gas flow rates, it is necessary to discretize the current data, and divide the data of each column into 18 intervals. The first value of the current rising edge and the current value greater than 100 A is taken as the first sampling point. We set different identifiers for the three shielding gas flow rates, for 25 L/min, the identifier is defined as 20000, for 10 L/min, the identifier is defined as 20001, for 0 L/min, the identifier is defined as 20002. The current data after discretization is shown in Table 2.

Table 2. Discrete transformed current data

Sampling point 1	Sampling point 2	Sampling point 3	...	Label 801
29	48	71	...	20000
29	47	71	...	20000
30	51	72	...	20000
...	...	...	...	....
27	49	70	...	20001
29	48	71	...	20001
26	45	71	...	20001
...	...	...	...	...
31	48	71	...	20002
27	46	69	...	20002
31	53	72	...	20002
...	...	...	...	...

In the specific mining classification process, we firstly mine the association rules under the shielding gas flow rate of 25 L/min. After data mining, the effective correlation points are analyzed and compared, as shown in Fig. 5.

```
------------------------------------------------Rules------------------------------------------------
                    9876,9916,9956,9976,9993,10013===>20000
                    9916,9956,9976,9993,10013,10034===>20000
```

Fig. 5. Effective correlation point under 25 L/min of shielding gas flow rate

The shielding gas flow rate of the mid-range is 10 L/min. After data mining, the effective correlation points are analyzed and compared, as shown in Fig. 6.

```
------------------------------------------------Rules------------------------------------------------
                    3389,3489,3968,4011,4793,5450===>20001
         3389,3489,3968,4011,4793,5450,9916,9956,9976,9993,10013,10034===>20001
```

Fig. 6. Effective correlation points under 107 L/min of shielding gas flow rate

The shielding gas flow rate of the low gear position is 0 L/min. After data mining, the effective correlation points are analyzed and compared, as shown in Fig. 7.

```
------------------------------------------------Rules------------------------------------------------
                    9876,9916,9956,9976,9993,10013===>20002
                    9916,9956,9976,9993,10013,10034===>20002
```

Fig. 7. Effective correlation point under 0 L/min of shielding gas flow rate

Thus, the shielding gas flow rate 10 L/min can be distinguished from the shielding gas flow rate 0 L/min and 25 L/min by the effective correlation points shown in Fig. 8, and the six correlation points are 170th, 175th, 199th, 201th, 240th and 273th, namely Rule 1. The current value intervals of these six points are derived, as shown in Table 3. However, Rule 1 cannot distinguish the shielding gas flow rate of 25 L/min and the shielding gas flow rate of 0 L/min, thus we need further analysis.

---Rules--

3389,3489,3968,4011,4793,5450===>20001

Fig. 8. Rule 1

Then, we should distinguish the shielding gas flow rate of 25 L/min and 0 L/min. According to the above method, the data of each column is re-divided into 8 intervals, and the obtained effective correlation points are as shown in Figs. 9 and 10 respectively. Therefore, the six associated points including 105th, 110th, 113th, 114th, 115th, and 131th that distinguish the shielding gas flow rate of 25 L/min and 0 L/min can be determined, namely Rule 2. As shown in Fig. 11. The current value intervals of these six points are derived, as shown in Table 3.

---Rules--

1046,1095,1126,1136,1146,1305===>20000
1046,1095,1126,1136,1146,1306,5045,5056,5066,5077,5088===>20000

Fig. 9. New effective correlation point under 25 L/min of shielding gas flow rate

---Rules--

5045,5056,5066,5077,5088===>20002

Fig. 10. New effective correlation point under 0 L/min of shielding gas flow rate

---Rules--

1046,1095,1126,1136,1146,1305===>20000

Fig. 11. Rule 2

Table 3. Current intervals of Rules 1 and 2

Sampling point	Rule 1 (Current (A))	Sampling point	Rule 2 (Current (A))
170th	334.2657–335.0456	105th	328.1497–329.6699
175th	334.2075–335.0205	110th	329.4093–331.0497
199th	334.6251–335.5847	113th	330.1341–331.5685
201th	334.5443–335.4582	114th	330.2790–331.6924
240th	334.6443–335.5454	115th	330.3020–331.7097
273th	334.5646–335.7040	131th	331.4386–333.7351

5.2   Testing and Verification of Different Shielding Gas Flow Rates on Line

After obtaining the classification rule, the accuracy of the rule needs to be verified on the ZYNQ board, and the current signals after the AD conversion are simultaneously judged by the Rules 1 and 2, that is to say, the CMT periods satisfying the associated points are respectively counted. Since the minimum confidence is set as 0.75 during data mining on the PC side. Therefore, for every 100 CMT cycles, if 75 CMT cycles can satisfy the associated points, they are considered to be in compliance with the rules. If it satisfies Rule 1, the shielding gas flow rate can be directly judged to be 10 L/min, and the LED 2 is lighted. If it satisfies Rule 2, it can be judged that the shielding gas flow rate is 25 L/min and the LED 1 is lighted. If the two rules are not satisfied, it can be judged that the shielding gas flow rate is 0 L/min and the LED 3 is lighted.

In the online welding experiment, after abundant experiments, the LED lights can be lighted correctly. In order to increase the accuracy of the conclusion, 15 channels of shielding gas flow rate are selected. The current of each welding is exported through the Gigabit Network port, and the confidence of the correlation points is calculated at PC. The confidence of Rule 1 and Rule 2 is shown in Table 4.

Table 4. Confidence of Rules 1 and 2

	Rule 1			Rule 2	
	25 L/min	10 L/min	0 L/min	25 L/min	0 L/min
1	0.3	0.87	0.01	0.98	0.03
2	0.53	0.88	0.01	0.96	0.01
3	0.46	0.87	0.01	0.97	0.01
4	0.5	0.89	0	0.99	0.01
5	0.56	0.89	0	0.99	0
6	0.6	0.91	0	0.98	0.01
7	0.57	0.90	0	0.99	0.01
8	0.58	0.87	0	0.99	0.01
9	0.53	0.87	0	0.99	0.03
10	0.56	0.86	0	0.99	0.01
11	0.57	0.89	0	0.98	0.01
12	0.61	0.89	0	0.99	0.01
13	0.57	0.81	0	1	0
14	0.61	0.78	0.01	0.99	0
15	0.58	0.86	0.01	0.99	0.01

In Table 4, the confidence of the current data with the shielding gas flow rate of 10 L/min is greater than the minimum confidence set by the Apriori algorithm (0.75), and the confidence of the current data of the shielding gas flow rate of 25 L/min and 0 L/min is less than the minimum confidence degree. That is to say Rule 1 is able to distinguish the shielding gas flow rate of 10 L/min from the shielding gas flow rate of 25 L/min and 0 L/min.

At the same time, the confidence of the current data with the shielding gas flow rate of 25 L/min is greater than the minimum confidence level set by the Apriori algorithm of 0.75, and the confidence of the current data of the shielding gas flow rate of 0 L/min is less than the minimum confidence. That is to say Rule 2 is able to distinguish the shielding gas flow rate of 25 L/min and 0 L/min.

In addition, similar shielding gas flow experiments were carried out with spectral data under TIG welding process. In the experiments, the shielding gas flow rates were 25 L/min, 10 L/min and 3 L/min respectively. Through data mining, three kinds of rules for shielding gas flow were found. The confidence levels of Rule 1 and Rule 2 are shown in Table 5. Rule 1 can distinguish the shielding gas flow rate of 25 L/min from the shielding gas flow rate of 10 L/min and 3 L/min. Rule 2 can the shielding gas flow rate of 25 L/min 3 L/min. The experimental results demonstrated the effectiveness of using data mining to distinguish shielding gas flow rate.

Table 5. Spectral confidence of Rules 1 and 2

	Rule 1			Rule 2	
	25 L/min	10 L/min	3 L/min	10 L/min	3 L/min
1	0.99	0.29	0	0.9	0
2	0.97	0.22	0	0.97	0
3	0.99	0.24	0	0.94	0
4	0.99	0.15	0	1	0.06
5	0.99	0.40	0	0.91	0
6	0.98	0.38	0	0.94	0.02
7	0.97	0.19	0	0.95	0.15
8	0.98	0.20	0	0.92	0.07
9	0.99	0.27	0	0.93	0
10	0.99	0.36	0	0.88	0.06
11	0.98	0.58	0.01	1	0.01
12	0.99	0.59	0.01	0.99	0.37
13	0.99	0.53	0	0.97	0.14
14	0.98	0.58	0	0.92	0.07
15	0.98	0.36	0	0.96	0.24

6 Conclusion

In this paper, in order to overcome the drawbacks of traditional welding quality detection methods, an online quality detection method based on Apriori data mining algorithm is proposed. The ZYNQ board is used as a carrier to distinguish the welding electrical signals under different shielding gas flow rates. The following conclusions are obtained:

(1) In the CMT process, the electrical signals have different rules under different shielding gas flow rates.

(2) A novel current data mining algorithm based on Apriori algorithm is proposed. The experimental results show that the algorithm can successfully identify the differences of current data under three different shielding gas flow rates, and the ZYNQ board is used to realize online identification of shielding gas flow rate.

(3) The conclusion of the experiment has been demonstrated by abundant experiments, and the current data is collected for offline verification, which has good accuracy and robustness.

Acknowledgement. This work was supported by the National Natural Science Foundation of China (Grant Nos. 61727802).

References

1. Wu, X., Zeng, Y.: Data mining of College Students' performance based on Apriori algorithm. J. Langfang Normal Univ. **19**(1), 31–36 (2019)
2. Liu, R., Yin, D., Zhao, X.: Design of current sampling system based on FPGA. Comput. Eng. **37**(10), 227–230 (2011)
3. Li, D., Song, Z., Ye, F.: GMAW defect online monitoring based on independent component analysis. J. Weld. **27**(3), 44–48 (2006)
4. Shi, Y., Nie, J., Health, et al.: Stability analysis of pulsed MIG welding of aluminium alloy based on arc voltage probability density. J. Weld. **31**(5), 13–16 (2010)
5. Gao, L., Xue, J., Zhang, W., et al.: CO2 arc welding monitoring based on the theory of normal repetition rate of weak periodic multi-channel. J. Weld. **32**(11), 29–32 (2011)
6. Gao, L., Xue, J., Chen, H., et al.: Quantitative evaluation of droplet transfer stability in arc welding based on autocorrelation analysis. J. Weld. **33**(5), 29–32 (2012)
7. Gao, L., Xue, J., Chen, H., et al.: Evaluation method of CO2 arc welding stability based on quantitative analysis of short-circuit time-frequency distribution. J. Weld. **7**, 43–46 (2013)
8. Hu, J., Wu, C., Jia, C.: Stability evaluation of underwater wet electrode arc welding process. J. Weld. **34**(5), 99–102 (2013)
9. Ahsan, M.R.U., Cheepu, M., Kim, T.H., et al.: Mechanisms of weld pool flow and slag formation location in cold metal transfer (CMT) gas metal arc welding (GMAW). Weld. World **61**(2), 1–11 (2017)
10. Chen, M., Dong, Z., Wu, C.: Current waveform effects on CMT welding of mild steel. J. Mater. Process. Technol. **243**, 395–404 (2017)
11. Wang, D., Zhang, Z., Liang, Z., et al.: Electric signals filtering of AC CMT welding based on wavelet analysis. Trans. China Weld. Inst. **35**(5), 17–20 (2014)
12. Rauma, K., Laakkonen, O., Luukko, J., et al.: Digital control of switch-mode welding machine using FPGA. In: IEEE Power Electronics Specialists Conference. IEEE (2006)
13. Wang, W.Q., Meng, Q.L., Niu, L.Y.: Study on CMT welding of stainless steel railway vehicle body. Adv. Mater. Res. **936**, 6 (2014)
14. Park, S.H., Synn, J., Kwon, O.H., et al.: Apriori-based text mining method for the advancement of the transportation management plan in expressway work zones. J. Supercomput. **74**(3), 1–16 (2017)
15. Zhen, L., Xuyou, W., Wei, W., et al.: Laser-CMT hybrid welding with argon as shielding gas for 304 stainless steel. Weld. Join. (2010)

Visual Tracking with Attentional Convolutional Siamese Networks

Ke Tan[1,2] and Zhenzhong Wei[1,2(✉)]

[1] School of Instrumentation and Optoelectronic Engineering,
Beihang University, Beijing, China
zhenzhongwei@buaa.edu.cn
[2] The Key Laboratory of Precision Opto-mechatronics Technology,
Ministry of Education, Beijing, China

Abstract. Recently Siamese trackers have drawn great attention due to their considerable accuracy and speed. To further improve the discriminability of Siamese networks for visual tracking, some deeper networks, such as VGG and ResNet, are exploited as backbone. However, high-level semantic information reduces the location discrimination. In this paper, we propose a novel Attentional Convolutional Siamese Networks for visual tracking (ACST), to improve the classical AlexNet by fusing spatial and channel attentions during feature learning. Moreover, a response-based weighted sampling strategy during training is proposed to strengthen the discrimination power to distinguish two objects with the similar attributes. With the efficiency of cross-correlation operator, our tracker can be trained end-to-end while running in real-time at inference phase. We validate our tracker through extensive experiments on OTB2013 and OTB2015, and results show that the proposed tracker obtains great improvements over the other Siamese trackers.

Keywords: Visual tracking · Siamese networks · Visual attentions

1 Introduction

Visual tracking is a one of the most important problems in computer vision due to its applications in diverse fields such as video monitoring, human-computer interactions, and industrial automation. Given only a bounding box of an arbitrary target in the first frame, the objective is to obtain the target region in the subsequent frames. Although great progress has been achieved in the past decades, it is still a herculean task to design a real-time high-performance tracker which can overcome all challenges including illumination, deformation, fast motion, occlusion and so on.

Recently, trackers based on Siamese networks [1,6,11,15,16,27,29,35,36] have achieved outstanding performances on various tracking datasets. In these

Supported by "the National Science Fund for Distinguished Young Scholars of China under Grant No. 51625501".

Y. Zhao et al. (Eds.): ICIG 2019, LNCS 11901, pp. 369–380, 2019.
https://doi.org/10.1007/978-3-030-34120-6_30

methods, tracking an arbitrary object is formulated as a similarity learning problem. To ensure tracking efficiency, the similarity function is learned in an initial offline phase and often fixed during online tracking. One of the representative works is the SiamFC [1], which feeds a candidate image and a much larger search image into a fully convolutional network with a cross correlation layer. Because of the translation invariance of the embedding function, it can compute the similarity at all translated sub-windows on a dense grid in a single evaluation. However, to satisfy the strict spatial translation invariance, SiamFC is restricted to use the no-padding AlexNet [14] as the backbone which is not powerful enough. To address this issue, some modern networks, such as VGG [22] and ResNet [10], are embedded into the Siamese framework. However, deeper networks are hard to train from limited data and high-level semantic information reduces the location discrimination. In RASNet [29], a spatial residual attention and a channel favored attention are exploited to enhance the discriminative capacity of cross-correlation layer, but it does not strength the ability to extract the underlying semantic features for the targeted objects. Inspired by [30], we improve the classical AlexNet by fusing spatial and channel attentions during feature learning.

Another issue is the imbalance of positive and negative samples caused by the dense spatial sampling. Although average weighted logistic loss is used in the SiamFC, it is unreasonable to assign the same weight to all negative samples. Zheng et al. [36] introduced an effective sampling strategy to control the sample distribution and make the model focus on the semantic distractors. Rather than a distractor-aware incremental learning phase in [36], we use the feed-forward response map to penalize the effects of simple and hard negative samples simultaneously, which can be seen as a spatial regularization (Fig. 1).

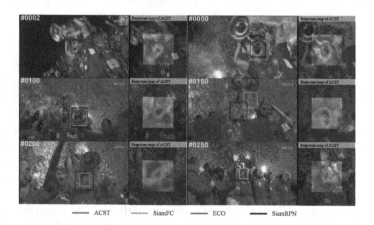

——— ACST   ——— SiamFC   ——— ECO   ——— SiamRPN

Fig. 1. Snapshots of the proposed tracker on sequence Soccer with comparison to SiamFC [1], ECO [5] and SiamRPN [16]. The right of each frame shows the response map produced by our tracker, which indicates the probability of target appears at every location.

Based on the above, we have developed an effective and efficient tracker, referred as ACST, where an improved attentional AlexNet is used as backbone and the learning phase is regularized by the response-based weighted sampling strategy. The model is trained with VID [21] and TrackingNet [19] and fine-tuned by ALOV [23] in an end-to-end manner. We evaluate our tracker on two benchmark datasets: OTB2013 [31] with 50 videos and OTB2015 [32] with 100 videos. Results show that our tracker performs favorably against a number of state-of-the-art trackers with the running speed over 30 frames per seconds.

To summarize, the main contributions of this work are listed below in three-fold:

- We improve the classical AlexNet by fusing spatial and channel attentions, which is embedded into the Siamese Network structure.
- We propose a response-based weighted sampling strategy to balance the training samples, which improve the discrimination ability to distinguish two objects with the similar attributes.
- A novel ACST tracker is proposed, which can be trained end-to-end and runs in real-time while achieves good performances.

The rest of the paper is organized as follows. Section 2 introduces the related works. The proposed architecture and training and tracking details are introduced in Sect. 3. While Sect. 4 presents the experimental results and Sect. 5 concludes this paper.

2 Related Work

In this section, we provide a brief review on methods closely related to this work.

2.1 CNNs in Visual Tracking

Recently, convolutional neural networks (CNNs) have made great breakthrough in many tasks of computer vision including visual tracking. A CNN is made up of several layers of convolution, activation, normalization and pooling operations. Existing trackers with CNNs can be roughly classified into two categories. One is that a CNN is used as a feature extractor which is combined with traditional machine learning methods [5,12,18,34]. [12] combines a pretrained CNN and online SVMs to find the target location directly from saliency map. In [18], features extracted from a pretrained deep CNN are exploited to train adaptive correlation filters (CF), which are then widely used in subsequent CF-based trackers [5,34]. However, these trackers suffer from slow speed and some are not particularly good compared to well-designed handcrafted trackers [25,26] in accuracy. The other is that CNNs are used for feature embedding in deep tracking networks [3,4,7,24,33]. [20] proposes a multi-domain CNN to identify target in each video domain, which is then improved by [13]. And recently Siamese networks-based trackers [1,6,11,15,16,27,29,35,36] have drawn great attentions due to the balance in speed and accuracy.

2.2   Siamese Trackers

Tao et al. [27] first apply Siamese networks into visual tracking where ROI pooling is used to evaluate the similarity of two regions. Then [11] trains a Siamese network to regress directly from two images to the location in the second image of the object shown in the first image. [1] exploits fully-convolutional Siamese networks with cross correlation operations to directly produce a response map. Although [1] achieves a good performance in both speed and accuracy, it still has a gap to the best online tracker. To further improve the performance of the Siamese tracking framework, a great deal of work has been done such as combing with RPN [15,16], using deeper networks [15,17,35], online learning [2, 8,28], augmented loss function [6,36] and so on.

3   The Proposed Tracker

In this section, we firstly introduce the architecture of proposed attentional convolutional Siamese network. Then a response-based weighted sampling strategy is developed to solve the unbalance of simple and hard samples. Finally, we present the online tracking pipeline of our ACST tracker.

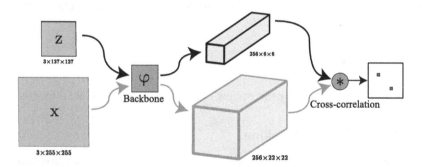

Fig. 2. The structure of fully-convolutional Siamese networks for visual tracking. The backbone extracts features from candidate and search image, which are then used to produce a response map with a cross-correlation layer.

3.1   Attentional Convolutional Siamese Network

We first review the Siamese network with fully-convolutional operations. A fully-convolutional network with integer stride k can be seen an embedding function φ satisfy

$$\varphi\left(L_{k\tau}\left(x\right)\right) = L_{\tau}\left(\varphi\left(x\right)\right) \tag{1}$$

where L_{τ} is the translation operator $\left(L_{\tau}x\right)\left[u\right] = x[u - \tau]$ for any translation τ. The advantage of fully-convolutional Siamese networks is that it can compute the similarity at all translated sub-windows on a dense grid in a single evaluation by a cross-correlation operator f by

$$f(z,x) = \varphi(z) * \varphi(x) + b\mathbb{1} \qquad (2)$$

where z and x are a candidate image and a much larger search image respectively, $b\mathbb{1}$ denotes a bias term at every location. Figure 2 shows the structure of Siamese network for visual tracking.

The discriminative ability of the above model depends largely on the backbone network φ which extracts features for the image. Inspired by [30], we exploit the attentional convolutional block to improve the classical AlexNet.

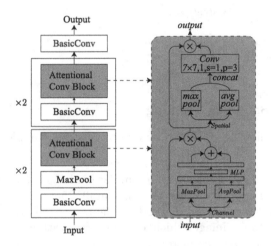

Fig. 3. The architecture of the proposed attentional AlexNet which is used as the backbone. The attentional convolutional block is illustrated in the right, where channel and spatial attention are applied to refine the input feature. Note that both max-pooling and average-pooling outputs are utilized in the channel attention with a shared Multi-Layer Perception (MLP), while the spatial attention concatenates two outputs that are pooled along the channel axis and forward them to a convolution layer.

As is shown in Fig. 3, the attentional convolutional block is made up of a basic convolution block and a dual attention module which operates along two separate dimensions, channel and spatial. Given an intermediate feature map, the output is obtained by sequentially multiplying the attention maps to it for adaptive feature refinement. Specifically, channel attention focuses on 'What' is meaningful while spatial attention focuses on 'Where' is an informative part.

We illustrate the Attentional Convolutional Siamese Network configuration in Table 1. In particularly, we modify the first 4 conv layers of no-padding AlexNet in [1] by the attentional convolutional block. Then we use a combination of cross correlation layers for calculating score maps. The generated maps indicate the similarity information between the candidate image and search image.

Table 1. Architecture of Attentional Convolutional Siamese Network which is modified from the no-padding AlexNet.

Layer		*Exemplar* output size	*Search* output size
Input		127×127	255×255
Conv1	$3, 96, 11 \times 11, stride = 3$	59×59	123×123
Pool1	$3 \times 3, stride = 2$	29×29	61×61
AttConvBlock1			
Conv2	$96, 256, 5 \times 5, stride = 1$	25×25	57×57
Pool2	$3 \times 3, stride = 2$	12×12	28×28
AttConvBlock2			
Conv3	$256, 384, 3 \times 3, stride = 1$	10×10	26×26
AttConvBlock3			
Conv4	$384, 384, 3 \times 3, stride = 1$	8×8	24×24
AttConvBlock4			
Conv5	$384, 256, 3 \times 3, stride = 1$	6×6	22×22
CrossCorrelation		17×17	

3.2   Weighted Sampling Training

We perform the following steps one by one to generate sample pairs and corresponding labels.

- Randomly choose two frames with a max interval N from a video and decode into 32-bit floating point raw pixel values in $[0, 255]$.
- Extract the exemplar image centered on the target following the way in [1].
- Extract the instance image which is translated within T pixels as well as scaled by a random stretch factor in $[1 - \alpha, 1 + \alpha]$.
- Calculate the labels for the training pair, which are belong to positive if they are within radius R of the target center c (accounting for the stride k)

$$y[u] = \begin{cases} 1, & if \ k \, \|u - c\| \leq R \\ 0, & otherwise \end{cases} \tag{3}$$

In [1], the average weighted logistic loss is adopted to train the network. However, assigning the same weight to all negative samples will reduce the ability to distinguish two objects with similar semantic information. Here we weight the training sample on the dense grid by the feed-forward response map, shown in Fig. 4. Given a real-valued score of a single exemplar-candidate pair v, we first calculate the sigmoid output $s[u] = sigmoid(v[u])$, which indicates the similarity of the training pair. Then the weight for each position is

$$w[u] = \begin{cases} \dfrac{s[u]}{2 \sum\limits_{u \in y^+} s[u]}, & if \ k \, \|u - c\| \leq R \\ \dfrac{s[u]}{2 \sum\limits_{u \in y^-} s[u]}, & otherwise \end{cases} \tag{4}$$

(a)   (b)   (c)   (d)

Fig. 4. Illustration of the weighted sampling strategy. (a) is the tracking frame. (b) is the candidate image extracted from the first frame. (c) is the search image with a response map added on it, which indicates the similarity between dense grid samples and candidate image. (d) is a hard sample with a similar attributes with candidate image, which is assigned more weight.

We choose the cross-entropy loss as the loss function, so the final weighted average loss is

$$loss(v, y) = -\sum_u w[u] (y[u] \ln (s[u]) + (1 - y[u]) \ln (1 - s[u])) \qquad (5)$$

Furthermore, Stochastic Gradient Descent with momentum is applied to obtain the network parameters.

3.3 Tracking

Given the first frame with the target annotated, we extract the exemplar image and feed it into the attentional convolutional network to obtain the filter kernel. Note that we do not update the kernel for speed consideration. When there comes a new frame, the target is searched around the previous position. To handle the scale variances, multi-scaled search patches are extracted as a mini-batch, which is used to input into the network and then calculate the cross-correlation with the kernel. The obtained response maps are up-sampled by bi-cubic interpolation and penalized by the scale factor. Meanwhile, a cosine window is added to the response maps to penalize large displacements. The peak of response maps indicates the position of the target and the scale is updated with a decay.

4   Experiments

In this section, we firstly provide details of training and tracking implementation. Then experiments on OTB2013 [31] and OTB2015 [32] are conducted to evaluate the proposed tracker and the results are compared with some state-of-the-arts. Finally, we show qualitative results on some challenging videos.

4.1 Implementation Details

During training phase, the weights of convolutional layers are initialized with Kaiming algorithm [9] and all bias are initialized to zero. For batch normalization layers, γ and β vectors are initialized to one and zero respectively. We first use VID [21] and TrackingNet [21] to train the model with a batch size 32. And SGD is applied with a warm-up learning rate increasing linearly for the first 5 epochs from 5×10^{-3} to 2.5×10^{-2} and decayed by 0.8576 for 15 epochs. Then the last two convolution layers of backbone are fine-tuned using ALOV [23] with a batch size 8. The learning rate of SGD is decayed by 0.8576 for 30 epochs. Moreover, the preprocessing steps are described in Sect. 3.2 with $N = 100$, $T = 64$, and $\alpha = 0.1$.

During tracking, the weight of cosine window is set to 0.31 and the upsampled factor of response map is set to 16. We search for the object over 3 scales $1.0375^{\{-1,0,1\}}$ and update the scale by a liner interpolation with a factor of 0.59. The proposed model is implemented in PyTorch 1.0 on a workstation with an Intel E5-1650 v4 CPU, 32G RAM, NVIDIA TITAN Xp GPU. Our tracker can run over 30 fps on OTB benchmark.

4.2 Experimental Validations

We evaluate the proposed tracker, referred as ACST, on OTB2013 and OTB2015 datasets which contain 50 and 100 videos respectively, with comparison to some state-of-the-art methods. We quantitatively evaluate trackers using center location error and overlap ratio. And for completeness, we also report qualitative results on some challenging videos.

Quantitative Analysis. We provide a comparison of ACST with 14 state-of-the-art trackers: MDNet [20], SCT [4], SiamFC [1], CREST [24], ECO [5], ADNet [33], SANet [7], SiamRPN [16], SiamTri [6], DaSiamRPN [36], ACT [3], SiamDW-FC [35], SiamDW-RPN [35] and SiamRPN++ [15]. Figure 5 shows the precision plots and success plots under one-pass evaluation (OPE) on OTB2013 and OTB2015 for all the 15 trackers which are ranked using the AUC (area under the curve) displayed in the legend.

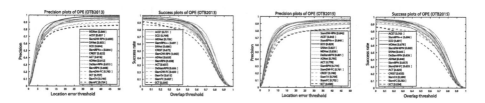

Fig. 5. Precision plots and Success plots under one-pass evaluation (OPE) on OTB2013 and OTB2015.

Qualitative Analysis. Figure 6 shows the tracking results of 4 methods: ECO [5], SiamFC [1], SiamRPN [16] and the proposed ACST on 6 challenging sequences. The ECO tracker performs well in sequences with illumination and fast motion (Singer2, DragonBaby) but fails when occlusion and rotation (MotoRolling, Freeman4). SiamRPN performs well in sequences with deformation, but fails when back-ground clutter (Tiger1). Our tracker performs well on all these videos, which validates the effectiveness.

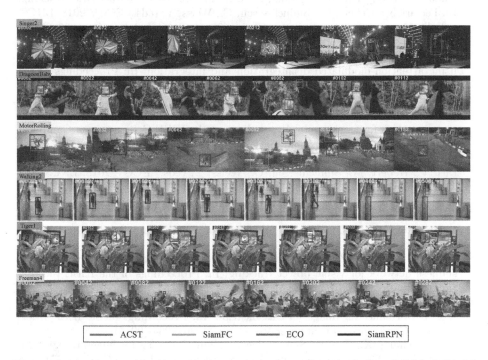

Fig. 6. Qualitative results of our ACST and 3 compared trackers: SiamFC [1], ECO [5] and SiamRPN [16].

5 Conclusion

In this paper, we improve the classical AlexNet with attention mechanism, which is used as the backbone of Siamese Networks for visual tracking. Furthermore, we enhance the ability of the proposed network to distinguish targets with similar attributes. Based on the above, we propose an ACST tracker, which achieves great performance on OTB2013 and OTB2015 datasets while running in real-time. In the future, we will make efforts to improve our tracker with RPN.

References

1. Bertinetto, L., Valmadre, J., Henriques, J.F., Vedaldi, A., Torr, P.H.S.: Fully-convolutional siamese networks for object tracking. In: Hua, G., Jégou, H. (eds.) ECCV 2016. LNCS, vol. 9914, pp. 850–865. Springer, Cham (2016). https://doi.org/10.1007/978-3-319-48881-3_56
2. Bhat, G., Danelljan, M., Van Gool, L., Timofte, R.: Learning discriminative model prediction for tracking. arXiv preprint arXiv:1904.07220 (2019)
3. Chen, B., Wang, D., Li, P., Wang, S., Lu, H.: Real-time 'Actor-Critic' tracking. In: Ferrari, V., Hebert, M., Sminchisescu, C., Weiss, Y. (eds.) ECCV 2018. LNCS, vol. 11211, pp. 328–345. Springer, Cham (2018). https://doi.org/10.1007/978-3-030-01234-2_20
4. Choi, J., Jin Chang, H., Jeong, J., Demiris, Y., Young Choi, J.: Visual tracking using attention-modulated disintegration and integration. In: Proceedings of the IEEE Conference on Computer Vision and Pattern Recognition, pp. 4321–4330 (2016)
5. Danelljan, M., Bhat, G., Shahbaz Khan, F., Felsberg, M.: ECO: efficient convolution operators for tracking. In: Proceedings of the IEEE Conference on Computer Vision and Pattern Recognition, pp. 6638–6646 (2017)
6. Dong, X., Shen, J.: Triplet loss in siamese network for object tracking. In: Ferrari, V., Hebert, M., Sminchisescu, C., Weiss, Y. (eds.) ECCV 2018. LNCS, vol. 11217, pp. 472–488. Springer, Cham (2018). https://doi.org/10.1007/978-3-030-01261-8_28
7. Fan, H., Ling, H.: SaNet: structure-aware network for visual tracking. In: Proceedings of the IEEE Conference on Computer Vision and Pattern Recognition Workshops, pp. 42–49 (2017)
8. Guo, Q., Feng, W., Zhou, C., Huang, R., Wan, L., Wang, S.: Learning dynamic siamese network for visual object tracking. In: Proceedings of the IEEE International Conference on Computer Vision, pp. 1763–1771 (2017)
9. He, K., Zhang, X., Ren, S., Sun, J.: Delving deep into rectifiers: surpassing human-level performance on ImageNet classification. In: Proceedings of the IEEE International Conference on Computer Vision, pp. 1026–1034 (2015)
10. He, K., Zhang, X., Ren, S., Sun, J.: Deep residual learning for image recognition. In: IEEE Conference on Computer Vision and Pattern Recognition, pp. 770–778 (2016)
11. Held, D., Thrun, S., Savarese, S.: Learning to track at 100 FPS with deep regression networks. In: Leibe, B., Matas, J., Sebe, N., Welling, M. (eds.) ECCV 2016. LNCS, vol. 9905, pp. 749–765. Springer, Cham (2016). https://doi.org/10.1007/978-3-319-46448-0_45
12. Hong, S., You, T., Kwak, S., Han, B.: Online tracking by learning discriminative saliency map with convolutional neural network. In: International Conference on Machine Learning, pp. 597–606 (2015)
13. Jung, I., Son, J., Baek, M., Han, B.: Real-time MDNet. In: Ferrari, V., Hebert, M., Sminchisescu, C., Weiss, Y. (eds.) ECCV 2018. LNCS, vol. 11208, pp. 89–104. Springer, Cham (2018). https://doi.org/10.1007/978-3-030-01225-0_6
14. Krizhevsky, A., Sutskever, I., Hinton, G.E.: ImageNet classification with deep convolutional neural networks. In: International Conference on Neural Information Processing Systems, pp. 1097–1105 (2012)
15. Li, B., Wu, W., Wang, Q., Zhang, F., Xing, J., Yan, J.: SiamRPN++: evolution of siamese visual tracking with very deep networks. arXiv preprint arXiv:1812.11703 (2018)

16. Li, B., Yan, J., Wu, W., Zhu, Z., Hu, X.: High performance visual tracking with siamese region proposal network. In: Proceedings of the IEEE Conference on Computer Vision and Pattern Recognition, pp. 8971–8980 (2018)

17. Li, Y., Zhang, X.: SiamVGG: visual tracking using deeper siamese networks. arXiv preprint arXiv:1902.02804 (2019)

18. Ma, C., Huang, J.B., Yang, X., Yang, M.H.: Hierarchical convolutional features for visual tracking. In: Proceedings of the IEEE International Conference on Computer Vision, pp. 3074–3082 (2015)

19. Müller, M., Bibi, A., Giancola, S., Alsubaihi, S., Ghanem, B.: TrackingNet: a large-scale dataset and benchmark for object tracking in the wild. In: Ferrari, V., Hebert, M., Sminchisescu, C., Weiss, Y. (eds.) ECCV 2018. LNCS, vol. 11205, pp. 310–327. Springer, Cham (2018). https://doi.org/10.1007/978-3-030-01246-5_19

20. Nam, H., Han, B.: Learning multi-domain convolutional neural networks for visual tracking. In: Proceedings of the IEEE Conference on Computer Vision and Pattern Recognition, pp. 4293–4302 (2016)

21. Russakovsky, O., et al.: Imagenet large scale visual recognition challenge. Int. J. Comput. Vis. **115**(3), 211–252 (2015)

22. Simonyan, K., Zisserman, A.: Very deep convolutional networks for large-scale image recognition. In: International Conference on Learning Representations (2015)

23. Smeulders, A.W.M., Chu, D.M., Rita, C., Simone, C., Afshin, D., Mubarak, S.: Visual tracking: an experimental survey. IEEE Trans. Pattern Anal. Mach. Intell. **36**(7), 1442–1468 (2014)

24. Song, Y., Ma, C., Gong, L., Zhang, J., Lau, R.W., Yang, M.H.: CREST: convolutional residual learning for visual tracking. In: Proceedings of the IEEE International Conference on Computer Vision, pp. 2555–2564 (2017)

25. Tan, K., Wei, Z.: Learning an orientation and scale adaptive tracker with regularized correlation filters. IEEE Access **7**, 53476–53486 (2019)

26. Tang, M., Yu, B., Zhang, F., Wang, J.: High-speed tracking with multi-kernel correlation filters. In: Proceedings of the IEEE Conference on Computer Vision and Pattern Recognition, pp. 4874–4883 (2018)

27. Tao, R., Gavves, E., Smeulders, A.W.: Siamese instance search for tracking. In: Proceedings of the IEEE Conference on Computer Vision and Pattern Recognition, pp. 1420–1429 (2016)

28. Valmadre, J., Bertinetto, L., Henriques, J., Vedaldi, A., Torr, P.H.: End-to-end representation learning for correlation filter based tracking. In: Proceedings of the IEEE Conference on Computer Vision and Pattern Recognition, pp. 2805–2813 (2017)

29. Wang, Q., Teng, Z., Xing, J., Gao, J., Hu, W., Maybank, S.J.: Learning attentions: residual attentional siamese network for high performance online visual tracking. In: IEEE Conference on Computer Vision and Pattern Recognition (2018)

30. Woo, S., Park, J., Lee, J.-Y., Kweon, I.S.: CBAM: convolutional block attention module. In: Ferrari, V., Hebert, M., Sminchisescu, C., Weiss, Y. (eds.) ECCV 2018. LNCS, vol. 11211, pp. 3–19. Springer, Cham (2018). https://doi.org/10.1007/978-3-030-01234-2_1

31. Wu, Y., Lim, J., Yang, M.H.: Online object tracking: a benchmark. In: IEEE Conference on Computer Vision and Pattern Recognition, pp. 2411–2418 (2013)

32. Yi, W., Jongwoo, L., Ming-Hsuan, Y.: Object tracking benchmark. IEEE Trans. Pattern Anal. Mach. Intell. **37**(9), 1834–1848 (2015)

33. Yun, S., Choi, J., Yoo, Y., Yun, K., Young Choi, J.: Action-decision networks for visual tracking with deep reinforcement learning. In: Proceedings of the IEEE Conference on Computer Vision and Pattern Recognition, pp. 2711–2720 (2017)
34. Zhang, T., Xu, C., Yang, M.H.: Learning multi-task correlation particle filters for visual tracking. IEEE Trans. Pattern Anal. Mach. Intell. **41**(2), 365–378 (2019)
35. Zhipeng, Z., Houwen, P., Qiang, W.: Deeper and wider siamese networks for real-time visual tracking. arXiv preprint arXiv:1901.01660 (2019)
36. Zhu, Z., Wang, Q., Li, B., Wu, W., Yan, J., Hu, W.: Distractor-aware siamese networks for visual object tracking. In: Ferrari, V., Hebert, M., Sminchisescu, C., Weiss, Y. (eds.) ECCV 2018. LNCS, vol. 11213, pp. 103–119. Springer, Cham (2018). https://doi.org/10.1007/978-3-030-01240-3_7

Enhanced Video Segmentation
with Object Tracking

Zheran Hong[1,2], Sheng Chen[1,2], Zhentao Tan[1,2], Qiankun Liu[1,2], Bin Liu[1,2(✉)],
and Nenghai Yu[1,2]

[1] School of Information Science and Technology,
University of Science and Technology of China, Hefei, China
`flowice@ustc.edu.cn`
[2] Key Laboratory of Electromagnetic Space Information,
Chinese Academy of Science, Beijing, China

Abstract. The high efficiency and superior performance of fully convolutional network (FCN) architecture makes it a recent trend that employing FCN in video object segmentation task. While these FCN-based methods usually ignore the motion information between frames, which may lead to similar object inference or background clutter issues. To deal with these, we propose to use tracking techniques to improve the performance of video object segmentation. The proposed algorithm performs video object segmentation and tracking simultaneously in a unified framework. After that, the motion information provided by initial tracking result is used to rejecting outliers in the segmentation mask caused by background complexities, such as similar object inference or background clutter issues. In return, the final segmentation result can be used to supervise the tracking result. In this iterative way, the performances of the both tasks are enhanced. Experimental results on the challenging benchmark demonstrate the effectiveness of our proposed method.

Keywords: Video segmentation · Object tracking · Unification framework · Background complexities

1 Introduction

Video object segmentation is a fundamental task in computer vision, which integrates the segmentation task with object tracking spirit. Compared to object detection or tracking, this task aims to find the exact object region at pixel-level. So video object segmentation is a complicated but realistic task for real applications, which shows a growing prospect in public security and surveillance technology and attracts increasingly more attentions. Recent methods [5,18] are nearly based on DCNN, and obtain encouraging score in some data sets, such as DAVIS [17] and SegTrack v2 [15]. However, there are still many difficulties to overcome.

© Springer Nature Switzerland AG 2019
Y. Zhao et al. (Eds.): ICIG 2019, LNCS 11901, pp. 381–392, 2019.
https://doi.org/10.1007/978-3-030-34120-6_31

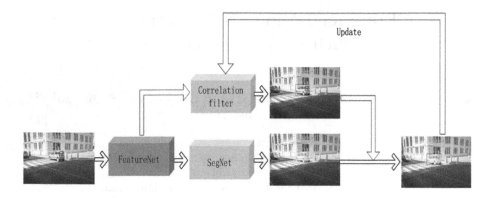

Fig. 1. The overall scheme of the proposed algorithm

The probability of objects and background that share similar appearances increases dramatically when the environment is complex. This makes the segmentation task be of greater difficulty and error-prone. The confusion between background and foreground is bound to lower the segmentation accuracy. But for videos there exists the relevance and continuity between frames, while many existing works [5,18,23] didn't take this characteristic into consideration. If the difference in two continuous frames is large or the foreground changes abruptly, some background regions may be easily blended into the foreground object area. To address this problem, a tracking based method is considered as an additional supplement in our work. Tracking results can provide a rough proposal bounding box in which the target locates, which helps to filter out many of the background pixels. Even if some regions are similar in appearance but when considered the motion similarity and the relationship among frames, many misclassifications could be eliminated. In turn, the segmentation mask will also provide strong evidence for the tracker when fast motion or occlusion happens.

In this paper, we propose a task-complementary algorithm which integrates a single object tracker to assist the segmentation network in bounding box level. As shown in Fig. 1, the tracker generates a bounding box (the left) which can be exploited as a candidate region for segmentation task. The segmentation network is first applied to generate output mask (the right) of the input frame, and at the same time, we use a correlation filter that shares the same CNN feature with the segmentation net to track the object. The segmentation results will be modified by the tracking output and used to adjust the parameters of the tracker as well.

2   Related Work

2.1   Video Object Segmentation

Semi-supervised video object segmentation aims to segment the foreground object with the knowledge of first frame. With the success of the DAVIS challenge [17], many recent segmentation methods based on CNNs have been pro-

(a) the tracking bounding box   (b) the segmentation mask

Fig. 2. The example of tracking result and segmentation result of the same frame

posed. All these methods could be roughly divided into two parts. One tends to regard video as a collection of images and process each frame independently. The most representative method for this is OSVOS [5], which use the famous image semantic segmentation network [6]. In order to focus on special object on each video sequence, OSVOS fine-tune the network with the first frame and its ground-truth. The other part realizes the temporal continuity between adjacent frames and tried to formulate it. MaskTrack [18] adds the mask from previous frame as a new input channel except RGB, and learns to extract features from static images. SegFlow [7] designs a network that jointly trains on object and optical flow. Bilateral filtering is used by VPN [13] to build a long range time correlation. And another technology for exploring the relationship between adjacent frames is RNN. Tokmakov *et al.* [20] propose to build a ?visual memory? in video with a convolutional recurrent unit. Although above works try to apply Interframe information, it is difficult to segment special objects when scene is complicated. With the advantage of object tracking, our method can handle such problem effectively.

2.2 Object Tracking

Object tracking is the task that predicts the location and scale of target which is given by a bounding box in the first frame of a video sequence, and it has been greatly developed in the last decade. Correlation filter, thanks to its high efficiency, has been widely used in object tracking. Bolme *et al.* developed MOSSE filter for tracking in [4] which can operate at 669 FPS. Though MOSSE only estimates the location of target, it can be extended to estimate changes in scale and rotation by filtering the log-polar transform of the input patch. But the training of MOSSE is conducted directly on the input image rather than features extracted from the image, and it is single channel which is not suitable for multi-channel features (deep features for example). A few years later, correlation filter was extended to multi-channel feature representations [3,11,12]. Dannelljan *et al.* proposed DSST tracker in [9], and the main contribution of DSST is the method to estimate the scale of target by computing the correlation scores in a scale pyramid representation. Danneljan also did some researches on correlation filter in [10] and [8]. C-COT in [10] focuses on the learning of continuous correlation filter by introducing the interpolation function, while ECO in [8] is the advanced version of C-COT, which learns and updates the correlation filter more efficiently and effectively. And the tracking performance is greatly improved by ECO.

For the remainder of the paper, we will introduce the overall architecture in detail in Sect. 3. Experimental results and analysis are given in Sect. 4. And our conclusion at the end.

3   Proposed Algorithm

Our proposed network can be divided into three parts, as shown in Fig. 2. The first part is a standard segmentation network which contains a CNN as encoder and a decoder network (noted as SegNet) to generate output masks. Secondly, we apply a correlation filter that shares the CNN feature extracted by the FeatureNet with the SegNet and works concurrently with it. Lastly, we fuse the outputs from the SegNet and the tracker to get the final result which is fed back into the correlation filter to get better results for both tasks.

3.1   Feature Extractor and Segmentation Network

We use OSVOS [5] which implements a VGG [19] based fully convolutional architecture as our base model for video object segmentation. The former 5 layers forms the feature network (FeatureNet) and the rest parts are the segmentation network (SegNet). The training process follows a general-to-specific manner. The entire network is pre-trained on ImageNet data set as a base model. Then it is further trained on DAVIS and results the parent model. Now the network is already capable to separate the foreground from an image. However it is still weak in segmenting certain object so far. Finally, when fine-tuned with the ground-truth mask of the first frame, the network learns to outline the given object from background scenes.

Another notable problem is that background pixels take up most space of images. This unbalance between foreground object and background scenes may finally incur more and more pixels in the frame be recognised as background. To avoid this, we applied a intra classes balancing cross entropy loss function.

$$L = -\beta \sum_{j \in Y_+} log P(y_j = 1|X)$$
$$- (1 - \beta) \sum_{j \in Y_-} log P(y_i = 0|X) \tag{1}$$

Where X is the input image, $y_i \in \{0,1\}$, $j = 1,...,|X|$ are 2 meaning labels by element. Y_+ and Y_- are positive and negative labeled pixels respectively, and

$\beta = \frac{|Y_-|}{|Y|}$. The probability $P(\cdot)$ is the result from the sigmoid activation of the output layer.

For the following post-processing procedure, many algorithms use Dense CRF [6,24] to refine the segmentation results. But these approaches are too costly in time and computation. In the proposed method, we apply edge detection method, HED [22], to detect overall contours all over the image and then we use Ultra metric Contour Map [2] to produce a superpixel representation [1] of the image. By voting of the most (over 50% experimentally), foreground regions (superpixels) are selected. With this procedure, we achieve similar results with CRF models but greatly accelerates the processing speed.

3.2 Tracker

We choose correlation filter method as our tracker for its high efficiency and robustness. Instead of using handcrafted features [9], we apply hierarchical convolutional features [16] as the input of the correlation filter. Since convolutional features can provide better performance and can be reused in the following segmentation network and the tracker at the meanwhile.

We briefly review the key principles of correlation filter. By exploiting the property of circulate matrix, the correlation filter can learn from a relatively large number of training samples effectively and perform fast tracking in the Fourier domain. Let x^l to be the feature map of l-th layer with size $M \times N \times D$, where M, N and D represents the width, height and the number of channels, respectively. Taking advantages of the property of cyclic matrix and befitting padding, all the circular shifts of $x_{m,n} \in \{0,1,...,M-1\} \times \{0,1,...,N-1\}$, are considered as training sample. Each training sample $x_{m,n}$ is assigned with a soft label $y_{m,n}$, which is generated by a Gaussian function and takes a value of 1 for the centred target, and smoothly decays to 0 for any other shifts. The goal of training is to find a function $f(z) = w^T z$ that minimizing the following cost:

$$w = \arg\min_w \sum_{m,n} |w * x_{m,n} - y_{m,n}|^2 + \lambda \|w\|^2 \tag{2}$$

where $w \cdot x_{m,n} = \sum_{d=1}^{D} w_{m,n,d}^T x_{m,n,d}$, and λ is a regularization parameter. In the Fourier domain, the learned filter for the d-th channel $(d \in \{1,...,D\})$ can be transformed into:

$$w^d = \mathcal{F}^{-1}\left(\frac{\mathcal{F}(y) \odot \mathcal{F}(\bar{x}^d)}{\sum_{i=1}^{D} \mathcal{F}(x^i) \odot \mathcal{F}(\bar{x}^i) + \lambda}\right) \tag{3}$$

Where $\mathcal{F}$ and $\mathcal{F}^{-1}$ represents the Fourier transformation and its inverse. The operation $\odot$ is the element-wise product, and the bar represents complex conjugation. Given an image patch in the new frame, the feature vector on the l-th layer is denoted as z, and its size is $M \times N \times D$. The score map $\hat{y}_l$ for the l-th correlation filter can be calculated as

$$\hat{y}_l = \mathcal{F}^{-1}\left(\sum_{d=1}^{D} \mathcal{F}(w^d) \odot \mathcal{F}(\bar{z}^d)\right) \tag{4}$$

The optimal position for the l-th correlation filter is obtained by searching the maximal value of the score map $\hat{y}_l$.

The tracking system works with the SegNet in a parallel manner. We apply hierarchical convolutional features to represent the object, which integrates low level features from lower layers of the CNN and high level features from upper layers' outputs. High level semantic features are helpful in handling appearance distortion problem, while low level features can be used to get accurate location. We choose to use the correlation filter to track the object for its high efficiency and superior performance. The specified tracking process is given as follows:

- Extracting features from layer conv3_3, conv4_3, and conv5_3 around the object according to the ground-truth in the first frame to train three correlation filters respectively.
- At time $t(t > 1)$, interpolating the features from these convolutional layers around the predicted location of time $t-1$. Feeding the features into the three correlation filters to get the output as score map.
- Updating the predicted location from upper filters, say conv5_3, to lower filters, like conv3_3. Taking outputs from the the upper filter as basis or constraint for lower filters. Note that $\arg\max_{m,n} f_l(m,n)$ represents the output location of the l-th correlation filter, so the output from the $l-1$-th correlation filter can be written as:

$$\arg \max_{m,n} f_{l-1}(m,n) + \gamma f_l(m,n)$$
$$s.t. \, |m - \hat{m}| + |n - \hat{n}| \leq r. \tag{5}$$

Where the constraint ensures the algorithm look for better results only in the upper layer. γ is a regularization term for former outputs. The predicted location can be calculated by optimizing Eq. (5).

- Updating three correlation filters with the tracking results.

3.3   Outputs Fusion

According to tracking bounding box, most of the irrelevant background pixels that lays outside the bounding box can be ignored naturally although some of them are close to the target region on appearance. The concrete implementing steps are given as follows:

- Firstly, as the rough segmentation results from SegNet are many foreground connected domains, we calculate bounding rectangles of each connected domain. If there exists any overlap region between two bounding rectangles, we cluster them into the same group. Going through the entire image we can get one bounding rectangle for every group.
- Secondly, we enlarge the box detected by the tracker by k times and calculate its overlap with the group bounding rectangles generated in the first step. If a group bounding rectangle have any overlap with the tracker bounding box we incorporate the segmented connected domains it contains into the foreground region. And the group bounding rectangles that lies apart are directly excluded.

With this mechanism, the segmentation network and the tracking correlation filter are complementary to each other while training. Ideally, the group bounding rectangle can completely covers the tracking bounding box, but actually the tracking results may usually deviate a lot from its ideal location. So, in order to measure the reliability of the tracker, we set a counting variable (cnt) that counts how many times the tracking bounding box drifts far away from the segmentation group bounding rectangle and lowers the IoU by a threshold (T_{min}). If cnt records over 4 times that the tracking box 'escapes' from the segmentation bounding rectangle, we think the tracking results is unreliable and reset the tracker. Which means if the tracking result deteriorates, we will retrack the object according to the bounding rectangle of the segmentation result at a certain middle frame of a video sequence.

4   Experiments and Results

The proposed algorithm shows distinct improvement when compared with common segmentation-only methods in complicated scenes. The experimental results in the following sections are obtained with Caffe and Matlab r2017b on 2 Nvidia Titan XP 12 GB GPUs.

4.1   Data Set and Parameter Settings

Our experiments are done mainly at DAVIS 2016 data set [17] which includes 50 videos from different scenes. The training set includes 30 of the videos and the rest are in the test set. One object in the video is selected to be the target and only the very first frame is labeled with binary mask as ground-truth. The videos are given in sequences of images with resolution at 480p and 1080p and we choose 480p in our experiments. DAVIS covers the main challenges in the video segmentation task including: background interference, similar objects, distortion, blur effect caused by motion, fast moving objects, low resolution, occlusion, scale variation and object that exceeds the screen etc.

For the SegNet, we set the parameters according to OSVOS [5]. For the tracker, the regularization coefficient γ are set to be 1, 0.5 and 0.02 for layer $conv4\_3$, $conv3\_3$ and $conv5\_3$ in our experiments. Statistics proves that threshold r shows little influence on results, so it is also acceptable to decide the final location by weight voting. For the output fusion part, scale ranges $k = 1.0$, threshold $T_{min} = 0.6$ experimentally.

4.2   Evaluation Metrics

i. Region Similarity, which is defined as the overlap or similarity between the predicted region and the ground-truth and denoted by J:

$$J = \frac{|M \bigcap G|}{|M \bigcup G|} \tag{6}$$

where G represents the ground-truth foreground and M means the predicted foreground.

ii. Contour Accuracy, which represents the output accuracy in edge level and denoted by F:

$$F = \frac{2P_c R_c}{P_c + R_c} \tag{7}$$

Where P_c and R_c are the precision and recall rate of the contour which is calculated from the predicted contour and the ground-truth mask.

iii. Temporal Stability, which measures the stability of the predicted foreground region and equals to the matching cost between frames at contiguous time. The concrete details about these evaluation metrics approaches are given in [17].

4.3   Results and Statistics

The comparison of the proposed algorithm with other related prevalent methods is shown in Table 1. Where M, O, and D represents mean, recall and decay respectively in first row. As illustrates in the table, the proposed algorithm outperforms other approaches and we attain 1.38% gain compared with our base model OSVOS [5] in the region similarity($J(M)$) which is the most significant accuracy quota. In other evaluation quotas , our algorithm also reaches the best or the state-of-the-art performance. We will further discuss the feasibility of our method at each video sequence of DAVIS data set.

In Table 2, we list the region similarity of state-of-the-art and our proposed methods at every video sequences in DAVIS data set. Our algorithm gets similar results with our baseline [5] at simpler sequences like Blackswan, Camel and Dog where the objects are discriminative against the background throughout the scenes. However, when the scene is complicated and heavily influenced by other moving objects, like Car-shadow, Horsejump-high and Libby, the tracker assistance enhances the segmentation output distinctively. Consequently, the tracker is proved to be helpful in promoting the robustness when handling the influence of similar objects and complex background.

We propose to use correlation filter to complement the segmentation network, which helps to exclude non-target regions. We give qualitative results on some challenging frames which include complex backgrounds, occlusions and similar objects in Fig. 3. Obviously, many regions that far part from the target are misclassified by the baseline model, but when guided by tracking bounding boxes the outlied green regions can be easily excluded.

On the other hand, the segmentation network can do a favor when the tracker fails to track some fast moving objects. As shown in Fig. 4, the tracker often crashes when the object has an abrupt motion or high speed like a drifting car or a excited dancer (right column). Nevertheless, the segmentation network is less affected in such circumstance. So we can use segmentation results to modify the bounding boxes generated by the tracker(left column). In left, we have a much more accurate bounding box than which in right.

Table 1. Comparison result with other state-of the-art methods

Methods	J(M) (%)	J(O) (%)	J(D) (%)	F(M) (%)	F(O) (%)	F(D) (%)	T(M)
MSK [18]	79.7	93.1	**8.9**	75.4	87.1	**9.0**	21.8
SFL [7]	76.1	90.6	12.1	76.0	85.5	10.4	18.9
CTN [14]	73.5	87.4	15.6	69.3	79.6	12.9	22.0
VPN [13]	70.2	82.3	12.4	65.5	69.0	14.4	32.4
PLM [23]	70.2	86.3	11.2	62.5	73.2	14.7	31.8
OFL [21]	68.0	75.6	26.4	63.4	70.4	27.2	22.2
OSVOS [5]	79.8	93.6	14.9	80.6	**92.6**	15.0	37.8
OURS	**80.9**	**93.8**	12.5	**81.9**	92.5	12.8	37.7

Table 2. Comparison result on averaged region similarity with other state-of the-art methods on each video in DAVIS.

Methods	OSVOS (%) [5]	MSK (%) [18]	SFL (%) [7]	OURS (%)
Blackswan	**94.2**	90.3	92.0	**94.2**
bmx-trees	55.5	57.5	45.7	**58.4**
breakdance	**70.8**	76.2	68.2	**70.8**
camel	**85.1**	80.1	79.1	**85.1**
car-roundabout	95.3	96.0	85.7	**95.9**
car-shadow	93.7	93.5	94.5	**95.7**
cows	**94.6**	88.2	90.6	**94.6**
dance-twirl	67.0	84.4	73.4	**67.9**
dog	**90.7**	90.9	93.0	**90.7**
drift-chicane	**83.5**	86.2	37.9	**83.5**
drift-straight	**67.6**	56.0	89.9	**67.6**
goat	**88.0**	84.5	86.1	**88.0**
horsejump-high	78.0	81.7	76.0	**86.7**
kite-surf	68.6	60.0	58.7	**69.1**
libby	80.8	77.5	70.0	**83.7**
motocross-jump	81.6	68.5	83.9	**82.7**
paragliding-launch	62.5	62.0	58.1	**62.7**
parkour	85.6	88.2	84.9	**88.3**
scooter-black	71.1	82.5	69.9	**71.4**
soapbox	80.9	89.9	83.7	**81.3**
Average	79.8	79.7	76.1	**80.9**

Fig. 3. Segmentation results with tracking (best viewed in color). Green masks are segmentation results from the baseline models, red boxes are the tracking results. If integrates the two methods, we can get a finer box (colored in green).

Fig. 4. Tracking modification by segmentation. Green regions are the segmentation results and red boxes are tracing results. (Color figure online)

5   Conclusion

In this paper, we propose to use object tracking mechanism as a complement for video segmentation, which works alternatively with the segmentation network to refine the final results. The experimental results shows that the proposed architecture is superior to the other state-of-the-art algorithms and much more robust in complicated scenes.

Acknowledgment. This work is supported by the National Natural Science Foundation of China (Grant No. 61371192), the Key Laboratory Foundation of the Chinese Academy of Sciences (CXJJ-17S044) and the Fundamental Research Funds for the Central Universities (WK2100330002).

References

1. Achanta, R., Shaji, A., Smith, K., Lucchi, A., Fua, P., Süsstrunk, S.: Slic superpixels. EPFL (2010)
2. Arbelez, P., Maire, M., Fowlkes, C., Malik, J.: Contour detection and hierarchical image segmentation. IEEE Trans. Pattern Anal. Mach. Intell. **33**(5), 898–916 (2011)
3. Boddeti, V.N., Kanade, T., Kumar, B.V.K.V.: Correlation filters for object alignment. In: 2013 IEEE Conference on Computer Vision and Pattern Recognition (CVPR), pp. 2291–2298. IEEE (2013)
4. Bolme, D.S., Beveridge, J.R., Draper, B.A., Lui, Y.M.: Visual object tracking using adaptive correlation filters. In: 2010 IEEE Conference on Computer Vision and Pattern Recognition (CVPR), pp. 2544–2550. IEEE (2010)
5. Caelles, S., Maninis, K.K., Pont-Tuset, J., Leal-Taixé, L., Cremers, D., Van Gool, L.: One-shot video object segmentation. In: CVPR 2017. IEEE (2017)
6. Chen, L.C., Papandreou, G., Kokkinos, I., Murphy, K., Yuille, A.L.: DeepLab: semantic image segmentation with deep convolutional nets, atrous convolution, and fully connected CRFs. IEEE Trans. Pattern Anal. Mach. Intell. **40**(4), 834–848 (2016)
7. Cheng, J., Tsai, Y.H., Wang, S., Yang, M.H.: SegFlow: joint learning for video object segmentation and optical flow. In: IEEE International Conference on Computer Vision, pp. 686–695 (2017)
8. Danelljan, M., Bhat, G., Khan, F.S., Felsberg, M.: ECO: efficient convolution operators for tracking. In: Proceedings of the 2017 IEEE Conference on Computer Vision and Pattern Recognition (CVPR), Honolulu, HI, USA, pp. 21–26 (2017)
9. Danelljan, M., Häger, G., Khan, F.S., Felsberg, M.: Discriminative scale space tracking. IEEE Trans. Pattern Anal. Mach. Intell. **39**(8), 1561–1575 (2017)
10. Danelljan, M., Robinson, A., Shahbaz Khan, F., Felsberg, M.: Beyond correlation filters: learning continuous convolution operators for visual tracking. In: Leibe, B., Matas, J., Sebe, N., Welling, M. (eds.) ECCV 2016. LNCS, vol. 9909, pp. 472–488. Springer, Cham (2016). https://doi.org/10.1007/978-3-319-46454-1_29
11. Galoogahi, H.K., Sim, T., Lucey, S.: Multi-channel correlation filters. In: 2013 IEEE International Conference on Computer Vision (ICCV), pp. 3072–3079. IEEE (2013)
12. Henriques, J.F., Carreira, J., Caseiro, R., Batista, J.: Beyond hard negative mining: efficient detector learning via block-circulant decomposition. In: 2013 IEEE International Conference on Computer Vision (ICCV), pp. 2760–2767. IEEE (2013)
13. Jampani, V., Gadde, R., Gehler, P.V.: Video propagation networks (2016)
14. Jang, W.D., Kim, C.S.: Online video object segmentation via convolutional trident network. In: IEEE Conference on Computer Vision and Pattern Recognition, pp. 7474–7483 (2017)
15. Li, F., Kim, T., Humayun, A., Tsai, D., Rehg, J.M.: Video segmentation by tracking many figure-ground segments. In: 2013 IEEE International Conference on Computer Vision (ICCV), pp. 2192–2199. IEEE (2013)

16. Ma, C., Huang, J.B., Yang, X., Yang, M.H.: Hierarchical convolutional features for visual tracking. In: IEEE International Conference on Computer Vision, pp. 3074–3082 (2016)
17. Perazzi, F., Pont-Tuset, J., McWilliams, B., Van Gool, L., Gross, M., Sorkine-Hornung, A.: A benchmark dataset and evaluation methodology for video object segmentation. In: Computer Vision and Pattern Recognition (2016)
18. Perazzi, F., Khoreva, A., Benenson, R., Schiele, B., Sorkine-Hornung, A.: Learning video object segmentation from static images. In: IEEE Conference on Computer Vision and Pattern Recognition, pp. 3491–3500 (2017)
19. Simonyan, K., Zisserman, A.: Very deep convolutional networks for large-scale image recognition. arXiv preprint arXiv:1409.1556 (2014)
20. Tokmakov, P., Alahari, K., Schmid, C.: Learning video object segmentation with visual memory, pp. 4491–4500 (2017)
21. Tsai, Y.H., Yang, M.H., Black, M.J.: Video segmentation via object flow. In: Computer Vision and Pattern Recognition, pp. 3899–3908 (2016)
22. Xie, S., Tu, Z.: Holistically-nested edge detection. In: IEEE International Conference on Computer Vision, pp. 1395–1403 (2016)
23. Yoon, J.S., Rameau, F., Kim, J., Lee, S., Shin, S., Kweon, I.S.: Pixel-level matching for video object segmentation using convolutional Neural Networks, pp. 2186–2195 (2017)
24. Zheng, S., et al.: Conditional random fields as recurrent neural networks, pp. 1529–1537 (2015)

Infrared and Visible Image Fusion Using NSCT and Convolutional Sparse Representation

Chengfang Zhang[1], Zhen Yue[2], Liangzhong Yi[1], Xin Jin[1], Dan Yan[1], and Xingchun Yang[1(✉)]

[1] Sichuan Police College, Luzhou 646000, Sichuan, China
zcf1838725417@163.com
[2] Beijing University of Posts and Telecommunications, Beijing 100876, China

Abstract. In this paper, a new infrared and visible image fusion method based on non-subsampled contourlet transform (NSCT) and convolutional sparse representation (CSR) is proposed to overcome defects in selecting the NSCT decomposition level, detail blur for the SR-based method, and low contrast for the CSR-based method. In the proposed method, NSCT is performed on source images to obtain the low-frequency NSCT approximation components and high-frequency NSCT detail components. Then, low-frequency NSCT approximation components are merged with the CSR-based method while the popular "max-absolute" fusion rule is applied for the high-frequency NSCT detail components. Finally, the inverse NSCT is performed over the low-pass fused result and high-pass fused components to obtain the final fused image. Three representative groups of infrared and visible images were used for fusion experiments to evaluate the proposed algorithm. More specifically, on the popular Leaves image, the objective evaluation metrics Q^{abf}, Q^e, and Q^p of the proposed method were 0.7050, 0.6029, and 0.7841, respectively; on the Quad image, Q^{abf}, Q^e, and Q^p were 0.6527, 0.4843, and 0.5169, respectively; and on the Kayak image, Q^{abf}, Q^e, and Q^p were 0.6882, 0.4470, and 0.5532, respectively. Compared with the fusion method based on NSCT and sparse representation, the objective evaluation metrics Q^{abf}, Q^e, and Q^p showed increases of 1.54%, 10.57%, 22.49% on average. These experimental results demonstrate that the proposed fusion algorithm provides state-of-the-art performance in terms of subjective visual effects and objective evaluation criteria.

Keywords: Infrared and visible image fusion · Contrast enhancement · Convolutional sparse representation · Decomposition level · Detail preservation · Non-subsampled contourlet transform

1 Introduction

Under optimal lighting conditions, visible images are capable of describing all of the visual information contained in a scene. However, the contrast of visible light images decreases rapidly in low-light conditions, which results in a loss of target information. Compared with visible light images, infrared images are highly effective in capturing

© Springer Nature Switzerland AG 2019
Y. Zhao et al. (Eds.): ICIG 2019, LNCS 11901, pp. 393–405, 2019.
https://doi.org/10.1007/978-3-030-34120-6_32

occluded or hidden heat source targets. However, infrared images are lacking in detail, and they cannot be used to capture the background information of a scene. By fusing infrared and visible light images, one may be able to highlight targets in a scene using the information provided by the infrared image, while retaining the contrast and detail of the visible light image. Consequently, the fusion of infrared and visible light images has already found widespread application in machine vision, target recognition, and military applications.

The progress of multiscale transform studies has led to the proposal of a litany of multiscale transform-based methods for the fusion of infrared and visible light images. In multiscale transform-based image fusion, multiscale decomposition is performed on the source images to obtain the sub-band coefficients of each scale, which may then be fused by a variety of image fusion rules. This approach has proven to be effective for image fusion. The most common multiscale image fusion methods are (1) the ratio of low-pass pyramid (RP) method [1]; (2) wavelet-based transforms such as discrete wavelet transform (DWT) [2], double-tree complex wavelet transform (DTCWT) [3], and stationary wavelet transform (SWT) [4]; and (3) multiscale geometry analyses such as curvelet transform (CVT) [5] and nonsubsampled contourlet transform (NSCT) [6]. However, images fused by the Laplacian pyramid transform are prone to blurring in certain areas and thus, a loss of detail and edge information. Although wavelet transforms are effective in capturing the local features of an image, the edges of the fused image can be blurry if the directional information carried by the captured signal is inadequate in quality. The curvelet transform is effective in preserving image edges and contours but is inadequate in presenting the information of certain areas of the fused image. The NSCT method is superior to the wavelet transform in terms of direction-ality, and it also addresses the lack of shift invariance in the contourlet transform and thus eliminates pseudo-Gibbs artifacts. However, the NSCT method performs poorly in terms of detail capture, which leads to a loss of contrast in the fused image.

To enable the capture of all salient features in source images, Yang et al. [7] used a fusion strategy in which images are represented with sparse "Max-L1" coefficients using a fixed dictionary of discrete cosine transforms (DCTs). This was the first attempt to use sparse representations (SR) in multifocus image fusion. In addition, it was experimentally demonstrated that multifocus image fusion is superior to conventional multiscale image fusion. However, a fixed DCT dictionary may not be sufficient to capture the fine detail of a signal (e.g., texture and edges). Furthermore, if the source images have different imaging mechanisms, naïve application of the "Max-L1" rule alone could lead to spatial inconsistencies in the fused image. Liu proposed a gener-alized image fusion framework based on the strengths of the multiscale transform and SR [8]. The effectiveness of this approach was found to be superior to conventional multiscale transform-based image fusion and simple SR-based image fusion. As the conventional SR approach does not account for correlations between the source ima-ges, numerous augmented SR algorithms have been proposed to address this flaw [10–13], which are based on distributed compressed sensing theory [9]. However, con-ventional SR and combined SR fusion algorithms both use the "sliding window" technique to process source images, and this patch-wise process disrupts the global structural features of an image whilst being computationally expensive. To address this issue, Liu incorporated the convolutional sparse representation (CSR) in image fusion for the first time, by using the shift-invariance of convolutional dictionaries. This approach eliminates the flaws of conventional SR such as segmentation-induced detail-blurring, ghosting artifacts, and misregistration [14].

Although CSR image fusion resolves the detail-blurring problem in image fusion, the fusion of infrared and visible light images via CSR results in areas of low contrast and brightness. To overcome the flaws of conventional SR and multiscale transform-based methods in the fusion of infrared and visible light images, we combined the NSCT approach with CSR to fuse infrared and visible light images. In our method, NSCT decomposition is first performed on the source images to obtain their low-frequency and high-frequency NSCT components. CSR and the "max-absolute" rule are then used to fuse the low-frequency and high-frequency NSCT components, respectively. Finally, the fused image is obtained by applying the fused coefficients in inverse NSCT. The effectiveness of the proposed algorithm was validated via a comparative experiment that included six multiscale transform-based algorithms (RP, DWT, DTCWT, SWT, CVT and NSCT) and three SR-based algorithms (SR, NSCT-SR, and CSR). This experimental validation was conducted using three common sets of infrared and visible light images. It was shown that our algorithm is capable of fusing the target information of infrared images with the background detail of visible light images. Furthermore, the fused images produced by our algorithm obtained excellent scores in subjective and objective measures. It was also shown that the fused images produced by our algorithm have higher levels of contrast and brightness than those produced by other image fusion algorithms.

2 Convolutional Sparse Representation

Although the conventional patch-based SR technique is superior to multiscale image fusion, the sparse coefficients produced by this approach are multivalued. Furthermore, they are not optimized with respect to the image as a whole. In CSR, the linear combination of a dictionary vector set is replaced by the sum of a set of convolutions with dictionary filters to calculate the SR of the entire signal. The maturation of CSR theory in recent years has led to its application in image super-resolution [16], image classification [17], and image denoising [18]. CSR may be thought of as a convolutional form of conventional SR. The basic principle of CSR is to treat a registered image as the convolution between a set of coefficient maps and a set of dictionary filters. The definition of the CSR model is shown in (1).

$$\arg\min_{\{a_m\}} \frac{1}{2} \left\| \sum_m d_m * a_m - x \right\|_2^2 + \lambda \sum_m \|a_m\|_1 \tag{1}$$

In (1), * denotes convolution. The alternating direction method of multipliers (ADMM) [19] is an effective method for solving the basis pursuit denoising (BPDN) problem; on this basis, Wohlberg used the convolutional form of BPDN (convolutional BPDN, or CBPDN) to solve (1). During this work, this approach was used to fuse the low-frequency NSCT approximation components.

3   The Proposed Image Fusion Algorithm

Based on NSCT image fusion, the max-absolute rule was used to fuse the high-frequency NSCT component whereas the "averaging" rule was used to fuse the low-frequency NSCT component. However, because visible light and infrared images have different imaging mechanisms, the averaging of low-frequency components could lead to a partial loss of energy and thus reduce the contrast of the fused image. To preserve detail, it is necessary to use a sufficiently high level of decomposition. However, an excessively high level of decomposition could increase noise or cause image misregistration. SR-based image fusion is constrained by inadequacies in the dictionary representation of the source signal and the sliding window technique, which leads to detail blurring in the fused image. As it is possible for a patch to be extremely bright in an infrared image but very dark in the corresponding visible light image, use of the "choose-max" rule in the base layer during CSR-based image fusion would produce fused images with lowered levels of contrast.

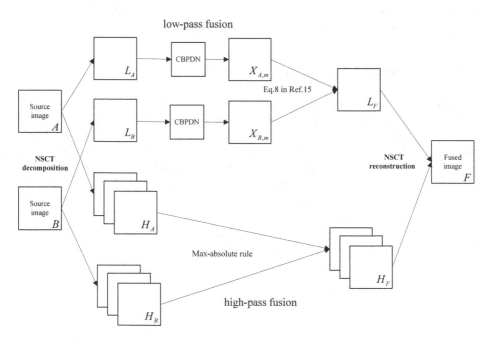

Fig. 1. Flow diagram of the proposed NSCT-CSR-based image fusion algorithm

Table 1. Flow of the proposed algorithm

Step 1: NSCT decomposition

NSCT decomposition is performed on a co-registered pair of source images, $\{A, B\}$. This yields the low-frequency NSCT components $\{L_A, L_B\}$ and high-frequency NSCT components $\{H_A, H_B\}$.

Step 2: Fusion of the low-frequency NSCT components

Firstly, the learning algorithm described in [14] is used to obtain a set of dictionary filters, d_m ($m \in \{1, ..., M\}$).

CSR decomposition is then performed on the low-frequency components L_k ($k = A, B$) via (1). This yields the set of coefficient maps for the low-frequency NSCT components of the source images, $A_{k,m}$, where $m \in \{1, ..., M\}$.

$$\arg\min_{A_{k,m}} \frac{1}{2} \left\| \sum_{m=1}^{M} d_m * A_{k,m} - L_k \right\|_2^2 + \lambda \sum_{m=1}^{M} \left\| A_{k,m} \right\|_1 \qquad (2)$$

Finally, the low-frequency fusion algorithm ((8) in [15]) is used to obtain the low-frequency NSCT component of the fused image, L_F.

Step 3: Fusion of the high-frequency NSCT components.

The high-frequency NSCT components are fused according to the "absolute-max" rule, thus producing the high-frequency NSCT component of the fused image, H_F.

Step 4: NSCT reconstruction.

The fused image is obtained by applying the inverse NSCT on L_F and H_F.

The flaws of the three aforementioned methods can be resolved using the framework proposed in this work for the fusion of infrared and visible light images. Firstly, the source images are decomposed into low-frequency and high-frequency components via NSCT decomposition. Low-pass fusion is then performed by applying the CSR fusion algorithm to the low-frequency NSCT components. The high-frequency NSCT components are fused by the "absolute-max" strategy, which enhances the detail of the fused image. The proposed approach reduces the difficulty of selecting an appropriate decomposition level for NSCT and enhances the contrast of the fused image. Furthermore, the fused image produced by this approach is significantly brighter than that of CSR-based image fusion. Figure 1 illustrates the flow of the NSCT-CSR framework proposed in this work for the fusion of visible light and infrared images. For simplicity, Fig. 1 only shows the fusion of two source images, A and B, although the proposed framework is readily extendable to the fusion of three or more images. Our algorithm consists of four steps: NSCT decomposition, fusion of the low-frequency NSCT components, fusion of the high-frequency NSCT components, and NSCT reconstruction. The steps employed in this algorithm is presented in Table 1.

4 Experimental Results and Analysis

The performance of the proposed algorithm was evaluated via the fusion of three common sets of visible light images and infrared images (Fig. 2). In addition, it was compared with nine other classic image fusion methods, which include three sparse domain-based image fusion algorithms (the SR-based method [7], NSCT-SR-based method, and CSR-based method [15]) and six multiscale transform-based image fusion algorithms (DWT, DTCWT, RP, CVT, SWT, and NSCT). The parameters of the multiscale transform-based image fusion algorithms were adopted from [8] and [12]. The parameters of the sparse domain-based image fusion algorithms were adopted from their respective publications. The NSCT decomposition was performed using "pyrexc" as the pyramid filter and "vk" as the directional filter. The NSCT decomposition levels (from coarsest to finest) are {2, 3, 3, 4}. The learning algorithms in [14] and [15] were used to train the dictionary filter (dm), and the size of dm was $8 \times 8 \times 32$. The value of λ in (1) and (2) was set to 0.01 to ensure that the experiment was fair.

(a)Leaves   (b)Quad   (c)Kayak

Fig. 2. Source images used in our experiments.

4.1 Source Images for Image Fusion

The three common sets of infrared and visible light images that were used in this experiment are shown in Fig. 2; the first and second row of images are the visible light images and infrared images, respectively. The infrared and visible light images were taken at the same time for the same scene. In the test images, the fusion targets include vegetation, cars, houses, street lights, and pedestrians. Having a variety of fusion targets helped to highlight the effectiveness of our algorithm in the fusion of infrared and visible light images.

4.2   Objective Evaluation Metrics

In addition to subjective evaluations based on the human eye, objective assessment metrics also play an important role in the evaluation of image fusion quality, because the performance of an image fusion algorithm is usually assessed by a number of quantitative metrics. In recent years, a variety of evaluation metrics have been proposed for assessing the effectiveness of image fusion. In a study conducted by Liu et al. [20] in which 12 commonly used fusion metrics were compared and examined, the metrics were classified into four categories: metrics based on information theory, metrics based on image features, metrics based on image structure similarity, and metrics based on human perception. Although numerous metrics have been proposed for the evaluation of image fusion quality, there is no single metric that can comprehensively evaluate all aspects of image fusion, as the imaging mechanisms of the source images can vary significantly. In most cases, the evaluation of image fusion quality requires the use of multiple evaluation metrics. In order to objectively assess the effectiveness of the proposed image fusion algorithm, three commonly used Q-series metrics (Q^{ABF} [20, 21], Q^e [20, 22], and Q^p [20, 23]) were used to assess the image fusion quality of our algorithm. A brief description of these metrics is provided below. Please note that A and B represent infrared and visible light source images with dimensions of $M \times N$, respectively.

4.3   Results of Image Fusion Experiment and Result Analysis

Fusion of the infrared and visible light images of "Leaves" by our algorithm and nine other image fusion algorithms is shown in Fig. 3. Figures 3(a) and (b) are the visible light and infrared source images of "Leaves." In the bracketed area of the "Leaves" image, it is shown that the visible light image (Fig. 3(a)) preserves the image information of the leaves but results in low target contrast, which makes it difficult to distinguish the target (a board). The infrared image (Fig. 3(b)) highlights the target but omits some of the leaves in the left side of the image. Figures 3(c)–(l) are the results produced by ten different image fusion algorithms. Here, it is shown that image fusion effectively allows the information contained by the visible light and infrared images to complement each other. Nevertheless, there are slight differences between the fused images in terms of fusion quality. For example, the DWT and DTCWT methods result in lower levels of contrast in the middle of the target board, as these images only emphasize the infrared information around the target. The fused image produced by the CVT method shows some "patchiness" in the target board. The target board is relatively dark in the fused images produced by the RP, SWT, and NSCT methods. Although the SR and NSCT-SR methods have preserved the leaves of the visible light image, the target is distorted in the fused images. Figure 3(l) shows that the fused image produced by our algorithm preserves the leaves and renders the target board with a high level of contrast. The fused image of our algorithm therefore exhibits a better level of visual efficacy than the fused images of the nine other image fusion algorithms.

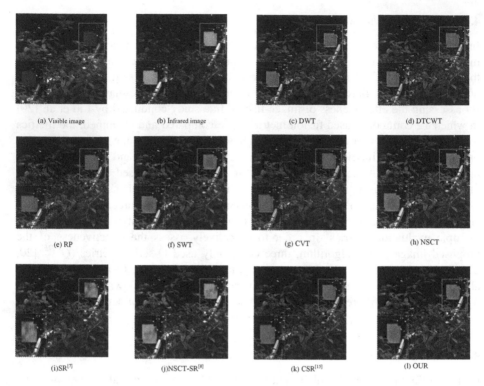

Fig. 3. Fusion results of the Leaves source images.

Figure 4 compares the results obtained by fusing the visible light and infrared images of "Quad." Figures 4(a) and (b) are the visible light and infrared source images of "Quad," respectively. In Fig. 4(a), the words on the advertisement board (top bracket) are clearly decipherable, but the pedestrians and cars on the street are difficult to see. Despite the lack of illumination, Fig. 4(b) clearly shows the pedestrians and cars on the street, but the information shown on the advertisement boards is missing in this image. Figures 4(c)–(l) show that the fusion algorithms have succeeded in fusing the visible light and infrared images. Nevertheless, there are differences between the fused images in certain details. In the fused image produced by the DWT method, the advertisement boards are distorted and the words on the advertisement boards are fuzzy. Furthermore, some patchiness is evident in this image. The fused images produced by the DTCWT, RP, and CVT methods have the lowest levels of contrast between the fused images, and the pedestrians are also quite dark in these images. In addition, the image produced by the CVT method shows some ghosting and dislocation in the advertisement boards and pedestrians. The fused images produced by the SWT and NSCT methods have brighter pedestrians and advertisement boards than those of the DTCWT, RP, and CVT methods, but they exhibit "excessive fusing" in certain areas. The fused images produced by the SR and NSCT-SR methods have high levels of contrast, and the pedestrians, advertisement boards, and cars are clearly distinguishable in these images. However, some fuzziness is evident in the upper left corner,

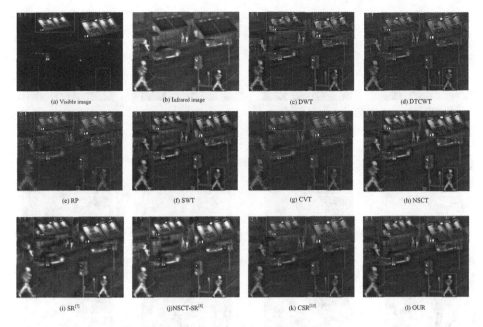

(a) Visible image   (b) Infrared image   (c) DWT   (d) DTCWT

(e) RP   (f) SWT   (g) CVT   (h) NSCT

(i) SR[7]   (j)NSCT-SR[8]   (k) CSR[15]   (l) OUR

Fig. 4. Fusion results of the Quad source images.

and the quality of image fusion is less than ideal in certain areas of these images. Furthermore, spatial discontinuities are present in the road surfaces. The CSR method and our method produced fused images that are superior to those of the SR and NSCT-SR methods. However, the CSR image exhibits a small degree of dislocation and ghosting, and it also has a slightly lower level of contrast than the image produced by our algorithm.

Figure 5 compares the results obtained by fusing the visible light and infrared source images of "Kayak." Figures 5(a) and (b) are the visible light and infrared source images of "Kayak," respectively. In the visible light image (Fig. 5(a)), the profile of the clocktower (top bracket) is clearly visible, but the rooftops (lower bracket) are fuzzy and poorly resolved. In the infrared image (Fig. 5(b)), the rooftops are clearly shown despite the lack of illumination, but the profile of the clocktower is indecipherable in this image. In Figs. 5(c)–(l), it is shown that the various image fusion algorithms have fused the visible light and infrared images of "kayak" to good effect. Nevertheless, the fused images differ slightly in certain details. In the fused images produced by the DWT, RP, and SR methods, the profile of the clocktower is not clear, and the fused image produced by the SR method displays some patchiness. The fused images produced by the SWT, CVT, and CSR methods are effective in preserving the profile of the clocktower, but they are generally lacking in contrast. The fused image produced by the NSCT-SR method has a high level of contrast, especially around the clocktower, but the image does exhibit a degree of patchiness. The fused images produced by the DTCWT and NSCT methods and our method have effectively preserved the profile of the clocktower. However, the DTCWT and NSCT images show a lower level of contrast in the rooftops than the image produced by our algorithm.

(a) Visible image

(b) Infrared image

(c) DWT

(d) DTCWT

(e) RP

(f) SWT

(g)CVT

(h) NSCT

(i) SR[7]

(j) NSCT-SR[8]

(k) CSR[15]

(l) OUR

Fig. 5. Fusion results of Kayak source images.

The objective evaluation metrics calculated for the fusion quality of the three aforementioned sets of infrared and visible light images are listed in Tables 2, 3 and 4. In terms of the "Leaves" images, our algorithm has the highest objective scores among the ten image fusion algorithms. Compared to the NSCT-SR method, the Q^{ABF}, Q^e, and Q^p scores of our method for the "Leaves" images are higher by 0.19%, 3.17%, and 3.72%, respectively. The Q^{ABF} and Q^e scores of our method with respect to the "Quad" image is superior to those of the nine other algorithms, and the Q^p score of our algorithm is only slightly lower than that of the CSR-based method. Compared to the NSCT-SR method, the Q^{ABF}, Q^e, and Q^p scores of our method for the "Quad" images are higher by 2.36%, 16.6%, and 30.45%, respectively. In terms of the "Kayak" image, the objective scores of our algorithm are superior to those of the nine other image fusion algorithms. Compared to the NSCT-SR method, the Q^{ABF}, Q^e, and Q^p scores of our method for the "Kayak" images are higher by 2.09%, 11.37%, and 33.32%, respectively. The objective scores indicate that our algorithm is highly effective in preserving edge information, angular information, and gradient information during image fusion, which helps to enhance detail levels in the fused images and highlight their salient features.

Table 2. Objective assessment of various methods for the 'Leaves' fused images

Metrics	Methods									
	DWT	DTCWT	RP	SWT	CVT	NSCT	SR [7]	NSCT-SR [8]	CSR [15]	OUR
Q^{ABF}	0.6336	0.6685	0.6029	0.6670	0.6298	0.6931	0.6738	0.7037	0.6979	**0.7050**
Q^e	0.5508	0.5681	0.5272	0.5625	0.5389	0.5712	0.5260	0.5812	0.5946	**0.6029**
Q^p	0.6800	0.7492	0.6282	0.7229	0.7009	0.7633	0.6895	0.7560	0.7831	**0.7842**

Table 3. Objective assessment of various methods for the 'Quad' fused images

Metrics	Methods									
	DWT	DTCWT	RP	SWT	CVT	NSCT	SR [7]	NSCT-SR [8]	CSR [15]	OUR
Q^{ABF}	0.5509	0.5505	0.3305	0.6068	0.4617	0.6442	0.5384	0.6377	0.6058	**0.6527**
Q^e	0.3575	0.4150	0.3782	0.4429	0.3688	0.4571	0.3219	0.4154	0.4765	**0.4843**
Q^p	0.3699	0.4633	0.2395	0.4625	0.4130	0.5082	0.2579	0.3963	**0.5582**	0.5169

Table 4. Objective assessment of various methods for the 'Kayak' fused images

Metrics	Methods									
	DWT	DTCWT	RP	SWT	CVT	NSCT	SR [7]	NSCT-SR [8]	CSR [15]	OUR
Q^{ABF}	0.6045	0.6194	0.5542	0.6441	0.5249	0.6789	0.6425	0.6742	0.6476	**0.6883**
Q^e	0.3301	0.3834	0.3494	0.3953	0.3329	0.4203	0.3737	0.4014	0.4254	**0.4470**
Q^p	0.3960	0.4688	0.3854	0.4995	0.4300	0.5331	0.3653	0.4150	0.5409	**0.5533**

5 Conclusion

In this work, we developed an image fusion algorithm that combines NSCT with CSR and applied it to the fusion of infrared and visible light images. Our CSR-NSCT approach effectively overcomes the detail blurring problem that plagues the conventional patch-based SR model. Furthermore, the CSR-NSCT method also resolves the weaknesses of the NSCT approach, which does not adequately capture the input signal and tends to produce fused images with low levels of contrast. In the proposed method, NSCT decomposition is first performed on the co-registered source images to obtain their low-frequency and high-frequency NSCT components. The low-frequency fusion coefficients are obtained via the CSR fusion algorithm, whereas the corresponding high-frequency NSCT component is obtained using the "absolute-max" rule. An experimental comparison was then conducted between our algorithm and nine other image fusion algorithms, which include six multiscale transform-based fusion algorithms (RP, DWT, DTCWT, SWT, CVT, and NSCT) and three sparse domain-based fusion algorithms (SR, NSCT-SR, and CSR). The fusion quality of our algorithm was validated by using three common sets of visible light and infrared images in the comparative experiment. It was experimentally demonstrated that our image fusion algorithm is highly effective in preserving detail in the fused image. Furthermore, the fused images produced by our algorithm are superior to those of other algorithms in terms of image contrast and brightness. In summary, the algorithm proposed in this work is superior to conventional algorithms for the fusion of visible light and infrared images in all objective and subjective measures.

Acknowledgment. The authors would like to thank the editors and anonymous reviewers for their detailed review, valuable comments and constructive suggestions. This work is supported by the National Natural Science Foundation of China (Grants 61372187), Research and Practice on Innovation of Police Station Work Assessment System (Grants 18RKX1034), Sichuan Science and Technology Program (2019YFS0068 and 2019YFS0069).

References

1. Toe, A.: Image fusion by a ratio of low-pass pyramid. Pattern Recogn. Lett. **9**(4), 245–253 (1989)
2. Li, H., Manjunath, B.S., Mitra, S.K.: Multi-sensor image fusion using the wavelet transform. In: IEEE International Conference on Image Processing, ICIP 1994, pp. 235–245. IEEE (2002)
3. Lewis, J.J., O'callaghan, R.J., Nikolov, S.G., et al.: Pixel- and region-based image fusion with complex wavelets. Inf. Fusion **8**(2), 119–130 (2007)
4. Beaulieu, M., Foucher, S., Gagnon, L.: Multi-spectral image resolution refinement using stationary wavelet transform. In: Geoscience and Remote Sensing Symposium, pp. 4032–4034 (2010)
5. Dong, L., Yang, Q., Wu, H., Xiao, H., Xu, M.: High quality multi-spectral and panchromatic image fusion technologies based on curvelet transform. Neurocomputing **159**(C), 268–274 (2015)
6. Wang, J., Li, Q., Jia, Z., et al.: A novel multi-focus image fusion method using PCNN in nonsubsampled contourlet transform domain. Optik – Int. J. Light Electron. Opt. **126**(20), 2508–2511 (2015)
7. Yang, B., Li, S.: Multifocus image fusion and restoration with sparse representation. IEEE Trans. Image Process. **15**(2), 3736–3745 (2006)
8. Liu, Y., Liu, S., Wang, Z.: A general framework for image fusion based on multi-scale transform and sparse representation. Inf. Fusion **24**, 147–164 (2014)
9. Duarte, M.F., Sarvotham, S., Baron, D.: Distributed compressed sensing of jointly sparse signals. In: Conference on Signals, Systems & Computers. IEEE (2006)
10. Haitao, Y.: Multimodal image fusion with joint sparsity model. Opt. Eng. **50**(6), 067007 (2011)
11. Zhang, Q., Fu, Y., Li, H.: Dictionary learning method for joint sparse representation-based image fusion. Opt. Eng. **52**(5), 7006 (2013)
12. Gao, Z., Zhang, C.: Texture clear multi-modal image fusion with joint sparsity model. Optik – Int. J. Light Electron. Opt. **130**, 255–265 (2017)
13. Yang, B., Li, S.: Visual attention guided image fusion with sparse representation. Optik – Int. J. Light Electron. Opt. **125**(17), 4881–4888 (2014)
14. Wohlberg, B.: Efficient algorithms for convolutional sparse representation. IEEE Trans. Image Process. **25**(1), 301–315 (2016)

15. Liu, Y., Chen, X., Ward, R.K.: Image fusion with convolutional sparse representation. IEEE Signal Process. Lett. **23**(12), 1882–1886 (2016)
16. Gu, S., Zuo, W., Xie, Q.: Convolutional sparse coding for image super-resolution. In: IEEE International Conference on Computer Vision, pp. 1823–1831 (2015)
17. Chen, B., Polatkan, G., Sapriro, G., et al.: Deep learning with hierarchical convolutional factor analysis. IEEE Trans. Pattern Anal. Mach. Intell. **35**(8), 1887–1901 (2013)
18. Zeiler, M.D., Krishnan, D., Taylor, G.W., et al.: Deconvolutional networks. In: Proceedings of IEEE Conference on Computer Vision Pattern Recognition, pp. 2528–2535 (2010)
19. Boyd, S., Parikh, N., Chu, E., et al.: Distributed optimization and statistical learning via the alternating direction method of multipliers. Found. Trends Mach. Learn. **3**(1), 1–122 (2010)
20. Liu, Z., Blasch, E., Xue, Z., et al.: Objective assessment of multiresolution image fusion algorithms for context enhancement in night vision: a comparative study. IEEE Trans. Pattern Anal. Mach. Intell. **34**(1), 94–109 (2011)
21. Xydeas, C.S., Petrovic, V.: Objective image fusion performance measure. Military Tech. Courier **56**(2), 181–193 (2000)
22. Piella, G., Heijmans, H.: A new quality metric for image fusion. In: International Conference on Image Processing, pp. 173–176 (2003)
23. Zhao, J., Laganiere, R., Liu, Z.: Performance assessment of combinative pixel-level image fusion based on an absolute feature measurement. Int. J. Innovative Comput. Inf. Control **3**(6), 1433–1447 (2006)

A Weakly Supervised Text Detection Based on Attention Mechanism

Lanfang Dong$^{(\boxtimes)}$ (iD), Diancheng Zhou (iD), and Hanchao Liu (iD)

School of Computer Science and Technology,
University of Science and Technology of China, Hefei, China
lfdong@ustc.edu.cn, {zdc0803,lhanchao}@mail.ustc.edu.cn

Abstract. In this paper, we propose a new method for natural image text detection under a weakly supervised data set. Currently, most of the text detection models are based on bounding box label training data. However, the cost of the bounding box label training data is very high. In order to solve this problem, we propose an attention mechanism that can be trained on image-level labels data and roughly identifies text regions via an automatically learned attentional map based on a convolutional neural network. There are three main steps: firstly, a VGG model is trained using image-level labels data to score the likelihood that a text region exists in the picture; secondly, the region of interest is extracted by means of the attention mechanism and the extracted region is evaluated using the network trained in the first step to getting the text region and finally, the text line is extracted in the text region using the MSER algorithm. Trained with the weakly supervised data which is only with image-level labels, our model can generate bounding boxes for the text line in the image. The results of our model are very close to those of the models using bounding box label training data on the text detection benchmark sets of MSRA-TD500, ICDAR2013, and ICDAR2015.

Keywords: Weakly supervised · Text detection · Attention mechanism · MSER algorithm

1 Introduction

Text information in images is of indispensable value in semantic visual understanding. Reading the text in a natural scene, however, compared to traditional OCR, is still a challenging problem. With which application of deep neural networks, the performance of text detection has been improved rapidly. However, training most of the detection models in the current research requires a large number of images that are labeled with bounding boxes. In fact, the cost of obtaining these strong supervised images is very high. And it is even impossible to get a lot of strong supervised data in some special tasks. On the other hand, the acquisition cost of weakly supervised image-level labels data which only marks whether the text is in or not in the image is far from lower than that of strong supervised data. Besides, the research on weakly supervised text

© Springer Nature Switzerland AG 2019
Y. Zhao et al. (Eds.): ICIG 2019, LNCS 11901, pp. 406–417, 2019.
https://doi.org/10.1007/978-3-030-34120-6_33

detection is still scarce at present. So developing a text detection model that can be trained with weakly supervised data is very valuable.

In recent years, people have found that human cognition of things is not a one-time focus on the entire scene, but gradually draws attention to different regions of the scene while extracting relevant information [1,2]. By quickly scanning the global image, humans obtain the target area that needs to be focused on and then invest more attention resources in this target area to obtain more attention targets. At present, the attention-based models have achieved good results on many challenging tasks, such as machine translation [1,3], question and answer system [4,5], image description [2], and so on. In these tasks, the attention mechanism reflects the excellent key area positioning ability. So we hope to use this excellent positioning ability to achieve the task of weakly supervised text detection.

Because the shape of the text is generally fixed, the size is moderate, and the colors of the continuous text fields are very similar. These features of the text allow the positioning ability of the attention mechanism to be fully utilized to perform the detection tasks we expect. And the results of our experiments have proved that the text detection model based on the attention mechanism has a very good effect in weakly supervised text detection task.

In this paper, we designed an attention-based text detection model that can be trained by images with only image-level labels data. We first trained a VGG network to classify whether texts are in the images or not. Than attention mechanism is applied to extract the interest regions and we use the trained VGG network to get the text regions. Finally, the MSER algorithm is applied to generate the bounding boxes of the text regions. In summary, our contributions are in two folds:

(1) We propose a text detection model under weakly supervised data. This is mainly achieved by training an evaluation model using image-level label data to evaluate the localization performance of another localization model.
(2) We apply the attention mechanism to the task of weakly supervised text detection and design a novel loss function for the training of the text detection model with an attention mechanism.

The remainder of this paper is organized as follows: In Sect. 2, we briefly review the previous related work. In Sect. 3, we describe the proposed method in detail, including the construction of the evaluation network based on weakly supervised data, the application of attention mechanisms and location the text line from the text region. Experimental results are described in Sect. 4. Finally, conclusive remarks and future work are given in Sect. 5.

2   Related Work

Deep learning technologies have significantly advanced the performance of text detection in the past years [6–9]. These approaches essentially work in a sliding window fashion, with two key developments: (i) they leverage deep features, jointly learned with a classifier, to enable the strong representation of text; (ii) sharing a convolutional mechanism was applied for reducing the computational cost remarkably. With these two improvements, a number of Fully Convolutional Network [11] based approaches have been proposed [8–10]. They compute pixel-wise semantic estimations of text or non-text, resulting in a fast text detector able to explore rich regional context information.

Recently, some methods cast the previous character based detection into direct text region estimation, avoiding multiple bottom-up post-processing steps by taking word or text-line as a whole. Tian et al. [12] modified Faster-RCNN [13] by applying a recurrent structure on the convolution feature maps of the top layer horizontally. The algorithm proposed by He et al. [14] was inspired from single shot multi box detector [15]. They both explored the framework from generic objects and convert to scene text detection by adjusting the feature extraction process to this domain specific task. However, these methods are based on bounding boxes, which need to be carefully designed in order to fulfill the requirements for training.

Methods of direct regression for inclined bounding boxes, instead of offsets to fixed bounding boxes, have been proposed recently. EAST [16] designed a fully convolutional network structure that outputs a pixel-wise prediction map for text/non-text and five values for every point of text region, i.e., distances from the current point to the four edges with an inclined angle. He et al. [17] proposed a method to generate arbitrary quadrilaterals by calculating offsets between every point of text region and vertex coordinates.

However, almost all text detection models are based on bounding box labels, but the high cost of bounding box labels in some cases has no way to obtain suitable data. In this paper, we train a text detection model only through image-level labels, which solves the problem that the text detection model relies on bounding box labels.

3   Methodology

The entire methodology consists of three main parts. First, we build and train a VGG network for classification and feature extraction. Then, the construction of attention maps based on the results of the VGG network. Finally, the MSER algorithm was used to locate the text block with the attention map.

3.1   Text/Non-text Classification Network

In order to get a text-detection model with only image-level labels, we have to classify whether the texts exist in the image. In the first step of our method, a VGG network [18] is trained to distinguish whether there is text in the image. This network will provide a discriminator for subsequent models for positioning accuracy evaluation.

The network, which is denoted as Discriminant Network, is inspired by VGG network [18]. However, there are several improvements in our networks to distinguish whether there is text in the image compared with the original VGG network. Because most of the text appears in lines, the shape is more like a flat rectangle, we adjusted the scale of the convolution kernel in some layers, 5×5 convolution kernels that are more suitable for extracting text features from complex backgrounds, we also use 1×2 max-pooling layer as in [19], which reserves more information along the horizontal axis and benefits the detection of narrow shaped. In order to make the effects of 5×5 convolution kernels and 1×2 max-pooling layer clear, we have designed four networks with similar architecture and the details of these networks are shown in Table 1. The experiments in Sect. 4 also prove that our adjustments are very effective.

Table 1. Network configurations (shown in columns). The detailed differences are shown in the contents of this section.

Network A	Network B	Network C	Network D
Input: 512×512 RGB images			
Conv - 3×3 - 64			
Conv - 3×3 - 64			
MaxPooling - k: 2×2 - s: 2×2			
Conv - 3×3 - 128			
Conv - 3×3 - 128			
MaxPooling - k: 2×2 - s: 2×2			
Conv - 3×3 - 256			
Conv - 3×3 - 256			
MaxPooling - k: 2×2 - s: 2×2		MaxPooling - k: 1×2 - s: 1×2	
Conv - 3×3 - 512	Conv - 3×3 - 512	Conv - 3×3 - 512	Conv - 3×3 - 512
Conv - 3×3 - 512	Conv - 5×5 - 512	Conv - 3×3 - 512	Conv - 5×5 - 512
Conv - 3×3 - 512	Conv - 3×3 - 512	Conv - 3×3 - 512	Conv - 3×3 - 512
MaxPooling - k: 2×2 - s: 2×2		MaxPooling - k: 1×2 - s: 1×2	
Conv - 3×3 - 512	Conv - 3×3 - 512	Conv - 3×3 - 512	Conv - 3×3 - 512
Conv - 3×3 - 512	Conv - 5×5 - 512	Conv - 3×3 - 512	Conv - 5×5 - 512
Conv - 3×3 - 512	Conv - 3×3 - 512	Conv - 3×3 - 512	Conv - 3×3 - 512
MaxPooling - k: 2×2 - s: 2×2		MaxPooling - k: 1×2 - s: 1×2	
FC - 4096			
FC - 512			
FC - 2			
Softmax			

In Table 1, 'Conv' stands for Convolutional layers, with kernel size and output channels presented. The stride and padding for convolutional layers are all set to '1'. For Maxpooling layers, 'k' means kernel size, and 's' represents stride.

It is worth mentioning that we are required to perform related operations on the input data of the model to make our model work. We will use the discriminator network in the subsequent steps to score the possibility of the presence of text in the picture, and the scored picture is generated by combining the processed picture with the attention map like Fig. 1(c), so we need to expand the input data by randomly smearing the training image make the picture more similar to the picture that combines the attention map. This method allows the discriminator to adapt to the pictures in the subsequent steps.

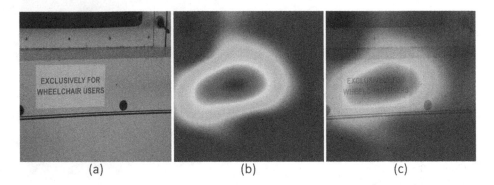

(a)   (b)   (c)

Fig. 1. The results of our method. (a) An input image; (b) The attention map, which is generated by the attention mechanism. Since the attention map is actually a matrix of the $[0, 1]$ interval, visualization is performed here; (c) Visualize the results of the combining the attention map with the input image.

3.2   Attention Mechanism

Our attention module is designed to learn rough spatial regions of text from the convolutional features automatically. It generates a pixel-wise probability heat map that indicates the text probability at each pixel location like Fig. 1(b). This probability heat map is referred to like the attention map which has an identical size of an input image and will be downsampled for each prediction layer.

We use the same convolution structure as that in Table 1, and inherit all the weight data from step one. This method of inheriting parameters is essentially a pretraining method that can make full use of the features of positioning that has been learned by convolutional neural networks.

The specific implementation of the attention mechanism is as follows: Denote by $L^s = \{l_1^s, l_2^s, \ldots, l_n^s\}$ the set of feature vectors extracted at a given convolutional layers s. Here, each l_i^s is the vector of output activations at the spatial location i of n total spatial locations in the layer. The global feature g is the feature of the fully connected layer of the previous model before the final output. We use the dot product between g and l_i^s as a measure of their compatibility:

$$c_i^s = \langle l_i^s, g \rangle, i \in \{1, \ldots, n\} \tag{1}$$

In this case, the relative magnitude of the scores would depend on the alignment between g and l_i^s in the high-dimensional feature space and the strength of activation of l_i^s. For each of one or more layers s, the set of compatibility scores $C(\hat{L}^s, g) = \{c_1^s, c_2^s, \ldots, c_n^s\}$, where $\hat{L}^s$, is the image of L^s under a linear mapping of the l_i^s to the dimensionality of g. The compatibility scores are then normalised by a softmax operation:

$$a_i^s = \frac{exp(c_i^s)}{\sum_j^n exp(c_j^s)}, i \in \{1, 2, \ldots, n\} \tag{2}$$

The normalised compatibility scores $A^s = \{a_1^s, a_2^s, \ldots, a_n^s\}$ is attention map which we will use it for location.

The attentional information is learned automatically in the training process, we can construct a loss function to describe whether the attention points are focusing on the right location. We have designed a new loss function by drawing on the idea of generative adversarial net [20].

$$Loss = -\lambda_{text}log(P_{text}) - \lambda_{non-text}log(1 - P_{non-text}) \tag{3}$$

λ_{text} and $\lambda_{non-text}$ is two constants used to adjust the attention mechanism, and we make it 0.5 in our model. P_{text} is a score for the possibility of text in the image which combined original image and attention map, we hope that this value is as large as possible. $P_{non-text}$ is similar to the definition of P_{text}, except that $P_{non-text}$ uses the attention map opposite to P_{text}. The meaning is the probability that there is text in the area that is not noticed, and we hope that this value is relatively small.

3.3   Bounding Boxes Generation

Although the text region detected by the attention map provides coarse localizations of text lines, they are still far from satisfactory. We borrowed from the methods in the paper [8] to further extract accurate bounding boxes of text lines.

At first, we extract the character components within the text blocks by MSER [21], since MSER is insensitive to variations in scales, orientations, positions, languages, and fonts. We use a constraint to reduce the wrong character components, the minimum area ratio of character components needs to be greater than 1% of the text region. After testing, in this way, most of the false components are excluded.

Then, we assume that text lines from the same text region have a substantially uniform spatial layout, and characters from one text line are in the arrangement of straight or near straight line. We use the statistical method to calculate the slope of the line passing through the most character components obtained by the MSER algorithm and record this slope as θ. In order to facilitate the calculation, the actual statistics are at intervals of $\frac{\pi}{24}$.

Finally, we combine the character components extracted by the MESR to text line candidate generation. We divide the components into groups. A pair of the components (A and B) within the text block α are grouped together if they satisfy the following conditions:

$$\frac{H(A)}{H(B)} \in \left[\frac{4}{5}, \frac{6}{5}\right] \tag{4}$$

$$O(A, B) - \theta \in \left[-\frac{\pi}{12}, \frac{\pi}{12}\right] \tag{5}$$

where $H(A)$ and $H(B)$ represent the heights of A and B, $O(A, B)$ represents the orientation of the pair.

For one group $\beta = \{c_i\}$, c_i is i-th character components, we draw a line l along the orientation θ passing the center of β. The point set ρ is defined as:

$$\rho = \{p_i\}, p_i \in l \cap \mathbb{B} \tag{6}$$

where $\mathbb{B}$ represents the boundary points of text region.

The minimum bounding box bb of β is computed as a text line candidate:

$$bb = \beta \cup \rho \tag{7}$$

where bb denotes the minimum bounding box that contains all points and components.

4 Experiments

Our methods are evaluated on three standard benchmarks, the MSRA-TD500 [22], ICDAR 2013 [23] and ICDAR 2015 [24]. The effectiveness of each proposed component is investigated by producing exploration studies. Full results are compared with the state-of-the-art performance on the three benchmarks.

4.1 Datasets

The following datasets are used in our experiments:

MSRA-TD500. The MSRA-TD500 dataset [22] is a multi-orientation text dataset including 300 training images and 200 testing images. The dataset contains text in two languages, namely Chinese and English. This dataset is very challenging due to the large variation in fonts, scales, colors and orientations. Here, we followed the evaluation protocol employed by [22], which considers both of the areas overlap ratios and the orientation differences between predictions and the ground truth.

ICDAR2013. The ICDAR2013 [23] consists of 229 training images and 233 testing images, with word-level annotations provided. It is the standard benchmark for evaluating near-horizontal text detection.

ICDAR2015. The ICDAR2015 [24] was collected by using Google Glass and it has 1,500 images in total: 1,000 images for training and the remained 500 images for testing. Different from the previous ICDAR competition, in which the text is well-captured, horizontal, and typically centered in images, these datasets focus on the incidental scene where text may appear in any orientation and any location with a small size or low resolution. This dataset is more challenging and has images with arbitrary orientations, motion blur, and low-resolution text. We evaluate our results based on the online evaluation system [24].

4.2   Implementation Details

The training of our model is divided into two steps: the first step is to train a classification network, and the second step is to train a regional extraction network based on the attention mechanism.

In the first step, we want to train a classification network to distinguish whether there is text in the image. However, the above three standard data sets are used for text detection, in which all pictures have the presence of text. So we use all the images of the above three datasets as a classification to represent images with text, and then randomly select the same number of images from the ImageNet [25] dataset as the classification without text. Our classification network is initialized by pretraining a model for ImageNet classification. The weights of the network are updated by using a learning rate of 10^{-3} for the first 100k iterations and 10^{-4} for the next 100k iterations, with a weight decay of 5×10^{-4} and a momentum of 0.9.

The results of our different networks are shown in Table 2. These results give strong evidence that the usage of 5×5 convolution kernels and 1×2 max-pooling layer can get better accuracies.

Table 2. Precision, Recall and F-measure of our networks in text/non-text classification task

Network	Precision	Recall	F-measure
A	0.81	0.86	0.83
B	0.83	0.85	0.84
C	0.86	0.89	0.87
D	**0.86**	**0.90**	**0.88**

It can be found that Network D works best, so in the subsequent steps we are all based on Network D. In the subsequent steps, we generate an attention map by combining the information of the fully connected layer with the information of the convolutional layer as described in Sect. 3.2. The weights of the network are updated by using a learning rate of 10^{-4} with a weight decay of 5×10^{-4} and a momentum of 0.9.

4.3   Experimental Results

MSRA-TD500. As shown in Table 3, although our method uses image-level labels, the results are still slightly better than other methods on MSRA-TD500 datasets. The proposed method achieves precision 0.81, recall 0.65 and f-measure 0.72. Compared to [26], our method obtains improvements on recall 0.02 and f-measure 0.01.

Table 3. Performance comparisons on the MSRA-TD500 dataset. The results are reported in the terms of Precision, Recall and F-measure

Method	Precision	Recall	F-measure
Yao et al. [22]	0.63	0.63	0.60
Yin et al. [27]	0.71	0.61	0.65
Kang et al. [28]	0.71	0.62	0.66
Yin et al. [26]	0.81	0.63	0.71
Proposed method	**0.81**	**0.65**	**0.72**

ICDAR2013. We also test our method on the ICDAR2013 dataset, which is the most popular for horizontal text detection. As shown in Table 4, the proposed method achieves 0.83, 0.73, 0.78 in precision, recall, and f-measure. Although the accuracy of our method is not as good as other state-of-the-art methods, it is worth emphasizing that our method is based on a weakly supervised data set. It is also very meaningful to train such a challenging result in a weakly supervised data set.

Table 4. Performance comparisons on the ICDAR2013 dataset.

Method	Precision	Recall	F-measure
SSD [15]	0.80	0.60	0.68
Yin et al. [26]	0.84	0.65	0.73
FASText [29]	0.84	0.69	0.77
TextBoxes [30]	0.86	**0.74**	0.80
Yin et al. [27]	0.88	0.66	0.76
CTPN [12]	**0.93**	0.73	**0.82**
Proposed method	0.82	0.71	0.76

ICDAR2015. As this dataset has been released recently for the competition in ICDAR2015, there is no literature to report the experimental result on it.

Table 5. Performance of different algorithms evaluated on the ICDAR2015 dataset. The comparison results are collected from ICDAR 2015 Competition on Robust Reading [24]

Method	Precision	Recall	F-measure
HUST_MCLAB [24]	0.44	0.38	0.41
AJOU [24]	0.47	0.47	0.47
Deep2Text-MO [24]	0.50	0.32	0.39
StradVision1 [24]	0.53	0.46	0.50
NJU-Text [24]	0.70	0.36	0.48
Yao et al. [10]	0.72	**0.57**	**0.64**
CTPN [12]	0.74	0.52	0.61
StradVision2 [24]	**0.77**	0.37	0.50
Proposed method	0.59	0.33	0.42

Therefore, we collect competition results [24] as listed in Table 5 for comprehensive comparisons. Our approach is less than ideal under this data set. After the analysis, we found that because the background of this data set is the most complicated and the size of the text is too small, the MSER algorithm will cause errors when extracting the bounding boxes from the attention area, affecting the final result.

The effectiveness and versatility of the proposed method can be proved by the above experiments. Besides the quantitative experimental results, several detection examples under various challenging cases of the proposed method on the MSRA-TD500 and ICDAR2013 datasets are shown in Fig. 2.

Fig. 2. Detection results by the proposed weakly supervised text detection model on the MSRA-TD500 and ICDAR2013 datasets.

5 Conclusion

In this paper, we have elaborately designed a model based convolutional neural network with an attention mechanism for weakly supervised text detection. In experiments, our model used image-level labels on MSRA-TD500, ICDAR2013, and ICDAR2015 datasets to compare the other state-of-the-art approaches which were trained in bounding box labels to show the validity and feasibility of our model for the text detection task. However, due to the limitations of the attention mechanism and the MSER algorithm. Our model does not have a beneficial effect on small text detection in complex backgrounds. This is also our future research direction.

References

1. Bahdanau, D., Cho, K., Bengio, Y.: Neural machine translation by jointly learning to align and translate. arXiv preprint arXiv:1409.0473 (2014)
2. Xu, K., Ba, J., Kiros, R., et al.: Show, attend and tell: neural image caption generation with visual attention. In: International Conference on Machine Learning, pp. 2048–2057 (2015)
3. Luong, M.T., Pham, H., Manning, C.D.: Effective approaches to attention-based neural machine translation. arXiv preprint arXiv:1508.04025 (2015)
4. Ye, Y., Zhao, Z., Li, Y., et al.: Video question answering via attribute-augmented attention network learning. In: Proceedings of the 40th International ACM SIGIR Conference on Research and Development in Information Retrieval, pp. 829–832 (2017)
5. Chen, Q., Hu, Q., Huang, J.X., et al.: Enhancing recurrent neural networks with positional attention for question answerin. In: Proceedings of the 40th International ACM SIGIR Conference on Research and Development in Information Retrieval, pp. 993–996 (2017)
6. Jaderberg, M., Vedaldi, A., Zisserman, A.: Deep features for text spotting. In: Fleet, D., Pajdla, T., Schiele, B., Tuytelaars, T. (eds.) ECCV 2014. LNCS, vol. 8692, pp. 512–528. Springer, Cham (2014). https://doi.org/10.1007/978-3-319-10593-2_34
7. Wang, T., Wu, D.J., Coates A., et al.: End-to-end text recognition with convolutional neural networks. In: Proceedings of the 21st International Conference on Pattern Recognition, pp. 3304–3308 (2012)
8. Zhang, Z., Zhang, C., Shen, W., et al.: Multi-oriented text detection with fully convolutional network. In: Proceedings of the IEEE Conference on Computer Vision and Pattern Recognition, pp. 4159–4167 (2016)
9. He, T., Huang, W., Qiao, Y., et al.: Accurate text localization in natural image with cascaded convolutional text network. arXiv preprint arXiv:1603.09423 (2016)
10. Yao, C., Bai, X., Sang, N., et al.: Scene text detection via holistic, multi-channel prediction. arXiv preprint arXiv:1606.09002 (2016)
11. Long, J., Shelhamer, E., Darrell, T.: Fully convolutional networks for semantic segmentation. In: Proceedings of the IEEE Conference on Computer Vision and Pattern Recognition, pp. 3431–3440 (2015)
12. Tian, Z., Huang, W., He, T., He, P., Qiao, Y.: Detecting text in natural image with connectionist text proposal network. In: Leibe, B., Matas, J., Sebe, N., Welling, M. (eds.) ECCV 2016. LNCS, vol. 9912, pp. 56–72. Springer, Cham (2016). https://doi.org/10.1007/978-3-319-46484-8_4
13. Ren, S., He, K., Girshick, R., et al.: Faster r-cnn: towards real-time object detection with region proposal networks. In: Advances in Neural Information Processing Systems, pp. 91–99 (2015)
14. He, P., Huang, W., He, T., et al.: Single shot text detector with regional attention. In: Proceedings of the IEEE International Conference on Computer Vision, pp. 3047–3055 (2017)

15. Liu, W., et al.: SSD: single shot multibox detector. In: Leibe, B., Matas, J., Sebe, N., Welling, M. (eds.) ECCV 2016. LNCS, vol. 9905, pp. 21–37. Springer, Cham (2016). https://doi.org/10.1007/978-3-319-46448-0_2

16. Zhou, X., Yao, C., Wen, H., et al.: EAST: an efficient and accurate scene text detector. In: Proceedings of the IEEE Conference on Computer Vision and Pattern Recognition, pp. 5551–5560 (2017)

17. He, W., Zhang, X.Y., Yin, F., et al.: Deep direct regression for multi-oriented scene text detection. In: Proceedings of the IEEE International Conference on Computer Vision, pp. 745–753 (2017)

18. Simonyan, K., Zisserman, A.: Very deep convolutional networks for large-scale image recognition. arXiv preprint arXiv:1409.1556 (2014)

19. Shi, B., Bai, X., Yao, C.: An end-to-end trainable neural network for image-based sequence recognition and its application to scene text recognition. IEEE Trans. Pattern Anal. Mach. Intell. **39**(11), 2298–2304 (2017)

20. Goodfellow, I., Pouget-Abadie, J., Mirza, M., et al.: Generative adversarial nets. In: Advances in Neural Information Processing Systems, pp. 2672–2680 (2014)

21. Neumann, L., Matas, J.: Real-time scene text localization and recognition. In: Proceedings of the IEEE Conference on Computer Vision and Pattern Recognition, pp. 3538–3545 (2012)

22. Yao, C., Bai, X., Liu, W., et al.: Detecting texts of arbitrary orientations in natural images. In: Proceedings of the IEEE Conference on Computer Vision and Pattern Recognition, pp. 1083–1090 (2012)

23. Karatzas, D., Gomez-Bigorda, L., Nicolaou, A., et al.: ICDAR 2013 competition on robust reading. In: 12th International Conference on Document Analysis and Recognition, pp. 1484–1493 (2013)

24. Karatzas, D., Gomez-Bigorda, L., Nicolaou, A., et al.: ICDAR 2015 competition on robust reading. In: 13th International Conference on Document Analysis and Recognition, pp. 1156–1160 (2015)

25. Deng, J., Dong, W., Socher, R., et al.: ImageNet: a large-scale hierarchical image database. In: IEEE Conference on Computer Vision and Pattern Recognition, pp. 248–255 (2009)

26. Yin, X.C., Pei, W.Y., Zhang, J., et al.: Multi-orientation scene text detection with adaptive clustering. IEEE Trans. Pattern Anal. Mach. Intell. **37**(9), 1930–1937 (2015)

27. Yin, X.C., Yin, X., Huang, K., et al.: Robust text detection in natural scene images. IEEE Trans. Pattern Anal. Mach. Intell. **36**(5), 970–983 (2014)

28. Kang, L., Li, Y., Doermann, D.: Orientation robust text line detection in natural images. In: Proceedings of the IEEE Conference on Computer Vision and Pattern Recognition, pp. 4034–4041 (2014)

29. Busta, M., Neumann, L., Matas, J.: Fastext: Efficient unconstrained scene text detector. In: Proceedings of the IEEE International Conference on Computer Vision, pp. 1206–1214 (2015)

30. Liao, M., Shi, B., Bai, X., et al.: Textboxes: a fast text detector with a single deep neural network. In: Thirty-First AAAI Conference on Artificial Intelligence (2017)

Proposal-Refined Weakly Supervised Object Detection in Underwater Images

Xiaoqian Lv, An Wang, Qinglin Liu, Jiamin Sun, and Shengping Zhang[✉]

School of Computer Science and Technology,
Harbin Institute of Technology, Weihai, China
xiaoqian.hit@gmail.com,wangan16@sina.com,qinglin.liu@outlook.com,
sunjiamin17@gmail.com,s.zhang@hit.edu.cn

Abstract. Recently, Convolutional Neural Networks (CNNs) have achieved great success in object detection due to their outstanding abilities of learning powerful features on large-scale training datasets. One of the critical factors of their success is the accurate and complete annotation of the training dataset. However, accurately annotating the training dataset is difficult and time-consuming in some applications such as object detection in underwater images due to severe foreground clustering and occlusion. In this paper, we study the problem of object detection in underwater images with incomplete annotation. To solve this problem, we propose a proposal-refined weakly supervised object detection method, which consists of two stages. The first stage is a weakly-fitted segmentation network for foreground-background segmentation. The second stage is a proposal-refined detection network, which uses the segmentation results of the first stage to refine the proposals and therefore can improve the performance of object detection. Experiments are conducted on the Underwater Robot Picking Contest 2017 dataset (URPC2017) which has 19967 underwater images containing three kinds of objects: sea cucumber, sea urchin and scallop. The annotation of the training set is incomplete. Experimental results show that the proposed method greatly improves the detection performance compared to several baseline methods.

Keywords: Underwater image · Weakly supervised · Object detection · Semantic segmentation

1 Introduction

Nowadays, aquaculture has become one of the most promising avenues for coastal fishermen by breeding marine products [13], especially high-quality marine products in sea floor, such as sea cucumbers, sea urchins and scallops. Underwater operation in the traditional aquaculture are mainly carried out by manual labor, which comes with low efficiency and high risk. Meanwhile, due to the development of artificial intelligence and the decrease of manufacturing costs, a huge

© Springer Nature Switzerland AG 2019
Y. Zhao et al. (Eds.): ICIG 2019, LNCS 11901, pp. 418–428, 2019.
https://doi.org/10.1007/978-3-030-34120-6_34

demand emerged for the application of underwater fishing robots, which are low-cost, reliable and affordable platforms for improving the efficiency of catching the marine products. Although underwater robots such as net cleaning robots have been widely used [13], the application of underwater fishing robots is still very challenging due to the difficulty of accurately detecting marine products in a complicated underwater environment.

With the development of Convolutional Neural Network (CNN), great improvements have been achieved on object detection on land, which are mainly divided into two categories: two-stage detectors and one-stage detectors. Two-stage detectors adopt a region proposal-based strategy, whose pipelines have two stages [3,5–7,9,16]. The first stage generates a set of category-independent region proposals, and the second stage classifies them into foreground classes or background. One-stage detectors does not separate detection proposal, making the overall pipeline single stage [8,10,12,14,15]. Although some methods without relying on region proposal have been proposed, region proposal-based methods possess leading accuracy on benchmarks datasets (e.g., PASCAL VOC [4], ILSVRC [19], and Microsoft COCO [11] datasets). Faster R-CNN [16] is one of the most well-known object detection framework, which proposed an efficient and accurate Region Proposal Network (RPN) to generate region proposals. Since then, these RPN-like proposals are standards for recent two-stage object detectors.

Existing object detectors heavily depend on a significant number of accurate annotated images [4,11,19]. The annotation of such benchmark datasets often cost too much time and labors. To reduce the cost of obtaining accurate annotation, some weakly supervised and semi-supervised object detection frameworks have been proposed over the past years. At present, Weakly supervised detection mainly focuses on image-level annotation instead of the bounding-box annotation [20,21,23]. Semi-supervised object detectors are trained by using few annotated data and massive unannotated data [1,2,17]. Nevertheless, the reduction of annotation cost is usually at the cost of degrading model accuracy. Though many promising ideas have been proposed in weakly supervised and semi-supervised object detection, they are still far from comparable to strongly supervised ones.

Unlike land images with common object categories, underwater images own the characteristics of image degradation and color distortion due to the absorption and scattering of light through water. Besides, objects in the underwater environment are usually small and tend to cluster. These reasons cause annotating underwater objects difficult and time-consuming particularly. Therefore, as shown in Fig. 1, missing partial annotations often occurs in underwater image datasets. Under these circumstances, the negative examples are generated not only from the background but also the unannotated foreground, which will misguide the training of detectors. Existing strongly and weakly supervised detection algorithms cannot achieve satisfied results in underwater object detection.

To solve this problem, we propose a proposal-refined weakly supervised object detection method, focusing on training detectors with incomplete annotated

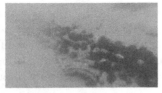
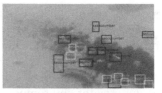

(a) Unannotated image.   (b) Groundtruth.

Fig. 1. Example of underwater image and corresponding groundtruth in URPC2017.

dataset. We discover that there are great differences between foreground and background in underwater images. Inspired by this, we design a weakly-fitted segmentation network to segment the foreground and background of an image by only using incomplete annotated detection dataset. Then, we use the segmentation map to control the generation of positive and negative examples when training the detection network, which is conducive to the generation of high-quality proposals. The proposed method does not restricted to a specific object detection framework. In fact, it can be incorporated into any advanced ones. Our experiments are carried out on the Underwater Robot Picking Contest 2017 dataset (URPC2017). Experiments show that the proposed method greatly improves the accuracy of object detection compared to several baseline methods.

2 The Proposed Method

2.1 Overview

In order to reduce the influence of the missed annotations of the training images, we design a weakly-fitted segmentation network to separate the foreground from background, and then utilize the results generated in the segmentation network to guide the generation of positive and negative examples in training of the detector. Figure 2 shows the overview of the proposed architecture. It consists of two stages, where the first stage is a weakly-fitted segmentation network and the second stage is a proposal-refined object detection network. The details of each part of our model are introduced in the following sections.

2.2 Weakly-Fitted Segmentation

To segment the foreground and background of an underwater image, we utilize the idea of U-Net [18], which consists of a contracting path to capture context information and an expanding path to guarantee the accuracy of localization. The traditional well-trained U-Net cannot accurately separate the foreground from background on our underwater images because there are a lot of unannotated foreground area in the training dataset. To address this problem, we propose two modifications: (1) As shown in Fig. 3, We design a light-weighted U-Net to reduce the ability to fit the training dataset. More specifically, we use

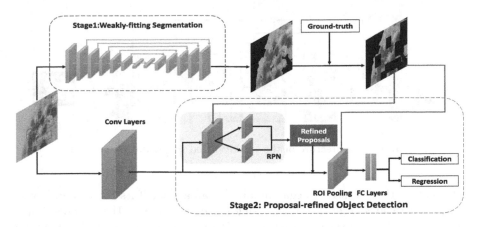

Fig. 2. The architecture of the proposed object detection network.

7 convolutional layers in downsampling and 6 deconvolutional layers in upsampling, which consist of 3×3 convolutions with stride 2 without double and halve the number of feature channels at each downsampling and upsampling step. The design of asymmetric convolutional layer can reduce the fitting degree of model to incomplete annotated training dataset. After that, the image size is restored by bilinear interpolation. (2) To segment the foreground as much as possible, the network is back-propagated via a modified MSE loss, denoted as

$$L(y, y^*) = \frac{1}{N} \sum_{i=0}^{N} (y_i^* - y_i)^2 + \lambda \frac{1}{N} \sum_{i=0}^{N} y_i^* (y_i^* - y_i) \tag{1}$$

where y is the output image of the weakly-fitted segmentation network, y^* is the ground-truth image generated by the bounding-box area of the underwater object detection dataset. i is the index of each pixel in an image, N is the number of pixels in an image. The value of y_i^* equals to 0 if it belongs to background while the value of y_i^* equals to 1 if it belongs to foreground. The term $y_i^* (y_i^* - y_i)$ means the last item is activated only for foreground. So, the last item can enlarge the loss, which takes the foreground as background influenced by the confusing of incomplete annotated datasets. Moreover, the two terms are normalized by N and weighted by a balancing parameter λ.

2.3 Proposal-Refined Object Detection

The quality of the proposals has great influence on the performance of object detection. Therefore various studies focus on region proposal generation [22, 24]. Among them, Region Proposal Network (RPN) proposed by Faster R-CNN [16] is the most influential method in recent years. Accordingly, we build our strategy based on the Faster R-CNN framework in this paper.

The architecture of Faster R-CNN can be divided into two parts: Region proposal network (RPN) and region-of-interest (ROI) classifier. For training RPN,

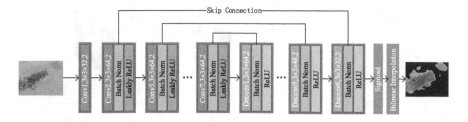

Fig. 3. The architecture of the weakly-fitted segmentation network.

traditional methods assign a negative label to an anchor if its Intersection-over-Union (IoU) ratio is lower than 0.3 for all ground-truth boxes. However, as shown in Fig. 4, many false negative examples which contain unlabeled objects will be generated due to the incomplete annotated dataset. It will affect the learning of RPN network directly. To address this problem, We add an input which is generated in the first stage to RPN and ROI classifier. When RPN and ROI classifier assigns a negative label to an anchor, it not only refers to the IoU for the ground-truth, but also the segmentation map.

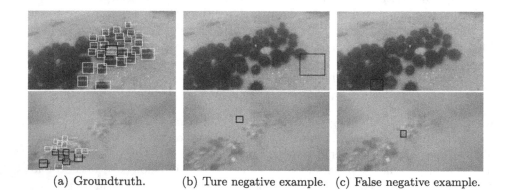

(a) Groundtruth.   (b) Ture negative example.   (c) False negative example.

Fig. 4. The instance of ture and false negative examples.

The specific steps are as follows: (1) Firstly, the foreground of underwater images can be obtained by the weakly-fitted segmentation network, which is denoted as S_1. Then, we subtract ground-truth boxes from S_1 to gain the unlabeled foreground region S_2. (2) For training RPN, the method of labeling positives is the same as traditional strategy. Nevertheless, assigning a negative label to an anchor needs to satisfy two conditions: (i) Its IoU ratio is lower than 0.3 for all ground-truth boxes. (ii) its IoU ratio is lower than or equal to β for S_2. Similarly, the generation of positive and negative examples is constrained by both ground-truth and segmentation map during the training of the ROI classifier.

By controlling the generation of negative examples, we can eliminate the false negative examples, thus provide more accurate positive and negative examples for training object detection network to generate high-quality proposals. Following [16], classification loss and bounding-box regression loss are computed for both the RPN and the RoI classifiers

$$L_{total} = L_{cls}^{rpn} + c^* L_{reg}^{rpn} + L_{cls}^{roi} + p^* L_{reg}^{roi} \tag{2}$$

where L_{cls} is the cross-entropy loss for classification, L_{reg} is the smooth L1 loss defined in [5] for regression, $c^* L_{reg}^{rpn}$ and $p^* L_{reg}^{roi}$ mean the regression loss activated only for positive anchors and non-background class proposals respectively. It is worth mentioning that although the proposed method is carried out on Faster R-CNN, it is applicable to other region proposal-based methods such as R-FCN [3], FPN [9], Mask R-CNN [7].

3 Experiment

3.1 Dataset and Metric

Our experiments are carried out on the Underwater Robot Picking Contest 2017 dataset (URPC2017), which contains 3 object categories (sea cucumber, sea urchin and scallop) with a total of 19967 underwater images. The dataset is divided into the train, val and test set, which have 17655, 1317 and 985 images respectively. In the dataset, the amount of complete annotated data is fewer than incomplete annotated data. We train our segmentation and detection network on the trainval set. The trainval set contains both complete and incomplete annotated images. The test set consists of accurate and complete annotated images. The dataset used to train the weakly-fitted segmentation network is generated from the bounding box area (see Fig. 5). Object detection accuracy is measured by mean Average Precision (mAP).

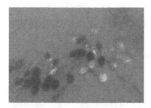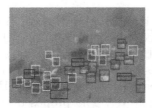

(a) Unannotated image.   (b) Groundtruth.   (c) Segmentation dataset.

Fig. 5. The generation of segmentation dataset.

3.2   Implementation Details

For the training of weakly-fitted Segmentation network, we use a learning rate of 0.0001 for 70k iterations and set $\lambda = 2$ which makes the two terms in Eq. 1 roughly equally weighted after normalization. For the training of the proposal-refined object detection network, we use Faster R-CNN as our baseline detection framework. The VGG16 pre-trained on ImageNet is used as the backbone architecture for feature extraction due to the small scale datasets. The initial learning rate is set to 0.0002 for the first 50k and then decrease to 0.00002 in the following 20k iterations. The momentum and weight decay are set to 0.9 and 0.0005, respectively. Other hyper-parameters are identical as those defined in [16].

3.3   Experimental Results

The Influence of IoU Threshold. We explore the influence of IoU threshold β of the segmentation map for detector. $\beta = 1$ is the baseline result of the original Faster R-CNN, which is not constrained by segmentation map when generating negative examples. As shown in Table 1, $\beta = 0.3$ outperforms other choices, which is 12.1% better than the baseline. It indicates that containing a part of object in the negative examples is beneficial to improve detection performance. When $\beta = 0$, detector will be trained on a large number of easily classified background examples, which is unuseful to improve detection accuracy. Consequently, we choose $\beta = 0.3$ for the following experiments.

Table 1. Comparison results with different IoU thresholds of segmentation map.

IoU thresholds (β)	0	0.3	0.5	0.8	1
mAP(%)	66.1	**69.7**	68.5	62.8	57.6

The Results of Weakly-Fitted Segmentation. Figure 6(c) shows the qualitative results of weakly-fitted segmentation: (a) is the input image, (b) is the segmentation result of U-Net. Obviously, Under the same experimental setting, U-Net cannot completely separate the foreground and background. Because the unannotated foreground area affects the ability of U-Net to distinguish foreground from background. However, The proposed weakly-fitted segmentation network can segment the foreground and background of an underwater images, including the unannotated region in the underwater object detection dataset. Because the proposed method reduce the fitting degree of model to training data and increase the penalty for regarding foreground as background.

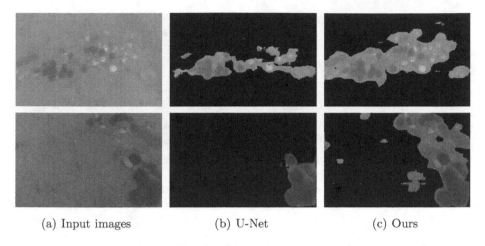

(a) Input images   (b) U-Net   (c) Ours

Fig. 6. Qualitative segmentation results on URPC2017 dataset.

The Results of Proposal-Refined Object Detection. To show how Faster R-CNN and proposal-refined detector improve during the learning, we plot mAP of the two detectors for different training iterations. As shown in Fig. 7, both detectors get improved at the beginning stage. But the proposal-refined detector always have a higher mAP than the Faster R-CNN, suggesting the effectiveness of the proposal-refined object detection network. Figure 8 shows the qualitative results of proposal-refined detector (top) compared with the benchmark Faster R-CNN (bottom). It can be seen that proposal-refined detector can detect more objects than the baseline framework, especially the small and challenging objects.

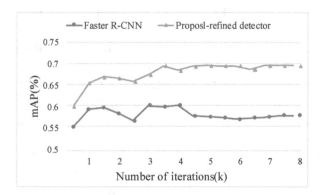

Fig. 7. The changes of mAP for Faster R-CNN and proposal-refined detector on URPC2017 dataset during the process of training.

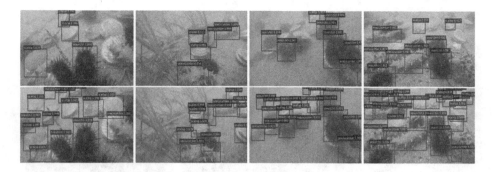

Fig. 8. Qualitative detection results on URPC2017. Top: the results of Faster R-CNN baseline model. Bottom: the results of proposal-refined detector.

Comparisons with the State-of-the-Arts. In this section, we present experimental results of our proposed method applied in other outstanding object detection networks: R-FCN [3], FPN [9], Mask R-CNN [7]. As shown in Table 2, our method improves the mAP of the original object detectors by about 10%, indicating the effectiveness and robustness of the proposal-refined weakly supervised object detection. By eliminating false negative examples, the proposed method can solve the problem of accuracy decrease caused by incomplete annotated dataset.

Table 2. Comparison results of different methods.

Method	Seaurchinn	Scallop	Seacucumber	mAP
Faster R-CNN	59.9	61.7	50.3	57.6
Ours	**72.4**	**77.7**	**58.3**	**69.7**
R-FCN	64.9	69.9	46.6	60.5
Ours	**73.8**	**76.9**	**55.4**	**68.7**
FPN	60.6	65.4	52.9	59.6
Ours	**69.7**	**77.0**	**56.0**	**67.5**
Mask R-CNN	61.8	68.5	51.6	60.7
Ours	**70.1**	**76.1**	**54.8**	**67.0**

4   Conclusion

In this paper, we propose a simple but efficient framework for object detection in underwater images with incomplete annotated dataset. Our proposal-refined weakly supervised object detection system is composed of two stages. The first stage is a weakly-fitted segmentation network that separates foreground from background. The second stage is the proposal-refined object detector that uses the segmentation map to generate high-quality proposals. Experiments show that the proposed method greatly improves the detection performance compared to several baseline methods. Through our method, we can not only reduce the cost of dataset annotation, but also offset the accuracy decrease caused by missed annotation. In addition, the idea of the proposed method can not only be applied to underwater object detection but also to other detect tasks with incomplete annotation.

References

1. Chen, G., Liu, L., Hu, W., Pan, Z.: Semi-supervised object detection in remote sensing images using generative adversarial networks. In: 2018 IEEE International Geoscience and Remote Sensing Symposium, pp. 2503–2506 (2018)
2. Choi, M.K., et al.: Co-occurrence matrix analysis-based semi-supervised training for object detection. In: 2018 25th IEEE International Conference on Image Processing, pp. 1333–1337 (2018)
3. Dai, J., Li, Y., He, K., Sun, J.: R-FCN: object detection via region-based fully convolutional networks. In: Advances in Neural Information Processing Systems, pp. 379–387 (2016)
4. Everingham, M., Gool, L.V., Williams, C.K.I., Winn, J., Zisserman, A.: The pascal visual object classes (voc) challenge. Int. J. Comput. Vis. **88**(2), 303–338 (2010)
5. Girshick, R.: Fast R-CNN. In: Proceedings of the IEEE International Conference on Computer Vision, pp. 1440–1448 (2015)
6. Girshick, R., Donahue, J., Darrell, T., Malik, J.: Rich feature hierarchies for accurate object detection and semantic segmentation. In: Proceedings of the IEEE Conference on Computer Vision and Pattern Recognition, pp. 580–587 (2014)
7. He, K., Gkioxari, G., Dollár, P., Girshick, R.: Mask R-CNN. In: Proceedings of IEEE International Conference on Computer Vision, pp. 2980–2988 (2017)
8. Law, H., Deng, J.: CornerNet: detecting objects as paired keypoints. In: Ferrari, V., Hebert, M., Sminchisescu, C., Weiss, Y. (eds.) Computer Vision – ECCV 2018. LNCS, vol. 11218, pp. 765–781. Springer, Cham (2018). https://doi.org/10.1007/978-3-030-01264-9_45
9. Lin, T.Y., Dollar, P., Girshick, R., He, K., Hariharan, B., Belongie, S.: Feature pyramid networks for object detection. In: Proceedings of the IEEE Conference on Computer Vision and Pattern Recognition, pp. 936–944 (2017)
10. Lin, T.-Y., Goyal, P., Girshick, R., He, K., Dollar, P.: Focal loss for dense object detection. IEEE Trans. Pattern Anal. Mach. Intell. (2018). https://doi.org/10.1109/TPAMI.2018.2858826
11. Lin, T.-Y., et al.: Microsoft COCO: common objects in context. In: Fleet, D., Pajdla, T., Schiele, B., Tuytelaars, T. (eds.) ECCV 2014. LNCS, vol. 8693, pp. 740–755. Springer, Cham (2014). https://doi.org/10.1007/978-3-319-10602-1_48
12. Liu, W., et al.: SSD: single shot multibox detector. In: Leibe, B., Matas, J., Sebe, N., Welling, M. (eds.) ECCV 2016. LNCS, vol. 9905, pp. 21–37. Springer, Cham (2016). https://doi.org/10.1007/978-3-319-46448-0_2
13. Naylor, R., Burke, M.: Aquaculture and ocean resources: raising tigers of the sea. Annu. Rev. Environ. Resour. **30**, 185–218 (2005)
14. Redmon, J., Divvala, S., Girshick, R., Farhadil, A.: You only look once: unified, real-time object detection. In: Proceedings of the IEEE Conference on Computer Vision and Pattern Recognition, pp. 779–788 (2016)

15. Redmon, J., Farhadi, A.: Yolo9000: better, faster, stronger. In: Proceedings of the IEEE Conference on Computer Vision and Pattern Recognition, pp. 6517–6525 (2017)
16. Ren, S., He, K., Girshick, R., Sun, J.: Faster R-CNN: towards real-time object detection with region proposal networks. In: Advances in Neural Information Processing Systems, pp. 91–99 (2015)
17. Rhee, P.K., Erdenee, E., Kyun, S.D., Ahmed, M.U., Jin, S.: Active and semi-supervised learning for object detection with imperfect data. Cogn. Syst. Res. **45**, 109–123 (2017)
18. Ronneberger, O., Fischer, P., Brox, T.: U-Net: convolutional networks for biomedical image segmentation. In: Navab, N., Hornegger, J., Wells, W.M., Frangi, A.F. (eds.) MICCAI 2015. LNCS, vol. 9351, pp. 234–241. Springer, Cham (2015). https://doi.org/10.1007/978-3-319-24574-4_28
19. Russakovsky, O., et al.: Imagenet large scale visual recognition challenge. Int. J. Comput. Vis. **115**(3), 211–252 (2015)
20. Tang, P., Wang, X., Bai, X., Liu, W.: Multiple instance detection network with online instance classifier refinement. In: Proceedings of the IEEE Conference on Computer Vision and Pattern Recognition, pp. 2843–2851 (2017)
21. Tang, P., et al.: Weakly supervised region proposal network and object detection. In: Ferrari, V., Hebert, M., Sminchisescu, C., Weiss, Y. (eds.) ECCV 2018. LNCS, vol. 11215, pp. 370–386. Springer, Cham (2018). https://doi.org/10.1007/978-3-030-01252-6_22
22. Uijlings, J.R.R., van de Sande, K.E.A., Gevers, T., Smeulders, A.W.M.: Selective search for object recognition. Int. J. Comput. Vis. **104**(2), 154–171 (2013)
23. Zhang, X., Feng, J., Xiong, H., Tian, Q.: Zigzag learning for weakly supervised object detection. In: Proceedings of the IEEE Conference on Computer Vision and Pattern Recognition, pp. 4262–4270 (2018)
24. Zitnick, C.L., Dollár, P.: Edge boxes: locating object proposals from edges. In: Fleet, D., Pajdla, T., Schiele, B., Tuytelaars, T. (eds.) ECCV 2014. LNCS, vol. 8693, pp. 391–405. Springer, Cham (2014). https://doi.org/10.1007/978-3-319-10602-1_26

Learning Cross Camera Invariant Features with CCSC Loss for Person Re-identification

Zhiwei Zhao, Bin Liu$^{(\boxtimes)}$, Weihai Li, and Nenghai Yu

Key Laboratory of Electromagnetic Space Information,
Chinese Academy of Sciences, School of Information Science and Technology,
University of Science and Technology of China, Hefei, China
zhaozhiwei1998@foxmail.com, {flowice,whli,ynh}@ustc.edu.cn

Abstract. Person re-identification (re-ID) is mainly deployed in the multi-camera surveillance scene, which means that learning cross camera invariant features is highly required. In this paper, we propose a novel loss named *Cross Camera Similarity Constraint loss* (CCSC loss), which makes full use of the camera ID information and the person ID information simultaneously to construct cross camera image pairs and performs cosine similarity constraint on them. The proposed CCSC loss effectively reduces the intra-class variance through forcing the whole network to extract cross camera invariant features, and it can be unified with identification loss in a multi-task manner. Extensive experiments implemented on the standard benchmark datasets including CUHK03, DukeMTMC-reid, Market-1501 and MSMT17 indicate that the proposed CCSC loss can bring a large performance boost on the strong baseline and it is also superior to other metric learning methods such as hard triplet loss and center loss. For instance, on the most challenging dataset CUHK03-Detect, Rank-1 accuracy and mAP are improved by **10.0%** and **10.2%** than the baseline respectively and simultaneously obtain a comparable performance with the state-of-the-art method.

Keywords: Person re-identification · CCSC loss · Cross camera

1 Introduction

Person re-identification (re-ID) has attracted close attention both in academic community and industry in recent years due to its great application prospects in many fields, such as video surveillance analysis, human-computer interaction, intelligent retail, etc. Given a query person-of-interest, person re-identification aims to retrieve all images that belong to the same person captured by multiple camera without view overlap at different time or scenarios.

Although the methods based on deep learning have brought great success to person re-ID [4,7,19,22], this field still faces many challenges. On the one hand, due to the extreme complexity of cross camera surveillance scenario, the image

© Springer Nature Switzerland AG 2019
Y. Zhao et al. (Eds.): ICIG 2019, LNCS 11901, pp. 429–441, 2019.
https://doi.org/10.1007/978-3-030-34120-6_35

pairs captured by different cameras for a specific person have dramatic variations in viewpoint, posture, background and illumination. As shown in Fig. 1, for this man, there is a large viewpoint variation between image pair captured by camera 2 and 4, an obvious posture change between image pair captured by camera 2 and 3, and a drastic change both in the background and illumination between image pair captured by camera 2 and 5. On the other hand, in a real large-scale surveillance scenario, the color of clothes and body shapes between different pedestrians may be very similar, which makes it difficult to distinguish even with the human eyes.

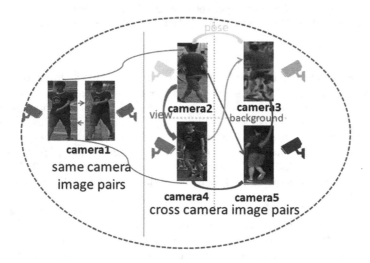

Fig. 1. In a multi-camera surveillance scenario, the image pairs of a pedestrian captured by different cameras have dramatic variations in viewpoint, posture, background and illumination, but the image pairs of a pedestrian captured by the same camera have very high similarity in appearance.

As mentioned above, many problems in person re-ID are caused by cross camera, so learn cross camera invariant feature representation is highly required. We notice that the few work in this field has utilized the camera ID information, which may play a very important role in the aspect of supervision. For instance, most existing re-ID methods based on deep metric learning such as constrastive loss [20] and triplet loss [3,15], which only consider the person ID information to construct positive and negative pairs but neglect useful camera ID information.

In this paper, we propose a novel loss named Cross Camera Similarity Constraint loss (CCSC loss), which takes full advantage of the camera ID information and person ID information to construct cross camera image pairs for every person and performs cosine similarity constraint on them. Specifically speaking, we first take each sample within a batch as the anchor, and then for each anchor, we select all the proper samples that have the same person ID but different camera

ID with the anchor to construct cross camera sample pair. Finally, we maximize the cosine similarity on all the cross camera sample pairs.

The proposed CCSC loss effectively alleviates a series of problems caused by cross camera, and it can be combined with identification loss in an multi-task learning framework. Compared with just using identification loss, the joint training of the CCSC loss and identification loss can bring significant performance improvements on the mainstream person re-ID benchmarks.

In summary, the contributions of this paper are two folds:

- We propose a novel loss named the CCSC loss, which explicitly utilizes the camera ID information and person ID information simultaneously to form cross camera sample pairs and performs similarity constraint on them. Through extensive experiments, we verify that the CCSC loss consistently improves the accuracy of the baseline over standard datasets, CUHK03, DukeMTMC- reid, Market-1501 and MSMT17.
- We fairly compare the proposed CCSC loss with hard triplet loss [7] and center loss [25]. Experiments show that when above losses are combined with identification loss respectively, the CCSC loss is not only superior to other losses in performance, but also makes it easier to train because it does not need to adjust additional hyperparameters. More importantly, it can achieve better performance with the help of hard triplet loss.

2   Related Work

With the tremendous success of deep learning in the field of computer vision, people abandoned hand-craft features for person re-identification [13,14], then deep learning based methods quickly dominate the person re-ID benchmarks. Recently deep re-ID methods mainly revolves around the following two lines.

One line is to find more discriminant and robust features to represent pedestrians, such as part-based methods [19,22,27], attention-based methods [12,17,21] and pose-guided methods [24,26,28]. Among them, the part-based methods achieve the state-of-the-art performance, which split a input feature map horizontally into a fixed number of strips and aggregate features from those strips. For example, Wang et al. [22] carefully design the multiple granularity network (MGN) to extract and utilize the global and local part features with multi-granularity for re-ID. However, MGN is very complex, and the computation cost of integrating all multi-branch feature vectors for testing is heavy, which is unrealistic for large-scale rapid person re-identification.

The other line is to find more effective metric learning method to make features of the same person more similar than those of different persons, such as constrastive loss [20], triplet loss [3,15], quadruplet loss [1] and center loss [25]. However, constrastive loss and triplet loss based methods suffer the common problem that they are prone to have a slow convergence speed and unstable performance in the circumstance of a large number of person identitiesnce and highly dependent on the sample's quality of the mini-batch in training. Although hard sample mining methods [7] effectively alleviates this problem, triplet loss still has extra margin hyperparameter to adjsut and it is not easy to train. Center loss, which simultaneously learns a center for deep features of each class and penalizes the euclidean distance between the deep features and their corresponding class centers. However, for center loss, each category center needs to be learned explicitly, and it does not fully take into account the rich variations of all the sample pairs. It is worth mentioning that none of the above metric learning methods have utilized the camera ID information. The proposed CCSC loss not only makes full use of the camera ID information to construct cross camera image pairs, but also takes full account of all the possible combinations of variations in appearance between image pairs. When above losses are all combined with identification loss respectively, compared with hard triplet loss and center loss, The proposed CCSC loss not only surpasses them in performance, but also make it easier to train.

3   Proposed Method

This section first introduces the structure of our model for person re-ID, and then explains the proposed CCSC loss in details.

3.1   Model Structure

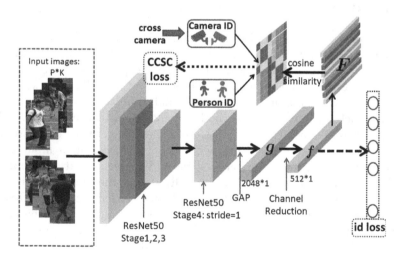

Fig. 2. The overall architecture of the proposed model. We adopt the modified ResNet-50 as the backbone for feature extraction, after that we use GAP (global average pooling) to get a 2048-dimensional feature vector g, then we employ a 1×1 convolutional layer to get a 512-dimensional feature vector f. Finally, the feature vectors of this batch $F = [f_1, f_2, \cdots, f_N]$ are fed into two branches, one branch is used to calculate the proposed CCSC loss and the other branch is implemented by a classifier to calculate the identification loss. We combine the CCSC loss and identification loss in multi-task manner to train the entire network.

ResNet-50 Baseline. We use ResNet-50 [6], which is widely used in person re-identification, as backbone for feature extraction. Notice that we change the down-sampling stride of ResNet-50 stage4 from 2 to 1 to enlarge the the spatial size of the feature map, just like recent works [19,22] have done. After stage 4 of ResNet-50, we use GAP (global average pooling) to get a 2048-dimensional feature vector g, then we employ a 1×1 convolutional layer, a batch normalization layer, and a ReLU layer to reduce the dimension of g to get a 512-dimensional feature vector f. Finally, the dimension-reduced feature vector f is fed into the classifier. Classifier is implemented by a fully-connected layer and a softmax layer. We denote the structure described above as ResNet-50 baseline. For simplicity, we will use baseline to refer to ResNet-50 baseline in the rest of this paper. For baseline, we only use identification loss (softmax loss) to optimize the whole network. The baseline achieves 91.4% Rank-1 accuracy and 77.9% mAP on the Market-1501 dataset, we believe that the innovation on a stronger baseline can better demonstrates the effectiveness of our method.

Our Approach. As shown in Fig. 2, on the basis of the baseline, our method only adds an extra branch to compute the proposed CCSC loss. Specifically speaking, we first randomly sample P identities and K images of per person to constitute a training batch, thus the batch size $N = P \times K$. After that we feed a batch of images to network to extract the feature vectors of this batch $F = [f_1, f_2, \cdots, f_N]$. Finally, F are fed into two branches, one branch is used to calculate the CCSC loss and the other branch is implemented by a classifier to calculate the identification loss. We combine these two types of losses in multi-task manner to train the entire network. Under this manner, our network not only has good property of distinguish different pedestrians, but also learns the cross camera invariant features, which greatly alleviates the problem of intra-class variation. In the test phase, the feature vector f is extracted for final distance metric.

3.2   Loss Function

The Proposed CCSC Loss. We follow the same batch sampling strategy with [7] to randomly sample P identities and K images of per person to constitute a training batch, thus in each mini-batch $\mathcal{N}$, we have $N = P \times K$ images. We denote the ith instance of pth person as s_i^p, denote the feature vector of s_i^p as f_i^p. As Eq. (1) illustrates, we use indicator function $I(s_i^p, s_j^p)$ to represent whether sample s_i^p and s_j^p come from different cameras, and camera(s_i^p) indicates the camera id of s_i^p.

Equation (2) shows the formulaic representation of the proposed CCSC loss, and diagram (Algorithm 1) shows a clear procedure to calculate the CCSC loss. Taking each sample $s \in \mathcal{N}$ as the anchor, we first select all the proper samples that have same person ID but different camera ID with anchor s, then we utilize them to construct cross camera sample pairs. At the same time, we compute the cosine similarity on all the cross camera sample pairs. For convenience of optimization, we transform the cosine similarity into the form of loss with the help of function: $T(x) = \frac{1}{1+x}$. Finally, we minimize the average loss on all the cross camera sample pairs.

$$
I(s_i^p, s_j^p) = \begin{cases} 1 & \text{if camera}(s_i^p) \neq \text{camera}(s_j^p); \\ 0 & \text{otherwise.} \end{cases}
\tag{1}
$$

$$
\mathcal{L}_{ccsc} = \frac{1}{M} \sum_{p=1}^{P} \sum_{i=1}^{K} \sum_{j=1}^{K} \frac{I(s_i^p, s_j^p)}{1 + \dfrac{(f_i^p)^T f_j^p}{\|f_i^p\|_2 \|f_j^p\|_2}}
\tag{2}
$$

$$
M = \sum_{p=1}^{P} \sum_{i=1}^{K} \sum_{j=1}^{K} I(s_i^p, s_j^p)
\tag{3}
$$

Algorithm 1. The procedure of calculating the proposed CCSC loss.

Input:

    A batch of feature vectors: $F = [f_1, f_2, \cdots, f_N]$, size: $N \times 512$;

    The corresponding person ID label vector: p, size: $N \times 1$;

    The corresponding camera ID label vector: c, size: $N \times 1$;

Output:

    The value of CCSC loss for this batch: $\mathcal{L}_{ccsc}$;

1: Calculating the cosine similarity matrix S, $S_{ij} = \frac{(f_i)^T f_j}{\|f_i\|_2 \|f_j\|_2}$, size: $N \times N$;

2: Converting similarity matrix S to loss matrix D with the help of function $T(\cdot)$ mentioned above, thus $D = T(S) = \frac{1}{1+S}$, element-wise operation;

3: Constructing the cross camera image pairs constraint flag matrix: $Mask$;

    Expanding person ID label vector p to matrix $P = [p, p, \cdots, p]$, size: $N \times N$;

    Expanding camera ID label vector c to matrix $C = [c, c, \cdots, c]$, size: $N \times N$;

    Calculating person constraint flag matrix: Mask-P $\Leftarrow (P == P^T)$;

    Calculating camera constraint flag matrix: Mask-C $\Leftarrow (C \neq C^T)$;

    $Mask \Leftarrow$ (Mask-P&Mask-C);

    Comment: Mask-P equals to 1 denotes that the corresponding image pair comes from the same person, Mask-C equals to 1 denotes that the corresponding image pair comes from different camera, so $Mask$ equals to 1 is the flag of the cross camera image pair that we desire.

4: Calculating the average loss on all the selected cross camera sample pairs.

    $\mathcal{L}_{ccsc} = \text{mean}(\ D[Mask == 1]\ )$;

5: **return** $\mathcal{L}_{ccsc}$;

Overall Loss. We regard the identification task as a multi-class classification problem, so identification loss actually refers to softmax loss in this paper. We combine identification loss with the proposed CCSC loss in multi-task manner like many works in this field [2,11] to train the entire network. The joint training of identification loss and the CCSC loss brings a significant improvement in performance. As illustrated in Eq. (4), the overall loss is the weighted sum of the proposed CCSC loss and identification loss, and λ is the balanced weight of the CCSC loss. In the experiments, we set $\lambda = 1.5$ for best performance.

$$\mathcal{L} = \mathcal{L}_{id} + \lambda\mathcal{L}_{ccsc} \tag{4}$$

4   Experiments

4.1   Datasets and Protocols

Datasets. We conduct extensive experiments on four public person re-identification benchmarks, *i.e.*, CUHK03 [10], DukeMTMC-ReID [16,31], Market-1501 [29] and MSMT17 [23]. Note that for CUHK03, according to whether it is manual or DPM labeling, it is divided into CUHK03-Label and CUHK03-Detect and we use the recently proposed new protocol in [32] for CUHK03.

Protocols. In the experiments, to evaluate the performances of re-ID methods, we report the cumulative matching characteristics (CMC) at Rank-1 and mean average precision (mAP) on four datasets.

4.2   Implementation Details

Training. In the training phase, the input images are resized to 384×128, then random horizontal flip, normalization and random erasing [33] are applied as data augmentation. We set $P = 16$ and $K = 4$ to construct mini-batch, thus batchsize $N = 64$. The backbone ResNet-50 is initialized from the ImageNet pre-trained model [5]. We use the Adam optimizer [9] to train the whole network. We set $\lambda = 1.5$ for best performance. We train 100 epochs in total. The learning rate warms up from 3.5e−5 to 3.5e−4 linearly in the first 5 epochs, then it is decayed to 3.5e−5 and 3.5e−6 at 35th and 55th epoch respectively. Our network is trained using 1 NVIDIA TITAN XP GPU and adopted Pytorch as the platform. Note that all comparative experiments adopted the same settings to ensure fairness.

Testing. In the testing phase, the input images are resized to 384×128, and only augmented with normalization. We extract feature vector f for test and use euclidean distance to rank.

4.3   Comparative Experiments Analysis

In order to prove the effectiveness of the proposed CCSC loss, a series of comparative experiments are conducted on all datasets we mentioned above.

Effectiveness of the Proposed CCSC Loss. As shown in Table 1, the proposed CCSC Loss can bring significant improvement to baseline on all four benchmarks. For instance, with the help of the proposed CCSC Loss, Rank-1 accuracy and mAP are improved by **10.0%** and **10.2%** respectively on the most challenging datasets CUHK03-Detect. It fully demonstrates the effectiveness of joint training with softmax loss and the CCSC loss. We can also see that the proposed CCSC loss is very effective for some datasets that have rich viewpoint variations, such as CUHK03, MSMT17 and DukeMTMC-ReID.

Figure 3 shows top-5 ranking results for some given query pedestrian images on CUHK03 and DukeMTMC-ReID dataset. For the first given pedestrian, the top-3 ranking results of the baseline are all mismatched due to the great similarity of clothing color and body shape, but the top-3 ranking results of our method for this pedestrian are all correct, even if there is a large change in angle of view. For the second query, even with serious occlusion problems and viewpoint changes, our methods can still find all the right results. The above results illustrate that our method is very effective for person re-ID in real complex scenes.

Table 1. Comparison of the proposed method (baseline + CCSC loss) with the baseline on four datasets. Rank-1 accuracy (%) and mAP (%) are shown.

Method	CUHK03-Label		CUHK03-Detect		DukeMTMC-reID		Market1501		MSMT17	
	Rank-1	mAP	Rank-1	mAP	Rank-1	mAP	Rank-1	mAP	Rank-1	mAP
baseline	63.8	60.8	60.4	56.6	82.8	66.9	91.4	77.9	69.0	40.7
baseline+ CCSC loss (**ours**)	73.6	70.8	70.4	66.8	85.0	69.8	92.1	81.8	72.7	45.4
increment ↑	**+9.8**	**+10.0**	**+10.0**	**+10.2**	**+2.2**	**+2.9**	**+0.7**	**+3.9**	**+3.7**	**+4.7**

Importance of Cross Camera Constraint. In this part, we illustrate the importance of cross camera constraint. As illustrated in Table 2, if we don't apply cross camera constraint when construct image pairs, i.e., we always set $I(s_i^p, s_j^p) = 1$, the re-ID performance significantly degraded on all datasets. As Fig. 1 shows, the image pairs come from the same camera have very high similarity, so we don't need to consider these simple sample pairs, because they are harmful to the optimization process of the whole network. The restriction of cross camera condition also reflects the idea of hard sample mining [7].

Table 2. The effect of whether to apply cross camera constraint or not.

Method	CUHK03-Label		CUHK03-Detect		DukeMTMC-reID		Market1501		MSMT17	
	Rank-1	mAP	Rank-1	mAP	Rank-1	mAP	Rank-1	mAP	Rank-1	mAP
With constraint	73.6	70.8	70.4	66.8	85.0	69.8	92.1	81.8	72.7	45.4
Without constraint	70.7	68.1	67.1	64.2	83.3	69.2	91.4	80.4	71.3	43.1
Reduction ↓	**−2.9**	**−2.7**	**−3.1**	**−2.6**	**−1.7**	**−0.6**	**−0.7**	**−1.4**	**−1.4**	**−2.3**

Comparison of the Proposed CCSC Loss with Hard Triplet Loss and Center Loss. In order to further prove the superiority of the proposed CCSC loss, we fairly compare the CCSC loss with several commonly used metric learning losses on all benchmarks. Table 3 shows the performance comparison on all datasets when given different combinations of losses. We can see that the joint training with softmax loss and CCSC loss is the best, whereas the joint training with softmax loss and hard triplet loss is the second, and the joint training with softmax loss and center loss is the worst, but both are better than the baseline model which only use softmax loss. Meanwhile, we can see from Table 3 that our method can achieve better performance when combined with hard triplet loss. Futhermore, the proposed CCSC loss not only surpasses other losses in performance, but also make it easier to train, because it has very few parameters to adjust.

Table 3. Comparison of the proposed CCSC loss with the hard triplet loss and center loss when based on the same baseline.

Method	CUHK03-Label		CUHK03-Detect		DukeMTMC-reID		Market1501		MSMT17	
	Rank-1	mAP	Rank-1	mAP	Rank-1	mAP	Rank-1	mAP	Rank-1	mAP
Softmax + CCSC (**ours**)	**73.6**	70.8	70.4	66.8	85.0	69.8	92.1	81.8	72.7	45.4
Softmax + triplet	**69.4**	65.7	65.5	62.2	84.0	69.4	91.8	80.2	72.0	44.5
Softmax + center	**66.1**	62.2	62.3	59.0	83.4	69.0	91.3	78.9	72.5	44.0
Softmax only (**baseline**)	**63.8**	60.8	60.4	56.6	82.8	66.9	91.4	77.9	69.0	40.7
Softmax + CCSC + triplet	**74.2**	71.0	71.8	68.6	84.5	71.5	92.2	82.7	72.4	47.5

Comparison with State-of-the-Art Methods. We compare our proposed method with state-of-the-art methods on all candidate datasets in Table 4. It can be clearly see that although our method only utilizes global feature, it still achieves comparable performance with BFE [4], which achieves the strongest

Table 4. The comparison with state-or-the-art methods on CUHK03, DukeMTMC-reID, Market1501 and MSMT17 datasets.

Method	CUHK03-Label		CUHK03-Detect		DukeMTMC-reID		Market1501		MSMT17	
	Rank-1	mAP	Rank-1	mAP	Rank-1	mAP	Rank-1	mAP	Rank-1	mAP
IDE [30]	22.2	21.0	21.3	19.7	67.7	47.1	72.5	46.0	–	–
SVDNet [18]	–	–	41.5	37.3	76.7	56.8	82.3	62.1	–	–
AlignedReID [26]	–	–	–	–	81.2	67.4	90.6	77.7	–	–
HA-CNN [12]	–	–	–	–	80.5	63.8	91.2	75.7	–	–
SPReID [8]	–	–	–	–	84.4	71.0	92.5	81.3	–	–
PCB [19]	–	–	61.3	54.2	81.9	65.3	92.4	77.3	–	–
PCB + RPP [19]	–	–	62.8	56.7	83.3	69.2	93.8	81.6	–	–
MGN [22]	68.0	67.4	66.8	66.0	88.7	78.4	95.7	86.9	–	–
BFE [4]	**75.4**	**71.2**	74.4	70.8	86.8	71.5	93.5	82.8	–	–
baseline	63.8	60.8	60.4	56.6	82.8	66.9	91.4	77.9	69.0	40.7
baseline+CCSC (**ours**)	73.6	70.8	70.4	66.8	85.0	69.8	92.1	81.8	72.7	45.4

performance on CUHK03 dataset at present. On Market1501 and DukeMTMC-reID dataset, although our approach is not as good as MGN in performance, the model complexity and computational cost are much lower than MGN, which is more suitable for large-scale rapid person re-identification.

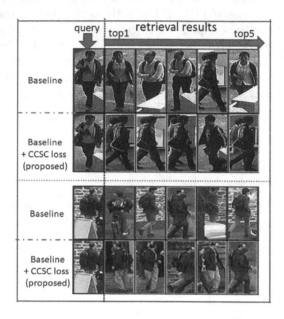

Fig. 3. Comparison of top-5 ranking results between our proposed method (base-line + CCSC loss) and baseline. The images with green borders belong to the same identity as the given query, and that with red borders do not. The images with blue borders represent query, best viewed in color. (Color figure online)

Visualization of the Feature Response Map. We believe that the proposed method did learns the cross camera invariant features. Figure 4 shows some feature response maps for some input pedestrian images, extracted from the last feature map before GAP. The brighter the area is, the more concentrated it is. We can clearly see that for query and gallery, which have a large change of view, our network is more concerned about some body areas that are keep unchanged cross cameras, such as the collar and sleeves of jacket, thighs, shoes and so on, which remain visible during the change of view.

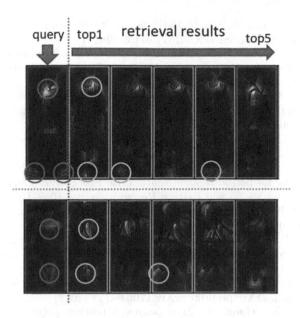

Fig. 4. Visualization is done by overlapping the intensity of corresponding last feature map onto the original images. The brighter the area is, the more concentrated it is, best viewed in color. The red circle and the yellow circle respectively point out the corresponding camera invariant regions for query and gallery. (Color figure online)

5   Conclusion

In this paper, we propose a novel loss named the CCSC loss to learn the cross camera invariant features for person re-identification. The proposed CCSC loss simultaneously utilizes the camera ID information and person ID information to construct cross camera sample pairs and performs cosine similarity constraint on them. The CCSC loss largely boost the performance of re-ID through the joint training with identification loss, and it is also superior than other metric learning losses in performance. Extensive experiments implemented on the standard benchmark datasets confirm the effectiveness of the proposed CCSC loss.

References

1. Chen, W., Chen, X., Zhang, J., Huang, K.: Beyond triplet loss: a deep quadruplet network for person re-identification. In: Proceedings of the IEEE Conference on Computer Vision and Pattern Recognition, pp. 403–412 (2017)
2. Chen, W., Chen, X., Zhang, J., Huang, K.: A multi-task deep network for person re-identification. In: Thirty-First AAAI Conference on Artificial Intelligence (2017)
3. Cheng, D., Gong, Y., Zhou, S., Wang, J., Zheng, N.: Person re-identification by multi-channel parts-based CNN with improved triplet loss function. In: Proceedings of the IEEE Conference on Computer Vision and Pattern Recognition, pp. 1335–1344 (2016)

4. Dai, Z., Chen, M., Zhu, S., Tan, P.: Batch feature erasing for person re-identification and beyond. arXiv preprint arXiv:1811.07130 (2018)
5. Deng, J., Dong, W., Socher, R., Li, L.J., Li, K., Fei-Fei, L.: ImageNet: a large-scale hierarchical image database. In: 2009 IEEE Conference on Computer Vision and Pattern Recognition, pp. 248–255. IEEE (2009)
6. He, K., Zhang, X., Ren, S., Sun, J.: Deep residual learning for image recognition. In: Proceedings of the IEEE Conference on Computer Vision and Pattern Recognition, pp. 770–778 (2016)
7. Hermans, A., Beyer, L., Leibe, B.: In defense of the triplet loss for person re-identification. arXiv preprint arXiv:1703.07737 (2017)
8. Kalayeh, M.M., Basaran, E., Gökmen, M., Kamasak, M.E., Shah, M.: Human semantic parsing for person re-identification. In: Proceedings of the IEEE Conference on Computer Vision and Pattern Recognition, pp. 1062–1071 (2018)
9. Kingma, D.P., Ba, J.: Adam: a method for stochastic optimization. arXiv preprint arXiv:1412.6980 (2014)
10. Li, W., Zhao, R., Xiao, T., Wang, X.: Deepreid: deep filter pairing neural network for person re-identification. In: Proceedings of the IEEE Conference on Computer Vision and Pattern Recognition, pp. 152–159 (2014)
11. Li, W., Zhu, X., Gong, S.: Person re-identification by deep joint learning of multi-loss classification. arXiv preprint arXiv:1705.04724 (2017)
12. Li, W., Zhu, X., Gong, S.: Harmonious attention network for person re-identification. In: Proceedings of the IEEE Conference on Computer Vision and Pattern Recognition, pp. 2285–2294 (2018)
13. Li, Z., Chang, S., Liang, F., Huang, T.S., Cao, L., Smith, J.R.: Learning locally-adaptive decision functions for person verification. In: Proceedings of the IEEE Conference on Computer Vision and Pattern Recognition, pp. 3610–3617 (2013)
14. Liao, S., Hu, Y., Zhu, X., Li, S.Z.: Person re-identification by local maximal occurrence representation and metric learning. In: Proceedings of the IEEE Conference on Computer Vision and Pattern Recognition, pp. 2197–2206 (2015)
15. Liu, H., Feng, J., Qi, M., Jiang, J., Yan, S.: End-to-end comparative attention networks for person re-identification. IEEE Trans. Image Process. **26**(7), 3492–3506 (2017)
16. Ristani, E., Solera, F., Zou, R., Cucchiara, R., Tomasi, C.: Performance measures and a data set for multi-target, multi-camera tracking. In: Hua, G., Jégou, H. (eds.) ECCV 2016. LNCS, vol. 9914, pp. 17–35. Springer, Cham (2016). https://doi.org/10.1007/978-3-319-48881-3_2
17. Si, J., et al.: Dual attention matching network for context-aware feature sequence based person re-identification. In: Proceedings of the IEEE Conference on Computer Vision and Pattern Recognition, pp. 5363–5372 (2018)
18. Sun, Y., Zheng, L., Deng, W., Wang, S.: SVDNet for pedestrian retrieval. In: Proceedings of the IEEE International Conference on Computer Vision, pp. 3800–3808 (2017)
19. Sun, Y., Zheng, L., Yang, Y., Tian, Q., Wang, S.: Beyond part models: person retrieval with refined part pooling (and a strong convolutional baseline). In: Ferrari, V., Hebert, M., Sminchisescu, C., Weiss, Y. (eds.) ECCV 2018. LNCS, vol. 11208, pp. 501–518. Springer, Cham (2018). https://doi.org/10.1007/978-3-030-01225-0_30
20. Varior, R.R., Haloi, M., Wang, G.: Gated siamese convolutional neural network architecture for human re-identification. In: Leibe, B., Matas, J., Sebe, N., Welling, M. (eds.) ECCV 2016. LNCS, vol. 9912, pp. 791–808. Springer, Cham (2016). https://doi.org/10.1007/978-3-319-46484-8_48

21. Wang, C., Zhang, Q., Huang, C., Liu, W., Wang, X.: Mancs: a multi-task attentional network with curriculum sampling for person re-identification. In: Ferrari, V., Hebert, M., Sminchisescu, C., Weiss, Y. (eds.) ECCV 2018. LNCS, vol. 11208, pp. 384–400. Springer, Cham (2018). https://doi.org/10.1007/978-3-030-01225-0_23

22. Wang, G., Yuan, Y., Chen, X., Li, J., Zhou, X.: Learning discriminative features with multiple granularities for person re-identification. In: 2018 ACM Multimedia Conference on Multimedia Conference, pp. 274–282. ACM (2018)

23. Wei, L., Zhang, S., Gao, W., Tian, Q.: Person transfer gan to bridge domain gap for person re-identification. In: Proceedings of the IEEE Conference on Computer Vision and Pattern Recognition, pp. 79–88 (2018)

24. Wei, L., Zhang, S., Yao, H., Gao, W., Tian, Q.: Glad: global-local-alignment descriptor for pedestrian retrieval. In: Proceedings of the 25th ACM International Conference on Multimedia, pp. 420–428. ACM (2017)

25. Wen, Y., Zhang, K., Li, Z., Qiao, Y.: A discriminative feature learning approach for deep face recognition. In: Leibe, B., Matas, J., Sebe, N., Welling, M. (eds.) ECCV 2016. LNCS, vol. 9911, pp. 499–515. Springer, Cham (2016). https://doi.org/10.1007/978-3-319-46478-7_31

26. Zhang, X., et al.: Alignedreid: surpassing human-level performance in person re-identification. arXiv preprint arXiv:1711.08184 (2017)

27. Zhao, L., Li, X., Zhuang, Y., Wang, J.: Deeply-learned part-aligned representations for person re-identification. In: Proceedings of the IEEE International Conference on Computer Vision, pp. 3219–3228 (2017)

28. Zheng, L., Huang, Y., Lu, H., Yang, Y.: Pose invariant embedding for deep person re-identification. IEEE Trans. Image Process. (2019)

29. Zheng, L., Shen, L., Tian, L., Wang, S., Wang, J., Tian, Q.: Scalable person re-identification: a benchmark. In: Proceedings of the IEEE International Conference on Computer Vision, pp. 1116–1124 (2015)

30. Zheng, L., Yang, Y., Hauptmann, A.G.: Person re-identification: past, present and future. arXiv preprint arXiv:1610.02984 (2016)

31. Zheng, Z., Zheng, L., Yang, Y.: Unlabeled samples generated by gan improve the person re-identification baseline in vitro. In: Proceedings of the IEEE International Conference on Computer Vision, pp. 3754–3762 (2017)

32. Zhong, Z., Zheng, L., Cao, D., Li, S.: Re-ranking person re-identification with k-reciprocal encoding. In: Proceedings of the IEEE Conference on Computer Vision and Pattern Recognition, pp. 1318–1327 (2017)

33. Zhong, Z., Zheng, L., Kang, G., Li, S., Yang, Y.: Random erasing data augmentation. arXiv preprint arXiv:1708.04896 (2017)

Tracker-Level Decision by Deep Reinforcement Learning for Robust Visual Tracking

Wenju Huang, Yuwei Wu$^{(\boxtimes)}$, and Yunde Jia

Beijing Laboratory of Intelligent Information Technology,
School of Computer Science, Beijing Institute of Technology,
Beijing 100081, People's Republic of China
{huangwenju,wuyuwei,jiayunde}@bit.edu.cn

Abstract. In this paper, we formulate the multi-tracker tracking problem as a decision-making task and train an expert by the deep reinforcement learning (DRL) to select the best tracker. Specifically, the expert takes the response map of the tracker as input and outputs a scalar to indicate the reliability of the tracker. With the DRL, the expert can make full use of complementary information among base trackers. Furthermore, under the guidance of the deep expert, base trackers update themselves adaptively to capture the changes of object appearance and prevent corruption. The experimental results on public tracking benchmarks demonstrate that the proposed method outperforms the state-of-the-art methods.

Keywords: Visual object tracking · Deep reinforcement learning · Tracker selection

1 Introduction

Visual object tracking is an important task in computer vision with a variety of applications, such as human-computer interaction, security and surveillance, traffic control, and so on [26,27]. Although many efforts have been made in the past decades, object tracking is still a challenging task. This is because there are various unfavorable factors in the tracking scenarios that a single tracker can't perfectly deal with. However, we observe that different trackers usually contain different and complementary information. For example, the tracker with features in the latter layers of deep convolutional neural networks (CNNs) captures strong semantic information, while the tracker with features in the earlier layers of CNNs captures more spatial details [17]. Therefore, exploiting the complementary information among different trackers provides a useful framework to deal with the challenge [18,25].

The existing multi-tracker based methods can be divided into two kinds. One mainly uses a fixed weight [13,23] or an attention model [7] to fuse different trackers. However, the fusion method requires most of its base trackers to

Y. Zhao et al. (Eds.): ICIG 2019, LNCS 11901, pp. 442–453, 2019.
https://doi.org/10.1007/978-3-030-34120-6_36

work correctly, which is difficult to achieve in complex tracking scenarios. The other uses a specific criterion to select the best tracker, such as the entropy minimization [29], the consistency degree between the different trackers [18,25], and the peak value of a response map [1]. However, these selection strategies lack foresight, because they only evaluate the performance of the trackers in the current frame, without considering that in the subsequent frames. The manual design methods are limited in the representation capability and can't take all the impact factors into account. Thus, these methods do not take full advantage of the complementary information among base trackers, and their results are suboptimal.

Recently, The advantage of model-free deep reinforcement learning (DRL) has been demonstrated on a range of challenging decision-making tasks [19]. DRL aims to learn a policy that maximizes the expected reward by interacting with the environment. This characteristic also makes the DRL competent for the task of tracker selection.

In this paper, we formulate the multi-tracker tracking problem as a decision-making task, in which an expert (called the agent in DRL) is trained by DRL to select the best base tracker to process the current frame. Our framework can be separated into the tracker pool part and the expert part. The tracker pool part consists of a series of base trackers and works as an environment, while the expert part is a deep network and works as an agent, as illustrated in Fig. 1. The tracker pool regresses the input image patch to a series of response maps. Then, these response maps are input into the expert network to estimate their value. Finally, the response map with the highest value will be taken to search the target and update base trackers for the next frame. The contributions of our method can be summarized as three-fold:

(1) We formulate the multi-tracker tracking problem as a decision-making task in which the tracker selection is executed by the deep expert. This impels the expert to consider the long-term expected performance of the tracker when making decisions, rather than the immediate performance, which makes the results more reliable.

(2) We train the deep expert offline by combining the supervised learning and deep reinforcement learning to estimate the reliability of base trackers. Benefiting from the CNNs, the expert has strong expressive ability to describe the complex relationship between the response map and the reliability of tracker.

(3) The deep expert guides base trackers to update themselves adaptively during online tracking to capture the changes of object appearance and prevent corruption.

The results on OTB-2013 [26], OTB-100 [27] and VOT2017 [15] benchmarks show that the proposed method takes full advantage of the complementary information among base trackers and outperforms the existing multi-tracker based methods.

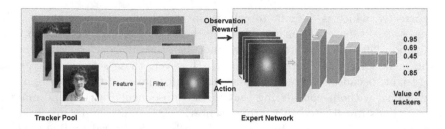

Fig. 1. The framework of the proposed method. The base tracker part consists of a series of correlation filter based trackers built in different feature spaces. The expert network observes the state offered by base trackers and outputs an action to select the best tracker.

2   Related Work

In this section, we briefly review the related works from two aspects: DRL based methods and multi-tracker based methods.

2.1   DRL Based Methods

Tracking an object always comes with a lot of decisions about the selection of tracker components. Such decision-making can be done through reinforcement learning. Yun et al. [28] focused on the motion model of the tracker and proposed ADNet in which a deep reinforcement network is used to generate a series of more and more precise candidates by iteration. Huang et al. [14] focused on feature extractor and proposed an early-stopping tracker, in which an agent is trained by reinforcement learning to select the cheap features for easy frames and the expensive deep features for challenging frames. In [6], an agent is trained to choose the best template for tracker from the template pool. Supancic et al. [24] focused on model update and formulated tracking as a partially observable decision-making process, in which the tracker must decide whether to update or not, or to reinitialize.

Different from the above methods, the proposed method in this paper focuses on the post-processor ensemble.

2.2   Multi-tracker Based Methods

Zhang et al. [29] introduced a multi-expert tracking framework (MEEM), in which an expert committee is constituted by a discriminative tracker and its former snapshots, and the best expert is selected according to the minimum entropy criterion to restore the tracking when there is a disagreement among the experts. Bai et al. [1] proposed a heterogeneous cue fusion method, where multi-templates, multi-features and multi-scales are integrated to further improve the discriminative ability of the filters. Otherwise, Wang et al. [25] proposed a multi-cue tracker, which integrates multi-features at the decision-level. Nam et al. [21]

proposed to deal with multi-modality in target appearances by modeling multiple CNNs in a tree structure.

However, the performances of the above methods are suboptimal because of the handcrafted fusion strategy. In this paper, we adopt the deep reinforcement network to dynamically select base trackers, which can maximize the role of each tracker.

3 Method

3.1 Definition of Expert Network

The goal of the agent in RL is to learn a policy function $\pi : \mathcal{S} \rightarrow \mathcal{A}$ to select the best tracker, where $\mathcal{S}$ is the state space and $\mathcal{A}$ is the action space. In our case, the action to select the based tracker does not output directly from the expert. Instead, we exploit expert to estimate the value of each base tracker according the value function it learns $V_{\pi} : \mathcal{S} \rightarrow \mathcal{V}$, where $\mathcal{V} \subset \mathbb{R}$ is the value space. Then the action to select the best tracker can be defined as

$$a = \arg \max_{i}(V_{\pi}(\mathbf{s}^{(i)}), a \in \mathcal{A}, \tag{1}$$

where $\mathbf{s}^{(i)} \in \{\mathbf{s}^{(1)}, \mathbf{s}^{(2)}, ..., \mathbf{s}^{(m)}\}$ is the state of base tracker.

State. The state $\mathbf{s}$ in our framework is a single response map of the CF-based tracker. Because the target output of the CF-based tracker is a Gauss distribution, the tracker will be more reliable if the output response map is more like Gauss distribution, such as has high peak and steep slope, as illustrated in Fig. 2a. The Peak-to-Sidelobe ratio (PSR) used in previous work for adaptive tracking [4] and fusion tracking [25] is also based on this property. However, the calculation of PSR only involves the peak value, mean value, and variance of the response map.

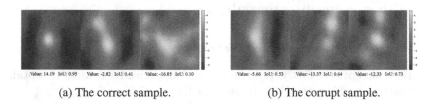

(a) The correct sample.   (b) The corrupt sample.

Fig. 2. Relationship between the response map, estimated value and Intersection-over-Union (IoU). (a) The correct sample. The high estimate value and IoU appear in map of Gaussian distribution. (b) The corrupt sample. The high IoUs appear in the response map of irregularity distribution, but the estimate values are still correct.

Value Function. Different from the general reinforcement learning methods which estimate a state-action value function (Q-function) to measure the value of executing a given action in a given state, we only estimate a state value function (Value function) to measure the value of a given state. Value function under a policy π is defined formally as

$$V_\pi(\mathbf{s}) = \mathbb{E}_\pi \left[\sum_{k=0}^{\infty} \gamma^k r_{t+k+1} \middle| \mathbf{s}_t = \mathbf{s} \right], \tag{2}$$

where r is the reward return from the environment and $\gamma \in [0,1]$ is a discounting rate. Because the reward to the agent depends on the action it takes, the value function is defined as the expected discounted return with respect to the policy.

The state space in our framework is enormous, which means we cannot expect to find the optimal value function using finite time and data. The usual way to solve this problem is to find a good approximate solution. The value function in (2) can be approximated by a deep network with parameter θ, which is called the expert network in this paper.

$$V_\pi(\mathbf{s}) \approx V(\mathbf{s}|\boldsymbol{\theta}). \tag{3}$$

The expert network can be solved by the gradient descent methodology, which will be described in Sect. 3.2.

Reward Function. The reward function $r(\mathbf{s}, a, \mathbf{s}')$ represents the consequences of the action a. The goal of the agent is to maximize the cumulative reward it receives. Since each state has a ground-truth and a predicted box, the reward function can be defined frame by frame as

$$r(\mathbf{s}, a, \mathbf{s}') = \begin{cases} 1, & IoU(\mathbf{b}, \mathbf{g}) >= 0.7 \\ 0, & 0.4 <= IoU(\mathbf{b}, \mathbf{g}) < 0.7 \\ -1, & IoU(\mathbf{b}, \mathbf{g}) < 0.4 \end{cases}, \tag{4}$$

where $IoU(\mathbf{b}, \mathbf{g}) = area(\mathbf{b} \cap \mathbf{g})/area(\mathbf{b} \cup \mathbf{g})$.

3.2   Offline Training of Expert Network

The expert network is firstly pre-trained by supervised learning to obtain a reliable initiation. Then the DRL is used for final training.

Pre-training. It is wildly known that the deep reinforcement learning network is hard in convergence due to the large state and action space. The usual methods to overcome it is to assign a reliable initiation to the deep network before training it with reinforcement learning [28]. Before applying reinforcement learning, we use supervised learning to pre-train our expert network offline. The output of our expert network can be approximated by the IoU between the predicting bounding box and the ground truth. Thus, the expert network can be pre-trained by minimizing the following loss,

$$\mathcal{L}_{sur} = \frac{1}{N} \sum_i (V(\mathbf{s}_i|\boldsymbol{\theta}) - y_i)^2, \tag{5}$$

where N is the number of training samples in a mini-batch, y is the IoU between the predict boundary box of the input response map and the ground truth.

Note that, the network trained by (5) can only output a coarse estimate of the state value, not the real estimate, for two reasons. One is that there are some noises in the corresponding relationship between IoU and the response map. For example, some response maps don't have the Gaussian-like distribution but have a high IoU, as shown in Fig. 2b. The other is that the IoU only represents the result of the current frame, but the value function focuses on not only the current return but the future return. The unequal relationship makes IoU fail to provide accurate guidance for network training. These two drawbacks in supervised learning can be well addressed by reinforcement learning.

Training with Deep Reinforcement Learning. After pre-training of supervised learning, we train our network with DRL. We use the temporal-difference based reinforcement learning method to train our expert network, which uses the current estimate value v instead of the true value v_π to formula the target as $r + \gamma v(\mathbf{s}')$. We adopt the strategies in [19] that uses a separate network for generating the targets and use an experience pool to remove data correlation. After randomly sampling N pairs of $(\mathbf{s}_i, a_i, r_i, \mathbf{s}'_i)$ samples from experience pool, the target of the training process is to minimize the following loss

$$\mathcal{L} = \frac{1}{N} \sum_i (r_i + \gamma V'(\mathbf{s}'_i|\boldsymbol{\theta}') - V(\mathbf{s}_i|\boldsymbol{\theta}))^2, \tag{6}$$

where V' with parameter θ' is the target network of V. Note that there are several temporary states $\{\mathbf{s}_t^{(1)}, \mathbf{s}_t^{(2)}, ..., \mathbf{s}_t^{(m)}\}$ generate at frame t (m is the number of based tracker). These states are estimated by the expert network and the state with the highest value will be selected as the final state s_t stored in the experience pool.

3.3   Online Tracking

In the online tracking, a search patch $\mathbf{p}_t$ is firstly cropped from the current frame F_t according to the previously estimated boundary box $\mathbf{b}_{t-1}$ and the pre-processing function $\mathbf{p}_t = \phi(\mathbf{b}_{t-1}, F_t)$. The patch $\mathbf{p}_t$ is then sent into the correlation filter base trackers and regressed into a series of response maps $S_t = \{\mathbf{s}_t^{(1)}, \mathbf{s}_t^{(2)}, ..., \mathbf{s}_t^{(m)}\}$. The values of these response maps are estimated by the expert network according to the equation (3). Next, the best tracker can be selected by the action defined in (1). Finally, the scale of the boundary box can be identified by the scale estimation method introduced in [9]. After process a frame, all base trackers are updated using the same template of the new target and share the same searching areas (ROI) in the next frame.

3.4   Adaptive Updating

In order to avoid the corruption of the tracker when online updating, we should carefully check the updated template. In our experiment, the value of the response map, which measures the quality of the tracking result, is a good adaptive update indicator. Thus, the learning rate of tracker update can be defined as

$$C = \left(\frac{\bar{V}(\mathbf{s}) - T_{min}}{T_{max} - T_{min}} \right) \cdot \eta', \tag{7}$$

where

$$\bar{V}(\mathbf{s}) = \begin{cases} T_{max}, & V(\mathbf{s}) > T_{max} \\ T_{min}, & V(\mathbf{s}) < T_{min} \\ V(\mathbf{s}), & otherwise \end{cases} \tag{8}$$

η' is the learning rate of model learning in the standard DCF, T_{max} and T_{min} are the upper and lower update boundary respectively.

4   Experiment

4.1   Implementation

Network Architecture. Because of the success of lightweight deep network in object tracking [22,28], we build our expert network with four convolution layers and three fully-connected layers (the last one is the output layer). The convolution layers have the same structure and initial parameters as the first three layers of VGG-M network [5]. The next three fully-connected layer has 256, 128 and 1 output units respectively. In addition to the output layer, each layer is initialized by the Xavier initializer and activated by the ReLU function.

Tracker Pool. Because of the good performance and high tracking speed of DCFs in object tracking, we take it as our base tracker. We implement our tracker pool following the setting in [25]. In order to capture low-level details and high-level semantic information meantime, the tracker pool consists of three types of feature: $\{Low, Middle, High\}$, where the low-level feature is the HOG feature, the middle-level feature is the conv4-4 layer of VGG-19, and high-level feature is the conv5-4 layer of VGG-19. Then these three types feature are optionally combined into $C_3^1 + C_3^2 + C_3^3 = 7$ kinds of base trackers. The hyperparameter setting of base trackers is consistent with [25].

Hyperparameters. To train our Expert Network offline, we use TC-128 [16] and UAV123 [20] as the training set, eliminating the same video in the test set. We randomly choose a video clip with a length less than 400 frames for each iteration.

The possibility of using the expert decision η is increased linearly from 0.5 to 0.9 over the first 100 thousand iterations, then is fixed. The reward decay rate γ

is set to 0.99. The target network is soft-updated with a learning rate of 0.001, while the evaluation network is optimized by the Adam and the learning rate is 0.001. The replay buffer size is 10^5 and the training batch size is 32. We train the expert network through 10,000 iterations of supervised learning and 250,000 iterations of reinforcement learning. In online tracking, the update boundary T_{max} and T_{min} is set to 8.0 and 2.0 respectively. Our algorithm is implemented in Python and runs with a single Titanx GPU.

4.2 Ablation Analysis

To verify the contributions of the different component of the proposed method, we implement and evaluate the proposed method with and without adaptive update, donated by Ours and Ours-NU respectively. We also compare them with the adaptive update version (MCCT) and the non-adaptive update version (MCCT-NU) of the baseline [25]. The ablation analyses mainly do on the OTB-2013 and OTB-100 datasets, using area-under-curve (AUC) of success plot and distance precision (DP) rate for as Metrics.

From the results in Table 1, we can observe that our methods significantly improve the performance of T_7, the best individual tracker, which illustrates the effectiveness of our method. When using our Expert Network only for tracker selection, Ours-NU outperforms T_7 with a gain of 3.4% in DP and 2.4% in AUC on OTB-2013, and 2.4% in DP and 1.6% in AUC on OTB100. When using our Expert Network for both tracker selection and adaptive update, the performance of the best base tracker is improved 3.6% in DP and 3.0% in AUC on OTB-2013, and 5.2% in DP and 3.7% in AUC on OTB100.

In comparison with the baseline, regardless of adaptive update, our Ours-NU method outperforms MCCT-NU with a gain of 1.5% in AUC and 2.3% in DP on OTB-2013, and 1.9% in AUC and 2.3% in DP on OTB-100. This means that our selection method can make better use of complementary information among base trackers than the handcrafted selection method in MCCT.

Table 1. Ablation analysis of the proposed method on the OTB-2013 [26] and OTB-2015 [27] datasets. The DP (@20px) and AUC scores are used for evaluation. The best result is marked in the **bold** font.

		T_7	MCCT-NU [25]	Ours-NU [25]	MCCT	Ours
OTB-2013	AUC	0.676	0.685	0.700	0.701	**0.706**
	DP	0.895	0.906	0.929	0.930	**0.931**
OTB-100	AUC	0.651	0.648	0.669	0.684	**0.688**
	DP	0.867	0.868	0.891	0.916	**0.919**

4.3   Tracker Selection VS Tracker Fusion

In order to compare the difference between the tracker selection and fusion strategy, we implement a tracker fusion strategy in our multi-tracker framework. The fusion weight of the ith tracker is defined as $w^{(i)} = V(s^{(i)}|\theta)/\sum_i V(s^{(i)}|\theta)$. Then, we can obtain the fusion response map by $R_f = \sum_i w^{(i)} R^{(i)}$. We evaluate these two version method on the 11 attributes of OTB-2013 [26] in terms of DP (@20px) and AUC scores. On the fast motion (FM), the in-plane rotation (IPR), the out-of-view (OV) and the background clutters (BC) these four attributes, the performance of tracker selection is obviously superior to the tracker fusion. Because in these attributes, the target changes dramatically, which makes most base trackers drift and affect the fusion results, as shown in Fig. 3. However, the tracker fusion outperforms tracker selection on the deformation (DEF), the out-of-plane rotation (OPR) and the low resolution (LR) these three attributes. Because most base trackers in these attributes have a good performance, so we can get a better performance with the fusion strategy, as shown in Fig. 3.

The above results show that the tracker fusion is suitable for simple tracking scenarios, while the tracker selection performs better in complex scenarios.

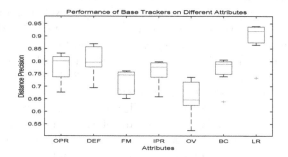

Fig. 3. The performance of base trackers on different attributes.

4.4   State-of-the-Art Comparisons

Evaluation on OTB Benchmarks. The proposed approach is compared with the state-of-the-art methods, including multi-tracker based methods MCCT [25], SCT4 [8], HDT [23], ACFN [7], MEEM [29], BranchOut [13], and other outstanding methods ADNet [28], CCOT [12], DSST [9], SRDCFdecon [11], CF2 [17], SiamFC [3], on both OTB-2013 and OTB-100 benchmarks by one-pass evaluation (OPE). The results of top ten trackers are shown in Fig. 4a and b. The comparison result shows that our method is better than the other trackers.

Figure 4a illustrates the precision and success plots of the OTB-2013 benchmark. The AUC score of the success plots of our method is 0.706, which outperforms the two existing best multi-tracker based trackers BranchOut and MCCT.

The improvement range is 0.4% and 0.5%, respectively. The DP (@20px) of the precision plot of our method is 0.931, which outperforms the MCCT, but is slightly worse than the BranchOut. Because base trackers of BranchOut are trained end-to-end, which are more superior than ours.

From the precision and success plots of OTB-100 benchmark illustrated in Fig. 4b, we can see that our method obtains the best results in both success and precision plots. Specifically, our method achieves an AUC score of 0.688 and a precision score of 0.919, which outperforms both BranchOut (0.678/0.917) and MCCT (0.684/0.916). The results of the OTB benchmark demonstrate the advantages of our selection strategy of multi-tracker.

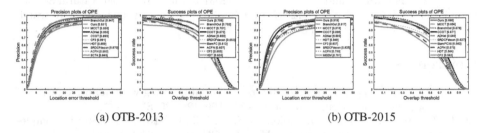

(a) OTB-2013   (b) OTB-2015

Fig. 4. The results of the OTB benchmark. The numbers in the legend are the precisions at 20 pixels and the area-under-curve scores, respectively for the precision plots and the success plots.

Evaluation on VOT2017 Benchmark. We also compare the proposed method with MCCT [25], SiamFC [3], CCOT [12], DSST [9], MEEM [29], SRDCF [10], Staple [2], UCT [30] on VOT2017. The expected average overlap (EAO), accuracy and failure times are used to evaluate these trackers, as shown in Tab. 2. We can see that our method obtains an EAO of 0.293, which outperforms the second best tracker MCCT with a gain of 2.3%. What's more, the average failure of our tracker is 18.29, which is the lowest among the above trackers. Although our method is slightly worse than MCCT in accuracy, it still has a comparable score of 0.5. The results of the VOT2017 benchmark also demonstrate the competitive performance of the proposed method.

Table 2. A comparison with the state-of-the-art trackers on VOT2017 dataset. The first and second best scores are highlighted in color. The ↑ represents the highest is the best, while the ↓ represents the lowest is the best.

Tracker	MCCT [25]	CCOT [12]	UCT [30]	MEEM [29]	SiamFC [3]	Staple [2]	SRDCF [10]	DSST [9]	Ours
EAO ↑	0.2703	0.2671	0.2058	0.1925	0.1880	0.1694	0.1189	0.0788	0.2932
Accuracy ↑	0.5283	0.4944	0.4921	0.4630	0.5029	0.5296	0.4903	0.3947	0.5000
Failures ↓	19.4526	20.4138	29.7991	33.6046	34.0259	44.0194	64.1136	95.5587	18.2966

5   Conclusions

In this work, we have formulated the multi-tracker tracking method as a decision-making task, where an expert is trained offline by deep reinforcement learning to select the best tracker for the current frame and guide the trackers to update. Formulating in decision-making process brings the foresight and strong representation to our method, which enables it to take full advantage of the complementary information among base trackers. The experimental results demonstrate the superiority of our method.

Acknowledgments. This work was supported in part by the Natural Science Foundation of China (NSFC) under Grant No. 61702037 Beijing Municipal Natural Science Foundation under Grant No. L172027, and Beijing Institute of Technology Research Fund Program for Young Scholars.

References

1. Bai, B., et al.: Kernel correlation filters for visual tracking with adaptive fusion of heterogeneous cues. Neurocomputing **286**, 109–120 (2018)
2. Bertinetto, L., Valmadre, J., Golodetz, S., Miksik, O., Torr, P.H.: Staple: complementary learners for real-time tracking. In: CVPR, pp. 1401–1409 (2016)
3. Bertinetto, L., Valmadre, J., Henriques, J.F., Vedaldi, A., Torr, P.H.S.: Fully-convolutional siamese networks for object tracking. In: Hua, G., Jégou, H. (eds.) ECCV 2016. LNCS, vol. 9914, pp. 850–865. Springer, Cham (2016). https://doi.org/10.1007/978-3-319-48881-3_56
4. Bolme, D.S., Beveridge, J.R., Draper, B.A., Lui, Y.M.: Visual object tracking using adaptive correlation filters. In: CVPR, pp. 2544–2550. IEEE (2010)
5. Chatfield, K., Simonyan, K., Vedaldi, A., Zisserman, A.: Return of the devil in the details: delving deep into convolutional nets. arXiv preprint arXiv:1405.3531 (2014)
6. Choi, J., Kwon, J., Lee, K.M.: Real-time visual tracking by deep reinforced decision making. Comput. Vis. Image Underst. **171**, 10–19 (2018)
7. Choi, J., Chang, H.J., Yun, S., Fischer, T., Demiris, Y., Choi, J.Y., et al.: Attentional correlation filter network for adaptive visual tracking. In: CVPR, vol. 2, p. 7 (2017)
8. Choi, J., Jin Chang, H., Jeong, J., Demiris, Y., Young Choi, J.: Visual tracking using attention-modulated disintegration and integration. In: CVPR, pp. 4321–4330 (2016)
9. Danelljan, M., Häger, G., Khan, F., Felsberg, M.: Accurate scale estimation for robust visual tracking. In: British Machine Vision Conference, Nottingham, 1–5 September 2014. BMVA Press (2014)
10. Danelljan, M., Hager, G., Shahbaz Khan, F., Felsberg, M.: Learning spatially regularized correlation filters for visual tracking. In: ICCV, pp. 4310–4318 (2015)
11. Danelljan, M., Hager, G., Shahbaz Khan, F., Felsberg, M.: Adaptive decontamination of the training set: a unified formulation for discriminative visual tracking. In: CVPR, pp. 1430–1438 (2016)
12. Danelljan, M., Robinson, A., Shahbaz Khan, F., Felsberg, M.: Beyond correlation filters: learning continuous convolution operators for visual tracking. In: Leibe, B., Matas, J., Sebe, N., Welling, M. (eds.) ECCV 2016. LNCS, vol. 9909, pp. 472–488. Springer, Cham (2016). https://doi.org/10.1007/978-3-319-46454-1_29

13. Han, B., Sim, J., Adam, H.: Branchout: regularization for online ensemble tracking with convolutional neural networks. In: ICCV, pp. 2217–2224 (2017)
14. Huang, C., Lucey, S., Ramanan, D.: Learning policies for adaptive tracking with deep feature cascades. In: IEEE International Conference on Computer Vision (ICCV), pp. 105–114 (2017)
15. Kristan, M., et al.: The visual object tracking vot2017 challenge results (2017)
16. Liang, P., Blasch, E., Ling, H.: Encoding color information for visual tracking: algorithms and benchmark. IEEE Trans. Image Process. **24**(12), 5630–5644 (2015)
17. Ma, C., Huang, J.B., Yang, X., Yang, M.H.: Hierarchical convolutional features for visual tracking. In: ICCV, pp. 3074–3082 (2015)
18. Meshgi, K., Oba, S., Ishii, S.: Efficient diverse ensemble for discriminative co-tracking. In: CVPR, pp. 4814–4823 (2018)
19. Mnih, V., et al.: Human-level control through deep reinforcement learning. Nature **518**(7540), 529 (2015)
20. Mueller, M., Smith, N., Ghanem, B.: A benchmark and simulator for UAV tracking. In: Leibe, B., Matas, J., Sebe, N., Welling, M. (eds.) ECCV 2016. LNCS, vol. 9905, pp. 445–461. Springer, Cham (2016). https://doi.org/10.1007/978-3-319-46448-0_27
21. Nam, H., Baek, M., Han, B.: Modeling and propagating CNNs in a tree structure for visual tracking. arXiv preprint arXiv:1608.07242 (2016)
22. Nam, H., Han, B.: Learning multi-domain convolutional neural networks for visual tracking. In: CVPR, pp. 4293–4302 (2016)
23. Qi, Y., et al.: Hedged deep tracking. In: CVPR, pp. 4303–4311 (2016)
24. Supancic III, J.S., Ramanan, D.: Tracking as online decision-making: Learning a policy from streaming videos with reinforcement learning. In: ICCV, pp. 322–331 (2017)
25. Wang, N., Zhou, W., Tian, Q., Hong, R., Wang, M., Li, H.: Multi-cue correlation filters for robust visual tracking. In: CVPR, pp. 4844–4853 (2018)
26. Wu, Y., Lim, J., Yang, M.H.: Online object tracking: a benchmark. In: CVPR, pp. 2411–2418 (2013)
27. Wu, Y., Lim, J., Yang, M.H.: Object tracking benchmark. T-PAMI **37**(9), 1834–1848 (2015)
28. Yoo, S., Yun, K., Choi, J.Y., Yun, K., Choi, J.: Action-decision networks for visual tracking with deep reinforcement learning. In: CVPR (2017)
29. Zhang, J., Ma, S., Sclaroff, S.: MEEM: robust tracking via multiple experts using entropy minimization. In: Fleet, D., Pajdla, T., Schiele, B., Tuytelaars, T. (eds.) ECCV 2014. LNCS, vol. 8694, pp. 188–203. Springer, Cham (2014). https://doi.org/10.1007/978-3-319-10599-4_13
30. Zhu, Z., Huang, G., Zou, W., Du, D., Huang, C.: Uct: learning unified convolutional networks for real-time visual tracking. In: ICCV, pp. 1973–1982 (2017)

Hierarchical Salient Object Detection Network with Dense Connections

Qing Zhang$^{(\boxtimes)}$, Jianchen Shi, Baochuan Zuo, Meng Dai, Tianzhen Dong, and Xiao Qi

Shanghai Institute of Technology, Shanghai 201418, China
zhangqing0329@gmail.com

Abstract. Fully convolutional neural networks (FCNs) have shown outstanding performance in many dense labeling tasks. FCN-like salient object detection models haven mostly developed lately. In the work, we propose a novel pixel-wise salient object detection network based on FCN by aggregating multi-level feature maps. Our model first makes a coarse prediction by automatically learning various saliency cues, including color and texture contrast, shapes and objectness. Then a densely connected feature extraction block is adopted to further extract rich features at each resolution. Moreover, skip-layer structure is introduced for providing a better feature representation and helping shallow side outputs locate salient objects. In addition, a weighted-fusion module is utilized to combine multi-level features. Finally, a fully connected CRF model can be optimally incorporated to improve spatial coherence and contour localization in the fused saliency map. The whole architecture works in a coarse to fine manner. Evaluations on five benchmark datasets and comparisons with 10 state-of-the-art algorithms demonstrate the robustness and efficiency of our proposed model.

Keywords: Salient object detection · Visual saliency detection · Deep learning · Feature extraction

1 Introduction

Salient object detection aims at modeling human visual attention mechanism to segment the most distinct regions or objects from the clutter backgrounds. It has received a great deal of attention in computer vision community because of its wide range of applications including video summarization [1], content-aware image cropping and resizing [3,4] and person re-identification [2].

Since the seminal approaches of Itti et al. [5] and Liu et al. [6] are reported, extensive visual saliency algorithms have been proposed to simulate human

This work is supported by Natural Science Foundation of Shanghai under Grant Nos. 19ZR1455300 and 19ZR1455200, Science and Technology Development Foundation of Shanghai Institute of Technology under Grant No. ZQ2018-23, and National Natural Science Foundation of China under Grant No. 61806126.

© Springer Nature Switzerland AG 2019
Y. Zhao et al. (Eds.): ICIG 2019, LNCS 11901, pp. 454–466, 2019.
https://doi.org/10.1007/978-3-030-34120-6_37

visual attention mechanism in images and videos. Traditional salient object detection methods [7–10] adopt heuristic priors and manually designed features which are usually considered as low-level information. These generic techniques are useful for keeping fine images structures. However, these models cannot generate satisfied predicted results and are less applicable to a wide range of problems in practice. For example, it is difficult to pop out the salient objects when the background and salient objects share similar attributes (See the first row of Fig. 1). Moreover, it might fail sometimes, when there are multiple salient objects (See the second row of Fig. 1).

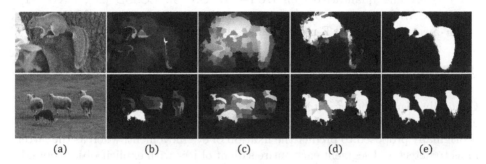

(a)   (b)   (c)   (d)   (e)

Fig. 1. Comparisons of results of different kinds of methods. For input images in (a), we show the salient object detection results of methods based on handcrafted features in (b) [10] and (c) [8], and salient object detection results of methods based on deep features in (d) [25] and (e) Ours.

In recent years, fully convolutional networks have shown powerful ability of feature representation and obtained impressive results in many dense labeling tasks including semantic segmentation [11,12], edge detection [14,15] and pose estimation [13]. Inspired by these achievements, researchers in the saliency detection community attempt to utilize its ability of adaptively extracting semantic features from raw images. These FCN-based models [16–18] have been successful in overcoming the disadvantages of handcrafted feature-based approaches and capturing high-level information about the objects and their clutter background, thus achieving better performance. However, although the saliency model using high-level information is superior, the low-level and mid-level features are also important in detecting salient objects. Therefore, it is a key and challenging issue to effectively and simultaneously aggregate multi-level saliency cues in a unified learning framework for capturing both the semantic objectness and detailed structure.

Motivated by these discussions, we propose a simple but effective salient object detection model for the pixel-wise saliency prediction task to simultaneously aggregate multi-level features to capture distinctive objectness and detailed information on complex images.

The main contributions are summarized as follows:

(1) A novel FCN-based saliency detection network model is proposed, which aggregates multi-level features as saliency cues. It performs image-to-image prediction and learns powerful and rich feature representations on complex images.
(2) We utilize the skip-layer scheme to guide low-level feature learning. With the help of deeper side information, shallower side outputs refine their predictions with more accurate location.
(3) The proposed model achieves state-of-the-art performance both quantitatively and qualitatively on DUT-OMRON [9], ECSSD [20], HKU [21], PASCAL-S [19] and SOD [34] benchmark datasets in terms of PR curves, F-measure, weighted F-measure and MAE scores.

2   Related Work

Generally, visual saliency detection approaches can be roughly classified into two categories: human fixation prediction and salient object detection. The former [5] is originally proposed to predict the fixation of eye movement, whereas the latter aims to detect and segment each entire salient object with explicit object boundaries from surroundings. Since this paper is focused on salient object detection based on deep learning, we will briefly review existing representative approaches for salient object detection.

2.1   Handcrafted Features Based Models

The majority of salient object detection approaches usually utilize handcrafted pixel/superpixel-level features, such as color, texture and orientation, by either local or global manner. The local based methods use rarity, contrast or distinctiveness of each pixel/region to capture the pixels/regions locally standing out from their surroundings, while the global based methods estimate the saliency of each pixel or region by using holistic priors of the entire image. Some researchers propose to build graphical models of superpixels to implicitly compute contrast [9,20]. They compute saliency by means of background, center, and compactness priors. However, traditional approaches, which mainly rely on handcrafted features, cannot describe semantic feature representation, therefore, they may fail to pop out salient objects in complex images.

2.2   Deep Neural Networks Based Models

Recently, deep learning based approaches, in particular the convolutional neural networks (CNNs), have been applied to detect salient objects and have improved the performance by a large margin. Wang et al. [23] propose one deep neural network to compute saliency score for each pixel in local context first, and then refine the saliency score for each object proposal over the global view with another network. Li et al. [21] predict saliency score of each superpixel by incorporating multi-scale features in a generic convolutional neural network. Zhao et al. [31] compute saliency by integrating global and local context into a deep learning based framework. Although these models achieve better results than traditional schemes, these models are very time-consuming due to the reason that they take segmented region as a basic unit to train a deep neural network for predicting saliency and the networks have to run many times for predicting saliency degree of all the superpixels in the image.

To remedy above problems, researchers prefer to adopt FCN-like model to detect saliency in a pixel-wise manner. Some researchers propose to use specific-level features for saliency prediction. For example, Lee et al. [25] propose to encode low-level distance map and high-level semantic features of deep CNNs. In [26], a network sharing features for segmentation and saliency tasks is proposed, and a graph Laplician regularized nonlinear regressor model is presented for refinement.

In contrary to these methods only use specific-level features, several works explore to integrate features from different side outputs and indicate that the features from all levels are potential saliency cues and are helpful for saliency prediction. The features from deep layers contain semantic information which is helpful for objectness, while the features from shallow layers contain rich detailed information which is helpful for explicit boundary in high-resolution prediction.

However, how to effectively and efficiently aggregate multi-level convolutional features remains challenging. To this end, several researchers make valuable attempts to solve this problem. Li et al. [27] combines a pixel-level fully convolutional stream and segmented-wise spatial pooling stream. The fully convolutional stream is a multi-scale fully convolutional network, which generates a saliency map with one eighth resolution of the raw input image by exploiting visual contrast across multiscale convolutional layers. Long et al. [11] introduce skip connections and adds high-level prediction layers to intermediate layers to generate pixel-wise prediction results at multiple resolutions. Liu et al. [16] design a two-stage deep network, in which a coarse global prediction is obtained by automatically learning various global structured saliency cues and another network is adopted to further refine the details of saliency maps via integrating local context information.

Though obvious achievement has been made by these deep learning based models in recent years, there is still a large room for improvement over the generic FCN-based models to uniformly highlight the entire salient objects and preserve the detailed boundaries against the cluttered background.

3 Proposed Model

Our proposed salient object detection model mainly consists of two stages: (1) a FCN-based deep network for multi-level features extraction and aggregation; and (2) a spatial coherence scheme for saliency refinement.

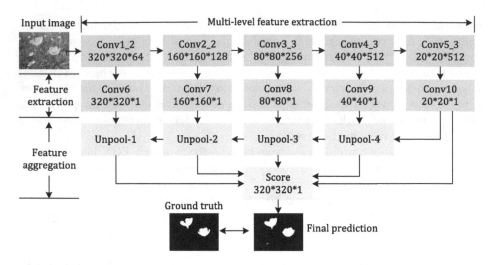

Fig. 2. The architecture of the proposed model. In the VGG-16 net, the names of the layers whose features are utilized are shown. The resolution of each step is also shown.

3.1   Network Architecture

To design a FCN-like network that is capable of accounting for both local and global context of an image and incorporating details from various resolutions, we develop a multi-scale deep convolutional neural network for learning discriminant saliency features (our mode is shown in Fig. 2). It consists of two components: feature extraction and aggregation.

Multi-level Feature Extraction. Our proposed model adopts VGG-16 net [28] (pre-trained over the ImageNet dataset for image classification) as our base network, and modifies it to meet our requirements. We retain its 13 convolutional layers, and remove the original 5th pooling layer and fully connected layers. Thus, the modified VGG-16 is composed of 5 groups of convolutional layers. For simplicity, we denote the third sub-layer in the fifth group of convolutional layer as $Conv5\_3$, and the other convolution layers in the VGG-16 is also denoted by this analogy. For an input image I with size $W \times H$, the modified VGGNet produces five feature maps f_i with decreasing spatial resolution by stride 2.

For each continuous feature f_i, $i \in \{5, 6, \ldots, 10\}$ extracted from VGG-16, we design a densely connected feature extraction block $Convi$. It utilizes a simple connectivity pattern: to preserve the feed-forward nature, each layer obtains additional inputs from all preceding layers and passes on its own feature maps to all subsequent layers, which is similar to DenseNet [24]. Figure 3 illustrates this layout schematically.

Features Aggregation. We obtain five feature maps with size different resolution from feature extraction blocks. The feature maps of deeper convolutional layers can accurately locate salient objects, while the feature maps generated by

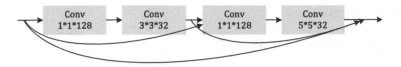

Fig. 3. Details of the feature extraction module.

shallower convolutional layers contain more details. To help the shallow side output contain more global properties, we refine these feature maps by skip-layer structure, namely, introducing the deeper side-output to its former shallower one. At each Unpool processing block, we combine features through summation. Moreover, we use a score module to integrate different maps and obtain a fused saliency map. To make the output maps of the features at different solutions have the same size for fusing, we use the deconvolutional layer for up-sampling. The strides of the last deconvolutional layers in the last four sides are respectively set to 2, 4, 8 and 16. And then, we combine features by concatenating them.

3.2 Spatial Coherence

To improve spatial coherence and achieve more accurate results, we adopt a pixel-wise saliency refinement model based on a fully connected conditional random field (CRF) [29] in the inference phase. This CRF model solves a binary pixel labeling problem, which is similar to our saliency prediction task, and employs the following energy function,

$$E(L) = -\sum_i logP(l_i) + \sum_{i,j} \theta_{ij}(l_i, l_j) \tag{1}$$

where L represents a binary label assignment for all pixels. $P(l_i)$ is the probability of pixel x_i with label l_i, which indicates the likelihood of pixel x_i being salient. Initially, $P(1) = S_i$ and $P(0) = 1 - S_i$, where S_i is the saliency score at pixel x_i from the fused saliency map S. $\theta_{i,j}(l_i, l_j)$ is a pairwise potential and defined as follows,

$$\theta_{ij} = \mu(l_i, l_j)[\omega_1 exp(-\frac{||p_i - p_j||^2}{2\sigma_\alpha^2}) - \frac{||I_i - I_j||^2}{2\sigma_\beta^2} + \omega_2 exp(-\frac{||p_i - p_j||^2}{2\sigma_\gamma^2})] \tag{2}$$

where $\mu(l_i, l_j) = 1$ if $l_i \neq l_j$, and zero otherwise. θ_{ij} involves two kernels. The first kernel depends on pixel positions p and pixel intensities I. This kernel makes nearby pixels having similar colors take similar saliency scores. Three parameters determine the degree of influence by color similarity and spatial relation, respectively. The second kernel is to remove small isolated regions. The parameters of ω_1, ω_2, σ_α^2, σ_β^2, σ_γ^2 are set to 3.0, 3.0, 60.0, 8.0 and 5.0 respectively in our experiments.

4   Experiments

4.1   Implementation Details

Our network is based on the publicly available Caffe library, an open source framework for CNNs training and testing. As mentioned above, we choose VGG-16 as our pre-trained model and fine-tune it for pixel-wise saliency prediction. We utilize the same training and validation sets as in [8]. The learning rate is set to 1e−9, the momentum parameter is 0.9, the weighted decay is set to 0.0005. The fusion weight in the feature integration module are all initialized with 0.2 in the training phase.

4.2   Datasets

We conduct evaluations on five widely used salient object benchmark datasets. DUT-OMRON is manually selected from more than 140,000 natural images, each of which has one or more salient objects and relatively complex backgrounds. As an extension of the Complex Scene Saliency Dataset (CSSD), ECSSD is obtained by aggregating the images from two publicly available datasets and the Internet. HKU contains 4447 images, most of which have low contrast and multiple salient objects. PASCAL-S is generated from the PASCAL VOC dataset with 20 object categories and complex scenes. SOD is more challenging with multiple salient object and background clutters in images.

4.3   Evaluation Metrics

We adopt the precision-recall (PR) curve to evaluate our proposed model. The precision and recall are computed by binarizing the saliency map with 256 thresholds, ranging from 0 to 255, and comparing the binary map with the ground truth. The PR curves demonstrate the mean precision and recall of saliency maps at different thresholds. We also use F-measure (F_β) and weighted F-measure (ωF_β) scores to comprehensively consider precision and recall. F_β is given by:

$$F_\beta = \frac{(1 + \beta^2) \cdot Precision \cdot Recall}{\beta^2 \cdot Precision + Recall} \tag{3}$$

where β is a balance parameter to weight the precision and recall, and β^2 is set to 0.3. Similar to F_β, ωF_β is computed with a weighted harmonic mean of $Precision^w$ and $Recall^w$: $F_\beta^w = \frac{(1+\beta^2) \cdot Precision^w \cdot Recall^w}{\beta^2 \cdot Precision^w + Recall^w}$.

Beside, we use the mean absolute error (MAE) to evaluate the average pixel-wise error between the saliency map and ground truth. It is defined as $MAE = \frac{1}{h \cdot w} \sum_{i=1}^{h} \sum_{j=1}^{w} |S_{ij} - G_{ij}|$ where S denotes the saliency map, G denotes the ground truth, and h and w denote the height and width of the image.

4.4 Performance Comparison with State of the Art

We compare our proposed approach with 10 state-of-the-art methods, including UCF [33], MTDS [26], LEGS [23], MDF [21], KSR [30], DRFI [8], SMD [10], ELD [25], MC [31], and ELE [32]. We use either the implementations or the saliency maps provided by the authors for fair comparison. Note that MC, UCF, ELD, MTDS, LEGS, MDF, KSR are deep learning based models.

Table 1. F_β and ωF_β scores of saliency maps produced by different approaches on DUT-OMRON, ECSSD, HKU, PASCAL-S, and SOD datasets (The top models are highlighted in bold. '-' denotes the saliency maps are not available).

Approach	DUT-OMRON		ECSSD		HKU		PASCAL-s		SOD	
	F_β	ωF_β	F_β	ωF_β	F_β	ωF_β	F_β	ωF_β	F_β	ωF_β
SMD	0.537	0.398	0.712	0.532	0.691	0.499	0.622	0.462	0.605	0.474
DRFI	0.555	0.374	0.732	0.567	0.722	0.502	0.613	0.446	-	-
ELE	0.575	0.525	0.755	0.720	0.699	0.655	0.652	0.604	-	-
LEGS	-	-	0.783	0.723	0.709	0.616	0.688	0.610	0.686	0.612
MC	-	-	0.797	0.750	0.759	0.700	0.692	0.628	0.589	0.391
KSR	0.591	0.493	0.782	0.675	0.747	0.638	0.703	0.610	0.668	0.579
MDF	0.596	0.499	0.749	0.643	0.764	0.641	0.648	0.557	0.697	0.601
ELD	0.614	0.564	0.817	0.773	-	-	0.721	0.659	-	-
MTDS	0.603	0.463	0.826	0.693	-	-	0.658	0.521	0.698	0.568
UCF	0.621	0.537	0.844	0.788	0.823	0.754	0.733	0.669	0.738	0.684
Ours	**0.660**	**0.615**	**0.862**	**0.851**	**0.868**	**0.845**	**0.747**	**0.719**	**0.759**	**0.759**

For quantitative evaluation, we show comparison results with PR curves and MAE scores in Figs. 4 and 5. And the comparisons of F_β and ωF_β are displayed in Table 1. We do not show the comparison of PR curves on DUT-OMRON due to the limited space. In terms of F_β, ωF_β and MAE scores, we can see that our model outperforms all other methods, especially on complex datasets. For the PR curves, our model also achieves a good performance on four datasets and is a little worse than UCF on ECSSD and PASCAL-S.

We show visual comparison in Fig. 6. We can see that our model not only detects and localizes salient objects accurately, but also preserves object details subtly. It can handle various complex situations well, including salient objects being small (row fourth and fifth), clutter backgrounds and salient objects (row first and sixth), backgrounds and salient objects sharing similar appearance (row second, third and fifth).

4.5 Evaluation on CRF Scheme

A fully connected CRF scheme is incorporated to further uniformly highlight the interior regions of salient object and preserve explicit contour in the saliency map

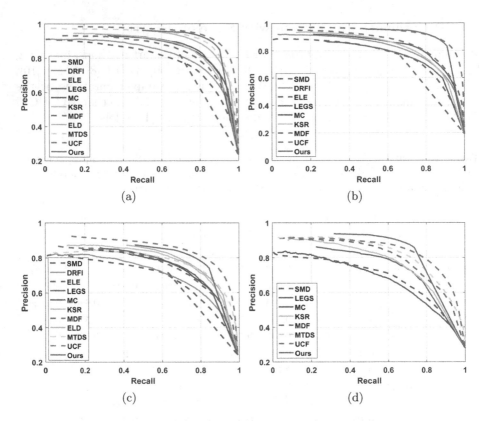

Fig. 4. PR curves of saliency maps produced by different approaches on four datasets. (a) ECSSD, (b) HKU, (c) PASCAL-S and (d) SOD.

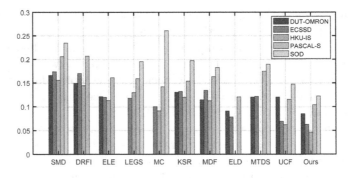

Fig. 5. MAE scores of the saliency maps produced by different models on five datasets. Lower is better.

from our proposed multi-scale FCN-like network. To validate its effectiveness, we have also evaluated the performance of our final saliency approach with and without (w/o) CRF scheme on five benchmark datasets in terms of F_β, ωF_β, and

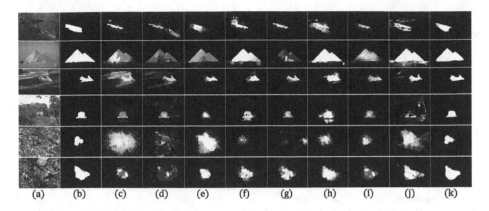

(a)   (b)   (c)   (d)   (e)   (f)   (g)   (h)   (i)   (j)   (k)

Fig. 6. Visual comparison results based on different models. (a) Input, (b) ground truth, (c) SMD, (d) DRFI, (e) LEGS, (f) MC, (g) MDF, (h) ELD, (i) MTDS, (j) UCF, and (k) Ours.

MAE scores. The results are displayed in Table 2, which shows that the CRF scheme improves the accuracy of our proposed model.

Table 2. Comparisons of our approach with and without(w/o) CRF scheme in terms of F_β, ωF_β, and MAE.

Datasets	Method	F_β	ωF_β	MAE
DUT-OMRON	Ours with CRF	0.6600	0.6152	0.0852
	Ours w/o CRF	0.6265	0.5753	0.0932
ECSSD	Ours with CRF	0.8621	0.8505	0.0627
	Ours w/o CRF	0.8299	0.8019	0.0730
HKU	Ours with CRF	0.8681	0.8454	0.0463
	Ours w/o CRF	0.8260	0.7897	0.0569
PASCAL-S	Ours with CRF	0.7465	0.7187	0.1041
	Ours w/o CRF	0.7180	0.6816	0.1127
SOD	Ours with CRF	0.7594	0.7589	0.1225
	Ours w/o CRF	0.7503	0.7303	0.1284

5   Conclusion

In this paper, we propose a simple but effective approach for pixel-wise salient object detection based on a fully convolutional network, which extracts multi-level features and utilizes the preceding information through a densely connected module. Moreover, the features from deeper layers are connected to the shallower ones by skip-layer structure for guiding the learning of shallower layers. Besides, a fusion layer is adopted to combine these rich features to generate a saliency map. In order to obtain more fine-gained saliency detection results, we introduce a saliency refinement scheme based on a fully connected CRF to further improve saliency performance. Experimental results demonstrate that our proposed approach achieves encouraging performance against 10 state-of-the-art methods on five benchmark datasets.

References

1. Simakov, D., Caspi, Y., Shechtman, E., Irani, M.: Summarizing visual data using bidirectional similarity. In: IEEE Conference on Computer Vision and Pattern Recognition, pp. 1–8 (2008)
2. Zhao, R., Ouyang, W., Wang, X.: Unsupervised salience learning for person re-identification. In: IEEE Conference on Computer Vision and Pattern Recognition, pp. 3586–3593 (2013)
3. Wang, W., Shen, J., Yu, Y., Ma, K.: Stereoscopic thumbnail creation via efficient stereo saliency detection. IEEE Trans. Vis. Comput. Graph. **23**(8), 2014–2027 (2017)
4. Wang, W., Shen, J., Ling, H.: A deep network solution for attention and aesthetics aware photo cropping. IEEE Trans. Pattern Anal. Mach. Intell. **41**, 1 (2018)
5. Itti, L., Koch, C., Niebur, E.: A model of saliency-based visual attention for rapid scene analysis. IEEE Trans. Pattern Anal. Mach. Intell. **20**(11), 1254–1259 (1998)
6. Liu, T., Sun, J., Zheng, N., Tang, X., Shum, H.: Learning to detect a salient object. In: IEEE Conference on Computer Vision and Pattern Recognition, pp. 1–8 (2007)
7. Perazzi, Y., Krahenbuhl, P., Hornung, H.: Saliency filters: contrast based filtering for salient region detection. In: IEEE Conference on Computer Vision and Pattern Recognition, pp. 733–740 (2012)
8. Wang, J., Jiang, H., Yuan, Z., Cheng, M., Hu, X., Zheng, N.: Salient object detection: a discriminative regional feature integration approach. Int. J. Comput. Vis. **123**(2), 251–268 (2017)
9. Yang, C., Zhang, L., Lu, H., Ruan, X., Yang, M.: Saliency detection via graph-based manifold ranking. In: IEEE Conference on Computer Vision and Pattern Recognition, pp. 3166–3173 (2013)
10. Peng, H., Li, B., Ling, H., Hu, W., Xiong, W., Maybank, S.: Salient object detection via structured matrix decomposition. IEEE Trans. Pattern Anal. Mach. Intell. **39**(4), 818–832 (2017)
11. Long, J., Shellhamer, E., Darrell, T.: Fully convolutional networks for semantic segmentation. In: IEEE Conference on Computer Vision and Pattern Recognition, pp. 3431–3440 (2015)
12. Noh, H., Hong, S., Han, B.: Learning deconvolution network for semantic segmentation. In: International Conference on Computer Vision, pp. 1520–1528 (2015)
13. Yang, W., Ouyang, W., Li, H., Wang, X.: End-to-end learning of deformable mixture of parts and deep convolutional neural networks for human pose estimation. In: IEEE Conference on Computer Vision and Pattern Recognition, pp. 3073–3082 (2016)
14. Xie, S., Tu, Z.: Holistically-nested edge detection. In: International Conference on Computer Vision, pp. 1395–1403 (2015)
15. Liu, Y., et al.: Richer convolutional features for edge detection. IEEE Trans. Pattern Anal. Mach. Intell. (2019). https://doi.org/10.1109/TPAMI.2018.2878849

16. Liu, N., Han, J.: DHSNet: deep hierarchical saliency network for salient object detection. In: IEEE Conference on Computer Vision and Pattern Recognition, pp. 678–686 (2016)
17. Zhang, P., Wang, D., Lu, H., Wang, H., Ruan, X.: Amulet: aggregating multi-level convolutional features for salient object detection. In: International Conference on Computer Vision, pp. 202–211 (2017)
18. Hou, Q., Cheng, M., Hu, X., Borji, A., Tu, Z., Torr, P.: Deeply supervised salient object detection with short connections. IEEE Trans. Pattern Anal. Mach. Intell. **41**(4), 815–828 (2019)
19. Li, Y., Hou, X., Koch, C., Rehg, J., Yuille, A.: The secrets of salient object segmentation. In: IEEE Conference on Computer Vision and Pattern Recognition, pp. 280–287 (2014)
20. Yan, Q., Xu, L., Shi, J., Jia, J.: Hierarchical saliency detection. In: IEEE Conference on Computer Vision and Pattern Recognition, pp. 1155–1162 (2013)
21. Li, G., Yu, Y.: Visual saliency based on multiscale deep features. In: IEEE Conference on Computer Vision and Pattern Recognition, pp. 5455–5463 (2015)
22. Zhu, W., Liang, S., Wei, Y., Sun, J.: Saliency optimization from robust background detection. In: IEEE Conference on Computer Vision and Pattern Recognition, pp. 2814–2821 (2014)
23. Wang, L., Lu, H., Yang, M.: Deep networks for saliency detection via local estimation and global search. In: IEEE Conference on Computer Vision and Pattern Recognition, pp. 3183–3192 (2015)
24. Huang, G., Liu, Z., Maaten, L., Weinberger, K.: Densely connected convolutional networks. In: IEEE Conference on Computer Vision and Pattern Recognition, pp. 2261–2269 (2017)
25. Lee, G., Tai, Y.W., Kim, J.: Deep saliency with encoded low level distance map and high level features. In: IEEE Conference on Computer Vision and Pattern Recognition, pp. 660–668 (2016)
26. Li, X., et al.: DeepSaliency: multi-task deep neural network mode for salient object detection. IEEE Trans. Image Process. **25**(8), 3919–3930 (2016)
27. Li, G., Yu, Y.: Deep contrast learning for salient object detection. In: IEEE Conference on Computer Vision and Pattern Recognition, pp. 478–487 (2016)
28. Simonyan, K., Zisserman, A.: Very deep convolutional networks for large-scale image recognition. In: International Conference on Learning Representations (2015)
29. Krahenbuhl, P., Koltun, V.: Efficient inference in fully connected CRFs with Gaussian edge potentials. In: Neural Information Processing Systems, pp. 109–117 (2011)
30. Wang, T., Zhang, L., Lu, H., Sun, C., Qi, J.: Kernelized subspace ranking for saliency detection. In: Leibe, B., Matas, J., Sebe, N., Welling, M. (eds.) ECCV 2016. LNCS, vol. 9912, pp. 450–466. Springer, Cham (2016). https://doi.org/10.1007/978-3-319-46484-8_27
31. Zhao, R., Ouyang, W., Li, H., Wang, X.: Saliency detection by multi-context deep learning. In: IEEE Conference Computer Vision and Pattern Recognition, pp. 1265–1274 (2015)
32. Xia, C., Li, J., Chen, X., Zheng, A., Zhang, Y.: What is and what is not a salient object? Learning salient object detector by ensembling linear exemplar regressors. In: International Conference on Computer Vision and Pattern Recognition, pp. 4399–4407 (2017)

33. Zhang, P., Wang, D., Lu, H., Wang, H., Yin, B.: Learning uncertain convolutional features for accurate saliency detection. In: International Conference on Computer Vision, pp. 212–221 (2017)
34. Movahedi, V., Elder, J: Design and perceptual validation of performance measures for salient object segmentation. In: International Conference on Computer Vision and Pattern Recognition, pp. 49–56 (2010)

Facial Expression Recognition Based on Group Domain Random Frame Extraction

Wenjun Zhou[1], Lu Wang[1], Yibo Huang[1], Linbo Qing[1,2(✉)],
Xiaohong Wu[1], and Xiaohai He[1]

[1] College of Electronics and Information Engineering, Sichuan University,
Chengdu 610065, China
qing_lb@scu.edu.cn
[2] Key Laboratory of Wireless Power Transmission of Ministry of Education,
Chengdu 610065, China

Abstract. Modeling the dynamic variation of facial expression from a sequence of images is a key issue in facial expression recognition. However, the analysis of complete sequence temporal information requires significantly computational power. To improve the efficiency, a dynamic frame sequence convolutional network (DFSCN) is proposed in this study. In the proposed DFSCN, an expression sequence simplification method is first proposed to reduce the sequence length and takes the reduced new sequence as the input of DFSCN. An adaptive weighted feature fusion method for spatiotemporal feature learning is then put forward in DFSCN. A still frame convolutional network (SFCN) is introduced for complementing the still appearance information and the fine-tuning of DFSCN. Finally, these two models are combined together by weighted fusion to enhance the performance. Two public-available databases, CK+ and Oulu-CASIA, are used to evaluate the performance of the proposed approach. Experimental results show that the proposed method can effectively capture the dynamic process of expression sequence and the recognition performance is superior to other state-of-the-art methods.

Keywords: Facial expression recognition · Sequence simplification · Adaptive weighted feature fusion · Dynamic Frame Sequence Convolutional Network · Still Frame Convolutional Network

1 Introduction

Automatic Facial Expression Recognition (FER) has become an attractive research topic in the field of computer vision, due to its significant role in numerous applications such as medical treatment [1], security monitoring [2] and many other human-computer interaction systems. Existing researches on FER can be divided into two categories depending on the type of data: dynamic video sequence-based and static image-based. Dynamic video sequence-based approaches can effectively extract useful temporal features from consecutive frames of input, whereas static image-based methods mainly focus on spatial information from the current single image. Extensive researches have demonstrated that the performance of sequence-based methods is usually better than

© Springer Nature Switzerland AG 2019
Y. Zhao et al. (Eds.): ICIG 2019, LNCS 11901, pp. 467–479, 2019.
https://doi.org/10.1007/978-3-030-34120-6_38

that of image-based one [3, 19] due to better exploiting of the dynamic spatial-temporal feature of facial expression. The expression recognition based on dynamic sequence usually proceeds from original sequence or the processed facial sequence, such as STM-ExpLet [3], DTAGN [4], PHRNN-MSCNN [5], FACRN-FGRN [6] and etc. Generally speaking, the research based on facial landmarks or other indirect information is more complicated than using the original sequence, because it requires special processing, and pretreatment may affect the recognition performance of the model. Therefore, it is a meaningful essay to use only the original sequence for facial expression recognition while ensuring high recognition accuracy.

There have been many pioneers in the research of expression recognition based on original expression sequence. Jung et al. [4] proposed a deep temporal appearance network (DTAN) to extract useful temporal features and achieved satisfied recognition rate. Huang et al. [6] introduced a facial appearance convolutional recurrent network (FACRN) to combine CNNs and RNNs [10] to learn characteristics from consecutive frames. In the above methods, the sequence is input into the network with original length and these networks can hardly process all the frames in one pass because of the varied length of sequence, then the efficiency of learning dynamic change of whole expression decreased [7]. Subsequently, Zhang et al. [18] proposed a new CNN architecture which imports a frame-to-sequence model based on the last few frames of the sequence for facial expression recognition, which fixed the problem of not being able to process all frames at once. Although this method reduces the original expression sequence to analyses the facial changes and achieves a certain recognition rate, it discards the frames in front of the sequence, that is to say the temporal correlation of the whole sequence is not fully considered. Therefore, it is of great value to enable the network to efficiently process expression sequences while preserving the temporal correlation of sequences, especially in the case of long sequences.

Usually, the sequence length of dynamic expression datasets, such as CK+ [13] and Oulu-CASIA [14], are varied. Meanwhile, as shown in Fig. 1, the variety of expression includes at least three stages for all expression databases, namely initial state, transition state and peak state. In each stage, the expression changes slightly or even almost unchanged, so the similarity of expression frames in the same stage is extremely high and the information contained are also similar. Based on the above analysis, if the redundant expression frames are eliminated and then the remaining sequences are used for expression recognition, the requirement for the network to process all sequences at once can be satisfied. And the time correlation of the sequence is also preserved.

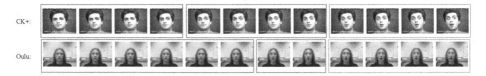

Fig. 1. Sequential expressions in two datasets. The expression of sequence can be divided into three parts: initial state (red box), transition state (green box) and peak state (pink box). (Color figure online)

In this work, we introduce a sequence simplification method to reduce the original expression sequence. Firstly, we propose a Dynamic Frame Sequence Convolutional Network (DFSCN). In our DFSCN model, we simplified the expression sequence to a fixed length and employ a convolutional structure to learn apparent characteristics for each extracted frame. Secondly, an Adaptive Weighted Feature Fusion (AWFF) algorithm is proposed in order to combine the previously obtained groups of spatial features to model the expression of the entire sequence. Considering frames still contain abundant spatial information, especially for frames with obvious expressions [19], a Still Frame Convolutional Network (SFCN) is also introduced to complement the spatial characteristics of DFSCN and fine-tune DFSCN. Finally, we use a score fusion approach to combine the DFSCN and SFCN together for final prediction. The contributions of this paper can be listed as follows:

- A deep network framework to extract temporal and spatial features of dynamic expression sequences for facial expression recognition is constructed.
- A new method to construct input sequences which can effectively shorten the length of the original sequence without losing global information of the sequence is proposed.
- A special feature fusion method is proposed to fuse the static features of different frames in the newly generated sequence, thus enhance the richness of temporal features.

This rest of this paper is structured as follows. Section 2 gives a detailed description of our DFSCN and SFCN. The performance of our proposed work compared with the state-of-art is evaluated in Sect. 3. Section 4 gives a conclusion of the whole paper.

2 Our Approach

Our proposed methodology is shown in Fig. 2. Our method consists of two kinds of networks: Dynamic Frame Sequence Convolutional Network (DFSCN) and Still Frame Convolutional Network (SFCN). Firstly, we introduce random frame extraction strategy and adaptive weighted feature fusion module in DFSCN to capture dynamic features of expression sequences. Secondly, SFCN is constructed to capture spatial features from still frames and fine-tune DFSCN. Finally, the two networks are integrated to boost the accuracy of facial expression recognition. The details of each component will be discussed in this section.

2.1 Image Pre-processing and Data Augmentation

Face Detection and Pre-processing. The expression frames in commonly used datasets, such as CK + and Oulu-CASIA, usually contain regions other than faces, which are helpless for expression recognition. Thus, we used the C++ library algorithm Dlib [15] for face detection and then cropped the detected faces. In addition, the obtained images were grayed and the size were normalized to $224 \times 224 \times 1$.

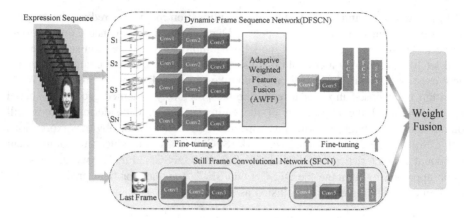

Fig. 2. Overall architecture of our method.

Data Augmentation. The existing facial expression datasets are relatively small, so the problem of overfitting is prone to occur during training. Therefore, various data augmentation techniques are required to increase the volume of data. The commonly used data enhancement methods include image rotation, noise addition and so on. Rotating image and increasing image brightness are used for data expansion in this paper. Image rotation is that the image revolves around a point and rotates at a certain angle clockwise or counterclockwise to form a new image. In this work, each image is rotated with seven angels, which can respond effectively to the change of slight. The image rotation formula is as follows:

$$
\begin{pmatrix} x \\ y \\ 1 \end{pmatrix} = \begin{pmatrix} cos\theta & -sin\theta & 0 \\ sin\theta & cos\theta & 0 \\ 0 & 0 & 1 \end{pmatrix} \begin{pmatrix} x_0 \\ y_0 \\ 1 \end{pmatrix}
\tag{1}
$$

where $P_0(x_0, y_0)$ is the pixel coordinate point of the original image, $P(x, y)$ is the pixel coordinate point of the rotated image, and θ is the angle of every rotation, $\theta \in \{-15°, -10°, -5°, 5°, 10°, 15°, 180°\}$. In addition, the brightness of datasets is increased by $5°$ and $15°$ respectively. Finally, we have ten times as much data as before.

2.2   Group Domain Frame Extraction

As shown previously in Fig. 1, the image sequence in public facial expression databases usually starts from a neutral face and gradually evolves into a peak expression. Thus, as shown in Fig. 1, in order to capture the information of each state and then mine the dynamic emotional changes of the whole sequence, we propose to sample a subset of frames representing the overall temporal dynamics of the sentiment sequence from the original whole sequence as the input of DFSCN. Same as [7] and [8], in this work, each expression sequence is split into N subsequences $\{S_1, S_2, \ldots, S_N\}$ of equal size, and one frame is selected randomly from each subsequence. Given the length of a sequence is L, the newly generated clip CL is formed for representative frame extraction as follows:

$$C_i = rand(S_i), i \in \{1, 2, \ldots, N\} \tag{2}$$

$$CL = \{C_1; C_2; \ldots; C_N\} \tag{3}$$

where C_i is the extracted frame from the subsequence S_i, $rand(\bullet)$ represents the random selection of a frame from the specified sequence, CL is the newly generated sequence. After the original sequence is sampled, we employ N CNNs to process the extracted frames independently in one process. Then, N sets of feature maps are obtained, all of which contain only static information (shown in Fig. 2).

2.3   Adaptive Weighted Feature Fusion

Up to this point, the frames selected from sequences are processed independently. In order to learn how facial expressions are made up of different appearances over time, we stack the multiple groups of features with AWFF method and feed them into a 2D-CNN (Conv4) for further study. The detailed implementation process is shown in Fig. 3. When each frame in CL is sent to the three CNNs (Conv1, Conv2, Conv3) respectively, we obtained N groups of features $\{G_1, G_2, \ldots, G_N\}$. For any network, the feature map of convolution layer h can be expressed as:

$$f_h = f(w_h * x_h + b_h) \tag{4}$$

where w_h is the connection weight between the upper layer and the next layer, x_h is the initial input for the hidden layer, and b_h is the bias of the hidden layer. $f(\bullet)$ is the activation function.

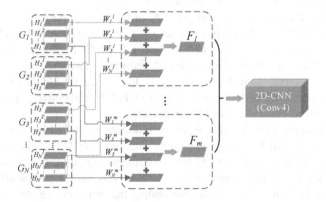

Fig. 3. Schematic diagram of spatial feature fusion.

In our AWFF model, the spatial features of the i-th ($i \in [1, N]$) channel include m feature maps, which are expressed as $G_i = \{H_i^1, H_i^2, \ldots, H_i^m\}$. When fusing the

N sets of features, a weight W_i^j will be assigned to the $j\text{-}th$ $(j \in [1, m])$ feature map in G_i. The fusion formula is as follows:

$$F_k = W_1^k * H_1^k + W_2^k * H_2^k + \ldots + W_N^k * H_N^k, k \in \{1, \ldots, m\} \tag{5}$$

$$P = \{F_1, F_2, \ldots, F_m\} \tag{6}$$

in which F_k is the $k\text{-}th$ feature fused from N group, P is the total feature output obtained after the fusion. In our work, the weight of W_i^j is updated adaptively to achieve the best fusion effect. Based on this method, we can more concisely combine the spatiotemporal information of the expression sequence.

2.4   The Detail of the Network Structure

In this paper, Our DFSCN model is designed as [Conv1(64) - Conv2(128) - Conv3 (256)] × N-AWFF - Conv4(512) - Conv5(512) - FC1(4096) - FC2(4096) -FC3(7/6), as shown in Fig. 2. The values in parentheses represent the total number of neurons used in the corresponding layer. For example, Conv2 (128) indicates that the number of convolution kernels of the second convolutional layer is 128. AWFF is designed to fuse the output of the N-channel [Conv1(64) - Conv2(128) - Conv3(256)] to learn the temporal features of the sequence. The weight information dimension in the AWFF module is $N \times m$. Besides, each convolution layer is sequentially followed by a max-pooling layer and Relu [22].

The biggest difference between DFSCN and SFCN lies in the feature processing method after the third convolution layer. For SFCN, the feature obtained after the third convolution layer is directly transmitted into the next convolution. However, for DFSCN, as shown in Fig. 2, N sets of features are obtained after N respective inputs, and then a method called adaptive weighted feature fusion is introduced to combine the features to get a new feature group and send it to Conv4.

SFCN network is designed to compensate for the spatial characteristics of DFSCN and fine-tune DFSCN as there is no relevant fine-tuning model for DFECN. The parameter settings of SFCN are similar to those of the DFSCN. We first train the SFCN network with the last frame of the original sequence, and then use parameters of Conv1, Conv2, Conv3, Conv4, Conv5, FC1, FC2, FC3 of SFCN to initialize the network layer parameters with the same name in DFSCN, respectively. Finally, the two models are fused to boost the recognition performance.

It is worth noting that the size of the convolution kernel of the five convolution layers in this paper is all 3 × 3, which is quite same with VGG [9]. There are two main reasons for this. Firstly, as the size of the convolution kernel increases, the number of parameters of the convolution kernel increases relatively, so the computational amount of convolution is bound to increase. Secondly, the network we designed must be moderate in depth and appropriate in parameters so as to avoid the problem of over-fitting caused by the small amount of data. Other than this, small receptive field is more capable of learning image features from details and distinguishable.

2.5   Model Fusion

In order to maximize the superiority of the two models, the DFSCN and SFCN are integrated by following fusion function [5]:

$$O(x) = \sum_{i=0}^{1} b_i(S_i(x) + P_i(x)) \tag{6}$$

where $P_i(x)$ $(0 < P_i(x) < 1)$ is the predicted probability of expressions in DFSCN and SFCN. $P_0(x)$ comes from the softmax layer of DFSCN while $P_1(x)$ comes from SFCN. $S_i(x)$ is sorted based on the prediction categories of expression. It can be expressed as:

$$S_i(x_1), \ldots, S_i(x_n) = Sorted(P_i(x_1), \ldots, P_i(x_n)) \tag{7}$$

where n refers to the total number of categories of expressions, $S_i(x) \in \{1, 2, \ldots, n\}$. In addition, b_i is a balance index for balancing different models. After a lot of comparative experiments, the value of b_i is set to 0.5 which can achieve the best performance as shown in Fig. 4.

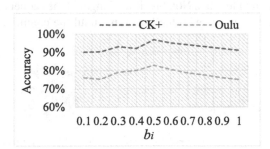

Fig. 4. Performance of our proposed work using the weighted fusion method with different values of b_i. We changed the value of b_i from 0 to 1 with interval of 0.01.

3   Experiment

In this section, we compare our models with some of the most advanced methods in facial expression recognition. This paper focuses on the emotion from neutral to peak. Thus, two widely used databases, namely CK+ and Oulu-CASIA, are used to assessing the performance of our approach. Details of our experimental results are given below.

3.1   Datasets

Description of CK+. The Extended Cohn-Kanade (CK+) database is one of the most extensively used databases for facial expression recognition. There are 593 video sequences from 123 subjects with a duration ranging from 9 to 60 frames in this database. Among these videos, 327 sequences are marked with seven typical emotional expressions (anger, contempt, disgust, fear, happiness, sadness, and surprise). Each expression sequence reflects the expression from neutral to emotional vertex. We adopt the most commonly used 10-fold validation method [4] to verify our experimental results.

Description of Oulu-CASIA. The Oulu-CASIA database contains 2880 expression sequences collected from 80 subjects. There are six types of emotions: anger, disgust, fear, happiness, sadness, and surprise. Similar to CK+ database, all of these sequences begins with a neutral expression and then gradually transit to the peak expression. We use 10-fold cross validation mentioned in the description of CK+.

3.2   Implementation Details

Depending on the characteristics of CK+ and Oulu-CASIA datasets, we set the value of N to 4. Figure 5 shows that more than 85% of sequences in these two datasets are between 10–30 in length. If these sequences are divided into fewer intervals, it may cause a loss of temporal features of sequences longer than 30. If the sequences are parted into more intervals, there will be more difficulties to network training. After comprehensive consideration, the sequences are divided into four groups, taking into account the sequences of different lengths. Notably, if the length of the sequence is not a multiple of $N = 4$, we pad the sequence with the last frame until the condition is satisfied.

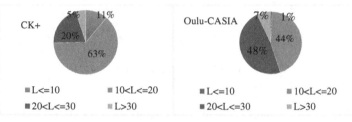

Fig. 5. Statistical analysis of sequence length in CK+ and Oulu-CASIA. L represents the length of the original sequence.

In this paper, experiments were carried out under the environment of tensorflow, a python-based deep learning framework. We firstly initialize the corresponding layer of SFCN by using the weights of VGG16 pre-trained on the ImageNet dataset [17]. When training our facial expression recognition network, we set the batch size of training to 16, the initial learning rate to 0.00001, and the training epoch to 300. For updating the network weight, we use Adam optimization algorithm [21]. Compared with the basic SGD algorithm [20], Adam can avoid local optimum and update faster. As for DFSCN, Adam optimization algorithm is also adopted. Meanwhile, we set the training batch size as 16, the initial learning rate as 0.00001, and the training epoch as 32, which means that the learning rate of 8 epochs decreased to 0.1 of the original learning rate.

It is important to note that we first train SFCN and then DFSCN. The model obtained from SFCN is not only used for classification, but for fine-tuning DFSCN. As shown in Fig. 2, the parameters obtained from the first three convolution layers of SFCN are loaded into the four single channels of DFSCN, and then the parameters of SFCN obtained from the fourth convolution layer to the last full connection layer are loaded into the feature fusion network of DFSCN. After all the two networks are trained, the fusion approaches mentioned in Sect. 3.5 is adopted for expression classification.

3.3 Experimental Results

In the experiments, we first evaluate the performance of each network separately, then the two streams are combined together by weight fusion to achieve complementary network performance. As for the comparison experiments, our model is mainly compared with hand-crafted methods and deep learning methods [3, 4, 6, 11, 12, 16, 23, 24].

Accuracies and Analysis. Tables 1 and 2 show the recognition accuracy of our model on each database, as well as the comparisons with other algorithms. From the two tables we can see that the performances of DFSCN and SFCN alone are not comparable to many other algorithms, while the recognition accuracy after fusion is higher than most of the advanced methods. In CK+, the accuracy of our model is higher than traditional algorithm, such as STM-ExpLet [3]. Compared with the recently proposed deep learning algorithm, the final accuracy of our model is slightly lower than that of DTAGN [4], but higher than that of other methods [6, 11, 16]. In Oulu-CASIA, our recognition accuracy even exceeds all the previous researches. Meanwhile, it can be clearly found that our method is better than hand-crafted methods [3, 11, 12, 23, 24], which is mainly attributed to the feature extraction ability of our network. Our model also has significant advantages over the deep learning approaches [4, 6], which is largely related to our frame-to-sequence network connectivity and the fine-tuning approach we use.

After the integrating our DFSCN and SFCN, the performance has reached the highest level, and this is shown in Fig. 6. In other words, the integration of DFSCN and SFCN improves the richness of network features, and the two channels complement each other. The combination of the two can boost the whole recognition accuracy, which proves the effectiveness of our method.

Table 1. Overall accuracy in CK+ database.

Method	Accuracy
HOG 3D [11]	91.94%
Cov3D [16]	92.30%
3DCNN [11]	85.90%
3DCNN-DAP [11]	92.40%
STM-ExpLet [3]	94.19%
DTAGN [4]	**97.25%**
FACRN-FGRN [6]	95.63%
SFCN	92.97%
DFSCN	95.41%
Fusion	**96.64%**

Table 2. Overall accuracy in Oulu database.

Method	Accuracy
HOG 3D [11]	70.63%
AdalLBP [23]	73.54%
Atlases [24]	75.52%
STM-ExpLet [3]	74.59%
3D SIFT [12]	75.83%
DTAGN [4]	81.46%
FACRN-FGRN [6]	76.50%
SFCN	78.13%
DFSCN	80.63%
Fusion	**83.13%**

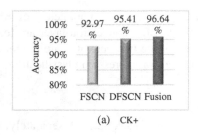 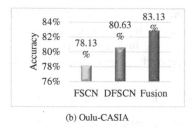

(a) CK+   (b) Oulu-CASIA

Fig. 6. Comparison of accuracy of three networks (SFCN, DFSCN and Fusion) on two databases. (a) CK + . (b) Oulu-CASIA.

Confusion Matrix. Table 3 gives the confusion matrices of our model on the two datasets, respectively. In CK+, our model achieved high recognition accuracies on six emotions except Contempt (Co). In our results, Contempt (Co) is easily misdiagnosed as Sad (Sa) due to the similarity of the expression variation between sad and contempt, especially the change of mouth, as shown in Fig. 7(a). As for Oulu-CASIA database, our model performed well in recognizing Fear (Fe), Disgust (Di), Sad (Sa), and other expressions except Anger (An). The main reason for the poor recognition effect of our model on Anger (An) may be that the volunteers in this dataset have similar facial expressions when expressing Anger (An), Disgust (Di) and Sad (Sa), as shown in Fig. 7(b), which makes the algorithm of this paper not distinguish enough. Moreover, the image quality of the Oulu-CASIA is not as clear as CK+, which may be another reason for the low recognition accuracy.

Table 3. Confusion matrix of our proposed method for two databases.

(a)   CK+ database

	An	Co	Di	Fe	Ha	Sa	Su
An	**96**	2	2	0	0	0	0
Co	6	**83**	0	0	0	11	0
Di	2	0	**98**	0	0	0	0
Fe	0	0	0	**96**	4	0	0
Ha	0	0	0	0	**100**	0	0
Sa	7	0	4	0	0	**89**	0
Su	0	1	0	0	0	0	**99**

(b)   Oulu-CASIA database

	An	Di	Fe	Ha	Sa	Su
An	**68**	15	3	3	12	0
Di	11	**78**	1	1	6	3
Fe	1	1	**81**	7	1	7
Ha	0	0	3	**97**	0	0
Sa	15	4	0	1	**80**	0
Su	0	0	4	0	1	**95**

(a) Two types of mood samples in Ck+   (b) The samples in Oulu

Fig. 7. Sample examples of two datasets.

4 Conclusion

In this paper, we presented a new deep network structure that provides a new approach for the effectively modeling of the dynamic changes of facial expression based on the image sequences through the combination of the static characteristics of sequences. Specially, the proposed DFSCN uses simplified sequences as the network input to study the dynamic variations of expression sequences with the combination of our adaptive weighted feature fusion method. To supplement static appearance information and fine-tune DFSCN, the SFCN is proposed to extract useful spatial information from the last frame of each sequence. These two networks capture dynamic and static information, respectively, at the same time, and complement each other to improve the performance of facial expression recognition. We evaluated our two models using two public-available datasets, CK+, and Oulu-CASIA, respectively. The experimental results demonstrate that the proposed methods have achieved the same accuracy as the state-of-the-art methods.

Acknowledgments. The authors would like to thank the anonymous reviewers for their comments. This work was supported by the National Natural Science Foundation of China (No. 61871278) and the Sichuan Science and Technology Program (No. 2018HH0143).

References

1. Lucey, P., Cohn, J., Lucey, S.: Automatically detecting pain using facial actions. In: International Conference on Affective Computing and Intelligent Interaction and Workshops, pp. 1–8. IEEE, Amsterdam (2009)
2. Cho, S., Kang, H.: Abnormal behavior detection using hybrid agents in crowded scenes. Pattern Recogn. Lett. **44**, 64–70 (2014)
3. Liu, M., Shan, S., Wang, R.: Learning expressionlets on spatio-temporal manifold for dynamic facial expression recognition. In: 2014 IEEE Conference on Computer Vision and Pattern Recognition (CVPR), pp. 1749–1751. IEEE, Columbus (2014)
4. Jung, H., Lee, S., Yim, J.: Joint fine-tuning in deep neural networks for facial expression recognition. In: IEEE International Conference on Computer Vision (ICCV), pp. 2983–2991. IEEE, Santiago (2015)

5. Zhang, K., Huang, Y., Du, Y.: Facial expression recognition based on deep evolutional spatial-temporal networks. IEEE Trans. Image Process. **2017**, 4193–4202 (2017)
6. Huan, Z., Shang, L.: Model the dynamic evolution of facial expression from image sequences. In: Phung, D., Tseng, V., Webb, G., Ho, B., Ganji, M., Rashidi, L. (eds.) PAKDD 2018, LNCS, vol. 10938, pp. 546–557. Springer, Cham (2018). https://doi.org/10.1007/978-3-319-93037-4_43
7. Jing, L., Yang, X., Tian, Y.: Video you only look once: overall temporal convolutions for action recognition. J. Vis. Commun. Image Represent. **52**, 58–65 (2018). S1047320318300233
8. Zolfaghari, M., Singh, K., Brox, T.: ECO: efficient convolutional network for online video understanding. In: Computer Vision 15th European Conference, pp. 1–7. ECCV, Munich (2018)
9. Simonyan, K., Zisserman, A.: Very deep convolutional networks for large-scale image recognition. Computer Science. **2014**, 1–14 (2014)
10. Graves, A.: Generating sequences with recurrent neural networks. Computer Science. **2013**, 1–8 (2013)
11. Klaser, A., Marszałek, M., Schmid, C.: A spatio-temporal descriptor based on 3D-gradients. In: Proceedings of the British Machine Vision Conference 2008, pp. 1–10. British Machine Vision Association, London (2008)
12. Scovanner, P., Ali, S., Shah, M.: A 3-dimensional sift descriptor and its application to action recognition. In: Proceedings of the 15th ACM International Conference on Multimedia, pp. 357–360. Association for Computing Machinery, Augsburg (2007)
13. Lucey, P., Cohn, J., Kanade, T.: The extended Cohn-Kanade dataset (CK+): a complete dataset for action unit and emotion-specified expression. In: 2010 IEEE Computer Society Conference on Computer Vision and Pattern Recognition-Workshops, pp. 94–101. IEEE, San Francisco (2010)
14. Taini, M., Zhao, G., Li, S.: Facial expression recognition from near-infrared video sequences. In: 2008 19th International Conference on Pattern Recognition, pp. 1–4. IEEE, Tampa (2008)
15. King, D.: Dlib-ml: a machine learning toolkit. J. Mach. Learn. Res. **10**(3), 1755–1758 (2009)
16. Sanin, A., Sanderson, C., Harandi, M., et al.: Spatio-temporal covariance descriptors for action and gesture recognition. In: 2013 IEEE Workshop on Applications of Computer Vision (WACV), pp. 103–110. IEEE, Clearwater (2013)
17. Krizhevsky, A., Sutskever, I., Hinton, G.: ImageNet classification with deep convolutional neural networks. In: Advances in Neural Information Processing Systems (NIPS), pp. 1097–1105. MIT Press, Lake Tahoe (2012)
18. Kuo, C., Lai, S., Sarkis, M.: A compact deep learning model for robust facial expression recognition. In: Proceedings of the IEEE Conference on Computer Vision and Pattern Recognition Workshops, pp. 2121–2129. IEEE, Salt Lake City (2018)
19. Li, S., Deng, W.: Deep facial expression recognition: a survey, pp. 1–25. arXiv preprint (2018)
20. Bottou, L.: Large-scale machine learning with stochastic gradient descent. In: Lechevallier, L., Saporta, G. (eds.) Proceedings of COMPSTAT 2010, pp. 177–186. Physica-Verlag HD, Heidelberg (2010). https://doi.org/10.1007/978-3-7908-2604-3_16
21. Kingma, D., Ba, J.: Adam: a method for stochastic optimization. In: International Conference on Learning Representations (ICLR), pp. 1–15 (2015)

22. Nair, V., Hinton, G.: Rectified linear units improve restricted Boltzmann machines. In: Proceedings of the 27th International Conference on Machine Learning (ICML), pp. 807–814 (2010)

23. Zhao, G., Huang, X., Taini, M., et al.: Facial expression recognition from near-infrared videos. In: Image and Vision Computing (IVC), pp. 607–619 (2011)

24. Guo, Y., Zhao, G., Pietikäinen, M.: Dynamic facial expression recognition using longitudinal facial expression atlases. In: Fitzgibbon, A., Lazebnik, S., Perona, P., Sato, Y., Schmid, C. (eds.) ECCV 2012. LNCS, pp. 631–644. Springer, Heidelberg (2012). https://doi.org/10.1007/978-3-642-33709-3_45

Blurred Template Matching
Based on Cascaded Network

Juncai Peng[✉], Nong Sang, Changxin Gao, and Lerenhan Li

National Key Laboratory of Science and Technology on Multispectral Information
Processing, School of Artificial Intelligence and Automation,
Huazhong University of Science and Technology, Wuhan, China
{jcpeng,nsang,cgao,lrhli}@hust.edu.cn

Abstract. Template matching is widely used in computer vision applications, but most matching methods simply assume the ideal images without real-word degradations, such as Gaussian blur. Traditional methods for blurred template matching either first resort to image deblurring and then perform template matching with the recovered image, or joint solve image deblurring and matching based on sparse expression prior. However, these methods always perform poor and the matching speed is slow. In this paper, we propose a blurred template matching method based on a cascaded network, which combines a coarse matching network and a fine matching network. The coarse matching network searches for a small image where the target matching position is located in the reference image, and then the fine matching network calculates the similarity between the blurred template image and all image patches in the small image, thus the matching position is the corresponding position of the image patch with the highest similarity. Extensive experiments demonstrate that our method significantly outperforms the state-of-art on the accuracy, speed and robustness.

Keywords: Blurred template matching · Image matching · Cascaded
network · Convolution network

1   Introduction

Template matching is finding the position of a template image in a reference image, and it is one of the fundamental techniques in a broad variety of computer vision applications, such as pattern recognition [1,2], image mosaic [3,4]. Different from object detection, template matching does not learn the features of specific objects and the template image may contain one, two or several objects. In general, there are two major classic template matching methods [5]: feature-based methods and pixel-based methods. The key point of feature-based methods is to extract robust feature vectors, and many feature extracting algorithms are proposed, including SUSAN [6], FAST [7], SIFT [8], SURF [9] and ORB [10]. Feature-based methods are resistant to illumination change and affine transformation, but it is difficult to extract robust feature vectors when the image is

Y. Zhao et al. (Eds.): ICIG 2019, LNCS 11901, pp. 480–492, 2019.
https://doi.org/10.1007/978-3-030-34120-6_39

heavy corrupted. Pixel-based methods make use of all pixels instead of feature vectors to find the image patches, which is the most similar to the template image. These methods always utilize different similarity measure, such as sum of squared differences, normalized cross correlation [11], increment sign correlation [12], selective correlation [13] and occlusion-free correlation [14]. Usually, pixel-based methods are superior to feature-based methods for image noise and occlusion. However, classic matching methods cannot tackle complex transformation. Recently, many improved methods are proposed to overcome real-life challenges. Dekel et al. [15] propose best-buddies similarity measure, Talmi et al. [16] introduce deformable diversity similarity measure and Kat et al. [17] introduce co-occurrence based similarity measure. Concurrently, researchers utilize deep learning network to compute the similarity of image patches. Han et al. [18] present a unified method named MatchNet, which consists of a deep convolutional network that extracts features from patches and a fully connected network that outputs a similarity between the extracted features. Zagoruyko et al. [19] also propose multiple neural network architectures to learn a general similarity function for comparing image patches.

In practical applications, the template image is inevitable to be blurred by Gaussian blur, but the above methods can not effectively tackle this case. For blurred template matching, a straightforward method is first utilizing image deblurring [20,21] to estimate the latent template image, and then performing template matching. With the help of image deblurring, the two-stage method relieves the effect of image blurring on matching, but it maybe suffer greatly from the deficiency of image deblurring. To avoid the problem, Shao et al. [5] propose a joint image deblurring and matching method (JRM-DSR), which utilizes the sparse representation prior to exploit the correlation between deblurring and matching. The method achieves deblurring and matching simultaneous, and these two tasks benefit greatly from each other. However, when the template image becomes more and more blurred, the matching accuracy of JRM-DSR will decrease dramatically. Besides, the optimization of JRM-DSR needs to solve the sparse representation of high-dimensional pixel vector, which results in slow matching speed. Moreover, once the reference image changes or the size of the template image changes, JRM-DSR has to reconstruct the image dictionary, which shows that the robustness of the method is not good.

In this paper, we propose a blurred template matching method based on a cascaded network. Adopting the coarse to fine matching strategy, the cascade network combines a coarse matching network and a fine matching network, both of which are derived from Siamese network. The framework of our method is shown in Fig. 1. Given a blurred template image and a clear reference image, the coarse matching network searches for a small image where the target matching position is located in the reference image, and then the fine matching network calculates the similarity between the blurred template image and all image patches in the small image, thus the matching position is the corresponding position of the image patch with the highest similarity. In brief, the main contributions of this paper are as follows:

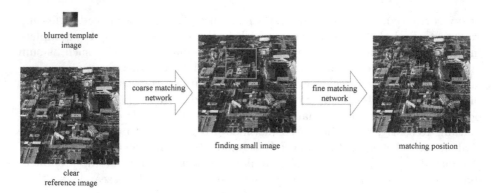

Fig. 1. The framework of cascaded network for blurred template matching. Given a blurred template image and a clear reference image, the coarse matching network search for the small image where the target matching position is located in the reference image, and then the fine matching network calculates the similarity between the blurred template image and all image patches in the small image to determine the matching position.

(1) We innovatively apply deep learning to blurred template matching and propose a new method based on a cascaded network. Different from the conventional blurred template matching methods, the proposed method directly learn feature vectors and similarity measurement from training data.
(2) Extensive experiments have been conducted. The results demonstrate the effectiveness and show the proposed method significantly outperforms the state-of-art in terms of matching accuracy, speed and robustness.

The remainder of this paper is organized as follows. Section 2 describes the proposed method and details the architecture of the cascaded network. Section 3 explains how to generate the training data and discusses the objective function. Section 4 presents the experimental results and analysis under different conditions. In the last of this paper, we conclude our work with a summary in Sect. 5.

2   Proposed Method and Network Architecture

Motivated by recent successes on learning features and similarity measure, we present a blurred template matching method based on the cascaded network, as shown in Fig. 1. The cascaded network contains a coarse matching network and a fine matching network. Given a blurred template image and a clear reference image, we first utilize the coarse matching network to search for a small image where the target matching position is located in the reference image. Afterwards, image patches of the same size as the template images are extracted in the small image, and the fine matching network calculates the similarity between the blurred template image and all image patches. Finally, we obtain the matching position in the reference image by the corresponding position of the image patch with the highest similarity. Generally speaking, the coarse matching network accelerates matching speed by reducing the search region in the reference image, and the fine matching network ensures matching accuracy by determining the matching position. Our method innovatively applies deep learning network to blurred template matching, and learns more robust feature vectors and more accurate similarity measurement, thus it greatly improves the matching accuracy. The computing time of the cascade network is short, and our method does not need any preparations for the change of template image and reference image, so the matching speed is very fast. Besides, our method is robust to the size of template image and reference image, and it directly outputs the matching position. In the next, we detail the architecture of the coarse matching network and the fine matching network.

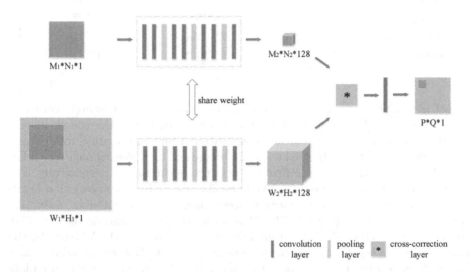

Fig. 2. The architecture of coarse matching network, which combines a fully-convolutional Siamese module and a cross-correction layer. A blurred template image and a clear reference image are input, and the coarse matching network searches for a small image where the target matching position is located in the reference image.

2.1 Coarse Matching Network

As for blurred template matching, we can utilize various type of Siamese network [18,19] to compute the similarity between the blurred template image and all image patches extracted in the reference image, while the size of image patch is the same as the template image. However, when the reference image size is large, the number of image patches is too big, and the time complexity of the straightforward method is too high to meet the speed requirement.

Inspired by the application of Siamese network in object tracking [22], we proposed a coarse matching network to search for the small image where the

Table 1. Layer parameters of fully-convolutional Siamese module. Layer type: C denotes convolution, MP denotes max-pooling. The output dimension of template image and reference image is *height × width × channel*.

Layer name	Layer type	Kernel size	Strides	Template image output dim.	Reference image output dim.
Conv1_1	C	3*3	1	50*50*64	350*350*64
Conv1_2	C	3*3	1	50*50*64	350*350*64
Pool1	MP	2*2	2	25*25*64	175*175*64
Conv2_1	C	3*3	1	25*25*128	175*175*128
Conv2_2	C	3*3	1	25*25*128	175*175*128
Pool2	MP	2*2	2	13*13*128	88*88*128
Conv3_1	C	3*3	1	13*13*256	88*88*256
Conv3_2	C	3*3	1	13*13*256	88*88*256
Pool3	MP	2*2	2	7*7*256	44*44*256
Conv4	C	3*3	1	7*7*128	44*44*128

target matching position is located in the reference image. The architecture of the coarse matching network is shown in Fig. 2. In short, the coarse matching network mainly contains a fully-convolutional Siamese module and a cross-correction layer. Given a blurred template image and a reference image, the fully-convolutional Siamese module first extracts the feature maps respectively. The Siamese module is influenced by VGGNet [23] and only has convolution layers and max-pooling layers, thus it can process images of different sizes. Assuming that the template image size is 50×50 and the reference image size is 350×350, the layer parameters of Siamese module are listed in Table 1. Afterwards, the cross-correction layer combines the feature maps of template image and reference image. Specifically, the cross-correction layer sets the feature maps of template image as kernels, and then performs convolution on each channel of the reference image's feature maps, as described in Eq. 1.

$$\mathbf{Y}_i = \mathbf{B}_i * \mathbf{I}_i, i = 1, 2, ..., 128 \tag{1}$$

where $\mathbf{B}$ denotes the feature maps of template image, $\mathbf{I}$ is the feature maps of reference image, and $\mathbf{Y}$ is the output. Generally, the operation of the cross-correlation layer is mathematically equivalent to utilize the inner product to independently evaluate the template image and each image patch in the reference image. Finally, we obtain a heat map by the output $\mathbf{Y}$ and a 1×1 convolution layer. The value of the heat map ranges from -1 to $+1$, and the pixel value greater than 0 denotes the small image where the target matching position is located in the reference image. Besides, we pad the convolution and pooling layers, so the output height and width are the same as input, and use ReLU as non-linearity for the convolution layers.

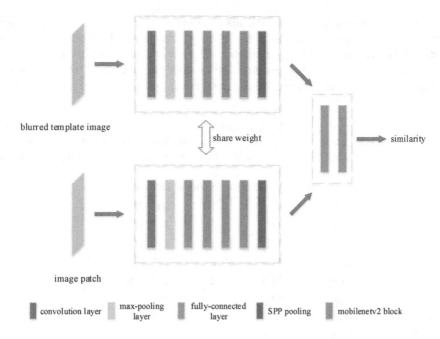

Fig. 3. The architecture of fine matching network, which contains a Siamese module and a metric module. Based on the matching result of the coarse matching network, the fine matching network calculates the similarity between the blurred template image and all image patches in the small image, and then outputs the matching position in the reference image.

2.2 Fine Matching Network

Based on the small image where the target matching position is located in the reference image, we propose a fine matching network to determine the final matching position. For the small image, we first extract the image patches of the same size as the template image. Afterward, we input the template image and an image patch into the fine matching network each time, and it calculates the similarity of the image pair. Therefore, the matching position is the corresponding position of the image patch with the highest similarity in the reference image.

The architecture of the fine matching network is shown in Fig. 3, and the layer parameters are listed in Table 2. As we can see, the fine matching network contains a Siamese module and a metric module. The Siamese module consists of convolution, max-pooling, three mobilenetv2 block [24] and SPP pooling [25]. Using the mobilenetv2 block, the Siamese module reduces a lot of computation and parameters. With the help of SPP pooling, it extracts fixed dimension features for the image of different size, thus we obtain two features with 640 dimensions. In the next, we concatenate the two features and pass it through the metric module, which consists of two fully-connected layers, and the similarity of two images is output. We also utilize ReLU as nonlinear activation function and set the output size is the same as the input. Different from these works [18,19], the proposed fine matching network calculates similarity based on the pixel deviation of two images, and the smaller the deviation, the higher the similarity. Moreover, the fine matching network is resistant to image blurring.

Table 2. Layer parameters of the fine matching network, where the size of blurred template image is 50×50. Layer type: C denotes convolution, MP denotes max-pooling, MB denotes mobilenetv2 block, SPP denotes SPP pooling, FC denotes fully-connected.

Layer name	Layer type	Strides	Output dim.
Conv1	C	1	50*50*32
Pool1	MP	2	25*25*32
Block1	MB	2	13*13*64
Block2	MB	2	7*7*128
Block3	MB	2	4*4*258
Block4	MB	1	4*4*128
Pool2	SPP	-	640
FC1	FC	-	256
FC2	FC	-	1

3   Training

In this section, we first discuss the training set of the coarse and fine matching network and then describe the objective functions.

The training data of the coarse matching network is a blurred template image, a reference image and a heat map. However, there are no standard datasets to train the network. Therefore, we choose 36000 images from MIT Places2 dataset [26] as clear reference images. For each reference image, we generate its blurred image by Gaussian blur kernel, and then randomly select a small image with random size from each blurred reference image as the blurred template image. Based on the blurred template image and the clear reference image, we can construct the corresponding heat map. In the heat map, the pixel value equal to 1 indicates the position of the template image in the reference image, otherwise the pixel value equal to -1. Besides, the size of the heat map changes with the reference image.

For the fine matching network, the training data is a blurred template image, a clear image with the same size of blurred template image and a label. When the two images is the same, the label equals to 1. Otherwise the label equals to 0. Given the 36000 images selected from MIT Places2 dataset, we also generate the blurred images of each reference image by Gaussian blur kernel, of which the standard deviation range from 1 to 5. Afterwards, a position in reference image

(x, y) and a image size (w, h) are randomly selected. On the blurred reference image, we extract the blurred template image I_1 with (x, y) as the center coordinate and (w, h) as the image size. On the clear reference image, we extract a clear image I_2 with the same center coordinate and image size, thus $(I_1, I_2, 1)$ is a positive training data. We extract another clear image I_3, of which the center coordinate is near to I_2, and combine the $(I_1, I_3, 0)$ into a negative training data. Therefore, we can obtain a large amount of training data by adopting the above method.

Given the training set, we train the coarse matching network and fine matching network in a strongly supervised manner. For the coarse matching network, the object function hopes the heat map and the ground truth will be the same as possible, so the logistic loss is adopted.

$$L = \frac{1}{P * Q} \sum_{i=0}^{P-1} \sum_{j=0}^{Q-1} log(1 + exp(-S(i, j) * G(i, j))) \tag{2}$$

where $S(i, j)$ means the value of predicted heat map, $G(i, j)$ means the value of ground truth.

In the small image where the target matching position is located in the reference image, only one image patch corresponds to the target matching position. However, there are many image patches that are close to the target image patch. In order to improve the matching accuracy of the fine matching network, we adopt the focal loss function [27]

$$L = -y(1 - p)^\gamma log(p) - (1 - y)p^\gamma log(1 - p) \tag{3}$$

where y is the training label, p is the predicted similarity, γ is the weight and we set $\gamma = 2$.

In the training and test stage, we use Nvidia GTX1080 in tensorflow and cuDNN library as usual. Adam with initial learning rate 0.0001 and weight decay 0.0003 is adopted to train the coarse matching network and fine matching network. Besides, the mini-batches is 8 and 128 for the two network.

4   Experiments

In this section, we first discuss the test image dataset. Besides, the experiments are carried out, and we compare our approach with other methods in terms of matching accuracy, speed and robustness.

In the experiments, the test reference images are six aerial images. We also apply Gaussian blur kernel to each reference image and randomly select 100 small images as the blurred template images. The size of reference images and template images are 600×600 and 50×50, respectively. Besides, the size of the Gaussian blur kernel is $(6\sigma + 1) \times (6\sigma + 1)$, where the σ denotes the standard deviation and range from 1 to 5. Therefore, the experiments are conducted with the reference image, blurred template image and the corresponding matching position. In order to quantify the matching accuracy, we adopt the Manhattan distance (MD) between the predicted matching position and the ground truth, and then obtain the percentage of test samples matched accurately under different MD.

4.1 Matching Accuracy

We apply our method to blurred template matching and examine the performance. The compared methods are as follows: (1) NCC [11] : template matching based on normalized correlation coefficient; (2) DNCC: utilize image deblurring to recover the latent template image and perform NCC; (3) JRM-DSR [5]: joint image deblurring and matching.

When the standard deviation of the Gaussian blur kernel is 3, the matching accuracy of the test dataset is listed in Table 3. The results show that the matching accuracy of DNCC is lower than NCC, which means image deblurring cannot help matching. We can observe that JRM-DSR has better performance than NCC and DNCC, because it combines image deblurring and matching. Obviously, our method has the highest matching accuracy. In details, the matching accuracy $\mathbf{P}_{md=5}$ for other method are 66.67%, 65.33% and 91.00%, but our method achieves 100.00% under the same conditions. In brief, our method significantly outperforms NCC, DNCC, and JRM-DSR in terms of matching accuracy.

Table 3. The matching accuracy of different methods, where the standard deviation of Gaussian blur kernel is 3.

Methods	$\mathbf{P}_{md=0}$	$\mathbf{P}_{md=1}$	$\mathbf{P}_{md=2}$	$\mathbf{P}_{md=3}$	$\mathbf{P}_{md=4}$	$\mathbf{P}_{md=5}$
NCC	47.00	61.33	65.50	66.33	66.50	66.67
DNCC	2.33	10.33	24.17	41.83	55.67	65.33
JRM-DSR	34.67	70.00	86.33	90.33	90.83	91.00
Ours	94.67	99.83	100.00	100.00	100.00	100.00

4.2 Matching Speed

We also examine the matching speed of the proposed method and compare it with other methods. In the experiment, we count the preparation time, computing time and total time of methods. The matching speed of different methods is listed in Table 4. We can see that it takes a lot of time for JRM-DSR to construct the image dictionary, but other methods do not require any preparations. The total time of NCC is the shortest, followed by our method, but the matching accuracy of NCC is low. Besides, the total time of DNCC and JRM-DSR is 4.52 and 1026.67 times that of our method, respectively. Therefore, while ensuring the highest matching accuracy, the matching speed of our method is very fast.

Table 4. The matching speed of different methods. The preparation time of JRM-DSR is the time spent in constructing the image dictionary. The computing time is the time from the start of matching to the end of matching, and the total time is the sum of preparation time and computing time.

Methods	Preparation time(s)	Computing time(s)	Total time(s)
NCC	0.00	0.26	0.26
DNCC	0.00	2.58	2.58
JRM-DSR	552.00	33.20	585.20
Ours	0.00	0.57	0.57

4.3 Robustness Analysis

Influence of the Standard Deviation of Gaussian Blur Kernel. Image blurring has a great influence on the matching accuracy, and the greater the standard deviation of Gaussian blur kernel, the more difficult the matching is. Therefore, experiments are carried out to demonstrate that our approach is robust to the standard deviation (σ) of Gaussian blur kernel. In the experiments, the σ range from 1 to 5 and the results are shown in Table 5. When the σ changes from 1 to 5, the matching accuracy of NCC, DNCC and JRM-DSR decreases by 67.33%, 76.67% and 51.50%, respectively. However, the matching accuracy of our method has always been 100%. Consequently, our method is more robust to the standard deviation of the Gaussian blur kernel.

Table 5. Image matching results comparison in terms of the standard deviation (σ) of Gaussian blur kernel, where the matching accuracy for $P_{md<=5}$

Methods	$\sigma = 1$	$\sigma = 2$	$\sigma = 3$	$\sigma = 4$	$\sigma = 5$
NCC	100.00	90.83	66.67	47.83	32.67
DNCC	99.50	93.67	65.33	25.67	23.33
JRM-DSR	100.00	98.83	91.00	72.00	48.50
Ours	100.00	100.00	100.00	100.00	100.00

Influence of Scale Variation. The experiment analyses the robustness of our method to template image size, and the matching accuracy of different methods are shown in Fig. 4. The results show that our method achieves the highest matching accuracy regardless of the size of the template image, while the matching accuracy of NCC and JRM-DSR varies greatly. Especially, when the template image size changes, JRM-DSR takes a lot of time to construct image dictionary or feature dictionary. However, our method does not have preparations and retraining, and it can directly process different reference images and

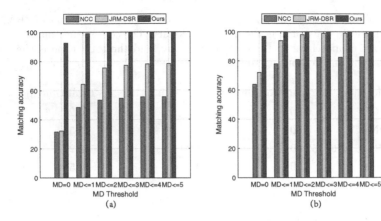

Fig. 4. Image matching results comparison in terms of the scale variation, where the standard deviation of Gaussian blur kernel is 3. (a) the size of template image is 40×40, (b) the size of template image is 60×60.

template images. Based on the above experimental results and analysis, it is obvious that the robustness of our method to scale variation is better than other methods.

5   Conclusions

In this paper, we have presented a blurred template matching method based on a cascaded network. Our method utilizes a coarse matching network to search for the small image where the target matching position is located, and then use a fine matching network to determine the final exact matching position in the reference image. The experimental results and analysis demonstrate its effectiveness on blurred template matching, and our method significantly outperforms the start-of-art in terms of matching accuracy, speed and robustness.

References

1. Ryan, M., Hanafiah, N.: An examination of character recognition on ID card using template matching approach. Proc. Comput. Sci. **59**, 520–529 (2015)
2. Boia, R., Florea, C., Florea, L., et al.: Logo localization and recognition in natural images using homographic class graphs. Mach. Vis. Appl. **27**(2), 287–301 (2016)
3. Brown, M., Lowe, D.G.: Automatic panoramic image stitching using invariant features. Int. J. Comput. Vis. **74**(1), 59–73 (2007)
4. Szeliski, R.: Image alignment and stitching: a tutorial. Found. Trends®Comput. Graph. Vis. **2**(1), 1–104 (2007)
5. Shao, Y., Sang, N., Gao, C., et al.: Joint image restoration and matching based on distance-weighted sparse representation. In: 2018 24th International Conference on Pattern Recognition (ICPR), pp. 2498–2503. IEEE (2018)

6. Smith, S.M., Brady, J.M.: SUSAN—a new approach to low level image processing. Int. J. Comput. Vis. **23**(1), 45–78 (1997)
7. Rosten, E., Drummond, T.: Machine learning for high-speed corner detection. In: Leonardis, A., Bischof, H., Pinz, A. (eds.) ECCV 2006. LNCS, vol. 3951, pp. 430–443. Springer, Heidelberg (2006). https://doi.org/10.1007/11744023_34
8. Lowe, D.G.: Distinctive image features from scale-invariant keypoints. Int. J. Comput. Vis. **60**(2), 91–110 (2004)
9. Bay, H., Tuytelaars, T., Van Gool, L.: SURF: speeded up robust features. In: Leonardis, A., Bischof, H., Pinz, A. (eds.) ECCV 2006. LNCS, vol. 3951, pp. 404–417. Springer, Heidelberg (2006). https://doi.org/10.1007/11744023_32
10. Rublee, E., Rabaud, V., Konolige, K., et al.: ORB: an efficient alternative to SIFT or SURF. ICCV. **11**(1), 2 (2011)
11. Aggarwal, J.K., Davis, L.S., Martin, W.N.: Correspondence processes in dynamic scene analysis. Proc. IEEE **69**(5), 562–572 (1981)
12. Kaneko, S., Murase, I., Igarashi, S.: Robust image registration by increment sign correlation. Pattern Recognit. **35**(10), 2223–2234 (2002)
13. Kaneko, S., Satoh, Y., Igarashi, S.: Using selective correlation coefficient for robust image registration. Pattern Recognit. **36**(5), 1165–1173 (2003)
14. Yoo, J.C., Ahn, C.W.: Image matching using peak signal-to-noise ratio-based occlusion detection. IET Image Proc. **6**(5), 483–495 (2012)
15. Dekel, T., Oron, S., Rubinstein, M., et al.: Best-buddies similarity for robust template matching. In: Proceedings of the IEEE Conference on Computer Vision and Pattern Recognition, pp. 2021–2029 (2015)
16. Talmi, I., Mechrez, R., Zelnik-Manor, L.: Template matching with deformable diversity similarity. In: Proceedings of the IEEE Conference on Computer Vision and Pattern Recognition, pp. 175–183 (2017)
17. Kat, R., Jevnisek, R., Avidan, S.: Matching pixels using co-occurrence statistics. In: Proceedings of the IEEE Conference on Computer Vision and Pattern Recognition, pp. 1751–1759 (2018)
18. Han, X., et al.: MatchNet: unifying feature and metric learning for patch-based matching. In: Proceedings of the IEEE Conference on Computer Vision and Pattern Recognition (2015)
19. Zagoruyko, S., Komodakis, N.: Learning to compare image patches via convolutional neural networks. In: Proceedings of the IEEE Conference on Computer Vision and Pattern Recognition (2015)
20. Xu, L., Zheng, S., Jia, J.: Unnatural L0 sparse representation for natural image deblurring. In: Proceedings of the IEEE Conference on Computer Vision and Pattern Recognition, pp. 1107–1114 (2013)
21. Pan, J., Sun, D., Pfister, H., et al.: Blind image deblurring using dark channel prior. In: Proceedings of the IEEE Conference on Computer Vision and Pattern Recognition, pp. 1628–1636 (2016)
22. Bertinetto, L., Valmadre, J., Henriques, J.F., Vedaldi, A., Torr, P.H.S.: Fully-convolutional siamese networks for object tracking. In: Hua, G., Jégou, H. (eds.) ECCV 2016. LNCS, vol. 9914, pp. 850–865. Springer, Cham (2016). https://doi.org/10.1007/978-3-319-48881-3_56
23. Simonyan, K., Zisserman, A.: Very deep convolutional networks for large-scale image recognition. arXiv preprint arXiv:1409.1556 (2014)
24. Sandler, M., Howard, A., Zhu, M., et al.: Mobilenetv 2: inverted residuals and linear bottlenecks. In: Proceedings of the IEEE Conference on Computer Vision and Pattern Recognition, pp. 4510–4520 (2018)

25. He, K., Zhang, X., Ren, S., et al.: Spatial pyramid pooling in deep convolutional networks for visual recognition. IEEE Trans. Pattern Anal. Mach. Intell. **37**(9), 1904–1916 (2015)
26. Zhou, B., Lapedriza, A., Khosla, A., et al.: Places: a 10 million image database for scene recognition. IEEE Trans. Pattern Anal. Mach. Intell. **40**(6), 1452–1464 (2018)
27. Lin, T.Y., Goyal, P., Girshick, R., et al.: Focal loss for dense object detection. In: Proceedings of the IEEE International Conference on Computer Vision, pp. 2980–2988 (2017)

A Method for Analyzing the Composition of Petrographic Thin Section Image

Lanfang Dong$^{(\boxtimes)}$ ⓘ and Zhongya Zhang

School of Computer Science and Technology,
University of Science and Technology of China, Hefei, China
lfdong@ustc.edu.cn, zhongya@mail.ustc.edu.cn

Abstract. The recognition and content analysis of the components in petrographic thin section image is a valuable study in geology. In this paper, we propose a two-stage method to segmentation and recognition of petrographic thin section image. In the first stage, we propose an image segmentation algorithm that can adaptively generate superpixel numbers based on SLIC algorithm. The algorithm is able to continually correct the number of superpixels in the iteration and then cluster the pixels of the image into superpixels by both color and spatial features. In the second stage, we designed a convolutional neural network and trained it with mineral grain images, which is then used to classify the superpixels obtained by first-stage. Finally, we count the categories and content of the components in the image based on the segmentation and classification results. We collected some images and invited geologists to label them for experimentation. The experimental results demonstrate the following: (1) Our proposed image segmentation algorithm is capable of dynamically generating the superpixels by the number of mineral grains in the image. (2) The CNN model we designed can accurately identify the categories of superpixel regions and has a small size. (3) The two-stage method is very effective in identifying the category of major components in an image and accurately estimating the content.

Keywords: Petrographic thin section image · Superpixels · Image segmentation · Image recognition · Component analysis

1 Introduction

The recognition and analysis of petrographic thin section plays an important role in the development of oil and gas resources. The traditional method of recognizing petrographic thin section is to cut and grind them into tens of micrometers of thin section, and then the researchers analyzes the image under the microscope [1]. Traditional methods require the knowledge of a researcher to complete, and the analysis of rock composition is cumbersome, time-consuming, and inaccurate. With the rapid development of digital image processing, image recognition technology has been widely used in petrographic thin section analysis. Segmentation and recognition of mineral grains in thin section images by image segmentation and recognition algorithms can accurately measure the categories and contents of major components in rock samples, thus better assisting petroleum development.

© Springer Nature Switzerland AG 2019
Y. Zhao et al. (Eds.): ICIG 2019, LNCS 11901, pp. 493–504, 2019.
https://doi.org/10.1007/978-3-030-34120-6_40

Because rock samples are usually formed by the bonding of mineral grains, there is a good idea for the analysis of the composition of petrographic thin section images. Using the image segmentation algorithm to segment the grains in the image and then identify each grain, the category and content of the main components in the thin section can be counted. In principle, almost all existing image segmentation algorithms can be used in the automatic segmentation of grain, such as region growing [2], ERS [3] and TurboPixel [4]. However, these algorithms are designed for ordinary scene images and may not be suitable for processing petrographic thin section image which contains large numbers of mineral grains. The superpixel segmentation algorithm is an image region over-segmentation algorithm. The image is divided into multiple irregular image blocks according to certain feature similarities. The current effective algorithm are LSC [5], SEEDS [6], SLIC [7] and so on. We believe that the sub-regions subdivided by the superpixels are similar to the contours of the mineral grains, so the superpixels obtained by the segmentation can be identified as a single grain. For example, Jiang et al. [8, 9] used a superpixel algorithm to segment and merge sandstone thin section images to obtain the contours of sandstone grains. However, the currently existing superpixel algorithm is very inefficient, or it is necessary to set the number of superpixels desired in advance. Therefore, we propose an algorithm that is fast and can dynamically adjust the number of superpixels based on the number of grains in the image.

After the image is segmented, the segmented sub-regions need to be identified. Since the crystal grains constituting the rock are generally colorless and transparent and have similar refractive indices, adjacent grains exhibit similar colors under a plane polarizing microscope, and it is sometimes difficult for the human eye to discern the difference. Unlike plane polarized light images, grains produce different interference colors under orthogonally polarized light [10]. For example, quartz and feldspar in sandstones require experts to combine orthogonal polarized and plane polarized images to distinguish them [11]. In recent years, convolutional neural networks have achieved remarkable success in a lots of computer vision tasks, such as image recognition [12, 13], object detection [14] and semantic segmentation [15]. Therefore, we have designed a CNN model that can combine the orthogonal polarized and plane polarized images to identify the sub-regions obtained by the previous segmentation. After identifying the category of each sub-regions, the area of each sub-regions of each category is counted, and the categories and contents of the major components in the image can be approximated.

In this paper, we have mainly completed the following work: (1) We propose a two-stage method to identify the major components of petrographic thin section. In the first stage, the image is segmented into over-segmented superpixel images. In the second stage, the CNN model is trained to identify each superpixel region to obtain the category of the mineral grain to which each superpixel region belongs. (2) We have enhanced the SLIC algorithm and proposed an algorithm that can generate superpixel adaptively based on the number of mineral grains in the image. For different images, our algorithm only needs to set a fixed K value, and the algorithm can dynamically adjust the number of superpixels to be segmented in multiple iterations. (3) We designed a CNN model that effectively classifies mineral grains that are common in petrographic thin section images. The CNN model can receive orthogonal polarized and plane polarized images as inputs, and can adapt to any size. (4) We created a dataset that included thousands of petrographic thin section images. The details of the dataset are described in Sect. 2.3.

2   Proposed Method

The compositional analysis of the petrographic thin section image is divided into two stage. In the first stage, we proposed the SLIC method of adaptive superpixel number (AS-SLIC). The AS-SLIC algorithm can adaptively generate multiple superpixels based on the number and characteristics of mineral grains in the image. In the second stage, the CNN is trained to identify the superpixels obtained by the previous stage. By identifying the category of mineral grains in each superpixel region, the approximate content of the major components in each thin section image can be calculated.

2.1   The SLIC of Adaptive Superpixel Number (AS-SLIC)

We first introduce the original SLIC algorithm. The original SLIC superpixels correspond to clusters in the CIELAB color space and spatial space. First, the algorithm needs to set two values m and K, m is a constant value, and K represents the number of superpixels that are desired to be divided. Then, the image is divided into K grids, and the center point of each grid is used as the initial cluster $C_k = [l_k, a_k, b_k, x_k, y_k]^T$. A pixel P_i color is represented in the CIELAB color space $[l, a, b]^T$, as shown in Eq. (1), the d_c is color distance between P_i and C_k. The $[x, y]^T$ represents the position of the pixel. As shown in Eq. (2), the d_s represents the spatial space distance between P_i and C_k. The D_{dist} computes the distance between P_i and C_k is measured in both color and spatial space, which is defined as Eq. (3). In the Eq. (3), $S = \sqrt{N/K}$, where N is the number of pixels of the input image. Since the initial superpixel approximates an $S \times S$ region, the search for similar pixels is done in a region $2S \times 2S$ around the cluster center. In one iteration, each pixel is assigned to the nearest cluster center, then the new cluster center is recalculated and the next iteration continues.

$$d_c = \sqrt{(l_i - l_k)^2 + (a_i - a_k)^2 + (b_i - b_k)^2} \tag{1}$$

$$d_s = \sqrt{(x_i - x_k)^2 + (y_i - y_k)^2} \tag{2}$$

$$D_{dist} = \sqrt{\left(\frac{d_c}{m}\right)^2 + \left(\frac{d_s}{S}\right)^2} \tag{3}$$

In the petrographic thin section image especially the sandstone thin section image, the mineral grains of different origins are different in size, so the number of grains in the image varies from tens to hundreds. The original SLIC algorithm needs to set the number of superpixels divided in advance. When the number of mineral grains in the image is large, setting a smaller K will cause a superpixel to contain multiple grains. Conversely, if the number of grains in the image is small, setting a larger K will cause over-segmentation. Both of the above cases are not conducive to subsequent classification. As shown in Fig. 2(a), we use the original SLIC algorithm and set K to 20. The results on the left Fig. 2(a) are acceptable, but because the grain area in the right of Fig. 2(a) is small, many superpixels contain multiple grain regions.

We proposed AS-SLIC algorithm takes into account the properties of mineral grains in petrographic thin section image. Most of the mineral grain composition is relatively single, so a single grain exhibits a uniform color in the image. Because of the optical properties of the crystals that make up the mineral grains, the differently oriented grain will show different colors under an orthogonal polarizing microscope. When a plurality of grain regions are included in one superpixel obtained by the segmentation, the superpixel will present a plurality of regions of different colors. So we can use the color histogram of all the pixels in the superpixel to determine whether the superpixel contains only a single grain. As shown in Fig. 1(a), this superpixel contains two grain regions, so the gray histogram of the superpixel has two peaks as shown in Fig. 1(b).

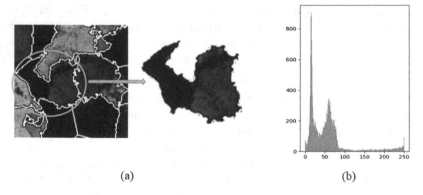

(a)   (b)

Fig. 1. (a) Is a superpixel that is not completely split. (b) Is the gray histogram of the (a).

Our method still needs to set K at the beginning, and K represents the number of cluster centers in the first iteration. Because the number of cluster centers will gradually increase during the iteration, the initial set K value is relatively small. In the original SLIC algorithm, once each pixel has been assigned to the appropriate cluster, then the mean vector of all the pixels belonging to each cluster are recalculated to update the cluster center. The difference between the algorithm we designed and the original SLIC lies in the update strategy of the cluster center. First, for each cluster center, the gray histogram H_k of all the pixels belonging to the cluster is counted. If the resulting gray histogram has only one peak, then it is likely that only one whole grain or a local region of one grain is included in the cluster. In this case, the average vector $V_k^i = [l_i, a_i, b_i, x_i, y_i]^T$ of all the pixels belonging to this cluster C_k is used as the new cluster center C_k', as shown in the Eq. (4). If the gray histogram has two or more peaks, the pixels belonging to this cluster C_k are from multiple grain regions. Therefore, it is necessary to split the cluster C_k into multiple cluster. It is assumed that there are T peaks in the gray histogram H_k, the corresponding gray value at the peak is represented by $Peak_i$, and V_k^j represents a vector of pixels in the cluster C_k whose gray value is equal to $Peak_i$. The previous cluster C_k is split into T new cluster according to peak. The calculation method of the new cluster C_k^i is as shown in the Eq. (5), where m represents the number of pixels in the cluster center C_k whose gray value is equal to $Peak_i$.

$$C_k' = \frac{1}{n}\sum\nolimits_{i=1}^{n} V_k^i, \mathrm{n} = |C_k| \tag{4}$$

$$C_k^i = \frac{1}{m}\sum\nolimits_{j=1}^{m} V_k^j, \mathrm{m} = |P_j|, P_j \in C_k \text{ and } P_j \text{ gray value is equal to } Peak_i \tag{5}$$

In order to avoid a superpixel splitting too many sub-regions, in each iteration we only select the two peaks containing the largest number of pixels in H_k to split. And in the experiment we set a maximum value K_{max}, when the number of cluster in the iteration is greater than K_{max}, the splitting strategy is no longer executed. We use L_2 norm shown in Eqs. (6) and (7) to compute the residual error R between the new cluster and previous cluster locations. The iteration is stopped when R is less than a certain threshold. The entire algorithm is shown in Table 1.

Table 1. The main steps of AS-SLIC algorithm

Algorithm 1.AS-SLIC

Input:I, the petrographic thin section image.
      K, the initial number of superpixels (The default value is 20).
Output: C, the initial segmentation results of I, $C = \{C_k\}_{k=1}^*$.

repeat
    for each cluster center C_k **do**
        for each pixel P_i in a $2S \times 2S$ region around C_k **do**
            Compute the distance D_{dist} between C_k and pixel P_i .
            Assign P_i to the best matching cluster C_k .
        end for
    end for
    for each cluster center C_k **do**
        Draw a gray histogram H_k of all the pixels in C_k.
        if The number of peaks in H_k is greater than 1 **then**
            Compute new cluster C_k^i by Eq.(5). Delete the cluster C_k .
            The new clusters C_k^i based on C_k is added to the clusters set C.
        else
            Compute new cluster centers C_k' .
            Set $C_k = C_k'$.
        end if
    end for
    Computer residual error R.
until $R \le threshold$

$$R = \frac{1}{|C|} \sum_{k=1}^{C} R_k \qquad (6)$$

$$R_k = \begin{cases} \left\| C_k' - C_k \right\|_2^2, & \text{if } C_k \text{ is not split.} \\ \frac{1}{T} \sum_{i=1}^{T} \left\| C_k^i - C_k \right\|_2^2, & \text{if } C_k \text{ is split.} \end{cases} \qquad (7)$$

2.2   Classification of Mineral Grains

Convolutional neural networks have been used for classification tasks for many years. Therefore, we designed a CNN model to identify the superpixel region obtained by the first stage segmentation. The orthogonal polarized image of the thin section focuses on the color features of the rock, and the plane polarized image focuses on the texture features of the rock. In order to make full use of the information in plane polarized and orthogonal polarized images, and to adapt to the superpixels of various sizes as input, we designed the CNN model to combine depthwise separable convolution [16] and spatial pyramid pooling [17].

Our CNN model architecture is shown in Table 2. The input to the model is two images, which are the orthogonal polarized image I_o and the plane polarized image I_p of the rock thin section. We expect the model to extract specific low-level semantic features from the two images, respectively. In the first input layer, the input I_o and I_p are group convolved with a $3 \times 3 \times 3 \times 32$ convolutional filter. The layer in front of the model uses a depth separable convolution, which is computationally intensive and leaves no information to communicate between the two images entered. After extracting the low-level semantic features from the two images, in order to combine the features in the two images to obtain the high-level semantic features of the rock sample as a whole, the 1×1 pointwise convolution is used for information exchange. All layers are followed by a ReLU nolinearity and batchnorm. Down sampling is handled with strided convolution in the depthwise convolutions.

Since the size of the superpixel obtained in the first stage is uncertain, and in order to prevent loss of features due to image scaling, we hope that the model can accept superpixel regions of any size or scale as input. Therefore, our model uses the strategy of spatial pyramid pooling to produce a fixed-size representation. There are two fully connected layers after the spatial pyramid pooling layer and feeds into a softmax layer for classification. To prevent overfitting, the dropout strategy is used and the value is set to 0.3. Finally, cross entropy loss is employed as the loss function to train the network. Counting convolutions and fully connected layers, our model has 28 layers.

2.3   Image Dataset

To verify the effect of our method on petrographic thin section images, we collected the images and invited geologists to mark them. First, we created the Petrographic Thin Section Image Dataset (PTSID). The PTSID includes images of 801 rock samples in 7 categories, each sample with a plane polarized image and a orthogonal polarized image. So there are 1602 images in the PTSID, and the resolution of each image is 1392 $\times$

Table 2. We designed the CNN Architecture. The "dw" represents a depthwise convolution, and "group" represents a group convolution. When the input resolution of the first layer of the model is 256 × 256, the convolution kernel and input size of each layer are as follows.

Type/Stride		Filter shape	Input size
Conv group/S = 2		$3 \times 3 \times 3 \times 32 \times 2$ group	$256 \times 256 \times 6$
Conv dw/S = 1		$3 \times 3 \times 32$ dw	$128 \times 128 \times 32$
Conv/S = 1		$1 \times 1 \times 32 \times 64$	$128 \times 128 \times 32$
Conv dw/S = 2		$3 \times 3 \times 64$ dw	$128 \times 128 \times 64$
Conv/S = 1		$1 \times 1 \times 64 \times 128$	$64 \times 64 \times 64$
2×	Conv dw/S = 1	$3 \times 3 \times 128$ dw	$64 \times 64 \times 128$
	Conv/S = 1	$1 \times 1 \times 128 \times 128$	$64 \times 64 \times 128$
Conv dw/S = 1		$3 \times 3 \times 128$ dw	$64 \times 64 \times 128$
Conv/S = 1		$1 \times 1 \times 128 \times 256$	$64 \times 64 \times 128$
Conv dw/S = 2		$3 \times 3 \times 256$ dw	$64 \times 64 \times 256$
2×	Conv/S = 1	$1 \times 1 \times 256 \times 256$	$32 \times 32 \times 256$
	Conv dw/S = 1	$3 \times 3 \times 256$ dw	$32 \times 32 \times 256$
Conv/S = 1		$1 \times 1 \times 256 \times 512$	$32 \times 32 \times 256$
Conv dw/S = 2		$3 \times 3 \times 512$ dw	$32 \times 32 \times 512$
2×	Conv/S = 1	$1 \times 1 \times 512 \times 512$	$16 \times 16 \times 512$
	Conv dw/S = 1	$3 \times 3 \times 512$ dw	$16 \times 16 \times 512$
Conv/S = 1		$1 \times 1 \times 512 \times 512$	$16 \times 16 \times 512$
Conv dw/S = 2		$3 \times 3 \times 512$ dw	$16 \times 16 \times 512$
Conv/S = 1		$1 \times 1 \times 512 \times 512$	$8 \times 8 \times 512$
Spatial pyramid pooling		2×2 pooling 1×1 pooling	$8 \times 8 \times 512$
FC1		2560×512	5×512
FC2		512×21	1×512
Softmax		–	1×21

1040 pixels. Each image is detailed with the category of sample, the category and content of the main components (the content is expressed as a decimal, and the sum of the components is 1). There are 500 images in the PTSID that are pixel-level annotations for typical mineral grains, with a total of 2126 mineral grains in 21 categories. To train the CNN model we designed, we also produced the Mineral Grain Classification Dataset (MGCD). The mineral grain region marked in the PTSID is first extracted to generate a sub-image whose size is equal to the minimum circumscribed rectangle of the grain contour. In addition to the grain region in the sub-image, the pixel values of the other regions are filled with 0. The sub-image is then rotated, scaled, and cropped to augment the dataset. In order to maintain data balance between categories, we rotate the categories with fewer images to rotate more angles to get more new images. After the above process, a total of 8150 mineral grain images in 21 categories were included in the MGCD.

3   Experiments

3.1   Superpixel Segmentation Experiment

We selected sandstone thin section images from PTSID to verify the performance of the AS-SLIC algorithm. The reason for selecting the sandstone image is that it contains a large number of grains of different sizes, which can reflect the characteristics of the algorithm adaptively generating the number of superpixels. Figure 2 shows a comparison of our AS-SLIC with the original SLIC. The initial K values of both algorithms are set to 20. As shown in the right column of Fig. 2, when the number of grains in the image is large, the superpixels segmented by the original SLIC contain multiple grains, which is not as good as our AS-SLIC. Obviously, our algorithm can generate different superpixel number segmentation results for different images.

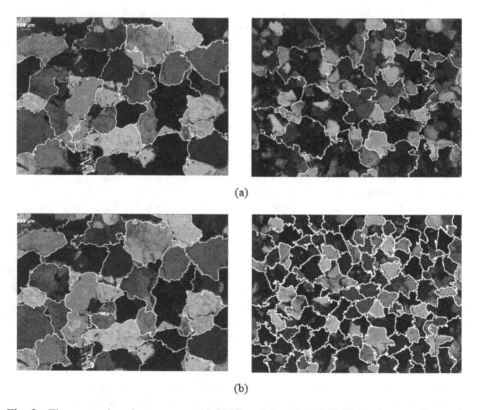

(a)

(b)

Fig. 2. The comparison between our AS-SLIC and the original SLIC. In the experiment, the initial K values of both algorithms were 20. The image in (a) is the result of segmentation using the SLIC, and (b) is the segmentation result of our AS-SLIC algorithm.

In terms of performance, we have found that majority of the image clusters do not change after 4 iterations. After 11 iterations, the residual error R of most images will be less than the threshold. Therefore, the time complexity of our algorithm in practical applications is close to that of the SLIC algorithm. We experimented with an image with a resolution of 1392 × 1040. The average computational complexity of our AS-SLIC and other existing superpixel algorithms is shown in Table 3. Because our algorithm needs to count the gray histogram in the iteration, and the number of clusters may increase, it takes more time than the original SLIC algorithm. However, compared to other classic superpixel algorithms such as GC [18], TP [4] and QS [19], our algorithm is still very fast.

Table 3. Time complexity compared to existing superpixel algorithms

Algorithm	GC [18]	TP [4]	QS [19]	SLIC [7]	Our AS-SLIC
Time (s)	77.6	203.6	70.3	**5.4**	7.1

3.2 Image Classification Result

Training Methods. Our model was built on the TensorFlow framework and was trained using the GTX 1080 GPU. We used the "Xavier" algorithm [20] to initialize the weights of all layers. The initial learning rate was 0.001 and reduced to $\sqrt{0.1}$ every 10 epochs. The training used stochastic gradient descent with 0.9 momentum. The batch size was set to 32 and the training "early stopping" (When the loss of the training set in five consecutive epochs is no longer reduced, the training is stopped) strategy is used. We use the "Multi-Size Training" strategy [17] mentioned in SPPNet for training. We resize the training set to four scales $s = \{224, 192, 160, 128\}$ and randomly selected one scale image for training in each epoch.

Results and Analyses. Our CNN model and other popular models were tested on the MGCD dataset. Table 4 compares our model to the VGG16 [21], ResNet [13], GoogLeNet [22] and MobileNet [16]. Overall, the ResNet34 model has the highest accuracy. This may be because the color and texture features in the thin section image are easily lost after multi-layer convolution, and the "shortcut connection" structure in the ResNet model makes the low-level information easier to forward propagation. Although the VGG model has the largest parameter, it is easier to overfit and the

Table 4. The experiment results of our model comparison to popular models

Network	MGCD test accuracy (%)		Number of parameters
	Top-1	Top-3	
VGG16	94.2	98.6	134 M
ResNet34	95.4	99.1	21.3 M
ResNet50	**96.1**	**99.4**	23.6 M
GoogLeNet	95.3	98.9	5.82 M
MobileNet	95.0	99.1	3.21 M
Ours	**95.4**	**99.2**	**2.74 M**

gradient disappears, so the performance is the worst. Our model is nearly as accurate as ResNet and is the smallest one of all models. Compared with the MobileNet and GoogLeNet models of the light weight, our model has the highest accuracy and the smallest model size. This shows that our model architecture is suitable for processing petrographic thin section images.

3.3   Analysis of Components

We summarize the results of the classification of each superpixel in the second stage, and the sum of the areas of the superpixel regions of each category is taken as the content of the mineral grains. As shown in Fig. 3, and 3(a) is a petrographic thin section image, and Fig. 3(b) is the segmentation and recognition result of Fig. 3(a). Each superpixel in the image covers an entire mineral grain or a part of a mineral grain, so the recognition result of the superpixel can be used as the category of the mineral grain. The different categories of mineral grains identified in the image are represented by different colors, and the percentage of each category of grains in the entire sample is counted. The results of the component analysis in this experiment were reviewed by several geologists. Experts believe that our analysis results are close to the results of manual analysis, and have a good application prospect, which provides a good solution for the automated analysis of petrographic thin section images in the future geology field.

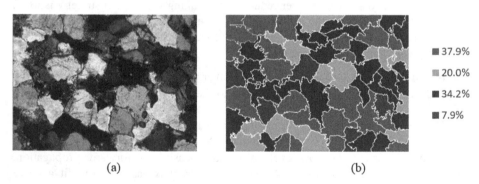

(a)   (b)

Fig. 3. Analysis of petrographic thin section image components. (a) is a petrographic thin section image. (b) is the segmentation and recognition result of (a).

3.4   Computation Cost

Our method is implemented using Python3.6 and Tensorflow1.12 on a PC with an GTX1080 GPU card (8 GB RAM) and Intel Core i7 3.6 GHz. The average processing time for an image with 1392 × 1040 resolution is 8.9 s. In the first-stage, since only the CPU was used for calculation, the average time was 7.1 s. In the second-stage, the superpixel region was identified using 1.7 s, and finally the statistics were consumed for 0.1 s. It can be seen that the first stage consumes a lot of time, and in our future research, implementing our superpixel segmentation algorithm as an algorithm that can be executed in parallel on the GPU can achieve faster execution efficiency.

4 Conclusion

In this paper, we propose a two-stage method for recognizing the category and content of the major components in the petrographic thin section image. We proposed an AS-SLIC algorithm, which adaptively adjusts the number of superpixels according to the gray histogram of the superpixel in each iteration. In the experiment, our AS-SLIC algorithm maintains the same speed as the original SLIC algorithm. When the number of grains in the image is large, it can adaptively increase the number of superpixels, and the generated superpixel is more suitable for the edge of the grain. We designed a CNN model to identify mineral grains that showed excellent performance in our MGCD dataset. Our model is a light weight network and has the same excellent classification results as other classic models such as ResNet. The results of our two-stage method for the analysis of petrographic thin section image were recognized and praised after evaluation by several geologists. Our method has important reference value for component analysis in other types of images, such as cancer cell analysis in medical images, remote sensing image analysis, and so on.

References

1. Tetley, M.G., Daczko, N.R.: Virtual Petrographic Microscope: a multi-platform education and research software tool to analyse rock thin-sections. Aust. J. Earth Sci. **61**(4), 631–637 (2014)
2. Izadi, H., Sadri, J., Mehran, N.A.: A new intelligent method for minerals segmentation in thin sections based on a novel incremental color clustering. Comput. Geosci. **81**, 38–52 (2015)
3. Liu, M.Y., Tuzel, O., Ramalingam, S., Chellappa, R.: Entropy rate superpixel segmentation. In: IEEE Conference on Computer Vision and Pattern Recognition (CVPR), pp. 2097–2104. IEEE (2011)
4. Levinshtein, A., Stere, A., Kutulakos, K.N., et al.: Turbopixels: fast superpixels using geometric flows. IEEE Trans. Pattern Anal. Mach. Intell. **31**(12), 2290–2297 (2009)
5. Li, Z., Chen, J.: Superpixel segmentation using linear spectral clustering. In: Proceedings of the IEEE Conference on Computer Vision and Pattern Recognition (CVPR), pp. 1356–1363. IEEE (2015)
6. Bergh, M.V.D., Boix, X., Roig, G., et al.: SEEDS: superpixels extracted via energy-driven sampling. In: Fitzgibbon, A., Lazebnik, S., Perona, P., Sato, Y., Schmid, C. (eds.) ECCV 2012. LNCS, vol. 7578, pp. 13–26. Springer, Heidelberg (2012). https://doi.org/10.1007/s11263-014-0744-2
7. Achanta, R., Shaji, A., Smith, K., et al.: SLIC superpixels compared to state-of-the-art superpixel methods. IEEE Trans. Pattern Anal. Mach. Intell. **34**(11), 2274–2282 (2012)
8. Jiang, F., Gu, Q., Hao, H.Z., et al.: Grain segmentation of multi-angle petrographic thin section microscopic images. In: IEEE International Conference on Image Processing (ICIP), pp. 3879–3883. IEEE (2017)
9. Jiang, F., Gu, Q., Hao, H.Z., et al.: Feature extraction and grain segmentation of sandstone images based on convolutional neural networks. In: 2018 24th International Conference on Pattern Recognition (ICPR), pp. 2636–2641. IEEE (2018)
10. Tarquini, S., Favalli, M.: A microscopic information system (MIS) for petrographic analysis. Comput. Geosci. **36**(5), 665–674 (2010)

11. Asmussen, P., Conrad, O., Gnther, A., et al.: Semi-automatic segmentation of petrographic thin section images using a seeded-region growing algorithm with an application to characterize wheathered subarkose sandstone. Comput. Geosci. **83**, 89–99 (2015)

12. Krizhevsky, A., Sutskever, I., Hinton, G.E.: ImageNet classification with deep convolutional neural networks. In: Proceedings of Advances in Neural Information Processing Systems, Lake Tahoe, pp. 1097–1105 (2012)

13. He, K., Zhang, X., Ren, S., et al.: Deep residual learning for image recognition. In: IEEE Conference on Computer Vision and Pattern Recognition (CVPR), Las Vegas, pp. 770–778 (2016)

14. Girshick, R., Donahue, J., Darrell, T., et al.: Rich feature hierarchies for accurate object detection and semantic segmentation. In: IEEE Conference on Computer Vision and Pattern Recognition (CVPR), Columbus, pp. 580–587 (2014)

15. He, K., Gkioxari, G., Dollár, P., et al.: Mask R-CNN. arXiv preprint arXiv:1703.06870 (2017)

16. Howard, A.G., Zhu, M., Chen, B., Kalenichenko, D., et al.: MobileNets: efficient convolutional neural networks for mobile vision applications. arXiv preprint arXiv:1704. 04861 (2017)

17. He, K., Zhang, X., Ren, S., Sun, J.: Spatial pyramid pooling in deep convolutional networks for visual recognition. IEEE Trans. Pattern Anal. Mach. Intell. **37**(9), 1904–1916 (2015)

18. Veksler, O., Boykov, Y., Mehrani, P.: Superpixels and supervoxels in an energy optimization framework. In: Daniilidis, K., Maragos, P., Paragios, N. (eds.) ECCV 2010. LNCS, vol. 6315, pp. 211–224. Springer, Heidelberg (2010). https://doi.org/10.1007/978-3-642-15555-0_16

19. Vedaldi, A., Soatto, S.: Quick shift and kernel methods for mode seeking. In: Forsyth, D., Torr, P., Zisserman, A. (eds.) ECCV 2008. LNCS, vol. 5305, pp. 705–718. Springer, Heidelberg (2008). https://doi.org/10.1007/978-3-540-88693-8_52

20. Glorot, X., Bengio, Y.: Understanding the difficulty of training deep feedforward neural networks. In: 13th International Conference on Artificial Intelligence and Statistics, vol. 9, pp. 249–256 (2010)

21. Simonyan, K., Zisserman, A.: Very deep convolutional networks for large-scale image recognition. arXiv preprint arXiv:1409.1556 (2014)

22. Szegedy, C., Liu, W., Jia, Y., et al.: Going deeper with convolutions. In: IEEE Conference on Computer Vision and Pattern Recognition (CVPR), pp. 1–9. IEEE (2015)

Curved Scene Text Detection Based on Mask R-CNN

Yuanping Zhu[✉] and Hongrui Zhang

College of Computer and Information Engineering,
Tianjin Normal University, Tianjin, China
zhuyuanping@tjnu.edu.cn

Abstract. Text detection in natural scenes has achieved good results in existing research methods. However, detecting the curved scene text is still a challenging task because of perspective distortion and variation of text scale. We proposed Mask-CSTD (Curved scene text detector based on Mask R-CNN), a detection model inspired by instance-aware semantic segmentation. Firstly, we utilized Mask instead of bounding box labels. Secondly, an integrated loss function was designed to detect curved text instances. The label loss satisfied the finer representation of curved text instances more efficiently. Finally, a Deformable NMS was proposed to improve heavily tilted, unnecessary overlap or severed text instances. Experimental result on curved benchmark dataset Total-Text achieves an F-measure of 82.41%, a highly-competitive performance, which exceeds the baseline by 2.46%. Moreover, results on MSRA-TD500 and ICDAR2015 verify the effectiveness of multi-oriented text detection.

Keywords: Scene text detection · Curved text · Arbitrary oriented

1 Introduction

Text detection has caused great attention in the field of computer vision. It has further been launched in the ICDAR series competition [1–3]. Particularly, detection attaches significance to image understanding because of the rich and high-level semantic information. Since the texts in natural scenes are mostly presented a large variety of text scale, size, font, color, and character spacing, it exhibits much more difficulties in the multiple artistic words. Furthermore, it is more complicated caused by the properties of curves, surfaces, and perspectives. In fact, the aspect of light and shadow varies significantly in the background brightness and contrast, which also greatly restrict the accuracy of text detection. The methods [4–7] adopt four different detection algorithms based on region feature extraction, image segmentation, classification regression in machine learning and multi-stage hybrid. The image-based segmentation of the existing models of text detection plays an important role in this paper. Perceptual instance segmentation has reached gratifying results in Mask R-CNN [8]. Inspired by it, we propose a novel method for curved text detection in natural scenes.

Recently, some text detectors obtain impressive performances. EAST [7] changes the multi-stage fusion approach, using an end-to-end framework on text detection tasks. RRPN [9] generates a rotating text area candidate box and calculates the rotation angle

© Springer Nature Switzerland AG 2019
Y. Zhao et al. (Eds.): ICIG 2019, LNCS 11901, pp. 505–517, 2019.
https://doi.org/10.1007/978-3-030-34120-6_41

of the text line to be verified in the border regression. FTSN [10] takes advantage of Mask-NMS to replace the standard NMS algorithm to filter candidate boxes. SegLink [11] cuts words into small blocks of text that are easier to detect and then predicts that adjacent links connect small blocks of text into words. STN-OCR [12] integrates the end-to-end model of image detection and recognition. The space transformation network is added to utilize for the affine transformation of the original input image. Mask TextSpotter [13] describes an end-to-end neural network for text spotting in natural scenes. TextSnake [14] proposes a flexible and versatile representation that can be employed to scene text of any shape. However, these methods are difficult to fuse curved factor in multi-oriented text detection algorithms. In this paper, based on the existing research and perceptual instance segmentation algorithm, we propose a method for curved text detection in natural scenes.

Our main contributions are summarized as follows:

- Inspired by the perspective of perceptual semantic instance segmentation, a method for detecting curved text in natural scenes is proposed.
- Adopting a multi-scale feature fusion model to fuse low-level and high-level semantic features of curved text semantics.
- DNMS is proposed to select the largest intersection and union of Mask regions to improve heavily tilted or unnecessary overlap text instances.
- An integrated loss serves to satisfy the finer representation of curved text instance more efficiently.

In this paper, we mainly describe the proposed method of curved scene text detection. Section 1 briefly reviews the related work for text detection. Section 2 discusses our proposed method in detail; Sect. 3 analyzes and demonstrates the experimental results; Sect. 4 briefly illustrates our future work.

2 Methodology

The key components of the proposed algorithm are the feature fusion in neural network and the curved text instance labeling. Our model is a deep neural network, which is suitable for curved text detection in natural scenes. The traditional RBOX and QUAD tags are visualized by Mask, and the curved scene text detection is inspired by the perceptual instance segmentation. Consequently, the detector is named Mask-CSTD (Curved Scene Text Detection Based on Mask R-CNN).

2.1 Pipeline

The architecture of Mask-CSTD is depicted in Fig. 1. The model follows the general idea of RCNN (Regions with CNN features) [15]. Firstly, extracting deep feature information of images through ResNet-101 [8] backbone network and using FPN (Feature Pyramid Networks) [8] fuses multi-scale feature. Secondly, using RPN (Region Proposal Network) [8] to generate multiple suggestion windows for each region of interest (ROI), using PSROI Align (Position Sensitive ROI Align) [16] to obtain location information, and superimposing on the fixed-size feature map. Finally, the classification network implements the prediction of the bounding box and the Mask generation through the classification layer.

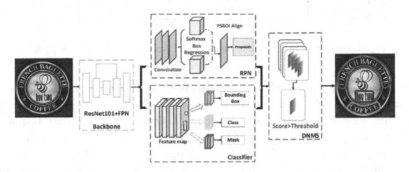

Fig. 1. An overview of the architecture.

2.2   Network Design

Text instances are different from general semantic objects, such as cats, dogs, and cars which have relatively strong semantic features. On the contrary, curved scene text shows a considerable difference within the category tremendously. The text in natural scenes has obvious background complexity, the variation of text scale, size, font, color, and character spacing, also presented curves, surfaces, and perspectives. However, due to the complexity of the curved text, traditional bounding-box cannot describe adaptability deformed text areas.

Illustrated in Fig. 2, the model consists of two parts: feature extraction module and feature fusion module. Basically, the backbone network consists of five stages. More specifically, a whole image is input into the model. Before region proposing, the ResNet stage 3 and Up-Sampling stage 4 combine the low-level and high-level features to form the fused feature map, and then the features extracted by the ResNet stage 5 are fused. The shared convolutional layer output feature is made use of in the RPN generation area suggestion window and generates multiple preset fixed aspect ratios bounding box for each position by means of a window sliding on the shared feature map. Mask is generated for each ROI and a fixed size is mapped in the spatial layout. Also, Mask position information is utilized to realize the pixel alignment of the convolution layer, and the PSROI Align layer is used to generate a fixed size feature map for each ROI. It is noted that we use the global average pooling for the classification branch and pixel-wise Softmax on Mask branch finally. It is worth mentioning that convolved pixels have spatially unaligned features, and the pixel-wise spatial mapping matching can better match the curved text regions in instance detection, so we take advantage of PSROI Align layer in feature aggregation. When the position information is manually introduced, the sensitivity of the deeper neural network to the position information of the object is effectively improved, and the accuracy of the detection of the little object of the detector can be heightened.

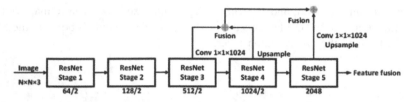

Fig. 2. Backbone network ResNet-101 and feature fusion network FPN.

The PSROI Align layer enhances the position information when the features are fused, which effectively improves the detection accuracy. In the pixel-wise feature fusion stage, RPN is merged using 3×3 conv to form the final feature fusion map, which is implemented by 1×1 conv layer softmax activation and 1×1 conv layer bounding-box regression. In the FPN, the design network of the Mask using the positive region selected by the ROI classifier is input, and the generated mask image size is 28×28 pixels.

2.3   Mask Generation

The Mask consists of a region surrounded by a number of point sets. Depicted in Fig. 3, the number of instances corresponds to the number of Masks and they are equal.

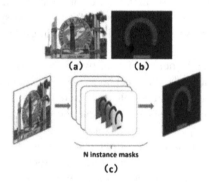

Fig. 3. Label generation process. (a) Real curved text; (b) Mask image; (c) Mask formation process.

$$Imask = \max \sum_{xi \in Cmask, yi \in Cmask} (\iint_{Mask} ds) \tag{1}$$

$$Cmask = \{(x1, y1), (x2, y2), \cdots, (xn, yn)\} \tag{2}$$

As shown in Formula (1) and (2), *Cmask* is the chain code of the mask contour; *Imask* is the area enclosed by the chain code, indicating the mask area. *s* represents the area.

As detailed in Table 1, our proposed method can accommodate almost any input of RBOX (e.g. MSRA-TD500) or QUAD (e.g. ICDAR2015) when the class label is embedded. For QUAD style tags, we choose the coordinate information to draw the contour of the Mask. For RBOX style tags, we calculate the center coordinates and then convert them to QUAD style tags. Diagrammed as Fig. 4, from left to right are the ground truth image, bounding box detection result, curved text instance image and the final multi-oriented result. Without loss of generality, it illustrates the flow of the Mask for curved text detection.

Table 1. Convert QUAD and RBOX to mask.

Geometry	Label	Description
QUAD	$(x1, y1, \cdots, x4, y4, text)$	$C = \{(xi, yi) | i \in \{1, 2, 3, 4\}\}$
RBOX	$(x, y, w, h, theat)$	$C = \{(xi, \theta), (yi, \theta)\}, 1 \leq i \leq 4, 0 \leq \theta \leq \pi$

Fig. 4. Curved text filtering result. (a) Ground truth; (b) bounding box result; (c) Mask image; (d) final curved result.

2.4 Loss Function

The loss can be interpreted as:

$$L = Lrpn + \lambda_1 Lmask\_map \tag{3}$$

$$Lrpn = \lambda_2 LBBox + \lambda_3 LCla \tag{4}$$

LCla is classification loss, *LBBox* denotes Bounding Box loss, *Lmask_map* illustrates Mask loss, *Lrpn* represents regional suggestion network loss and weight $\lambda_1, \lambda_2, \lambda_3$ is regularization constant coefficient.

Region Proposal Network Loss. Cross-entropy loss function has significant performance in classification problems, and we utilize it in curved scene text detection. To optimize the balance between the real Mask and the category confidence box, we use the Smooth L1 loss introduced in [17], depicted by the Formula (5).

$$smoothL1(x) = \begin{cases} 0.5x^2 & |x| < 1 \\ |x| - 0.5 & otherwise \end{cases} \tag{5}$$

As illustrated in Formula (6), we add the offset relative to default boxes and optimize them with a scale-normalized Smooth L1 loss. $yBBox$ is the ground truth and $yBBox'$ interprets as the predicted. Actually, they could typically be of any shape.

$$L_{BBox} = SmoothL1(yBBox - yBBox') \tag{6}$$

Mask Loss. One difficulty for curved scene text detection is how to define the contour and background. It is a challenge to fully generate accurate textual geometric predictions. If the L1 or L2 loss is directly used for regression, the larger curved contour regions will be guided. Consequently, we utilize IoU to measure the overlap rate of the corresponding pixel in the Mask area and choose Smooth L1 to optimize them. To prevent loss bias towards larger, we use a logarithmic fit.

$$Lmask\_map = Llabel + Lmask \tag{7}$$

$Lmask$ uses cross-entropy as loss function, $Llabel$ illustrates as follow.

$$Llabel = SmoothL1(\log IoU(R, R')) \tag{8}$$

Where R' represents the predicted Mask area and R illustrates the ground truth. When $R = R'$, we determine that the Mask fits the curved text area perfectly.

Classification Loss. After the Softmax regression, if MSE (Mean Square Error loss) is used, the partial derivative calculation will guide the loss bias towards abnormal. Both ends are flat and close to 0 during the calculation. Accordingly, the cross-entropy loss is used to fit the curved scene text. When the error is large, the weight update is fast. In contrast, the weight is finer and slower, which is consistent with the loss trend of the network.

$$LCla = class\_loss(Y, Y') = -\frac{1}{n}\sum_x [y \ln y' + (1 - y) \ln(1 - y')] \tag{9}$$

Where x is the sample, n represents the total number of samples, y' illustrates the expected output, and y is the actual result.

2.5   DNMS

Deformable Non-Maximum Suppression (DNMS) helps to obtain the final result. We utilize DNMS to filter overlapped and severed text instances. Depicted in Fig. 5, we generate a minimum contour for each curved scene text instance covered by the mask after DNMS. It mainly modifies the intersection and union when computing Mask IoU instead of bounding box IoU.

Fig. 5. Intermediate results of generating a minimum contour covered by the mask. From left to right, input images, Mask results, final results and ground truth images.

Diagrammed in Fig. 6, it is still a challenge to generate accurate textual geometric predictions to filter overlapped and severed text instances. We calculate the maximal union of different detected Masks corresponding to the same instance. Consequently, we propose a modified NMS [18] called DNMS, a method computation to Mask maximum union instead of bounding boxes, to handle such situations.

Fig. 6. Curved scene text filtering results. From left to right are: NMS results, DNMS results and the Mask ground truth images.

$$Si = \begin{cases} Si & iou(SM, Mi) < Nt \\ 0 & iou(SM, Mi) \geq Nt \end{cases} \tag{10}$$

Illustrated in Formula (10), M is the list of initial detection Mask, S contains corresponding detection scores, Nt is the NMS threshold, which is set to $Nt = 0.5$ in our experiment.

3 Experiment

We conduct experiments to verify the effectiveness of Mask-CSTD on multi-datasets: Total-Text [19], MSRA-TD500 [20] and ICDAR2015 [2]. Particularly, we simply utilize official training images without any extra data augmentation. To further to assess the significance of our proposed algorithm on curved scene text detection, we mainly experiment on Total-Text by picking curved text instances.

3.1   Datasets and Evaluation Protocol

Total-Text [19]. It consists of 1555 images in natural scenes, divided into 1225 training samples and 300 test images. It has been more than three different text directions: horizontal, arbitrary direction and curved. Most of them are mainly curved. This dataset contains multiple languages both Chinese and English. We use the hand-craft to generate real Mask. It also aims at solving the arbitrary-shaped text detection problem.

MSRA-TD500 [20]. The dataset is used for text detection in any direction. It contains Chinese and English, a total of 500 natural scene images. For the training set of 300, the test set of 200, marked in units of behavior rather than words. Each image is completely marked, difficult to identify with difficult labels, text box labels with RBOX style annotations.

ICDAR2015 [2]. The dataset in the data samples provided in the ICDAR competition and is taken from real scene images. It is widely used to benchmark multi-oriented text detectors. It contains 1500 images: the training image set contains 1000 images for the training label set, and the test image set contains 500 images. Text box labels with QUAD style annotation.

Evaluation Protocol. We utilize the standard evaluation protocol depending on Precision, Recall, F-measure. They are defined as below:

$$Precision = \frac{TP}{TP + FP} \tag{11}$$

$$Recall = \frac{TP}{TP + FN} \tag{12}$$

$$F - measure = \frac{2 \times Precison \times Recall}{Preciosn + Recall} \tag{13}$$

Where TP, FP and FN illustrated as the number of correctly text instances detection, incorrect detection, and missing text instances, respectively.

3.2   Training

In order to decline the training time, we use GPU NVIDIA GTX1080. Experimental environment configuration as follow: Ubuntu 16.04, Python 3.6.2, Keras 2.1.3, Tensorflow 1.4.0.

The learning rate is 0.001 initially, the loss function weight coefficient $\lambda 1, \lambda 2, \lambda 3$ defaults set to 1.0, and RPN ANCHOR is $(32, 64, 128, 256, 512)$. We pre-trained the model on the COCO dataset and fine-tuned the model parameters.

3.3   Results

The precise description of the curved scene text instance is the essential difference between our method and the other approach. It can predict the shape of the curved text instance and display the description accurately. We attribute this accurate method describing the instance segmentation to Mask. The Mask mechanism is a spatial mapping of positional information. It has two distinct advantages: (1) intuitive visualization of spatial mapping; (2) multi-scale fusion feature integration is obvious. Shown in Figs. 7 and 8, it handles the text instances in some challenge scenarios, such as non-uniform illumination, varying curved orientation and perspective distortion.

(a)   (b)   (c)

Fig. 7. Results on: (a) Total-Text, (b) MSRA-TD500, (c) ICDAR2015.

Fig. 8. More detection details on Total-Text. From left to right are the baseline result, our result and ground truth.

As Mask-CSTD models are trained on the datasets Total-Text, MSRA-TD500 and ICDAR2015 separately, the experimental results of our method compared with those of the state-of-the-art approaches are diagrammed in Tables 2, 3 and 4. For the Total-Text dataset, the performance of our method even exceeding the FTSN [10]. The result is a precision of 83.02%, a recall of 81.81%, and an F-measure of 82.41% under the evaluation protocol. In Fig. 9, it contains the detail of the evaluation protocol on Total-Text. As illustrated, we learn about the count of the evaluation protocol value in each range on Total-Text. Compared with the substantial performance gains over the published works confirm the effectiveness of our approach on curved scene text detection. Specifically, compared with the baseline model, our method achieves an improvement of about 2.46%. For evaluating our algorithm comprehensively, we utilize our model to detect horizontal or non-curved text instance on MSRA-TD500 and ICDAR2015 datasets. As depicted in Tables 3 and 4, all the experiments exhibit an increase in the F-measure.

Table 2. Results on Total-Text.

Algorithm	Precision	Recall	F-measure
TextSnake [14]	82.7	74.5	78.4
TextField [22]	81.2	79.9	80.6
FTSN [10]	**84.7**	78.0	81.3
Baseline(Mask R-CNN [8])	81.69	78.28	79.95
Ours	83.02	**81.81**	**82.41**

Table 3. Results on MSRA-TD500.

Algorithm	Precision	Recall	F-measure
RRPN [9]	82.0	69.0	75.0
EAST [7]	87.3	67.4	76.1
PixelLink [21]	83.0	73.2	77.8
TextField [22]	87.4	75.9	81.3
FTSN [10]	**87.6**	77.1	82.0
IncepText [23]	87.5	79.0	**83.0**
Baseline(Mask R-CNN [8])	73.67	67.07	70.22
Ours	85.48	**79.68**	82.48

Table 4. Results on ICDAR2015.

Algorithm	Precision	Recall	F-measure
RRPN [9]	84.0	77.0	80.0
EAST [7]	83.3	78.3	80.7
TextField [22]	84.3	80.5	82.4
FTSN [10]	88.6	80.0	84.1
IncepText [23]	**93.8**	**87.3**	**90.5**
Baseline(Mask R-CNN [8])	85.99	74.32	79.73
Ours	87.24	82.35	84.72

Fig. 9. The count of the evaluation protocol value in each range on Total-Text test datasets.

Furthermore, to further demonstrate the ability of the proposed Mask-CSTD, we present more results on the Total-Text dataset. As displayed in Fig. 10, our model still achieves good results when the curved text is mixed with horizontal text. In natural scenarios, such as non-uniform illumination, curved, multi-oriented and irregular text instance is not special. Consequently, in natural scenes, curved and horizontal text instance mixed is also a commonly seen artistic-style text. There is still a challenge for text detection is that the color, sizes, large character spacing of text images vary tremendously in natural scenes. Due to its irregular text length and multi-lingual text lines, detecting curved text is also very challenging. As can be seen, Mask-CSTD handles these difficult scenarios available. This demonstrates the validity of the proposed Mask-CSTD.

Fig. 10. Results on the Total-Text: curved, horizontal mixed detection examples.

3.4 Error Analysis

As displayed in our experiments, Mask-CSTD performs well in most scenarios. However, it still fails to handle some special situations. Depicted in Fig. 11, the limitation of the proposed method is that: (a) when the color contrast is not obvious, although the image centralization has been completed, our model has not achieved the expected effect; (b) in the small curved text detection task, the feature sequence covers a small number of these samples and so the text area detect inaccurately occurs; (c) it may give false predictions for text instances when they take only a special portion in the training datasets.

Fig. 11. Some failure examples on the Total-Text. (a) Incorrect detection; (b) missing detections for little text instance; (c) false detection.

4 Conclusions

Inspired by the perceptual instance segmentation, we propose a curved scene text detector Mask-CSTD. We utilize a deep neural network and Mask spatial alignment mapping to predict curved text regions and use multi-scale feature fusion networks to integrate multi-level semantic information. By designing a reasonably effective integral loss function, the detector positioned image area and text boundaries more precisely. The improved DNMS helps to improve the results. The experimental results on the curved scene text dataset Total-Text demonstrate that Mask-CSTD achieves effectiveness substantially. ICDAR2015 and MSRA-TD500 verify the effectiveness for multi-oriented text detection. Furthermore, there are still a small number of inaccurate and missed detection in our algorithm. In the future, curved scene text detection in distant images is still a difficult problem. We also interested in combining the networks of detection and recognition into an end-to-end framework.

Acknowledgment. This work was supported by the National Natural Science Foundation of China (Grant No. 61602345, 61703306) and the Natural Science Foundation of Tianjin (Grant No. 18JCYBJC85000, 16JCQNJC00600).

References

1. Karatzas, D., Shafait, F., Uchida, S., et al.: ICDAR 2013 robust reading competition. In: 12th International Conference on Document Analysis and Recognition, vol. 1, pp. 1484–1493. IEEE, Washington, America (2013)
2. Karatzas, D., Gomez-Bigorda, L., Nicolaou, A., et al: ICDAR 2015 robust reading competition. In: 13th International Conference on Document Analysis and Recognition, vol. 1, pp. 1156–1160. IEEE, Nancy, France (2015)
3. Iwamura, M., Morimoto, N., Tainaka, K., et al.: ICDAR2017 robust reading challenge on omnidirectional video. In: 14th International Conference on Document Analysis and Recognition, vol. 1, pp. 1448–1453. IEEE, Kyoto, Japan (2017)
4. Liu, Y., Jin, L.: Deep matching prior network: toward tighter multi-oriented text detection. In: 2017 IEEE Conference on Computer Vision and Pattern Recognition on Proceeding, pp. 3454–3461. IEEE, Honolulu, Hawaii (2017)
5. He, T., Huang, W., Qiao, Y., et al.: Accurate text localization in natural image with cascaded convolutional text network. arXiv preprint arXiv:1603.09423 (2016)
6. He, W., Zhang, X.Y., Yin, F., Liu, C.L.: Deep direct regression for multi-oriented scene text detection. In: the IEEE International Conference on Computer Vision on Proceedings, pp. 745–753. IEEE, Venice, Italy (2017)
7. Zhou, X., Yao, C., Wen, H., et al.: EAST: an efficient and accurate scene text detector. In: the IEEE conference on Computer Vision and Pattern Recognition on Proceedings, pp. 2642–2651. IEEE, Hawaii (2017)
8. He, K., Gkioxari, G., Dollr, P., et al.: Mask R-CNN. In: the IEEE international conference on computer vision on Proceedings, pp. 2961–2969. IEEE, Venice, Italy (2017)
9. Ma, J., Shao, W., Ye, H., et al.: Arbitrary-oriented scene text detection via rotation proposals. IEEE Trans. Multimedia **20**(11), 3111–3122 (2018)

10. Dai, Y., Huang, Z., Gao, Y., et al.: Fused text segmentation networks for multi-oriented scene text detection. In: 24th International Conference on Pattern Recognition on Proceedings, pp. 3604–3609. IEEE, Beijing, China (2018)

11. Shi, B., Bai, X., Belongie, S.: Detecting oriented text in natural images by linking segments. In: the IEEE Conference on Computer Vision and Pattern Recognition on Proceedings, pp. 2550–2558. IEEE, Honolulu, Hawaii (2017)

12. Bartz, C., Yang, H., Meinel, C.: STN-OCR: a single neural network for text detection and text recognition. Archives of Computational Methods in Engineering. pp. 1–22 (2017)

13. Lyu, P., Liao, M., Yao, C., et al.: Mask TextSpotter: an end-to-end trainable neural network for spotting text with arbitrary shapes. In: 15th European Conference on Computer Vision, pp. 71–88. IEEE, Munich, Germany (2018)

14. Long, S., Ruan, J., Zhang, W., et al.: Textsnake: a flexible representation for detecting text of arbitrary shapes. In: the European Conference on Computer Vision on Proceedings, pp. 20–36. IEEE, Munich, Germany (2018)

15. Girshick, R., Donahue, J., Darrell, T., et al.: Region-based convolutional networks for accurate object detection and segmentation. IEEE Trans. Pattern Anal. Mach. Intell. $38(1)$, 142–158 (2016)

16. https://github.com/afantideng/R-FCN-PSROIAlign. Accessed 2 July2018

17. Ren, S., He, K., Girshick, R., et al.: Faster R-CNN: towards real-time object detection with region proposal networks. In: Advances in neural information processing systems on Proceedings, pp. 91–99. IEEE, Canada (2015)

18. Neubeck, A., Gool, L.V.: Efficient non-maximum suppression. In: International Conference on Pattern Recognition on Proceedings, pp. 850–855. IEEE, Cancun, Mexico (2006)

19. Ch'ng, C.K., Chan, C.S.: Total-text: a comprehensive dataset for scene text detection and recognition. In: 14th IAPR International Conference on Document Analysis and Recognition on Proceedings, pp. 935–942. IEEE, Kyoto, Japan (2017)

20. Noh, H., Hong, S., Han, B.: Learning deconvolution network for semantic segmentation. In: IEEE International Conference on Computer Vision on Proceedings, pp. 1520–1528. IEEE, Venice, Italy (2015)

21. Deng, D., Liu, H., Li, X., et al.: PixelLink: detecting scene text via instance segmentation. In: National Conference on Artificial Intelligence, pp. 6773–6780 (2018)

22. Xu, Y., Wang, Y., Zhou, W., et al.: TextField: learning a deep direction field for irregular scene text detection. arXiv preprint arXiv: 1812.01393 (2018)

23. Yang, Q., Cheng, M., Zhou, W., et al.: IncepText: a new inception-text module with deformable PSROI pooling for multi-oriented scene text detection. In: International Joint Conference on Artificial Intelligence, pp. 1071–1077 (2018)

Semantic Inference Network for Human-Object Interaction Detection

Hongyi Liu, Lisha Mo, and Huimin Ma[✉]

Department of Electronic Engineering, Tsinghua University, Beijing, China
{liuhy18,mls18}@mails.tsinghua.edu.cn, mhmpub@tsinghua.edu.cn

Abstract. Recently many efforts have been made to understand the scenes in images. The interactions between human and objects are usually of great significance to scene understanding. In this paper, we focus on the task of detecting human-object interactions (HOI), which is to detect triplets $< human, verb, object >$ in challenging daily images. We propose a novel model which introduces a semantic stream and a new form of loss function. Our intuition is that the semantic information of object classes is beneficial to HOI detection. Semantic information is extracted by embedding the category information of objects with pre-trained BERT model. On the other hand, we find that the HOI task suffers severely from extreme imbalance between positive and negative samples. We propose a weighted focal loss (WFL) to tackle this problem. The results show that our method achieves a gain of 5% compared with our baseline.

Keywords: Human-object interaction · Visual relationship detection · Word embedding

1 Introduction

In the task of general object detection, many efficient frameworks and novel methods have been proposed in recent years. Most previous detection work focused on object instances individually, while the latest researches further considered the relationships between objects. For scene understanding and world cognition, machine vision is expected to learn to mine and establish the connections between objects. Moreover, the interaction between human and objects is a kind of relationship, which contains more effective information and is of more significance in practice.

As a subtask of visual relationship detection, human-object interactions (HOI) aims to detect all interaction triplets $< human, verb, object >$. The HOI detection task attempts to bridge the gap between object detection and image understanding. Besides, it also makes sense to other high-level vision tasks, such as image captioning [1,2] and visual question answering (VQA) [3], etc.

In this paper, we introduce a semantic stream and propose a weighted focal loss to assist in the detection of human-object interactions. Our core idea is that

© Springer Nature Switzerland AG 2019
Y. Zhao et al. (Eds.): ICIG 2019, LNCS 11901, pp. 518–529, 2019.
https://doi.org/10.1007/978-3-030-34120-6_42

the semantic information embedded in the category labels of objects is of great benefit to HOI detection. On the other hand, we find that HOI task suffers from extreme imbalance between positive and negative samples. Therefore, we put forward a weighted focal loss based on focal loss in [4] to tackle this problem.

We validate the effectiveness of our model on Verbs in COCO (V-COCO) dataset [5]. Our results show that both semantic stream and weighted focal loss contribute a lot to the task, and our model achieves a gain of 5% compared to the baseline iCAN [6].

2   Related Work

Object Detection. Both one-stage [7,8] and two-stage [9–11] object detection frameworks have improved by a wide margin in the last few years. Faster R-CNN, a typical representative of two-stage object detection, utilizes Region Proposal Network and ROI pooling operation to improve the efficiency and accuracy of object detection. In our work, Faster R-CNN framework is used to detect persons and objects in images. With these detected bounding boxes, human-object interactions can be inferred for each pair of person and object instances.

Visual Relationship Detection. Recently, a series of work has focused on the detection of visual relationships [12–14]. Visual relationship detection aims to detect objects and the relationships between these objects simultaneously. The relationships include spatial relationships, preposition terms, verbs and actions. Nevertheless, limited data samples and large numbers of relationship types lead to the difficulty of the task. Therefore, some methods were proposed to use language priors to assist in the detection of visual relationships, which is instructive for our work. In comparison to visual relationship detection, our task merely concentrates on the interactions between human and objects. The interactions between human and objects are more fine-grained, therefore resulting in new challenges.

Human-Object Interactions. Human-object interactions are similar to visual relationships, but face the challenge of more specific interaction types (e.g., hold, carry, cut, hit). This challenge demands a more intrinsic understanding of the image.

Recently, many papers have put forward rather effective methods to tackle this problem. Chao et al. [15] believed that spatial locations between human and objects are of great help to the detection of human-object interactions. A CNN-based interaction pattern was proposed to extract the spatial features between these two bounding boxes. Gkioxari et al. [16] introduced an action-specific density map to predict the locations of target objects. Then an interaction branch was proposed to predict the score for each action based on human and object appearances. Gao et al. [6] put forward an instance-level attention module to learn the contextual information around persons and objects.

Input Image   Object Detection   HOI Detection

Fig. 1. Results of human-object interaction. Given an input image, our method detects the human and object instances and the interactions between them.

Based on these progresses in the HOI detection, our work focuses on different aspects. Since the HOI detection evaluation does not require the correct classification of object categories, most work does not take the category information of objects into account. However, it is evident that the category information of objects is beneficial for the HOI detection. In addition, each pair of human and objects needs to be considered in the HOI task. However, most human-object pairs are not interactive at all. There are only one or two interactions among 26 interactions in the interactive pairs of a single image. Therefore, we are confronted with a problem of extreme imbalance between positive and negative samples. To solve above problems, we use semantic word embedding to utilize the category information and a weighted focal loss to tackle the imbalance.

3   Method

3.1   Algorithm Overview

To detect a human-object interaction, all the persons and objects in the image have to be detected firstly (denoted as b_h and b_o respectively). In this work, we choose Faster R-CNN to carry out object detection. Then the interactions between human and objects will be identified through the semantic stream, the visual stream and the spatial stream.

Training. Since each person can perform several interactions with an object, human-object interaction detection is a multi-label classification task. For instance, a person can 'hold' and 'hit' a baseball bat simultaneously. Therefore, we have to predict the interaction score s^a for each interaction by applying binary sigmoid classifiers. Accordingly, binary sigmoid cross-entropy is applied in this task.

Inference. Different from training, the object detection outputs a confidence score for all human and object bounding boxes (denoted as s_h and s_o respectively) during inference. Besides, the interaction score predicted by our algorithm is denoted as s^a. Consequently, the final HOI score $S^a_{h,o}$ for human-object pair (b_h, b_o) is represented as:

$$S^a_{h,o} = s_h \cdot s_o \cdot s^a \tag{1}$$

For those interaction categories that are irrelevant to objects (e.g., walk, run, smile), the form of the final HOI score is updated as:

$$S^a_h = s_h \cdot s^a \tag{2}$$

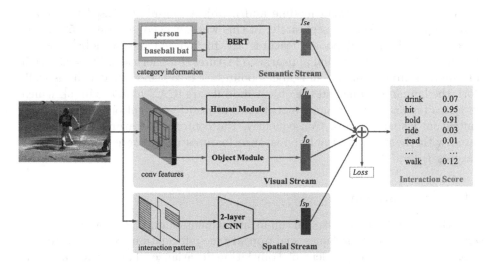

Fig. 2. Overview of semantic inference network. Our proposed model mainly consists of three streams: (1) a semantic stream utilizing semantic information embedded in categories of objects. (2) a visual stream that extract visual appearance of human and objects. (3) a spatial stream to encode the spatial relationship between human and object bounding boxes. Then the features from three streams are fused together to make final predictions.

3.2 Semantic Stream

It is clear that the category information of objects can assist in the detection of human-object interactions. For example, a person can 'eat' bananas but cannot 'ride' bananas. On the other hand, different categories of objects may share similar semantic meanings, thus resulting in similar preference of interactions. For example, bananas and apples are both likely to link with the interaction 'eat'.

However, the implicit semantic information is hard to discover solely based on visual appearance. In order to solve this problem, a semantic stream was introduced to utilize the category information of objects. More specifically, the pre-trained BERT model, which has an excellent performance in the field of natural language processing (NLP), was used to embed the category information. The 768-D feature vectors generated by the BERT are followed by a fully connected layer to get the feature encodings f_{Se} of human and objects. Due to the excellence of BERT model, we can explore the implicit semantic information between similar categories of objects. The utilization of semantic knowledge will contribute a lot to the ultimate performance of our model.

3.3 Visual Stream

For visual stream, a human module and a object module are used to extract visual appearance features separately. In this work, we choose ResNet-50 as our backbone, and the RoI pooled features are further fed into a global average pooling layer to get the visual appearance features. Furthermore, the contextual information around human and object bounding boxes is of great help as well. Both the human module and the object module are iCAN modules in [6], which performs instance-level channel attention and position attention to extract the contextual visual features. Consequently, the contextual features are concatenated with the original visual appearance features to generate the final visual features f_H and f_O.

3.4 Spatial Stream

Even though the semantic stream can utilize semantic information and the visual stream can provide abundant visual appearance information to predict interactions, CNN faces the problem of translation invariance. That is to say, it is hard for CNN to learn the spatial relationships between persons and objects. Nevertheless, it is evident that the relative position is beneficial for the HOI detection.

The interaction pattern in [15] is used to encode the spatial relationship in our model, which is represented by a two-channel binary image. As only the spatial relationship are focused, pixel values are neglected to exploit information of bounding box locations. More specifically, the first channel has value 1 where the pixel is inside the human bounding box and value 0 elsewhere, while the second channel has value 1 where the pixel is inside the object bounding box and value 0 elsewhere. Then, a 2-layer CNN is applied to extract spatial features f_{Sp}, which are further fused with another two streams to constitute the final features.

3.5 Weighted Focal Loss

Since the above three streams extract different aspects of information, the above features generated by each stream are concatenated to obtain a better prediction performance. As stated in Sect. 3.1, our task is a multi-label classification problem. Given a human-object pair, binary sigmoid classifiers predict the score of every interaction category. The interaction is considered positive if the score exceeds the threshold.

In general, there are several persons and objects in an image. Each bounding box of person and object forms a human-object pair. Then, it needs to predict the interaction scores for all human-object pairs. However, most human-object pairs do not interact at all. In addition, even in those interactive samples, only one or few interactions are positive among 26 categories. Therefore, the number of positive samples is much smaller than the number of negative samples. Under such circumstances, the model is actually facing the problem of extremely imbalanced samples. As there are too many negative samples, the loss of negative samples is considerably large, even though the loss of each negative sample is small. In contrast, the loss of positive samples has an insignificant effect on the total loss, which has no benefit to the convergence of the model.

In existing models, binary cross entropy (CE) is selected as the loss. The normal binary cross entropy loss can be represented as:

$$CE(p, y) = -y \log(p) - (1 - y) \log(1 - p) \tag{3}$$

In formula 3, p is the score of prediction and y is the ground truth label.

Considering the imbalance problem, we use focal loss in [4] which introduces a weighted factor α and a modulating factor γ. α increases the significance of minority class and decreases the significance of majority class, while γ reduces the loss of easy samples.

$$FL(p, y) = -\alpha y (1 - p)^\gamma \log(p) - (1 - \alpha)(1 - y)p^\gamma \log(1 - p) \tag{4}$$

Besides, there exists some interactions that are similar with others. For example, 'hold' and 'carry' interactions are similar in semantic stream, visual stream and spatial stream. To some extent, it is hard to distinguish them. However, usually only one of these two interactions occurs in a human-object pair. It is probable that our model outputs positive predictions for both interactions when one interaction occurs, resulting in a false prediction. Besides, it is hard for the convergent model to rectify this type of hard samples, which in turn will affect the precision of major samples which have already been well-recognized. To reduce the impact of this type of false prediction, we introduce a penalty term for hard samples based on focal loss, so that the loss of these hard samples has less significance. The weighted focal loss can be represented as:

$$WFL(p, y) = \begin{cases} FL(p, y)(1 - \beta \frac{|p-y|-\theta}{1-\theta}) & , if \ |p - y| > \theta \\ FL(p, y) & , if \ |p - y| \leq \theta \end{cases} \tag{5}$$

The parameters are set to $\alpha = 0.25, \gamma = 2, \theta = 0.8, \beta = 0.5$ in our work.

4   Datasets and Evaluation Metrics

We use V-COCO (Verbs in COCO) dataset to evaluate the performance of our model. V-COCO is a subset of MS COCO that only has images with human-object interactions. There are 5400 images including 8431 person instances in the trainval set, and 4946 images including 7768 person instances in the test set. V-COCO is annotated with 26 different interactions, and each person has 2.9 interactions simultaneously on average.

There are two types of Average Precision (AP) metrics in HOI detection. The first one is the AP of the triplet $< human, verb, object >$, which is called average role AP. A triplet is judged as true positive if: (i) the detected bounding boxes of human and objects both have the Intersection of Union (IoU) higher than 0.5 with the ground truth, and (ii) the predicted interaction is consistent with the ground truth interaction. The second evaluation metric is the AP of the pair $< human, verb >$, which is called average agent AP. Agent AP has no demand for localizing the object.

In V-COCO, there are three different types of interactions among 26 interactions. Five interactions (point, run, stand, smile, walk) are not interacted with any objects, therefore only evaluated with agent AP. Three interactions (cut, hit, eat) are annotated with different types of objects. For instance, 'eat' + 'spoon' which includes the instrument means eat with a spoon, and 'eat' + 'apple' which includes the object means eat an apple. Other interactions are interacted with objects and annotated with one type of objects.

5   Experiments

5.1   Implementation Details

Our network was based on Faster R-CNN from Detectron [11] with a ResNet-50 backbone. The Faster R-CNN model pre-trained on ImageNet is used to detect human and objects. Only human bounding boxes with score higher than 0.8 and object bounding boxes with score higher than 0.4 are reserved for subsequent implementation.

In the semantic stream, the category information is embedded to semantic features by the BERT module, which is an excellent NLP model developed by Google. For each human-object pair, the pre-trained BERT model by [17] can output a 768-D embedding feature of human and object categories. Then a fully connected layer is applied to get a 1024-D feature f_{Se}.

In the visual stream, the iCAN modules in [6] are responsible for extracting the visual appearance features. More specifically, a 7×7 feature of human or object regions extracted by RoI pooling is fed into a global average pooling (GAP) to obtain a 2048-D feature f_{inst}. Channel attention operation and position attention operation are followed to extract a 1024-D contextual visual features $f_{context}$. Finally, the output of human/object iCAN module is the concatenation of visual appearance feature f_{inst} and contextual visual features $f_{context}$, which can be represented as f_H and f_O respectively.

In the spatial stream, given a $64 \times 64 \times 2$ interaction pattern, a 2-layer CNN is applied to extract a 5408-D spatial feature f_{Sp}. Finally, the features from three streams are concatenated together to make the final predictions.

We train our model on V-COCO trainval set with a learning rate 0.001, a momentum of 0.9, a weight decay of 0.0005, and a 0.96 factor for reducing the learning rate. It takes 14 h to train on a single NVIDIA TITAN Xp and 1 h to test on V-COCO test set.

5.2   Qualitative Analysis

Our model has been tested on V-COCO dataset. The detection results of human-object interactions are displayed in Figs. 3 and 4. Different colors represent different bounding boxes of human and objects. The color of the interaction results represents which bounding box is interacted with. Figure 3 shows that our network can detect $<human, verb, object>$ triplets in a wide range of scenes. Figure 4 shows that multiple interactions between human and objects can be detected correctly at the same time.

Fig. 3. Results of HOI detections in simple scenes. Our model can detect human-object interactions in a wide range of scenes. (Color figure online)

Fig. 4. Results of HOI detections in complex scenes. Our model can detect multiple human-object interactions simultaneously. (Color figure online)

5.3   Quantitative Analysis

As stated in Sect. 4, we mainly use AP_{agent} and AP_{role} to evaluate our model. Our model is based on iCAN [6] and our reimplementation result is shown in Table 1. It has an AP_{agent} of 65.32 and AP_{role} of 44.67. By utilizing semantic information and weighted focal loss, our model makes an improvement of 1.59 points in AP_{agent} and 2.13 points in AP_{role}. As AP_{role} is the more concerned target, it can be observed that our method gains an improvement of 5% relatively.

Table 1. Detailed results on V-COCO test set. We compare our model with baseline and bold the better results.

	Baseline (iCAN)		Ours	
	AP_{agent}	AP_{role}	AP_{agent}	AP_{role}
Hold	62.06	24.81	**67.16**	**31.65**
Sit	**70.86**	**27.09**	65.90	20.52
Ride	80.68	63.91	**80.73**	**65.57**
Look	60.32	16.80	**63.37**	**27.86**
Jump	**57.63**	**52.01**	55.70	49.73
Lay	30.30	23.37	**30.41**	**25.01**
Talk-on-phone	75.06	50.96	**75.31**	**53.44**
Carry	61.42	34.35	**66.21**	**36.92**
Throw	47.39	42.17	**48.88**	**42.20**
Catch	58.26	**46.71**	**58.52**	45.68
Work-on-computer	**78.87**	**62.41**	77.91	59.73
Ski	79.31	42.50	**84.02**	**51.35**
Surf	91.73	79.45	**92.35**	**79.77**
Skateboard	94.02	**83.77**	**94.75**	80.39
Drink	40.81	27.84	**49.92**	**39.48**
Kick	76.16	63.73	**77.59**	**66.36**
Read	47.01	23.09	**49.21**	**34.45**
Snowboard	84.47	71.56	**87.40**	**74.35**
Point	0.70	/	**1.13**	/
Run	70.44	/	**74.38**	/
Stand	84.59	/	**88.35**	/
Smile	58.18	/	**59.64**	/
Walk	48.62	/	**57.27**	/
Cut (object)	**70.59**	36.80	68.02	**37.02**
Cut (instrument)		36.80		**38.82**
Eat (object)	**81.42**	**37.75**	76.72	32.73
Eat (instrument)		6.60		**7.33**
Hit (object)	**87.36**	42.43	80.87	**49.20**
Hit (instrument)		75.12		**75.43**
Mean AP	65.32	44.67	**66.91**	**46.80**

It can be observed from the experimental results that the categories with significant decline are 'sit' and 'eat', while the categories with significant improvement are 'hold', 'look', 'ski', 'drink' and 'read', where a float more than 5% is considered to be a significant change. The decline of 'sit' suggests that our network is more concerned with obvious interactions of limbs, and the bounding box is not sufficient to provide detailed posture information. Because of the ambiguity and indistinguishability of some interactions, there can be a trade-off between 'eat' and other subtle actions such as 'drink' and 'look'. Eventually, our model achieved performance improvements of most categories, with a relatively small cost.

5.4   Ablation Analysis

We evaluate the effectiveness of semantic stream and weighted focal loss in Tables 2 and 3.

With vs. Without Semantic Stream. We use BERT model to embed the category information into a 768-D semantic feature. Table 2 shows that utilizing semantic features contributes a lot to the performance of human-object interactions detection.

Table 2. Ablation study on V-COCO test set about the semantic stream.

	AP_{agent}	AP_{role}
w/o semantic stream	65.81	45.91
w semantic stream	**66.91**	**46.80**

Binary Cross Entropy vs. Focal Loss vs. Weighted Focal Loss. Table 3 proves that our weighted focal loss can make progress by solving the problem of imbalanced samples and hard samples. The weighted focal loss performs better than focal loss because it reduces the loss of hard samples which might mislead our model.

Table 3. Ablation study on V-COCO test set about the form of loss.

	AP_{agent}	AP_{role}
Binary cross entropy	65.92	45.78
Focal loss	66.30	46.47
Weighted focal loss	**66.91**	**46.80**

6   Conclusions

In order to understand the image scenes and mine the visual relationship better, we propose Semantic Inference Network for human-object interaction detection, which implements a semantic stream to introduce the category information and a weighted focal loss to tackle the imbalance between positive and negative samples. Our method gains a boost of 5% than the baseline. The ablation experiments have validated the effectiveness of our semantic stream and weighted focal loss.

Acknowledgement. This research is supported by The National Key Basic Research and Development Program of China (No. 2016YFB0100900) and National Natural Science Foundation of China (No. 61773231).

References

1. Li, Y., Ouyang, W., Zhou, B., Wang, K., Wang, X.: Scene graph generation from objects, phrases and region captions. In: Proceedings of the IEEE International Conference on Computer Vision, pp. 1261–1270 (2017)
2. Xu, D., Zhu, Y., Choy, C.B., Fei-Fei, L.: Scene graph generation by iterative message passing. In: Proceedings of the IEEE Conference on Computer Vision and Pattern Recognition, pp. 5410–5419 (2017)
3. Anderson, P., et al.: Bottom-up and top-down attention for image captioning and visual question answering. In: Proceedings of the IEEE Conference on Computer Vision and Pattern Recognition, pp. 6077–6086 (2018)
4. Lin, T.-Y., Goyal, P., Girshick, R., He, K., Dollár, P.: Focal loss for dense object detection. In: Proceedings of the IEEE International Conference on Computer Vision, pp. 2980–2988 (2017)
5. Gupta, S., Malik, J.: Visual semantic role labeling. Computer Science (2015)
6. Gao, C., Zou, Y., Huang, J.-B.: iCAN: instance-centric attention network for human-object interaction detection. In: British Machine Vision Conference (2018)
7. Redmon, J., Divvala, S., Girshick, R., Farhadi, A.: You only look once: unified, real-time object detection. In: Proceedings of the IEEE Conference on Computer Vision and Pattern Recognition, pp. 779–788 (2016)
8. Liu, W., et al.: SSD: single shot multibox detector. In: Leibe, B., Matas, J., Sebe, N., Welling, M. (eds.) ECCV 2016. LNCS, vol. 9905, pp. 21–37. Springer, Cham (2016). https://doi.org/10.1007/978-3-319-46448-0_2
9. Ren, S., He, K., Girshick, R., Sun, J.: Faster R-CNN: towards real-time object detection with region proposal networks. In: Advances in Neural Information Processing Systems, pp. 91–99 (2015)
10. He, K., Gkioxari, G., Dollár, P., Girshick, P.: Mask R-CNN. In: Proceedings of the IEEE International Conference on Computer Vision, pp. 2961–2969 (2017)
11. Girshick, R., Radosavovic, I., Dollár, P., He, K., Gkioxari, G.: Detectron (2018)
12. Dai, B., Zhang, Y., Lin, D.: Detecting visual relationships with deep relational networks. In: Proceedings of the IEEE Conference on Computer Vision and Pattern Recognition, pp. 3076–3086 (2017)
13. Zhuang, B., Liu, L., Shen, C., Reid, C.: Towards context-aware interaction recognition for visual relationship detection. In: Proceedings of the IEEE International Conference on Computer Vision, pp. 589–598 (2017)

14. Liang, K., Guo, Y., Chang, H., Chen, X.: Visual relationship detection with deep structural ranking. In: Thirty-Second AAAI Conference on Artificial Intelligence (2018)
15. Chao, Y.W., Liu, Y., Liu, Y., Zeng, H., Deng, J.: Learning to detect human-object interactions. In: 2018 IEEE Winter Conference on Applications of Computer Vision (WACV), pp. 381–389. IEEE (2018)
16. Gkioxari, G., Girshick, R., Dollár, P., He, K.: Detecting and recognizing human-object interactions. In: Proceedings of the IEEE Conference on Computer Vision and Pattern Recognition, pp. 8359–8367 (2018)
17. Xiao, H.: Bert-as-service (2018). https://github.com/hanxiao/bert-as-service

Disentangled Representation Learning for Leaf Diseases Recognition

Xing Wang, Congcong Zhu, and Suping Wu[✉]

School of Information Engineering, Ningxia University, Yinchuan 750021, China
wx_nxu@163.com, ccz_nxu@163.com, wspg123@163.com, pswuu@nxu.edu.cn

Abstract. Plant disease detection plays an important role in agricultural production and ecological protection. However, it is always a challenge to detect the severity of plant diseases in multi-species and multi-disease conditions. Unlike most existing classification methods which are difficult to solve multi-properties detection, we propose a disentangled representation interactive network (DRIN), which disentangles the global features of each plant leaf and learns the discriminative representation of multiple sub-properties, including plant species, disease types and disease severity. To achieve it, the disentangled representation network transform the joint probability into the conditional probability through the information interaction between the sub-properties. Moreover, data filtering was introduced to reduce the error messages in property interactions. Experimental results demonstrate the effectiveness of our DRIN on the plant disease detection dataset.

Keywords: Representation learning · Plant disease detection · Information interaction · Data filtering

1 Introduction

With the continuous development of computer technology, the research on intelligent identification of plant diseases has made good progress. The main challenge existing is the fact that it can only detect specific plant diseases or multiple plant diseases, but cannot distinguish the severity of the disease. It is important to distinguish the severity of the disease, for it represents different strategies in dealing with the disease (e.g. the medicine quantity).

There are many traditional methods for detecting plant diseases. Tan et al. [1] established a multi-layer BP neural network model by calculating the leaf chromaticity value to realize the identification of soybean leaf diseases. Tian et al. [2] used the support vector machine (SVM) recognition method to extract the color and texture features of grape diseased leaves, and achieved a better recognition effect than the neural network. Wang et al. [3] extracted the color, shape,

This work was supported in part by the Ningxia Key Research and Development Plan Major Projects under Grant 2016BZ0901 and in part by the Research and Innovation Foundation of First-class Universities in Western China under Grant ZKZD2017005.

© Springer Nature Switzerland AG 2019
Y. Zhao et al. (Eds.): ICIG 2019, LNCS 11901, pp. 530–540, 2019.
https://doi.org/10.1007/978-3-030-34120-6_43

texture and other features of leaf lesions, combined with environmental informa-tion, and used discriminant analysis method to identify the types of cucumber lesions. Zhang et al. [4] also extracted the color, shape and texture features of the diseased spots after the spots were segmented, and then identified five kinds of corn leaves by the k-nearest neighbor (KNN) classification algorithm. In the literatures above, specific plant image features were extracted and combined with traditional classification methods to identify diseases. Although the meth-ods above have achieved good recognition effects, they are not able to completely represent plant disease information due to specific features. For instance, some certain diseased leaves may appear in other features (e.g. powder) rather than disease spots, which makes the segmentation more difficult and has a negative impact on the recognition effect. Moreover, the number of test samples selected by the methods above is limited, i.e. the selected leaves are only from one plant, and these methods are only limited to the identification of the same plant leaf diseases.

In recent years, convolutional neural network [5] has been widely used in the field of image recognition (such as handwritten font recognition [6], face recog-nition [7,8] and object detection [9,10]) without relying on specific features. In general recognition, convolutional neural network models such as AlexNet [11], GoogLeNet [12] and ResNet [13] have achieved good results. An increas-ing number of scholars applied these models to image recognition in a narrow sense. [14,15] used convolutional neural network to conduct relevant studies on plant leaf classification. Sladojevic et al. [16], Brahimi et al. [17] and Amara et al. [18] applied the convolutional neural network to the identification of plant leaf diseases, and improved the model CaffeNet and AlexNet respectively with fine-tuning methods, achieving good identification results. The literatures above proved that convolutional neural network is feasible to identify plant leaf dis-eases. In the field of agriculture, it is far from enough to just identify the type of plant diseases, which did not consider collaborative predictions of species, disease types and disease severity. Therefore, modeling the interaction between mutil-properties is critical.

To address the above-mentioned challenges, in this paper, we propose a dis-entangled multi-representation interactive network to identify various plant dis-eases. Different from the existing detection methods of plant diseases, which can only detect one or more diseases, the DRIN adopts the learning methods of property decomposition, property interaction, property fusion and data fil-tering, aiming at predicting the plant disease severity and improving classifica-tion accuracy. Specifically, the DRIN disentangle the total plant property into three sub-properties: species, disease types and disease severity. In the process of sub-property prediction, the interactive information was added between sub-properties and fuse the information of sub-property species into sub-property disease and disease severity respectively. At the same time, we integrate the information of sub-property diseases into the severity of sub-property diseases. After that, we fuse the information of multiple sub-properties into the final result by means of property fusion. Finally, in order to ensure the effective information

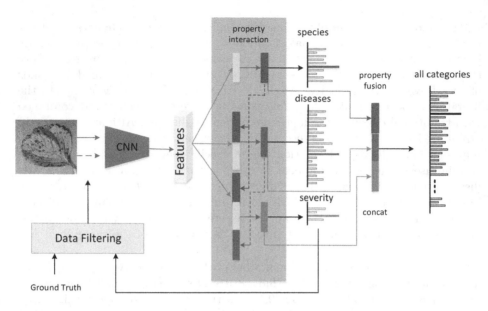

Fig. 1. Schematic diagram of the network structure of our method. We first disentangle the total plant property into three sub-properties: species, disease types and disease severity. In the process of sub-property prediction, we add the interactive information between sub-properties and fuse the information of sub-property species into sub-property disease and disease severity respectively. At the same time, we integrate the information of sub-property diseases into the severity of sub-property diseases. After that, we fuse the information of multiple sub-properties into the final result by means of property fusion. Finally, in order to ensure the effective information transfer in property interaction, we introduce data filtering to reduce error information in the process of property interaction.

transfer in property interaction, data filtering was introduced to reduce error messages in the process of property interaction. Figure 1 shows the structure of our framework.

The core contributions are summarized as follows:

- For the first time, we use feature learning in plant disease prediction. To achieve it, multi-branch network was proposed to disentangle the global information of plant leaves. Therefore, this method avoids the problem of difficult optimization in joint prediction of multiple sub-properties.
- We propose a property interaction to change the joint probability into the conditional probability through the information interaction between sub-properties. For error information, data filtering was introduced to reduce it in the process of property interaction.

2   Our Approach

In this paper, multi-branch network was introduced to disentangle the global information of plant leaves. This method avoids the problem of difficult optimization in joint prediction of multiple sub-properties. In addition, the DRIN put forward the property interaction method to transform the joint probability into conditional probability. Furthermore, a data filtering was introduced to reduce the property interaction error messages.

2.1   Multi-properties Interactive Networks

Detection of disease severity is a complex classification problem in multi-species and multi-disease conditions, and the network needs to be able to simultaneously predict species, disease and disease severity. The traditional network can only detect these three properties jointly, which leads to increase fitting complexity of the model. In order to solve this problem, we transform the joint prediction task into multiple sub-property classification tasks.

Given an input image I, we first put the image into the convolutional layer of pre-training to extract the features of plant diseases. The extracted deep representations are denoted as $F = W_c * I$, where $*$ denotes a set of operations of convolution, pooling and activation, and W_c denotes the overall parameters. Then the extracted global features are sent to the disentangle representational network. In the disentangle representational network, the three branch networks learn the higher-order feature expressions of species, disease types and disease severity respectively. This clearer expression of higher-order features contributes to more accurate detection tasks. Finally, the three sub-properties are fused to obtain the final result. The extracted feature F obtained species feature F_s, disease feature F_d and disease severity feature F_l through species decomposition function f_s, disease types decomposition function f_d and disease severity decomposition function f_l, respectively. Therefore, the formula to formalize the property decomposition is as follows:

$$F_s = f_s(F), F_d = f_d(F), F_l = f_l(F) \tag{1}$$

The disentangled representation sub-network has a natural hierarchical relationship among multiple sub-properties. For example, it's easier to identify a disease with a known species. Similarly, it is easier to deduce the severity in the presence of a known species and disease. There is no doubt that this natural hierarchy can provide prior information in multi-properties tasks. Therefore, we make use of the probabilistic formulation over the variables including the image I, species S, disease D and severity of plant disease L. In accordance with the natural relation of the interaction between multiple sub-properties, the joint probability and conditional probability under multiple properties can be written as:

$$p(S, D, L, I) = p(L|S, D, I)p(D|S, I)p(S|I)p(I) \tag{2}$$

$$p(S, D, L|I) = \frac{p(S, D, L, I)}{p(I)} = \underbrace{p(S|I)}_{\text{Species}} \cdot \underbrace{p(D|S, I)}_{\text{Disease}} \cdot \underbrace{p(L|S, D, I)}_{\text{Severity}} \tag{3}$$

We have carried out the information interaction between the sub-properties in the disentangled representation sub-network which approximately transforms the joint probability into conditional probability.

After the property decomposition module, there is a very important fusion module, which fuses the results of decomposing sub-properties to obtain the final predicted results. There are intersections and differences between species and disease (for example, two widely different species may share a disease, and two similar species may not have the same one). Thus, there is a possibility that the species sub-network and disease sub-network predict two outcomes those are absolutely impossible to coexist (like citrus is free from powdery mildew). Therefore, we introduce the concept of mutual supervision to constrain sub-properties fusion. When two incompatible results appear, the result with the highest confidence in two properties is taken as the prediction guide. In the other sub-properties, a high confidence result that coexists with the previous sub-property is selected as the prediction result. By using the mutual supervision information among the sub-properties, the dependence among the branch networks can be well restrained.

2.2 Data Filtering

One of the most challenging aspects of this task is the intersection of species, disease and severity. There are difficult samples that are similar in certain property. As shown in Fig. 2, it can be seen that there are hard samples in the data set that cannot be distinguished by people. This means that when some samples are highly similar in one property, it will confuse the judgment in other sub-properties. For example, when the sample prediction is wrong, it means that the condition is wrong in the conditional probability of the disentangled representation network, so the network will predict the probability under the wrong condition. The two problems above will affect the gradient descent direction of the network and make it difficult for the network to converge to the optimal value.

To solve this problem, we propose a method of data filtering. As shown in Fig. 3, the network will conduct two iterations. The first iteration uses all the data in batch size to train the network. After that, the data will be filtered once and the network gradient direction under correct conditional probability will be optimized again.

During the data filtering phase, a batch size sample disease severity label was used to test the consistency with the predicted results. When the label is consistent with the predicted result of the network, the sample will be fully used to train the network. When the label is inconsistent with the network prediction result, the confidence of the positive class in the prediction result will be taken as

Fig. 2. The pictures in the first line show the general plant diseases severity. The second line shows the severity of the plant disease. From left to right are Citrus Greening June, Pepper Scab, Grape Leaf Blight Fungus and Peach_Bacterial Spot.

the weight value of loss calculation for this sample. When the label of a sample is significantly different from the predicted result of the network, the influence of it as an error condition on the classification in other sub-properties will be greater. Therefore, it will produce false conditional probability to affect the prediction accuracy. In the second iteration, our network reduce the contribution of this sample to network optimization. Through data filtering, data will describe the difference between properties more accurately, and more robust data distribution will guide the conditional probability more accurately between sub-properties.

3   Experiments

To evaluate the effectiveness of the DRIN, we firstly conducted the main experimental results on the 2018 Global AI Challenge Plant Disease Datasets[1]. Next, we present details on evaluation data set, protocol and experimental analysis, respectively.

3.1   Evaluation Dataset and Protocol

The dataset has 61 classifications (according to "species-disease-degree"), 10 species, 27 diseases (24 of which are classified into general and severe), 10 health classifications, and a total of 47,393 pictures. Each picture contains a leaf of one plant, and the leaf occupies the main position of the picture. The dataset is randomly divided into four sub-data sets: training (70%), validation (10%), test A (10%) and test B (10%). Among them, the training set has 32,739 pictures,

---

[1] https://challenger.ai/competition/pdr2018.

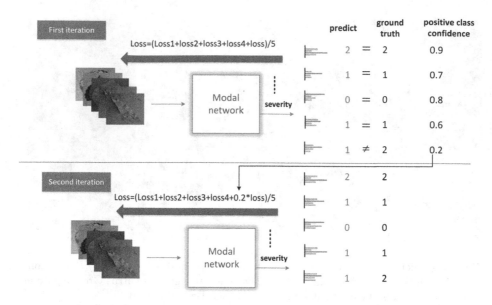

Fig. 3. Schematic diagram of data filtering structure. In the data filtering phase, the batch size is assumed to be five. When the five photos are iterated for the first time, the final loss value is equal to the average value of the sum of each loss. When the predicted result of the first iteration is inconsistent with the label, the loss of the second iteration is multiplied by the confidence of the predicted result of the positive class; When the predicted results of the first iteration are consistent with the label, the loss of the second iteration remains unchanged.

the validation set has 4,982 pictures, the test set A has 4,959 pictures, and the test set B has 4,957 pictures. Since the labels for test set A and test set B are not publicly available in the dataset, We mix the two data sets of training and validation, and finally randomly select 10% of them as the test set and the rest as the training set.

3.2   Implementation Details

Datasets are unevenly distributed across the data distribution. In the training dataset, there are only two images of tomato scab, one general and one serious, which cannot be trained to achieve good result, so we delete these two images. For other data, the training set was perturbed by randomly rotation, horizontal and vertical flip to increase the data to solve the problem of data imbalance. At the same time, we scale the image to 224×224 pixels for training. For optimization, Adam optimizer was used with a learning rate begins at 0.0001 and decays 0.9 after each 20 epochs. The batch size is set as 40.

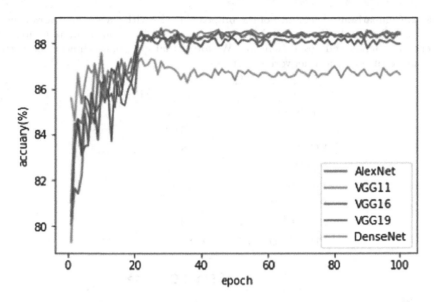

Fig. 4. The accuracy curve of classical network architecture in plant disease data set. From these results, the performance of VGG19 surpasses that of other classic network structures.

3.3   Results and Analysis

Plant disease dataset was trained in the model DRIN. To evaluate the methods, we use the classical network structure Alexnet, VGG11 [20], VGG16, VGG19 and DenseNet [19] to classify all the categories. The accuracy curve shown in Fig. 4 is obtained. From these results, the performance of VGG19 surpasses that of other classic network structures.

In order to prove the effectiveness of the method, we select VGG19 as our basic network and design three groups of experiments. Specifically, we put an image into the VGG19 network, and then enter three different full-connection layers to decompose its property into three sub-properties (species, disease types, and disease severity). There are interactions between sub-properties in the full-connection layers, and finally the three sub-properties was fused to obtain the final result. At the same time, in order to prove the effectiveness of the data filtering, we design two sets of experiments. Specifically, The former is the basis of the first set of experiments. We cancel the interaction between sub-properties of full connection and increase data filtering. The latter is based on the first set of experiments, the way of data filtering was added. The prediction accuracy of experimental results is shown in Table 1.

As can be seen from Table 1 that VGG19 has the best performance in the classical deep learning network structure, which is far superior to other deep learning network structures. The experimental results show that when property decomposition, property interaction and property fusion are used in the method, the performance is improved by 0.08%, which proves the effectiveness

Table 1. Comparisons of accuracy of the proposed DRIN with classical network framework. PD, DF, PI and PF respectively represent property decomposition, data filtering, property interaction and property fusion. We found that our approach achieved better performance than classical network structures.

Methods	Accuracy(%)
AlexNet	88.42
DenseNet	86.62
VGG11	88.39
VGG16	87.92
VGG19	88.47
PD+PI+PF(ours)	88.55
PD+DF+PF(ours)	88.44
PD+PI+DF+PF(ours)	88.86

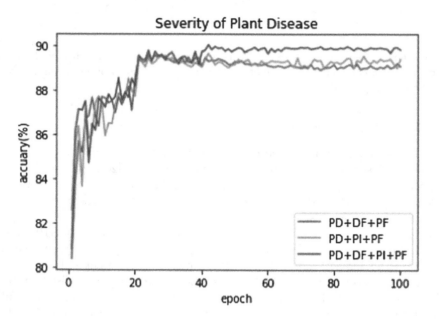

Fig. 5. The curve of the accuracy of each group in predicting the severity of plant diseases was shown. PD, DF, PI and PF respectively represent property decomposition, data filtering, property interaction and property fusion. We found that our methods with PD, DF, PI, and PF performed better than others.

of the property interaction in our experiment. When adding data filtering to the method, the performance is improved to 88.86% and increased by 0.29%. At the same time, we do a counter example experiment in which data filtering is added when there was no property interaction in the experiment, so the accuracy of the experiment was reduced. This experiment proves that the proposed data filtering method reduces the weight of error message interaction in property interaction, and thus proves the effective performance of data filtering. Since it is the most difficult part to predict the severity of diseases in the sub-property results, we have drawn the curve of the prediction accuracy of plant disease severity, as shown in Fig. 5.

4   Conclusion

We propose a disentangled representational interactive network to solve the problem of predicting plant disease in the case of multiple plants and multiple diseases. Our method consists of property decomposition, property interaction, property fusion and data filtering. Property decomposition is to disentangle the entire property into three sub-properties. Property interaction refers to the information transfer between sub-properties. Property fusion is the fusion of three sub-properties into final result. Data filtering reduces the transmission of error messages between sub-properties. Experimental results demonstrate the effectiveness of our method for plant disease detection. In future work, we will use metric learning to further improve our performance.

References

1. Feng, T., Xiaodan, M.: The method of recognition of damage by disease and insect based on laminae. J. Agric. Mechan. Res. **6**(2009), 41–43 (2009)
2. Youwen, T., Tianlai, L., Chenghua, L., Zailin, P., Guokai, S., Bin, W.: Method for recognition of grape disease based on support vector machine. Trans. Chin. Soc. Agric. Eng. **23**(6), 175–180 (2007)
3. Wang, X., Zhang, S., Wang, Z., Zhang, Q.: Recognition of cucumber diseases based on leaf image and environmental information. Trans. Chin. Soc. Agric. Eng. **30**(14), 148–153 (2014)
4. Zhang, S., Shang, Y., Wang, L., et al.: Plant disease recognition based on plant leaf image. J. Anim. Plant Sci. **25**(3 Suppl. 1), 42–45 (2015)
5. Zeiler, M.D., Fergus, R.: Visualizing and understanding convolutional networks. In: Fleet, D., Pajdla, T., Schiele, B., Tuytelaars, T. (eds.) ECCV 2014. LNCS, vol. 8689, pp. 818–833. Springer, Cham (2014). https://doi.org/10.1007/978-3-319-10590-1_53
6. LeCun, Y., Bottou, L., Bengio, Y., Haffner, P., et al.: Gradient based learning applied to document recognition. Proc. IEEE **86**(11), 2278–2324 (1998)
7. Sun, Y., Wang, X., Tang, X.: Deep learning face representation from predicting 10,000 classes. Proceedings of the IEEE conference on computer vision and pattern recognition. 1891–1898 (2014)
8. Wen, Y., Zhang, K., Li, Z., Qiao, Y.: A discriminative feature learning approach for deep face recognition. In: Leibe, B., Matas, J., Sebe, N., Welling, M. (eds.) ECCV 2016. LNCS, vol. 9911, pp. 499–515. Springer, Cham (2016). https://doi.org/10.1007/978-3-319-46478-7_31
9. Ren, S., He, K., Girshick, R., Sun, J.: RCNN Faster. [n.d.]. Towards Real-Time Object Detection with Region Proposal Networks ([n. d.])

10. Girshick, R., Donahue, J., Darrell, T., Malik, J.: Rich feature hierarchies for accurate object detection and semantic segmentation. Proceedings of the IEEE Conference on Computer Vision and Pattern Recognition, pp. 580–587 (2014)
11. Krizhevsky, A., Sutskever, I., Hinton, G.E.: ImageNet classification with deep convolutional neural networks. In: Advances in Neural Information Processing Systems, pp. 1097–1105 (2012)
12. Szegedy, C., et al.: Going deeper with convolutions. In: Proceedings of the IEEE Conference on Computer Vision and Pattern Recognition, pp. 1–9 (2015)
13. He, K., Zhang, X., Ren, S., Sun, J.: Deep residual learning for image recognition. In: Proceedings of the IEEE Conference on Computer Vision and Pattern Recognition, pp. 770–778 (2016)
14. Gong, D.X., Cao, C.R.: Plant leaf classification based on CNN. Comput. Modernization **4**(2014), 12–15 (2014)
15. Zhang, S., Huai, Y.: Leaf image recognition based on layered convolutions neural network deep learning. J. Beijing For. Univ. **38**(9), 108–115 (2016)
16. Sladojevic, S., Arsenovic, M., Anderla, A., Culibrk, D., Stefanovic, D.: Deep neural networks based recognition of plant diseases by leaf image classification. Comput. Intell. Neurosc. **2016**, 11 (2016)
17. Brahimi, M., Boukhalfa, K., Moussaoui, A.: Deep learning for tomato diseases: classification and symptoms visualization. Appl. Artif. Intell. **31**(4), 299–315 (2017)
18. Amara, J., Bouaziz, B., Algergawy, A., et al.: A deep learning based approach for banana leaf diseases classification. In: BTW (Workshops), pp. 79–88 (2017)
19. Simonyan, K., Zisserman, A.: Very deep convolutional networks for large-scale image recognition. arXiv preprint arXiv:1409.1556 (2014)
20. Huang, G., Liu, Z., Van Der Maaten, L., Weinberger, K.Q.: Densely connected convolutional networks. In: Proceedings of the IEEE Conference on Computer Vision and Pattern Recognition, pp. 4700–4708 (2017)

A Noise Robust Batch Mode Semi-supervised and Active Learning Framework for Image Classification

Chaoqun Hou[✉], Chenhui Yang, Fujia Ren, and Rongjie Lin

Fujian Key Laboratory of Sensing and Computing for Smart Cities,
School of Information Science and Engineering,
Xiamen University, 422 Siming Road South, Xiamen 361005, Fujian, China
chaoquncql@foxmail.com

Abstract. Supervised learning with convolutional neural networks has made a great contribution to computer vision largely due to massive labeled samples. However, it is far from adequate available labeled samples for training in many applications. Realistically, annotation is a tedious, time consuming, and costly task while a strong need for specialty-oriented knowledge and skillful expert. Therefore, in order to take full advantage of limited resources to observably reduce the cost of annotation, we propose a noise robust batch mode semi-supervised and active learning framework which named NRMSL-BMAL. When querying labels in an iteration, firstly, a convolutional autoencoder cluster based batch mode active learning strategy is used for querying worthy samples from annotation experts with a cost. Then, a noise robust memorized self-learning is successively used for extending training samples without any annotation cost. Finally, these labeled samples are added to the training set for improving the performance of the target model. We perform a thorough experimental evaluation in image classification tasks, using datasets from different domains, including medical image, natural image, and a real-world application. Our experimental evaluation shows that NRMSL-BMAL is capable to observably reduce the annotation cost range from 44% to 95% while maintaining or even improving the performance of the target model.

Keywords: Active learning · Semi-supervised learning · Convolutional autoencoder cluster · Continuously fine-tuning · Image classification

1 Introduction

Image classification is a challenging task in computer vision and pattern recognition with a long history. Significantly, convolutional neural networks (CNNs), which is one of the most successful deep learning models, have achieved groundbreaking results owing to massive labeled samples and computational power in the recent few years. Unfortunately, in many real-world applications, there has

© Springer Nature Switzerland AG 2019
Y. Zhao et al. (Eds.): ICIG 2019, LNCS 11901, pp. 541–552, 2019.
https://doi.org/10.1007/978-3-030-34120-6_44

been far from adequate existing labeled samples for training [25]. Moreover, especially in medical image analysis, annotation is a tedious, time consuming and costly task while a strong need for specialty-oriented knowledge and skillful annotation expert [29]. Therefore, the motivation of this paper is how to take full advantage of limited resources to minimize the need for human annotation while maintaining or even slightly improving the performance of the model.

Semi-supervised learning (SSL) is a kind of method that attempts to exploit the availability of unlabeled samples for the performance improvement of the model. Self-training [28] (we called "self-learning, SL") is a classic SSL algorithm. Firstly, SL starts with a model that pre-trained on a smaller set from target domain labeled samples. Subsequently, SL attempts to add those relatively certain samples and its respective labels to training set from unlabeled sample pool. The model is re-trained with the incremental training set until reach the mark. Obviously, there is a drawback of SL that noisy labels shall seriously degenerate the performance by itself. Moreover, most learning algorithms, including deep learning, are sensitive to noisy labels. To remit the noisy label issue, [6] proposed a self-paced manifold regularization framework which was inspired by the learning principle of human. And [4] designed a two-stage approach to train AlexNet and ResNet respectively from noisy labels in a semi-supervised manner without any prior knowledge or distribution of noisy labels. Furthermore, lots of existing methods [12, 18, 22] were proposed to address noisy labels.

In this paper, we propose a noise robust memorized self-learning algorithm named NRMSL. Firstly, we attempt to memorize the prediction information of the model in each iteration, and apply it into SSL procedure for reducing noisy labels. Subsequently, we apply a self-error-correcting method, which is inspired by [18], to improve the anti-noise ability of the target model.

Alternatively, active learning (AL) is another similar field which could greatly improve the performance of the target model with fewer labeled samples thanks to some heuristic strategies, such as uncertainty [1, 5, 13]. There are three classical settings [25], including membership query synthesis, stream-based sampling, and pool-based sampling. We focus on pool-based sampling which is the most popular setting recently years. However, in real-world applications, parallel labeling environment may be available that means single mode AL queries in serial may be inefficient. Fortunately, batch mode active learning (BMAL)[2, 25] is exactly the way to remit the parallel labeling problem by selecting more than one sample in each iteration. We consider overlap in information content among those most informative samples through combining uncertainty strategy and convolutional autoencoder (CAE) cluster. Furthermore, based on a mass of researches, we have summarized various related work and listed them in Table 1.

Obviously, there is an essential difference between SL and AL. SL directly obtains label through model while AL querying for label by annotation experts who also called oracle. In other words, AL assurances annotated samples by experts, leading to notable performance improvements, but costly. On the contrary, SL obtains label without the intervention of human annotators while along

Table 1. Various related researches about BMAL and SSL-AL in recent years.

Research of active learning	Relevant literature
Hand-craft strategy	[1,5,13]
Combining hand-craft strategies	[2,3,9,17,20,29]
Cluster active learning	[21,27]
Semi-supervised active learning	[10,19,23]

with noisy labels. A tandem certainty-based AL and self-learning method was proposed in [10], considering both the pool-based scenario and the stream-based scenario. In addition, [10] have summarized overview of previous work combining AL and SSL. [19] is the first work exploring combinations of AL and SL algorithm in 1998. However, those work mentioned above just skillfully combined AL and SL seldom considering noisy labels. According to these important observations, we propose a framework named NRMSL-BMAL to combine NRMSL and BMAL naturally.

Contributions of this paper can be summarized as follows. (1) We propose a memorized self-learning (MSL) algorithm to reduce noisy labels owing to the memorized information of historical model's prediction. Furthermore, to reduce the impact of noise labels, we apply a noisy label self-adjusting method in an effective way to MSL, named NRMSL. (2) We propose a framework for taking full advantage of limited resources that interactively integrated NRMSL, CAE cluster based BMAL and other technologies, such as transfer learning and continuously fine-tuning. (3) We perform a thorough experimental evaluation in image classification task and show that NRMSL-BMAL can reduce annotation cost range from 44% to 95%, using datasets from different domains as detailed in Sect. 3.1. (4) We verify that these samples queried by NRMSL-BMAL are representative.

2   Proposed Framework

2.1   NRMSL-BMAL Framework

NRMSL-BMAL framework includes the following four cores. (1) Transfer learning: NRMSL-BMAL initializes with a pre-trained model that source domain can be changed flexibly, and ImageNet [24] was used in this paper. (2) Then, CAE cluster based BMAL method is used for querying those most uncertain and diverse samples from unlabeled sample pool. (3) NRMSL algorithm is successively used for extending training samples without any annotation cost. (4) Fine-tuning the target model continuously by these incremental labeled samples. Finally, these four parts were synergistically integrated into a framework.

In another point of view, as we can see from Fig. 1, directly perceived through the senses of NRMSL-BMAL framework. It works in an incremental way, improving the performance of the target model with increasing labeled samples. In each

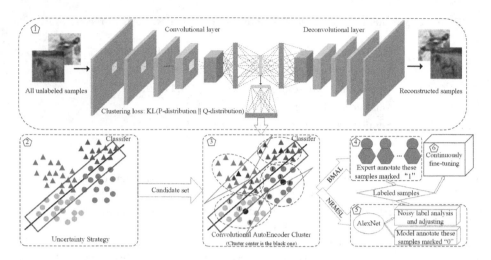

Fig. 1. Diagram of NRMSL-BMAL framework combines some deep learning components cooperatively. To simplify the discussion, this diagram was designed for binary classification, although it is also available for multi-classification. The pentagon shape shows the core procedures. These dotted circles and ellipses of 2nd and 3rd pentagon mark different clusters while these solid lines circles and triangles represent different categories, and the heavy line is the classifier. Besides, for example, these solid lines circles and triangles with letter "1" and "0" represent the selected samples for BMAL and NRMSL respectively.

iteration, firstly, we calculate the corresponding prediction probability value from the softmax layer of each unlabeled sample without real labels. Secondly, an uncertainty indicator, such as entropy, BvSB [13], least confidence, and smallest margin [25], will be calculated according to these probability value. In this paper, we use entropy for binary classification and BvSB for multi-classification. Thirdly, CAE cluster is the next step to gather similar samples into a predefined set of clusters as shown in the 3rd pentagon of Fig. 1. Furthermore, we innovatively apply a self noisy label adjusting method into NRMSL procedure as detailed in [18], using a self-error-correcting softmax loss to adaptively switch between the noisy label or the max-activated neuron and a Bernoulli distribution to decide whether to select the max-activated label (see details in Sect. 2.2). Finally, fine-tuning AlexNet [15] by those incremental labeled samples obtained respectively from experts and the current model. Experts annotate those most uncertain and diverse samples while the current model annotates those relatively certain and diverse samples cooperatively.

2.2 Noise Robust Memorized Self Learning

Self-learning was used for extending training samples without any annotation cost in NRMSL-BMAL framework. The core challenge of self-learning is to mitigate noise samples. NRMSL is exactly the way to mitigate the problem, which containing two-stage.

The first stage is to reduce the generation of noise samples. Firstly, we design a structure for storing historical predictive information. Then, we use two constraints to filter samples. Those two constraints respectively are (1) the prediction results of latest m-times must be consistent and (2) the score of latest m-times must be within the preset range ($score_{min} < score < score_{max}$).

		current model		C_t
initialization	unlabeled samples	model-0	dog	$C_{t-0} = 1$
iteration-1	labeled samples / unlabeled samples	model-1	dog	C_{t-1}
iteration-2	labeled samples / unlabeled samples	model-2	dog	C_{t-2}
	⋮	⋮	⋮	⋮
iteration-N	labeled samples / unlabeled samples	model-N	cat	C_{t-N}

Fig. 2. Diagram of the noisy label adjust. These processed samples are from the self-learning procedure.

The second stage is to improve the anti-noise ability of the target model which is the core of NRMSL. Considering that the performance of the target model will be improved with the increase of the labeled samples, as shown in Fig. 2. We allow those labeled samples, which comes from self-learning procedure, to modify the label with a certain probability. Besides, inspired by [18], we used the formula 1 as polynomial confidence policy to elaborately cooperate with the target model in each iteration:

$$C_t = C_0 * \left(1 - \frac{t}{T}\right)^\lambda, \tag{1}$$

where C_0 denotes the initial confidence, t denotes the current iteration of training, T denotes the total number of iterations. Subsequently, for each labeled sample from self-learning, a random value r between 0 and 1 will be generated to decide whether or not to adjust the label of this sample. If $r > C_t$ then change the label of this sample as $\tilde{y}$, where $\tilde{y}$ is the predicted result of the current model, otherwise do nothing. Note that, as the iteration t increases, more and more labeled samples will be added into the training set, and the value of C_t decreases. Therefore, as shown in Fig. 2, the further the iteration, the performance of the target model will get better, and the probability of adjusting the label will be higher.

2.3 CAE Cluster Based BMAL

As we can see from the 2nd pentagon of Fig. 1, for probabilistic classification models, we used entropy to measure the uncertainty of samples, and the entropy score of samples are defined as:

$$EntropyScore = -\sum_{i=1}^{C} p_i \times log(p_i), \qquad (2)$$

where C is the number of categories, p_i is the predicted probability value of the corresponding sample. Uncertainty measure captures the informativeness of samples without true labels. For example, as a binary classification task, the predicted probability value of sample$_1$ and sample$_2$ respectively are $\{p_{11} = 0.5, p_{12} = 0.5\}$ and $\{p_{21} = 0.1, p_{22} = 0.9\}$. Then the entropy score of sample$_1$ and sample$_2$ respectively are 1 and 0.47. So, sample$_1$ with higher scores will be the prioritized target sample.

However, as shown in Fig. 3(b), it would generate much redundant information when we need to query more than one sample in an iteration. In other words, the information of those samples provide for the model is similar. Therefore, as shown in Fig. 3(c), the clustering algorithm is used for querying those representative samples.

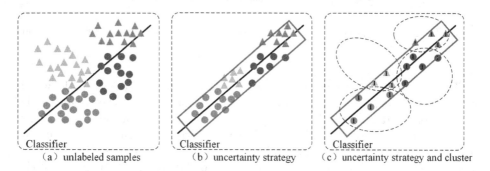

(a) unlabeled samples   (b) uncertainty strategy   (c) uncertainty strategy and cluster

Fig. 3. Diagram of batch mode active learning: combining uncertainty strategy and cluster. The closer the sample is to the classifier, the higher the uncertainty will be. Distance between samples reflects the similarity.

As mentioned, the core challenge of BMAL is to reduce duplicate information of queried samples. CAE cluster, as shown in the 1st pentagon from Fig. 1, plays a role both in querying diverse samples and being able to deal with more complex images. We design a CAE network which encoder and decoder are composed of convolutional, deconvolutional, fully connected and leaky ReLU layers. The main loss is added by reconstruction loss and clustering loss, as detailed in [7,8,26]. CAE cluster is pre-trained in an unsupervised way with unlabeled sample pool and fine-tuned with a subset selected by uncertainty strategy. Subsequently, we apply the CAE cluster not only in BMAL procedure but also in NRMSL procedure, as shown in the red rectangles and green triangles respectively of the 3rd pentagon from Fig. 1. Finally, we try to select these diverse samples from each cluster. Besides, in many real-world applications, the number of querying samples is usually larger than clusters. In other words, we need more than one sample from each cluster so that we use the cosine distance to measure the similarity between different samples, as shown in these solid lines circles and triangles with letter "1" and "0" from Fig. 1. Moreover, we can deal with more complex images owing to the parameters sharing of convolution, such as single-multi packing dataset which discussed in Sect. 3.1.

The overall NRMSL-BMAL algorithm is given in Algorithm 1.

Algorithm 1. Algorithm of NRMSL-BMAL

Input: Unlabeled sample pool, U; Validation set, V; Training set, L; Labeled set by experts, L_1; Labeled set by self-learning, L_2; Uncertainty strategy, Q; Pre-trained AlexNet, target model; Pre-trained CAE cluster with U, CAE-C; Target validation accuracy, $target\_acc$; Validation accuracy of current model, $active\_acc$; Iteration times, t; Number of BMAL, k; constraint conditions, CC.

Output: Model of achieving target accuracy; Labeled samples by experts.

1: **initial:** $L = \varnothing$, $L_1 = \varnothing$, $L_2 = \varnothing$, $t = 0$, $m = 10$.
2: **repeat**
3:     $t + = 1$, $L = L_1 \cup L_2$
4:     **if** $L \neq \varnothing$ **then**
5:         training target model according to L and update $active\_acc$
6:     **end if**
7:     **if** $L_2 \neq \varnothing$ **then**
8:         adjust each sample from L_2 according to NRMSL
9:     **end if**
10:     predict each sample from U and calculate score according to target model
11:     descending order U by score and select $3 * k$ samples by Q
12:     select k from these $3 * k$ samples by CAE-C, denoted as X_1
13:     query label of X_1 by experts, $L_1 = L_1 \cup X_1$, $U = U/X_1$
14:     **if** $t >= m$ **then**
15:         ascending order U by score and select $3 * k$ samples by CC
16:         select k from these $3 * k$ samples by CAE-C, denoted as X_2
17:         query label of X_2 by current target model, $L_2 = L_2 \cup X_2$, $U = U/X_2$
18:     **end if**
19: **until** $active\_acc >= target\_acc$ **or** empty(U)

---

3   Simulation Experimental Results

In this section, we performed a thorough experimental evaluation in image classification tasks, including five different datasets to evaluate the effectiveness of the proposed method and framework. We keep the same parameters to train with all labeled samples and marked the best validation accuracy as the target accuracy of NRMSL-BMAL. PyTorch is used to develop our framework, with support of 1080Ti GPU and Ubuntu system. We repeat the experiment 5 times and record the mean value as the final evaluation result for each dataset, where a random 15%~20% and 10% of annotated samples is divided into validation set and a test set respectively.

3.1   Dataset

NRMSL-BMAL is available for both binary classification and multi-classification task. We evaluate on two multi-classification datasets including (1) **MNIST** [16]: handwritten digital images with 28 × 28 pixels, (2) **CIFAR10** [14]: real object images with 32 × 32 × 3 pixels, and three binary classification datasets including (1) Open Access Series of Imaging Studies (**OASIS**): it is a project aimed at making neuroimaging data sets of the brain freely available to the scientific community and a subset was available in [11] that we used in this paper, (2) **Dog-Cat**: it is a competition of Kaggle, (3) **Image attribute from Tmall**: it is a binary classification task with attributes including single-packing and multi-packing. And we completed the experiment as an intern in Tmall. Particularly, cause of the real-world demand (single-multi packing classification), two annotation experts have participated in our experimental procedure.

3.2   Full Training with AlexNet

We train AlexNet with all labeled samples for each dataset respectively. It is necessary to make clear that our uppermost goal is to reduce the cost of annotation while maintaining the performance of the target model. Therefore, as shown in Table 2, we set the best validation accuracy of full training as the target accuracy. Besides, relevant parameters are optimizer = Adam, pre-trained dataset = ImageNet, classifier learning rate $= 1e - 3$, feature learning rate $= 1e-4$, batch size $= 32$.

Table 2. Result of full training with AlexNet. Abbreviations are Tra: Training, val: Validation, num: number, acc: accuracy.

Dataset	Tra num	Val num	Tra acc	Best val acc
Cat-dog	25000	10000	99.59%	97.05%
OASIS	5120	1280	99.71%	99.06%
Single-Multi	55551	8000	95.20%	92.89%
MNIST	55000	10000	99.47%	99.12%
CIFAR-10	50000	10000	97.22%	90.10%

3.3   Annotation Cost of NRMSL-BMAL

NRMSL: We compare our NRMSL method with the pure SL method. As shown in Fig. 4, the cause of noisy labels and the uninformative labeled samples, SL barely improved the performance of the target model after several iterations. On the contrary, owing to the memorized information and noisy label self-adjusting

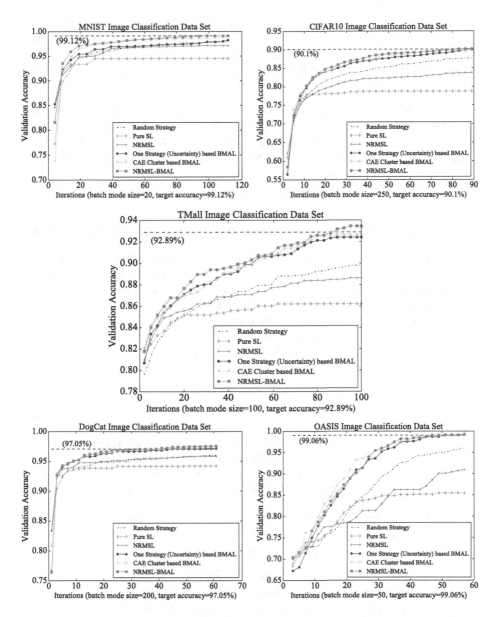

Fig. 4. Experiments on random strategy, pure self earning, NRMSL, one strategy based BMAL, CAE cluster based BMAL and NRMSL-BAML. The abscissa and the ordinate indicate the number of iterations and the validation accuracy respectively

method, NRMSL relatively improved the accuracy of the target model. Especially in MNIST and Dog-Cat dataset.

BMAL: Our purpose in this part is to verify the effectiveness of applying CAE cluster method into BMAL. As shown in Fig. 4, considering the duplicate information of samples, CAE cluster based BMAL outperforms one strategy based BMAL. And both of these two BMAL methods significantly reduced annotation cost compared with random strategy.

NRMSL-BMAL: As shown in Fig. 4, NRMSL-BMAL is significantly superior to random strategy. Because of those extended training samples and noisy label adjusted from NRMSL, NRMSL-BMAL outperforms CAE cluster based BMAL. On the single-multi set from TMall, NRMSL-BMAL is not only significantly reduced the cost of annotation, but also improve the performance of the target model (the final validation accuracy is higher than target validation accuracy).

3.4   The Representativeness of Samples Queried by NRMSL-BMAL

To verify the representativeness of those labeled samples by experts (L_1 in Algorithm 1), the statistical accuracy of remaining unselected samples (U/L_1) was evaluated according to the final model in Table 3. As it shows, the accuracy of remaining unselected samples is almost close to 100%, which means those selected samples could represent those unselected samples. In other words, those unselected samples are hardly enough to improve the target model, so that we can reduce the annotation cost of these samples. Besides, the unlabeled sample pool of OASIS is smaller than others, which is the main reason why its much lower saved rate. NRMSL-BMAL could be more effective in a larger scale unlabeled sample pool.

Table 3. Experiments on the representativeness of selected samples. **Expert** means the number of labeled samples from experts, **Saved Rate** is equal to *one* minus *"Expert"* divided by *"the number of all full training samples"*, **Remaining accuracy** means the predicted accuracy of those samples that are not annotated by experts.

Dataset	Expert	Saved rate	Remaining accuracy
Cat-dog	6100	75.60%	99.04% (18718/18900)
OASIS	2850	**44.34%**	99.56% (2250/2270)
Single-multi	10000	82.00%	98.04% (44658/45551)
MNIST	2240	**95.93%**	99.62% (52558/52760)
CIFAR-10	22250	55.50%	99.06% (27489/27750)

4   Conclusion

We have designed and implemented NRMSL-BAML framework to remit the critical problem: how could we take full advantage of limited resources to minimize the need for human annotation while maintaining the performance of the target model for image classification. It naturally combined NRMSL and CAE cluster based BMAL method. We have evaluated in five different image classification datasets, demonstrating that NRMSL-BMAL could reduce annotation cost range from 44% to 95% for different dataset mentioned in Sect. 3.1. It is worth noting that NRMSL-BMAL could prevent partly self-learning procedure from skewing the target model on account of the high quality labeled samples from experts and noisy label self-adjusting method. Moreover, we are eager to apply it to more real-world applications, such as single-multi packing classification from Tmall. However, it could not be neglected that the time complexity of NRMSL-BAML still needs to be improved, because of the CAE cluster procedure and a large number of training epochs. Therefore, the tradeoff between time complexity and annotation cost is also a critical problem in future works.

References

1. Campbell, C., Cristianini, N., Smola, A., et al.: Query learning with large margin classifiers. In: Proceedings of International Conference on Machine Learning, pp. 111–118 (2000)
2. Cardoso, T.N.C., Silva, R.M., Canuto, S., Moro, M.M., Gonçalves, M.A.: Ranked batch-mode active learning. Inf. Sci. **379**, 313–337 (2017)
3. Chiu, S.C., Jin, Z., Gu, Y.: Active learning combining uncertainty and diversity for multi-class image classification. IET Comput. Vis. **9**(3), 400–407 (2015)
4. Ding, Y., Wang, L., Fan, D., Gong, B.: A semi-supervised two-stage approach to learning from noisy labels. arXiv preprint arXiv:1802.02679 (2018)
5. Freund, Y., Seung, H.S., Shamir, E., Tishby, N.: Selective sampling using the query by committee algorithm. Mach. Learn. **28**(2–3), 133–168 (1997)
6. Gu, N., Fan, M., Meng, D.: Robust semi-supervised classification for noisy labels based on self-paced learning. IEEE Sig. Process. Lett. **23**(12), 1806–1810 (2016)
7. Guo, X., Gao, L., Liu, X., Yin, J.: Improved deep embedded clustering with local structure preservation. In: International Joint Conference on Artificial Intelligence, pp. 1753–1759 (2017)
8. Guo, X., Liu, X., Zhu, E., Yin, J.: Deep clustering with convolutional autoencoders. In: Liu, D., Xie, S., Li, Y., Zhao, D., El-Alfy, E.S. (eds.) Neural Information Processing. ICONIP 2017, vol. 10635, pp. 373–382. Springer, Cham (2017). https://doi.org/10.1007/978-3-319-70096-0_39
9. Guo, Y., Schuurmans, D.: Discriminative batch mode active learning. In: Proceedings of the International Conference on Neural Information Processing Systems, NIPS 2007, pp. 593–600. Curran Associates Inc., USA (2007)
10. Han, W., et al.: Semi-supervised active learning for sound classification in hybrid learning environments. PLoS ONE **11**(9), e0162075 (2016)
11. Hon, M., Khan, N.M.: Towards alzheimer's disease classification through transfer learning. In: International Conference on Bioinformatics and Biomedicine, pp. 1166–1169 (2017)

12. Jindal, I., Nokleby, M., Chen, X.: Learning deep networks from noisy labels with dropout regularization. In: IEEE International Conference on Data Mining, pp. 967–972. IEEE (2016)
13. Joshi, A.J., Porikli, F., Papanikolopoulos, N.: Multi-class active learning for image classification. In: Proceedings of the IEEE Conference on Computer Vision and Pattern Recognition, pp. 2372–2379 (2009)
14. Krizhevsky, A., Hinton, G.: Learning multiple layers of features from tiny images. Technical Report (2009)
15. Krizhevsky, A., Sutskever, I., Hinton, G.E.: ImageNet classification with deep convolutional neural networks. In: Proceedings of the International Conference on Neural Information Processing Systems, pp. 1097–1105 (2012)
16. LeCun, Y., Bottou, L., Bengio, Y., Haffner, P.: Gradient-based learning applied to document recognition. Proc. IEEE **86**(11), 2278–2324 (1998)
17. Li, X., Guo, Y.: Adaptive active learning for image classification. In: Proceedings of the IEEE Conference on Computer Vision and Pattern Recognition, pp. 859–866 (2013)
18. Liu, X., Li, S., Kan, M., Shan, S., Chen, X.: Self-error-correcting convolutional neural network for learning with noisy labels. In: IEEE International Conference on Automatic Face & Gesture Recognition, pp. 111–117 (2017)
19. McCallumzy, A.K., Nigamy, K.: Employing EM and pool-based active learning for text classification. In: Proceedings of International Conference on Machine Learning, pp. 359–367 (1998)
20. Patra, S., Bruzzone, L.: A batch-mode active learning technique based on multiple uncertainty for SVM classifier. IEEE Geosc. Remote Sens. Lett. **9**(3), 497–501 (2012)
21. Patra, S., Bruzzone, L.: A cluster-assumption based batch mode active learning technique. Pattern Recogn. Lett. **33**(9), 1042–1048 (2012)
22. Patrini, G., Rozza, A., Menon, A.K., Nock, R., Qu, L.: Making deep neural networks robust to label noise: a loss correction approach. In: Proceedings of the IEEE Conference on Computer Vision and Pattern Recognition, pp. 2233–2241 (2017)
23. Rong, C., Cao, Y.F., Hong, S.: Multi-class image classification with active learning and semi-supervised learning. Acta Automatica Sinica **37**(8), 954–962 (2011)
24. Russakovsky, O., et al.: Imagenet large scale visual recognition challenge. Int. J. Comput. Vis. **115**(3), 211–252 (2015)
25. Settles, B.: Active learning. Synth. Lect. Artif. Intell. Mach. Learn. **6**(1), 1–114 (2012)
26. Xie, J., Girshick, R., Farhadi, A.: Unsupervised deep embedding for clustering analysis. In: Proceedings of International Conference on Machine Learning, pp. 478–487 (2016)
27. Xiong, S., Azimi, J., Fern, X.Z.: Active learning of constraints for semi-supervised clustering. IEEE Trans. Knowl. Data Eng. **26**(1), 43–54 (2014)
28. Yarowsky, D.: Unsupervised word sense disambiguation rivaling supervised methods. In: Proceedings of Annual Meeting of the Association for Computational Linguistics, pp. 189–196 (1995)
29. Zhou, Z., Shin, J., Zhang, L., Gurudu, S., Gotway, M., Liang, J.: Fine-tuning convolutional neural networks for biomedical image analysis: Actively and incrementally. In: Proceedings of the IEEE Conference on Computer Vision and Pattern Recognition, pp. 4761–4772 (2017)

Salient Object Detection via Distribution of Contrast

Xiaoming Huang[✉]

Computer School, Beijing Information Science & Technology University,
Xiaoying East Road 12, Haidian District, Beijing 100101, China
Huangxm0556@163.com

Abstract. In human vision system, the contrast of image regions to their surroundings is highly sensitive. Motivated by this intuition, many contrast based salient object detection methods have been proposed in recent years. In these previous methods, saliency for one region is generally measured by sum of contrast from its surrounding. We find that the spatial distribution of contrast is one important cue of saliency. Salient region usually shows high contrast in the most of directions, while background region present high contrast in a few directions. Inspired by this phenomenon, we propose one salient object detection method via distribution of contrast. In proposed method, input image will be segmented into superpixels first, for each superpixel, contrast between this superpixel and its surrounding will be computed in every direction, relative standard deviation (RSD) of contrast in all directions is measured as cue of saliency. Experimental results on four benchmark dataset demonstrate the proposed method performs well when against the state-of-the-art methods.

Keywords: Salient object detection · Contrast · Distribution · Relative standard deviation

1 Introduction

Salient object detection aims to simulate the human visual system for detecting pixels or regions that most attractive. The detection result can be used for numerous computer vision task such as image classification [1, 2], object detection and recognition [3, 4], image compression [5], and image segmentation [6, 7].

Motivated by increasing application demand, a number of algorithms have been proposed [8–25]. These methods can be divided into two categories: bottom-up (stimulus-driven) [8–23] methods and top-down (goal-driven) [24, 25] methods. Most bottom-up detection methods rely on low level visual cues such as color, intensity, and orientation, the main advantage of such methods is fast and need not specific prior knowledge. On the contrary, top-down methods usually learn models from training examples with manually labeled ground truth. Based on the supervised learning frame work, such methods require domain specific prior knowledge. A comprehensive survey of saliency detection can be found in [26, 27], and a quantitative comparison of different methods was provided in [28, 29].

© Springer Nature Switzerland AG 2019
Y. Zhao et al. (Eds.): ICIG 2019, LNCS 11901, pp. 553–565, 2019.
https://doi.org/10.1007/978-3-030-34120-6_45

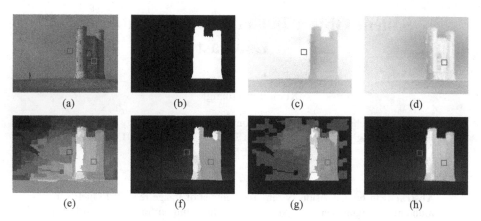

Fig. 1. Motivation of proposed method. (a) Input image with two marked target pixels, one foreground pixel in the center of red box and one background pixel in the center of blue box. (b) Saliency ground truth. (c) (d) Contrast from entire image to two target pixels, higher contrast is shown in darker green. The sum of contrast for two target pixels is comparable, while distribution differ greatly. (e) Saliency computed by sum of contrast [18] that two target pixels show comparable saliency. (f) Proposed saliency computed by distribution of contrast. (g) (h) Final result of (e) (f) with post-processing. (Color figure online)

Recently, works belonging to the bottom-up methods have made significant progress due to simple and fast implementation. One representative research field is based on the contrast between image pixels or regions and their surroundings. These contrast based methods can be roughly divided into local methods [14, 16, 30] and global methods [17, 18, 20]. Local contrast-based methods consider the contrast between the pixels or regions and their local neighborhoods, whereas global contrast-based methods consider contrast relationships over the entire image.

In all of these contrast based methods, saliency is measured by sum of contrast between the pixels or regions and their surrounding. We find that spatial distribution of contrast is one important cue of saliency. Figure 1(a) is one source image with two marked target pixels (one foreground pixel in center of red box and one background pixel in center of blue box). The spatially weighted contrast from entire image to these two target pixels are shown in Figs. 1(c) and 1(d), darker green indicate higher contrast. We find that the sum of contrast in Fig. 1(c) and the sum of contrast in Fig. 1(d) is comparable, but the spatial distribution of contrast differ greatly. If the target pixel is regarded as center of view, high contrast in Fig. 1(c) is concentrated in some directions (mainly in right-bottom), while Fig. 1(d) show high contrast in almost every directions. If saliency is simply defined as the sum of contrast from entire image (such as global contrast method [18]), the saliency of these two target pixels is comparable. This result can be seen from saliency result of global contrast method [18] in Fig. 1(e). In our vision system, object with high contrast in more directions usually have higher saliency. Inspired by this phenomenon, we propose one saliency detection method via distribution of contrast in this paper. Figure 1(f). is the proposed saliency result without any post-processing, the target pixel in red box shows greatly high saliency than the target pixel in blue box, this result is more close to the ground-truth result Fig. 1(b). Final saliency result of Figs. 1(e) and (f) with post-processing are also shown in Figs. 1 (g) and (h).

In proposed method, input image will be segmented into superpixels first, for each superpixel, contrast between this superpixel and its surrounding will be computed in every direction, relative standard deviation (RSD) of contrast in all directions is measured as cue of saliency. The main flowchart is shown in Fig. 2.

We evaluated the proposed approach on four datasets. Experimental results demonstrate the proposed method performs well when against the state-of-the-art methods.

The remainder of this paper is organized as following. Section 2 contains a review of contrast based saliency detection methods. Proposed saliency detection method and experimental results are described in Sects. 3 and 4. Finally, Sect. 5 is conclusion and future work of this paper.

2 Related Work

In human vision system, the contrast of image regions to their surroundings is highly sensitive. Many contrast based saliency detection methods have been proposed in recent years. These contrast based methods can be roughly divided into local methods and global methods.

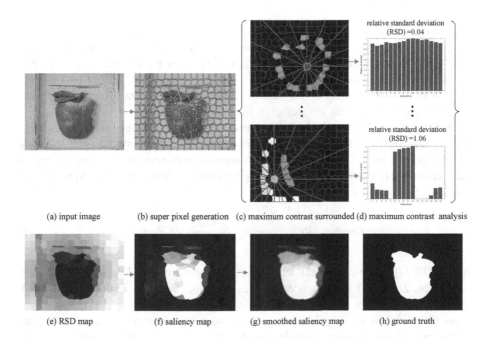

(a) input image   (b) super pixel generation   (c) maximum contrast surrounded (d) maximum contrast analysis

(e) RSD map   (f) saliency map   (g) smoothed saliency map   (h) ground truth

Fig. 2. Main flowchart of proposed method.

Local contrast based methods investigate the rarity of image regions with respect to local neighborhoods. Itti et al. [14] use a difference of Gaussians approach to extract multi-scale color, intensity, and orientation information from images. This information was then used to define saliency by calculating center-surround differences. Based on this work, Harel et al. [15] propose a bottom-up visual saliency model to highlight conspicuous parts and permit combination with other importance maps. Liu et al. [31] propose multi-scale contrast by linearly combining contrast in a Gaussian image pyramid.

Global contrast based methods evaluate saliency of pixels or regions using contrast relationships over the entire image. Zhai and Shah [32] define pixel-wise saliency as the pixel's contrast to all other pixels. Achanta et al. [17] present a frequency tuned algorithm that defines pixel saliency as its color difference from the average image color. This simple but very fast approach usually failed in complex natural images. In [18], a region-based saliency algorithm is introduced by measuring the global contrast between the target region with respect to all other regions in the image, the saliency of one region is defined as the sum of contrast to all other regions. By avoiding the hard decision boundaries of superpixels, a soft abstraction is proposed in [33] to generate a set of large scale perceptually homogeneous regions using histogram quantization and a global Gaussian Mixture Model (GMM), such approach provides large spatial support and can more uniformly highlight the salient object.

3 Proposed Detection Method

3.1 Main Flowchart

Since directly computing pixel-level contrast is computationally expensive, we use superpixels to represent the image. Currently existing edge-preserving superpixel segment algorithms mainly include [34–36], in this paper we adopt the SLIC algorithm [36] for its high efficiency and apply it to over-segment the input image I into K (e.g., $K = 200$ in this work) parts which denoted as R_i ($i = 1..K$). One source image and superpixel generation result are shown in Figs. 2(a) and (b).

The region contrast is usually defined as color distance between target region and other regions. Besides contrast, spatial relationships are also important in human vision system. One near region with high contrast is usually more sensitive than one far region with comparable contrast. The spatially weighted contrast between region R_i and R_j can be defined as following:

$$Contrast(R_i, R_j) = exp(\frac{D_s(R_i, R_j)}{-\delta_s^2})D(R_i, R_j) \tag{1}$$

where $D(R_i, R_j)$ is the color Euclidean distance between region R_i and R_j in CIE-Lab color space, $D_s(R_i, R_j)$ is the space Euclidean distance between region R_i and R_j, δ_s controls the strength of spatial distance weighting.

As discussed in introduction part, region with high contrast in more directions usually have higher saliency. In order to estimate saliency of each region R_i, we first need to calculate contrast in every direction, then we need one measure to represent the distribution of contrast in all directions.

We first considerate contrast from other regions to region R_i in every direction. In order to reduce computational cost, we divide all surrounding regions of target region R_i into N directions. Direction from region R_j to R_i can be determined by coordinate of region center. Although there are many regions in each direction, human usually pay more attention to the region with maximum contrast. This maximum contrast can be regarded as the contrast in this direction. The maximum contrast of region R_i in direction n can be computed as following:

$$MaxContrast_i(n) = \max_{R_j \text{ at direction } n \text{ of } R_i} (Contrast(R_i, R_j)) \tag{2}$$

One example of the maximum contrast surrounded in each direction is shown in Fig. 2(c). Blue radical line indicate all directions, darker green indicate higher contrast, lighter green or even white indicate lower contrast.

Since region with high contrast in more directions usually have higher saliency. One measure to represent the distribution of maximum contrast surrounded is relative standard deviation (RSD) which is determined as following:

$$RSD_i = \frac{Cov(MaxContrast_i)}{mean(MaxContrast_i)} \tag{3}$$

where $Cov(MaxContrast_i)$ and $mean(MaxContrast_i)$ are the variance and average of maximum contrast surrounded. If one region show high contrast in most of directions, RSD value will be lower, one example is shown in top row of Fig. 2(d). On the contrary, if this region show high contrast only in a few directions, RSD value will be higher, one example is shown in bottom row of Fig. 2(d). RSD map of all superpixels is shown in Fig. 2(e), we can find that foreground regions show lower RSD and background regions show higher RSD.

According to explanation of RSD, we can define one decreasing function to estimate saliency:

$$S_i = e^{-\lambda \cdot RSD_i} \tag{4}$$

where λ is parameter to control saliency. Example of saliency map of all superpixels is shown in Fig. 2(f).

3.2   Saliency Smoothing

In order to reduce noisy saliency results we use a smoothing procedure to refine the saliency value. We first smooth saliency at region level as following:

$$S_i' = \sum_{j=1}^{K} S_j \cdot W_j \cdot \exp(-\frac{D_s(R_i, R_j)}{\delta_{smooth\_space}^2} - \frac{D(R_i, R_j)}{\delta_{smooth\_color}^2}) \tag{5}$$

where W_j means weight of region R_j defined by the number of pixels, $D(R_i, R_j)$ and $D_s(R_i, R_j)$ are the color and space distance between region R_i and R_j, δ_{smooth_space} and δ_{smooth_color} controls the strength of smoothing in spatial and color domain.

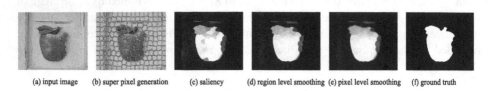

(a) input image   (b) super pixel generation   (c) saliency   (d) region level smoothing   (e) pixel level smoothing   (f) ground truth

Fig. 3. Saliency smoothing.

Superpixel segmentation usually results in discontinuous saliency at boundary of superpixel. In order to get refined saliency, we smooth saliency at pixel level with guided-filter [37, 38]:

One proposed saliency map is shown in Fig. 3(c), region level smoothed saliency and pixel level smoothed saliency is shown in Figs. 3(d) and (e).

4   Experimental Results

4.1   Datasets

We evaluate the proposed method on four benchmark datasets. The first MSRA-10000 dataset [18, 33] consists of 10000 images, each of which has an unambiguous salient object with pixel-wise ground truth labeling. Since this widely used dataset include all images of ASD-1000 dataset [17], the evaluation on ASD-1000 dataset [17] is omitted in this paper. The second ECSSD-1000 [10] contains more salient objects under complex scenes and some images come from the challenging Berkeley-300 dataset. The third DUT-OMRON dataset [23] contains 5166 challenging images with pixel-wise ground truth annotated by five users. The final PASCAL-S dataset [42] constructed on the validation set of recent PASCAL VOC segmentation challenge. It contains 850 natural images with multiple complex objects and cluttered backgrounds. Unlike the traditional benchmarks, the PASCAL-S is believed to eliminate the dataset design bias (e.g., center bias and color contrast bias).

4.2   Evaluation Metrics

(1)   Precision-recall (PR) curve

For a saliency map S, we can convert it to a binary mask M and compute *Precision* and *Recall* by comparing M with ground-truth G:

$$Precision = \frac{|M \cap G|}{|M|}, \quad Recall = \frac{|M \cap G|}{|G|} \tag{6}$$

where $|\cdot|$ represent the number of non-zero entries in the mask.

One common way to bipartite S is to use a fixed threshold which changes from 0 to 255. On each threshold, a pair of precision/recall scores are computed, and are finally combined to form a precision-recall (PR) curve to describe the model performance at different situations.

(2) **F-measure**

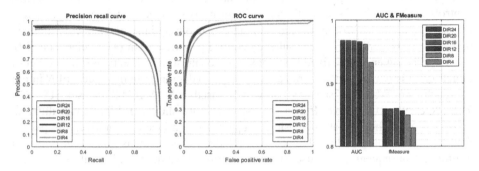

Fig. 4. Number of contrast directions.

The F-measure is the overall performance measurement computed by the weighted harmonic of precision and recall:

$$F_\beta = \frac{(1 + \beta^2) \, \mathrm{Pr}ecision \times \mathrm{Re}call}{\beta^2 \, \mathrm{Pr}ecision + \mathrm{Re}call} \tag{7}$$

As suggested by many salient object detection works, where β^2 is set to 0.3 to raise more importance to the precision value.

F-Measure can be computed with adaptive or fixed threshold. We use a fixed threshold which changes from 0 to 255, the resulted PR curve can be scored by its maximal F_β, which is a good summary of the detection performance.

(3) **Receiver operating characteristic (ROC) curve**

Similar to *Precision* and *Recall*, false positive rate (*FPR*) and true positive rate (*TPR*) also can be computed when saliency map S is converted to a binary mask M with a set of fixed threshold:

$$TPR = \frac{|M \cap G|}{|G|}, \quad FPR = \frac{|M \cap \overline{G}|}{|\overline{G}|} \tag{8}$$

where $\overline{G}$ represent the opposite of ground-truth.

The ROC curve is the plot of *TPR* versus *FPR* by a fixed threshold which changes from 0 to 255.

(4) **AUC**

This metric represents the area under the ROC curve and can effectively reflect the global properties of different algorithms.

4.3   Parameter Setting

We set the number of superpixels $K = 200$ in all the experiments. Besides, there are four parameters in the proposed algorithm are empirically chosen: δ_s in (1) which controls the strength of spatial distance weighting is selected as 1.6, δ_{smooth_space} and δ_{smooth_color} in (5) which controls the strength of smoothing in spatial and color domain is selected as 0.2 and 16, and the saliency control parameter λ in (4) is selected as 6.0.

Since the saliency is derived from maximum contrast in N directions, the number of directions N is one important parameter. In order to demonstrate the effects of N in our approach, we evaluated saliency on widely used MSRA-10000 dataset [18, 33]. The PR curve, ROC curve, F-measure and AUC are shown in Fig. 4 with different values of N. It shows that, increasing N up to 16 significantly improves the performance, but it makes little difference beyond this. This result is consistent with our intuition. Consequently, we use 16 as the optimal value for N in all subsequent experiments.

4.4   Comparison with State-of-Art

The proposed algorithm is compared with both the classic and newest state-of-the-arts: SR [16], FT [17], RC [18], SF [20], GS [21], GC [33], MR [23], MC [39], MAP [40], MBD [41], Water [43]. To evaluate these methods, we either use results provided by authors or run their implementations based on the available codes or software.

Figure 5 shows quantitative comparison (including PR curve, ROC curve, F-measure and AUC measure) between the proposed method and some previous methods. On MSRA-10000 [18], [33] dataset, our method outperforms all state-of-the-arts including Water [43], MBD [41], and MC [39] which were the top-performing methods for saliency detection. For the DUTOMRON [23] and PASCAL-S [42] dataset, our method outperform the most of previous method except MC [39]. We present comparable F-measure with MC [39] but obviously better PR curve, ROC curve and AUC measure. On the ECSSD [10] dataset, only MAP [40] demonstrate comparable F-measure with our approach but slight worse PR curve, ROC curve and AUC measure. Our method outperforms all of other methods. Note that the RC [18] and GC [33] method is also based on region contrast while neglect distribution of contrast, the proposed method significantly outperforms these two methods on all datasets. This experimental results prove that distribution of contrast is important cue of saliency.

Figure 6 shows a few saliency maps of the evaluated methods. In first row of Fig. 6, one flower is located at center of input image, sum of contrast to petal region is larger than sum of contrast to pistil region, but petal and pistil both have high contrast

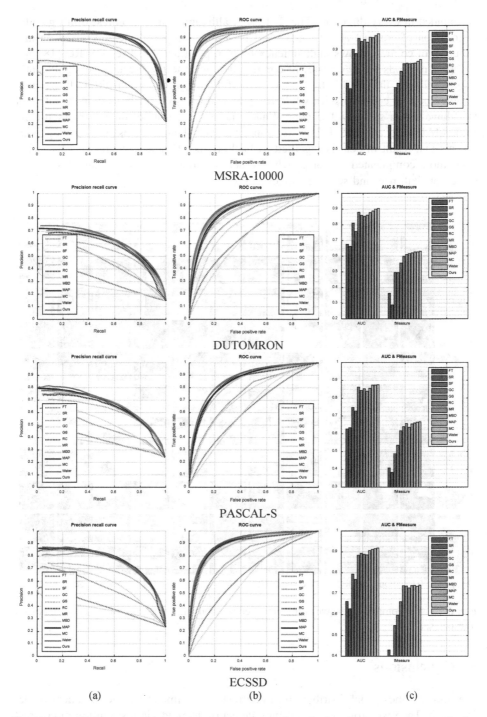

Fig. 5. Quantitative comparison between the proposed method and some previous methods. (a) precision-recall curves. (b) ROC curves. (c) F-Measure and AUC.

surrounded in every direction. That is to say, although sum of contrast differ greatly, distribution of contrast at petal and pistil is similar. Proposed method based on distribution of contrast shows high saliency both in petal and pistil region. RC [18] and GC [33] methods based on sum of contrast demonstrate that petal have significantly higher saliency than pistil. In second row of Fig. 6, proposed method shows high saliency both in coat and skirt, while RC [18] and GC [33] methods demonstrate that skirt have higher saliency than coat. What's more, some state-of-the-arts method (MC [39], MAP [40], and MR [23]) only can effectively handle cases with homogenous objects that result in coat is regarded as background. Besides, our model can tackle even more complicated scenarios while other methods failed, two examples is presented at sixth row and seventh row of Fig. 6.

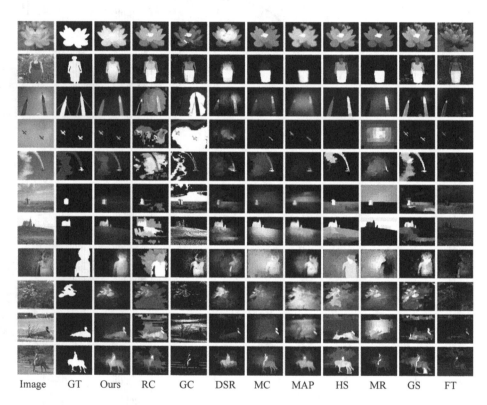

Image   GT   Ours   RC   GC   DSR   MC   MAP   HS   MR   GS   FT

Fig. 6. Visual comparison between the proposed method and some state-of-the-art methods on four datasets.

5   Conclusions

We find that the spatial distribution of contrast is one important cue of salient object detection. Then we propose one saliency detection method via distribution of contrast in this paper. Experimental results on four benchmark dataset demonstrate the proposed method performs well when against the state-of-the-art methods. Next, optimal algorithms will be considered.

References

1. Siagian, C., Itti, L.: Rapid biologically-inspired scene classification using features shared with visual attention. IEEE Trans. Pattern Anal. Mach. Intell. **29**(2), 300–312 (2007)
2. Sharma, G., Jurie, F., Schmid, C.: Discriminative spatial saliency for image classification. In: Proceedings of the IEEE Conference on Computer Vision and Pattern Recognition, pp. 3506–3513, June 2012
3. Walther, D., Rutishauser, U., Koch, C., Perona, P.: Selective visual attention enables learning and recognition of multiple objects in cluttered scenes. Comput. Vis. Image Underst. **100**(1–2), 41–63 (2005)
4. Alexe, B., Deselaers, T., Ferrari, V.: Measuring the objectness of image windows. IEEE Trans. Pattern Anal. Mach. Intell. **34**(11), 2189–2202 (2012)
5. Guo, C., Zhang, L.: A novel multiresolution spatiotemporal saliency detection model and its applications in image and video compression. IEEE Trans. Image Process. **19**(1), 185–198 (2010)
6. Wang, L., Xue, J., Zheng, N., Hua, G.: Automatic salient object extraction with contextual cue. In: Proceedings of IEEE International Conference Computer Vision, pp. 105–112, November 2011
7. Jung, C., Kim, C.: A unified spectral-domain approach for saliency detection and its application to automatic object segmentation. IEEE Trans. Image Process. **21**(3), 1272–1283 (2012)
8. Shen, X., Wu, Y.: A unified approach to salient object detection via low rank matrix recovery. In: Proceedings of the IEEE Conference on Computer Vision and Pattern Recognition, pp. 853–860, June 2012
9. Jiang, P., Ling, H., Yu, J., Peng, J.: Salient region detection by UFO: Uniqueness, focusness and objectness. In: Proceedings of IEEE International Conference of Computer Vision, pp. 1976–1983, December 2013
10. Yan, Q., Xu, L., Shi, J., Jia, J.: Hierarchical saliency detection. In: Proceedings of IEEE Conference Computer Vision Pattern Recognition, pp. 1155–1162, June 2013
11. Li, X., Lu, H., Zhang, L., Ruan, X., Yang, M.-H.: Saliency detection via dense and sparse reconstruction. In: Proceedings of IEEE International Conference of Computer Vision, pp. 2976–2983, December 2013
12. Liu, R., Cao, J., Lin, Z., Shan, S.: Adaptive partial differential equation learning for visual saliency detection. In: Proceedings of IEEE Conference Computer Vision Pattern Recognition, pp. 3866–3873, June 2014
13. Li, N., Ye, J., Ji, Y., Ling, H., Yu, J.: Saliency detection on light field. In: Proceedings of IEEE Conference Computer Vision Pattern Recognition, pp. 2806–2813 (2014)
14. Itti, L., Koch, C., Niebur, E.: A model of saliency-based visual attention for rapid scene analysis. IEEE Trans. Pattern Anal. Mach. Intell. **20**(11), 1254–1259 (1998)
15. Harel, J., Koch, C., Perona, P.: Graph-based visual saliency. In: Proceedings Advanced Neural Information Processing System, pp. 545–552 (2006)
16. Hou, X., Zhang, L.: Saliency detection: a spectral residual approach. In: Proceedings of IEEE Conference on Computer Vision and Pattern Recognition, pp. 1–8, June 2007

17. Achanta, R., Hemami, S., Estrada, F., Süsstrunk, S.: Frequency-tuned salient region detection. In: Proceedings of IEEE Conference on Computer Vision and Pattern Recognition, pp. 1597–1604, June 2009

18. Cheng, M.-M., Zhang, G.-X., Mitra, N.J., Huang, X., Hu, S.-M.: Global contrast based salient region detection. In: Proceedings of IEEE Conference on Computer Vision and Pattern Recognition, pp. 409–416, June 2011

19. Jiang, H., Wang, J., Yuan, Z., Liu, T., Zheng, N., Li, S.: Automatic salient object segmentation based on context and shape prior. In: Proceedings of British Machine Vision Conference, pp. 1–12 (2011)

20. Perazzi, F., Krahenbuhl, P., Pritch, Y., Hornung, A.: Saliency filters: contrast based filtering for salient region detection. In: Proceedings of IEEE Conference Computer Vision Pattern Recognition, pp. 733–740, June 2012

21. Wei, Y., Wen, F., Zhu, W., Sun, J.: Geodesic saliency using background priors. In: Proceedings of 12th European Conference on Computer Vision, pp. 29–42 (2012)

22. Xie, Y., Lu, H., Yang, M.-H.: Bayesian saliency via low and mid level cues. IEEE Trans. Image Process. **22**(5), 1689–1698 (2013)

23. Yang, C., Zhang, L., Lu, H., Ruan, X., Yang, M.-H.: Saliency detection via graph-based manifold ranking. In: Proceedings of IEEE Conference Computer Vision Pattern Recognition, pp. 3166–3173, June 2013

24. Kanan, C., Tong, M.H., Zhang, L., Cottrell, G.W.: SUN: top-down saliency using natural statistics. Vis. Cognit. **17**(6–7), 979–1003 (2009)

25. Yang, J., Yang, M.-H.: Top-down visual saliency via joint CRF and dictionary learning. In: Proceedings of IEEE Conference of Computer Vision Pattern Recognition, pp. 2296–2303, June 2012

26. Borji, A., Itti, L.: State-of-the-art in visual attention modeling. IEEE Trans. Pattern Anal. Mach. Intell. **35**(1), 185–207 (2013)

27. Borji, A., Cheng, M.M., Jiang, H., Li, J.: Salient object detection: a survey, arXiv preprint. http://arxiv.org/pdf/1411.5878.pdf

28. Borji, A., Sihite, D.N., Itti, L.: Salient object detection: a benchmark. In: Proceedings of 12th European Conference Computer Vision, pp. 414–429 (2012)

29. Borji, A., Cheng, M.M., Jiang, H., Li, J.: Salient object detection: a benchmark. IEEE Trans. Image Process. **24**(8), 5706–5722 (2015)

30. Goferman, S., Zelnik-Manor, L., Tal, A.: Context-aware saliency detection. In: Proceedings of IEEE Conference Computer Vision Pattern Recognition, pp. 2376–2383, June 2010

31. Liu, T., et al.: Learning to detect a salient object. IEEE Trans. Pattern Anal. Mach. Intell. **33**(2), 353–367 (2011)

32. Zhai, Y., Shah, M.: Visual attention detection in video sequences using spatiotemporal cues. In: Proceedings of 14th Annual ACM International Conference Multimedia, pp. 815–824 (2006)

33. Cheng, M.-M., Warrell, J., Lin, W.-Y., Zheng, S., Vineet, V., Crook, N.: Efficient salient region detection with soft image abstraction. In: Proceedings of IEEE International Conference of Computer Vision, pp. 1529–1536, December 2013

34. Felzenszwalb, P.F., Huttenlocher, D.P.: Efficient graph-based image segmentation. Int. J. Comput. Vis. **59**(2), 167–181 (2004)

35. Shi, J., Malik, J.: Normalized cuts and image segmentation. IEEE Trans. Pattern Anal. Mach. Intell. **22**(8), 888–905 (2000)

36. Achanta, R., Shaji, A., Smith, K., Lucchi, A., Fua, P., Süsstrunk, S.: SLIC superpixels compared to state-of-the-art superpixel methods. IEEE Trans. Pattern Anal. Mach. Intell. **34**(11), 2274–2282 (2012)

37. He, K., Sun, J., Tang, X.: Guided image filtering. In: Proceedings of the IEEE European Conference Computer Vision, pp. 1–14 (2010)
38. He, K., Sun, J., Tang, X.: Guided Image Filtering. IEEE Trans. Pattern Anal. Mach. Intell. **35**(6), 1397–1409 (2013)
39. Jiang, F., Kong, B., Adeel, A., Xiao, Y., Hussain, A.: Saliency detection via bidirectional absorbing markov chain. In: Ren, J., et al. (eds.) BICS 2018. LNCS (LNAI), vol. 10989, pp. 495–505. Springer, Cham (2018). https://doi.org/10.1007/978-3-030-00563-4_48
40. Sun, J., Lu, H., Liu, X.: Saliency region detection based on markov absorption probabilities. IEEE Trans. Image Process. **24**(5), 1639–1649 (2015)
41. Zhang, S.A.J., Sclaroff, S., Lin, Z., Shen, X., Price, B., Mech, R.: Minimum barrier salient object detection at 80 FPS. In: Proceedings of International Conference on Computer Vision, pp. 1404–1412, December 2015
42. Li, Y., Hou, X., Koch, C., Rehg, J., Yuille, A.: The secrets of salient object segmentation. In: CVPR, vol. 5 (2014)
43. Huang, X., Zhang, Y.: Water flow driven salient object detection at 180 FPS. Pattern Recognit. **76**, 95–107 (2018)

FVCNN: Fusion View Convolutional Neural Networks for Non-rigid 3D Shape Classification and Retrieval

Yan Zhou, Fanzhi Zeng, Jiechang Qian$^{(\boxtimes)}$, Yang Xiang, and Zhijian Feng

Department of Computer Science, Foshan University, Foshan 528000, China
512502487@qq.com

Abstract. Most 3D shape classification and retrieval algorithms were based on rigid 3D shapes, deploying these algorithms directly to non-rigid 3D shapes may lead to poor performance due to complexity and changeability of non-rigid 3D shapes. To address this challenge, we propose a fusion view convolutional neural networks (FVCNN) framework to extract the deep fusion features for non-rigid 3D shape classification and retrieval. We first propose a projection module to transform the non-rigid 3D shape into a 2D view plane. We then propose a feature coding module to extract the new scale invariance heat kernel signature (NS) feature and structural relationship (SR) feature of the 3D shape, which are used as the pixel values on the projection points of the corresponding vertices to generate two views, respectively. Finally, we propose a fusion module based on CNNs to extract the view-based features, which are fused to extract the deep fusion features as the 3D shape descriptors. The experiments on standard dataset SHREC show that our method outperforms the state-of-the-art methods on non-rigid 3D shape classification and retrieval.

Keywords: Non-rigid 3D shape · Fusion view ·
Convolutional neural networks · Projection module ·
Feature coding module

1 Introduction

With the development of computer graphics, 3D shapes play an important role in many domains with a wide range of applications. Accurate 3D shape classification and retrieval are necessary. In recent years, with remarkable advances in deep learning, various network structures have been proposed for 3D shape classification and retrieval, such as 3D ShapeNets [35], PointNet [3], Rotation-Net [12]. Quite significantly, view-based methods have the best performance so far. Using deep learning schemes to extract view descriptor typically refers to exploiting well-established models, such as VGGNet [28], GoogLeNet [32], and ResNet [9]. Although deep learning methods for 2D views have been well investigated [19,36], the structural relationship and the local details of the 3D

© Springer Nature Switzerland AG 2019
Y. Zhao et al. (Eds.): ICIG 2019, LNCS 11901, pp. 566–581, 2019.
https://doi.org/10.1007/978-3-030-34120-6_46

shapes are still unexplored. If these methods are directly deployed to non-rigid 3D shapes classification and retrieval, it may lead to poor performance because the non-rigid 3D shapes are more complex and changeable. Comparing with the rigid 3D shapes, such as desk, chair and bed, the structure of non-rigid 3D shapes is more complex, such as ant, cat and human. In addition, the non-rigid 3D shapes have varieties of postural changes, which leads to different shapes that can be very similar in one posture. Therefore, the classification and retrieval of non-rigid 3D shapes are more difficult and challenging.

To tackle this issue, we propose here a FVCNN framework. It contains three function modules: projection module, feature coding module and descriptor generation module, as shown in Fig. 1. First, the projection module follows the principles and laws of the human visual system, constructing an efficient coordinate system to project the vertices on the observable region of the 3D shape onto a 2D plane. Then a feature coding module is used to extract the NS feature and SR feature as the pixel of the 2D plane, so that two kinds of views are generated. Since the views contain the local details and structural relationship features of the 3D shapes, the views can comprehensively describe the 3D shapes if the content features are explored efficiently. Finally, we use an efficient CNNs fusion module to extract the features of the two views, which are fused to further extract the deep fusion features as the 3D shape descriptors. We evaluate the proposed method on SHREC and the experiment results show that it outperforms state-of-the-art methods.

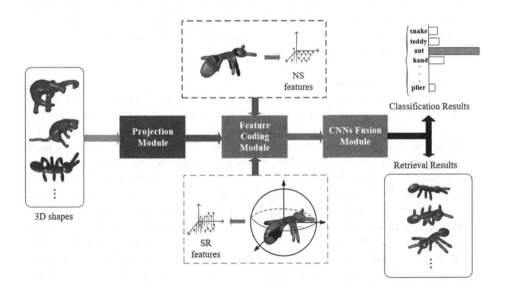

Fig. 1. A classification and retrieval framework for Non-rigid 3D shapes based on fusion view.

Our main contributions of this paper are as follows:

- We propose a projection and feature encoding module to generate the NS and SR views, which contain the local details and structural relationships of the 3D shape, allowing the content features to be explored efficiently such that the views can comprehensively describe the 3D shapes.
- We develop a CNNs fusion module to extract the features from the views, then fuse the features and extract the deep fusion features as the 3D shapes descriptors. Through deep fusion features, the expression ability of the shape descriptor can be improved, and the limitations of single feature are overcome.

2 Related Works

In earlier research on 3D shape classification and retrieval, the features of 3D shapes can be divided into five categories: statistic features [21,30], view features [4,24], topological features [2,16], function transformation features [5,17] and fusion features [27,40]. The key issues with these features are the their weaknesses in descriptive ability and meanwhile expensive in computation time to calculate. It has been generally realized that obtaining efficient features is the key to the classification and retrieval of 3D shapes. Recently, deep learning has been applied in many fields and achieved satisfying results. The deep feature in a 3D shape is a comprehensive one integrating the characteristics of all aspects of the object [8,15,18,22,33,37,39,41]. Introducing deep learning into 3D shape classification and retrieval has been a hot research topic recently. There are two categories of approaches for the CNN-based 3D shape classification: one is voxel-based and the other 2D image-based.

(1) voxel-based approaches

Charles et al. [3] introduced a hierarchical neural network called Point-Net++ which applies PointNet recursively on a nested partition in the input point set. By exploiting the metric space distances, the network can learn local features by increasing contextual scales. The experimental results show that PointNet++ can learn deep point set features efficiently and robustly. Luciano et al. [41] brought forth a deep learning framework for efficient 3D shape classification by geodesic moments. It uses a two-layer stacked sparse autoencoder to learn deep features from geodesic moments by training the hidden layers individually in an unsupervised fashion followed by a softmax classifier. Ren et al. [23] developed a new definition about 2D multilayer dense representation (MDR) for 3D volumetric data to extract concise informative shape description. As a result, a novel adversarial network is designed to train a set of CNN, recurrent neural network (RNN) and an adversarial discriminator. The method improved the efficiency and effectiveness of 3D volumetric data processing.

(2) 2D image-based approaches

Bai et al. [1] presented a real-time 3D shape retrieving engine GIFT based on the projection image of 3D shapes, which combines GPU acceleration and Inverted File (Twice). As a result, this method achieved ultra-high time efficiency where every retrieval task can be finished within one second. Sinha et al. [29] adopted an approach by converting the 3D shape into a geometry image so that standard CNNs can be used to learn 3D shapes directly. By projecting and cutting the spherical parameterized shape, the original 3D shape is transformed into a flat and regular geometric image. Based on the geometric image, the shape descriptor is extracted by CNNs. Shi et al. [26] introduced a rotation-invariant deep representation for 3D shape classification and retrieval known as DeepPano that verifies the rotation invariance of the representation. A variant of CNN is specifically designed to learn the deep representations directly from such views. Different from a typical CNN, a row-wise max-pooling layer is inserted between the convolution and fully-connected layers, making the learned representation invariant to the rotation around a principle axis. Su et al. [31] proposed a new CNN architecture that combines information groups to provide better recognition performance from multiple views of 3D graphics to single compact shape descriptor. The same structure can also be used to identify the hand-drawn sketches of human bodies accurately.

3   Methodology

3.1   Overview

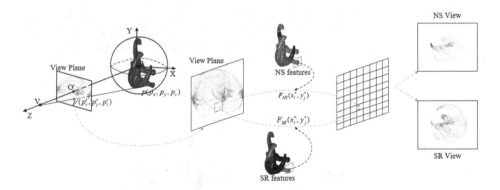

Fig. 2. Schematic diagram of visual views generation.

In human visual system, people recognize an object by observing its local details and structural relationships. Motivated by this observation, we propose a novel method to generate views. The views are formed by projecting the vertices in the visual area of the 3D shape onto a 2D plane where the object features labeled with pixel values, as shown in Fig. 2. We extract the features from the views, and then fuse them to extract the deep fusion features as the descriptors of a 3D shape through the CNNs fusion module. The architecture of the method contains the following main steps as illustrated in Fig. 1.

Step. 1: Projection module for non-rigid 3D shape
In order to ensure the consistency of the extracted features, the 3D shape is preprocessed by the method described in [34] that can eliminate the influence caused by rotation and translation. The 3D shape is eventually surrounded by a sphere. A projection module *similar to* a visual imaging system is established, and the partial vertices of the 3D shape are then projected onto the 2D plane.

Step. 2: Feature coding module for the views
In this step, we propose a feature coding module. The features of the 3D shape such as NS and SR are coded as the pixel values on the view plane from which the views are generated.

Step. 3: CNNs fusion module for non-rigid 3D shape
A feature extractor combining view-pooling and CNNs is developed to extract the 3D shape descriptors. The module can be iteratively updated by training until the number of iterations reaches a given threshold or the performance of the module convergence.

We will describe the design and analysis details for each key part of the model in the following sections.

3.2   Projection Module for Non-rigid 3D Shape

We shall first define a sphere with a radius r which surrounds the 3D shape, and establish a coordinate system, as shown in Fig. 3. The center of the sphere is defined as the coordinate origin O(0,0,0) which also is the centroid of the 3D shape. The Z axis is the long spindle of the coordinates. The view plane can then be set up as perpendicular to the Z axis with the size of $h \times h$ and its

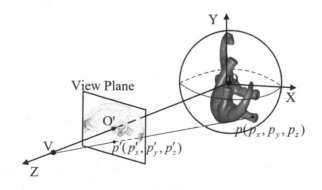

Fig. 3. Schematic diagram of the projection module.

center at O'(0,0,d). At the viewpoint V, we can observe the 3D shape through the view plane where the partial vertices are projected onto. In order to achieve this, the coordinate of the viewpoint can be determined by $V(0,0,\alpha)$ where $\alpha = dr/(r-h/2)$ according to the theory of similar triangles. Using the projection function $E_{pro} : R^3 \rightarrow R^3$ as shown in Eq. 1, we can calculate the coordinates of point p' on the view plane by $p' = F_{pro}(p)$ where the point $p(p_x, p_y, p_z)$ is the vertex on the 3D shape and the point $p'(p'_x, p'_y, p'_z)$ is the projection point.

$$
\begin{cases}
p'_x = (1 - \dfrac{d}{\alpha})p_x \\[2mm]
p'_y = (1 - \dfrac{d}{\alpha})p_y \\[2mm]
p'_z = d
\end{cases}
\tag{1}
$$

3.3 Feature Coding Module for the Views

Discrete Grid Division of the View Plane. We set S_h and S_v as the horizontal and vertical step length such that $n_h = h/S_h$ and $n_v = h/S_v$ are the division number of the view plane. According to the step-length, the 2D plane can then be divided into the areas $A_{ij}, i = 1, 2, \ldots, n_h, j = 1, 2, \ldots, n_v,$ according to

$$
A_{ij} = \left\{ (x,y,z) \left| \begin{matrix} (i-1)S_h \leq x < iS_h \\ (j-1)S_v \leq x < jS_v \\ z = d \end{matrix} \right. \right\}
\tag{2}
$$

The center point (x_i^*, y_j^*, d) of the area A_{ij} can be calculated as

$$
\begin{cases}
x_i^* = (i-1)S_h + \dfrac{1}{2}S_h \\[2mm]
y_j^* = (j-1)S_v + \dfrac{1}{2}S_v
\end{cases}
\tag{3}
$$

In order to simplify the expression, a local filter function can be defined as

$$
I_{A_{ij}}(p) = \begin{cases} 1, & if \quad p' = F_{pro}(p) \in A_{ij} \\ 0, & else \end{cases}
\tag{4}
$$

The pixel value of each center point of A_{ij} can finally be calculated accordingly

$$
F_{NS}(x_i^*, y_j^*) = \underset{p}{avg}(H_{NS}(p)I_{A_{ij}}(p))
\tag{5}
$$

$$
F_{SR}(x_i^*, y_j^*) = \underset{p}{max}(H_{SR}(p)I_{A_{ij}}(p))
\tag{6}
$$

We obtain two categories of view: NS view such as $(x_i^*, y_j^*, F_{NS}(x_i^*, y_j^*))$ and SR view such as $(x_i^*, y_j^*, F_{SR}(x_i^*, y_j^*))$. $H_{NS}(p)$ and $H_{SR}(p)$ are the feature coding functions as described in the following section.

Feature Coding for the Views. We design two coding functions to code the NS and SR features as the pixel value of the non-rigid 3D shape. The views developed this way that contain not only the local shape features of the object observed, but also the positional relationship between the features. The pixel values reflect geometric features such as structural relationship, local details and topological structure of the 3D shape. So the features based on views are aligned into the comprehensive descriptors of the 3D shape.

(1) NS features as the pixel value for the views

The NS features [38] of a 3D shape can be used to describe the local structure and details. A view with its pixel values constructed with NS features is named NS view here. By optimizing the heat kernel signature (HKS) features, we can obtain the NS features on the vertices of the 3D shape according to

$$H_{NS}(p) = F[\frac{d}{d\tau}logK_{\beta\tau}(p)] \quad and \quad K_{\beta\tau}(p) = \sum_{i\leq0} e^{-\Lambda_i\beta_\tau}\Phi_i^2(x) \tag{7}$$

where Λ_i and Φ_i are the eigenvalues and eigenfunctions of the discrete Laplace-Beltrami operator, β is a constant, and $\tau \in [lb(t_{min}), lb(t_{max})]$ in which t_{min} and t_{max} are the critical time values beyond which NS features of the 3D shape no longer change.

As shown in Fig. 4, the same colors of vertices indicate that their NS features are similar. The NS features have isometric invariance and robustness under small perturbation such as small topological change or noise.

Fig. 4. The NS feature of the man models

As shown in Fig. 5(a) and (b) or (c) and (d) are the same 3D models with different scales, but their NS features are similar. Although the scale is changed, the NS features are still robust. For different types of 3D shapes, such as (a) and (c) or (b) and (d), their NS features are distinctly different.

Fig. 5. The NS features of the house and man models.

Fig. 6. The model for extracting the SR features.

(2) SR features as the pixel value for the views

The minimum circumferential sphere enclosing the 3D shape is adopted as shown in Fig. 6. According to Eq. 8, we can obtain the pixel value H_{SR} to describe the global structural features on the vertex p of the 3D shape.

$$H_{SR}(p) = (cos\theta + cos\varphi)dis(Op) \tag{8}$$

where (θ, φ, r) is the spherical coordinates of the vertex p of the 3D shape, $dis(Op)$ is the distance between point O and point p.

3.4   CNNs Fusion Module for Non-rigid 3D Shape

In this section, we develop two CNNs: the convolutional neural networks based on traditional networks (CNNs-T) and the convolutional neural networks based on ResNet (CNNs-R), as shown in Figs. 7 and 8, respectively. For each CNN, the input data has two categories of views obtained in the previously defined modules.

Motivated by [10], we define a composite function of three consecutive operations for each block of each CNN: batch normalization (BN) [11], followed by a rectified linear unit (ReLU) [6] and a 3×3 convolution (Conv). We train two kinds of views in CNN_1 to extract the features of the corresponding views, and then fuse the features at the view-pooling layer and input them to CNN_2 for further extraction.

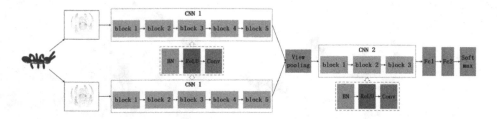

Fig. 7. The convolutional neural networks based on traditional networks (CNNs-T).

CNNs-T. The training process of CNNs-T is illustrated in Fig. 7. Traditional convolutional feed-forward networks use the output of the l_{th} layer as the input to the $(l + 1)_{th}$ layer [13].

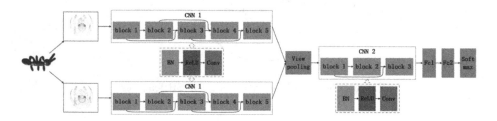

Fig. 8. The convolutional neural networks based on ResNet (CNNs-R).

CNNs-R. The training process of CNNs-R is illustrated in Fig. 8. In ResNet [9], a skip-connection is added, bypassing the non-linear transformation by an identity function. The output of the l_{th} layer is used as the input to the $(l+1)_{th}$ layer and $(l + 2)_{th}$ layer. The advantage of ResNet is that the gradient can flow directly from the later layers to the earlier layers by the identity function.

View-Pooling Layer. View-pooling layers are closely related to max-pooling layers, and the only difference is that the pooling operations are carried out in three dimensions.

Implementation Details. There are five blocks for the CNN_1, and each block has the same number of layers. For a 3×3 convolutional layer, each side of the input is zero-padded by one pixel for the purpose of fixing the feature-map size. At the end of the CNN_1, a view-pooling is performed and CNN_2 is attached, so it forms three parts. At the end of the CNN_2, two fully-connected layers and one softmax classifier are used. In addition, the numbers of the feature-map for each block are 32, 32, 64, 64, 64, 128, 128 and 256.

In our experiments, we use the above two network structures to extract the descriptors of the 3D shapes and implement the classification and retrieval of 3D shapes.

4   Experiment Result and Analysis

All algorithms proposed in this work are implemented and tested using Matlab2017b on a PC with the following specifications, CPU: Intel(R) Core(TM) i9-7960X 2.80 GHz, GPU: NVIDIA GeForce GTX1080TI, RAM: 16 GB DDR4, OS: Windows10 SP1 of 64 bits.

4.1   Dataset

We evaluate our method on the SHREC [14] database of watertight meshes. SHREC contains 600 3D shapes from 30 categories, among them 480 and 120 3D shapes are used for training and testing, respectively. We randomly select 30 shapes from 30 categories, are shown in Fig. 9.

Fig. 9. 30 selected 3D shapes from SHREC database.

4.2   The NS and SR Features of the 3D Shapes

Based on the feature coding module in Sect. 3.3, we extract the NS features and SR features of 30 selected 3D shapes, respectively. Using the color represents the NS features and SR features of the 3D shapes, as shown in Fig. 10(a) and (b). These color blocks reflect the local details and structural relations of the 3D shapes, and the similarity between the NS features and SR features from different categories is low.

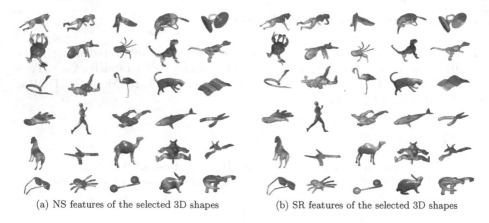

| (a) NS features of the selected 3D shapes | (b) SR features of the selected 3D shapes |

Fig. 10. The NS features (a) and SR features (b) of the selected 3D shapes.

4.3 The Examples of NS and SR Views for Non-rigid 3D Shapes

Based on the projection and feature coding module described in Sect. 3, we extracted the views of the 3D shapes from 30 categories to analyze their expression capabilities, as shown in Fig. 11(a) and (b), the NS views and SR views correspond to the 3D shapes in Fig. 9. We can find that the views reflect well the local details and structural relations of the 3D shapes and the similarities between the views of different categories are low. Through these two views, the 3D shapes can be described efficiently and the similarities and differences between shapes can be distinguished.

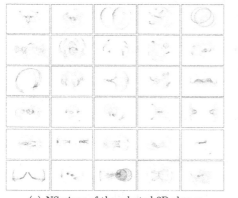
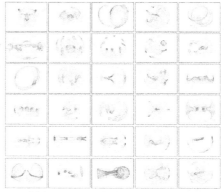
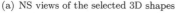

| (a) NS views of the selected 3D shapes | (b) SR views of the selected 3D shapes |

Fig. 11. The NS views (a) and SR views (b) of the selected 3D shapes.

4.4 Non-rigid 3D Shape Classification and Retrieval Efficiency Analysis

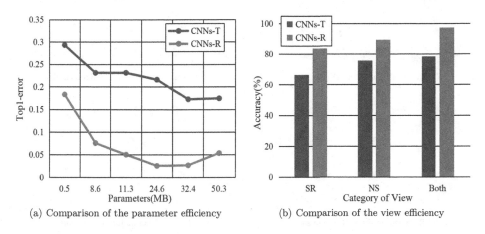

(a) Comparison of the parameter efficiency   (b) Comparison of the view efficiency

Fig. 12. The comparison of the parameter (a) and the view (b) efficiency.

The Comparison of CNNs-T and CNNs-R. The results in Fig. 12(a) show that CNNs-R utilize parameters more efficiently, consistently outperforming CNNs-T in reducing top1-errors when they both have the same parameters. Moreover, CNNs-R also explores the view features more effectively as it delivers better performance in accuracy using the same visual view (e.g., 83.76% vs 66.35%, 89.29% vs 75.68%, 97.44% vs 78.41%), as shown in Fig. 12(b).

The Retrieval Results of CNNs-T and CNNs-R. In the experiment, 3D man and ant shapes are chosen as the query example shapes. We compared the 3D shape retrieval performance of CNNs-R and CNNs-T, and the results are shown in Fig. 13. In Fig. 13(a) and (b), we can see that all shapes in retrieval results are relevant. Although the 3D ant shape can be complex and many forms of deformation exist, CNNs-R delivered good retrieval performance. In contrast, the retrieval results obtained from CNNs-T have one irrelevant retrieval in the 3D man shape and two irrelevant ones in the 3D ant shape, as shown in Fig. 13(c) and (d). These results verify the superior performance of the proposed CNNs-R approach.

Comparative Analysis with the State of the Art Methods. We now compare our methods with state-of-the-art approaches, including Zer [20], LFD [4], SN [35], Conf [7], Sph [25], Geometry Image [29]. The results of non-rigid 3D shape classification and retrieval are summarized in Table 1 and in Fig. 14. We

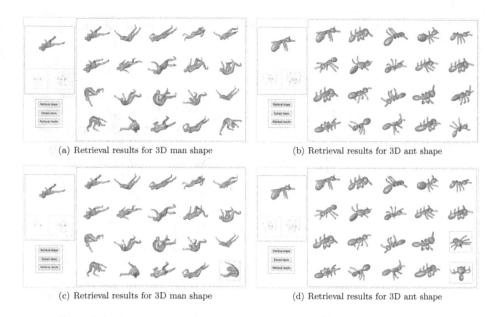

(a) Retrieval results for 3D man shape   (b) Retrieval results for 3D ant shape

(c) Retrieval results for 3D man shape   (d) Retrieval results for 3D ant shape

Fig. 13. The retrieval results.

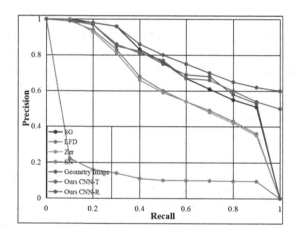

Fig. 14. Precision-recall curves among different algorithms on the SHREC-11 dataset.

can see that both CNNs-T and CNNs-R have better performance. CNNs-T delivered classification accuracy and MAP retrieval reaching 82.7% and 76%, respectively, which is 4% higher than Geometry Image [29] in MAP retrieval. Moreover, CNNs-R has the best performance of them all with the classification accuracy of 97.4% and MAP retrieval of 81% which are 0.8% and 9% higher than Geometry Image [29], respectively.

Table 1. Comparison results among different algorithms on the SHREC dataset.

Method	Classification (Accuracy)	Retrieval (MAP)
Zer [20]	50.8%	64%
LFD [4]	65.8%	65%
SN [35]	48.4%	13%
Conf [7]	85.0%	45%
Sph [25]	82.5%	66%
Geometry Image [29]	96.6%	72%
Ours CNNs-T	82.7%	76%
Ours CNNs-R	97.4%	81%

5   Conclusion

In this paper, we bring forward a FVCNN framework for classifying and retrieving non-rigid 3D shapes. Firstly, we propose a projection module to transform the non-rigid 3D shape onto a 2D view plane and a feature coding module to extract the NS features and SR features of the 3D shape. And then the NS views and SR views are generated by using the NS features and SR features as the pixel values, respectively, which are able to express the 3D shapes efficiently. Finally, we propose a CNNs fusion module to extract the view-based features and fuse them to extract the deep fusion features as the 3D shape descriptors. The method in this paper use neural network architecture and outperformed a more traditional non-learning based approach, these is still much space for improvement.

In the future we wish to build upon these insights for generative models of 3D shape with encoded views instead of traditional images. An future direction is to consider integrating the discriminative power of view-based approaches and the robustness approaches reasoning more locally with geometry.

Acknowledgement. This work is partially supported by the following projects in China: the National Natural Science Foundation of China (no. 61602116), Natural Science Foundation of Guangdong Province (no. 2017A030313388), Engineering Technology Research Center of Foshan City (no. 2017GA00015, 2016GA10156), Engineering Technology Research Center of Guangdong Province (no. G601624), and Special Fund for Science and Technology Innovation of Foshan City (no. 2015AG10008).

Compliance with Ethical Standard. We confirm that there are no potential conflicts of interest; also, the work involves no human participants and/or animals. All authors consent to the submission of the paper.

References

1. Bai, S., Bai, X., Zhou, Z., Zhang, Z., Latecki, L.J.: Gift: a real-time and scalable 3D shape search engine. In: 2016 IEEE Conference on Computer Vision and Pattern Recognition (CVPR), pp. 5023–5032, June 2016. https://doi.org/10.1109/CVPR.2016.543

2. Barra, V., Biasotti, S.: 3D shape retrieval using kernels on extended Reeb graphs. Pattern Recogn. **46**(11), 2985–2999 (2013)
3. Charles, R.Q., Su, H., Kaichun, M., Guibas, L.J.: Pointnet: deep learning on point sets for 3D classification and segmentation. In: 2017 IEEE Conference on Computer Vision and Pattern Recognition (CVPR), pp. 77–85, July 2017. https://doi.org/10.1109/CVPR.2017.16
4. Chen, D., Tian, X., Shen, Y., Ouhyoung, M.: On visual similarity based 3D model retrieval. Comput. Graph. Forum **22**(3), 223–232 (2010)
5. Daras, P., Zarpalas, D., Tzovaras, D., Strintzis, M.G.: Efficient 3-D model search and retrieval using generalized 3-D radon transforms. IEEE Trans. Multimedia **8**(1), 101–114 (2006)
6. Glorot, X., Bordes, A., Bengio, Y.: Deep sparse rectifier neural networks. In: International Conference on Artificial Intelligence and Statistics, pp. 315–323 (2011)
7. Gu, X., Wang, Y., Chan, T.F., Thompson, P.M.: Genus zero surface conformal mapping and its application to brain surface mapping. IEEE Trans. Med. Imaging **23**(8), 949–958 (2003)
8. Han, X., Wu, Z., Wu, Z., Yu, R., Davis, L.S.: VITON: an image-based virtual try-on network. CoRR abs/1711.08447 (2017). arxiv:1711.08447
9. He, K., Zhang, X., Ren, S., Sun, J.: Deep residual learning for image recognition. In: 2016 IEEE Conference on Computer Vision and Pattern Recognition (CVPR), pp. 770–778, June 2016. https://doi.org/10.1109/CVPR.2016.90
10. He, K., Zhang, X., Ren, S., Sun, J.: Identity mappings in deep residual networks. In: Leibe, B., Matas, J., Sebe, N., Welling, M. (eds.) ECCV 2016. LNCS, vol. 9908, pp. 630–645. Springer, Cham (2016). https://doi.org/10.1007/978-3-319-46493-0_38
11. Ioffe, S., Szegedy, C.: Batch normalization: accelerating deep network training by reducing internal covariate shift, pp. 448–456 (2015)
12. Kanezaki, A.: Rotationnet: Learning object classification using unsupervised viewpoint estimation. CoRR abs/1603.06208 (2016). arxiv:1603.06208
13. Krizhevsky, A., Sutskever, I., Hinton, G.E.: Imagenet classification with deep convolutional neural networks. In: International Conference on Neural Information Processing Systems, pp. 1097–1105 (2012)
14. Laga, H., Schreck, T., Ferreira, A., Godil, A.: Shrec'11 track: shape retrieval on non-rigid 3D, pp. 79–88 (2012). Editors, I.P., Meshes, W., Lian, Z., Godil, A., Bustos, B., Daoudi, M
15. Leng, B., Liu, Y., Yu, K., Zhang, X., Xiong, Z.: 3D object understanding with 3D convolutional neural networks. Inf. Sci. **366**, 188–201 (2016)
16. Li, C., Hamza, A.B.: Symmetry Discovery and Retrieval of Nonrigid 3D Shapes Using Geodesic Skeleton Paths. Kluwer Academic Publishers, Dordrecht (2014)
17. Lian, Z., et at.: A comparison of methods for non-rigid 3D shape retrieval. Pattern Recogn. **46**(1), 449–461 (2013)
18. Lin, T.Y., Goyal, P., Girshick, R., He, K., Dollar, P.: Focal loss for dense object detection. IEEE Trans. Pattern Anal. Mach. Intell. **PP**(99), 2999–3007 (2017)
19. Liu, B., Jing, L., Li, J., Yu, J., Gittens, A., Mahoney, M.W.: Group collaborative representation for image set classification. Int. J. Comput. Vis. **4**, 1–26 (2018)
20. Novotni, M., Klein, R.: Shape retrieval using 3D Zernike descriptors. Comput.-Aided Des. **36**(11), 1047–1062 (2004)
21. Osada, R., Funkhouser, T., Chazelle, B., Dobkin, D.: Shape distributions. ACM Trans. Graph. **21**(4), 807–832 (2002)
22. Patricia, M., Daniela, S.: Multi-objective optimization for modular granular neural networks applied to pattern recognition. Inf. Sci. **460–461**, 594–610 (2018)

23. Ren, M., Niu, L., Fang, Y.: 3D-a-nets: 3D deep dense descriptor for volumetric shapes with adversarial networks. CoRR abs/1711.10108 (2017). arxiv:1711.10108

24. Sang, M.Y., Kuijper, A.: View-based 3D model retrieval using compressive sensing based classification. In: International Symposium on Image and Signal Processing and Analysis, pp. 437–442 (2011)

25. Shen, L., Makedon, F.: Spherical mapping for processing of 3D closed surfaces. Image Vis. Comput. **24**(7), 743–761 (2006)

26. Shi, B., Bai, S., Zhou, Z., Bai, X.: Deeppano: deep panoramic representation for 3-D shape recognition. IEEE Sig. Process. Lett. **22**(12), 2339–2343 (2015)

27. Shih, J.L., Chen, H.Y.: A 3D model retrieval approach using the interior and exterior 3D shape information. Multimedia Tools Appl. **43**(1), 45–62 (2009)

28. Simonyan, K., Zisserman, A.: Very deep convolutional networks for large-scale image recognition. arXiv:1409.1556

29. Sinha, A., Bai, J., Ramani, K.: Deep learning 3D shape surfaces using geometry images. In: Leibe, B., Matas, J., Sebe, N., Welling, M. (eds.) ECCV 2016. LNCS, vol. 9910, pp. 223–240. Springer, Cham (2016). https://doi.org/10.1007/978-3-319-46466-4_14

30. Sipiran, I., Bustos, B., Schreck, T.: Data-Aware 3D Partitioning for Generic Shape Retrieval. Pergamon Press, Inc., Oxford (2013)

31. Su, H., Maji, S., Kalogerakis, E., Learned-Miller, E.: Multi-view convolutional neural networks for 3D shape recognition. In: IEEE International Conference on Computer Vision, pp. 945–953 (2015)

32. Szegedy, C., et al.: Going deeper with convolutions. In: 2015 IEEE Conference on Computer Vision and Pattern Recognition (CVPR), pp. 1–9, June 2015. https://doi.org/10.1109/CVPR.2015.7298594

33. Tang, H., Xiao, B., Li, W., Wang, G.: Pixel convolutional neural network for multi-focus image fusion. Inf. Sci. **433–434**, 125–141 (2018)

34. Vranic, D.: 3D model retrieval. University of Leipzig (2004)

35. Wu, Z., et al.: 3D shapenets: a deep representation for volumetric shapes. In: 2015 IEEE Conference on Computer Vision and Pattern Recognition (CVPR), pp. 1912–1920, June 2015. https://doi.org/10.1109/CVPR.2015.7298801

36. Xu, C., Govindarajan, L.N., Zhang, Y., Cheng, L.: Lie-x: Depth image based articulated object pose estimation, tracking, and action recognition on lie groups. CoRR abs/1609.03773 (2016). arxiv:1609.03773

37. Yuan, J., Li, W., Zhang, Z., Fleet, D., Shotton, J.: Guest editorial: human activity understanding from 2D and 3D data. Int. J. Comput. Vis. **118**(2), 113–114 (2016)

38. Zeng, F., Qian, J., Zhou, Y., Yuan, C., Wu, C.: Improved three-dimensional model feature of non-rigid based on HKS. In: Qiu, M. (ed.) SmartCom 2017. LNCS, vol. 10699, pp. 427–437. Springer, Cham (2018). https://doi.org/10.1007/978-3-319-73830-7_42

39. Zhou, Y., Yuan, C., Zeng, F., Qian, J., Wu, C.: An object detection algorithm for deep learning based on batch normalization. In: Qiu, M. (ed.) SmartCom 2017. LNCS, vol. 10699, pp. 438–448. Springer, Cham (2018). https://doi.org/10.1007/978-3-319-73830-7_43

40. Zhou, Y., Zeng, F.: 2D compressive sensing and multi-feature fusion for effective 3D shape retrieval. Inf. Sci. **409–410**, 101–120 (2017)

41. Zhou, Y., Zeng, F., Qian, J., Han, X.: 3D shape classification and retrieval based on polar view. Inf. Sci. **474**, 205–220 (2019)

Densenet-Based Multi-scale Recurrent Network for Video Restoration with Gaussian Blur

Liyou Wu[1(⊠)], Nong Sang[1], Junyan Yang[2], Lihong Jing[2], Changxin Gao[1], and Lerenhan Li[1]

[1] National Key Laboratory of Science and Technology on Multispectral Information Processing, Huazhong University of Science and Technology, Wuhan, China
861010729@qq.com
[2] Infrared Detection Technology Research & Development Center of CASC, Shanghai Aerospace Control Technology Institute, Shanghai, China

Abstract. Video cameras are now commonplace and available, and videos can be obtained almost everywhere at anytime. However, due to turbulence or thermal effects of air, blurring occurs during image acquisition. Removing these artifacts from the blurry recordings is a highly ill-posed problem as neither the sharp image nor the blur kernel is known. Propagating information between multiple consecutive blurry observations can help restore the desired sharp video. In this work, we propose an efficient approach to produce a significant amount of realistic training data and introduce a novel multi-scale recurrent network architecture to deblur frames taking temporal information into account. The experimental results demonstrate the effectiveness of the proposed method.

Keywords: Deep learning · Video deblurring · DenseNet · Multi-scale recurrent network

1 Introduction

Videos captured by cameras usually contain blurring effects caused by turbulence or thermal effects of air. As for removing blur caused by turbulence or thermal effects, the kernel is usually modeled as a stationary Gaussian function. A reconstruction of the sharp frame from a blurry observation is a highly ill-posed problem, denoted as blind or non-blind deconvolution depending on whether blur information is known or not.

Early image deblurring is a non-blind deconvolution algorithm. [1, 20, 21] are classic non-blind deconvolution algorithms, but these algorithms are not high quality due to simple modeling. Since the blurred kernel of the blurred image is mostly unknown, Gaussian blurred image restoration is mainly a single frame restoration. It is usually obtained by estimating the Gaussian blur kernel of the blurred image and then using the deconvolution algorithm to obtain a sharp image. Pan et al. [2] present a simple and effective blind image deblurring method based on the dark channel prior. Yan and Shao [3] and Schuler et al. [4] constructed a neural network to estimate the

© Springer Nature Switzerland AG 2019
Y. Zhao et al. (Eds.): ICIG 2019, LNCS 11901, pp. 582–594, 2019.
https://doi.org/10.1007/978-3-030-34120-6_47

blur kernel s by training network. However, most Gaussian-image deblurring approaches are focused on single image restoration. But in video deblurring, the reconstruction process can make use of additional data from neighboring frames. However, recent network architectures for multi-frame and video deblurring [5–8] are primarily directed to eliminate motion blur. To alleviate the ill-posedness of the problem [9], one might take multiple observations into account. Thus, views of a static scene, each of which is differently blurred, serve as inputs [12–14]. Quite recently, Wieschollek et al. in [5] introduce an end-to-end trainable neural network architecture for multi-frame deblurring. The related task of deblurring of videos has been approached by Su et al. [6] Their approach uses the U-Net architecture [10] with skip connection to directly regress the sharp image from an input burst.

So in order to combine the multi-frame Gaussian blurred image information to achieve better recovery, in this paper, we propose a new multi-scale recurrent encoder-decoder network. In the network, we combine the spatial residual connections and introduce recurrent connections between different scale networks. At the same time, in order to combine multi-frame blurred image information, we use a cyclic network to train the model. In brief, the main contributions of this paper are as follows:

(1) We present a novel recurrent network architecture – Densenet-based multi-scale recurrent network (DMSRDBN) – for Gaussian blurred video restoration.
(2) We present a novel method for the efficient generation of a vast number of blurry/sharp video sequence pairs.
(3) Extensive experiments have been conducted. The results prove the validity and show that the proposed method is significantly better than the current mainstream video deblurring method.

The remainder of the paper is organized as follows. Section 2 explains how to generate the training data and describes the proposed method and details the architecture of the network. Section 3 explains the loss function and training process. Section 4 presents the experimental results and analysis under different conditions. In the last of this paper, we conclude our work with a summary in Sect. 5.

2   Method

Overview. In our approach, by building a fully convolutional neural network deblurs a frame I using information from previous frames $I_{-1}, I_{-2}, \ldots$ in an iterative, recurrent way. Combining a previous (blurry) frame improves the current prediction for I step by step. Thus, the whole Densenet-based multi-scale recurrent network (DMSR network) consists of several deblur blocks. We share network weights through feature extraction and recurrent connections. At the inference time, the inputs can have arbitrary spatial dimensions as long as the processing of a minimum of two frames fits on the GPU.

2.1 Generating Realistic Ground-Truth Data

Training the neural network to predict a sharp frame of a blurry input requires realistic training data for these two aligned versions of each video frame: a blurred version used as input and a related sharp version used as a ground truth. Obtaining this data is challenging because any recorded sequence may suffer from the described blurring effect itself. Recent work [11] to obtain video data through online media, and then generate the required motion blur by means of averaging synthesis provides a good idea for this paper.

Training Data. With the rapid development of Internet technology, digital multimedia technology has penetrated every aspect of people's lives due to the popularity of video capture devices such as smartphones and video cameras. Today's various social software and platforms, such as Weibo, YouTube, etc., upload a large number of video resources every day, ranging from short films to professional videos up to 8k resolution. From this source, we collect video with a resolution of 4k–8k and a frame rate of 30fps–60fps. Video content includes movies, sports events, daily life and, other videos.

Consider such a video with frames $(f_t)_{t=1,2,...T}$. For each frame at time t, we add Gaussian blur to the original sharp frame.

$$b_t = k_t \otimes f_t$$

$$k_t = abs(\sigma \sin(\omega t)) \tag{1}$$

Where b_t denotes the blurred frame produced at time t, f_t denotes the sharp frame at time t, $\otimes$ denotes convolution operation, k_t denotes a sinusoidal convolution kernel, where σ denotes the standard deviation of the Gaussian blur kernel, used to control the degree of blur, ω is used to control the sine function period, $abs(\cdot)$ denotes absolute operation.

For practical purposes, we set $\sigma = 5$, $\omega = 0.3$. For all video parts that passed our sharpness test (5 h in total) we produce a ground-truth video and blurry version. To add variety to the training data we crop random parts from the frames and resize them to 128×128px. Figure 1 shows a few random examples from our training dataset.

2.2 Network Architecture

To combine multi-frame blurred image information, we propose that the entire network uses a cyclic training method to process continuous input pairs. Consider a single deblur step with a current prediction I of shape [H,W,C] and blurry observation I_{-1}. Inspired by the work of Nah et al. [16] and the recent success of dense connections [15] we use an encoder-decoder architecture in each deblur block. At the same time, we add recurrent connections based on existing multi-scale networks to share weights between different scale networks, thereby reducing the number of parameters of the overall network and enhancing the learning ability of the network. Finally, we built a Densenet-based multi-scale recurrent network(DMSRDBN). See Fig. 2. Hereby, the network only consists of convolution and transpose convolution layers with batchnorm [17]. We applied the ReLU activation to the input of the convolution layers, as proposed in [18].

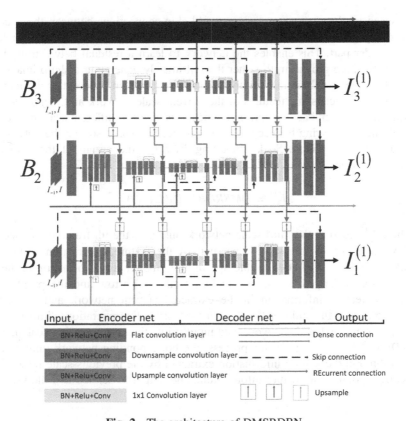

Fig. 1. Snapshot of the training process. Each triplet shows the input with synthetic blur (left), the current network prediction (middle), and the associated ground-truth (right).

Fig. 2. The architecture of DMSRDBN

For raw input frames (I, I_{-1}), first downsample them and then concat them into a 6-channel input $B_k(I, I_{-1})$, where $k = 1, 2, 3$, representing the input of three different scale networks, and the subscript k indicates from large to small. At three scales, the sampling factor between each scale is $1/2$. Thus, the inputs to the three scale networks are $B_1(I, I_{-1})$, $B_2(I, I_{-1})$, $B_3(I, I_{-1})$. The overall network adopts the training idea from coarse to fine. The input is first sent to the first convolutional layer, and the number of channels is expanded. Then enter encoder net part, each encoder net consists of a downsampling convolutional layer (■■■), a Dense Block, and a convolutional layer (▭) with a convolution kernel size of 1×1. The downsampling layer halves the spatial dimension with stride 2 and doubles the effective number of channels $[H, W, C] \rightarrow [H/2, W/2, 2C]$, where H, W denote the spatial size of the feature map, and C denotes the number of channels of the feature map. Then the output of the downsample convolutional layer is sent to the Dense Block. Since the output of the Dense Block will concat the output of the previous convolutional layer, the number of channels of the feature map is large, so a convolutional layer (▭) with a kernel size of 1×1 merges the stitched feature maps. In the subsequent decoder net section, the upsample convolution layer (■■■) inverts the effect of the downsampling $[H, W, C] \rightarrow [2H, 2W, C/2]$. We use a filter size of $3 \times 3 / 4 \times 4$ for all flat convolution/upsample convolution layers. The resulting output is a three-channel color image.

To speed up the training process, we add skip-connections between the encoding and decoding part. With this, we add the extracted features from the encoder to the related decoder part. This enables the network to learn a residual between the blurry input and the sharp ground-truth rather than ultimately generating a sharp image from scratch. To better combine the feature information between different scale networks, the algorithm extracts some features of the current scale network and merges with the next scale network, and adds recurrent connections between different scale networks to make information sharing between networks more. Sufficient, so that the network can gain more learning ability. So by inputting $B_3(I, I_1)$, the output of the network is defined as follows:

$$I_3^{(1)} = DMSRDBN(B_3(I, I_{-1})) \tag{2}$$

Where $I_3^{(1)}$ denotes the third scale network output of the ith frame combined with the previous frame. At the same time, part of the feature information in the scale network is extracted, and after double upsampling, the feature information in the next scale network is merged. As shown in Fig. 1, the red arrow and the purple arrow indicate the feature information in the extracted current network and are merged through the feature fusion layer (■■■) after the upsampling operation. Similarly, input $B_2(I, I_{-1})$ for the next-scale network, and input it into the corresponding scale network $DMSRDBN(\cdot)$ for training. In this process, the corresponding network output can be obtained through the feature information extracted by the previous-scale network and the recurrent connection feature information. The output is defined as follows:

$$I_2^{(1)} = DMSRDBN\left(B_2(I, I_{-1}), e^{3\uparrow}, r^{3\uparrow}\right) \tag{3}$$

Where $B_2(I, I_{-1})$ denotes the input of the second scale network, $e^{3\uparrow}$ denotes the feature of the decoder network in the extracted previous scale network, $r^{3\uparrow}$ denotes the recurrent feature in the extracted previous scale network, and $\uparrow$ denotes upsampling operation. The final output of the network is defined as follows:

$$I_1^{(1)} = DMSRDBN\left(B_1(I, I_{-1}), e^{2\uparrow}, r^{2\uparrow}\right) \tag{4}$$

Hence, the network is fully-convolutional and therefore allows for arbitrary input size. Please refer to Table 1 for more details.

Table 1. Network Sepcification. The output dimesion of feature map is height $\times$ width $\times$ channel.

layer		filter	stride	Output shape
	$A_{0,1}$	3 x 3 x 32	2	H x W x 32
	$C_{1,1}$	3 x 3 x 32	2	H/2 x W/2 x 32
	$C_{1,2}$	1 x 1 x 32	1	H/2 x W/2 x 32
	$C_{1,3}$~$C_{1,5}$	3 x 3x 32	1	H/2 x W/2 x 32
	$C_{1,6}$	1 x 1 x 32	1	H/2 x W/2 x 32
	$C_{1,7}$	1 x 1 x 32	1	H/2 x W/2 x 32
	$C_{2,1}$	3 x 3 x 64	2	H/4 x W/4 x 64
	$C_{2,2}$	1 x 1 x 64	1	H/4 x W/4 x 64
	$C_{2,3}$~$C_{2,5}$	3 x 3 x 64	1	H/4 x W/4 x 64
	$C_{2,6}$	1 x 1 x 64	1	H/4 x W/4 x 64
	$C_{2,7}$	1 x 1 x 64	1	H/4 x W/4 x 64
	$C_{3,1}$	3 x 3 x 128	2	H/8 x W/8 x 128
	$C_{3,2}$	1 x 1 x 128	1	H/8 x W/8 x 128
	$C_{3,3}$~$C_{3,5}$	3 x 3 x 128	1	H/8 x W/8 x 128
	$C_{3,6}$	1 x 1 x 128	1	H/8 x W/8 x 128
	$C_{3,7}$	1 x 1 x 128	1	H/8 x W/8 x 128
	$C_{4,1}$	4 x 4 x 64	1/2	H/4 x W/4 x 64
	$C_{4,2}$~$C_{4,4}$	3 x 3 x 64	1	H/4 x W/4 x 64
	$C_{4,5}$	1 x 1 x 64	1	H/4 x W/4 x 64
	$C_{4,6}$	1 x 1 x 64	1	H/4 x W/4 x 64
	$C_{5,1}$	4 x 4 x 32	1/2	H/2 x W/2 x 32
	$C_{5,2}$~$C_{5,4}$	3 x 3 x 32	1	H/2 x W/2 x 32
	$C_{5,5}$	1 x 1 x 32	1	H/2 x W/2 x 32
	$C_{5,6}$	1 x 1 x 32	1	H/2 x W/2 x 32
	$C_{6,1}$	4 x 4 x 32	1/2	H x W x 32
	$C_{6,2}$	3 x 3 x 6	1	H x W x 6
	$C_{6,3}$	3 x 3 x 3	1	H x W x 3

3 Training

We train the model on our dataset. We produce a ground-truth video and blurry version (5 h in total), so we used 3 h of it for training and 2 h for the test. To account for the effect of vanishing gradients, we force the output $I_1^{(n)}$ of each deblur block to match the sharp ground-truth $I^{(gt)}$ in the corresponding loss term L_n (see Fig. 3).

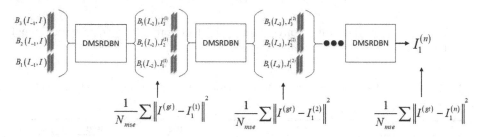

Fig. 3. Training flowchart of DMSRDBN

Network Loss. The Densenet-based multi-scale recurrent network to downsample the input image to get the network output from coarse to fine. Thus, for each iteration, the algorithm uses the mean square loss (MSE) between the output of the largest-scale network and the original sharp frame of the blurred frame of the first frame as a loss function. The loss function is defined as follows:

$$L_n = \frac{1}{N_{mse}} \sum \left\| I^{(gt)} - I_1^{(n)} \right\|^2 \tag{5}$$

Where $I_1^{(n)}$ denotes maximum scale network output combining the previous n frames of blurred images information, $I^{(gt)}$ denotes the original sharp image corresponding to the I frame blurred image, N_{mse} denotes the total number of pixels of the output image. Finally, the loss of the n frames of blurred images is summed to obtain the final loss function of the cyclic network. The final loss is defined as follows:

$$L_{total} = \sum L_n \tag{6}$$

Training Process. The complete network obtains the final output by downsampling the input image to obtain multiple scales and then inputting different scales to train the network according to the process from coarse to fine. At the same time, the network uses the cyclic network to train. Figure 3 shows the network training process.

For the i-th frame and the first n frames in the blurred video, the frame I and the previous frame I_{-1} are downsampled into an image of size B_k, where $k = 1, 2, 3$. Then concat them to a 6-channel input image $B_k(I, I_{-1})$. For the first two frames of blurred images, the output result $\left(I_3^{(1)}, I_2^{(1)}, I_1^{(1)}\right)$ are obtained by the coarse to fine training method, where the subscripts represent the different scale networks built, and the superscript n denotes the combined n-frame blurred image information. Similarly, the multi-scale network output obtained is combined with the frame I_{-2} to obtain the output of the first two frames of blurred image information. Thereby, the network output combining the previous n frames of blurred image information can be obtained, is defined as follows:

$$I_k^{(n)} = Net_{DMSRDBN}\left(B_k\left(I_{-n}, I_k^{(n-1)}\right), e^{k+1\uparrow}, r^{k+1\uparrow}\right) \qquad (7)$$

Where $I_k^{(n)}$ denotes the output of networks of different scales, the subscript k denotes different scale networks built, and the superscript $(n), n = 1, 2, 3, \ldots, n$ denotes combined n-frame blurred image information. $B_k(\cdot), k = 1, 2, 3$ denote the different scale inputs of DMSRDBN. e denotes the feature of the decoder network in the extracted previous scale network, r denotes the recurrent feature in the extracted previous scale network, and $\uparrow$ denotes upsampling operation. Therefore, the network implements a video deblurring algorithm, and the final network output is defined as follows:

$$I_1^{(n)} = Net_{DMSRDBN}\left(B_1\left(I_{-n}, I_k^{(n-1)}\right), e^{2\uparrow}, r^{2\uparrow}\right) \qquad (8)$$

Training details. We use ADAM [19] for minimizing the total loss $L_{total} = \sum_{n=1}^{4} L_n$ for sequences of 5 inputs. We leave the optimizer's default parameters ($\beta_1 = 0.9$, $\beta_2 = 0.999$) unchanged and use 5e-3 as the initial learning rate. We implemented our model with tensorflow library. All the following experiments were performed in a desktop with i7-6700 K CPU and NVIDIA GTX TitanX GPU.

4   Experiments

In this section, we evaluate the performance of our model on the previously created dataset. Our test dataset contains 1 h of sharp and blurred video versions. We first compare the results with the mainstream Gaussian deblurring method and then compare the results with the mainstream video deblurring method. At the same time, we use the PSNR and SSIM to measure the final restoration effect.

4.1   Compare with Gaussian Deblurring Method

We apply our method to Gaussian blurred image restoration. We compare with the current mainstream deblurring methods. At the same time, we compare the Gaussian deblurring method to compare the average of PSNR and SSIM on the test set to compare their restoration effects.

Figure 4 shows the restoration results of a partial Gaussian blurred image. We can see that the mainstream method [2] obtains the final deblurring result by alternately iteratively solving the blurred kernel and the clear latent image, and the restored image has a strong sharpening effect, and our algorithm can achieve better recovery results, with more detailed restoration effects on the edges.

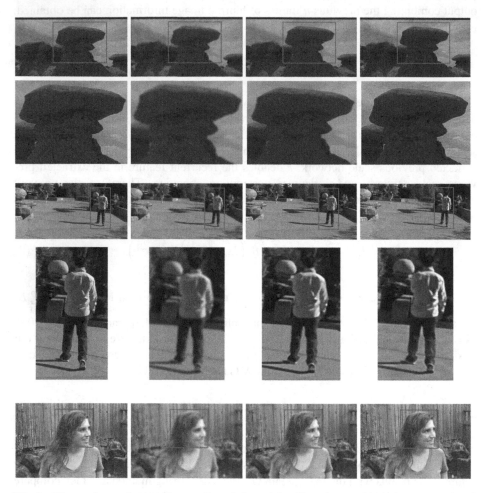

Fig. 4. The results on the test dataset. From left to right: Sharp image, blurry image, results of [2], the result of the proposed method.

From the statistical data of Table 2, we can see that our method has better results in the PSNR and SSIM evaluation indicators than the method [2], which shows that the method of restoring the target frame by combining multi-frame Gaussian blurred image information is effective and robust against many common natural scenes.

Table 2. Quantitative deblurring performance comparison on the test dataset.

Measure	[2]	Ours
PSNR	25.5	**28.2**
SSIM	0.735	**0.853**

4.2 Compare with Video Deblurring Method

We apply our method to Gaussian blurred video restoration. Also, to verify that the proposed algorithm has good deblurring effect for Gaussian blurred video, we will compare it with the current mainstream video deblurring method [11].

Fig. 5. The results on the test dataset. From left to right: Sharp image, Blurry image, results of [11], the results of the proposed method.

Figure 5 shows some of the experimental results of the mainstream video deblur-ring method [11] and the deblurring method in this paper. Since the current mainstream video deblurring algorithm is mainly aimed at the restoration of motion blurred images, this paper focuses on the restoration of Gaussian blurred video. We can see that the current mainstream video deblurring algorithm does not recover well for Gaussian blurred video. The effect of the reconstructed video frame has an obvious distortion effect, and the proposed DMSRDBN network can achieve better recovery effect for Gaussian blurred video, which proves the effectiveness of the proposed algorithm.

The mainstream video deblurring method is for motion blur. From the statistics of PSNR and SSIM evaluation indicators in Table 3, we can see that when we use the current mainstream video deblurring method to recover Gaussian blurred video, we will get a weak recovery effect. At present, the mainstream video deblurring method is not suitable for Gaussian blurred video restoration, and our proposed method has better restoration effect for Gaussian blurred video. PSNR and SSIM have distinct advantages compared with method [11], which sufficiently proves the method phase. Compared with the mainstream video deblurring algorithm, it is more suitable for Gaussian blurred video restoration.

Table 3. Quantitative video deblurring performance comparison on the test dataset.

Measure	[11]	Ours
PSNR	24.6	**28.2**
SSIM	0.661	**0.853**

5 Conclusion

Our proposed model can use the upper and lower frame sequence information of Gaussian blurred video to get a clear restored video. We fuse the feature information between different deblur blocks and realize the weight sharing between multi-scale networks through recurrent connections, which not only reduces the parameter quantity of the network but also enhances the learning ability of the network. Finally, our experimental results and analysis prove the effectiveness of the Gaussian blurred video restoration method. Our method is superior to the current mainstream technology in terms of restoration effect and robustness.

References

1. Helstrom, C.W.: Image restoration by the method of least squares. JOSA **57**(3), 297–303 (1967)
2. Pan, J., Sun, D., Pfister, H., Yang, M.H.: Blind image deblurring using dark channel prior. In: Proceedings of the IEEE Conference on Computer Vision and Pattern Recognition, pp. 1628–1636 (2016)
3. Yan, R., Shao, L.: Blind image blur estimation via deep learning. IEEE Trans. Image Process. **25**(4), 1910–1921 (2016)

4. Schuler, C.J., Hirsch, M., Harmeling, S., Schölkopf, B.: Learning to deblur. IEEE Trans. Pattern Anal. Mach. Intell. **38**(7), 1439–1451 (2016)
5. Wieschollek, P., Schölkopf, B., Lensch, H.P.A., Hirsch, M.: End-to-End learning for image burst deblurring. In: Lai, S.-H., Lepetit, V., Nishino, K., Sato, Y. (eds.) ACCV 2016. LNCS, vol. 10114, pp. 35–51. Springer, Cham (2017). https://doi.org/10.1007/978-3-319-54190-7_3
6. Su, S., Delbracio, M., Wang, J., Sapiro, G., Heidrich, W., Wang, O.: Deep video deblurring for hand-held cameras. In: Proceedings of the IEEE Conference on Computer Vision and Pattern Recognition, pp. 1279–1288 (2017)
7. Noroozi, M., Chandramouli, P., Favaro, P.: Motion deblurring in the wild. In: Roth, V., Vetter, T. (eds.) GCPR 2017. LNCS, vol. 10496, pp. 65–77. Springer, Cham (2017). https://doi.org/10.1007/978-3-319-66709-6_6
8. Chakrabarti, A.: A neural approach to blind motion deblurring. In: Leibe, B., Matas, J., Sebe, N., Welling, M. (eds.) ECCV 2016. LNCS, vol. 9907, pp. 221–235. Springer, Cham (2016). https://doi.org/10.1007/978-3-319-46487-9_14
9. Hasinoff, S.W., Kutulakos, K.N., Durand, F., Freeman, W.T.: Time-constrained photography. In: 2009 IEEE 12th International Conference on Computer Vision, pp. 333–340. IEEE (2009)
10. Ronneberger, O., Fischer, P., Brox, T.: U-Net: convolutional networks for biomedical image segmentation. In: Navab, N., Hornegger, J., Wells, W.M., Frangi, A.F. (eds.) MICCAI 2015. LNCS, vol. 9351, pp. 234–241. Springer, Cham (2015). https://doi.org/10.1007/978-3-319-24574-4_28
11. Wieschollek, P., Hirsch, M., Scholkopf, B., Lensch, H.: Learning blind motion deblurring. In: Proceedings of the IEEE International Conference on Computer Vision, pp. 231–240 (2017)
12. Hirsch, M., Sra, S., Schölkopf, B., Harmeling, S.: Efficient filter flow for space-variant frame blind deconvolution. In: 2010 IEEE Computer Society Conference on Computer Vision and Pattern Recognition, pp. 607–614. IEEE (2010)
13. Sroubek, F., Milanfar, P.: Robust multichannel blind deconvolution via fast alternating minimization. IEEE Trans. Image Process. **21**(4), 1687–1700 (2011)
14. Zhu, X., Šroubek, F., Milanfar, P.: Deconvolving PSFs for a better motion deblurring using multiple images. In: Fitzgibbon, A., Lazebnik, S., Perona, P., Sato, Y., Schmid, C. (eds.) ECCV 2012. LNCS, vol. 7576, pp. 636–647. Springer, Heidelberg (2012). https://doi.org/10.1007/978-3-642-33715-4_46
15. Huang, G., Liu, Z., Van Der Maaten, L., Weinberger, K.Q.: Densely connected convolutional networks. In: Proceedings of the IEEE Conference on Computer Vision and Pattern Recognition, pp. 4700–4708 (2017)
16. Nah, S., Hyun Kim, T., Mu Lee, K.: Deep multi-scale convolutional neural network for dynamic scene deblurring. In: Proceedings of the IEEE Conference on Computer Vision and Pattern Recognition, pp. 3883–3891 (2017)
17. Ioffe, S., Szegedy, C.: Batch normalization: accelerating deep network training by reducing internal covariate shift. In: International Conference on Machine Learning, pp. 448–456 (2015)
18. He, K., Zhang, X., Ren, S., Sun, J.: Deep residual learning for image recognition. In: Proceedings of the IEEE Conference on Computer Vision and Pattern Recognition, pp. 770–778 (2016)
19. Kingma, D.P., Ba, J.: Adam: a method for stochastic optimization. arXiv preprint arXiv:1412.6980 (2014)

20. White, R.L.: Image restoration using the damped Richardson-Lucy method. In: Instrumentation in Astronomy VIII, vol. 2198, pp. 1342–1349. International Society for Optics and Photonics (1994)
21. Wiener, N.: Extrapolation, Interpolation, and Smoothing of Stationary Time Series: with Engineering Applications. Technology Press, Cambridge (1950)

Large Kernel Spatial Pyramid Pooling
for Semantic Segmentation

Jiayi Yang[ID], Tianshi Hu[ID], Junli Yang[⊠][ID], Zhaoxing Zhang, and Yue Pan

Beijing University of Posts and Telecommunications, Beijing, China
{markyang,ertyoii,yangjunli}@bupt.edu.cn

Abstract. Spatial pyramid pooling is growing to become an important component in the network for semantic segmentation. However, it faces a dilemma of using larger kernels for better global context and computation cost. Recent architectures like ASPP have tried to solve this problem by using atrous convolution to keep large reception field while reducing the cost. However, atrous convolutions bring new problems like the "gridding effect", and the large gap between convolutional points make it hard to extract features of small or narrow objects. Inspired by the idea of stacking small filters to simulate large kernels, we propose a Large Kernel Spatial Pyramid Pooling to address both sufficient receptive field while maintaining efficiency. Our approach is evaluated on PASCAL VOC 2012 dataset and Road Extraction Challenge dataset, and achieved better results than competing architectures.

Keywords: Semantic segmentation · Spatial Pyramid Pooling

1 Introduction

Since the design of Spatial Pyramid Pooling (SPP) [4], using parallel convolution to extract features of different sizes has been a trend in computer vision [2, 10]. SPP is a parallel connected convolution module located after the network backbone to extract multi-scale features. Semantic Segmentation has also been using the idea of SPP in many of the modern architectures. PSPNet was one of the first to use them.

SPP is a parallel connected convolution module located after the network backbone to extract multi-scape features. It usually splits the feature map into four parallel channels with different sizes of convolution kernel on each channel. Each parallel convolution outputs feature map of the same size and is concatenated at the end of the SPP module, the overall structure is shown in Fig. 1. As different kernel size results in different sizes of receptive fields.

One of the limitations of SPP is that to extract features of different scales requires different size of large convolutional kernels, which will introduce many parameters to the network. While it may not affect much in tasks like classifications, using large kernel can result in a huge increase of training and inference time in semantic segmentation, due to the pixel-wise segmentation output [5]

© Springer Nature Switzerland AG 2019
Y. Zhao et al. (Eds.): ICIG 2019, LNCS 11901, pp. 595–605, 2019.
https://doi.org/10.1007/978-3-030-34120-6_48

and back propagation. This limits the size of kernel used in SPP module and further limits its ability to extract global context.

DeepLab series [1,2] were introduced to alleviate this problem. In their proposed network, a modified version of Spatial Pyramid Pooling was introduced to solve the problem of computation cost limiting the kernel size used in Spatial Pyramid Pooling. The core innovation of ASPP is that they replaced traditional convolution network with Atrous Convolution [9]. With the same receptive field, ASPP greatly reduced the number of parameters compared to traditional SPP. However, atrous convolution of ASPP brings out another problems.

In this paper, we propose an improved SPP architecture, called Large Kernel Spatial Pyramid Pooling, to deal with the problems mentioned above:

- large convolution kernel in SPP module of introduces significant amount of additional parameters.

- Atrous convolution fails to extract local context information.

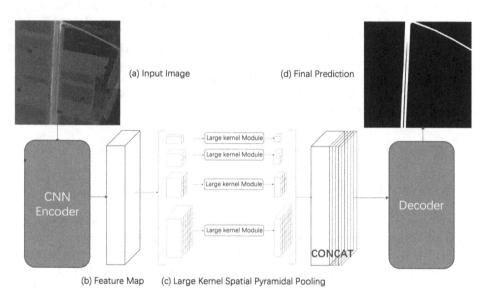

Fig. 1. A network structure with Large Kernel Spatial Pyramid Pooling. It receives feature map extracted from the network backbone. Output of the module fed into the decoder to recover pixel resolution.

To solve these two problems, we describe a novel Large Kernel module inspired by [6,8]. This new module increases the reception field of the kernel by using a combination of depthwise separable convolution [3] and Global Context Network [6], thereby extracts dense features with large reception field while keeping the number of parameters in an acceptable range. This module can be replace any other neural network that uses SPP or SPP-like module to achieve better performance.

2   Related Works

In this work, we mainly focus on the Spatial Pyramid Pooling module of the network, and pay less attention on the backbone of the network.

Spatial Pyramid Pooling: Spatial pyramid pooling layer [4] has been successfully applied to deep convolutional networks to deal with problem of classification and object detection. The SPP layer can pool the features extracted at variable scales and produce fixed-length outputs, which can be used as input of fully-connected layers. What's more, the SPP uses multi-level spatial bins. These advantages of SPP layer help to remove the requirement of fixed-size image input, which means we do not need to crop or wrap the image at the risk of image containing incomplete object or unwanted distortion.

PSPNet: PSPNet [10] works well on scene parsing, which is based on scene segmentation. It is used to predict the label and location for each element. PSPNet make good use of global scene category clues by using global pyramid pooling features. It embeds difficult scenery context features and adopts an effective optimization strategy based on deeply supervised loss. With this improvements PSPNet can do better prediction.

DeepLabv3+: DeepLabv3+ [2] is an effective model in semantic segmentation. It extends the DeepLabv3 model by adding a decoder module to improve the result of semantic segmentation on the boundaries of object. The encoder-decoder structure of DeepLabv3+ can control the resolution of extracted encoder features. With this improvement, DeepLabv3+ can perform better on semantic segmentation along object's boundaries.

Large Kernel Matters: Large kernels play an important role when we perform the classification and localization tasks simultaneously. Large kernel enables the densely connections between feature maps and per-pixel classifiers to make sure the models are invariant to different transformations in classification task.

Rethinking Inception: Inception architecture has much lower computational cost, however, the inception architecture is very complex. To trade off between the computational cost an the complexity, they point out 4 rules: First, we should avoid representational bottlenecks, especially in early stage of the network; Secondly, higher dimensional representations are easier to process locally in the network; Thirdly, we can do spatial aggregation over lower dimensional embeddings; Fourthly, we should balance between the width and depth of the network.

3   Proposed Model

In this section, we start from analyzing spatial pyramid pooling and propose our Large Kernel Spatial Pyramid Pooling (LKSPP).

3.1 Spatial Pyramid Pooling

Spatial Pyramid Pooling is first introduced in [4] to extract multiple features in the task of object detection, and was soon implemented in the task of Semantic Segmentation by PSPNet. In their work, observations were made that traditional FCN often suffer from problems like mismatching relationship, confusion categories and inconspicuous classes. These problems are mainly related to missing of global context information for different sizes of receptive fields.

To alleviate this problem, Pyramid Pooling Module was introduced. Pyramid Pooling Module accepts extracted feature maps from the network backbone and performs convolution on different scales to extract global context information. However, convolution kernels cannot be as large as desired due to the exponentially growing parameters and computation loads. Atrous Spatial Pyramid Pooling was designed to alleviate this problem. Compared with traditional Spatial Pyramid Pooling, Atrous Convolution was introduced to increase the receptive filed while remaining the same computational cost.

3.2 Large Kernel Spatial Pyramid Pooling

To overcome the shortcomings of SPP and ASPP, we introduce Large Kernel Spatial Pyramid Pooling, which proves to find a better balance between computation cost and effectiveness. The structure of Large Kernel Spatial Pyramid Pooling is illustrated in Fig. 2 Each atrous convolution kernel is replaced with the large kernel modulereference Large Kernel Matters. Large Kernel Module was originally inspired by [6,8], which uses a two channel convolution of $n*1+1*n$ and $1*n+n*1$ followed by batch normalization layer on each channel to avoid the affect of covariate shifting.

Further more, we use depthwise separable convolution instead of traditional convolution to further reduce our computation cost. The computation cost can be calculated with the following equation: The overall structure consists of a network backbone followed by a 4-channel parallel LKSPP module. It collects dense features of multiple sizes and the output feature maps are concat together for the decoder to recover the resolution for prediction, see Fig. 1.

3.3 Model Size Control

An important feature of LKSPP is that it significantly reduces the parameters needed in the network compared to traditional Spatial Pyramid Pooling.

The parameters existing in a traditional SPP module can be given by the following equation:

$$V_{conv}(K) = K^2 * C * D \tag{1}$$

Where C stands for the number of input feature maps and D for the number of output feature maps. K represents the height and width of the kernel, assuming that they are of the same length.

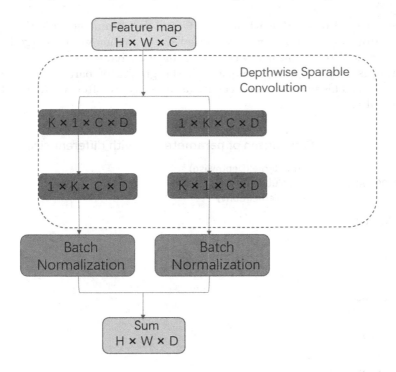

Fig. 2. Structure of Large Kernel Spatial Pyramid Pooling. The convolutional kernels used are depthwise separable convolution followed by batch normalization on each channel. This structure simulates a traditional $K * K$ convolution kernel. The input and output feature maps are C and D respectively.

Parameters can be significantly reduced by using Large Kernel module:

$$V_{lk}(K) = (K * 1 * C * D) * 4 \qquad (2)$$

To reduce further reduce parameters, we replace convolutional kernel used in LKSPP with depthwise separable convolution. $K * C$ describes the parameters of the depthwize convolution. Note that in Xception it is described as $K^2 * C$, but since we use a kernel size of $K * 1$ or $1 * K$, the kernel size is reduced to $K * C$. The following $C * D$ describes the pointwise convolution with $1 * 1$ convolution kernel. The scale factor 4 represents the two kernels used in the two different channels each in Large Kernel.

$$V_{lk}(K) = (K * C + C * D) * 4 \qquad (3)$$

The total equation is given by Eq. 4, where $k_1, k_2, ..., k_n$ represents the kernel size of each parallel channel with a total of n channels. In case of our standard LKSPP, we use a combination of $[k_1, k_2, k_3, k_4] = [1, 4, 8, 16]$.

$$V_{total} = \sum_{k=k_1}^{k_n} V(K) \qquad (4)$$

It is not hard to see that when given C and D, parameters traditional convolution grows exponentially with its kernel size K. When we use Large Kernel, The parameters grows linearly with K, significantly reducing parameters when the kernel is very large, Fig. 3 illustrates the growth of parameters with the increase of K. This property is especially useful since we often set K very big in SPP modules.

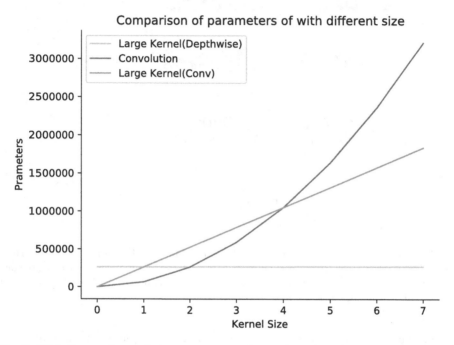

Fig. 3. This figure shows the parameters used in a kernel with different kernel size K. Traditional convolution increases notably after the kernel grows bigger than 2. We assume that the input and output feature map has a depth of 256.

4   Experiment

We firstly evaluate our model, which is pretrained on Pascal VOC 2012 dataset, on Road Extraction Challenge semantic segmentation dataset which contains one foreground road class and one background class. The dataset contains 6226 training, 1243 evaluation and 1101 testing pixel-wise annotated images of size 1024×1024. We train the network on the training set and report the intersection over union (IoU) of the road class.

Then, in order to further prove the effectiveness of our proposed model, we employ another Xception65, which is pretrained on ImageNet-1k dataset, as the network backbone and evaluate this model on augmented [7] Pascal VOC

2012 benchmark, which contains 10582 training, 1449 validation and 1456 testing images of 21 classes. And the performance is evaluated in terms of mean intersection over union (mIoU) across all 21 classes.

4.1 Training Setting

The training setting is very similar to [1,2], but there are some changes. We discuss in details in this subsection.

Learning Rate: we use a base learning rate $= 0.01$ with 'poly' learning rate policy which multiplies the base with $(1 - \frac{iter}{iter_{max}})^{power}$ where power is 0.9 as set by [1,2].

Batch Size and Crop Size: as constrained by the computation power, we strike a balance between large batch size and more information on each image. And after some experiments, we choose batch size $= 3$ with crop size $= 513$ during all the training and testing.

Output Stride: recall that output stride is the ratio of input image spatial resolution to final output resolution [1] and [2] shows that the smaller output stride the better result, so we adapt the smallest output stride possible and during out training and testing the output stride $= 16$.

Data Augmentation: during training, we apply HSV and spatial data augmentation as proposed by [HSV transfer] and when evaluating or visualizing, we follow [1,2] method scaling the input images by 1.75, 2.0 and 2.25 and randomly left-right flipping those images.

Besides, we implement Nesterov momentum optimizer with momentum $=$ 0.9 and weight decay 4e-5. All trainings are implemented on 2 NVIDIA GTX 1080 using tensorflow 1.8.

4.2 Road Extraction Challenge Dataset

Baseline Model: We use DeepLabv3+ with Xception65 as backbone, which is pretrained on Pascal VOC 2012, to train our baseline model. Specifically, the network has an encoder where the Xception65 backbone resides with an ASPP module whose atrous rate is [6, 12, 18] and a simple yet effective decoder. With all training settings modified, except using data augmentation, as indicated by our training protocol section, we alter the evaluation function from calculating mean IoU to IoU since in this dataset there is only one road class matters. After training the network for 30000 steps, we observe that the loss converges and the result IoU 57.6% is recorded in the first row of Table 1.

Adding New Loss Function: The second row of Table 1 is our experiment with a new loss function inspired by [11]. The performance increase from 57.6% to 58.4%.

Table 1. Road Extraction Challenge test set result of different models

Model	IoU
Baseline	57.6%
Baseline+new loss function	58.4%
Baseline+new loss function+data augmentation	61.5%
Large Kernel SPP+new loss function+data augmentation	64.6%

Adding Data Augmentation: The third row of Table 1 shows the result of adding both HSV and spatial transformation and test time augmentation. The evaluation result increase another 3% to 61.5.

Ablation Study: We have also examined the effect of different atrous rates of ASPP module on the network performance. Specifically, we compare different combinations of training and evaluating atrous rate as shown in Table 2.

Analysis: Here, as introduced in the previous section, the atrous rate in DeepLabv3+ model represents how much context information can be obtained from the atrous convolution operation. When the training atrous rate is set, we can see from Table 2 that the larger evaluation atrous rate the better IoU and when the evaluation atrous rate is set, we can also conclude that the larger training atrous rate the better IoU result. It can be seen that atrous rate has strong relationship with the network performance. Therefore, we come up with using large kernel in the ASPP module.

Table 2. Effect of different atrous rate on training and evaluation using baseline DeepLabv3+ model

Model	Train atrous rate	Eval atrous rate	TTA	IoU
DeepLabv3+	[2,4,8]	[3,6,9]		56.7%
DeepLabv3+	[2,4,8]	[6,12,18]		57.3%
DeepLabv3+	[2,4,8]	[24,28,32]		57.9%
DeepLabv3+	[2,4,8]	[24,28,32]	✓	60.4%
DeepLabv3+	[4,8,12]	[3,6,9]		57.0%
DeepLabv3+	[4,8,12]	[6,12,18]		57.8%
DeepLabv3+	[4,8,12]	[24,28,32]		58.1%
DeepLabv3+	[4,8,12]	[24,28,32]	✓	60.6%
DeepLabv3+	[6,12,18]	[3,6,9]		57.3%
DeepLabv3+	[6,12,18]	[6,12,18]		57.6%
DeepLabv3+	[6,12,18]	[24,28,32]		58.7%
DeepLabv3+	[6,12,18]	[24,28,32]	✓	61.7%

Large Kernel Spatial Pyramid Pooling: We use the DeepLabv3+ having the best performance (with data augmentation and the new loss function added) as a starting point. By modifying all dilated convolutions in ASPP module to Large Kernel convolution, we achieve a 3% performance gain to an IoU of 64.6%. We also compare the solo effect of using Large Kernel Spatial Pyramid Pooling without data augmentation. The results are recorded in the fouth row of Table 1 and we present the inference result of both baseline and LK SPP model in the following Fig. 4.

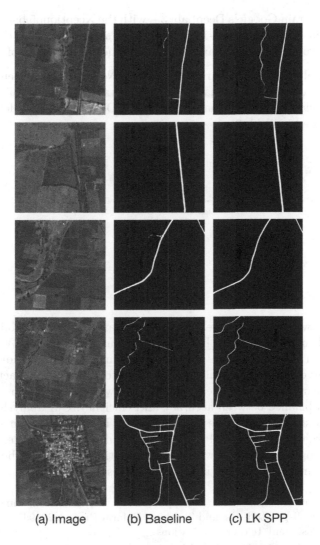

      (a) Image   (b) Baseline   (c) LK SPP

Fig. 4. Visualization result of road extraction. (a) is the original images to be inferred. (b) is the visualization result of the baseline model and (c) is the inference result of our Large Kernel Spatial Pyramid Pooling model.

4.3   Pascal VOC 2012 Dataset

We compare our proposed model with a baseline model on another dataset to prove that the proposed model is truly effective. And due to our limited computation resources, we understand that we cannot reproduce the score that Google's team had made using DeepLabv3+ on Pascal VOC 2012 benchmark. Therefore, rather than fine-tuning the Pascal VOC 2012 pretrained model, which will bring very few improvement, we use ImageNet-1k pretrained model as basic backbone for comparison.

Baseline: We train Google's DeepLabv3+ with the Xception65 backbone, which is now pretrained on ImageNet-1k dataset, on Pascal VOC 2012 dataset. After 30000 steps, we find the loss start to converge and get an mIoU of 38.1%.

Large Kernel Spatial Pyramid Pooling: Next, we implement our Large Kernel mechanism on the original ASPP module and achieve an mIoU of 46.0%, which is significantly higher than the original DeepLabv3+ model. The mean IoU are recorded in Table 3.

Table 3. Result on Pascal VOC 2012 dataset

Model	mIoU
Baseline	38.1%
LKSPP	46.0%

5   Conclusion

In this paper, we propose a novel method for Spatial Pyramid Pooling to alleviate the problem gridding effect brought by Atrous Convolution. We carried out experiments on popular datasets in both natural image segmentation and special tasks like road extraction to evaluate its ability to extract to extract multi-scale features while maintaining efficiency. Results shows that our proposed structures found a good balance between accuracy and efficiency. Our analysis and experiment results shows that Large Kernel Spatial Pyramid Pooling is promising in many tasks containing small objects like road extraction of remote sensing images. We hope our module with implementation details can be adopted to other models and help them to achieve better results.

Acknowledgments. This work is supported by the Research Innovation Fund 201811062 for College Students, and Teaching Reform Project 2019JY-A05 of Beijing University of Posts and Telecommunications.

References

1. Chen, L.C., Papandreou, G., Schroff, F., Adam, H.: Rethinking atrous convolution for semantic image segmentation. arXiv preprint arXiv:1706.05587 (2017)
2. Chen, L.-C., Zhu, Y., Papandreou, G., Schroff, F., Adam, H.: Encoder-decoder with atrous separable convolution for semantic image segmentation. In: Ferrari, V., Hebert, M., Sminchisescu, C., Weiss, Y. (eds.) ECCV 2018. LNCS, vol. 11211, pp. 833–851. Springer, Cham (2018). https://doi.org/10.1007/978-3-030-01234-2_49
3. Chollet, F.: Xception: deep learning with depthwise separable convolutions. In: Proceedings of the IEEE Conference on Computer Vision and Pattern Recognition, pp. 1251–1258 (2017)
4. He, K., Zhang, X., Ren, S., Sun, J.: Spatial pyramid pooling in deep convolutional networks for visual recognition. IEEE Trans. Pattern Anal. Mach. Intell. 37(9), 1904–1916 (2015)
5. Long, J., Shelhamer, E., Darrell, T.: Fully convolutional networks for semantic segmentation. In: Proceedings of the IEEE Conference on Computer Vision and Pattern Recognition, pp. 3431–3440 (2015)
6. Peng, C., Zhang, X., Yu, G., Luo, G., Sun, J.: Large kernel matters-improve semantic segmentation by global convolutional network. In: Proceedings of the IEEE Conference on Computer Vision and Pattern Recognition, pp. 4353–4361 (2017)
7. Sun, T., Chen, Z., Yang, W., Wang, Y.: Stacked U-Nets with multi-output for road extraction. In: 2018 IEEE/CVF Conference on Computer Vision and Pattern Recognition Workshops (CVPRW), pp. 187–1874. IEEE (2018)
8. Szegedy, C., Vanhoucke, V., Ioffe, S., Shlens, J., Wojna, Z.: Rethinking the inception architecture for computer vision. In: Proceedings of the IEEE Conference on Computer Vision and Pattern Recognition, pp. 2818–2826 (2016)
9. Yu, F., Koltun, V.: Multi-scale context aggregation by dilated convolutions. arXiv preprint arXiv:1511.07122 (2015)
10. Zhao, H., Shi, J., Qi, X., Wang, X., Jia, J.: Pyramid scene parsing network. In: Proceedings of the IEEE Conference on Computer Vision and Pattern Recognition, pp. 2881–2890 (2017)
11. Zhou, L., Zhang, C., Wu, M.: D-linknet: linknet with pretrained encoder and dilated convolution for high resolution satellite imagery road extraction. In: Proceedings of the IEEE Conference on Computer Vision and Pattern Recognition Workshops, pp. 182–186. IEEE (2018)

Weighted Feature Pyramid Network for One-Stage Object Detection

Xiaobo Tu and Yongzhao Zhan[(⊠)]

School of Computer Science and Telecommunication Engineering,
Jiangsu University, Zhenjiang, China
yzzhan@ujs.edu.cn

Abstract. One-stage object detection methods have attracted much attention for their high speed performance compared with two-stage methods. But one-stage methods under performs with small object detection. Feature Pyramid Network (FPN) was widely used to deal with this problem for its multi-scale feature present ability. However there still remains a few problems that are not considered in FPN, which results in limited improvement in detector performance. We note that FPN does not take the weight and scale distribution between different levels of feature maps into account when merging high-level feature maps and low-level feature maps. We present a network named Weighted Feature Pyramid Network (WFPN) to address these problems. Our experimental results on PASCAL VOC and MS COCO show that WFPN can significantly improve the detector performance, especially on small object detection.

Keywords: Object detection · Weighted Feature Pyramid Network

1 Introduction

In recent years, deep learning has significantly pushed the development of object detection. There are a large quantity of deep learning based methods for detection task [7,10–13,21,22]. On the whole, these methods can be classified into two-stage and one-stage methods. The first stage of two-stage methods is proposing some candidate bounding boxes [3]. The following network classification objects that is indicated by those bounding boxes, regression on each bounding box is used to accelerate the intersection over union (IoU) of predicted box and ground truth box. One-stage methods do not need candidate bunding boxes, they use neural network regression to calculate objects location and class.

Two-stage methods such as R-CNN [5], Fast R-CNN [4], Faster R-CNN [16] and other R-CNN series methods obtain high accuracy but low speed on many public detection datasets such as PASCAL VOC 2007 and COCO. One-stage methods such as YOLO series [7,12–14], SSD [11] reach a very high FPS compared with two-stage methods. But there is a big gap between one-stage and two-stage methods for mAP of public detection datasets. One-stage methods

© Springer Nature Switzerland AG 2019
Y. Zhao et al. (Eds.): ICIG 2019, LNCS 11901, pp. 606–617, 2019.
https://doi.org/10.1007/978-3-030-34120-6_49

remove the proposal process that two-stage methods use to generate candidates for bounding box. Such measures dramatically accelerate the speed of methods but reduce the accuracy at the same time. This is largely because one-stage methods suffer from the small object detection accuracy [15].

Small objects occupy fewer pixels on image, therefore repeated CNN convolution operations on features of small object which could even be one pixel make it difficult for the regression network to recognize such small object on the feature map. There are numerous researches for this issue. Feature Pyramid Network (FPN) [10] that is inspired by feature pyramids built upon image pyramids were heavily used in the era of one-stage methods. FPN use a bottom-up pathway, a top-down pathway, and lateral connections upon several layers output feature map of CNN to build a feature pyramid with high-level semantics. But FPN consider the low-level feature as important as high-level feature for detection task. Detection task can be divided into classification and location tasks. High-level feature maps have strong semantic information which can benefits classification task, whereas low-level feature maps have high resolution that can contribute to location task [6]. But is the importance of those two subtasks equal for the detection task?

In order to figure out this problem, we consider giving high-level feature maps and low-level feature maps different weights when building feature pyramid. We speculate that the weights of these two features may have different effects on the detection task of large objects and small objects. The bounding box of small objects is small, so the network does not need much location information to perform regression on the predicted bounding box. In contrast, the feature of small objects occupies so few pixels on the whole feature map that it may increase the classification loss. Based on the above reasons, we give high weight to high-level feature map that contain more semantic information to benefit classification. The large ground truth bounding box of large objects may lead to high location loss. On the other hand, their features on the whole feature map is usually big enough to provide class information needed for the classification task. So we give a small weight to low-level feature map that contain more location information.

We verified our assumptions on the PASCAL VOC 2007, 2012 and MS COCO dataset. The results demonstrate that our proposed WFPN can significantly improve the detection performance, especially small object detection.

Our contributions are summarized as follows:

- High-level feature maps and low-level feature maps have different importance when merging different level feature map for multi-scale object detection. Since high-level feature map contains strong semantic information and is more important than low-level feature map.
- Batch normalization before feature merging improves the effect of feature representation. Since high-level and low-level feature maps have different scale distribution.

- Experimental results on PASCAL VOC and MS COCO demonstrate the Weighted Feature Pyramid Network (WFPN) can significantly improve the performance of one shot object detection network.

2 Related Work

Small object detection at vastly different scales is a fundamental challenge in object detection task and many researchers have made efforts to this end. In this section, we take an overview of some methods that attempts to enhance small object detection performance on multi-scale detection task.

Featurized Image Pyramid. Hand-craft features were heavily used to extract features of image prior to the proposal of AlexNet in 2012. Feature image pyramid that built upon image pyramids from the basis of a standard solution is a widely used method to deal with multi-scales object detection. This method firstly resize image to different resolutions to get an image pyramid, and then extract the hand-engineered features from the image pyramid to establish the featurized image pyramid. Featurized image pyramid has used extensively not only on object detection but also on human pose estimation and other computer vision tasks.

Feature Pyramid Network. With the development of deep learning, ConvNets show the powerful capabilities in feature extraction. But extracting features from image pyramid to build a feature pyramid by ConvNets need a huge computing and storage resources. Tsung-Yi Lin et al. proposed Feature Pyramid Network (FPN) [10] to exploit the inherent multi-scale, pyramidal hierarchy of deep convolutional networks to construct feature pyramids with marginal extra cost. FPN use a top-down architecture with lateral connections to build high-level semantic feature maps at all scales. The top-down pathway uses nearest neighbor upsample by a factor of 2 to resize the upper layer feature map to the low layer feature map size [17]. And the lateral connections combine the upper layer feature map with bottom-up feature map (which undergoes a 1×1 convolutional layer to reduce channel dimensions equal to the upper layer feature map) by element-wise addition. FPN was widely used in the area of multi-scale object detection.

YOLOv3 is a state-of-art one stage object detector proposed by Joseph Redmon et al. and uses a specially designed backbone network named darknet-53 [14]. And YOLOv3 draws on the idea of FPN, which uses final three stage of feature maps to build an FPN structure to predict the bounding boxes and categories of objects. However, like FPN, YOLOv3 also does not consider the influence of feature maps from different level on the detection performance.

3 Weighted Feature Pyramid Network

3.1 The Entire Framework

ConvNets have almost replaced hand-engineered image feature extraction in computer vision for their powerful feature extraction capabilities [9,18,19]. It is well known that high-level feature maps in ConvNets contain strong semantic information compared to low-level feature map [1]. FPN aims to improve object detector performance by merging high-level and low-level features, which is a widely used method in the area of object detection research. But FPN did not consider the different importance of high-level and low-level features for detection task. In this paper, we are committed to research the effects of different weights for high-level features and low-level features on the detection task. In light of the fact that YOLOv3 is a state-of-art one stage object detector which used the FPN idea, we adopt the YOLOv3 to bulid our detection framework. The whole framework as shown in the Fig. 1 composed of darknet-53 for feature extraction and WFPN for feature fusion.

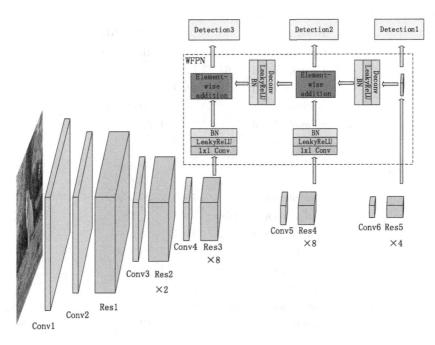

Fig. 1. We use the outputs of Res3, Res4 and Res5 to build WFPN. Feature fusion module consists of a top-down pathway (which composed of a deconvolutional layer, a ReLU layer and a batch normalization layer) and a lateral connection (which composed of a 1×1 convolutional layer, a ReLU layer and batch normalization layer). Merged feature map serves as the input of detection layer and another fusion module top-down pathway.

3.2 Feature Extraction

Same as the YOLOv3, we use darknet-53 [14] as the backbone to extract feature maps from input images. The structure of darknet-53 is shown in Table 1. We call the combination of two convolutional layers and one residual layer in the box as a residual block. In this paper, we use the feature maps that is output by the final three residual blocks to build the feature pyramid network.

Table 1. Darknet-53 structure. The **Convolutional** means a set composed of a 2-D convolutional layer, a batch normalization layer and a LeakyReLU layer.

	Type	Filters	Size	Output
	Convolutional	32	3×3	416×416
	Convolutional	64	$3 \times 3/2$	208×208
1x	Convolutional	32	1×1	
	Convolutional	64	3×3	
	Residual			208×208
	Convolutional	128	$3 \times 3/2$	104×104
2x	Convolutional	64	1×1	
	Convolutional	128	3×3	
	Residual			104×104
	Convolutional	256	$3 \times 3/2$	52×52
8x	Convolutional	128	1×1	
	Convolutional	256	3×3	
	Residual			52×52
	Convolutional	512	$3 \times 3/2$	26×26
8x	Convolutional	256	1×1	
	Convolutional	512	3×3	
	Residual			26×26
	Convolutional	1024	$3 \times 3/2$	13×13
4x	Convolutional	512	1×1	
	Convolutional	1024	3×3	
	Residual			13×13

3.3   Feature Fusion

Feature fusion module contains a top-down pathway, a lateral connection and a element-wise addition block. Top-down pathway consists of an upsample layer, a Leaky-ReLU layer and a batch normalization layer [8]. Upsample layers upsamples the upper layer feature map by a factor of 2 to make sure it can be merged with lower layer feature map. Since it has been proven that deconvolutional performs better than linear upsample methods, we use deconvolutional layer for upsampling [20]. We deem that different layer feature maps show different scale distribution, it is essential to normalize features before fusion. So we add a ReLU layer and a batch normalization [8] layer to deal with this problem. Lateral connection do 1x1 convolution on the lower layer feature map to adopt its channels unified to be 256, which is the minimum of final four residual block output channel. As with the top-down pathway, a ReLU layer and a batch normalization layer are used to handle the different scale distribution problem. Then we fuse the output of top-down pathway and the lateral connection by an element-wise addition block. We consider that upper layer feature map contains more strong semantic information which is beneficial to classification. So we give a high weight to the output of the fusion modules top-down pathway when merging the feature map by element-wise addition. The fusion module can be formulize as:

$$
\begin{aligned}
F_c &= Comb(F_h, F_l) \\
&= w_h * BN(ReLU(DeConv(F_h))) + w_l * BN(ReLU(DeConv(F_l)))
\end{aligned}
\tag{1}
$$

where

$$
BN(x) = \gamma \frac{x - \mu}{\sqrt{\sigma^2 + \varepsilon}} + \beta
\tag{2}
$$

$$
ReLU(x) = max(0, x)
\tag{3}
$$

$$
Deconv(x) = x \oplus K
\tag{4}
$$

in the above, F_c is the feature map that fuse F_h (which is the input of fusion module top-down pathway) and F_l (which is the input of fusion module lateral connection). And w_h is the weight of high level feature map, that is, the upper layer feature map or merged upper layers feature map. w_l is the weight of low layer feature map. γ and β is learnable parameters of network, μ is the mean of mini-batch, σ^2 is the variance of mini-batch. k is the filter of previous convolutional layer.

4   Experiments

4.1   Datasets

To verify the effectiveness of the proposed WFPN, we conduct experiments on two widely used datasets, namely, PASCAL VOC and MS COCO. PASCAL VOC have over 10000 color images in 20 classes. We split the data into 50% for training/validation and 50% for testing. The distributions of images and objects by class are approximately equal across the training/validation and test sets. MS COCO is a large-scale object detection dataset which have over 330000 images in 80 object categories. Compared with PASCAL VOc, MS COCO have more small objects and more objects per image, and most of the objects are not centered, which is more in line with daily environment, so detection on COCO is more difficult.

4.2 Training Mechanism

We fine tune the Network based on pretrained YOLOv3 weights provided by [14]. We start training with learning rate of 10^{-3} and desent it by a factor of 0.1 if the validation loss does not decent for two consecutive epochs. We stop our training if validation loss does not decent for ten consecutive epochs.

4.3 Ablation Experiments

We perform ablation experiments on PASCAL VOC 2007 and 2012 benchmarks datasets and on MS COCO 2015 benchmark dataset to analy in detail each components of our network. We first replace each component based on same hyper parameters step-by-step to analyse the real effects of each component. On these experiments, we give same weights to high-level and low-level feature map. In practice, we set w_h and w_l equal to 1. We show these in experiments below.

Fusion Methods. We first consider three different element-wise operations to merge the output feature maps of top-down pathway and lateral connection. The results in the Table 2 shows that element-wise addition perform well than element-wise maximum and element-wise product, and element-wise product perform worst. This situation can be understood from the information flow during training. In the forward propagation phase, addition enables network to take full advantage of the information from two branches complementary without losing any information. And in the backpropagation phase, it can equally distribute the gradient to each branches. For the element-wise maximum, the network only use one branch information which has high values during the forward propagation phase and only routes the gradient to the higher input branch. The element-wise product assigns a small gradient to the high input branch and a large gradient to the low input branch, which makes the network hard to converge. Therefore, element-wise addition has a better performance on feature map fusion, and element-wise perform worst. For these reasons, we choose element-wise addition to merge feature maps.

Table 2. The efforts of fusion methods. The detection mAP(%) of three different fusion methods on above datasets shows that element-wise addition perform better.

Dataset	ADDITION	MAX	PRODUCT
VOC07	**80.7**	79.2	78.6
VOC12	**79.8**	78.9	78.2
COCO15	**38.5**	37.8	37.3

Activation and Batch Normalization. We assume that feature map of different level have different scale distributions, so we add a Leaky-ReLU and a batch normalization layer in the end of the top-down pathway and the lateral connection. We verify the necessity of adding these two layers through the ablation experiments. The results in the Table 3 shows that adding a ReLU and batch normalization layer can effectively improve the detection effect. Hence, we use a Leaky-ReLU layer and a batch normalization layer to solve the problem of distribution diverseness.

Table 3. The efforts of Leaky-ReLU and Batch Normalization layer. **TD**:Top-down pathway, **LC**: Lateral Connection, **with BN**: means there add a Leaky-ReLU layer and a batch normalization layer after original module.

Fusion module		VOC07	VOC12	COCO15
TD	LC	79.3	78.3	35.1
TD	LC with BN	78.8	78.6	34.8
TD with BN	LC	79.1	78.9	34.6
TD with BN	LC with BN	**81.2**	**79.6**	**38.2**

4.4   Weight Selection

The ablation experiments proved the rationality and necessity of our network design. On this basis, we discuss the distribution of different level feature map weights when feature map merging.

Firstly, we fixed w_l to focus on the efforts of w_h to the network. We trained the network on PASCAL VOC 2012, and evaluates it on the PASCAL VOC 2007 benchmark. As the results shown in the Fig. 2(a), when we set w_l equal to 0.8, we can get a better mAP in most cases. It is worth noting that almost all curves have an upward trend when the value of w_h are 1 to 1.8, and begin decline after 1.8 except the curve with w_l choose as 0.4. We can deduce two conclusions from this situation. An obvious one is that it is a fine choise to let w_h equal to 1.8 to get a better detector. Another one is that it is not a good idea to assign a very low value of w_l, otherwise a serious imbalance between high-level feature and low-level feature may lead to poor detection AP.

Then we fixed w_h to observe the influence of w_l to the final detection performance. The results shown in the Fig. 2(b) shows again that 1.8 is an appropriate value of w_h and the best choice of w_l maybe 0.8.

4.5   YOLOv3 with WFPN

Finally, we set w_l equal to 0.8 and w_h equal to 1.8, and use the feature maps which is output from the final three residual blocks of darknet-53 to build a

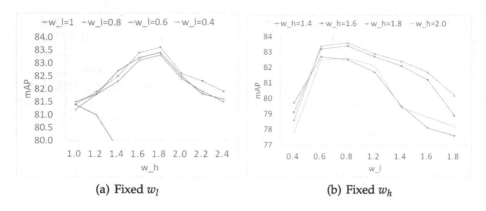

(a) Fixed w_l (b) Fixed w_h

Fig. 2. Each weights efforts

weighted FPN (for simplicity, we call it WFPN-YOLOv3). We conducted a comparative test with YOLOv3, SSD and DSSD on the PASCAL VOC 2007+2012 benchmark and MS COCO 2015 benchmark. The results is shown in the Tables 4 and 5.

As can be seen from the Table 4, our network performs better on most categories, especially on the "small" categories such as bird, boat, bottle and plant etc. Comparied with YOLOv3, our network achieves a 3.2% improvement on bird class, a 5.7% improvement on boat class, a 6.1% on bottle class and a 5.7% improvement on plant class. WFPN-YOLOv3 performance on other classes also gets a little improvement or nearly the same with YOLOv3.

Results on Table 5 shows that our proposed WFPN-YOLOv3 achieves 44% mAP on MS COCO, which outperforms the YOLOv3 by 4%. WFPN-YOLOv3 gets a 7% improvement from YOLOv3 on small object AP, a 3% improvement on middle object AP and gets the same AP on large object AP. On PASCAL VOC dataset, WFPN-YOLOv3 only gets a 2.1% improvement from YOLOv3, but it gets a 4% improvement from YOLOv3 on the MS COCO dataset. The performance can be attributed to MS COCO having more small objects and more objects per image.

Experimental results above demonstrate that our WFPN which gives high level feature map a higher weight, have a good performance on small object (Fig. 3).

Table 4. Evaluation results on PASCAL VOC 07+12

Model	mAP	aero	bike	bird	boat	bottle	bus	car	cat	chair	cow	table	dog	horse	mbike	person	plant	sheep	sofa	train	tv
WFPN-YOLOv3	83.6	91.8	89.2	85.4	77.4	71.6	89.3	84.5	91.1	66.7	89.1	76.0	91.5	91.6	88.9	89.8	68.2	84.5	72.6	89.4	82.6
YOLOv3 [14]	81.5	90.0	88.6	82.2	71.7	65.5	85.5	84.2	92.9	67.2	87.6	70.0	91.2	90.5	90.0	88.6	62.5	83.8	70.7	88.8	79.4
SSD [11]	80.0	90.7	86.8	80.5	67.8	60.8	83.6	85.5	93.5	63.2	85.7	64.4	90.9	89.0	88.9	86.8	57.2	85.1	72.8	88.4	75.9
DSSD [2]	81.2	86.6	86.2	82.6	74.9	62.5	89.0	88.7	88.8	65.2	87.0	78.7	88.2	89.0	87.5	73.7	51.1	86.3	87.6	85.7	83.7

Fig. 3. Qualitative results on PASCAL VOC. The left column is the test result of YOLOv3, and the right column is the test result of WFPN-YOLOv3.

Table 5. Evaluation results on MS COCO 15

Model	AP	AP50	AP75	APS	APM	APL
WFPN-YOLOv3	0.44	0.63	0.43	0.25	0.42	0.48
YOLOv3 [14]	0.40	0.59	0.39	0.18	0.39	0.48
SSD [11]	0.39	0.55	0.36	0.20	0.38	0.47
DSSD [2]	0.41	0.57	0.38	0.23	0.40	0.46

5 Conclusion

In this paper, we have presented a Weighted Feature Pyramid Network (WFPN) which considers the different importance between high-level feature maps and low-level feature maps when feature pyramid network is used to merge feature maps. WFPN assigns higher weight to upper layer feature maps since the upper layer feature maps with stronger semantic information are more useful for object detection. Moreover, WFPN introduces batch normalization to solve the problem of different scale distributions between high-level feature maps and low-level feature maps when FPN is performing feature fusion. The experimental results on PASCAL VOC and MS COCO demonstrate that WFPN significantly improves the small object detection effect of detector, while the large object detection effect are not reduced.

Acknowledgments. This work was supported by the National Natural Science Foundation of China (No. 61672268) and the Primary Research & Development Plan of Jiangsu Province (No. BE2015137).

References

1. Cai, Z., Fan, Q., Feris, R.S., Vasconcelos, N.: A unified multi-scale deep convolutional neural network for fast object detection. In: Leibe, B., Matas, J., Sebe, N., Welling, M. (eds.) ECCV 2016. LNCS, vol. 9908, pp. 354–370. Springer, Cham (2016). https://doi.org/10.1007/978-3-319-46493-0_22
2. Fu, C.-Y., Liu, W., Ranga, A., Tyagi, A., Berg, A.C.: DSSD: deconvolutional single shot detector (2017)
3. Ghodrati, A., Diba, A., Pedersoli, M., Tuytelaars, T., Van Gool, L.: DeepProposals: hunting objects and actions by cascading deep convolutional layers. Int. J. Comput. Vis. **124**(2), 115–131 (2017)
4. Girshick, R.: Fast R-CNN. In Proceedings of the IEEE International Conference on Computer Vision, vol. 2015, pp. 1440–1448 (2015)
5. Girshick, R., Donahue, J., Darrell, T., Malik, J.: Rich feature hierarchies for accurate object detection and semantic segmentation. In The IEEE Conference on Computer Vision and Pattern Recognition (CVPR), June 2014
6. He, K., Zhang, X., Ren, S., Sun, J.: Deep residual learning for image recognition. In The IEEE Conference on Computer Vision and Pattern Recognition (CVPR), June 2016

7. Huang, R., Pedoeem, J., Chen, C.: YOLO-LITE: a real-time object detection algorithm optimized for non-GPU computers. In: Proceedings - 2018 IEEE International Conference on Big Data, Big Data 2018, pp. 2503–2510 (2019)
8. Ioffe, S., Szegedy, C.: Batch normalization: accelerating deep network training by reducing internal covariate shift (2015)
9. Krizhevsky, A., Sutskever, I., Hinton, G.E.: Imagenet classification with deep convolutional neural networks. In: Pereira, F., Burges, C.J.C., Bottou, L., Weinberger, K.Q. (eds.) Advances in Neural Information Processing Systems, vol. 25, pp. 1097–1105. Curran Associates Inc. (2012)
10. Lin, T.Y., Dollár, P., Girshick, R., He, K., Hariharan, B., Belongie, S.: Feature pyramid networks for object detection. In: Proceedings - 30th IEEE Conference on Computer Vision and Pattern Recognition, CVPR 2017, pp. 936–944 (2017)
11. Liu, W., et al.: SSD: single shot multibox detector. In: Leibe, B., Matas, J., Sebe, N., Welling, M. (eds.) ECCV 2016. LNCS, vol. 9905, pp. 21–37. Springer, Cham (2016). https://doi.org/10.1007/978-3-319-46448-0_2
12. Redmon, J., Divvala, S., Girshick, R., Farhadi, A.: Unified, real-time object detection, you only look once (2015)
13. Redmon, J., Farhadi, A.: YOLO9000: better, faster, stronger. In: Proceedings - 30th IEEE Conference on Computer Vision and Pattern Recognition, CVPR 2017, pp. 6517–6525 (2017)
14. Redmon, J., Farhadi, A.: Yolov3: an incremental improvement. CoRR, abs/1804.02767 (2018)
15. Ren, J., et al.: Accurate single stage detector using recurrent rolling convolution. CoRR, abs/1704.05776 (2017)
16. Ren, S., He, K., Girshick, R.B., Sun, J.: Faster R-CNN: towards real-time object detection with region proposal networks. CoRR, abs/1506.01497 (2015)
17. Rusk, N.: Deep learning. Nat. Methods **13**(1), 35 (2015)
18. Simonyan, K., Zisserman, A.: Very deep convolutional networks for large-scale image recognition. Computer Science (2014)
19. Szegedy, C., et al.: Going deeper with convolutions. In: 2015 IEEE Conference on Computer Vision and Pattern Recognition (CVPR), pp. 1–9, June 2015
20. Woo, S., Hwang, S., Kweon, I.S.: StairNet: top-down semantic aggregation for accurate one shot detection. In: Proceedings - 2018 IEEE Winter Conference on Applications of Computer Vision, WACV 2018, pp. 1093–1102 (2018)
21. Zhang, S., Wen, L., Bian, X., Lei, Z., Li, S.Z.: Single-shot refinement neural network for object detection. In: Proceedings of the IEEE Computer Society Conference on Computer Vision and Pattern Recognition, pp. 4203–4212 (2018)
22. Zhu, C., He, Y., Savvides, M.: Feature selective anchor-free module for single-shot object detection (2019)

TA-CFNet: A New CFNet with Target Aware for Object Tracking

Jiejie Zhao and Yongzhao Zhan[✉]

School of Computer Science and Communication Engineering,
Jiangsu University, Zhenjiang, China
yzzhan@ujs.edu.cn

Abstract. A new network constructed by the combination of Siamese network and correlation filter has achieved enormous popularity because Siamese networks obtain high accuracy and correlation filters provide amazing speed. How to tackle boundary effects problems brought by filters, to fuse CF layers with multi-layers of CNN is an essential problem of the new network. Most papers deal with the problem by simply adding cosine window to every image. However, if the target is too small for bounding box to include background information or if target is too big for bounding box to loss partial information. In this paper, a new CFNet with target aware (TA-CFNet) for object tracking is proposed. TA-CFNet intergrates current target position and feature weight map to form target likelihood matrix. This target likelihood matrix is used to optimize and update the correlation filter, so that the template object of the deep tracking network is framed as accurately as possible. Experimental results on OTB benchmarks for visual tracking demonstrate that our proposed method outperforms other trackers in deep learning.

Keywords: Object-tracking · Target-likelihood matrix · Correlation filter · Siamese network

1 Introduction

Visual tracking aims to locate targets which are usually marked at the first frame, meanwhile it needs to record the identities and trajectories of target at every frame in a video [18]. The technology of visual tracking is widely applied in video surveillance, autonomous navigation, medical diagnosis and so on. But the task of visual tracking is challenging because it needs to solve many problems such as occlusion, deformation, background disorder and scale change.

In the past two decades, the technology of tracking has made great progress. The earliest tracking method was particle filter based [21]. Then methods about contour model [16], meanshift and camshift appeared. In particular, the method using deep learning has achieved satisfactory results. These works are aimed to train a two-category deep network and extract features by using already trained networks. At present, there are four main representative tracking methods based

© Springer Nature Switzerland AG 2019
Y. Zhao et al. (Eds.): ICIG 2019, LNCS 11901, pp. 618–630, 2019.
https://doi.org/10.1007/978-3-030-34120-6_50

on deep learning: tracking with stacked denoising autoencoder [24]; tracking with multi-domain convolutional neural networks [20]; tracking with fully convolutional Siamese networks [2] and tracking with deep regression networks [11]. These methods can achieve robust visual tracking by learning efficient features with deep learning, but most of them encounter speed bottlenecks.

Correlation filters can achieve high speed [12] by utilizing Fast Fourier Transform for calculation. MOSSE algorithm [15] first applied correlation filter to tracking with the speed of 600–700fps. Subsequently, many methods have been proposed to improve correlation filter such as C-COT [8] and ECO [5].

Nowadays, methods based on correlation filters and deep learning have become mainstream tracking methods. Method about tracking with fully convolutional Siamese networks always considers the first frame as template. Some researchers have suggested improving template precision which is very important in the process of tracking. One approach [23] wants to update template branch by introducing correlation filter which is less time consuming. However, introducing correlation filter to deep networks can bring negative effects. Since correlation filter needs to formulate circulation matrix which leads to boundary effects, restricting the target area.

Some networks based on correlation filters try to solve the problems of boundary effects. Discriminative Correlation Filter (DCF) is a representative method utilizing CF. Then many methods such as SRDCF [7], CSK [13], C-COT, BACF [9] proposes measures to settle boundary effects. These works either enlarge the search area which can't guarantee the target box exactly matching the target area or introduce cosine box to blur the value of the bounding box which may lead to the loss of target information.

To suppress boundary effects, our work adds the target likelihood matrix which is about relationship of target, background in the rectangle box and background in other area of the frame to the first branch of CFNet. We focus on target to ensure the integrity and accuracy of target in box when learning a filter.

End-to-end training is popular in the tracking community. We integrate feature extraction network and correlation filter with target likelihood matrix network which has no need to finish learning work independently. The entire network includes not only forward propagation but also back propagation. With Deep layers feedbacking mistakes to shallow layer, extraction features layer and correlation filter with target likelihood matrix layer are adjusted together.

Overall, the major contributions of our work are as follows:

(1) We introduce target likelihood matrix to correlation filter in deep network when training filter. This module is more specific consist of target position and feature weight map. It guides filter to give high weight to target while reducing the emphasis on backgrounds inside bounding box.
(2) After introducing the target likelihood matrix, our work derives the end-to-end closed solution of the template-correlation filter and establish its forward and backward propagation expressions throughout the whole network.

(3) Experimental results are obtained on typical tracking benchmark datasets such as OTB2013, OTB-50, OTB-100. Our method achieves a superior performance compared with some other trackers.

2   Related Work

The tracking community leads a fashion of deep learning for visual tracking. In this section, the most relevant tracking algorithms especially based on Siamese networks or correlation filters are presented.

Deep Trackers Based on Siamese Network. Recently, Siamese networks have obtained enormous popularity because of their simplicity and excellent performance. The most representative one is SiamFC [2] with two identical branches sharing parameters. Subsequently, many methods are introduced to Siamese networks which expect to improve both precision and efficiency, most of them modify the branches of Siamese network. Also, double branches cannot satisfy the requirement of their works. One method [9] fuses three branches into one branch and takes it as a template branch. In addition, method [17,29] divides every branch into two branches and fuse the result of all branches.

Trackers Based on Correlation Filter. Correlation filters are widely used owing to their combination of kernal skills, cycle structure, multi-channel and deep feature. CSK [13] adds cycle matrix to MOSSE which brings a huge boost in speed. KCF [12] adds multi-channel to CSK which improves both speed and accuracy. DCF, SRDCF [7] adds features extracted by CNN to correlation filter which improves the robustness of trackers. SRDCF adds spatial regular part that needs to be carefully adjusted parameters to achieve regularization weight. STC [28] adds context when learning filter bringing fast and robust effect. CCOT [8] integrates multi-resolution features in continuous domain. Many studies have considered methods to improve the performance of filters.

Siamese Network + Correlation Filters. Correlation filters have great advantages in efficiency because they can utilize circulating samples to accelerate calculation in frequency domain, therefore many papers [3,4,23] add correlation filters to Siamese networks. Most of them add correlation filter to the first branch of Siamese network intending to get reliable and expressive template.

However, it suffers from boundary effects. That boundary effects [11] produces erroneous samples that can make the ability of classifier to become weak and has a serious impact on tracking performance.

Generally, the problem of boundary effects gets more and more attention of trackers based on correlation filters. Many measures are taken to deal with the problem. For example, MOSSE filter suppresses boundary effects by cosine window multiplying with every pixel which makes the value of the bounding box smaller compared with center pixels. Then CFLB [10] expands target area and fills up edge area with zero, this process however needs constant iteration bringing massive calculation. Also, SRDCF adds spatial regularization which supresses the response of the background area. BACF [9] introduces the likelihood of foreground and background, taking awareness of background.

Most of papers combining Siamese network with correlation filters only simply process boundary effects by just introducing a cosine box. Recently, many methods take the background information into consideration. However, our work introduces target likelihood matrix, giving more attention to the target to instruct correlation filter to learn.

3   Methods

Before the detailed description of our work, we first review the details of CFNet (Sect. 3.1). Then we give on overview of the whole framework of our tracking method (TA-CFent) (Sect. 3.2), presenting target likelihood matrix (Sect. 3.3) and then formulating special correlation filter (Sect. 3.4). In Sect. 3.5, we introduce the process of training and tracking.

3.1   Siamese Network with Correlation Filter

Siamese network includes two branches corresponding to inputs, it extracts features from inputs simultaneously by CNN and obtains a score map according to the cross-correlation operation of two branches results. Correlation filter is added to the first branch of Siamese network to make current template combine with the previous template to ensure the accuracy of the template during the tracking process. The architecture can be described as Eq. 1:

$$h_{\rho,s,b}(x^{'}, z^{'}) = sw(f_\rho(x^{'})) \star f_\rho(z^{'}) + b \tag{1}$$

where function f represents feature extraction, function w represents CF module. ρ, s, b are parameters learned during training process. ρ controls the extraction of CNN. s, b make the score map more suitable for logistic regression.

CFNet constructs a correlation filter and combines with quadratic regularization to find the optimal w. The whole solution process is obtained by back-propagation in the Fourier domain.

3.2   Our Overall Network Framework

Our overall network framework is shown in Fig. 1. We introduce target likelihood module into CFNet. The two parallel branches with respective inputs (training and test image), next connected with extract-feature module. This module can contain two or even more convolutional layers and share parameters with two branches. It is however not correct that the more the number of convolutional layers is, the high the accuracy of tracking is. The CF layer is connected after the extract-feature module. At last, score map is obtained by taking the result of training branch as convolution kernel and having a convolution operation on the result of test branch. The target is located at the area with the highest score in the score map. Features extracted from last frame and target location calculated from score map are simultaneously used to calculate target likelihood matrix which is fed into correlation filter to train and update the optimal filter.

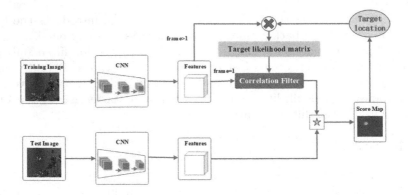

Fig. 1. Our overall network framework (TA-CFNet).

3.3   Target Likelihood Matrix

The purpose of target likelihood matrix is to emphasize on target in bounding box by giving high weights to them and low weights to background pixels in bounding box and zero weights to pixels outside bounding box. This work is more specific compared with other works [7,9] which only regard bounding box as a whole. Both location and value of every pixel is taken into consideration by target likelihood matrix. Therefore, the solution to target likelihood matrix is comprised of serval steps: calculate weight map of the whole frame, calculate spatial prior map related to location on the basis of bounding box and lastly consolidate two maps. The entire implementation process is shown in Fig. 2.

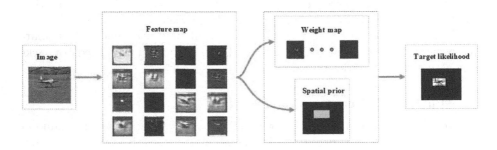

Fig. 2. The whole process of obtaining target likelihood map. Impose spatial prior map on weight map of feature map.

(1) Calculate weight map for every feature map as Eq. 2:

$$\beta_{ijk} = \frac{\ln(1 + \exp(x_{ijk})) + \varepsilon}{\sum\limits_{i} \sum\limits_{j} (\ln(1 + \exp(x_{ijk})) + \varepsilon)} \tag{2}$$

where k feature maps with the size $i \times j$, ε is quite small to guarantee the establishment of the formula, β_{ij} denotes the weight of different area in frame.

(2) Calculate spatial prior map

As is shown in Fig. 2, in different feature maps, target area is not always the highest value, so not every feature map focus on target, hence spatial prior map is introduced. The spatial prior map β_{f_b} is relatively simple only with 0 and 1 padding, within the target bounding box which is mapped to the feature space is 1 and the rest of the area is 0.

(3) Calculate target likelihood matrix as:

$$\beta = \beta_{f\_b} \times \beta_{ijk} \tag{3}$$

The target likelihood matrix is composed of the weight map and the spatial prior map, target has high values while background has lower values. When training a filter, areas with 0 value help reduce calculations and filter focuses on target which effectively mitigates boundary effects.

We selected four different categories of video to visualize the results of target likelihood. Figure 3 visually shows the contrast between the original image and the target likelihood.

Fig. 3. Visualization of the original and the target likelihood.

3.4 Formulation for Correlation Filter with Target Likelihood Matrix

Correlation filter is responsible for template updating by combing a new template in each frame of video with the previous template in a moving average. The optimal w is the value when the inner product between feature map x of input image and it is closest to the reponse value y which denotes the signal of every domain of a frame. This work of finding the optimal w can be solved by defining an optimization problem combined with secondary regularization. The construction of formulas is as Eq. 4:

$$\arg\min_{w} \frac{1}{2n} \left\| (\beta w) \star x - y \right\|^2 + \frac{\lambda}{2} \left\| w \right\|^2 \tag{4}$$

where $x \in R^{m \times m}$, signal $y \in R^{m \times m}$, correlation filter $w \in R^{m \times m}$ is required to be solved in this work, $\beta \in R^{m \times m}$ is target likelihood matrix appear in Sect. 3.3, n denotes the number of samples, λ is regularization parameter.

Both convolution operation $(*)$ and cross-correlation $(\star)$ operation appear in our work. The difference between two operations is whether to rotate the convolution kernel 180° or not. And the result of them makes a difference.

The solution of w is to formulate a Lagrangian expression as Eq. 5. Here $r = X^T \beta w - y$.

$$L(w, r, v) = \frac{1}{2n} \left\| r \right\|^2 + \frac{\lambda}{2} \left\| w \right\|^2 + v^T (r - X^T \beta w + y) \tag{5}$$

After the operation of partial derivatives and equations, w must satisfy a set of equations as follows:

$$\begin{cases} k = x \star x + n \frac{\lambda}{\beta \beta^T} \delta \\ k * \alpha = \frac{1}{\beta \beta^T} y \\ w = \alpha \star x \end{cases} \tag{6}$$

where k and α both are introduced to better express w. k is a regularized kernel matrix and α is a scaled double variable.

To accelerate the speed of calculation, this work puts the above equations into the frequency domain. The cross-correlation function of x and x in frequency domain is equal to the conjugate of the signal x in the frequency domain multiplied by the signal x in the frequency domain. Therefore, equations above can be transformed as follows:

$$\begin{cases} \hat{k} = (\hat{x}^* \circ \hat{x}) + \frac{\lambda n}{\beta \beta^T} I \\ \hat{\alpha} = \frac{1}{\beta \beta^T} \hat{k}^{-1} \circ \hat{y} \\ \hat{w} = \hat{\alpha}^* \circ \hat{x} \end{cases} \tag{7}$$

where $\hat{\ }$ denotes Fourier transform signals, $*$ denotes the conjugate of signals, $\circ$ denotes Element multiplication which simplifies operation beneficial to speed.

Because this work considers correlation filter as a separate layer of network, it refers to the basis of neural networks forward-back propagation. In order to train a better network, we need to derive back propagation expressions. First, transform the equals of the solution of w into differential equations as the left of Eq. 8. Similar to Eq. 7, this work puts the process of back-propagation into the Fourier domain to accelerate speed as the right of Eq. 8.

$$\begin{cases} dk = (dx \star x + x \star dx) \\ dk * \alpha + k * d\alpha = \frac{1}{\beta \beta^T} dy \\ dw = d\alpha \star x + \alpha \star dx \end{cases} \rightarrow \begin{cases} d\hat{k} = (d\hat{x}^* \circ \hat{x} + \hat{x}^* \circ d\hat{x}) \\ d\hat{\alpha} = \hat{k}^{-1} \circ (\frac{1}{\beta \beta^T} d\hat{y} - d\hat{k} \circ \hat{\alpha}) \\ d\hat{w} = d\hat{\alpha}^* \circ d\hat{x} \end{cases} \tag{8}$$

Last, with the map of $dx \rightarrow dk$; $dk, dy \rightarrow d\alpha$; $dx, d\alpha \rightarrow dw$, we can calculate their inner product. Back propagation expression of w is derivated as:

$$\begin{cases} \nabla_\alpha \hat{l} = \hat{x} \circ \nabla_w \hat{l}^* \\ \nabla_y \hat{l} = \frac{1}{\beta \hat{\beta}^T} \hat{k}^{-*} \circ \nabla_\alpha \hat{l} \\ \nabla_k \hat{l} = -\hat{k}^{-*} \circ \hat{\alpha}^* \circ \nabla_\alpha \hat{l} \\ \nabla_x \hat{l} = 2\hat{x} \circ \mathrm{Re} \left\{ \nabla_k \hat{l} \right\} + \hat{\alpha}^* \circ \nabla_w \hat{l} \end{cases} \quad (9)$$

3.5 Training and Tracking

Training. During training process, we construct a network with two branches including adding CNN layers, CF layer, crop layer and corr layer after loading metadata about datasets. Every layer has own function and forward and back propagation. Then loss function is set on the response map by adding loss layer and weighting. Last, we directly call the cnn_train_dag API to start training datasets. During the training process, each iteration generates a network model structure and saves the corresponding parameters.

Tracking. The overall algorithm of tracking is summarized in Algorithm 1. We call two branches of TACFNet as TIB (template image branch) and SIB (search image branch). I denotes the number of all frames of a video.

Algorithm 1. CFNet with Target Aware (TA−CFNet) tracking algorithm
1   Import model, parameter, test video
2   Obtain first frame
3   Initialize TIB, SIB
4   **funtion** TACFNet_tracker
5   The first frame as template is forwarded through TIB
5   **for**$(i = 2; i \leq I; i ++)$
6   sample the ith frame in SIB according to $(i-1)$th frame
7             $\hat{x}'$: the i frame in SIB is forwarded
8   response map = template $\star$ $\hat{x}'$
9   calculate location from response map
10   resample the ith frame in TIB and forward propagation
11   obtain β_{f_b}, β_{ijk}
12   obtain target likelihood matrix β as Eq. 3
13   update template w
14   count score and speed
15   **end function**

4 Experimental Results

In this section, we use experiments to validate the performance of TA-CFNet. Implementation details and results are also presented. We evaluate the proposed tracker on OTB benchmarks [25,26] including OTB-2013, OTB-50, OTB-100. At last, we give a detailed analysis of the OTB-100.

4.1 Data Sets

Train Set: ILSVR2015 [22] which contains 4417 sequences is selected as our training set. It contains 30 basic classes that takes occlusion, deformation, average target numbers and many more into account. Before training, we divide the set into five folders and crop each frame according to ground truth. When training the network, network files are saved with repeated iterations finished.

Test Set: All the experiments are running on the challenging tracking datasets: OTB-2013, OTB-50, OTB-100. The sequences OTB benchmarks include are with a wide variety of tracking challenging. The AUC (area under curve) is mainly considered to rank tracking algorithms.

All of OTB benchmark also adopt two standard metrics: precision plot and success plot. The precision plot measures the percentage of frames whose CLE (Center Location Error; a distance between the center of tracked target and ground truth) is within 20 pixels threshold. The success plot can be obtained by evaluating the success rate at different IoU (intersection over union; a distance between the tracked and ground truth bounding boxes) thresholds We obtain all experimental data by the standard OTB toolkit.

Implementation Details: 100 different epoches are used to achieve stable and excellent network performance. The model performs best when the number of epoch is 80. During online tracking, we set three scales to 1.0410, 0.9805, 0.615. And the learning rate of template is 8e-3.

4.2 Overall Performance

In this subsection, we compare proposed tracker TA-CFNet with correlation filter and Siamese network based trackers on the OTB-2013, OTB-50 and OTB-100 datasets. The compared trackers include SRDCF [7], Staple [1], CFNet [23], SiamFC [2], DSST [6], SAMF [27], LCT [19], CNN-SVM [14]. In the following, we present and analyse the results.

Table 1. Comparison on different kinds trackers by OTB benchmark.

Tracker	OTB-2013		OTB-50		OTB-100	
	AUC	Prec	AUC	Prec	AUC	Prec
SRDCF	0.626	0.838	0.539	0.732	0.598	0.789
Staple	0.6	0.793	0.507	0.784	0.578	0.784
SiamFC	0.608	0.809	0.516	0.692	0.582	0.771
CFNet	0.610	0.822	0.538	0.723	0.582	0.776
TA-CFNet	**0.633**	**0.849**	**0.552**	**0.749**	**0.605**	**0.798**

Table 1 shows the AUC score and precision plot on OTB benchmark. In the table, we choose two kinds trackers, SRDCF and Staple based on correlation filters while SiamFC and CFNet based on double branch network. The proposed tracker TA-CFNet achieves the best performance on OTB-2013(AUC score: 0.633), OTB-50(AUC score: 0.552) and OTB-100(AUC score: 0.605). TA-CFNet is the perfection of the CFNet, the result is improved compared to CFNet which indicates that our work is meaningful.

Also, our work presents a comparison with 8(SRDCF [7], Staple [1], CFNet [23], SiamFC [2], DSST [6], SAMF [27], LCT [19], CNN-SVM [14]) different trackers of our proposed tracker. Although the proposed tracker TA-CFNet wins second place in precision plots of OPE of OTB2013, OTB-50 and OTB-100, second only to CNN-SVM. It obtains the best performance in success plots of OPE better than SVM-CNN. On the whole, our tracker is better than other trackers. This is because success plots are preferred over the precision plots, since precision only uses the bounding box locations, and ignores the size or overlap.

4.3   Analysis of 11 Attributes on OTB-100 in Different Sceneries

In this section, we analyze the proposed method on the OTB-100 datasets. Since OTB-100 contains more videos than OTB-2013, OTB-50. TA-CFNet achieves an improvement of 2.3% in success plot and 2.2% in precision plot of OTB-100 compared with CFNet. This is because target likelihood exploits semantic information of target to deal with variations and scale changes. But speed is 11 fps lower than CFNet on GPU.

Table 2. Comparison of 11 attributes success plot on OTB-100.

A T	Staple	LCT	SAMF	DSST	SiamFC	CFNet	CNN-SVM	SRDCF	TA-CFNet
IV	0.576	0.538	0.523	0.530	0.566	0.519	0.532	0.599	**0.612**
SV	0.520	0.494	0.488	0.477	0.556	0.550	0.552	0.527	**0.558**
OCC	**0.562**	0.503	0.535	0.471	0.527	0.509	0.539	0.550	0.545
DEF	**0.577**	0.507	0.498	0.443	0.510	0.500	0.572	0.531	0.547
MB	0.534	0.519	0.525	0.465	0.544	0.555	0.541	**0.594**	0.579
FM	0.550	0.560	0.514	0.468	0.570	0.576	0.542	0.601	**0.604**
IPR	0.547	0.545	0.547	0.499	0.564	0.559	0.506	0.544	**0.573**
OPR	0.556	0.537	0.549	0.489	0.545	0.532	0.486	0.544	**0.567**
OV	0.468	0.429	0.480	0.383	0.500	0.483	0.464	0.460	**0.516**
BC	0.554	0.536	0.525	0.517	0.520	0.554	0.545	0.583	**0.602**
IR	0.396	0.399	0.471	0.382	**0.609**	0.525	0.362	0.514	0.565

Also, we analyze 11 different attributes of OTB-100. These attributes include   illumination   variation(IV),   scale   variation(SV),   occlusion(OCC), deformation(DEF), motion blur(MB),fast motion(FM), in-plane rotation(IPR), out-of-plane rotation(OPE), out-of-view(OV), background clutter(BC) and low resolution(IR). As shown in Table 2, TA-CFNet achieves the best results in 7 of 11 attributes, which can be attributed to the target likelihood matrix for correlation filter, demonstrating the robustness of proposed tracker under challenging scenarios.

5 Conclusion

In this work, we propose an end-to-end target-aware for correlation filter framework (TA-CFNet) for visual tracking. We incorporate target likelihood matrix with correlation filter to obtain more precise information of template during the process of updating the template each time. A optimization procedure is designed on correlation filter and derives expressions for forward and back propagation. Experiments on OTB-2013, OTB-50, OTB-100 show that our approach obtains favorable performance. Future work will focus on the improvement of search image branch by attention mechanism and add other data sets for verification.

Acknowledgments. This work was supported by the National Natural Science Foundation of China (No. 61672268) and the Primary Research & Development Plan of Jiangsu Province (No. BE2015137).

References

1. Bertinetto, L., Valmadre, J., Golodetz, S., Miksik, O., Torr, P.H.S.: Staple: complementary learners for real-time tracking. In Computer Vision & Pattern Recognition (2016)
2. Bertinetto, L., Valmadre, J., Henriques, J.F., Vedaldi, A., Torr, P.H.S.: Fully-convolutional siamese networks for object tracking. In: Hua, G., Jégou, H. (eds.) ECCV 2016. LNCS, vol. 9914, pp. 850–865. Springer, Cham (2016). https://doi.org/10.1007/978-3-319-48881-3_56
3. Choi, J., Jin Chang, H., Jeong, J., Demiris, Y., Young Choi, J.: Visual tracking using attention-modulated disintegration and integration. In: Computer Vision and Pattern Recognition (2016)
4. Choi, J., Jin Chang, H., Yun, S., Fischer, T., Young Choi, J.: Attentional correlation filter network for adaptive visual tracking. In: IEEE Conference on Computer Vision and Pattern Recognition (2017)
5. Danelljan, M., Bhat, G., Shahbaz Khan, F., Felsberg, M.: Eco: efficient convolution operators for tracking. In: Proceedings of the IEEE Conference on Computer Vision and Pattern Recognition, pp. 6638–6646 (2017)
6. Danelljan, M., Häger, G., Khan, F., Felsberg, M.: Accurate scale estimation for robust visual tracking. In British Machine Vision Conference, Nottingham, 1–5 September 2014. BMVA Press (2014)
7. Danelljan, M., Häger, G., Shahbaz Khan, F., Felsberg, M.: Learning spatially regularized correlation filters for visual tracking (2016)
8. Danelljan, M., Robinson, A., Shahbaz Khan, F., Felsberg, M.: Beyond correlation filters: learning continuous convolution operators for visual tracking. In: Leibe, B., Matas, J., Sebe, N., Welling, M. (eds.) ECCV 2016. LNCS, vol. 9909, pp. 472–488. Springer, Cham (2016). https://doi.org/10.1007/978-3-319-46454-1_29

9. Kiani Galoogahi, H., Fagg, A., Lucey, S.: Learning background-aware correlation filters for visual tracking (2017)
10. Kiani Galoogahi, H., Sim, T., Lucey, S.: Correlation filters with limited boundaries. In: Computer Vision & Pattern Recognition (2015)
11. Held, D., Thrun, S., Savarese, S.: Learning to track at 100 FPS with deep regression networks. In: Leibe, B., Matas, J., Sebe, N., Welling, M. (eds.) ECCV 2016. LNCS, vol. 9905, pp. 749–765. Springer, Cham (2016). https://doi.org/10.1007/978-3-319-46448-0_45
12. Henriques, J.F., Caseiro, R., Martins, P., Batista, J.: High-speed tracking with kernelized correlation filters. IEEE Trans. Pattern Anal. Mach. Intell. **37**(3), 583–596 (2015)
13. Henriques, J.F., Caseiro, R., Martins, P., Batista, J.: Exploiting the circulant structure of tracking-by-detection with kernels. In: Fitzgibbon, A., Lazebnik, S., Perona, P., Sato, Y., Schmid, C. (eds.) ECCV 2012. LNCS, vol. 7575, pp. 702–715. Springer, Heidelberg (2012). https://doi.org/10.1007/978-3-642-33765-9_50
14. Hong, S., You, T., Kwak, S., Han, B.: Online tracking by learning discriminative saliency map with convolutional neural network. In: International Conference on International Conference on Machine Learning (2015)
15. Hsu, L.C., Chen, H.M.: On optimizing scan testing power and routing cost in scan chain design. In: International Symposium on Quality Electronic Design (2006)
16. Kass, M., Witkin, A., Terzopoulos, D.: Snakes: active contour models. Int. J. Comput. Vis. **1**(4), 321–331 (1988)
17. Li, B., Yan, J., Wu, W., Zhu, Z., Hu, X.: High performance visual tracking with Siamese region proposal network. In: Proceedings of the IEEE Conference on Computer Vision and Pattern Recognition, pp. 8971–8980 (2018)
18. Luo, W.: Multiple object tracking: a literature review. arXiv preprint arXiv:1409.7618 (2014)
19. Ma, C., Yang, X., Zhang, C., Yang, M.H.: Long-term correlation tracking. In: Computer Vision & Pattern Recognition (2015)
20. Nam, H., Han, B.: Learning multi-domain convolutional neural networks for visual tracking. In: Proceedings of the IEEE Conference on Computer Vision and Pattern Recognition, pp. 4293–4302 (2016)
21. Nummiaro, K., Koller-Meier, E., Van Gool, L.: Object tracking with an adaptive color-based particle filter. In: Van Gool, L. (ed.) DAGM 2002. LNCS, vol. 2449, pp. 353–360. Springer, Heidelberg (2002). https://doi.org/10.1007/3-540-45783-6_43
22. Russakovsky, O., et al.: Imagenet large scale visual recognition challenge. Int. J. Comput. Vis. **115**(3), 211–252 (2014)
23. Valmadre, J., Bertinetto, L., Henriques, J., Vedaldi, A., Torr, P.H.S.: End-to-end representation learning for correlation filter based tracking (2017)
24. Wang, N., Yeung, D.-Y.: Learning a deep compact image representation for visual tracking. In: Advances in Neural Information Processing Systems, pp. 809–817 (2013)
25. Wu, Y., Lim, J., Yang, M.-H.: Online object tracking: a benchmark. In: Proceedings of the IEEE Conference on Computer Vision and Pattern Recognition, pp. 2411–2418 (2013)
26. Wu, Y., Lim, J., Yang, M.H.: Object tracking benchmark. IEEE Trans. Pattern Anal. Mach. Intell. **37**(9), 1834–1848 (2015)

27. Li, Y., Zhu, J.: A scale adaptive kernel correlation filter tracker with feature integration. In: Agapito, L., Bronstein, M.M., Rother, C. (eds.) ECCV 2014. LNCS, vol. 8926, pp. 254–265. Springer, Cham (2015). https://doi.org/10.1007/978-3-319-16181-5_18

28. Zhang, K., Zhang, L., Liu, Q., Zhang, D., Yang, M.-H.: Fast visual tracking via dense spatio-temporal context learning. In: Fleet, D., Pajdla, T., Schiele, B., Tuytelaars, T. (eds.) ECCV 2014. LNCS, vol. 8693, pp. 127–141. Springer, Cham (2014). https://doi.org/10.1007/978-3-319-10602-1_9

29. Zheng, Z., Wei, W., Wei, Z., Yan, J.: End-to-end flow correlation tracking with spatial-temporal attention (2017)

Multi-view Similarity Learning
of Manifold Data

Rui-rui Wang[1], Si-bao Chen[1,2(✉)], Bin Luo[1], and Jian Zhang[2]

[1] Key Lab of Intelligent Computing and Signal Processing of Ministry of Education,
School of Computer Science and Technology, Anhui University, Hefei 230601, China
sbchen@ahu.edu.cn
[2] Peking University Shenzhen Institute, Shenzhen 518057, China

Abstract. In recent years, multi-view learning methods have developed rapidly where graph-based approaches have achieved good performance. Usually, these learning methods construct information graph for each view or fuse different views into one graph. In this paper, a novel multi-view learning model that learns one similarity matrix for all views named Multi-view Similarity Learning (MSL) is proposed, where adaptive weights are learned for each view. The multi-view similarity learning method is further extended to kernel space. Experiments of classification, clustering and semi-supervised classification on different real-world datasets show the effectiveness of the proposed method.

Keywords: Laplacian Eigenmaps · Multi-view learning · Similarity learning

1 Introduction

Usually, in the field of machine learning and pattern recognition, the dimension of data is very high, so how to map high-dimensional data to low-dimensional data and preserve the topology of data is an important issue we are studying now. Then, some classical linear dimensionality reduction methods are proposed, such as Principle Component Analysis (PCA) [5], Linear Discriminant Analysis (LDA) [13], Multidimensional Scaling (MDS) [4]. But the linear dimension reduction method can not represent the manifold structure of data well, so in recent years, many nonlinear dimensionality reduction methods are proposed, such as isometric feature mapping (ISOMAP) [14], local linear embedding (LLE) [8], and Laplacian Eigenmaps (LE) [1]. Laplacian Eigenmaps is a nonlinear dimensionality reduction method for manifold data learning, it can well preserve the nonlinear structure of data space.

This work was supported in part by NSFC Key Projects of International (Regional) Cooperation and Exchanges under Grant 61860206004 and in part by Shenzhen Science & Research Project under Grant JCYJ20170817155854115.

However, in many real world applications, the representation of actual data is not a single form, but can have many forms of expression, such as person's different angles of faces, which can collect different information from different angles. Usually, for each thing, we observe things from different angles that cannot be observed from another angle, so in order to understand more comprehensive information, we need to observe the problem from multiple angles. Nowadays, we can get different feature from one dataset, we think that compared with single feature, multi-view features can get better results. So for classification, clustering and semi-supervised, we think compared with the single-view similarity learning algorithm, multi-view similarity learning can have the higher accuracy.

Today, in many areas of science, such as pattern recognition, computer vision, genetics, and data mining, we can more easily get data that contain heterogeneous features from samples from different perspectives. In visual data, images can be represented by different descriptors. For example, gray features, gabor features [10], local binary patterns (LBP) features [12], and so on. In image processing, the gabor function is a linear filter for edge extraction. Local binary patterns (LBP) were first proposed as an effective texture description operator, and have been widely used due to their excellent rendering ability for image local texture features. LBP features have significant advantages such as gray invariance and rotation invariance. Because LBP features are simple to calculate and have good effects, LBP features have been widely used in many fields of computer vision.

One of the methods to solve the multi-view problem is to connect vectors from different perspectives into a vector and then on the cascaded vector, directly apply the single view clustering algorithm. However, this connection results in overfitting on small samples and has no meaning to the multi-view problem. Our solution to multi-view is to learn a similarity matrix by adding weights to each view. The weights can be updated through each iteration. Our method can be used for classification, clustering and semi-supervised classification.

The k-nearest neighbor (k-NN) [7] classification algorithm is one of the simplest machine learning algorithms. And k is a very important parameter in the k-NN algorithm. The selection of the k value will affect the classification result of the sample to be classified. However, it is hard to choose a proper neighbor number k beforehand, because if the value of k is too small, the model is easily interfered by noisy data, and if k is too large, the prediction ability of the model is greatly weakened. Generally we use cross-validation to select an appropriate k value.

2   Learning Multi-view Similarity in Laplacian Eigenmaps

In this section, we first introduce a nonlinear dimensionality reduction method LE. And then simply illustrate our multi-view similarity learning algorithms.

2.1 Laplacian Eigenmaps

For the success of the graph-based approach, preserving local manifold structures is an important factor. Laplacian Eigenmaps (LE) is a graph-based dimensionality reduction algorithm and it constructs the relationship between data uses a local angle.

Given a set of data points $\{X_1, X_2, \cdots, X_n\}$, denote data matrix $X \in \mathcal{R}^{n \times p}$, where n is the number of data points and p is the dimension of features, LE pursues their low dimensional representation $Y_1, Y_2, \cdots, Y_n \in \mathcal{R}^q (q < p)$, which constructs a weighted graph with n points as nodes, and a set of weighted edges connecting neighboring points. If the two data instances i and j are very similar, then i and j should be as close as possible after dimensionality reduction.

The steps are as follows:

1. **Constructing the Adjacency Graph:** Construct neighborhood graph $\mathcal{G}$ through k-nearest neighbors algorithm, k is a preset value. Given n data points $\{X_1, X_2, \cdots, X_n\}$. Nodes i and j are connected if X_i is among k nearest neighbors of X_j or X_j is among k nearest neighbors of X_i.
2. **Choosing the weights:** Choose edge weights using heat kernel or simply set edge weight to be 1 if connected and 0 otherwise, or we can get similarity matrix S with heat kernel by:

$$S_{ij} = \exp\left\{-\frac{d_{ij}^2}{2r}\right\} \tag{1}$$

where $d_{ij} = \|X_i - X_j\|$, $r > 0$ is a suitable constant.
3. **Eigenmaps:** Calculate the eigenvectors and eigenvalues of the Laplacian matrix L by:

$$Lv = \lambda Dv \tag{2}$$

where D is diagonal matrix and its entries are row sums of S, $D_{ii} = \sum_j S_{ij}$, Laplacian matrix $L = D - S$. We omitting the eigenvector v_0 and use the next q eigenvectors for embedding in q-dimensional Euclidean space: $X_i \mapsto Y_i = (v_1(i), v_2(i), \cdots, v_q(i))^\top$.

2.2 Learning New Multi-view Similarity

For multi-view data, the representation $X^1, X^2, \cdots, X^m$ is the data matrix for each view. $X^v \in R^{n \times p^v} (v = 1, 2, \cdots, m)$, where n is the number of data and p^v is the feature dimension of the v-th view. For graph-based methods, each view can build a similar graph and maximize performance by itself. The similarity between two data points in 1 does not reflect the local popular structure of manifold data, so we add a locally linear reconstruction to sample point by its neighbor points. Then for each view, we propose an effective method is to combine these views with the appropriate weights $w_v(v = 1, 2, \cdots, m)$, so our objective function can be written as

$$\min_{S,w} \sum_{v=1}^{m} w_v^2 \|X^v - X^v S\|_F^2 \tag{3}$$
$$s.t. \ S_{ij} = S_{ji} \geq 0, S_{ii} = 0$$

If the distance between sample points are larger, the corresponding reconstruction weight is smaller, and vice versa. We limited reconstruction weight S_{ij} is non-negative, and reconstruction weight is symmetry. Therefore, between sample points we add linear reconstruction constraints to learn the new similarity, so the objective function can be written as

$$\min_{S,w} \sum_{v=1}^{m} w_v^2 \|X^v - X^v S\|_F^2 + \alpha_v \|S - S_0^v\|_F^2 \tag{4}$$
$$s.t. \ S_{ij} = S_{ji} \geq 0, S_{ii} = 0$$

The L_1 paradigm can produce relatively sparse solutions, and has the ability to select features. It is useful when solving high-dimensional feature space. Then the minimization of the final objective function of our learning new similarity turns into

$$\min_{S,w} \sum_{v=1}^{m} (w_v^2 \|X^v - X^v S\|_F^2 + \alpha_v \|S - S_0^v\|_F^2) + \beta \|S\|_1 \tag{5}$$
$$s.t. \ S_{ij} = S_{ji} \geq 0, \ S_{ii} = 0$$

2.3   Learning Weight for Each View

Where each view shares the same similarity matrix, so we can get a more accurate similarity matrix by adding appropriate weights to each view. We want the distance between the data points in the same class to be as small as possible, so the objective function of weight can be written as

$$\min \sum_{v=1}^{m} w_v^2 d_v \tag{6}$$
$$s.t. \ \Sigma_{v=1}^{m} w_v = 1.$$

The Lagrange function of Eq. (6) can be written as

$$\min \sum_{v=1}^{m} w_v^2 d_v - \lambda \left(\sum_{v=1}^{m} w_v - 1 \right). \tag{7}$$

Taking the derivative of Eq. (7) and setting the derivative to zero, we get the iterative formula for w_v is

$$w_v = \frac{(d_v)^{-1}}{\sum_{v=1}^{m} (d_v)^{-1}} \tag{8}$$

where $d_{ij} = \|X_i - X_j\|$.

3   Algorithms and Analyses

In this section, we respectively analyze the multi-view similarity learning algorithm for mix-signed data and non-negative data.

Algorithm 1. Algorithm of Learning Multi-view Similarity for Mix-Signed Data

---

Input: $X = \{X^1, X^2, \cdots, X^m\}$, $X^v \in R^{n \times p^v}$, $v \in \{1, 2, \cdots, m\}$, positive tuning parameter α and β.

Output: An n-by-n similarity matrix S among n training samples.

Initial The weight for each view, $w_v = \frac{1}{v}$, for compute initial similarity matrix S_0^v, with its elements being heat kernels $S_0^v{}_{ij} = exp\{-d_v^2{}_{ij}/2r\}$.

Set $S_{ij}^{(0)} = 1$, $S_{ii}^{(0)} = 0 (i, j = 1, 2, \cdots, n)$, $t = 0$.

repeat

For each $i, j = 1, 2, \cdots, n$, update $S_{ij}^{(t+1)}$ as in (9).

Update weight by $w_v = \frac{(d_v)^{-1}}{\sum_{v=1}^{m}(d_v)^{-1}}$.

$t = t + 1$.

until $t >$ MaxIterNum.

return $S^{(t)}$.

3.1   Algorithm for Mix-Signed Data

When the elements of the cell array X are mixed with symbols (some are positive and some are negative), we learn the similarity matrix S by the Eq. (9) iterative update formula. Algorithm 1 summarizes the overall similarity learning iterative update algorithm for mix-signed data. In this algorithm, we set $Q = \sum_{v=1}^{m}(w_v^2 X^{v\top} X^v)$. And in addition to the input cell array X, there are two tuning parameters α and β. In practice, we found that our algorithm is robust to both parameters α and β, so in all experiments in this paper, we only set $\alpha = \beta = 1$.

Theorem 1. *The objective function in Eq. (5) monotonically decreases (ie, does not increase) under the update rule Eq. (9) of Algorithm 1.*

$$S_{ij}^{(t+1)} = S_{ij}^{(t)} \sqrt{\frac{[S^{(t)}Q^- + Q^- S^{(t)}]_{ij} + 2[Q^+ + \alpha_v S^v)]_{ij}}{[S^{(t)}Q^+ + Q^+ S^{(t)} + 2\sum_{v=1}^{m} \alpha_v S^{(t)}]_{ij} + 2Q_{ij}^- + \beta)}} \qquad (9)$$

For proof of Theorem 1, refer to article [2].

3.2   Algorithm for Nonnegative Data

For nonnegative data, we propose a more efficient multi-view similarity learning algorithm to learn the similarity matrix S as in Eq. (10), and we also set $Q = \sum_{v=1}^{m}(w_v^2 X^{v\top} X^v)$. We summarize the multi-view similarity algorithm for nonnegative data in Algorithm 2. And we only set $\alpha = \beta = 1$ in all the experiments of this paper.

Algorithm 2. Algorithm of Learning Multi-view Similarity for Nonegative Data

Input: $X = \{X^1, X^2, \cdots, X^m\}$, $X^v \in R^{n \times p^v}$, $v \in \{1, 2, \cdots, m\}$, positive tuning parameter α and β.
Output: An n-by-n similarity matrix S among n training samples.
Initial The weight for each view, $w_v = \frac{1}{v}$, for compute initial similarity matrix S_0^v, with its elements being heat kernels $S_0^v{}_{ij} = exp\{-d_v^2{}_{ij}/2r\}$.
Set $S_{ij}^{(0)} = 1$, $S_{ii}^{(0)} = 0 (i, j = 1, 2, \cdots, n)$, $t = 0$.
repeat
    For each $i, j = 1, 2, \cdots, n$, update $S_{ij}^{(t+1)}$ as in (10).
    Update weight by $w_v = \frac{(d_v)^{-1}}{\sum_{v=1}^{m}(d_v)^{-1}}$.
    $t = t + 1$.
until $t >$ MaxIterNum.
    return $S^{(t)}$.

Theorem 2. *For nonnegative data, the objective function in Eq. (5) decreases monotonically (i.e. it is non-increasing) under the update rule Eq. (10) in Algorithm 2.*

$$S_{ij}^{(t+1)} = S_{ij}^{(t)} \sqrt{\frac{2[Q + \alpha_v S^v)]_{ij}}{[S^{(t)}Q + QS^{(t)} + 2\sum_{v=1}^{m}\alpha_v S^{(t)}]_{ij} + \beta)}} \tag{10}$$

For proof of Theorem 2, refer to article [2].

4   Learning Multi-view Similarity in Kernel Spaces

The role of the kernel function is to imply a mapping from low-dimensional space to high-dimensional space $\mathcal{F}$, then in $\mathcal{F}$ space, we learn new similarity for data.

For each view, we use a nonlinear map: $\phi : \mathcal{R}^{p^v} \rightarrow \mathcal{F}^v$, $X_i^v \rightarrow \phi(X_i)^v$, the mapped data matrix is $\phi(X)^v = [\phi(X_1)^v, \phi(X_2)^v, \cdots, \phi(X_n)^v]$. So the minimization objective function becomes

$$\min_{S,w} \sum_{v=1}^{m}(w_v^2\|\phi(X)^v - \phi(X)^v S\|_F^2 + \alpha_v\|S - S_0^v\|_F^2) + \beta\|S\|_1 \tag{11}$$

In this paper, four kernel functions are mainly used, include linear kernel, gaussian kernel, cosine kernel and polynomial kernel. In implementation, the mapping ϕ does not need to be computed explicitly. By choosing a proper kernel function k, The ϕ mapping and $\mathcal{F}$ space can determined implicitly by the dot product between two mapped data samples $\phi(X_i)^v$ and $\phi(X_j)^v$ in $\mathcal{F}$ space by

$$k(X_i^v, X_j^v) = (\phi(X_i)^v \cdot \phi(X_j)^v) \tag{12}$$

Note that most of kernel functions are nonnegative, such as gaussian kernel and cosine kernel. So we replacing $\phi(X)^{v\top}\phi(X)^v$ with kernel matrix K^v in the iterative updating (10) for non-negative data, then we can get

$$S_{ij}^{(t+1)} = S_{ij}^{(t)} \frac{2[\sum_{v=1}^{m}((w_v^2 K^v) + \alpha_v S_0^v)]_{ij}}{[S^{(t)} \sum_{v=1}^{m}(w_v^2 K^v) + \sum_{v=1}^{m}(w_v^2 K^v)S_0^v + 2\sum_{v=1}^{m}\alpha_v S_0^v]_{ij} + \beta)} \tag{13}$$

We replacing (10) in Algorithm 2 for nonnegative data with (13), we can obtain the multi-view similarity learning algorithm in kernel space.

5   Experiments

In this section, we first introduce the data set we used. And then we will perform the proposed method on many benchmark datasets and compare it with other related graph-based multi-view learning methods. In the following experiments, we will learn new multi-view similarity matrix with Algorithm 2.

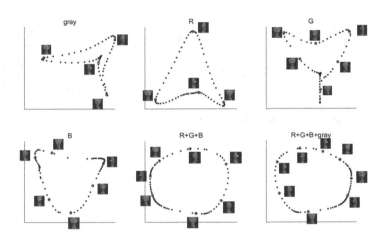

Fig. 1. Images of toy tiger mapped into the embedding space described by the two coordinates of MSL. Different angles of tiger are shown next to circled points in different parts of the space.

5.1   Brief Description of Data Sets

ORL[1] data set include 400 images with 40 classes. We extract three visual features from each image: gray feature with dimension $4,096$, gabor feature with dimension $2,560$, and LBP feature with dimension $3,776$.

AR[2] data set include $3,120$ images with 120 class. We extract three visual features from each image: gray feature with dimension $2,000$, gabor feature with dimension $3,200$, and LBP feature with dimension $4,720$.

---

[1] http://www.cl.cam.ac.uk/research/dtg/attarchive/facedatabase.html.
[2] http://www.pudn.com/Download/item/id/2427991.html.

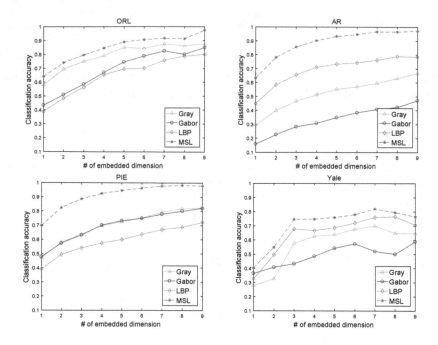

Fig. 2. Classification accuracy of MSL at four datasets.

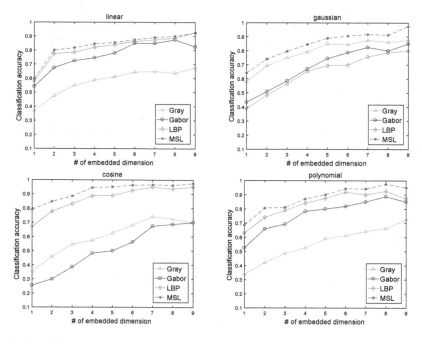

Fig. 3. Classification accuracy of MSL with different kernel functions at ORL dataset.

Multi-view Similarity Learning of Manifold Data   639

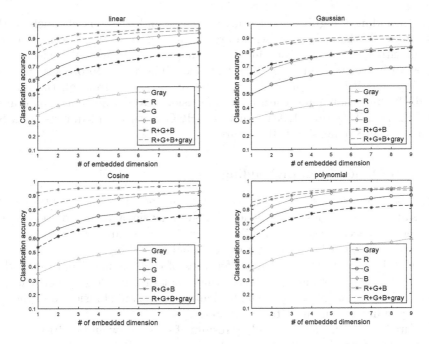

Fig. 4. Classification accuracy of MSL with different kernel functions at COIL-100 dataset.

PIE[3] data set is a face data set, we select its subset pose27 that include 1, 440 images with 20 class. We extract three visual features from each image: gray feature with dimension 1, 024, gabor feature with dimension 640, and LBP feature with dimension 944.

Yale[4] data set include 166 images with 15 class. We extract three visual features from each image: gray feature with dimension 4, 096, gabor feature with dimension 2, 560, and LBP feature with dimension 3, 776.

COIL20[5] data include 1, 440 images with 20 class. We extract three visual features from each image: gray feature with dimension 1, 024, gabor feature with dimension 640, and local binary pattern (LBP) with dimension 944.

Handwritten numerals[6] (HW) data set is comprised of 2, 000 data points for 0 to 9 digit classes, 200 data points for each class. We extract three visual features from each image: gray feature with dimension 256, gabor feature with dimension 160, and LBP feature with dimension 236.

The COIL100[7] data set contains 7, 200 colorized images corresponding to 100 different objects in 72 different viewpoints. We extracted 7 features from each

---

3 http://www.cs.cmu.edu/afs/cs/project/PIE/MultiPie/Multi-Pie/Home.html.
4 http://vision.ucsd.edu/content/yale-face-database.
5 http://www.cs.columbia.edu/CAVE/software/softlib/coil-20.php.
6 https://archive.ics.uci.edu/ml/datasets/Multiple+Features.
7 http://www.cs.columbia.edu/CAVE/software/softlib/coil-100.php.

image, including the gray scale features features and other six channels in the RGB and HSV channels, all of which are characterized by 16, 384 dimensions.

Caltech101[8] data set is containing 101 categories of images. We select 1, 474 images with 7 classes, include Dolla-Bill, Face, Garfield, Motorbikes, Snoopy, Stop-Sign and Windsor-Chair. Six features are extracted from all the images, include 48 dimension gabor feature, 40 dimension wavelet moments, 254 dimension CENTRIST feature, 1, 984 dimension HOG feature, 512 dimension GIST feature, and 928 dimension LBP feature.

5.2   Low Dimensional Embedding

In this part, we use one class of COIL-100 datasets, this class containing 72 images at different angles, we embedding high-dimensional images data into low-dimensional space. We can see in Fig. 1, these 72 images are rotated at different angles, so when embedding to a two-dimensional space, the closer to a circle. We extracted the gray feature and RGB features of each channel, and found that the performance after fusion is better than that of a single channel features. Figure 1 showing the results of a single channel and channel fusion, and there is a corresponding picture next to each point.

Figure 1 shows the 2-D embedding results of single-view and multi-view with each row corresponding to one manifold benchmark. From the figure, we can see our method MSL is more robust and can effectively find the proper low-dimensional embedding.

5.3   Classification of MSL

In this section, in order to validate the performance of the proposed method, we apply our method into multi-view classification. We used the indicator accuracy (ACC) [6] to evaluate the performance of the algorithm on four benchmark datasets. The four datasets we used are ORL, Yale, AR, and PIE. Each database is randomly divided into a training set and a test set, with different numbers of images being used for training.

The neighbor number k in computing heat kernel similarity of MSL are tuned such that MSL reach its best classification performance. We set the parameters $\alpha = \beta = 1$. The classification accuracy is computed by the nearest neighbor classifier.

To test the performance of the multi-view similarity learning in kernel spaces computed in (13), we test the classification accuracy of single-view and multi-view new similarity learning method with different kernel functions on ORL database. In this paper, four kernel functions (linear, polynomial, cosine, and Gaussian) are adopted in the experiments. We test MSL with tenfold cross-validation [3] as different number of embedded dimension is chosen. Figure 2 shows the classification accuracy of the single-view feature on the four data sets and the classification accuracy after the fusion of the three features. Figure 3

---

[8] http://www.vision.caltech.edu/feifeili/Datasets.htm.

Table 1. Clustering accuracy of MSL on two datasets

Database	Yale	Caltech101
SC(1)	0.3675	0.3460
SC(2)	0.3313	0.4480
SC(3)	0.3375	0.5290
SC(4)	–	0.6070
SC(5)	–	0.6720
SC(6)	–	0.5910
MVSC	0.6050	0.7250
MLAN	0.6325	0.7800
MSL	**0.6506**	**0.8433**

shows the classification performance variations of MSL with different kernel functions, and the classification performance variations of single-view and multi-view. And from the Fig. 3, We can see that classification performance of multi-view is better than classification performance of single-view on any kernel. From the Fig. 4, we extracted the fours features include gray feature and features of the RGB three channels of the picture and then calculate the classification accuracy of each channel, and classification accuracy of fusion with multiple channels, we can see the classification accuracy after fusion is higher than the classification accuracy of a single channel.

5.4 Clustering of MSL

In this section, we compared our method with other two multi-view learning methods, one is Multi-view Spectral Clustering (MVSC) [9] and the other is Multi-view Learning with Adaptive Neighbours (MLAN) [11], Table 1 show the clustering result in terms of accuracy of different method in two different datasets, SC is single-view feature. We can see our clustering results are better than the other two methods.

5.5 Semi-supervised Classification of MSL

In this section, we compared our method with MLAN [11] method, and in terms of semi-supervised classification, we choose the front 20% data as labeled sample, Table 2 show the semi-supervised classification performance of different method in three different datasets, we can see that our method is better than other method.

Table 2. Semi-supervised classification accuracy of MSL on three datasets

Database	ORL	COIL20	HW
SC(1)	0.5500	0.7207	0.9250
SC(2)	0.7750	0.8690	0.8040
SC(3)	0.7813	0.5474	0.7630
SC(4)	–	–	0.6930
SC(5)	–	–	0.6910
SC(6)	–	–	0.4730
MLAN	0.8313	0.9716	0.9760
MSL	**0.8700**	**0.9920**	**0.9813**

6   Conclusion and Remarks

In this paper, we introduce a novel multi-view similarity learning method named MSL, and our method can preserves the manifold structure of the data. For multi-view learning, our method can automatically learns weights for each view. The experimental results on real world benchmark data sets demonstrate that the classification accuracy of multi-view features is higher than that of single-view feature. And in clustering and semi-supervised classification, the accuracy of our method is higher than that of other multi-view methods. These experimental results demonstrate the effectiveness of our method.

References

1. Belkin, M., Niyogi, P.: Laplacian eigenmaps for dimensionality reduction and data representation. Neural Comput. **15**, 1373–1396 (2003)
2. Chen, S., Ding, C.H.Q., Luo, B.: Similarity learning of manifold data. IEEE Trans. Cybern. **45**(9), 1744–1756 (2015)
3. Chik, Z., Aljanabi, Q.A., Kasa, A., Taha, M.R.: Tenfold cross validation artificial neural network modeling of the settlement behavior of a stone column under a highway embankment. Arab. J. Geosci. **7**(11), 4877–4887 (2014)
4. Cox, M.A.A., Cox, T.F.: Multidimensional scaling. J. R. Stat. Soc. **46**(2), 1050–1057 (2001)
5. Debruyne, M., Verdonck, T.: Robust kernel principal component analysis and classification. Adv. Data Anal. Classif. **4**(2–3), 151–167 (2010)
6. Diebold, F.X., Mariano, R.S.: Comparing predictive accuracy. J. Bus. Econ. Stat. **13**(1), 134–144 (1995)
7. Guo, G., Wang, H., Bell, D., Bi, Y., Greer, K.: KNN model-based approach in classification. In: Meersman, R., Tari, Z., Schmidt, D.C. (eds.) OTM 2003. LNCS, vol. 2888, pp. 986–996. Springer, Heidelberg (2003). https://doi.org/10.1007/978-3-540-39964-3_62
8. Hsieh, P., Yang, M., Gu, Y., Liang, Y.: Classification-oriented locally linear embedding. IJPRAI **24**(5), 737–762 (2010)

9. Kumar, A., III, H.D.: A co-training approach for multi-view spectral clustering. In: Proceedings of the 28th International Conference on Machine Learning, ICML 2011, Bellevue, Washington, USA, 28 June–2 July 2011, pp. 393–400 (2011)
10. Liu, C., Wechsler, H.: Independent component analysis of gabor features for face recognition. IEEE Trans. Neural Netw. 14(4), 919–928 (2003)
11. Nie, F., Cai, G., Li, X.: Multi-view clustering and semi-supervised classification with adaptive neighbours. In: Proceedings of the Thirty-First AAAI Conference on Artificial Intelligence, pp. 2408–2414 (2017)
12. Ojala, T., Pietikäinen, M., Mäenpää, T.: Multiresolution gray-scale and rotation invariant texture classification with local binary patterns. IEEE Trans. Pattern Anal. Mach. Intell. 24(7), 971–987 (2002)
13. Wang, X., Tang, X.: Dual-space linear discriminant analysis for face recognition. In: 2004 IEEE Computer Society Conference on Computer Vision and Pattern Recognition, pp. 564–569 (2004)
14. Yang, M.-H.: Discriminant isometric mapping for face recognition. In: Crowley, J.L., Piater, J.H., Vincze, M., Paletta, L. (eds.) ICVS 2003. LNCS, vol. 2626, pp. 470–480. Springer, Heidelberg (2003). https://doi.org/10.1007/3-540-36592-3_45

Salient Points Driven Pedestrian Group Retrieval

Xiao-Han Chen[1,2,4] and Jian-Huang Lai[1,3,4](✉)

[1] School of Data and Computer Science, Sun Yat-sen University,
Guangzhou 510006, People's Republic of China
chenxh45@mail2.sysu.edu.cn, stsljh@mail.sysu.edu.cn
[2] Faculty of Mathematics and Computer Science, Guangdong Ocean University,
Zhanjiang 524088, People's Republic of China
[3] School of Information Science and Technology, XinHua College,
Sun Yat-sen University, Guangzhou 510006, People's Republic of China
[4] Guangdong Key Laboratory of Information Security Technology,
Sun Yat-sen University, Guangzhou 510006, People's Republic of China

Abstract. Groups are the primary constituent units of crowd and the study on groups can help us better understand the collective phenomena in public area. In this paper, collection of stable individuals with some social relationship in public area, called group, is selected as the research object, and a novel task of pedestrian group retrieval is introduced. Different from the individual person matching, groups often show high aggregation due to their inherent characteristics, individuals in the group are more occluded. Therefore, the performance of individual person based detection and matching will be affected. At the same time, group matching also needs to handle difficulties like variations in the shape and ordering of people within the group. We then design a salient points driven framework for pedestrian group retrieval across non-overlapping cameras. The work focuses on the problems of overall appearance characteristics extraction of a deformable pedestrian collection and matching of groups at varying scales. Experiments on Pedestrian-Groups dataset demonstrate the effectiveness of our proposed framework for Pedestrian Group retrieval.

Keywords: Salient points · Pedestrian group · Group retrieval · Group entire descriptor

1 Introduction

With increasing need for public safety, crowd management in public area with the computer vision technology is essential to improve the management ability of security departments. In the past decade, crowd analysis has been a active field and widely studied and applied in crowd event detection [2,24], crowd counting [21,25] and segmentation [1,15]. Nevertheless, as one of the major constituent units of crowd, the group is also the important research object in public safety

© Springer Nature Switzerland AG 2019
Y. Zhao et al. (Eds.): ICIG 2019, LNCS 11901, pp. 644–656, 2019.
https://doi.org/10.1007/978-3-030-34120-6_52

management field. It contains the information that facilitates the understanding about collective phenomena, which raises great interest from the researchers on the study of group, such as group detection [10,12,19] or group activity recognition [11,16,22]. However, the understanding of group remains challenging, especially the cross-camera group retrieval, which is still few studies.

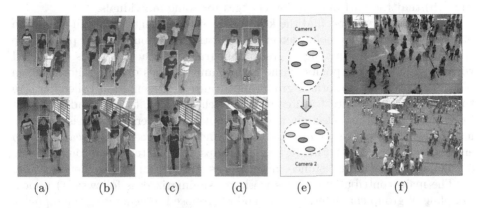

(a)   (b)   (c)   (d)   (e)   (f)

Fig. 1. The challenges in group retrieval. (a)–(d) The two rows of pictures are collected from different cameras, and those in the one column are of the same group obtained from two cameras. Changes could be seen in relative location, distance, and way of occlusion of individuals (the rectangles of the same color in every column indicate the same person in different camera scene). (e) Group may modify their configuration on movement. (f) Pedestrian Groups in real public area (Color figure online)

The major goal of group retrieval is to search and match the same pedestrian collective from different non-overlapping cameras images. As defined in [19], pedestrian group is a cluster of members who tend to move together for a sustained period of time. They usually have some kind of social relationships, e.g. friends or family members. In here, we mainly focus on small groups composed of several pedestrians. The study on group retrieval is of practical value for security management, such as anomaly crowd source tracing or group movement route detection. Therefore, group retrieval has become one of the urgent issues for the security departments. This paper conducts an exploration into this less studied issue and proposes a salient points driven framework for pedestrian group retrieval.

Compared with individual re-identification, group re-identification confronts the following challenges. The first challenge is that the relative position and distance of individuals may change with the movement, and the group structure will also change accordingly. This makes it impossible to measure the similarity through the global matching method. As in Fig. 1(a) and (d), we can see that there are significant changes in the relative locations of the individuals in the group. In addition, the individuals in Fig. 1(b) and (c) are also becoming farther or closer from each other while in motion, and the shape of the group also changes accordingly. The second challenge concerns the inconsistent scales in the matching of different group due to the difference in individual quantity. The third challenge results from the highly aggregated individuals in the group, this causes them to highly occlude each other. The individual-based matching method turns out to be increasingly hard to conduct in such case. Furthermore, the occlusion varies in form and is largely random. For instance, it can be observed from Fig. 1(b) that the part of occlusion changes for some individuals.

In this study, we first find the key points of the salient feature in the group image by using the key point detection algorithm. Combining with perspective transformation, the sampling of group appearance characteristics is conducted with these salient key points as the center, and a series of non-uniform image blocks are obtained to form a group appearance representation collection. We then extract appearance features vectors such as color, texture and structure for each block in the collection, and carry out the clustering analysis for these features with normalization process, the obtained cluster centers are called group entire descriptor (GED). Finally, for two groups to be matched, we computer the optimal GED matching distance, and use it as a metric for groups identification.

The main contributions of this work are summarized as follows: (1) a local sampling of group characteristics method is proposed to generate group appearance representation collection in dealing with the deformation caused by group movement; (2) a salient points driven framework for group retrieval is designed to solve the issue concerning group matching in different camera; and (3) we also introduce a new cross cameras group dataset *Pedestrian-Groups* for evaluating our proposed methods; it contains various situations such as occlusion each other, group shape and relative position or distance changes.

2   Related Works

Group detection and group activity recognition have attracted a great deal of attention of many researchers [10–12,16,22]. A context-aware parameter-free (MPF) framework is proposed to detect groups [12]. In this framework, feature points are detected and clustered by using the motion information. Approach in [10] proposes an instant group motion refining framework based on group motion. Our proposed method also detects feature points at first, but we use the motion clues for denoise processing. In the methods of group activity recognition [11,16,22], the representation of group activities often based on individual actions or pairwise interactions. In contrast to these approaches, we use a sub-individual way to construct a group entire descriptor.

About the retrieval problems, single person re-identification has been well studied in surveillance video analysis. In general, these approaches can be categorized into two classes: hand-crafted algorithms [5,13] and deep learning methods which benefit from the use of large data [20,23]. One of the crucial tasks of person re-identification is to extract the description of color and textural information of a person image. For this purpose, a person image is usually divided according to a certain principle. In [13], an image is equally divided into six horizontal stripes, and an effective feature representation called Local Maximal Occurrence (LOMO) is proposed. A method adopts perceptual principles to find two horizontal axes of asymmetry to isolate three main body regions [5]. In [23], for learning different body parts representations, part bounding boxes are obtained by an unsupervised person part generation procedure. In our framework, we fetch a series of image blocks according to the detected feature points.

Perspective normalization is an important step in crowded scene analysis. Abnormal crowd behavior detection [2,3], person detection and tracking [18] and crowd counting [25] have considered the effects of perspective in their proposed methods. To obtain perspective map, the approach in [25] randomly selects several adult pedestrians, and then labels them from head to toe. Similarly, we conduct perspective estimation by selecting and artificially labeling pedestrians in the group image to acquire their heights.

3   Extraction of Group Appearance Representation

When we are identifying groups, we generally first pay our attention to some discriminating local parts that can distinguish it from others instead of directly giving an overall matching. Then, those characteristics will be found from some individuals of one group. The more significant regions that can be matched at another group, the higher possibility that they are the same group. In this process, the entirety of group is decomposed, and some blocks with distinguishable features are selectively re-collected as the basis for final matching. Inspired by such intuition, a series of salient points in group image are first extracted, then local sampling processing is performed to divide a complete group into some units of smaller granularity in order to provide useful clues for the subsequent group matching. In the following, how to develop local appearance sampling on group as driven by those salient points, and acquire the distinguishable group appearance representation collection are going to be discussed.

3.1   Finding the Salient Points Associated to Group

Groups usually have no regular shape, the grid cell division on the images may easily introduce some unrelated information. Therefore, interest point detection is combined with noise reduction by motion information to find out the significance points on the group image. In this paper, Harris detector is employed to extract the salient points [6], while those points usually exist on the locations with critical information. In Fig. 2(a), we provide the detection result of the significance points in the scenario, from which it could be observed that most of the key points are located on the individuals and their locations contain abundant characteristics of the group images. Apart from that, it is also found that some points extracted in such a case may fall into the areas outside the moving groups just like those on the door of background in Fig. 2(a), and those noise points must be eliminated. In view of those, the noise points are removed on using the moving information of optical flow with two continue frames, and then the target salient points of the group are obtained. As shown in Fig. 2(b), these salient points have a higher correlation with the group.

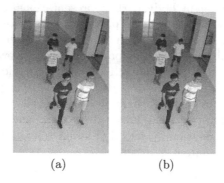

(a)   (b)

Fig. 2. The detection result of salient points associated to group (marked with green circles). (a) The points are extracted with detection operator; some noise points resided outside the group area. (b) The points in the group area after being treated with a noise removal process. (Color figure online)

3.2   Local Sampling of Group Based Salient Points

After obtaining a series of salient points of group, local appearance sampling is performed as per those points. With points as the centers, some square image blocks with certain width are extracted to represent the characteristics of the group. However, it is noticed that under the effect of perspective distortion, the group pixel size in the scenario changes with the distance from the cameras. This causes the lack of uniformity in image scale. If the sampling process is carried out with the same size in the scenario, the scale of the body parts contained in the image blocks would be inconsistent. To cope with those problems, reasonable perspective transformation should be conducted towards the size of the image blocks. As in [25], to ensure the extracted image blocks could fit the change in individual size as much as possible, the size of sampling window is determined by its location coordinates and scale factor. Assume P is a point in the set of salient points of a group at location (x_P, y_P), its sampling width w with the image block is defined as

$$w = (\eta_1 + \eta_2 \cdot y_P)/\rho, \tag{1}$$

where η_1, η_2 are scale factors, and ρ is the height coefficient, value of ρ means the width of image block w is $1/\rho$ of that of a pedestrian image height, ρ is set to 7. The scale factors η_1, η_2 can be calculated by estimating the camera perspective, and each camera is to be estimated for one time. It is assumed that the height of an individual varies linearly from near to far in camera scene. Several images taken by one camera are randomly selected and artificially labeled to acquire the pixel height from head to toe of K individuals, the height of the ith individual is supposed to be h_i, and the coordinates of its central point is denoted (x_i, y_i). After that, the scale factors η_1 and η_2 can be calculated by

$$\min_{\eta_1, \eta_2} \sum_{i=0}^{K} (\eta_1 + \eta_2 \cdot y_i - h_i)^2. \tag{2}$$

Figure 3(a) presents the marking results of one image using bounding boxes. The height of the yellow box is the pixel value of the individual height, whereas the green cross in the middle of each box is the position of central point.

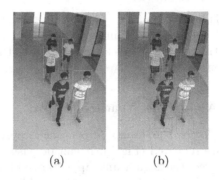

(a)   (b)

Fig. 3. The local appearance sampling of group. (a) The height of the yellow box is the pixel value of the resulting individual height and green cross in the middle corresponding its location coordinates. (b) The sampling box size is adjusted adaptively according to its location after a perspective transformation (marked in green). (Color figure online)

According to the definition of Eq. (1), image blocks with non-uniform size are next extracted from every point in the salient points set acquired in Sect. 3.1 according to its location. The bounding boxes in Fig. 3(b) illustrate the local appearance sampling results of a group, and the sampling box size is adjusted adaptively according to the specific location. It can be seen from the figure that the size of image blocks is inconsistent due to the varying distances, while the scale of body part in image blocks is close to each other. Although the size of the sampling block is approximate, this kind of perspective transformation can effectively normalize each image block in scale. This is helpful for the subsequent group matching. In Fig. 4, the comparison of the extracted group appearance image blocks is further presented. To facilitate the comparison, we select some image blocks containing person's head; the first row in Fig. 4 is the result extracted in the non-uniform block sampling method as described in this paper, while the second row is the result of the extraction in a uniform width. As shown in Fig. 4, the scale of person's head in the images of first row appears to have a better consistency.

In addition, as the salient points of significance obtained from Sect. 3.1 are usually too densely. When the Euclidean distance between the central points of

Fig. 4. The comparison of the group local appearance image blocks extraction results with non-uniform and uniform in row 1 and row 2, respectively; the scale of person's head in the images of first row appears to have a better consistency.

the two image blocks is lower than τ, $\tau = 10$, one of them is randomly reserved in order to reduce the redundancy of image blocks and computation as well. Finally, all of these extracted image blocks are resized to the same size (40×40 pixels), and used to construct a group appearance representation collection.

3.3   Extraction of Appearance Feature Vectors

When the image blocks in the group appearance representation collection obtained in Sect. 3.2 are reviewed (Fig. 5), it can be easily observed that those blocks have well reserved the significant appearance characteristics of their group. By extracting the non-uniform appearance blocks of the group, the deformed collective composed of the pedestrian individuals is disassembled into some units of sub-individual granularity; they retain the local characteristics of the group so as to overcome the impact of deformation.

Fig. 5. The group appearance representing image blocks. They can preserve the significant appearance characteristics of group well.

Then, the appearance feature vectors of all the image blocks of each group will be extracted. We use a sliding window of size of 10×10 with an overlapping step of 5 pixels to the 40×40 image block and extract such characteristic information as color, texture. Color is an important characteristic for depicting the appearance of group. The Retinex algorithm is firstly used to enhance each image block [7,8], so that the image lightness and color under different cameras could be more consistent; we then extract a 512×49 dimensions HSV histogram vector. Next, the scale invariant local ternary pattern (SILTP) descriptor is applied to obtain the illumination invariant texture feature [14]. The SILTP histograms of two scales (radius: 3 and 5) are extracted separately to get texture feature vectors with the dimension of $81 \times 2 \times 49$. Finally, a 576-dimensional histogram of oriented gradients (HOG) vectors is extracted to gain the structural characteristics of the images [4]. Those results constitute all the feature vectors of the group appearance representation collection.

4   Group Matching

The acquisition of group appearance representation collection enables a complete group to be represented with a set of image units of smaller granularity. However, due to the difference in scenario or group, the number of appearance blocks obtained by each group image is different. Therefore, the appearance feature vectors extracted from the image blocks cannot be directly applied to the matching of groups. We solved the problem in two steps. Firstly, clustering analysis is carried out with the feature vectors of every group by a fixed class number, and the second step is to view the clustering centers resulting from previous step as the group entire descriptor for the final matching.

4.1   Construction of Group Entire Descriptor

Assume $\boldsymbol{R}$, $\boldsymbol{S}$, $\boldsymbol{T}$ are the HSV, HOG and SILTP features matrixes of image blocks in all group appearance representation collections, respectively. Each row in those matrixes is a feature vector of one image block. We then denote the vectors $\boldsymbol{m}_{hsv}$, $\boldsymbol{m}_{hog}$, $\boldsymbol{m}_{siltp}$ as the mean and $\boldsymbol{s}_{hsv}$, $\boldsymbol{s}_{hog}$, $\boldsymbol{s}_{siltp}$ as the standard deviation of each column of matrixes.

For the ith group C_i, there are L image blocks in its group appearance representation collection, and its color, texture and structure features sets can be given by $HSV_i = \{\boldsymbol{hsv}_i^1, \ldots, \boldsymbol{hsv}_i^l, \ldots, \boldsymbol{hsv}_i^L\}$, $HOG_i = \{\boldsymbol{hog}_i^1, \ldots, \boldsymbol{hog}_i^l, \ldots, \boldsymbol{hog}_i^L\}$ and $SILTP_i = \{\boldsymbol{siltp}_i^1, \ldots, \boldsymbol{siltp}_i^l, \ldots, \boldsymbol{siltp}_i^L\}$, where $\boldsymbol{hsv}_i^l$, $\boldsymbol{hog}_i^l$ and $\boldsymbol{siltp}_i^l$ indicate the HSV, HOG and SILTP feature vectors of the lth image block, respectively. We next conduct the normalization process by Eqs. (3)–(5) on feature vectors $\boldsymbol{hsv}_i^l$, $\boldsymbol{hog}_i^l$, $\boldsymbol{siltp}_i^l$, and obtain the results as $\boldsymbol{r}_i^l$, $\boldsymbol{s}_i^l$, $\boldsymbol{t}_i^l$, respectively.

$$r_i^l = (\boldsymbol{hsv}_i^l - \boldsymbol{m}_{hsv}) \oslash \boldsymbol{s}_{hsv}, \tag{3}$$

$$s_i^l = (\boldsymbol{hog}_i^l - \boldsymbol{m}_{hog}) \oslash \boldsymbol{s}_{hog}, \tag{4}$$

$$t_i^l = (\boldsymbol{siltp}_i^l - \boldsymbol{m}_{siltp}) \oslash \boldsymbol{s}_{siltp}, \tag{5}$$

where the operator $\oslash$ is the Hadamard division, which denotes element-wise division. A new feature fusion vector $\boldsymbol{hhs}_i^l$ is then obtained by vector concatenation, $\boldsymbol{hhs}_i^l = (\boldsymbol{r}_i^l, \boldsymbol{s}_i^l, \boldsymbol{t}_i^l)$, and new features set of all image blocks of C_i is denoted as $HHS_i = \{\boldsymbol{hhs}_i^1, \ldots, \boldsymbol{hhs}_i^l, \ldots, \boldsymbol{hhs}_i^L\}$ accordingly. The k-means clustering analysis with N classes is performed on HHS_i, clustering centers of group C_i are then achieved and denoted as $GED_i = \{\boldsymbol{ged}_i^1, \ldots, \boldsymbol{ged}_i^n, \ldots, \boldsymbol{ged}_i^N\}$, where $\boldsymbol{ged}_i^n$ is the nth clustering center vector. Through this processing, the set of N vectors, namely GED, is derived for group C_i.

As clustering to the same number of classes, the difference in the number of image block resulting from the scale difference of groups has been transformed to have the same matching dimensions.

4.2   Optimal Group Entire Descriptors Matching

For a pair of groups to be matched, the ith group C_i and the jth group C_j, it is necessary to compute the final general distance between them. Suppose GED_i and GED_j indicate the GED of C_i and C_j, respectively. The bipartite graph matching method is adopted to find the optimal GED matching distance between C_i and C_j at a minimum cost. The distance between each element of GED_i and GED_j is firstly computed by the Bhattacharyya distance to generate a cost matrix $\boldsymbol{D}_{ij}$; then a bipartite graph $G = (V, E)$ (in which $V = GED_i \bigcup GED_j$) for GED_i and GED_j is constructed, and it is assumed that they are both disjoint sets. In Graph G, every edge $e \in E$ means the cost (distance) between two vertices from GED_i and GED_j, respectively, namely certain value in $\boldsymbol{D}_{ij}$. Therefore, the optimal matching of bipartite graph G can be determined through solving of Hungarian algorithm [9], and the total cost corresponding to the optimal matching can also be figured out and denoted as $D(GED_i, GED_j)$, that is GED distance between groups C_i and C_j. It means the criteria for measuring the similarity of groups.

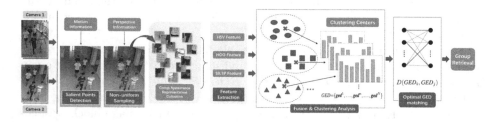

Fig. 6. Overview of the proposed framework for pedestrian group retrieval.

5   Experiments

To evaluate our proposed method, the experiments are conducted on Pedestrian-Groups dataset. In experiments, the clustering parameter N is set to 11. And the framework of our proposed approach is shown in Fig. 6.

5.1   Pedestrian-Groups Dataset

Groups retrieval is one of the new and less studied issue in crowd video analysis field. In order to verify the method proposed in this paper, a new dataset, the

Pedestrian-Groups dataset is constructed. In this dataset, there are 120 images of 30 groups, and two parts of data are collected from each group under two non-overlapping cameras. For the purpose of obtaining the motion information, there are two continuous image frames of one pedestrian group under each camera. In each group, the number of pedestrian individuals ranged from 2 to 5. The dataset is inclusive of such situations as the changing shape of groups, the inter-occlusion and the change in relative location or distance between different individuals.

5.2   Experimental Results

The method proposed in this paper is evaluated on Pedestrian-Groups dataset. The results are shown by the Cumulative Matching Characteristics (CMC) curve. The CMC curve can be associated with the ratio of the probe group found in the top r matches in the group image gallery. In Fig. 7, we can see that the proposed method can effectively identify group across camera, and the matching correctness of rank 1 identification rate is 70.0%. Furthermore, we use single person re-identification based match method as a comparison baseline, which first extracts individuals in group by pedestrian detection method [17] and leverages the person re-dentification approach [13] to compute the optimal match cost as a group pair distance. As shown in Fig. 7, person re-identification based match method has lower results than ours for the reason of the highly aggregation feature of group.

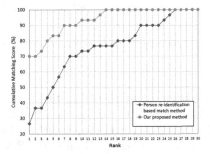

Fig. 7. The CMC curves on the Pedestrian-Groups dataset by comparing the proposed method to the person re-identification based match method.

The effect of perspective transformation-based sampling method proposed in this paper is analyzed by comparing the experimental results from the sampling window of non-uniform and uniform in the group appearance image blocks extraction. As shown in Fig. 8(a), the rank 1 identification rate of non-uniform block extraction method is 70.0% and of the uniform method is 43.3%, it increases by 26.7%, which indicates the perspective transformation processing is conducive to the improvement of the identification effect. For the number of clusters, we conduct the experiments on varying the value of N. Figure 8(b) shows how the number of clusters affects the rank 1 identification rate. It shows that the performance obtains higher precision when N is more than 8. We fix $N = 11$ in our other experiments.

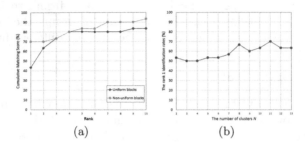

(a)   (b)

Fig. 8. (a) Results of the uniform and the non-uniform sampling window. (b) Comparing the results of rank 1 identification rates at different number of clusters N.

Table 1. Comparison of different feature representation methods (%); concatenation means that all of the features are directly concatenated without normalization

Features	Rank 1	Rank 10	Rank 20
HSV	66.67	100.00	100.00
HOG	20.00	60.00	83.33
SILTP	40.00	80.00	96.67
Concatenation	40.00	83.33	93.33
Normalization	**70.00**	93.33	100.00

A comparative evaluation of using the different feature representation methods is carried out. Table 1 presents the evaluation results based on features of HSV, HOG, and SILTP. Although the rank 1 identification rate achieves 66.7% by using just HSV features, combining HOG and SILTP features can increase to 70%, it shows that the combined feature can improve the performance of group identification in rank 1. At last, the results also indicate that treating the direct concatenating of feature without conducting the normalization process has failed to generate a significant effect on matching.

6   Conclusion

In this paper, local sampling method is adopted to decompose the deformative group and a group retrieval framework driven by salient points is proposed. Experimental evaluations are conducted on our Pedestrian-Groups dataset. The experimental results suggest our method can effectively identify the group across cameras. As a preliminary exploration in group matching, we will extend the research on terms of group characteristic representation and modeling of group matching. Meanwhile, group detection or segmentation may be combined to apply the group retrieval to some more complicated practical scenarios in our future work.

Acknowledgments. This work was supported by National Key Research and Development Program of China (2016YFB1001003), the NSFC (61573387).

References

1. Ali, S., Shah, M.: A lagrangian particle dynamics approach for crowd flow segmentation and stability analysis. In: IEEE Conference on Computer Vision and Pattern Recognition, CVPR 2007, pp. 1–6. IEEE (2007)
2. de Almeida, I.R., Cassol, V.J., Badler, N.I., Musse, S.R., Jung, C.R.: Detection of global and local motion changes in human crowds. IEEE Trans. Circ. Syst. Video Technol. **27**(3), 603–612 (2017)
3. Chen, X.H., Lai, J.H.: Detecting abnormal crowd behaviors based on the Div-Curl characteristics of flow fields. Pattern Recogn. **88**, 342–355 (2019)
4. Dalal, N., Triggs, B.: Histograms of oriented gradients for human detection. In: IEEE Computer Society Conference on Computer Vision and Pattern Recognition, CVPR 2005, vol. 1, pp. 886–893. IEEE (2005)
5. Farenzena, M., Bazzani, L., Perina, A., Murino, V., Cristani, M.: Person re-identification by symmetry-driven accumulation of local features. In: 2010 IEEE Conference on Computer Vision and Pattern Recognition (CVPR), pp. 2360–2367. IEEE (2010)
6. Harris, C., Stephens, M.: A combined corner and edge detector. In: Alvey Vision Conference, vol. 15, pp. 10–5244. Citeseer (1988)
7. Jobson, D.J., Rahman, Z.U., Woodell, G.A.: A multiscale retinex for bridging the gap between color images and the human observation of scenes. IEEE Trans. Image Process. **6**(7), 965–976 (1997)
8. Jobson, D.J., Rahman, Z.U., Woodell, G.A.: Properties and performance of a center/surround retinex. IEEE Trans. Image Process. **6**(3), 451–462 (1997)
9. Kuhn, H.W.: The Hungarian method for the assignment problem. Naval Res. Logist. Q. **2**(1–2), 83–97 (1955)
10. Li, N., Zhang, Y., Luo, W., Guo, N.: Instant coherent group motion filtering by group motion representations. Neurocomputing **266**, 304–314 (2017)
11. Li, X., Choo Chuah, M.: SBGAR: semantics based group activity recognition. In: Proceedings of the IEEE International Conference on Computer Vision, pp. 2876–2885 (2017)
12. Li, X., Chen, M., Nie, F., Wang, Q.: A multiview-based parameter free framework for group detection. In: AAAI, pp. 4147–4153 (2017)
13. Liao, S., Hu, Y., Zhu, X., Li, S.Z.: Person re-identification by local maximal occurrence representation and metric learning. In: Proceedings of the IEEE Conference on Computer Vision and Pattern Recognition, pp. 2197–2206 (2015)
14. Liao, S., Zhao, G., Kellokumpu, V., Pietikäinen, M., Li, S.Z.: Modeling pixel process with scale invariant local patterns for background subtraction in complex scenes. In: 2010 IEEE Conference on Computer Vision and Pattern Recognition (CVPR), pp. 1301–1306. IEEE (2010)

15. Lin, W., Mi, Y., Wang, W., Wu, J., Wang, J., Mei, T.: A diffusion and clustering-based approach for finding coherent motions and understanding crowd scenes. IEEE Trans. Image Process. **25**(4), 1674–1687 (2016)
16. Qi, M., Qin, J., Li, A., Wang, Y., Luo, J., Van Gool, L.: stagNet: an attentive semantic RNN for group activity recognition. In: Ferrari, V., Hebert, M., Sminchisescu, C., Weiss, Y. (eds.) ECCV 2018. LNCS, vol. 11214, pp. 104–120. Springer, Cham (2018). https://doi.org/10.1007/978-3-030-01249-6_7
17. Redmon, J., Farhadi, A.: Yolov3: An incremental improvement. arXiv (2018)
18. Rodriguez, M., Laptev, I., Sivic, J., Audibert, J.Y.: Density-aware person detection and tracking in crowds. In: 2011 IEEE International Conference on Computer Vision (ICCV), pp. 2423–2430. IEEE (2011)
19. Shao, J., Loy, C.C., Wang, X.: Learning scene-independent group descriptors for crowd understanding. IEEE Trans. Circ. Syst. Video Technol. **27**(6), 1290–1303 (2017)
20. Shen, Y., Li, H., Xiao, T., Yi, S., Chen, D., Wang, X.: Deep group-shuffling random walk for person re-identification. In: Proceedings of the IEEE Conference on Computer Vision and Pattern Recognition, pp. 2265–2274 (2018)
21. Shi, Z., et al.: Crowd counting with deep negative correlation learning. In: Proceedings of the IEEE Conference on Computer Vision and Pattern Recognition, pp. 5382–5390 (2018)
22. Shu, T., Todorovic, S., Zhu, S.C.: Cern: confidence-energy recurrent network for group activity recognition. In: IEEE Conference on Computer Vision and Pattern Recognition, vol. 2 (2017)
23. Yao, H., Zhang, S., Hong, R., Zhang, Y., Xu, C., Tian, Q.: Deep representation learning with part loss for person re-identification. IEEE Trans. Image Process. **28**, 2860–2871 (2019)
24. Yuan, Y., Feng, Y., Lu, X.: Structured dictionary learning for abnormal event detection in crowded scenes. Pattern Recogn. **73**, 99–110 (2018)
25. Zhang, C., Li, H., Wang, X., Yang, X.: Cross-scene crowd counting via deep convolutional neural networks. In: Proceedings of the IEEE Conference on Computer Vision and Pattern Recognition, pp. 833–841 (2015)

Small Object Detection on Road by Embedding Focal-Area Loss

Zijie Wang[1], Jianwu Fang[1,2](✉), Jian Dou[1], and Jianru Xue[1]

[1] Institute of Artificial Intelligence and Robotics, Xian Jiaotong University,
Xi'an, China
fangjianwu@chd.edu.cn , jrxue@mail.xjtu.edu.cn
[2] School of Electronic and Control Engineering, Chang'an University, Xi'an, China

Abstract. In recent years, with the continuous popularity of deep learning, the research on artificial intelligence has boosted the progress of many new applications, such as the autonomous driving. At present, the detection methods of vehicles, pedestrians and other objects in the self-driving technology have been investigated numerously, but there is no good solution for the detection of small objects such as stones on road. However, small targets on road seriously affects the stability of automated vehicle system. Therefore, it is important to carry out the detection of small targets on road.

This paper designs a focal-area loss function which is learned by focusing the area change of small targets. The contribution of small object is weighted more in learning. We embed this focal-area loss into a newly proposed Scale Normalization for Image Pyramids (SNIP). Exhaustive experiments on Lost And Found (LAF) dataset show that our method can significantly boost the performance of state-of-the-art.

Keywords: Autonomous driving · Small object detection · Focal Loss · Area constraint

1 Introduction

The detection of small objects on road is of great significance for autonomous driving research. This is because small objects on road, such as bricks and stones, have a major safety impact on the stability of high-speed cars. Automated vehicles are mainly based on the sensors to evaluate the overall state of the car. Sudden car bumps may cause a serious error occurred in the sensor data acquisition, which led to an error in the evaluation of the overall state of the vehicle by the automatic driving system. It may eventually caused the entire autonomous driving system to malfunction. Therefore, for the field of automatic driving, the detection of small objects on road is very important for the safety of self-driving cars.

Because of the limited resolution and information of small objects in images, detection of small objects is a really challenging task. As for small objects on

© Springer Nature Switzerland AG 2019
Y. Zhao et al. (Eds.): ICIG 2019, LNCS 11901, pp. 657–665, 2019.
https://doi.org/10.1007/978-3-030-34120-6_53

road, when they are combined with large-scale targets for detection, the scale range is very large. Besides, the color of small objects on road are usually grey, such as the stones. So it is easy to be confused with the color of the road itself and treated as noise.

For the definition of small targets, as shown in Table 1 [8], the Microsoft COCO [12] dataset considers the small target to be a target with a target frame of less than 32×32 pixels.

Table 1. COCO dataset target scale standard.

	Min rectangle area	Max rectangle area
Small object	0×0	32×32
Medium object	32×32	96×96
Large object	96×96	$\infty \times \infty$

In this paper, we improve the detection method of small targets on road based on Scale Normalization for Image Pyramids [19]. This method is based on Faster R-CNN [18] and uses the idea of image pyramid to scale the image. Ground truth in a certain range is selected for images of different scales, and then chips [19] of uniform size are selected for images of each scale. Finally, the ground truth is input to Faster R-CNN [18] for the detection of small targets. In addition, this paper designed a loss function –focal-area loss with the scale change fitting of target ground truth. This kind of loss function can improve the importance of the small target in the classification loss, so as to improve the detection accuracy of the small object detection. According to the Lost And Found dataset [15] released by Peter Pinggera et al., a comparison was conducted between the network with and without the focal-area loss. Based on the experiments, the superiority of the proposed method is proved.

2   Related Work

Researchers have proposed many methods to detect small objects [1,2,4,9]. These methods have achieved considerable improvement by using contextual information. Besides, enlarging the small regions has also been used to fit the features of the network pretrained [7]. Some methods exploit structural associations between objects of different scales. For example, the generation network of the GAN [6] network converts the weak representation of the small target into a super-resolution representation [10], and enhances the feature representation of the small object to make it similar to the large object.

Apart from the traditional computer vision method, 3D laser is also adopted to detect the small objects by tracing the sparse point cloud [16].

As this paper studies the small objects on road, we proposed a new loss function detection method based on SNIP [19] with combination of ground truth area due to the difference with general objection detection.

3   Brief Introduction of SNIP

Image pyramid is one of the traditional strategies to improve the accuracy of object detection. By upsampling the image, the resolution of the image can be improved, so as to improve the accuracy of object detection. On the other hand, the image pyramid also has a very serious disadvantage, that is, the calculation amount is increased.

To this end, Bharat Singh et al. [19] proposed a small target detection and recognition algorithm based on image pyramid combined with Faster R-CNN [18] named SNIP [19]. The motivation in SNIP [19] is that the use of image pyramids is beneficial to improve the accuracy of object detection, while at the same time discarding the shortcomings of excessive image pyramid calculation. SNIP [19] reduces the amount of computation by ignoring large-sized targets in high-resolution images and small-sized targets in low-resolution images. On the other hand, due to the pyramid operation of the image, the resolution of the small target is improved, and finally the accuracy of the small target detection is improved. SNIP [19] not only avoids the large calculation of image pyramid, but also achieves good results for small target detection. Specifically, the input of SNIP [19] is the original image, and the output is the subimages sampled from the image, which are called chips [19]. In fact, chips are a series of fixed size of windows arranged at constant intervals on a certain scale of an image. Suppose the size of chips is 512×512, For an image generated by image pyramid, assuming the size of $W_i \times H_i$, $\frac{W_i}{32} \times \frac{H_i}{32}$ chips with size of 512×512 can be obtained by sliding window with step size of 32.

As shown in SNIP [19] structure Fig. 1, the targets in the image are divided into three categories: large target, medium target and small target. After passing through the RPN [18] network, the images of different scales generated by the image pyramid get three feature maps of different sizes. Obviously, the top feature maps have the largest size, so the large target in the original image will become larger after being enlarged, which is not conducive to detection. However, the size of small targets in the original image will increase after being enlarged, which is conducive to detection. In this way, those small objects which originally were difficult to detect become easy to detect through the zooming in of image pyramid.

In conclusion, the top feature map goes with the smallest anchors, the middle one goes with the medium ones, and the bottom one goes with the biggest ones. In this way, feature maps of each scale are good at detecting large, medium and small targets in the image, respectively, and the detection frame of each feature maps only detects anchor in a specific scale, ignoring anchor not in a specific scale. In this way, SNIP [19] not only detects the small targets in the image through the image pyramid processing, but also does not ignore the targets of other scales because of focusing on detecting the small targets. Moreover, each feature map [5] only detects anchor with a specific size, which avoids the disadvantage of large calculation amount of traditional image pyramid, and avoids the increase of calculation amount and calculation time caused by image scale enlargement.

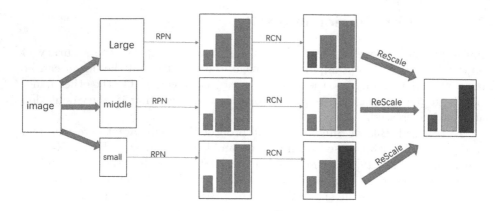

Fig. 1. The network structure of SNIP.

4   Modification of SNIP

Problems in SNIP. SNIP [19] is not a target detection and recognition algorithm, but a sampling strategy for input images. The results of sampling will be inputed into the target detection and recognition algorithm. In addition, another contribution of SNIP [19] is to solve the time consuming issue of traditional image pyramid. However, although the SNIP [19] method has all the above advantages and achieved good results in the Microsoft COCO dataset [12], it only considers the preprocessing of images. The chips obtained from the image pyramid sampling for Faster R-CNN [18] network was directly input into the Faster R-CNN [18] network after NMS [14], and the original Faster R-CNN [18] network was directly used instead of optimization. However, due to the small size of small objects on road, such as stones, bricks, tires, etc, relative to the size of the car, pedestrians, etc, the scale distribution range in the pictures of the dataset is too large. Therefore, it is necessary to optimize Faster R-CNN [18] network, so as to achieve higher accuracy in the detection of small targets.

Focal Loss. This paper draws on Focal Loss [11] proposed by Tsung-Yi Lin et al. Focal Loss is proposed for one stage algorithms with lower detection accuracy than two stage algorithms, and solves the problem of sample class imbalance in training samples. Object detection algorithms can be divided into two categories: the two stage detection algorithm and the one stage detection algorithm. The former refers to the detection algorithm that requires region proposal like Faster R-CNN [18] and RFCN [3], which can achieve a high accuracy rate but at a low speed. Although the speed can be increased by reducing the number of proposals or reducing the resolution of the input image, the speed is not improved in a qualitative way. The latter refers to detection algorithms like YOLO [17] and SSD [13], which do not require region proposal but direct regression. Such algorithms are fast but not as accurate as the former ones. In view of this, we propose a loss function for focusing the area priori of small target, named focal-area loss.

The Focal Loss [11] can be defined as:

$$FL(p_t) = -\alpha_t(1 - p_t)^\gamma log(p_t),\qquad(1)$$

where p_t is defined as below:

$$p_t = \begin{cases} p & \text{if y=1} \\ 1 - p & \text{otherwise,} \end{cases}\qquad(2)$$

Where $y = \{-1, 1\}$ and 1 represents target label and -1 represents background class, $p \in [0,1]$ is the probability, α_t and γ are two constant.

The equilibrium factor α_t is designed to balance the uneven distribution of positive and negative examples. The factor γ is used to reduce the loss of easily classified positive or negative examples, making the loss function more focused on difficult samples that are prone to be misclassified.

Focal-Area Loss. As for small objects detection on road, we also found that in the images of the Lost And Found dataset [15], the number of small objects is far less than the number of large objects, and the area of large objects is far greater than the one of small objects, but these are not reflected in the network of SNIP. However, based on the definition of COCO, the maximum size of the small target is 32×32, while the scale of most large targets is more than 100×100, the area is usually dozens of times than the area of small targets. Therefore, a loss function focusing the target area variable is needed to make the large target less important to than the small target. By referring to Focal Loss [11], We designed a new loss and called it focal-area loss.

The focal-area loss is defined as:

$$F(p_t) = -(\frac{1}{log(Area)})^\alpha (1 - p_t)^\gamma log(p_t)\qquad(3)$$

where p_t is the same meaning as the one in the Focal Loss, $Area$ means the maximum ground truth area corresponding to an anchor. Therefore, for large objects, due to the area of its ground truth is relatively large, the $\frac{1}{log(Area)}$ coefficient is relatively small and vice versa for small objects. In this way, the contribution of the small object to the classification loss function is boosted, and the contribution rate of the large target to the classification loss function is reduced.

5 Experimental Results

The dataset used in this experiment is the Lost And Found dataset [15]. The Lost And Found dataset [15], which addresses unpredictable microscopic obstacles on the road. The dataset uses a binocular camera to vision data. The dataset includes 112 stereo video sequences with 2104 annotated frames. The dataset contains seven broad categories of objects, namely Miscellaneous,

Counter hypotheses, Standard objects, Random hazards, Random non-hazards, "Emotional" hazards, and Humans. Obstacles in the dataset include cardboard boxes, bumpers, tires, children's toy cars, children, etc., which are basically small objects commonly found on traffic roads to meet the needs of the experiment. Since the dataset is collected by a binocular camera, and this experiment does not involve the depth problem of the image object, so it is sufficient to be used only by any of the binocular cameras. In this experiment, a total of 2239 images in the dataset were used, and 1990 of them were randomly selected as the training dataset, and 249 were left as the test dataset.

Evaluation Index. We evaluate and analyze the test results by using mAP and PR curve. mAP's full name is mean Average Precision, the average value of AP of all categories. mAP is often used in objection detection to measure the detection accuracy. PR curve is the curve between precision and recall. The more the curve is skewed to the right and above, the better the precision of model detection is. And mathematics proves that AP is the area under the PR curve. The specific experimental results are shown in the figure and table below.

Experimental Results. Figure 2 shows the different detection results of the same picture for SNIP [19] with and without focal-area loss. The above three pictures, from left to right, are the original pictures, the detection results of the SNIP [19] not using focal-area loss and the detection results of SNIP [19] using focal-area loss. It can be seen that the SNIP [19] using focal-area loss can detect better results than the original SNIP [19].

Fig. 2. Comparison of some experimental results from the left to the right, the detection results of the SNIP without focal-area loss and the detection results of SNIP with focal-area loss.

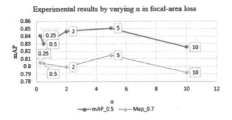

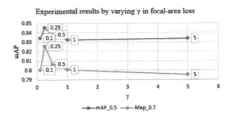

Fig. 3. Varying α in focal-area loss.   **Fig. 4.** Varying γ in focal-area loss.

The experimental quantitative results are show in Figs. 3 and 4.

The other experimental conditions were epoch $= 7$, batch size $= 2$. In Fig. 3 the γ is 2 and learning rate is 0.015; In Fig. 4 the α is 5, learning rate is 0.015. It can be seen from the above Figs. 3 and 4 that when other conditions remain unchanged, the detection result is the best when $\alpha = 5$. This also applies when $\gamma = 0.25$. Therefore, we can consider that when $\alpha = 5$ and $\gamma = 0.25$, the detection result of SNIP with focal-area loss is the best. In addition to adjusting α and γ, we also adjusted the learning rate. The mAP of the detection results under different learning rates is shown in the Table 2.

Table 2. Experiment by varying learning rate.

Learning rate	SNIP without focal-area loss		SNIP with focal-area loss	
	$mAP_{0.5}$	$mAP_{0.7}$	$mAP_{0.5}$	$mAP_{0.7}$
0.01	0.8368	0.8140	**0.8437**	0.8073
0.02	0.8422	0.8035	**0.8567**	0.8258
0.0225	0.8414	0.8070	**0.8565**	0.8288
0.025	0.8416	0.8127	**0.8570**	0.8012
0.03	0.8366	0.8064	**0.8434**	0.8061

Apart from the figures and table above, we also made Fig. 5 of the PR curve of SNIP with focal-area loss and SNIP without focal-area loss under the conditions of $\alpha = 5$, $\gamma = 0.25$ and learning rate $= 0.0225$. In Fig. 5, The top two pictures use focal-area loss, while the bottom two does not. Obviously, under different ovthresh, the area surrounding the lower left of the PR curve with focal-area loss is skewed to the up right than that without focal-area loss. This also proves that focal-area loss can improve the performance of SNIP in detecting small targets.

As we can see from the experimental data, the influence of α on the final detection accuracy is small, while the influence of γ on the final detection accuracy is large. In addition, it is obvious that when the value of α is 5, the value of

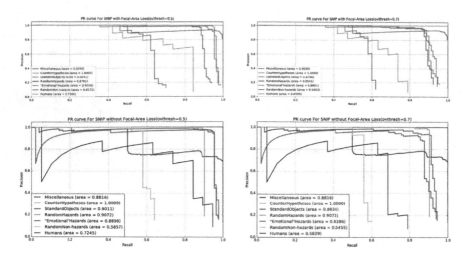

Fig. 5. Comparison of PR curve

γ is 0.25, and the learning rate is 0.0225, the final detection accuracy improves the most compared with the original SNIP [19] algorithm. Specifically, the detection result after changing the loss function is that $mAP_{0.5}$ is 85.65% and $mAP_{0.7}$ is 82.88%. $mAP_{0.5}$ and $mAP_{0.7}$ of the original SNIP [19] object detection algorithm without changing the loss function are 84.14% and 80.70% respectively. It is obvious that $mAP_{0.5}$ is improved by 1.51% and $mAP_{0.7}$ by 2.18%, which is greatly improved compared with the original SNIP [19] algorithm. The results show that the proposed loss function focal-area loss based on target ground truth area is scientific and feasible. It is verified that the detection accuracy of target detection network can be improved by adjusting the contribution of target ground truth area to the loss function.

6   Conclusion

At present, the detection algorithms for pedestrian and vehicle have achieved high accuracy. However, due to the complexity of detection of small objects on road, the research on detection of small targets on road has been progressing slowly. Based on SNIP [19], this paper came up with a new loss function focal-area loss to improve the detection accuracy of small objects on road. The new network classification Loss function embeded the ground truth area of the detection target to the loss function. The algorithm has been successfully applied in Lost and Found dataset. Compared with the original SNIP [19], SNIP [19] using focal-area loss has a higher detection accuracy than the original SNIP [19] in almost all objects.

Acknowledgement. This work is supported by the National Natural Science Foundation of China (No. 61751308, 61773311 and 61603057), and China Postdoctoral Science Foundation (No. 2017M613152).

References

1. Bell, S., Zitnick, C.L., Bala, K., Girshick, R.B.: Inside-outside net: detecting objects in context with skip pooling and recurrent neural networks. In: Computer Vision and Pattern Recognition, pp. 2874–2883 (2016)
2. Chen, C., Liu, M.-Y., Tuzel, O., Xiao, J.: R-CNN for small object detection. In: Lai, S.-H., Lepetit, V., Nishino, K., Sato, Y. (eds.) ACCV 2016. LNCS, vol. 10115, pp. 214–230. Springer, Cham (2017). https://doi.org/10.1007/978-3-319-54193-8_14
3. Dai, J., Li, Y., He, K., Sun, J.: R-fcn: object detection via region-based fully convolutional networks. In: Neural Information Processing Systems, pp. 379–387 (2016)
4. Fu, C., Liu, W., Ranga, A., Tyagi, A., Berg, A.C.: Dssd : Deconvolutional single shot detector. arXiv Computer Vision and Pattern Recognition (2017)
5. Girshick, R.B., Donahue, J., Darrell, T., Malik, J.: Rich feature hierarchies for accurate object detection and semantic segmentation. In: Computer Vision and Pattern Recognition, pp. 580–587 (2014)
6. Goodfellow, I.J., et al.: Generative adversarial nets, pp. 2672–2680 (2014)
7. Hu, P., Ramanan, D.: Finding tiny faces. In: Computer Vision and Pattern Recognition, pp. 1522–1530 (2017)
8. Kisantal, M., Wojna, Z., Murawski, J., Naruniec, J., Cho, K.: Augmentation for small object detection. arXiv Computer Vision and Pattern Recognition (2019)
9. Kong, T., Yao, A., Chen, Y., Sun, F.: Hypernet: towards accurate region proposal generation and joint object detection. In: Computer Vision and Pattern Recognition, pp. 845–853 (2016)
10. Li, J., Liang, X., Wei, Y., Xu, T., Feng, J., Yan, S.: Perceptual generative adversarial networks for small object detection. In: Computer Vision and Pattern Recognition, pp. 1951–1959 (2017)
11. Lin, T., Goyal, P., Girshick, R.B., He, K., Dollar, P.: Focal loss for dense object detection. In: International Conference on Computer Vision, pp. 2999–3007 (2017)
12. Lin, T.-Y., et al.: Microsoft COCO: common objects in context. In: Fleet, D., Pajdla, T., Schiele, B., Tuytelaars, T. (eds.) ECCV 2014. LNCS, vol. 8693, pp. 740–755. Springer, Cham (2014). https://doi.org/10.1007/978-3-319-10602-1_48
13. Liu, W., et al.: SSD: single shot multibox detector. In: Leibe, B., Matas, J., Sebe, N., Welling, M. (eds.) ECCV 2016. LNCS, vol. 9905, pp. 21–37. Springer, Cham (2016). https://doi.org/10.1007/978-3-319-46448-0_2
14. Neubeck, A., Van Gool, L.: Efficient non-maximum suppression, vol. 3, pp. 850–855 (2006)
15. Pinggera, P., Ramos, S., Gehrig, S.K., Franke, U., Rother, C., Mester, R.: Lost and found: detecting small road hazards for self-driving vehicles. In: Intelligent Robots and Systems, pp. 1099–1106 (2016)
16. Razlaw, J., Quenzel, J., Behnke, S.: Detection and tracking of small objects in sparse 3d laser range data. arXiv Robotics (2019)
17. Redmon, J., Divvala, S.K., Girshick, R.B., Farhadi, A.: You only look once: unified, real-time object detection. In: Computer Vision and Pattern Recognition, pp. 779–788 (2016)
18. Ren, S., He, K., Girshick, R.B., Sun, J.: Faster r-cnn: Towards real-time object detection with region proposal networks. IEEE Trans. Pattern Anal. Mach. Intell. **39**(6), 1137–1149 (2017)
19. Singh, B., Davis, L.S.: An analysis of scale invariance in object detection - snip. In: Computer Vision and Pattern Recognition, pp. 3578–3587 (2018)
20. Singh, B., Najibi, M., Davis, L.S.: SNIPER: efficient multi-scale training. In: Neural Information Processing Systems, pp. 9310–9320 (2018)

Multi-scale Feature and Spatial Relation Inference for Object Detection

Tianyu Zhou$^{(\boxtimes)}$, Zhenjiang Miao, and Jiaji Wang

Beijing Jiaotong University, Haidian District, Beijing 100044, China
{12120360,zjmiao,12112069}@bjtu.edu.cn

Abstract. Real scene images contain rich contextual visual cues. How to use contextual information is a fundamental problem in object recognition task. Contextual information mainly includes visual relevance and spatial relations. In this work we propose a context model that utilize both kinds of context. Our model is based on common detection framework Faster R-CNN. A multi-scale feature module extract features from different scales, and use a gated bi-directional structure to control message pass. This module collects contextual information from relevant region and different scales. A spatial relation inference module builds object spatial relations through a GRU structure, which can learn effective object spatial state. By special designing to adapt to CNN structure, both modules work well in the integrated detection model. And the whole model can be trained end-to-end. Experiments prove the effectiveness of our context model. Our method achieves 76.82% mAP on PASCAL VOC 2007 dataset, which outperforms Faster R-CNN by 3.5%.

Keywords: Object detection · Contextual information

1 Introduction

Object detection is a classic recognition task in the field of computer vision. The goal of object detection is to use bounding boxes to localize the region of every target object instance in an image, and to recognize object category of each region. Most of the current object detection methods use deep convolutional neural networks (DCNNs) [1], which aim to extract deep visual features of the image and build efficient detectors to generate accurate target regions. While these methods have made great breakthroughs and achieved good performance in many detection benchmarks, they only use visual features in local regions around object instances independently. Usually there are often close relationships between object instances in a same scene [2]. In other words, rich contextual information in an image can contact all independent object instances together. Taking good advantage of this information can help us infer the category of each target region more accurately.

Utilizing contextual information can improve the recognition result, which has been proved by many studies [2–4]. There are two main types of contextual

© Springer Nature Switzerland AG 2019
Y. Zhao et al. (Eds.): ICIG 2019, LNCS 11901, pp. 666–675, 2019.
https://doi.org/10.1007/978-3-030-34120-6_54

information between objects: visual relevance and spatial relations. There are two types of method for modeling the corresponding contextual information. One is to incorporate global scene visual features or local visual features around object regions [2,5–7],The other to describe spatial relations by building a structured model [8–13]. The goal of this paper is to utilize both types of contextual information, and build a better context model in detection task.

To achieve this goal, we use both multi-scale visual features and object relationships separately. In order to make better use of contextual visual information, we introduce multi-scale visual feature extraction. Different scales of contextual information may contain different object relationships. However, directly combined features may not be the most effective. To preserve the useful parts of multi-scale features, we use a gated bi-directional network (GBD-Net) [7,14]. This network passes messages between different scales of features, which can learn the visual relevance of neighbor regions. To model object spatial relations we are inspired by structure inference network (SIN) [15]. Structure inference network is a graphical model to infer object state, which takes objects as nodes and relationships between them as edges. It uses a structured manner to learn some spatial and semantic relationships between different visual concepts, e.g. a man is riding a horse, birds are in the sky. Both of the two modules can be used in a typical detection framework such as Faster R-CNN.

The significance of this paper is to combine two types of contextual information modeling methods and build an integrated model. Through the experiments on PASCAL VOC2007, we validate the effectiveness of our approach, and rich context cues can help on boosting the performance detection frameworks.

2 Related Work

Object Detection Pipeline. Most current object detection systems use deep convolutional neural networks (DCNNs) to extract deep visual features. The mainstream detection framework pipelines can be grouped into one of two types: region-based pipelines and unified pipelines. Region based framework was proposed by Girshick et al. [16] in the name of R-CNN. It regards detection task as region proposal and region classification task, which first generates many candidate boxes those may contain object instances, and then classify these regions using deep CNN structure. R-CNN framework has achieved the state-of-art detection performances on the mainstream benchmarks, and continued to be improved by many studies, e.g. Fast R-CNN [17], Faster R-CNN [18]. Although region-based approaches have achieved outstanding detection performances, they could be computationally expensive which may not suitable for some mobile devices. The other type of approaches is called unified pipelines, which use CNN to predict bounding box offsets and category probabilities directly from the whole images. The representative methods are YOLO [19] and SSD [20]. Unified pipelines are fast enough for real-time detection but with a little loss of accuracy. While for salient objects detection these methods work well, their performances are limited in complex scenes.

Contextual Visual Features. It is natural to believe that objects coexisting in the same scene usually have strong visual correlations. Many studies have validated the role of contextual visual features in object recognition tasks in early years [2–4]. With the widespread use of DCNN methods in the field of computer vision, many researchers tried to use CNN feature representations to utilize contextual information. It is well known that DCNNs can learn hierarchical feature representations, which implicitly integrate contextual information [5]. The value of contextual information based on DCNN representations can be further explored. MR-CNN [21] uses some additional net branches to extract features inside or around original object proposal, in order to obtain richer contextual representation. ION [6] uses skip pooling to extract multiple scales of information, and uses spatial recurrent neural networks to integrate contextual information outside the region of interest. Zeng *et al.* [7,14] proposed a gated bi-directional CNN (GBD-Net) to incorporate different scales of features of regions around objects, and pass effective message between different levels. From these methods, the effective extraction and integration of multi-region and multi-scale information is the key to build contextual visual features.

Spatial Relation Inference. The construction of an object spatial relation inference model is usually a combination of a graphical model and CNN structure [8–13,15]. Deng *et al.* [12] proposed structure inference machines, which use recurrent neural networks (RNNs) for analyzing relations in group activity recognition. Xu *et al.* [13] proposed a graph inference model which uses conditional random fields (CRF) and RNN structure to generate scene graph from an image. SIN [15] proposes a graph structure inference model, which can learn the representation of spatial relationships between objects during training stage, and infer object state during detection stage.

Our work is mainly based on the most effective detection methods at present, and proposes a detection model that effectively integrates two types of context modeling methods. This model makes full use of both visual relevance and spatial relations, which can further improve detection performance than original approaches.

3   Method

We use Faster R-CNN as our base detection framework, and added a multi-scale visual feature module and a spatial relation inference module. The multi-scale visual feature module uses different scales of local feature and whole image feature as contextual information. Different scales of features are incorporated through a gated bi-directional network (GBD-Net). The spatial relation inference module uses the scheme in structure inference network (SIN) to represent spatial relationships between objects. The whole model can be trained end-to-end. We explain the details of each part in the following subsections.

3.1   Main Framework

The main structure of our model consists of a base detection framework, a multi-scale feature module and a spatial relation inference module. We adopt Faster R-CNN as our base detection framework because it is a standard region based detection model. It first generates some candidate regions by region proposal network (RPN) on an image, and then classifies these regions through deep CNN structure. As either of the multi-scale feature module and the spatial relation inference module needs region proposals as input, so a region based framework is necessary. The whole framework of our detection model are shown in Fig. 1.

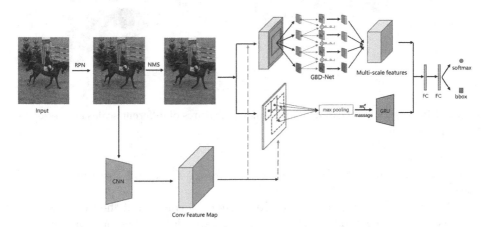

Fig. 1. The framework of the whole detection model. The base model is Faster R-CNN. It first generates region proposals using RPN, and calculate features through CNNs, shown in the bottom of the figure. NMS method selects best boxes and input them to the two context modules. The multi-scale feature module is shown in the top part, which integrate multi-scale features through GBD-Net structure. The spatial relation inference module is in the middle of the figure, which learns object spatial relations through a GRU.

3.2   Multi-scale Feature Module

This module is connected behind RPN. After RPN generates a large number of region proposals from an image, we use Non-Maximum Suppression (NMS [22]) to choose proper number of ROIs (Region of Interest). Multi-scale feature module takes these ROIs as input. For each ROI $r_0 = (x, y, w, h)$, we choose several scales of region with the same center location (x, y). In this paper we use three scales, with the parameters $\lambda_1 = 0.8$, $\lambda_2 = 1.2$ and $\lambda_3 = 1.8$, so the three regions $r_i = (x, y, \lambda_i w, \lambda_i h)$. We crop the size of r_i from the output feature map of the last convolutional layer f_1 as our local contextual features. We then use the whole feature map as global contextual features f_g. All fi are resized to 7

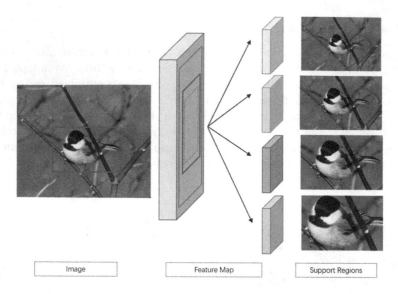

| Image | Feature Map | Support Regions |

Fig. 2. Different scales of candidate region. The features of different scales are obtained by roi-pooling.

$\times$ 7 $\times$ 512, which is the same size as f_g. Figure 2 illustrates different scales of features.

These multiple scales of features give richer information of the target object, but also may bring redundancy or noisy information. Different from directly concatenate all these features together, we use a gated bi-direction structure [7] to control information flow. For region r_i, we use h_i^0 in $\{h_1^0, h_2^0, h_3^0, h_4^0\}$ to represent the original input features f_1, f_2, f_3, f_g in different scales, h_i^1 and h_i^2 are middle features, and h_i^3 is the output features. The detailed calculation process can be represented as follows:

$$h_i^1 = \sigma(h_i^0 \times w_i^1 + b_i^{0,1}) + G(h_{i-1}^0, w_{i-1,i}^g + b_{i-1,i}^g) \cdot \sigma(h_i^0 \times w_{i-1,i}^1 + b_i^1), \quad (1)$$

$$h_i^2 = \sigma(h_i^0 \times w_i^2 + b_i^{0,2}) + G(h_{i+1}^0, w_{i+1,i}^g + b_{i+1,i}^g) \cdot \sigma(h_i^0 \times w_{i+1,i}^2 + b_i^2), \quad (2)$$

$$h_i^3 = \sigma(cat(h_i^1, h_i^2) \times w_i^3 + b_i^3), \quad (3)$$

$$G(x, w, b) = sigm(x \times w + b) \quad (4)$$

h_i^1 and h_i^2 represent two directions of feature integration, and h_i^3 is the concatenation of h_i^1 and h_i^2. $G(x, w, b)$ is a sigmoid function as the gate to control effective messages and prevent noisy information. More implementation details can refer to [7]. With the help of multi-scale feature module, the detection model can get optimized features with rich contextual information from different scales, which for region classification and bounding box regerssion.

3.3   Spatial Relation Inference Module

In this module, we model object relations as a graph $G = (V, E)$. Each node $v \in V$ represents an object, and each edge $e \in E$ represents the spatial relationships between a pair of objects. We define the spatial relationships in the similar form in SIN method [15]. Different from SIN, our module only encodes message from object spatial relationships, but no message from scene. That is because the form of message from scene is similar to the feature form in multi-scale visual feature module, so we drop this part to avoid repetitive work. Each ROI is a node in G, which can be represented as $v_i = (x_i, y_i, w_i, h_i, s_i, f_i)$, where (x_i, y_i) is the center location, wi and hi are the width and height, s_i is the area of v_i, and fi is the convolutional feature vector of v_i. The spatial relationship of v_i and v_j can be represented as edge R_{ij}:

$$R_{ij} = [w_i, h_i, s_i, w_j, h_j, s_j, \frac{x_i - x_j}{w_j}, \frac{y_i - y_j}{h_j},$$
$$\frac{(x_i - x_j)^2}{w_j{}^2}, \frac{(y_i - y_j)^2}{h_j{}^2}, \log\left(\frac{w_i}{w_j}\right), \log\left(\frac{h_i}{h_j}\right)] \tag{5}$$

The form of R_{ij} is a set of relationships between v_i and v_j including location, shape, orientation and other type of spatial relationships. Thus we get the content of each node and edge in G. Similar to our multi-scale visual feature module, we use a RNN structure to learn spatial relations during training process. In this module we use GRU [23] to learn and remember long-term information. GRU is a lightweight but efficient memory cell to enhance useful spatial relation state, and ignore useless state. It uses an aggregation function to fuse messages from object spatial relationships into a meaningful representation. Implementation details of GRU can refer to [23]. For each node v_i, message from object spatial relationships mei comes from an integration of all other nodes. The integrated message can be calculated as:

$$m_i^e = \max_{j \in V} pooling(e_{j \to i} * f_j), \tag{6}$$

$$e_{j \to i} = relu(W_p R_{ij}) * tanh(W_v * [f_i, f_j]). \tag{7}$$

W_p and W_v are weight matrixes. Maxpooling is to integrate important messages from all other nodes. Visual feature vector $[f_i, f_j]$ present visual relationship. Detailed calculation process is mostly same as the form in SIN [15]. After training on different kinds of images, the final hidden state of GRU can learn effective spatial relation state representations, and finally help to predict the region category and the offsets of bounding box.

4   Experimental Results

4.1   Implementation Details

We use Faster R-CNN with a VGG-16 model [24] pre-trained on ImageNet [1] as base detection framework, all parameters are same as the original paper [18].

Specially, we use NMS [25] during training and testing process, which chooses top 100 rated proposals as ROIs. We use PASCAL VOC Dataset to train and test our method. In each set of experiment, we use VOC 07 trainval and VOC 12 trainval as training set, and VOC 07 test as test set. We compared the detection results of the baseline model, baseline model with either single module in our paper, and with both modules. In addition, we also compared the relevant methods under the same conditions. For all experiments, we use dynamic learning rate: $5e^{-4}$ as initial for 80 K iterations and $5e^{-5}$ for the rest 50 K iterations. We choose $1e^{-4}$ as weight decay and 0.9 as momentum.

4.2   Results

The results of mAP of all compared methods are shown in Table 1. The upper part of Table 1 shows the results of baseline and methods in relevant papers. The lower part of Table 1 shows the results of our methods, where the effect of each module can be seen. SIN edge is the structure inference network method with edge GRU part only, which is the same as our spatial relation inference module. The result of GBD-Net is our implementation because the original method uses different base net and parameters. Table 2 shows detailed detection precision of our model on each category.

Table 1. Object detection mAP (%) on VOC 2007 test.

Method	mAP
Faster R-CNN	73.20
SIN edge	74.87
SIN	76.00
GBD-Net	76.32
Faster R-CNN +MSF	75.26
Faster R-CNN +SRI	74.73
Faster R-CNN +MSF+SRI	76.82

From the experimental results, we can see the significance of contextual information. In our methods, both multi-scale feature module and spatial relation inference module obviously improve detection accuracy comparing to the baseline. The two modules work well together and bring more accuracy improvement. These results benefit from reasonable context model design, which makes full use of the advantages of network structure, and is very suitable for end-to-end training.

Table 2. Object detection mAP (%) of each category on VOC 2007 test.

	Faster R-CNN	Ours
Aeroplane	76.5	**81.05**
Bicycle	79.0	**82.10**
Bird	70.9	**73.75**
Boat	65.5	**69.37**
Bottle	52.1	**58.51**
Bus	83.1	**87.03**
Car	84.7	**89.04**
Cat	86.4	**88.70**
Chair	52.0	**59.43**
Cow	81.9	**84.27**
Table	65.7	**71.51**
Dog	84.8	**88.53**
Horse	84.6	**87.82**
Motorbike	77.5	**82.31**
Person	76.7	**79.84**
Plant	**38.8**	37.96
Sheep	73.6	**78.14**
Sofa	**73.9**	73.67
Train	83.0	**86.46**
Tv	72.6	**76.81**
mAP	73.2	**76.82**

5 Conclusion

In this paper, we study the contextual information modeling methods in detection task. Contextual information is mainly divided into visual relevance and spatial relations. In order to adapt to the common CNN detection structure, we propose a context model with two special modules to two types of contextual information. The multi-scale feature module incorporates features in different scales, where a gated bi-directional structure helps to select useful context messages. The spatial relation inference module uses a graphical model to learn and predict object spatial relations. Both modules are designed to adapt to a CNN structure, which can be trained end-to-end. And our method achieved good performance on PASCAL VOC dataset.

Acknowledgments. This work is supported by the NSFC 61672089, 61273274, 61572064, and National Key Technology R&D Program of China 2012BAH01F03.

References

1. Krizhevsky, A., Sutskever, I., Hinton, G.E.: Imagenet classification with deep convolutional neural networks. In: Advances in Neural Information Processing Systems, pp. 1097–1105 (2012)
2. Divvala, S.K., Hoiem, D., Hays, J.H., Efros, A.A., Hebert, M.: An empirical study of context in object detection. In: 2009 IEEE Conference on Computer Vision and Pattern Recognition, pp. 1271–1278. IEEE (2009)
3. Torralba, A.: Contextual priming for object detection. Int. J. Comput. Vis. **53**(2), 169–191 (2003)
4. Oliva, A., Torralba, A.: The role of context in object recognition. Trends Cogn. Sci. **11**(12), 520–527 (2007)
5. Zeiler, M.D., Fergus, R.: Visualizing and understanding convolutional networks. In: Fleet, D., Pajdla, T., Schiele, B., Tuytelaars, T. (eds.) ECCV 2014. LNCS, vol. 8689, pp. 818–833. Springer, Cham (2014). https://doi.org/10.1007/978-3-319-10590-1_53
6. Bell, S., Lawrence Zitnick, C., Bala, K., Girshick, R.: Inside-outside net: detecting objects in context with skip pooling and recurrent neural networks. In: Proceedings of the IEEE Conference on Computer Vision and Pattern Recognition, pp. 2874–2883 (2016)
7. Zeng, X., Ouyang, W., Yang, B., Yan, J., Wang, X.: Gated bi-directional CNN for object detection. In: Leibe, B., Matas, J., Sebe, N., Welling, M. (eds.) ECCV 2016. LNCS, vol. 9911, pp. 354–369. Springer, Cham (2016). https://doi.org/10.1007/978-3-319-46478-7_22
8. Li, Y., Tarlow, D., Brockschmidt, M., Zemel, R.: Gated graph sequence neural networks (2015). arXiv preprint arXiv:1511.05493
9. Hu, H. , Zhou, G.-T., Deng, Z., Liao, Z., Mori, G.: Learning structured inference neural networks with label relations. In: Proceedings of the IEEE Conference on Computer Vision and Pattern Recognition, pp. 2960–2968 (2016)
10. Battaglia, P., Pascanu, R., Lai, M., Rezende, D.J., et al.: Interaction networks for learning about objects, relations and physics. In: Advances in Neural Information Processing Systems, pp. 4502–4510 (2016)
11. Jain, A., Zamir, A.R., Savarese, S., Saxena, A.: Structural-RNN: deep learning on spatio-temporal graphs. In: Proceedings of the IEEE Conference on Computer Vision and Pattern Recognition, pp. 5308–5317 (2016)
12. Deng, Z., Vahdat, A., Hu, H., Mori, G.: Structure inference machines: recurrent neural networks for analyzing relations in group activity recognition. In: Proceedings of the IEEE Conference on Computer Vision and Pattern Recognition, pp. 4772–4781 (2016)
13. Xu, D., Zhu, Y., Choy, C.B., Fei-Fei, L.: Scene graph generation by iterative message passing. In: Proceedings of the IEEE Conference on Computer Vision and Pattern Recognition, pp. 5410–5419 (2017)
14. Zeng, X., et al.: Crafting GBD-Net for object detection. IEEE Trans. Pattern Anal. Mach. Intell. **40**(9), 2109–2123 (2018)
15. Liu, Y., Wang, R., Shan, S., Chen, X.: Structure inference net: object detection using scene-level context and instance-level relationships. In: Proceedings of the IEEE Conference on Computer Vision and Pattern Recognition, pp. 6985–6994 (2018)
16. Girshick, R., Donahue, J., Darrell, T., Malik, J.: Rich feature hierarchies for accurate object detection and semantic segmentation. In: Proceedings of the IEEE Conference On Computer Vision And Pattern Recognition, pp. 580–587 (2014)

17. Girshick, R.: Fast R-CNN. In: Proceedings of the IEEE International Conference on Computer Vision, pp. 1440–1448 (2015)
18. Ren, S., He, K., Girshick, R., Sun, J.: Faster R-CNN: towards real-time object detection with region proposal networks. In: Advances in Neural Information Processing Systems, pp. 91–99 (2015)
19. Redmon, J., Divvala, S., Girshick, R., Farhadi, A.: You only look once: unified, real-time object detection. In: Proceedings of the IEEE Conference on Computer Vision and Pattern Recognition, pp. 779–788 (2016)
20. Liu, W., et al.: SSD: single shot multibox detector. In: Leibe, B., Matas, J., Sebe, N., Welling, M. (eds.) ECCV 2016. LNCS, vol. 9905, pp. 21–37. Springer, Cham (2016). https://doi.org/10.1007/978-3-319-46448-0_2
21. Gidaris, S., Komodakis, N.: Object detection via a multi-region and semantic segmentation-aware CNN model. In: Proceedings of the IEEE International Conference on Computer Vision, pp. 1134–1142 (2015)
22. Felzenszwalb, P.F., Girshick, R.B., McAllester, D., Ramanan, D.: Object detection with discriminatively trained part-based models. IEEE Trans. Pattern Anal. Mach. Intell. **32**(9), 1627–1645 (2009)
23. Cho, K., Van Merriënboer, B., Bahdanau, D., Bengio, Y.: On the properties of neural machine translation: Encoder-decoder approaches (2014). arXiv preprint arXiv:1409.1259
24. Simonyan, K., Zisserman, A.: Very deep convolutional networks for large-scale image recognition (2014). arXiv preprint arXiv:1409.1556
25. Everingham, M., Van Gool, L., Williams, C.K.I., Winn, J., Zisserman, A.: The pascal visual object classes (voc) challenge. Int. J. Comput. Vis. **88**(2), 303–338 (2010)

Deep Stacked Bidirectional LSTM Neural Network for Skeleton-Based Action Recognition

Kai Zou, Ming Yin$^{(\boxtimes)}$, Weitian Huang, and Yiqiu Zeng

School of Automation, Guangdong University of Technology, Guangzhou, China
`yiming@gdut.edu.cn`

Abstract. Skeleton-based action recognition has made great progress recently. However, many problems still remain unsolved. For example, the representations of skeleton sequences learned by most of the existing methods lack spatial structure information and detailed temporal dynamics features. To this end, we propose a novel Deep Stacked Bidirectional LSTM Network (DSB-LSTM) for human action recognition from skeleton data. Specifically, we first exploit human body geometry to extract the skeletal modulus ratio features (MR) and the skeletal vector angle features (VA) from the skeletal data. Then, the DSB-LSTM is applied to learning both the spatial and temporal representation from MR features and VA features. This network not only leads to more powerful representation but also stronger generalization capability. We perform several experiments on the MSR Action3D dataset, Florence 3D dataset and UTKinect-Action dataset. And the results show that our approach outperforms the compared methods on all datasets, demonstrating the effectiveness of the DSB-LSTM.

Keywords: Deep learning · Skeleton-based action recognition · Bidirectional LSTM

1 Introduction

In computer vision, human action recognition plays a fundamental and import role. It has many important applications including video surveillance, human-computer interaction, game control, sports video analysis and so on [1].

Action recognition is a challenging task in the computer vision community. According to the type of the input data, the existing approaches can be grossly divided into two categories: RGB video based action recognition and 3D skeleton data based action recognition [2]. The RGB video based action recognition

The Project was supported in part by NSF China (No. 61876042), Science and Technology Planning Project of Guangdong Province (No.2017A010101024), and in part by Special Funds for Scientific and Technological Innovation and Cultivation of Guangdong University Students, China (No. pdjh2019b0153).

Y. Zhao et al. (Eds.): ICIG 2019, LNCS 11901, pp. 676–688, 2019.
https://doi.org/10.1007/978-3-030-34120-6_55

methods mainly focus on modeling spatial and temporal representation from RGB frames and temporal optical flow. Despite RGB video based methods have achieved promising results [1,3], there still exit some limitation, *e.g.*, background clutter, illumination changes, and so on. Moreover, the spatial appearance only contains 2D information that is hard to capture all the action information, and the optical flow generally is computing intensive. On the other hand, Johansson et al. [4] have recognized that 3D skeleton sequences can effectively represent the dynamics of human actions. Moreover, skeleton sequence does not contain color information and is not affected by the limitations of RGB video. Such robust representation allows to model more discriminative temporal characteristics about human actions. Beside, the skeleton sequences can be captured by the Kinect [5] and the advanced human pose estimation algorithms [6]. In the last decades, skeleton-based human action recognition has attracted more and more attention [1,2]. In this paper, we focus on the problem of skeleton based action recognizing.

For skeleton based action recognition, the existing methods explore different models to learn spatial and temporal features of skeleton sequences. There has been a lot of existing methods applying relative joints coordinates, Vemulapalli et al. [8] utilized rotations and translations to represent the 3D geometric relationships of body parts in Lie group, which can overlook absolute movements of skeleton joints. Recently, there is a growing trend toward Long Short-Term Memory (LSTM) networks based methods, Shahroudy et al. [9] introduce a part-aware LSTM network to further improve the performance of the action recognition. Despite the great improvement in performance, there exist two urgent problems needed to be solved [7]. First, human behavior is accomplished in coordination with each part of the body. It is very difficult to capture the high-level spatial structural information within each frame if directly feeding the concatenation of all body joints into networks. Second, these methods utilize LSTM to directly model the overall temporal dynamics of skeleton sequences. The hidden representation of the final RNN is used to recognize the actions. For the task of action recognition, the current prediction depends not only on the past but also on the expectations of the future. However, the last hidden representation cannot completely contain the detailed temporal dynamics of sequences.

In this paper, we propose a novel Deep Stacked Bidirectional LSTM Network (DSB-LSTM) for this task, which can effectively solve the above challenges. Figure 1 shows the overall pipeline of our model. Firstly, we propose a feature extraction method to capture modulus ratio spatial features and vector angle spatial features within each frame. Then we concatenate this spatial features and normalize them using the mask layer. Next, we apply base on deep stacked Bidirectional LSTM layers to model spatial-temporal features. Finally, a fully connected layer and a softmax layer are performed on the obtained representation to classify the actions. The main contributions of this work are summarized as follows:

- We propose a novel DSB-LSTM network for skeleton-based action recognition, which is able to effectively capture discriminative spatiotemporal information.
- Model is performed extensive experiments to demonstrate the effectiveness. Experimental results on three action recognition datasets consistently demonstrate the effectiveness of the proposed model.

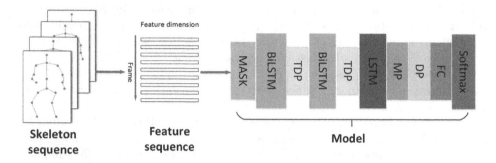

Fig. 1. The architecture of the proposed Deep Stacked Bidirectional LSTM Network (DSB-LSTM). MASK layer fills the sequence of different frames to normalized frame number. TDP layer strengthen the robustness of the model. MP layer enhances representation of temporal features. BiLSTM denote bidirectional LSTM.

2   Related Work

Human action recognition based on skeleton data has received a lot of attention, due to its effective representation of motion dynamics. Traditional skeleton-based action recognition methods mainly focus on designing hand-crafted features [8,10,11]. Vemulapalli et al. [4] represent each skeleton using the relative 3D rotations between various body parts. The relative 3D geometry between all pairs of body parts is applied to represent the 3D human skeleton in [8].

Recent works mainly learn human action representations with deep learning networks. Du et al. [1] divide human skeleton into five parts according to the human physical structure, and then separately feed them into a hierarchical recurrent neural network to recognize actions. A spatial-temporal attention network learns to selectively focus on discriminative spatial and temporal features in [11]. Zhang *et al.* [12] present a view adaptive model for skeleton sequence, which is capable of regulating the observation viewpoints to the suitable ones by itself. The works in [13] further show that learning discriminative spatial and temporal features is the key element for human action recognition. A hierarchical CNN model is presented in [14] to learn representations for joint co-occurrences and temporal evolutions. A spatial-temporal graph convolutional network (ST-GCN) is proposed for action recognition in [13]. Each spatial-temporal graph convolutional layer constructs spatial characteristics with a graph convolutional operator, and models temporal dynamic with a convolutional operator. Compared with ST-GCN, Si et al. [11] apply graph neural networks to capture spatial structural information and then use LSTM to model temporal dynamics. Despite the significant performance improvement in [11], it ignores the co-occurrence relationship between spatial and temporal features. In this paper, we propose a novel DSB-LSTM network that can not only effectively extract discriminative spatial and temporal features but also explore the relationship between spatial and temporal domains.

3   Method

In this section, we first review some necessary backgrounds. Then, we introduce features extracted from the skeleton sequence. Finally, a novel deep DSB-LSTM is proposed of action recognition are discussed.

3.1   Preliminaries

Recurrent neural networks (RNN) have an internal state to exhibit dynamic temporal behavior, which make them naturally suitable for supervised sequence labelling. They map an input sequence to another output sequence, and can process sequences with arbitrary length. The advanced RNN architecture of LSTM which can learn long-range dependencies. The hidden state representation h_t of an unit at each time step t is updated by:

$$f_t = \sigma(W_f \cdot [h_{t-1}, x_t] + b_f),$$

$$i_t = \sigma(W_i \cdot [h_{t-1}, x_t] + b_i),$$

$$C_t = f_t * C_{t-1} + i_t * tanh(W_c * [h_{t-1}, x_t] + b_c), \tag{1}$$

$$o_t = \sigma(W_o \cdot [h_{t-1}, x_t] + b_o),$$

$$h_t = o_t * tanh(C_t),$$

where x_t denotes the input, and i_t, f_t, o_t denote the internal representations correspond to the input gate, forget gate and output gate, respectively. All the matrices W are the connection weights and all the variables b are biases. The gates are used to determine when the input is significant enough to remember, when it should continue to remember or forget the value, and when it should output the value.

Due to past and future contexts are important for sequence labelling, for the task of action recognition, the current prediction depends not only on the past but also on the expectations of the future. Bidirectional LSTM neuron (BiLSTM)[15] elegantly combine both forward and backward dependencies by using two separate recurrent hidden layers to present the input sequence. By using BiLSTM, the output sequence at each time step provides complete historical and future contexts. BiLSTM hidden state representation h_t is update by:

$$h_t^l = f_h^{(l)}(h_t^{(l-1)}, h_{t-1}^l), \tag{2}$$

where $h_t^{(l)}$ is the hidden state of the l-th level at time step t, and $f_h^{(l)}$ is nonlinear function of the BiLSTM unit. When $l = 1$, the state is computed using x_t instead of $h_t^{(l-1)}$.

3.2   Skeleton Sequence Feature Extraction

A human body can be represented by a stick figure called human skeleton, which consists of line segments linked by joints, and the motion of joints can provide the key to motion estimation and recognition of the whole figure. Given a human subject, the skeleton data involves two geometric constraints. First, as a bone length is constant, the distance and direction between two adjacent points along a connected segment is fixed. Second, Fixed range of rotation angles at each joint. Based on above observations, the skeleton data conveys two types of information: Skeleton sequence modulus ratio, skeleton sequence vector angles. To make the most of the ability of deep networks to learn representations from raw data, extracted MR and VA features should have excellent spatiotemporal. The details are presented as follows.

Skeleton Sequence Modulus Ratio: Skeleton data contains the positions of joints in a frame. The relative position of the joints will vary with body type, the captured joints are based on the position of the sensor. In order to overcome the influence of the subject body type and sensor's coordinate system, we uses the joints modulus ratio to characterize of the relative change of the skeletal point position. We sets the subject body's hip joint as the new coordinates, converts others joints coordinate on the sensor's coordinate into the new coordinate. The resulting new coordinate calculation formula is shown in formula (3) by:

$$f = p_n - p_0, \qquad (n = 1, 2, 3, ..., N), \tag{3}$$

Where p_0 is coordinates of the hip joint and p_n denotes coordinates of others. f denotes vector of others joints to hip. Feature vector of each frame $f^t = [f_x^t, f_y^t, f_z^t]$, all sequences are characterized by $F = [f^1, f^2, ..., f^t, ..., f^T]$. In order to balance the effects of subject body difference, we normalize each feature vector using the distance from the head to the hip.Where h is distance from the head to the center of the hip, finally modulus ratio by,

$$\overline{f} = f/h. \tag{4}$$

Skeleton Sequence Vector Angles: The joints drive the interconnected joint movements, causing the angle of the interconnected joint to change, thereby Skeleton sequence vector angles indirectly reflecting human action. In order to obtain the skeletal sequence vector angle, we need to construct the skeletal structure vector from the skeletal data.Base on geometric constraints, we first constructed 22 structure vectors are shown in Table 1. The t on Table 1 $V_{lshoulder\_to\_lelbow}$ denotes a vector from left shoulder to left elbow. For example,the t frame of the action sequence left shoulder point $p_{lshoulder}^t = (x_1^t, y_1^t, z_1^t)$, left elbow point $p_{lelbow}^t = (x_2^t, y_2^t, z_2^t)$, left wrist point $p_{lwrist}^t = (x_3^t, y_3^t, z_3^t)$. $V_{lshoulder\_to\_lelbow}$ and $V_{lelbow\_to\_lwrist}$ calculation formulas are as :

$$
\begin{aligned}
V_{lshoulder\_to\_lelbow}^t &= (x_1^t - x_2^t, y_1^t - y_2^t, z_1^t - z_2^t), \\
V_{lelbow\_to\_lwrist}^t &= (x_2^t - x_3^t, y_2^t - y_3^t, z_2^t - z_3^t),
\end{aligned}
\tag{5}
$$

Table 1. The definition of the structure vectors.

Number	Vector	Number	Vector
0	$V_{lshoulder\_to\_lelbow}$	11	$V_{mspine\_to\_rhip}$
1	$V_{lelbow\_to\_lwrist}$	12	$V_{mspine\_rhip\_lhip}$
2	$V_{rshoulder\_to\_relbow}$	13	$V_{neck\_to\_lshoulder}$
3	$V_{relbow\_to\_rwrist}$	14	$V_{neck\_to\_rshoulder}$
4	$V_{lhip\_to\_lknee}$	15	$V_{mspine\_to\_lelbow}$
5	$V_{lknee\_to\_lankle}$	16	$V_{mspine\_to\_lhand}$
6	$V_{rhip\_to\_rknee}$	17	$V_{mspine\_to\_relbow}$
7	$V_{rknee\_to\_rankle}$	18	$V_{mspine\_to\_rhand}$
8	$V_{mspine\_to\_neck}$	19	$V_{mspine\_to\_head}$
9	$V_{mspine\_to\_lshoulder}$	20	$V_{head\_to\_rhand}$
10	$V_{mspine\_to\_rshoulder}$	21	$V_{head\_to\_lhand}$

Table 2. The calculated angles from the structure vectors.

Number	Angle	Number	Angle
0	$\theta_{neck\_rshoulder\_relbow}$	10	$\theta_{rshoulder\_mspine\_rhip}$
1	$\theta_{rshoulder\_relbow\_rwirst}$	11	$\theta_{mspin\_rhip\_rknee}$
2	$\theta_{neck\_lshoulder\_lelbow}$	12	$\theta_{rhip\_rknee\_rankls}$
3	$\theta_{lshoulder\_lelbow\_lwrist}$	13	$\theta_{lshoulder\_mspin\_lhip}$
4	$\theta_{neck\_rshoulder\_mspine}$	14	$\theta_{mspin\_lhip\_lknee}$
5	$\theta_{rshoulder\_relbow\_mspin}$	15	$\theta_{lhip\_lknee\_lankle}$
6	$\theta_{relbow\_rwirst\_mspin}$	16	$\theta_{neck\_mspin\_rhip\_rknee}$
7	$\theta_{neck\_lshoulder\_mspin}$	17	$\theta_{neck\_mspin\_rknee\_rankle}$
8	$\theta_{lshoulder\_lelbow\_mspin}$	18	$\theta_{neck\_mspin\_lhip\_lknee}$
9	$\theta_{lelbow\_lwrist\_mspin}$	19	$\theta_{neck\_mspin\_lknee\_lankle}$

Then, calculated angle by the structure vectors are shown in Table 2: As is shows in Table 2. For example, $\theta_{lshoulder\_lelbow\_lwrist}$ denotes angle formed by vector $V_{lshoulder\_to\_lelbow}$ and $V_{lelbow\_to\_lwrist}$. each Skeleton sequence vector angles is obtained by the following cosine theorem:

$$\theta_{lshoulder\_lelbow\_lwrist} = \frac{V_{lshoulder\_to\_lelbow} \cdot V_{lelbow\_to\_lwrist}}{|V_{lshoulder\_to\_lelbow}||V_{lelbow\_to\_lwrist}|}, \tag{6}$$

where the $V_{lshoulder\_to\_lelbow} \neq 0$ and $V_{lelbow\_to\_lwrist} \neq 0$.

3.3   Deep Stacked Bidirectional LSTM Neural Network

The DSB-LSTM for action recognition is shown in Fig. 1. Feeding Skeleton sequence modulus ratio and angle as the inputs. The backbone of our network consists of two BiLSTM layers and a LSTM layer due to its excellent performance for classification. A temporal max pooling (MP) layer along the time axis is placed on top of the last LSTM layer to obtain a time invariant vector representation of the sequence. After that, dropout (DP) is employed and a fully-connected (FC) layer with softmax activation is used to classify actions. In particular, to facilitate feature learning and improve model robustness, we introduce two novel layers: masking mechanism (MASK) layer and temporal dropout (TDP) layer. The details are described as follows.

Masking Mechanism: In reality, the length of the skeleton sequences collected in sensor is different. For the LSTM-based prediction problem, if the number of input data frames time series data contains missing/null values and inconsistent, the LSTM based model will fail due to null values cannot be computed during the training process. If the missing values are set as zero,or some other predefined values, the training and testing results will be highly biased. Thus, we adopt a masking mechanism to overcome the potential missing values problem.

Figure 2 demonstrates the details of the masking mechanism. BiLSTM cell denotes a BiLSTM layer. A mask value, ϕ, is pre-defined, which normally is 0 or Null. We employ the maximum of frames in sequence as the standard, sequence that not reach the maximum number of frames is processed with missing values. For an input MR features and VA features series data X_T, if x_t is the missed element, which equals to ϕ, the training process at the t-th step will be skipped, and thus, the calculated cell state of the $(t\text{-}1)$-th step will be directly input into the $(t+1)$-th step. In this case, the output of t-step also equals to ϕ, which will be considered as a missing value. Similarly, we can deal with input data with consecutive missing values using the masking mechanism.

Temporal Dropout: Skeletons collected by sensor may not always be reliable due to noise and occlusion. To address this problem, We adopt approach based on dropout [16], which improves model robustness. As is shown Fig. 3(a), for the standard dropout, each hidden unit is randomly omitted from the network with a probability of p_{drop} during training. For testing, all activations are used and $1 - p_{drop}$ is multiplied to account for the increase in the expected bias. Temporal

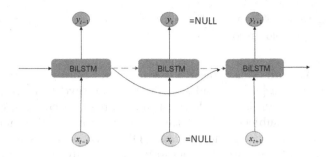

Fig. 2. Masking layer for time series data with missing values

dropout is slightly different from the standard dropout. Given the $T \times d$ matrix representation of a sequence, where T is the length of the sequence and d is the feature dimension, we only perform T dropout trials and extend the dropout value across the feature dimension. This technique is inspired by the spatial dropout to process the convolution feature 4D tensor [17]. We modify it for 3D tensor and apply it for feature learning from sequences. As shown in Fig. 3(b), the temporal dropout is performed before the bidirectional LSTM layer.

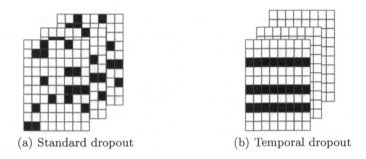

(a) Standard dropout   (b) Temporal dropout

Fig. 3. Black grids of feature map denotes the dropout part.

4   Experiment

we have evaluated our proposed model on three benchmark datasets: MSR Action3D dataset, Florence 3D Action, UTKinect-Action. The analysis of experimental results confirms the effectiveness of our model for skeleton-based action recognition.

4.1   Datasets

MSR Action3D Dataset: This dataset was captured using a depth sensor like Kinect. It consists of 20 actions performed by 10 subjects for two or three times.

Altogether, there are 557 valid action sequences, and each frame in a sequence is composed of 20 skeleton joints [8].

Florence 3D Action: This dataset was collected at the University of Florence using a Kinect camera [5]. It includes 9 actions: arm wave, drink from a bottle, answer phone, clap, tight lace, sit down, stand up, read watch, bow. Each action is performed by 10 subjects several times for a total of 215 sequences. The sequences are acquired using the OpenNI SDK, with skeletons represented by 15 joints instead of 20 as with the Microsoft Kinect SDK. The main challenges of this dataset are the similarity between actions, the human object interaction, and the different ways of performing a same action [13].

UTKinect-Action: This dataset was captured using a single stationary Kinect. It consists of 10 actions performed by 10 different subjects, and each subject performed every action twice. Altogether, there are 199 action sequences, and the 3D locations of 20 joints are given. This is regarded as a challenging dataset because of variations in the view point and high intra-class variations [8].

4.2   Results and Comparisons

MSR Action3D Dataset: MSR Action3D dataset follow the standard protocol provided in [8]. In this standard protocol, the dataset is divided into three action sets such as Action Set1 (AS1), Action Set2 (AS2) and Action Set3 (AS3). In Experiments, We use the samples of subjects 1, 3, 5, 7, 9 for training and the samples of subjects 2, 4, 6, 8, 10 for testing. The results of experiments as shown in Table 3 where MR and VA denote the skeleton sequence modulus ratio and angles, MR+VA denote combination of the two features. As shown in Table 3, the addition of MR and SM into LSTM makes the average accuracy increase by 5.31% and 8.09%, respectively, which indicates that our feature representation is very useful on this dataset. The DSB-LSTM are around 3.5% higher than the previous method [20]. The confusion matrices on the dataset are shown in Fig. 4 (a),(b),(c). We can see that the misclassifications. For example in Fig. 4(a), the action "Pick up&Throw" is often misclassified to "Bend" while the action "Forward-Punch" is misclassified to "TennisServe". Actually, "Pick up&Throw" just has one more "Throw" move than "Bend", and the "throw" move often holds few frames in the sequence. So it is very difficult to distinguish these two actions.

Florence 3D Action: Florence 3D Action dataset is follow the protocol [21], the dataset is benchmarked by leave-one-subject-out cross-validation. The results of experiments as shown in left on Table 4. As shown in Table 4, the DSB-LSTM achieves the best accuracy of 97.46% on the Florence 3D Action dataset. Where MR and VA denote the skeleton sequence modulus ratio and angles, MR+VA denote combination of the two features. As shown in Table 4, the using the

Table 3. The experimental results comparison on the MSR Action3D dataset.

Method	Year	AS1	AS2	AS3	Ave.
Bag of 3D [8]	2010	72.9	71.9	79.2	72.7
Hoj 3D [18]	2012	**87.9**	**85.5**	63.5	79.0
Eigen joints [19]	2012	74.5	76.1	**96.4**	82.3
Two-layer AP+HMM [20]	2017	83.8	75.0	93.7	84.17
LSTM	–	70.48	71.43	72.07	71.33
LSTM+MR	–	82.27	79.11	92.55	84.64
LSTM+VA	–	79.54	74.84	86.89	80.42
LSTM+MR+VA	–	84.89	80.20	93.24	86.11
DSB-LSTM+MR+VA	–	87.10	80.20	95.85	**87.72**

MR+SM feature is 2.17% and 3.24% higher than the single feature MR, SM. The DSB-LSTM are around 3.5% higher than the previous method [20]. The confusion matrices on Florence 3D Action dataset are shown in Fig. 4(d).

UTKinect 3D Action: UTKinect-Action dataset is follow the protocol [13], in which half of the subjects are used for training and the remaining are used for testing. The first 5 subjects are used for training while the last 5 subjects are used for testing. As shown in right on Table 4, the propose DSB-LSTM achieves the best accuracy of 95.96% on the UTKinect 3D Action dataset. The confusion matrices on UTKinect 3D Action dataset are shown in Fig. 4(e).

Table 4. Experimental result comparison on the Florence dataset and UTKinect dataset.

Florence 3D			UTKinect 3D		
Method	Year	Acc.	Method	Year	Acc.
Multi-part Bag-of Poses [21]	2013	82.00	STIPs [26]	2013	80.8
Motion trajectories [22]	2017	87.04	Primitive Pretrained [27]	2015	82.83
Elastic Function Coding [23]	2015	89.67	ST-LSTM+Trust Gate [28]	2016	95.0
Covariange descriptors [24]	2016	86.13	Geometric feature [28]	2017	**95.96**
Mining keyLatent SVM [25]	2017	87.00	HKC+MKL+MLE [29]	2018	94.95
LSTM+MR	–	94.95	LSTM+MR	–	91.92
LSTM+VA	–	93.88	LSTM+VA	–	78.08
LSTM+MR+VA	–	97.12	LSTM+MR+VA	–	93.13
DSB-LSTM+MR+VA	–	**97.46**	DSB-LSTM+MR+VA	-	**95.96**

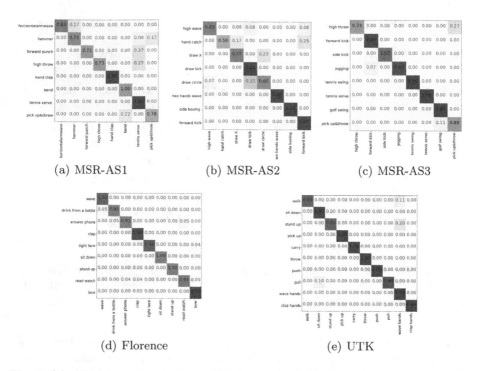

(a) MSR-AS1   (b) MSR-AS2   (c) MSR-AS3

(d) Florence   (e) UTK

Fig. 4. (a), (b), (c) are the result of MSR dataset is divided into three action sets, (c), (d) are the confusion matrices of Florence dataset and UTKinect dataset

5   Conclusion

In this paper, we propose an Deep Stacked Bidirectional LSTM Network(DSB-LSTM) for skeleton-based action recognition, which achieves promising results than the existing methods. We first extract modulus ratio features and angle features from the skeletal sequence. Then, DSB-LSTM is presented to effectively capture discriminative spatiotemporal feature. The success of our approach can be explained by introduction of masking layer and temporal dropout layer, such that impose action recognition precision and stronger generalization capability. Nevertheless, in the future we will try the combination of skeleton sequence and object appearance to promote the performance of human action recognition.

References

1. Simonyan, K., Zisserman, A., Ghahramani, Z., et al.: Two-stream convolutional networks for action recognition in videos. In: Advances in Neural Information Processing Systems, vol. 27, pp. 568–576 (2014)
2. Si, C.Y., Chen, W.T., Wan, W., et al.: An attention enhanced graph convolutional LSTM network for skeleton-based action recognition. In: The IEEE Conference on Computer Vision and Pattern Recognition (2019)

3. Wang, L.M., Xiong, Y., Wang, Z., et al.: Temporal segment networks: towards good practices for deep action recognition. In: European Conference on Computer Vision, pp. 20–36 (2016)

4. Johansson, G.: Visual perception of biological motion and a model for its analysis. Percept. Psychophys. **4**(2), 201–211 (1973)

5. Zhang, Z.: Microsoft kinect sensor and its effect. IEEE Multimed. **19**(2), 4–10 (2012)

6. Cao, Z., Simon, T., Wei, S.E., et al.: Realtime multi-person 2d pose estimation using part affinity fields. In: The IEEE Conference on Computer Vision and Pattern Recognition, pp. 7291–7299 (2017)

7. Si, C., Jing, Y., Wang, W., Wang, L., Tan, T.: Skeleton-based action recognition with spatial reasoning and temporal stack learning using temporal sliding LSTM networks. In: The European Conference on Computer Vision, pp. 103–118 (2018)

8. Vemulapalli, R., Arrate, F., Chellappa, R.: Human action recognition by representing 3d skeletons as points in a lie group. In: The IEEE Conference on Computer Vision and Pattern Recognition, pp. 588–595 (2014)

9. Shahroudy, A., Liu, J., Ng, T.T., et al.: Ntu rgb+d: A large scale dataset for 3d human activity analysis. In: The IEEE Conference on Computer Vision and Pattern Recognition, pp. 1010–1019 (2016)

10. Wang. J., Liu, Z., Wu, Y., et al.: Mining actionlet ensemble for action recognition with depth cameras. In: The IEEE Conference on Computer Vision and Pattern Recognition, pp. 1290–1297 (2012)

11. Hussein, M.E., Torki, M., Gowayyed, M.A., El-Saban, M.: Human action recognition using a temporal hierarchy of covariance descriptors on 3d joint locations. In: International Joint Conference on Artificial Intelligence (2013)

12. Wenjun, Z., Junliang, X., et al.: View adaptive recurrent neural networks for high performance human action recognition from skeleton data. In: Proceedings of the IEEE International Conference on Computer Vision (2017)

13. Yan, S., Xiong Y., Lin, D., et al.: Spatial temporal graph convolutional networks for skeleton-based action recognition. In: International Joint Conference on Artificial Intelligence (2018)

14. Li, C., Zhong, Q., Xie, D. et al.: Co-occurrence feature learning from skeleton data for action recognition and detection with hierarchical aggregation. In: International Joint Conference on Artificial Intelligence (2018)

15. Schuster, M., Paliwal, K.K.: Bidirectional recurrent neural networks. Trans. Signal Process. **45**(11), 2673–2681 (1997)

16. Hinton, G.E., Srivastava,N., Krizhevsky, A. et al.: Improving neural networks by preventing co-adaptation of feature detectors, arXiv preprint arXiv:1207.0580, (2012)

17. Tompson, J., Goroshin, R., Jain, A., et al.: Efficient object localization using convolutional networks. In: The IEEE Conference on Computer Vision and Pattern Recognition, pp. 648–656 (2015)

18. Xia, L., Chen, C., Aggarwal, J.K., et al.: View invariant human action recognition using histograms of 3D joints. In: The IEEE International Conference on Computer Vision and Pattern Recognition, pp. 20–27 (2012)

19. Yang, X., Tian, Y.: Eigen joints based action recognition using Naitive-Bayes-nearest-neighbor. In: The IEEE International Conference on Computer Vision and Pattern Recognition, pp. 14–19 (2012)

20. Yuan, M., Chen, E., Gao, L.: Posture selection based on two-layer ap with application to human action recognition using HMM. In: IEEE International Symposium on Multimedia, pp. 359–364 (2017)

21. Seidenari, L., Varano, V., Berretti, S., et al.: Recongnizing actions from depth cameras as weakly aligned multi-part bag-of-poses. In: the IEEE International Conference on Computer Vision and Pattern Recognition Workshops (2013)
22. Ding, W.W., Liu, K., Li, G.: Human action recognition using spectral embedding to similarity degree between postures. In: Visiual Communications and Image Processing, Chengdu (2017)
23. Anirudh, R., Turaga, P., Su, J., Srivastava, A.: Elastic functional coding of human actions: from vector-fields to latent variables. In: the IEEE International Conference on Computer Vision and Pattern Recognition, pp. 3147–3155 (2015)
24. Youssef, C.: Spatiotemporal representation of 3d skeleton joints-based action recognition using modified spherical harmonics. Pattern Recogn. Lett. **83**, 32–41 (2016)
25. Li, X., Liao, D., Zhang, Y.: Mining key skeleton poses with latent SVM for action recognition. Appl. Comput. Intell. Soft Comput. **2017**, 11 (2017)
26. Zhu, Y., Chen, W., Guo, G.: Fusing spatiotemporal features and joints for 3d action recognition. In: The IEEE International Conference on Computer Vision and Pattern Recognition, pp. 486–491 (2013)
27. Liu, J., Shahroudy, A., Xu, D.: Spatio-temporal lstm with trust gates for 3d human action recognition. In: European Conference on Computer Vision, pp. 816–833 (2016)
28. Zhang, S., Liu, X., Xiao, J.: On geometric features for skeleton-based action recognition using multilayer lstm networks. In: IEEE Winter Conference on Applications of Computer Vision, pp. 148–157 (2017)
29. Ghorbel, E., Boonaert, J., Boutteau, R., et al.: An extension of kernel learning methods using a modified Log-Euclidean distance for fast and accurate skeleton-based human action recognition. Comput. Vis. Image Understand. **175**, 32–43 (2018)

Insect Recognition Under Natural Scenes Using R-FCN with Anchor Boxes Estimation

Hong-Wei Pang[1,2], Peipei Yang[1,2], Xiaolin Chen[3], Yong Wang[3],
and Cheng-Lin Liu[1,2(✉)]

[1] National Laboratory of Pattern Recognition, Institute of Automation,
Chinese Academy of Sciences, Beijing, China
{ppyang,liucl}@nlpr.ia.ac.cn
[2] University of Chinese Academy of Sciences, Beijing, China
panghongwei17@mails.ucas.ac.cn
[3] Key Laboratory of Zoological Systematics and Evolution, Institute of Zoology,
Chinese Academy of Sciences, Beijing, China
xlchen@ioz.ac.cn, 530468931@qq.com

Abstract. Insect species recognition is an important application of computer vision in zoology and agriculture. Most of existing methods resort to hand-crafted features and traditional classifiers, which usually give poor accuracy and apply only to elaborately taken full-size pictures. In this paper, we focus on a more challenging case where the images are taken in the wild with complex backgrounds, and propose to use a deep learning based detection model to deal with it. It exploits multi-class object detection to eliminate interferences from complex backgrounds, while taking advantages of deep learning to significantly improve the performance of recognition. After evaluating several popular detection methods, R-FCN is selected as the base model. To further improve its performance, we introduce a clustering algorithm for estimation of the anchor boxes instead of using predefined ones. The experimental results on a dataset of insect images collected in the wild prove the effectiveness of our proposed method in improving both accuracy and speed.

Keywords: Multi-class object detection · R-FCN · Deep learning · Insect recognition · Natural scenes

1 Introduction

Many species of insects are damaging to vegetables, fruits or other crops, which might have negative impact on agriculture and related international trade [1]. This requires rapid and accurate species identification where expertise is often lacking. Automatic or semi-automatic identification of insects are greatly needed for diagnosing causes of damage and quarantine protocols for the economically relevant species. Some computer-aided systems have been implemented in recent

© Springer Nature Switzerland AG 2019
Y. Zhao et al. (Eds.): ICIG 2019, LNCS 11901, pp. 689–701, 2019.
https://doi.org/10.1007/978-3-030-34120-6_56

decades for identifying harmful species [2–6], and the intelligent recognition approaches for insects living in natural scenes have also made some progresses recently [7–9].

For example, Favret et al. [4] used a sparse processing technique and support vector machine to successfully recognize the specimens. Wang et al. [5] proposed to use Gabor surface features in automated identification to further improve the recognition accuracy. Deng et al. [7] proposed to recognize the insect living in natural scene using natural statistics (SUN) model. In spite of the obvious progresses in insect recognition, existing approaches suffer from the following two limitations.

First, these methods apply to almost only elaborately taken pictures. Earlier approaches extract hand-crafted features and then predict the label of an image with a traditional classifier. They are effective for the images with clean background such as specimens, but liable to fail on the images taken in the wild due to their complex backgrounds. They can neither deal with the case where there are more than one species in an image. Recently, there appear a few works applying detection techniques to insect recognition [7,9,10]. However, in these works, the images are always taken elaborately, where the insects occupy most of the space in the image with a simple background as shown in Fig. 1. Up to now, we have not seen a method that recognizes insects freely taken in real natural scenes with complex backgrounds, which seriously limits its applicability.

Fig. 1. The images are almost taken elaborately, where the insects occupy most of the space in the image with a simple background.

Second, with the breakthrough of deep learning technology, computer vision made considerable progress and has been successfully applied in many fields. The deep learning has been the state-of-the-art technique for both image classification and object detection. Meanwhile, most of existing insect recognition approaches still resort to traditional classifiers with hand-crafted features. Thus, they cannot take full advantages of deep learning in its high accuracy and robustness.

In this paper, we propose an insect recognition for images taken freely under natural scenes by exploiting the technique of deep learning based multi-class object detection. Our approach exceeds previous works in the following aspects. (1) We focus on direct recognition of the images taken freely under natural scenes (Fig. 2), which are obviously different from the ones appearing in previous works

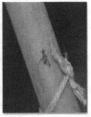

Fig. 2. The images in our dataset are taken in the wild more freely. Without prepro-cessing, insects occupy only small space of the whole images with complex backgrounds.

(Fig. 1). The backgrounds are much more complex and the insects occupy only small space in the images. This makes the recognition much more difficult. (2) We exploit multi-class object detection with deep learning for insect recognition. On one hand, the multi-class object detection model discovers the area of insect to be recognized directly from the image, rather than recognizing the whole image, and thus can effectively eliminate the interferences from complex natural backgrounds. It also easily solve the problem of multiple objects in one image. On the other hand, it exploits deep learning that is the state-of-the-art method in computer vision to significantly improve the recognition accuracy. Further-more, we evaluate several currently popular deep learning detection methods on our dataset and select R-FCN that gives the best performance as our base model. (3) We introduce the technique of estimating the anchor boxes by a clus-tering algorithm into R-FCN. It supplies more appropriate anchor boxes to the detection algorithm, which improves the recognition accuracy and reduces the training time.

The rest of this paper is organized as follows. Section 2 introduces the dataset for insect recognition in the wild and the selection of the base detection method. Section 3 introduces our insect recognition based on R-FCN with estimation of anchor boxes. Section 4 details the experimental setting and results, and Sect. 5 concludes this paper.

2 Dataset Construction and Selection of Basic Model

In this section, we will give a introduction to the dataset for insect recognition under natural scenes. Then we evaluate three object detection methods on this dataset, and select the one giving the best performance to be our base model.

2.1 Dataset Construction

As stated in Sect. 1, we aim at recognizing the insects from images taken under natural scenes. To obtain the insect recognition model, we first build the dataset. There are 19 species of insects to be recognized, and the images are collected by researchers of Key Laboratory of Zoological Systematics and Evolution, Institute of Zoology, Chinese Academy of Sciences. They travel to Yunnan Province, Xinjiang Province, Hebei Province, and suburb of Bejing City in China for data acquisition. The images are taken freely in the wild and there are often complex backgrounds. Some examples of the data are shown in Fig. 2.

We then build our dataset using the collected images. There are a total of 4,538 samples in 19 classes in our dataset. Each sample is an image of insect whose resolution is 5000 × 4000. Every class corresponds to an insect species to be recognized, and the names of the species are listed in Table 1. For each image, we annotate the objects with bounding boxes, and assign their labels to them. For convenience of evaluation, we divide the data into a training set and a test set with a ratio of 3 : 1. The number of samples in each class and the division of training/test set are also listed in Table 1.

Table 1. The names, numbers of samples, and the division of training/test set for all species in our dataset.

No.	Species	Images	Train	Test
1	Bemisia tabaci	120	90	30
2	Bactrocera dorsalis	295	221	74
3	Phenacoccus solenopsis	27	20	7
4	Vinsonia stellifera	180	135	45
5	Rhyncophorus ferrugineuss	338	253	85
6	Lissorbqptrus oryzqphilus	404	303	101
7	Leptinotarsa decemlineata	412	309	103
8	Bactrocera correcta	195	146	49
9	Anoplophora sp	333	250	83
10	Agrilus mali	307	230	77
11	Aleurodes proletella	189	142	47
12	Bactrocera cucurbitae	121	91	30
13	Batracomorphus pandaru	224	168	56
14	Carpomya vesuviana	499	374	125
15	Ceroplastes rubens	152	114	38
16	Cydia pomonella	273	205	68
17	Eulecanium rugulosum	92	69	23
18	Hyphantria cunea	213	159	54
19	Stilprotia salicis	164	123	41

2.2   Selection of the Base Model

The most straightforward way to implement the insect recognition is to use a classifier to predict its class from an image of the insect [11–16]. However, since our images are collected freely in the wild, the insects to be recognized usually occupy only a small proportion in the images and there exist complex backgrounds. This obviously affects the recognition accuracy, because the classifier is incapable to well distinguish the insect to recognize and the background. Considering that our insect recognition may be used in various applications, where the backgrounds of images are various and different from the samples in dataset, the recognition accuracy will further deteriorate in practical usage.

One way to eliminate the impact of the background is to cut the region of the insect from the image and then apply the classification. It is evidently avoid the interferences from background. However, users have to cut the region of insect before recognition, which makes it inconvenient to use. In this paper, we propose to use the multi-class object detection to recognize the insects under complex backgrounds. It is capable to detect the object to recognize directly from the image, giving both the class of the object and its location. It can effectively facilitate the insect recognition while obtaining high accuracy.

To select a proper detection model for our insect recognition, we evaluate three currently popular detection methods on our dataset: Faster R-CNN [17], YOLO [18], and R-FCN [19]. Among them, Faster R-CNN and R-FCN are two-stage methods which include a process of Region Proposal Network (RPN), while YOLO belongs to one-stage methods that directly predict class and location of object on the feature map.

We evaluate the three methods above on our insect recognition dataset and the results are given in Fig. 4. In our experiments, the R-FCN with ResNet-101 outperforms other methods. On one side, as a two-stage detection method, R-FCN benefits from the RPN that can improve the accuracy of detection especially for small objects such as some insects in our dataset. On the other side, although there is also a region proposal process in Faster R-CNN, the direct full connection of feature map after the ROI pooling layer is not an optimal choice. Full convolution networks such as GoogLeNet [20] and ResNet [21] have proved better performance and adaptation to different scales of the image. Besides, the number of channels in Faster R-CNN ROI pooling layer is very large, resulting in a large amount of calculations for the fully connected layer. In contrast, R-FCN without the fully connected layer significantly reduces the calculation. In summary, the R-FCN is appropriate for our insect recognition and thus we select it as the basic model.

3   Insect Recognition Using R-FCN with Anchor Boxes Estimation

In this section, we will propose our insect recognition in detail. Firstly, a brief introduction to the base model R-FCN is given. Then we propose to estimate the anchor boxes by clustering of bounding boxes instead of predefined ones in RPN of R-FCN. It effectively improve its performance on small objects while speeding up the detection process. Finally, the algorithm of anchor boxes generation is presented in detail.

3.1   A Brief Introduction to R-FCN

R-FCN is a region-based, fully convolutional networks for object detection [19]. It proposes the position-sensitive score maps to address a dilemma between translation-invariance in image classification and translation-variance in object detection, which makes it possible to adopt fully convolutional image classifier backbones in object detection. R-FCN usually costs less calculations while keeping high accuracies.

The framework of R-FCN is shown as Fig. 3. The first part is a convolutional network (ResNet-101 is used here) that is used to extract the convolutional features from the images. After feature extraction, a region proposal network (RPN) is exploited to generate region of interest (ROI) on the feature map. At the same time, a k^2-dimensional position-sensitive score map is obtained from the feature

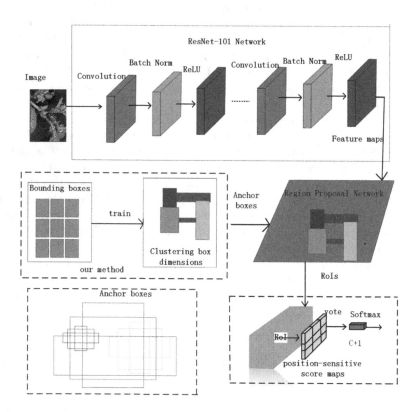

Fig. 3. Proposed framework comprised of R-FCN and the clustering bounding boxes in the region proposal network.

map for classification, while a $4k^2$-dimensional position-sensitive score map is for regression. Finally, the class and position information can be obtained by a position-sensitive ROI pooling.

3.2   Improve the Detection by Anchor Boxes Estimation

In the original version of R-FCN, the RPN generates ROIs from feature map using 9 predefined anchor boxes of different scales and shapes. The predefined anchor boxes work well in most practical cases. However, in the problem of insect detection, some objects are so small that the predefined anchor boxes cannot cover their shapes. This makes these insects apt to be missed by the detector.

To solve this problem, we can use a set of more appropriate anchor boxes instead of the predefined ones. In Yolov2 [18], the authors propose to set the scale and shape of the anchor boxes according to data. To be specific, the bounding boxes are collected from the training data, and then clustered into several clusters. The scales and shapes of the bounding boxes in cluster centers are used to set the anchor boxes. We also introduce this technique into R-FCN in our insect recognition. Using the anchor boxes estimated in this way, the region proposal network can generate ROIs that adapt better to the data, which improves the performances of detection, especially on small objects. Besides, the algorithm can obtain comparable performances with less anchor boxes, and this efficiently reduces the calculation.

3.3   The Anchor Boxes Estimation Algorithm

The detailed anchor boxes estimation algorithm based on clustering of bounding boxes is given in Algorithm 1. In this algorithm, each bounding box in training sample is represented as a point in 2-dimensional space, where the two axes correspond to the width and height of the box. Then these points are clustered using the k-means clustering algorithm [22]. After initialization, the update of the cluster of each point and the center are updated alternately. The criteria to determine the closest point is not the Euclidean distance as traditional k-means algorithm because it always generates greater errors for large boxes. Instead, the distance is measured according to the IOU score as [18]

$$d(box, centroid) = 1 - IOU(box, centroid)$$

Using this algorithm, a set of appropriate anchor boxes are selected, which improves the detection performances and accelerates the calculation.

Algorithm 1. Anchor Boxes Estimation Algorithm By Clustering

Input: The set of samples $D = \{\mathbf{x}_1, \mathbf{x}_2, \ldots, \mathbf{x}_m\}$, where $\mathbf{x}_i = [x_i^w, x_i^h]^T$, k.
  m is the total number of samples, k is the number of clusters, and x_i^w, x_i^h represents
  the width and height of the bounding box of the i-th sample.
Output: $C = \{\mathbf{c}_1, \mathbf{c}_2, \ldots, \mathbf{c}_k\}$
1: Randomly select k samples from D to be the initial values of $\{\boldsymbol{\mu}_1, \boldsymbol{\mu}_2, \ldots, \boldsymbol{\mu}_k\}$;
2: **repeat**
3:     $C_i = \emptyset$ $(1 \leq i \leq k)$
4:     **for** $j = 1, 2, \ldots m$ **do**
5:         Compute the overlap between edges of the j-th box and the i-th center:
        $L^w = \min(x_j^w, \mu_i^w)$, $L^h = \min(x_j^h, \mu_i^h)$
6:         Compute their intersection: $S_I = L^w \times L^h$
7:         Compute their union: $S_U = x_j^w \times x_j^h + \mu_i^w \times \mu_i^h - S_I$
8:         Compute IOU: $S_{ji} = S_I / S_U$
9:         Compute the distance between $\mathbf{x}_j$ and $\boldsymbol{\mu}_i$: $d_{ji} = (1 - S_{ji})$
10:        Assign $\mathbf{x}_j$ to the cluster with the closest mean:
        $\lambda_j = \arg\min_{i \in \{1,2,\ldots,k\}} d_{ji}$, $C_{\lambda_j} = C_{\lambda_j} \cup \{\mathbf{x}_j\}$
11:    **end for**
12:    **for** $i = 1, 2, \ldots k$ **do**
13:        Calculate new means: $\boldsymbol{\mu}_i' = \frac{1}{|C_i|} \sum_{\mathbf{x} \in C_i} \mathbf{x}$
14:        **if** $\boldsymbol{\mu}_i' \neq \boldsymbol{\mu}_i$ **then**
15:            update $\boldsymbol{\mu}_i = \boldsymbol{\mu}_i'$
16:        **end if**
17:    **end for**
18: **until** all $\boldsymbol{\mu}_i$'s keep unchanged

---

4   Experiments

4.1   Data Preparation and Experimental Settings

In this paper, we evaluate our proposed method on the data of insect recognition taken in the wild. The dataset has been proposed in Sect. 2. Since the original images are of unnecessarily high resolution, we downsample the images to be of resolution 600×600 before training or recognition.

Our experiments are performed a NVIDIA Titan X GPU. In the experiment, the learning rate for ResNet-101 is set to 0.001, the maximum number of iteration is set to $110,000$, the momentum is set to 0.9, and the weight decay is set to 0.0005. The mean average precision (MAP) [23] is used to evaluate the performance of algorithms, which reflects both the recall and precision of the detection algorithm. In our experiment, if the ratio of the intersection of a ground-truth bounding box and a predicted bounding box to the union of them is greater than 0.5, it is regarded as a good prediction.

4.2   Experimental Results

The detection results are shown in Figs. 4 and 5 in MAP. Comparing the accuracies and the running time of the detection methods to the ones with clustering of

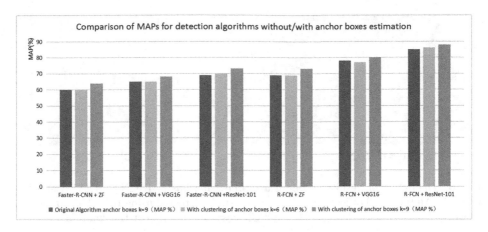

Fig. 4. Comparison of MAPs for original detection algorithms and the ones with anchor boxes estimation, where both 6 and 9 clusters are used in the estimation.

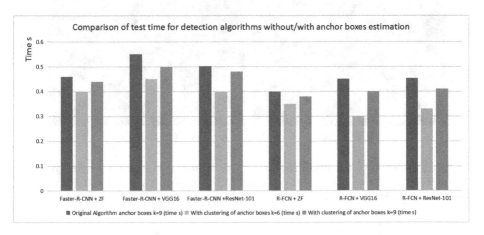

Fig. 5. Comparison of test time for original detection algorithms and the ones with anchor boxes estimation, where both 6 and 9 clusters are used in the estimation.

anchor boxes estimation, we obtain the following observations. (1) When we use 6 anchor boxes obtained by clustering in the detection, the algorithms obtain almost the same accuracies as the ones using 9 predefined anchor boxes. With less anchor boxes, the test time is obviously reduced. Thus clustering of anchor boxes is helpful to reduction of the calculations while keeping comparable accuracies. (2) When we use the same number of anchor boxes obtained by clustering as the original algorithms, the performances of detection is improved. This shows that a set of more appropriate anchor boxes are obtained by clustering.

In the following, we will illustrate some detection results directly in the images. For brevity of illustration, we use R-FCN-Cl to represent R-FCN with anchor boxes estimation using the clustering algorithm. Figure 6 illustrates the

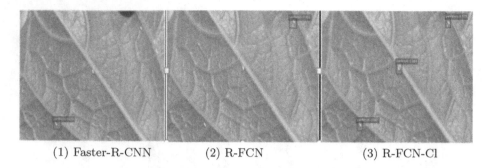

| (1) Faster-R-CNN | (2) R-FCN | (3) R-FCN-Cl |

Fig. 6. The detection results of *Bemisia tabaci* using Faster R-CNN, R-FCN, and R-FCN-Cl. All the small objects have been successfully detected using R-FCN-Cl method. However, the Faster-R-CNN and the R-FCN easy missing detect the small objects.

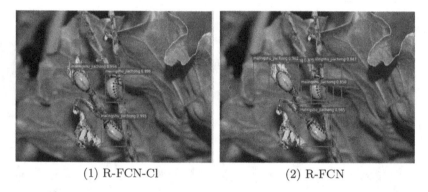

| (1) R-FCN-Cl | (2) R-FCN |

Fig. 7. Comparison of detection results on *Leptinotarsa decemlineata* using R-FCN-Cl and R-FCN. There are several obviously false detection given by R-FCN, while R-FCN-Cl gives better results.

detection results of *Bemisia tabaci* using Faster R-CNN, R-FCN, and R-FCN-Cl. The objects of *Bemisia tabaci* are relatively small in the images, which are therefore apt to be missed by the detector. Figure 7 compares the detection results of *Leptinotarsa decemlineata* given by R-CNN and R-CNN-Cl. There are less false retrievals in the result of R-CNN-Cl.

Figure 8 compares the detected bounding boxes of *Anoplophora sp* given by Faster R-CNN, R-FCN, and R-FCN-Cl, which are shown in red, blue, and green borders. From the image, we see that R-FCN-Cl gives the most appropriate box for the object. Besides, we also show some detection results of other insect species in Fig. 9. In these images, the objects are all correctly detected and recognized even under complex backgrounds. It proves the effectivity of our proposed method.

We also evaluate the performance of our method as an insect recognizer. To be specific, we construct a multi-label classification problem: the labels of an image are defined to be all the insect species that appear in it. Then, we

Fig. 8. The detection results of *Anoplophora sp* using Faster R-CNN, R-FCN, and R-FCN-Cl. The red, blue, and green lines represent bounding boxes given by Faster R-CNN, R-FCN, and R-FCN-Cl, respectively. (Color figure online)

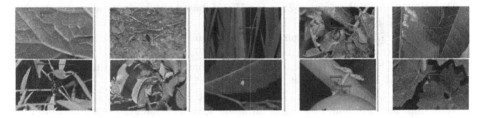

Fig. 9. Illustration of the detection results on some other insect species using R-FCN-Cl.

apply the detection algorithm to the image and assign all detected insect species to it as its labels. A sample is correctly classified if and only if all its labels are exactly matched to the ground truth. In our experiments, we obtain a recognition accuracy of 98.5%.

5 Conclusion

In this paper, we propose to use multi-class object detection based on deep learning to solve the problem of insect recognition under natural scenes. On one side, using detection technique, it is capable to accurately recognize the insects in the images even taken freely under complex backgrounds, and easily deal with the case where there are multiple species of insects to be recognized in one image. The positions of the objects in the image are also given by the algorithm. On the other hand, we take advantages of deep learning on object detection to significantly improve the performance of insect recognition, and simplify the recognition process.

We select the R-FCN method for multi-object detection in our insect recognition system after evaluating some popular methods. To further improve the detection accuracy and speed, we propose to design anchor boxes adaptively by clustering. We build a dataset for insect recognition in the wild, where the images are collected under natural scenes. The experimental results show that our method work well on the real data, and anchor boxes estimation by clustering is effective to improve both detection accuracy and speed.

Acknowledgement. This work is supported by the National Natural Science Foundation of China (NSFC) Grants 61721004, 61836014, the Beijing Municipal Science and Technology Project grant Z181100008918010, and National Key R&D Program of China grant 2017YFC1200602.

References

1. White, I.M., Elson-Harris, M.M., et al.: Fruit Flies of Economic Significance: Their Identification and Bionomics. CAB International (1992)
2. Hassan, S.N.A., Rahman, N., Zaw, Z.: Vision based entomology: a survey. Int. J. Comput. Sci. Eng. Sur. **5**, 19–31 (2014)
3. MacLeod, N.: Automated Taxon Identification in Systematics: Theory Approaches and Applications. CRC Press, Boca Raton (2007)
4. Favret, C.R., Sieracki, J.M.: Machine vision automated species identification scaled towards production levels. Syst. Entomol. **41**, 133–143 (2016)
5. Wang, J.N., Chen, X.L., Hou, X.W., Zhou, L.B., Zhu, C.D., Ji, L.Q.: Construction, implementation and testing of an image identification system using computer vision methods for fruit flies with economic importance (Diptera: Tephritidae). Pest Manag. Sci. **73**(7), 1511–1528 (2017)
6. Wang, L., Huang, L., Yang, H., Gao, L., et al.: Developing and testing of image identification system for Bactrocera spp. Plant Quar. (Shanghai) **27**(5), 29–36 (2013)
7. Deng, L., Wang, Y., Han, Z., Yu, R.: Research on insect pest image detection and recognition based on bio-inspired methods. Biosyst. Eng. **169**, 139–148 (2018)
8. Ebrahimi, M., Khoshtaghaza, M., Minaei, S., Jamshidi, B.: Vision-based pest detection based on svm classification method. Comput. Electron. Agric. **137**, 52–58 (2017)
9. Hu, Z., Liu, B., Zhao, Y.: Agricultural robot for intelligent detection of pyralidae insects. In: Agricultural Robots-Fundamentals and Applications. IntechOpen (2018)
10. Zhong, Y., Gao, J., Lei, Q., Zhou, Y.: A vision-based counting and recognition system for flying insects in intelligent agriculture. Sensors **18**(5), 1489 (2018)
11. Gassoumi, H., Prasad, N.R., Ellington, J.J.: Neural network-based approach for insect classification in cotton ecosystems. In: International Conference on Intelligent Technologies, pp. 13–15 (2000)
12. Asefpour Vakilian, K., Massah, J.: Performance evaluation of a machine vision system for insect pests identification of field crops using artificial neural networks. Arch. Phytopathol. Plant Prot. **46**, 1262–1269 (2013)
13. Wang, J., Lin, C., Ji, L., Liang, A.: A new automatic identification system of insect images at the order level. Knowl.-Based Syst. **33**, 102–110 (2012)

14. Ding, W., Taylor, G.: Automatic moth detection from trap images for pest management. Comput. Electron. Agric. **123**, 17–28 (2016)
15. Wang, J., Ji, L., Liang, A., Yuan, D.: The identification of butterfly families using content-based image retrieval. Biosyst. Eng. **111**(1), 24–32 (2012)
16. Sun, Y., et al.: A smart-vision algorithm for counting whiteflies and thrips on sticky traps using two-dimensional fourier transform spectrum. Biosyst. Eng. **153**, 82–88 (2017)
17. Ren, S., He, K., Girshick, R., Sun, J.: Faster R-CNN: Towards real-time object detection with region proposal networks. In: Advances in Neural Information Processing Systems, pp. 91–99 (2015)
18. Redmon, J., Farhadi, A.: Yolo9000: better, faster, stronger. In: Proceedings of the IEEE Conference on Computer Vision and Pattern Recognition, 7263–7271 (2017)
19. Dai, J., Li, Y., He, K., Sun, J.: R-FCN: Object detection via region-based fully convolutional networks. In: Advances in Neural information Processing Systems, PP. 379–387 (2016)
20. Szegedy, C., et al.: Going deeper with convolutions. In: Proceedings of the IEEE Conference on Computer Vision and Pattern Recognition. pp. 1–9 (2015)
21. He, K., Zhang, X., Ren, S., Sun, J.: Deep residual learning for image recognition. In: Proceedings of the IEEE Conference on Computer Vision and Pattern Recognition. (2016) 770–778
22. Hartigan, J.A., Wong, M.A.: Algorithm as 136: A k-means clustering algorithm. J. Royal Stat. Soc. Ser. C (Appl. Stat.) **28**(1), 100–108 (1979)
23. Everingham, M., Van Gool, L., Williams, C.K., Winn, J., Zisserman, A.: The pascal visual object classes (voc) challenge. Int. J. Comput. Vis. **88**(2), 303–338 (2010)

A Full-Reference Image Quality Assessment Model Based on Quadratic Gradient Magnitude and LOG Signal

Congmin Chen and Xuanqin Mou[(✉)]

Institute of Image Processing and Pattern Recognition,
Xi'an Jiaotong University, Xi'an, China
chencongmin@stu.xjtu.edu.cn, xqmou@mail.xjtu.edu.cn

Abstract. Image quality assessment aims at estimating the subject quality of images and builds models to high efficiently evaluate the perceptual quality of the image for many applications. Because the human visual system (HVS) is highly sensitive to structural information, various image features have been studied and widely applied in IQA metrics design. Previous work has validated that the image gradient magnitude and the Laplacian of Gaussian (LOG) operator are efficient structural features in IQA tasks. Most of the IQA metrics work capably only when the distorted image is totally registered with the reference image, and perform poorly on images even with small translations. In this paper, we suggested an FR-IQA method with a simple combination of the gradient magnitude and the LOG signals, which obtains satisfied performance in evaluating image quality while considering the shift-invariance property for not well-registered reference and distortion image pair. Experimental results show that the proposed model works robustly on three large scale subjective IQA databases which contain a variety of distortion types and levels, stays in the state-of-the-art FR-IQA models and achieves the best performance in terms of weighted average score over the three databases. Furthermore, we proved that the proposed model performs better in translation-invariance test compared with the competitors.

Keywords: Image quality assessment (IQA) · Full reference (FR) ·
Gradient magnitude · Laplacian of Gaussian · Translation-invariance

1 Introduction

Since the distortion of information existed during the operation of image transmission, compression, restoration, etc., it is a significant procedure to evaluate the quality of digital images. In most cases, human beings are the conclusive observers who provide the ranks of image quality. Although the subjective score from the observation of human beings is able to estimate the image quality, automatic algorithms are absolutely much more convenient and economic in most practical applications. Image Quality Assessment (IQA), which is able to build models to compute the objective score in accordance with the subjective grading of distorted images, has been widely applied to imitate the observation results of the human visual system (HVS). Among different

© Springer Nature Switzerland AG 2019
Y. Zhao et al. (Eds.): ICIG 2019, LNCS 11901, pp. 702–713, 2019.
https://doi.org/10.1007/978-3-030-34120-6_57

metrics, IQA models can be classified into three types: full reference (FR) metrics, which provide reference image, no reference (NR) or blind IQA is effective where the pristine reference image is not available, and reduced reference (RR) IQA methods is employed when partial information of the source image is provided. This paper focuses on FR-IQA methods in which the reference image is assumed to be available and the quality to be perfect.

The traditional metrics such as the peak signal-to-noise ratio (PSNR) and the mean squared error, which are operated on the intensity domain of the image, are not well consistent with the subjective score of human judgement. The structural similarity (SSIM) index [1] is designed on the assumption that the HVS is sensitive to local structures and is able to capture the structural information when evaluating the quality of the visual signal. The multi-scale SSIM (MS-SSIM) [2], with an extension to single-scale algorithm, and the information weighted SSIM (IW-SSIM) [3], with consideration of the different types of local regions, bring better results than the original algorithm. The information fidelity criteria (IFC) [4] predicts the image quality by computing the information shared between the distorted images and reference images using the image fidelity measurement, and has been extended to a more efficient measurement named visual information fidelity (VIF) [5]. Another IQA metric based on the fact that HVS understands an image according to its low-level features is the feature-similarity (FSIM) [6] index, which uses the phase congruency (PC) to measure the significance of local structure and treats it as the primary feature. In consideration of the sensitivity of the image gradients to image distortions, the gradient magnitude similarity deviation (GMSD) [7] metric makes use of the variation of gradient based local quality map and shows good performance in image quality prediction. [8] shows comprehensive evaluations of different FR-IQA methods.

Image gradient is an elementary feature in IQA field, since it is sensitive to the variance of the intensity and is able to extract local structures, which is the main target of the HVS. As is well known, gradient is the first-order derivative information of the image. So what does the second-order derivative know about the image quality? Obviously, the second-order derivative of a Gaussian function turns out to be an LOG (Laplacian of Gaussian) filter, which is proved to be approximate to the decorrelation mechanism of the retinal ganglion receptive field [9, 10]. Since the perceptual distortion can be sensed by extracting the difference information between image structures, the LOG operator shows the ability to capture the structural information. Actually, the LOG signal has been proved to be highly effective in FR, RR, and blind IQA model developments [11–15]. Particularly, [13] proved that GM and LOG show great performance in Blind-IQA with training method.

Since the LOG signal and gradient magnitude both work well for quality estimating in previous works, we make use of the mathematical methods to find further relationship between the structural information in the LOG signal and gradient magnitude of images. Based on the idea that the LOG signal and gradient magnitude are efficient features and can reflect the image quality, we are motivated to find a new metric using these two features which performs more stably and robustly when small spatial translation occurs in images. In this paper, we propose a simple combination of the gradient magnitude and the LOG signals, which we proved to be stable in the quality estimating process. With less cost of time, the proposed model shows good performance on the commonly used databases. Meanwhile, it is proved to perform well on translation-invariance property when small translation of pixels occurs.

2   Methods

2.1   LOG Signal and Gradient Magnitude

Laplacian filters are derivative filters used in edge detection. Since the derivative filters are sensitive to noise and discrete points, a procedure of smoothing and denoising by using a Gaussian filter is commonly used before applying the Laplacian operator. This two-step process is called the Laplacian of Gaussian operation. The LOG operator takes the second derivative of the given signals. Where the image is basically uniform, the LOG will give zero. Wherever a change occurs, the LOG will give a positive response to the darker side and a negative response to the lighter side, while the point on the edge itself turns out to be zero.

The LOG filter that we used in this paper is defined as:

$$\nabla^2 G(x,y,\sigma) = -\frac{1}{\pi\sigma^4}\left(1 - \frac{x^2+y^2}{2\sigma^2}\right)e^{-\frac{x^2+y^2}{2\sigma^2}} \tag{1}$$

where the variables x and y denote the coordinate of the input image, parameter σ represents the scale factor of the LOG filter. In our experiment, we select the scale factor of the LOG operator as 0.5, which performs better in the evaluation process of our method. A scaling factor can be used on the filter to restrict the range of the LOG values. We use $L_{R,\sigma}$ to denote the transformation results with the parameter σ of the reference images, while $L_{D,\sigma}$ to denote the distorted signals.

The image gradient, which is defined as the root mean square of image directional gradients along two orthogonal directions, which is usually computed by convolving an image with a linear filter. In order to smooth the image, a Gaussian filter is commonly used before the convolving process. The first-order derivative of Gaussian filter on horizontal direction and vertical direction is defined as:

$$\frac{\partial G}{\partial x} = \left(-\frac{1}{2\pi\sigma^4}\right)xe^{-\frac{x^2+y^2}{2\sigma^2}} \tag{2}$$

$$\frac{\partial G}{\partial y} = \left(-\frac{1}{2\pi\sigma^4}\right)xe^{-\frac{x^2+y^2}{2\sigma^2}} \tag{3}$$

We convolve the reference image with the two directional derivative filters to yield the horizontal and vertical gradient images $d_{R,x}$ and $d_{R,y}$. The gradient magnitude of reference images is computed as follows:

$$D_{R,\sigma} = \sqrt{d_{R,x}^2 + d_{R,y}^2} \tag{4}$$

where σ denotes the parameter of scale in the Gaussian filter. The gradient magnitude of distorted images can be produced in the same way as:

$$D_{D,\sigma} = \sqrt{d_{D,x}^2 + d_{D,y}^2} \tag{5}$$

For simplicity of computation we use divisive normalization method to remove contrast variation in the image in a large scale, before which we adjust the range of LOG signals and gradient magnitudes with a ratio coefficient. The divisive normalization processes are shown as following equations:

$$U_{R,\sigma}(i,j) = \frac{L_{R,\sigma}(i,j)}{\sqrt{G_{2\sigma}(i,j) * [D_{R,\sigma}^2(i,j) + L_{R,\sigma}^2(i,j)]} + c_0} \tag{6}$$

$$V_{R,\sigma}(i,j) = \frac{D_{R,\sigma}(i,j)}{\sqrt{G_{2\sigma}(i,j) * [D_{R,\sigma}^2(i,j) + L_{R,\sigma}^2(i,j)]} + c_0} \tag{7}$$

where $G_{2\sigma}$ denotes scale factor of the employed large scale Gaussian filter, and c0 is an adjustable constant which makes the denominator not to be zero. In the same way, $U_{D,\sigma}$ and $V_{D,\sigma}$ can be produced from distorted images.

Since GM and LOG are relatively complemented in image expressing [13], here we design a simple combination of $U_{R,\sigma}$ and $V_{R,\sigma}$ to express the information of reference images, which is defined as:

$$r_{R,\sigma}(i,j) = \sqrt{U_{R,\sigma}^2(i,j) + V_{R,\sigma}^2(i,j)} \tag{8}$$

where σ denotes the scale factor of the Gaussian filter. In the same way, $r_{D,\sigma}$ is defined as:

$$r_{D,\sigma}(i,j) = \sqrt{U_{D,\sigma}^2(i,j) + V_{D,\sigma}^2(i,j)} \tag{9}$$

which is generated from distorted images.

2.2   The Proposed FR-IQA Model

We use a similarity computation to ascertain the structural difference between distorted and reference images. The similarity map is defined as:

$$Q(i,j) = \frac{2r_{R,\sigma}(i,j)r_{D,\sigma}(i,j) + c_1}{r_{R,\sigma}^2(i,j) + r_{D,\sigma}^2(i,j) + c_1} \tag{10}$$

where c_1 is a positive constant that supplies numerical stability.

Equation (10) measures the local similarity at each image pixel. To yield the overall evaluation score of the image, a polling strategy shall be introduced to integrate the similarities of all image pixels. Average pooling is a general way to produce the general estimate of the image quality [20, 21]. Based on this, in this study we propose the first FR-IQA metric named as the mean of the quadratic gradient magnitude and LOG signal, abbreviated as mQGL, as shown in Eq. (11). In the same time, standard deviation pooling has been validated to be a more efficient method in synthesizing for gradient similarity based IQA method [7], here we accordingly propose the second FR-IQA metric named as the standard deviation of the quadratic gradient magnitude and LOG signal, abbreviated as sQGL, as shown in Eq. (12).

$$mQGL = \frac{1}{N} \sum_{i,j} Q(i,j) \tag{11}$$

$$sQGL = \sqrt{\frac{1}{N} \sum_{i,j} (Q(i,j) - mean(Q))^2} \tag{12}$$

Note that the value of mQGL means the average similarity between reference and distorted images, which gives higher score to higher image quality, while sQGL indicates the difference between reference and distorted images, and gives higher score to the larger distortion level and lower image quality.

3   Experimental Results

3.1   Test Database

We evaluate the performance of the IQA models on three publicly accessible IQA databases: LIVE [16], CSIQ [17] and TID2013 [18]. The LIVE database consists of 779 distorted images created from 29 reference images with 5 types of distortions: JPEG compression, JPEG2000 compression, white noise, Gaussian blur and simulated fast fading. The CSIQ database contains 866 subject-rated distorted images which are created from 30 reference images with 6 different types of distortions: JPEG compression, JPEG2000 compression, additive white noise, additive pink Gaussian noise, Gaussian blur, and global contrast decrements. The Difference Mean Opinion Score (DMOS) values are provided as part of the LIVE database and CSIQ database. For each image, higher DMOS value means higher distortion and lower image quality in the subjective evaluation. The TID2013 database is the largest database of the commonly used databases which is intended for evaluation of full-reference image visual quality assessment metrics. It contains 3000 distorted images, generated from 25 reference images with 24 types of distortions at 5 levels. This database covers the most types of distortions among all existed databases. The human subjective score has been given as Mean Opinion Score (MOS) in TID2013 database, where higher MOS value means higher subjective image quality. The distortion types in these databases reflect a broad range of image impairments, and are widely applied in IQA research.

3.2 IQA Performance

One of the most commonly used performance metrics, SROCC (Spearman rank-order correlation coefficient) index, which can measure the monotonicity of the computed result and the subjective score, is employed to evaluate the performance of the proposed IQA model, as well as the competitors. We investigate the results of the model scores for each image from the three benchmark databases, which include both pristine images and distorted images, to compare with the value of DMOS or MOS value given from the databases. The SROCC regarding to the model scores versus the human subjective opinion scores on the different databases are shown in Table 1. The top three models for each database are shown in boldface. The competitors are selected for the representation of their experimental evaluation results on the three IQA databases mentioned in the preceding paragraph, which have been reported in their original literature, or the publically achievement of their source code of the proposed model which are given on website.

Table 1. Performance comparison of different FR-IQA models on three benchmark databases. The top three models for each database are shown in boldface.

SROCC	LIVE (779 images)	CSIQ (866 images)	TID2013 (3000 images)	Weighted average
PSNR	0.8756	0.8058	0.6394	0.7100
SSIM [1]	0.9479	0.8756	0.7417	0.8012
MS-SSIM [2]	0.9513	0.9133	0.7859	0.8374
IW-SSIM [3]	0.9567	0.9213	0.7779	0.8346
IFC [4]	0.9259	0.7671	0.5390	0.6463
VIF [5]	**0.9636**	0.9195	0.6770	0.7703
FSIM [6]	**0.9634**	0.9240	**0.8022**	**0.8519**
NLOG-MSE	0.9405	0.9259	0.7734	0.8299
NLOG-COR	0.9429	**0.9308**	0.7772	0.8336
GMSD [7]	**0.9603**	**0.9570**	**0.8044**	**0.8590**
RFSIM [24]	0.9438	0.9292	0.7744	0.8317
mQGL	0.9520	0.9217	0.7895	0.8414
sQGL	0.9574	**0.9541**	**0.8108**	**0.8621**

As shown in Table 1, the proposed sQGL method ranks 4th on LIVE database, 2nd on CSIQ, and 1st on TID2013 database. With stable performance across the three databases, especially on the largest TID2013 database, the proposed sQGL model is significantly better than all the other competitors. Model working well on TID2013 is much more significant since the TID2013 database is the largest database among the three commonly used databases. Especially, sQGL achieves the best performance in terms of weighted average score over the three databases. It has stable performance on different databases, and is efficient on different types of distorted images. According to the experimental results, average pooling to the quality map is not so valid as standard deviation pooling. Standard deviation pooling shows more efficient property in the proposed QGL model.

Table 2 shows the performance of the proposed model and the competitors on each individual distortion type, where the top three models for each distortion type are shown in boldface.

Table 2. Performance comparison of the IQA models on each individual distortion type in terms of SROCC. The top three models for each type are shown in boldface.

	Distortion	PSNR	SSIM	MS-SSIM	IW-SSIM	IFC	VIF	FSIM	GMSD	NLOG-MSE	NLOG-COR	mQGL	sQGL
LIVE	JP2K	0.8954	0.9614	0.9654	0.9653	0.9100	0.9683	**0.9717**	**0.9711**	0.9499	0.9515	0.9668	**0.9705**
	JPEG	0.8809	0.9764	0.9793	**0.9809**	0.9440	**0.9842**	**0.9834**	0.9782	0.9610	0.9629	0.9801	0.9803
	WN	**0.9854**	0.9694	0.9731	0.9671	0.9377	0.9845	0.9652	0.9737	**0.9877**	**0.9880**	0.9582	0.9685
	GB	0.7823	0.9517	0.9584	**0.9722**	0.9649	**0.9722**	**0.9708**	0.9567	0.9440	0.9470	0.9535	0.9650
	FF	0.8907	**0.9556**	0.932⊦	0.9443	**0.9644**	**0.9652**	0.9499	0.9416	0.9127	0.9148	0.9497	0.9500
CSIQ	AWN	0.9363	0.8974	0.9471	0.9377	0.8460	0.9571	0.9262	**0.9676**	**0.9663**	**0.9664**	0.9557	**0.9656**
	JPEG	0.8882	0.9546	0.9622	**0.9664**	0.9395	**0.9705**	0.9654	0.9651	0.9483	0.9475	0.9596	**0.9661**
	JP2K	0.9363	09606	**0.9691**	0.9681	0.9262	0.9672	0.9685	**0.9717**	0.9503	0.9481	0.9590	**0.9728**
	PGN	0.9338	0.8922	0.9330	0.9057	0.8279	**0.9509**	0.9234	0.9502	**0.9588**	**0.9594**	0.9379	0.9461
	GB	0.9289	0.9609	0.9720	**0.9781**	0.9593	**0.9747**	**0.9729**	0.9712	0.9519	0.9519	0.9548	0.9722
	Contrast	0.8622	0.7922	**0.9521**	**0.9540**	0.5416	0.9361	0.9420	0.9040	0.9238	0.9264	0.9356	**0.9422**
TID2013	AWGN	**0.9291**	0.8671	0.8645	0.8438	0.6611	0.8994	0.8973	**0.9462**	0.9251	0.9245	0.9094	**0.9404**
	ANMC	**0.8984**	0.7726	0.7729	0.7514	0.5351	0.8299	0.8207	**0.8684**	0.8414	0.8414	0.8280	**0.8641**
	SCN	0.9198	0.8515	0.8543	0.8166	0.6601	0.8834	0.8749	**0.9350**	0.9242	**0.9250**	0.8939	**0.9284**
	MN	0.5416	0.7767	0.8014	0.8063	0.6732	**0.8642**	0.8013	0.7075	**0.8271**	**0.8298**	0.8026	0.7674
	HFN	**0.9141**	0.8634	0.8603	0.8553	0.7405	0.8972	0.8983	**0.9162**	0.9001	0.8993	0.8930	**0.9116**
	IMN	**0.8968**	0.7503	0.7628	0.7281	0.6407	0.8536	0.8072	0.7637	**0.8799**	**0.8763**	0.7983	0.7360
	QN	0.8808	0.8657	0.8705	0.8467	0.6282	0.7853	0.8719	**0.9049**	**0.8917**	0.8912	0.8659	**0.9016**
	GB	0.9149	0.9667	0.9672	**0.9701**	0.8906	0.9649	0.9550	0.9113	**0.9694**	**0.9705**	0.9663	0.9545
	DEN	0.9480	0.9254	0.9267	0.9152	0.7779	0.8910	0.9301	**0.9525**	**0.9488**	0.9478	0.9343	**0.9481**
	JPEG	0.9189	0.9200	0.9265	0.9186	0.8356	0.9191	0.9324	**0.9507**	**0.9553**	0.9469	0.9367	**0.9475**
	JP2K	0.8840	0.9468	0.9504	0.9506	0.9077	0.9516	0.9576	**0.9657**	**0.9614**	0.9598	0.9602	**0.9647**
	JGTE	0.7685	**0.8493**	0.8475	0.8387	0.7425	0.8409	0.8463	0.8403	0.8117	0.8143	**0.8614**	0.8544
	J2TE	0.8883	0.8828	0.8888	0.8656	0.7769	0.8760	0.8912	0.9136	**0.9371**	**0.9344**	0.9076	**0.9161**
	NEPN	0.6860	0.7821	0.7968	0.8010	0.5736	0.7719	0.7917	**0.8140**	0.7509	0.7554	**0.8045**	**0.8187**
	Block	0.1552	0.5720	0.4800	0.3716	0.2413	0.5306	0.5489	**0.6625**	0.5926	0.6148	**0.6384**	**0.6591**
	Mean shift	0.7672	0.7752	**0.7906**	0.7833	0.5522	0.6275	0.7530	0.7351	**0.7993**	**0.8009**	0.7192	0.7159
	Contrast	0.4403	0.3775	0.4633	0.4592	-0.180	**0.8385**	0.4686	0.3235	0.4654	0.4677	**0.4849**	0.3404
	CCS	0.0944	-0.414	-0.410	-0.420	-0.403	-0.310	**-0.275**	**-0.295**	-0.317	-0.342	-0.382	-0.313
	MGN	**0.8905**	0.7803	0.7785	0.7727	0.6142	0.8468	0.8469	**0.8886**	0.8678	0.8676	0.8332	**0.8693**
	CN	0.8411	0.8566	0.8527	0.8761	0.8160	0.8946	0.9120	**0.9298**	**0.9277**	**0.9245**	0.8959	0.9212
	LCN	0.9145	0.9057	0.9067	0.9037	0.8180	0.9203	**0.9466**	**0.9629**	0.9339	0.9310	0.9399	**0.9623**
	CQD	**0.9269**	0.8542	0.8554	0.8401	0.6006	0.8414	0.8759	0.9102	**0.9176**	**0.9062**	0.8910	0.9092
	Chr. abr.	**0.8873**	0.8775	0.8784	0.8681	0.8209	0.8848	0.8714	0.8530	**0.8872**	**0.8902**	0.8815	0.8609
	Sampling	0.9042	0.9461	0.9482	0.9474	0.8884	0.9352	0.9565	**0.9683**	**0.9579**	0.9573	0.9573	**0.9648**

As is shown in the table, the proposed sQGL model performs well on most distortion types. The experimental results also reveals that the proposed method pooling with standard deviation works robustly on different image distortion types across the three databases.

In our experiment, the scale factor σ of the Gaussian filter is selected as 0.5, which shows the best performance in image quality estimating, and the scale factor for the divisive normalization Gaussian filter is 2σ. In the translation-invariance experiment, we compare the performance of the model for $\sigma = 0.5$ and $\sigma = 1$ with other methods since the larger scale works more stably on the translation-invariance property. The constant $c0$ in the divisive normalization process is selected as 1, while the constant $c1$ for the quality map computation is selected as 0.0009, which shows the best property in the experimental performance of the proposed model.

3.3   Test on Translation-Invariance Property

We transform the reference images in LIVE database by spatial shift with the range from 0.1 to 2 pixels on horizontal direction, and investigate the SROCC between the image subjective scores and model scores computed by the proposed model, the GMSD metric, and the SSIM metric, as shown in Fig. 1. The scale factor of the Gaussian filters are selected as 0.5 and 1 separately.

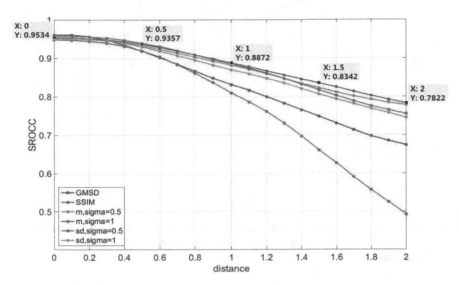

Fig. 1. Comparison of the SROCC values for several metrics on LIVE database along with the spatial translation distance by pixel. The curves from top to bottom in the legend are: the GMSD method, the SSIM method, the proposed model by average pooling with scale factor 0.5 and 1 for Gaussian filter, the proposed model by standard deviation pooling with scale factor 0.5 and 1 for Gaussian filter. Five discrete points on the purple line are marked on the curve. (Color figure online)

Experimental results show that the proposed model performs better than the competitors when small spatial translation occurs in the image, and the SROCC value of the average pooling method declines more slowly than the result computed by standard deviation pooling with same scale factor of the Gaussian filter. We mark the purple curve (average pooling, $\sigma = 1$) with five specific discrete points from translation distance 0 to 2 pixels in the figure. Although the marked SROCC is not the highest on the point with zero displacement, the value falls the most slowly among these metrics. When the reference image is shifted by two pixels on horizontal direction, the SROCC value of the proposed model holds 0.7822 while the GMSD drops below 0.7, and SSIM falls below 0.5. This comparison indicates that the proposed model ignores the spatial position of an edge and catch the information of the texture feature in a local area, thus tends to be shift-invariant when small spatial shift occurs in images.

The comparisons on CSIQ and TID2013 databases are shown in Figs. 2 and 3 separately.

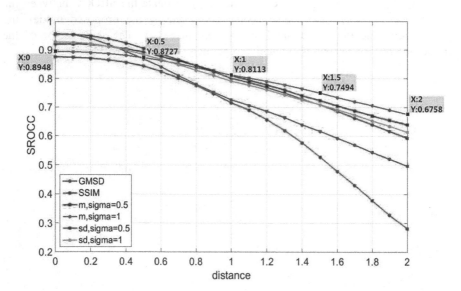

Fig. 2. Comparison of the SROCC values for several metrics on CSIQ database along with the spatial translation distance by pixel. Five points on the purple line are marked on the curve. (Color figure online)

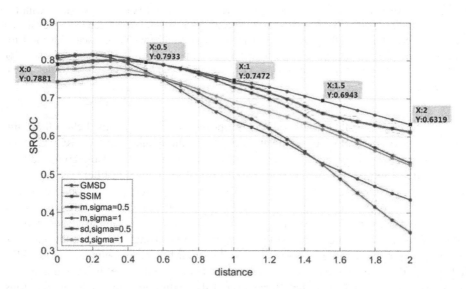

Fig. 3. Comparison of the SROCC values for several metrics on TID2013 database along with the spatial translation distance by pixel. Five points on the purple line are marked. (Color figure online)

On CSIQ database, the SROCC of the proposed model holds 0.8113 when the translation distance is one pixel on the horizontal direction, while the GMSD and SSIM method falls below 0.75. When the reference image is translated by two pixels, the SROCC value of the proposed model holds 0.6758, while the GMSD drops to 0.5, and SSIM falls below 0.3. Note that the average pooling method shows better performance than the standard deviation based pooling in our computational model.

The comparison on TID2013 database also gives the result that the proposed model works better on the translation-invariance property. The SROCC of the proposed method holds 0.7472 when the translation distance is one pixel along the horizontal direction, while the GMSD and SSIM method falls below 0.7. When the reference image is translated by two pixels, the SROCC value of the proposed model holds 0.6319, while the GMSD drops below 0.5, and SSIM falls below 0.45. The average pooling method still shows better performance than the standard deviation pooling method in the proposed model.

These comparisons validate that the proposed average based metric works stably on translation-invariance property, and is significantly better than the competitors on the three databases when small translation is given to the reference image. This result shows more practical significance since images obtained in many applications suffer dithering or movement.

4 Discussion and Conclusion

In this paper, we proposed an FR-IQA model based on the local structure represented by the root of the sum of squared LOG signal and gradient magnitude, which has been proved to be an efficient feature in quality evaluation. The similarity map helps capture the local quality of images, and we validated that compared with average pooling method in our computation, the overall image quality computed by standard deviation pooling is highly relevant to subjective image quality. Compared with the state-of-the-art FR methods, the proposed sQGL model performs well and more stably across the three benchmark databases, and achieves the best performance in terms of weighted average score over the three databases. sQGL metric also shows stable result on different types of distortions separately on these databases. According to our experimental comparison, the result validated that the combination strategy of LOG signal and gradient magnitude is efficient in IQA applications.

Another contribution in our work is that the proposed model shows good performance and robust property on images with small spatial translations compared with other metrics, and this model does not cost much time to achieve in computational process. Since the images obtained from different ways suffer different dithering and movement so that the distorted image is not always totally matched with the reference image, the translation-invariance property is more efficient and significant in most practical applications. Furthermore, translation-invariance property would be significant in discriminating different types of local textures, for which we will investigate in the future.

Acknowledgment. This work was supported in part by the National Natural Science Foundation of China (NSFC, No. 61571359) and the National Program on Key Research Project (No. 2016YFA0202003).

References

1. Wang, Z., Bovik, A.C., Sheikh, H.R., Simoncelli, E.P.: Image quality assessment: from error visibility to structural similarity. IEEE Trans. Image Process. **13**(4), 600–612 (2004)
2. Wang, Z., Simoncelli, E.P., Bovik, A.C.: Multiscale structural similarity for image quality assessment. In: Proceedings of IEEE 37th Conference Rec. Asilomar Conference on Signals, System Computers, vol. 2, pp. 1398–1402 (2003)
3. Wang, Z., Li, Q.: Information content weighting for perceptual image quality assessment. IEEE Trans. Image Process. **20**(5), 1185–1198 (2011)
4. Sheikh, H.R., Bovik, A.C., de Veciana, G.: An information fidelity criterion for image quality assessment using natural scene statistics. IEEE Trans. Image Process. **14**(12), 2117–2128 (2005)
5. Sheikh, H.R., Bovik, A.C.: Image information and visual quality. IEEE Trans. Image Process. **15**(2), 430–444 (2006)
6. Zhang, L., Zhang, L., Mou, X., Zhang, D.: FSIM: a feature similarity index for image quality assessment. IEEE Trans. Image Process. **20**(8), 2378–2386 (2011)
7. Xue, W., Zhang, L., Mou, X., et al.: Gradient magnitude similarity deviation: a highly efficient perceptual image quality index. IEEE Trans. Image Process. **23**(2), 684–695 (2014)
8. Zhang, L., Zhang, L., Mou, X., Zhang, D.: A comprehensive evaluation of full reference image quality assessment algorithms. In: Proceedings of 19th IEEE ICIP, pp. 1477–1480, October 2012
9. Simoncelli, E.P., Olshausen, B.A.: Natural image statistics and neural representation. Annu. Rev. Neurosci. **24**, 1193–1216 (2001)
10. Croner, L.J., Kaplan, E.: Receptive fields of P and M ganglion cells across the primate retina. Vis. Res. **35**(1), 7–24 (1995)
11. Zhang, M., Mou, X., Zhang, L.: Non-shift edge based ratio (NSER): an image quality assessment metric based on early vision features. IEEE Signal Process. Lett. **18**(5), 315–318 (2011)
12. Shao, W., Mou, X.: Edge patterns extracted from natural images and their statistics for reduced-reference image quality assessment. In: Proceedings of IS&T/SPIE Electronic Imaging, California, USA, vol. 8660 (2013)
13. Xue, W., Mou, X., Zhang, L., Bovik, A.C.: Blind image quality assessment using joint statistics of gradient magnitude and Laplacian features. IEEE Trans. Image Process. **23**(11), 4850–4862 (2014)
14. Mou, X., Xue, W., Chen, C., Zhang, L.: LoG acts as a good feature in the task of image quality assessment. In: Proceedings of IS&T/SPIE Electronic Imaging, California, USA, vol. 9023 (2014)
15. Chen, C., Mou, X.: A reduced-reference image quality assessment model based on joint-distribution of neighboring LOG signals. In: Proceedings of IS&T Electronic Imaging, no. 18, pp. 1–8 (2016)
16. Sheikh, H.R., Wang, Z., Cormack, L., Bovik, A.C.: Live image quality assessment database release 2 (2005). http://live.ece.utexas.edu/research/quality
17. Larson, E.C., Chandler, D.M.: Most apparent distortion: full-reference image quality assessment and the role of strategy. J. Electron. Imaging **19**(1), 011006 (2010)

18. Ponomarenko, N., et al.: Color image database TID2013: peculiarities and preliminary results. In: Proceedings of 4th European Workshop on Visual Information Processing, Paris, France, pp. 106–111 (2013)
19. Zhang, L., Zhang, L., Mou, X.: RFSIM: a feature based image quality assessment metric using Riesz transforms. In: Proceedings of IEEE International Conference on Image Processing, Hong Kong (2010)
20. Wang, Z., Shang, X.: Spatial pooling strategies for perceptual image quality assessment. IEEE International Conference on Image Processing, pp. 2945–2948, September 2006
21. Moorthy, A.K., Bovik, A.C.: Visual importance pooling for image quality assessment. IEEE J. Special Topics Signal Process 3, 193–201 (2009)

A Universal Fusion Strategy for Image Super-Resolution Jointly from External and Internal Examples

Wei Wang[1], Xuesen Shang[1], Wenming Yang[1(✉)], Canrong Zhang[2], and Qingmin Liao[1]

[1] Department of Electronic Engineering, Graduate School at Shenzhen, Tsinghua University, Shenzhen, China
yang.wenming@sz.tsinghua.edu.cn
[2] Research Center for Modern Logistics, Graduate School at Shenzhen, Tsinghua University, Shenzhen, China

Abstract. The validity of learning-based image super-resolution is largely limited by supporting dataset. Neither external-based nor internal-based super-resolution methods can perform well in real applications such as medical endoscopic images. This paper studies the strategy of joint learning of two kinds of methods. We first build sub-dictionaries and study the corresponding mapping matrices on the respective samples. Due to the consistency of learning strategies, we establish joint mapping matrices based on the distance between the input low-resolution image patches and the dictionary atoms in the reconstruction phase. We adopt the nearest neighbor strategy and the weighted joint strategy to obtain the new mapping matrix. The high-resolution image is reconstructed by the new mapping model. The experiments prove the effectiveness of our strategy.

Keywords: Super-resolution · External examples · Internal examples · Medical endoscopic images · Joint learning

1 Introduction

Single image super-resolution (SISR) aims to recover a high-resolution (HR) image from the input low-resolution (LR) image via complex linear or nonlinear models. The SR problem arises in many practical applications, such as medical imaging and video applications [13,21]. It is a classical problem in low-level computer vision and has attracted a lot of research attention. In recent years, numerous approaches have been proposed to solve this problem. In general, SR algorithms can be divided into three categories: interpolation-based methods [8,12], reconstruction-based methods [4,14], and learning-based methods [16,17,21].

Interpolation-based methods [8,12], such as the bilinear method and bicubic method, are efficient, but tend to generate oversmoothing images. Another class

© Springer Nature Switzerland AG 2019
Y. Zhao et al. (Eds.): ICIG 2019, LNCS 11901, pp. 714–723, 2019.
https://doi.org/10.1007/978-3-030-34120-6_58

of SR approach is based on reconstruction [4,14]. These methods estimate an HR image by enforcing some reasonable assumptions or prior knowledge to it. However, the high frequency details in images are not reconstructed very well [1].

The most popular method for SR now is the third category, which is known as learning-based method. These approaches usually assume that the lost high-frequency details in LR images can be predicted by the learned information from training set, which consists of a large set of LR patches and HR patches. These methods attempt to capture the co-occurrence prior between LR and HR image patches. Inspired by compressed sensing, Yang *et al.* [21] adopted sparse representation to solve SR problem. Timofte *et al.* [16] proposed an anchored neighborhood regression (ANR) method, which learned a sparse dictionary and utilized the sparse dictionary atoms for ridge regression, while, its refined variant, A+ [17], utilized the neighborhood taken from the training pool of samples for each sparse dictionary atom. Deep learning has also been adopted to address SR problem. Dong *et al.* [5] proposed a super-resolution convolutional neural network (SRCNN) for SR. Kim *et al.* [11] presented a very deep networks for super-resolution.

According to the ways of extracting training examples, learning-based SR method can be split into two classes. One uses an external database of natural images [3,5,11,16,17,20–22] and the other utilizes a database obtained from the input LR image itself [2,6,7].

The external example-based methods are based on the assumption that the mapping model between LR and HR image patches can be learned from an external database. The methods above are almost external example-based SR models. The internal example-based methods assume that the patches in a natural image tend to redundantly recur many times inside the image, both within the same scale, as well as across different scales [7]. Bevilacqua *et al.* [2] generated a double pyramid of recursively scaled and interpolated images, thus built a dictionary from the input LR image itself.

External example-based SR methods and internal example-based SR methods both have their own advantages and disadvantages, for example, some features of medical endoscopic images can not be well represented by widely used training set. Therefore, we can jointly train the model to get better medical image super-resolution results. Wang *et al.* [18] defined two loss functions using sparse coding-based external exmaples, and epitomic matching based on internal examples. Timofte [15] proposed a method, which fused A+ [17] and CSCN [19] as a new image feature, and applied the anchor strategy for SR. However, both of them adopted two different SR strategies and are based on the results of reconstruct HR image patches. In this paper, We propose novel joint SR to adaptively integrate the merits of both external-based and internal-based SR methods. What's more, we can fuse the mapping matrices in training phase, and thus obtain fusion matrices.

The remainder of the paper is as follows: Sect. 2 details the universal fusion strategy for SR. Section 3 shows the experimental results. Conclusion follows in Sect. 4.

2   Proposed Method

In the external example-based SR methods, we cannot guarantee that any input image patch can be matched and expressed by a limited set of external database. When dealing with some textures which are missing in the external database, the SR results may be oversmoothing and product serious noise. Internal strategy can handle this situation. But it can not perform well when the image has some patches that rarely recur. So it is reasonable that jointly learning for SR from external and internal examples.

However, there are lots of different SR methods. With different SR methods, it is hard to identify that the result of the final improvement is from whether the two different SR approaches or the combination of two different example selection strategies. For the purpose of getting a universal conclusion, we adopt the same strategy, A+ [17], based on external examples and internal examples respectively, to obtain a joint SR model. In this way, the improvement only depends on a combination of samples.

2.1   Training Model

We adopt the same training strategy with A+ to obtain the mapping matrix for each anchor point.

In external example-based A+ method, we apply K-SVD to get a sparse dictionary $\mathbf{A}_e$. Each atom of the dictionary is regarded as an anchor point. We search N_e nearest neighbors in the training set to conduct a sub-dictionary pair $\{\mathbf{D}_{He}^{ke}, \mathbf{D}_{Le}^{ke}\}_{ke=0}^{N_e}$ for each anchor point.

As for internal example-based A+ method, we adopt the double pyramid method to get the internal database. As shown in Fig. 1, we regard the input LR image $\mathbf{Y}$ as an HR training image. The other HR training images are generated by scaling down the LR input image $\mathbf{Y}$ with small factor p_i. So the HR training set is denoted as $\{\mathbf{Y}_H^i\}_{i=0}^{N_s}$, and N_s is the number of generated HR images. The LR image training set is conducted by scaling down each HR image with factor s, which is the same with the factor in reconstruction step. We also rotate and flip the input LR image for data augmentation. Then, we can conduct an HR and LR patch set for training. With the training set obtained, a similar sparse dictionary $\mathbf{A}_i$ is learned by K-SVD. For each anchor point in sparse dictionary, we also conduct the sub-dictionary pair $\{\mathbf{D}_{Hi}^{ki}, \mathbf{D}_{Li}^{ki}\}_{ki=0}^{N_i}$, N_i is the number of anchor points in internal model.

2.2   Mapping Model

In this paper, we adopt the ridge regression for learning the mapping matrix. We take the external example-based method as example, the regression is formulated as:

$$\mathbf{w} = \arg\min_{\mathbf{w}} \|\mathbf{y}_l - \mathbf{D}_{Le}^{ke}\mathbf{w}\|_2^2 + \lambda\|\mathbf{w}\|_2^2, \tag{1}$$

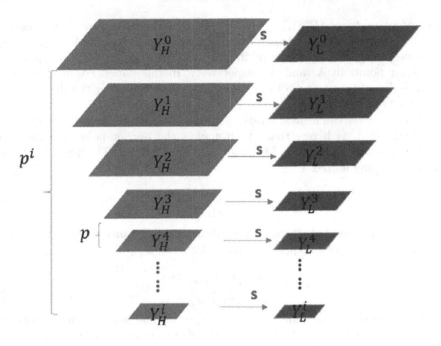

Fig. 1. The strategy of generating training set by the input image.

where $\mathbf{y}_l$ is an input LR patch. $\mathbf{D}_{Le}^{ke}$ is the corresponding sub-dictionary of $\mathbf{y}_l$, and ke is the index and is depended on the distance between anchor point and LR patch $\mathbf{y}_l$. $\mathbf{w}$ is the representation of $\mathbf{y}_l$ on sub-dictionary $\mathbf{D}_{Le}^{ke}$.

Equation 1 has a closed-form solution:

$$\mathbf{w} = \left(\mathbf{D}_{Le}^{ke}{}^T \mathbf{D}_{Le}^{ke} + \lambda \mathbf{I}\right)^{-1} \mathbf{D}_{Le}^{ke}{}^T \mathbf{y}_l, \tag{2}$$

Thus, we can get the corresponding HR image patch $\mathbf{y}_h$ using the same coefficient on HR sub-dictionary $\mathbf{D}_{He}^{ke}$:

$$\mathbf{y}_h = \mathbf{D}_{He}^{ke} \mathbf{w}, \tag{3}$$

We can obtain the mapping matrix $\mathbf{P}_e^{ke}$:

$$\mathbf{P}_e^{ke} = \mathbf{D}_{He}^{ke} \left(\mathbf{D}_{Le}^{ke}{}^T \mathbf{D}_{Le}^{ke} + \lambda \mathbf{I}\right)^{-1} \mathbf{D}_{Le}^{ke}{}^T, \tag{4}$$

The mapping matrix $\{\mathbf{P}_i^{ki}\}_{ki=0}^{N_i}$ in internal example-based method also can be computed in the same way.

2.3 Fusion Model and Image SR Reconstruction

In this stage, the input LR image are divided into overlapped image patches $\{\mathbf{y}_i\}_{i=0}^N$. The underlying HR image patches are noted as $\{\mathbf{x}_i\}_{i=0}^N$. Once we get the mapping matrices $\{\mathbf{P}_e^{ke}\}_{ke=0}^{N_e}$ and $\{\mathbf{P}_i^{ki}\}_{ki=0}^{N_i}$, we need to fuse them based on the distance between input LR patch and anchor point in $\mathbf{A}_e$ and $\mathbf{A}_i$ respectively.

We denote $\mathbf{d}_e$ and $\mathbf{d}_i$ as the minimum distance between LR input patch $\mathbf{y}_i$ and anchor points in $\mathbf{A}_e$ and $\mathbf{A}_i$ respectively. In this paper, cosine distance is chosen as distance metric. So the greater the value, the closer the distance. We attempt two joint strategies.

The first one we call nearest strategy. For each input LR patch $\mathbf{y}_i$, we compare $\mathbf{d}_e$ with $\mathbf{d}_i$. If $\mathbf{d}_e$ is bigger than $\mathbf{d}_i$, it means the anchor point generated by external example-based method is closer than internal one. Thus, we choose the external mapping matrix $\mathbf{P}_e^{ke}$.

$$\mathbf{P}^k = \begin{cases} \mathbf{P}_e^{ke} & if \quad d_e > d_i \\ \mathbf{P}_i^{ki} & else, \end{cases} \tag{5}$$

The other is weighted strategy. According to the distance $\mathbf{d}_e$ and $\mathbf{d}_i$, we give different weights to two mapping matrices $\mathbf{P}_e^{ke}$ and $\mathbf{P}_i^{ki}$.

$$\mathbf{P}^k = w_1 \mathbf{P}_e^{ke} + w_2 \mathbf{P}_i^{ki}$$
$$s.t. \quad w_1 + w_2 = 1, \tag{6}$$

where, w_1 and w_2 are weights that balance the two mapping matrices. Since the bigger the value of $\mathbf{d}_e$ and $\mathbf{d}_i$, the closer the distance, the corresponding weight should also be bigger than another. Thus, if $\mathbf{d}_e$ is bigger than $\mathbf{d}_i$, w_1 should also be bigger than w_2. We apply a simple weighted strategy to our model:

$$w_1 = \frac{d_e}{d_e + d_i}$$
$$w_2 = \frac{d_i}{d_e + d_i}, \tag{7}$$

Once the fusion mapping matrix is got, we directly use it to reconstruct the underlying HR image patch $\mathbf{x}_i$.

$$\mathbf{x}_i = \mathbf{P}^k \mathbf{y}_i, \tag{8}$$

The desired HR image $\mathbf{X}$ is reconstructed by merging all the HR image patches $\{\mathbf{x}_i\}_{i=0}^{N}$, and averaging the overlapping regions between the adjacent patches.

3   Experimental Results

In this section, we first compare the proposed method with external A+ method and internal A+ method to evaluate the validity of fusion strategy. We also compare it with several representative SISR methods, including external-based methods ScSR [21], Zeyde's [22], A+[17], internal-based method SelfEx [9] and deep method SRCNN [5]. All the experiments are carried out in the Matlab (R2016a) environment. For fair comparison, the external example-based methods are all trained on 91-image dataset [21]. The peak signal-to-noise ratio (PSNR) and structural similarity (SSIM) are applied to evaluate the quality of SR reconstruction and results are listed in Tables 1 and 3. We use three testing set (Set5, Set14 and B100) for SR evaluation.

3.1   Implementation Details

We convert RGB color space into YCbCr color space and apply the proposed algorithm on luminance channel (Y) and up-sample color channels (CbCr) by interpolation since human vision is much more sensitive to illuminance changes. The magnification factor is 3. The size of LR and HR image patch is 5×5 with overlapping 4 pixels. The features of LR images are the first and second order derivatives of the patches. The features of HR images are the residual between ground truth and the interpolated LR images, and represent the lost high frequency details. The number of generated HR images, N_s is 19, which means there are 20 HR images (including the input image itself). We also make a data argumentation for training. We rotate the image in 64 angles, and there is a 5.625° difference in each angle. The size of sparse dictionary in external part is 2048, i.e. N_e is 2048. N_i, the size in internal part, is 1024. The regularization parameter, λ, is set as 0.01.

3.2   Quality Evaluation

Table 1 shows average performance of fusion using two different strategies. Compared with external A+ SR method and internal A+ SR method, both joint methods can improve the SISR result, indicating the effectiveness of the method. And nearest strategy defeats weighted strategy, which impels us to use it in the rest of the experiment.

Table 2 shows the PSNR results on Set5. Our method achieves the best performance on most test images. We also compare the proposed method (with nearest strategy) with some state-of-the-art SR methods on Set14 and BSD100. Table 3 shows the average PSNR and SSIM results for up-scaling factor 3. Our method outperforms external, internal, and deep-based methods on all datasets. The average SSIM also performs best, revealing that our reconstructed results achieve best structural similarity with the ground truth. We also collect some medical endoscopic images for visual comparison, as shown in Fig. 2. We can see that our method (with nearest strategy) recovers more visually pleasing results with fewer artifacts, more accurate details and sharper edges.

Table 1. Average performance in PSNR and SSIM using nearest strategy and weighted strategy on BSD100. Up-scaling factor: 3

Method	PSNR	SSIM
External A+	28.31	0.7851
Internal A+	28.26	0.7831
Weighted strategy	28.34	0.7853
Nearest strategy	28.35	0.7856

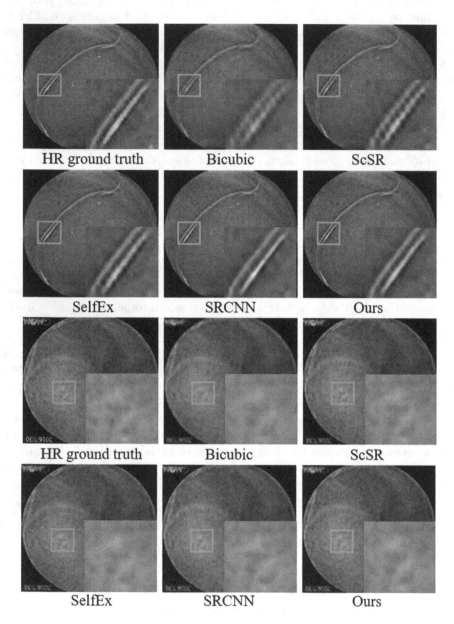

Fig. 2. Results of medical endoscopic images.

Table 2. Comparison on PSNR with different methods on test images Set5. Upscale factor: 3.

Images	Bicubic	SCSR [21]	SRCNN [5]	LANR [10]	Ours (nearest strategy)
Baby	33.92	34.29	35.01	35.23	**35.25**
Bird	32.58	34.11	34.91	34.82	**35.51**
Butterfly	24.04	25.58	**27.58**	26.14	27.50
Head	32.89	33.17	33.55	33.68	**33.79**
Woman	28.57	29.94	30.92	30.50	**31.26**
Average	30.40	31.42	32.39	32.07	**32.66**

Table 3. Benchmark SISR results. Average PSNR/SSIM for scale factor ×3 on datasets Set14 and BSD100. **Bold** represent the best performance.

Algorithm	Set14		BSD100	
	PSNR	SSIM	PSNR	SSIM
Bicubic	27.54	0.7736	27.15	0.7364
ScSR [21]	28.12	0.8055	27.82	0.7744
Zeyde's [22]	28.67	0.9075	27.87	0.7695
A+[17]	29.13	0.8188	28.18	0.7808
SelfEx [9]	28.95	0.8200	28.29	0.7846
SRCNN [5]	29.00	0.8145	28.21	0.7800
Ours	**29.21**	**0.8211**	**28.35**	**0.7856**

4   Conclusion

External-based and internal-based super-resolution methods both have their own advantages. This paper studies the strategy of joint learning of two kinds of methods and propose a universal fusion strategy for super-resolution. We utilize the strategy, which is the same with A+ [17], to obtain a external sub-dictionaries and internal sub-dictionaries. Then, we use the nearest strategy and weighted strategy to fuse the external and internal mapping matrices. The high-resolution image is reconstructed by the new mapping model. The experiments prove the effectiveness of our strategy.

Acknowledgment. This work was supported by the Natural Science Foundation of China (Nos. 61471216 and 61771276), the National Key Research and Development Program of China (No. 2016YFB0101001) and the Special Foundation for the Development of Strategic Emerging Industries of Shenzhen (Nos. JCYJ20170307153940960 and JCYJ20170817161845824).

References

1. Baker, S., Kanade, T.: Limits on super-resolution and how to break them. IEEE Trans. Pattern Anal. Mach. Intell. **24**(9), 1167–1183 (2002)
2. Bevilacqua, M., Roumy, A., Guillemot, C., Morel, M.L.A.: Single-image super-resolution via linear mapping of interpolated self-examples. IEEE Trans. Image Process. **23**(12), 5334–5347 (2014)
3. Chang, H., Yeung, D.Y., Xiong, Y.: Super-resolution through neighbor embedding. In: 2004 Proceedings of the 2004 IEEE Computer Society Conference on Computer Vision and Pattern Recognition, CVPR 2004, vol. 1, p. I. IEEE (2004)
4. Dai, S., Han, M., Xu, W., Wu, Y., Gong, Y.: Soft edge smoothness prior for alpha channel super resolution. In: 2007 IEEE Conference on Computer Vision and Pattern Recognition, pp. 1–8. IEEE (2007)
5. Dong, C., Loy, C.C., He, K., Tang, X.: Image super-resolution using deep convolutional networks. IEEE Trans. Pattern Anal. Mach. Intell. **38**(2), 295–307 (2016)
6. Freedman, G., Fattal, R.: Image and video upscaling from local self-examples. ACM Trans. Graph. (TOG) **30**(2), 12 (2011)
7. Glasner, D., Bagon, S., Irani, M.: Super-resolution from a single image. In: 2009 IEEE 12th International Conference on Computer Vision, pp. 349–356. IEEE (2009)
8. Hou, H., Andrews, H.: Cubic splines for image interpolation and digital filtering. IEEE Trans. Acoust. Speech Signal Process. **26**(6), 508–517 (1978)
9. Huang, J.B., Singh, A., Ahuja, N.: Single image super-resolution from transformed self-exemplars. In: Proceedings of the IEEE Conference on Computer Vision and Pattern Recognition, pp. 5197–5206 (2015)
10. Jiang, J., Ma, X., Chen, C., Lu, T., Wang, Z., Ma, J.: Single image super-resolution via locally regularized anchored neighborhood regression and nonlocal means. IEEE Trans. Multimed. **19**(1), 15–26 (2017)
11. Kim, J., Kwon Lee, J., Mu Lee, K.: Accurate image super-resolution using very deep convolutional networks. In: Proceedings of the IEEE Conference on Computer Vision and Pattern Recognition, pp. 1646–1654 (2016)
12. Li, X., Orchard, M.T.: New edge-directed interpolation. IEEE Trans. Image Process. **10**(10), 1521–1527 (2001)
13. Park, S.C., Park, M.K., Kang, M.G.: Super-resolution image reconstruction: a technical overview. IEEE Signal Process. Mag. **20**(3), 21–36 (2003)
14. Sun, J., Xu, Z., Shum, H.Y.: Image super-resolution using gradient profile prior. In: 2008 IEEE Conference on Computer Vision and Pattern Recognition, CVPR 2008, pp. 1–8. IEEE (2008)
15. Timofte, R.: Anchored fusion for image restoration. In: 2016 23rd International Conference on Pattern Recognition (ICPR), pp. 1412–1417. IEEE (2016)
16. Timofte, R., De Smet, V., Van Gool, L.: Anchored neighborhood regression for fast example-based super-resolution. In: Proceedings of the IEEE International Conference on Computer Vision, pp. 1920–1927 (2013)
17. Timofte, R., De Smet, V., Van Gool, L.: A+: adjusted anchored neighborhood regression for fast super-resolution. In: Cremers, D., Reid, I., Saito, H., Yang, M.-H. (eds.) ACCV 2014. LNCS, vol. 9006, pp. 111–126. Springer, Cham (2015). https://doi.org/10.1007/978-3-319-16817-3_8
18. Wang, Z., Yang, Y., Wang, Z., Chang, S., Yang, J., Huang, T.S.: Learning super-resolution jointly from external and internal examples. IEEE Trans. Image Process. **24**(11), 4359–4371 (2015)

19. Wang, Z., Liu, D., Yang, J., Han, W., Huang, T.: Deep networks for image super-resolution with sparse prior. In: Proceedings of the IEEE International Conference on Computer Vision, pp. 370–378 (2015)
20. Yang, J., Wang, Z., Lin, Z., Cohen, S., Huang, T.: Coupled dictionary training for image super-resolution. IEEE Trans. Image Process. **21**(8), 3467–3478 (2012)
21. Yang, J., Wright, J., Huang, T.S., Ma, Y.: Image super-resolution via sparse representation. IEEE Trans. Image Process. **19**(11), 2861–2873 (2010)
22. Zeyde, R., Elad, M., Protter, M.: On single image scale-up using sparse-representations. In: Boissonnat, J.-D., et al. (eds.) Curves and Surfaces 2010. LNCS, vol. 6920, pp. 711–730. Springer, Heidelberg (2012). https://doi.org/10.1007/978-3-642-27413-8_47

Unsupervised Optic Disc Segmentation for Cross Domain Fundus Image Based on Structure Consistency Constraint

XueSheng Bian, Cheng Wang$^{(\boxtimes)}$ (iD), Weiquan Liu, and Xiuhong Lin

Fujian Key Laboratory of Sensing and Computing for Smart Cities,
School of Information Science and Engineering, Xiamen University,
Xiamen 361005, China
xbc0809@gmail.com, cwang@xmu.edu.cn, wqliu1026@163.com, xhlinxm@gmail.com

Abstract. Deep convolution neural networks (DCNNs) are playing critical roles in various computer vision tasks, though they also suffer from many problems. The performance of DCNNs will severely degrade when facing gaps between the training set and test set. It is quite common in the field of medical image analysis where domain shift between images acquired from different devices or patients. A feasible solution to this problem is collecting more labeled data from the testing set for fine-tuning DCNNs to improve the generalization ability. Unfortunately, medical image labeling is complicated and time-consuming, which makes fine-tuning hard to apply. In this paper, we propose a novel network model which makes full use of the existing pre-trained network models, effectively reducing domain shift and restraining performance degradation. This method dramatically reduces the dependence on labeled data and achieve better results even if trained with a small amount of unlabeled data. We transform the testing images (target domain) with adversarial learning mechanism and make them look similar to training data (source domain) that can be segmented directly by the pre-training model trained on the training set. In addition, we introduce additional structural consistency constraints to suppress the distortion during forwarding propagation when the model is trained with fewer samples, which ensure the structure of the generated image is consistent with the input image. We validate our method in retinal fundus image segmentation task. Experiments show that the proposed method suppresses the degradation of model performance caused by domain migration, and achieves almost the same segmentation performance as the original training data.

Keywords: DCNN · Domain adaption · Adversarial learning

1 Introduction

Deep convolution neural network (DCNN) has achieved great success in various computer vision tasks, such as image recognition [17], target detection [14],

© Springer Nature Switzerland AG 2019
Y. Zhao et al. (Eds.): ICIG 2019, LNCS 11901, pp. 724–734, 2019.
https://doi.org/10.1007/978-3-030-34120-6_59

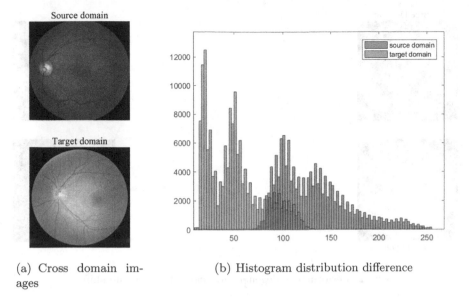

(a) Cross domain images

(b) Histogram distribution difference

Fig. 1. The appearance and Histogram distribution difference between cross domain images.

image segmentation [3], etc. Trained by a large amount of labeled data, this learning-based approach outperforms traditional methods in performance. However, encountering some new samples which are different from training data, the pre-trained model often lacks generalization. In the field of medical image analysis, a variety of different devices are used to capture medical images of patients. Because of the different imaging mechanism, these images collected by different devices have great differences in appearance, which is called domain gap, as shown in Fig. 1. These images are called cross-domain images. In this paper, the existing images with annotations are regarded as the source domain and the new images without annotations are seen as the target domain. Our goal is to make full use of the existing segmentation network trained in the source domain to enhance its generalization ability in the target domain.

In this case, some supervised learning method based on fine-tuning is proposed [12]. However, fine-tuning method requires labeling data in different image domains, which is time-consuming and expensive. In this paper, we propose an unsupervised image segmentation network based on synchronous domain adaption on image space and output space. This method only requires a small amount of unpaired data sampled from different image domains, and need not to annotated data. We use image synthesis method to make domain adaptation in image space and reduce the gap between training images and new images (testing images). In order to alleviate the problem of geometric distortion when training the transformation network with a small amount of data, we propose a structural consistency constraint to effectively suppress image distortion. Obviously,

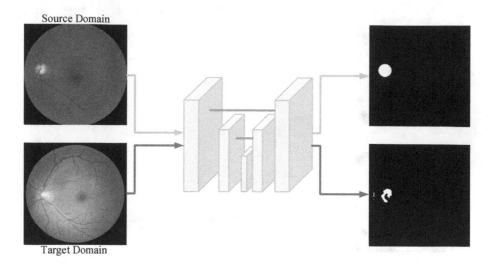

Fig. 2. Performance degeneration caused by domain shift

even though the appearances of source and target domains differ greatly, their semantics information is similar, thus, we adopt domain adaptation in the output space to further improve the segmentation performance.

In order to verify the partitioning performance of the proposed network, we validated the proposed algorithm on Retinal Fundus Glaucoma Challenge of MICCAI 2018 [1]. In terms of segmentation accuracy, the unsupervised network proposed in this paper is equivalent to other supervised learning methods.

The contributions of this work are as follows: (1) we proposes an unsupervised method for fundus image segmentation, which can make full use of the existing pre-training model, effectively reduce the dependence on training data, and also achieve the same segmentation effect as supervised learning. (2) A structural consistency constraint is proposed, which effectively suppress the geometric distortion during image conversion and ensure the structural consistency. (3) Image space and output space are synchronously adapted to effectively improve the segmentation performance of the model.

2   Related Work

Optic Disc Segmentation. Recently, since the convolutional neural network (CNN) has been proposed for image classification [10], there are variants of CNNs have been successfully applied in various computation vision tasks [11,13,18]. For the optic disc segmentation task, most of these methods achieving state-of-the-art performance are mainly based on deep learning. Fu [5], Zilly [21], Sedai [16] and Almotiri [2] propose various segmentation networks based on supervised learning and effectively overcome the challenges of optic disc segmentation, such as the unbalanced of the foreground and background, the blurred boundary, etc.

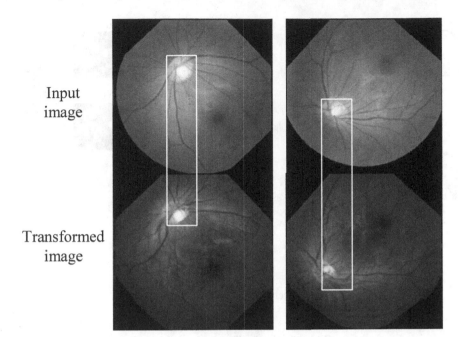

Input
image

Transformed
image

Fig. 3. Geometric distortion

These networks work well on a specific dataset, however, if the testing data are significantly different from the training data, the performance will be degraded severely, like the bad result shown in Fig. 2.

Adversarial Learning. After the Generative Adversarial Network (GAN) [6] was proposed by Goodfellow, adversarial learning mechanism achieves great success in image generation task and domain adaption methods. GAN consists of two parts: a generator and a discriminator. The generator aims to generate a realistic sample to fool discriminator, and the discriminator tries to distinguish whether the input is generated by the generator or not. Moreover, CycleGan [20] makes it easy to achieve unpaired image translation.

Domain Adaption. To deal with domain shift between the source domain and the target domain, many domain adaption methods are proposed which are usually divided into three categories. (1) Domain adaption on image space. Huo [8] proposed a segmentation method based on image generation, it first converted the source domain images to the target domain images, then the generated images are trained with the annotation of the corresponding source domain images. However, this method strongly relied on the quality of image generation. (2) Domain adaption on feature space. Dou [4] first mapped the source domain images and the target domain images into the sharing feature subspace by using the adversarial training strategy, then performed the segmentation on the latent

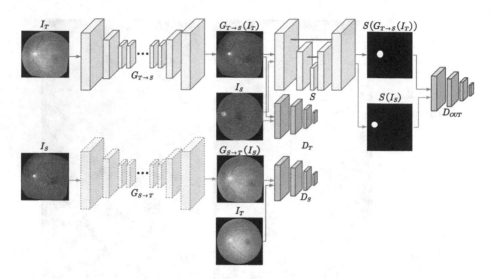

Fig. 4. The architecture of proposed network

feature representation. This method effectively reuses the pre-trained model. However, since only the features are adapted and the weight of the encoder is not shared, the spatial structure consistency cannot be guaranteed, which causes geometric distortion easily, which is shown in Fig. 3. (3) Domain adaption on output space. Tsai [19] proposed an adaptive method based on output space, which made full use of the semantic similarity between the source domain and the target domain images. However, the performance cannot be promised when the source domain and target domain images are too different.

3   Method

The proposed model consists of three components: Image generator G, performing image style transformation between two domains; Image discriminators D, distinguishing whether the input is from specific domain; Segmentation network S, predicting the category of each pixel. We use the Unet [15] pre-trained on the source domain as our segmentation and never update its weight. The architecture of our network is shown in Fig. 4.

3.1   Domain Adaption on Image Space

In order to take advantage of the performance of the pre-trained model on the training set, we need to generate a realistic target domain image close to images in source domain. Here, we utilize CycleGAN as the backbone to create image generator, which is full convolutional Resnet [7]. The discriminators are standard AlexNet [9]. Given two group of images $I_S, I_T \in R^{H \times W \times 3}$ sampled from the source domain and the target domain respectively. We feed I_T into the generator $G_{T\rightarrow S}$ to get new images $I_{T\rightarrow S}$ which are similar to images in the target domain. The corresponding generator $G_{S\rightarrow T}$ perform the inverse process. D_T, D_S are domain discriminators and D_{OUT} is used to tell semantic consistency between source domain and target domain in output space. The generators and discriminators are optimized alternately. The objective is formulized as:

$$
\begin{aligned}
\mathcal{L}_{GAN}\left(G_{S\rightarrow T}, D_T, I_S, I_T\right) = {} & E_{x\sim I_T}\left[\log D_T(x)\right] \\
& + E_{y\sim I_S}\left[\log\left(1 - D_T\left(G_{S\rightarrow T}(y)\right)\right)\right]
\end{aligned}
\tag{1}
$$

$$
\begin{aligned}
\mathcal{L}_{GAN}\left(G_{T\rightarrow S}, D_S, I_T, I_S\right) = {} & E_{y\sim I_S}\left[\log D_S(y)\right] \\
& + E_{x\sim I_T}\left[\log\left(1 - D_T\left(G_{T\rightarrow S}(x)\right)\right)\right]
\end{aligned}
\tag{2}
$$

To keep the semantic consistency between input and output, we expect that the generated image should have the ability of reconstructing the input. Cycle consistency loss is integrated to maximize the correlation between input image and generated image, which is described as:

$$
\mathcal{L}_{cycle}\left(G_{S\rightarrow T}, G_{T\rightarrow S}, S\right) = E_{x\sim I_T}\left[\left\|G_{S\rightarrow T}\left(G_{T\rightarrow S}(x)\right) - x\right\|_1\right]
\tag{3}
$$

$$
\mathcal{L}_{cycle}\left(G_{S\rightarrow T}, G_{T\rightarrow S}, S\right) = E_{y\sim I_S}\left[\left\|G_{T\rightarrow S}\left(G_{S\rightarrow T}(y)\right) - x\right\|_1\right]
\tag{4}
$$

The cycle consistency loss cannot promise the structure remain unchanged when the network is trained with small dataset. Considering the awareness that source domain image and target domain images have greater similarity in the gradient space, we propose an additional structure consistency loss with gradient constraint to suppress distortion during forward propagation. The structure consistency loss is as follows:

$$
\mathcal{L}_{structure}\left(G_{T\rightarrow S}, S\right) = E_{x\sim I_T}\left[\left\|gradient\left(G_{T\rightarrow S}(x)\right) - gradient\left(x\right)\right\|_1\right]
\tag{5}
$$

$$
\mathcal{L}_{structure}\left(G_{S\rightarrow T}, S\right) = E_{y\sim I_S}\left[\left\|gradient\left(G_{S\rightarrow T}(y)\right) - gradient\left(y\right)\right\|_1\right]
\tag{6}
$$

3.2   Domain Adaption on Output Space

Although the image generators output similar images to the target domain, the cross domain gap cannot be completely eliminated. We perform domain adaption on output space to force $S(G_{T\rightarrow S(I_T)})$ share semantic information with $S(I_S)$. The constrain is written as:

$$
\begin{aligned}
\mathcal{L}_{GAN}\left(G_{T\rightarrow S}, D_{out}, S, I_S, I_T\right) = {} & E_{y\sim I_S}\left[\log D_{OUT}(S(y))\right] \\
& + E_{x\sim I_T}\left[\log\left(1 - D_{OUT}\left(S\left(G_{S\rightarrow T}(x)\right)\right)\right)\right]
\end{aligned}
\tag{7}
$$

The training objective for proposed network is extended as:

$$\begin{aligned}
\mathcal{L}_{total} = \ &\lambda_1 * (\mathcal{L}_{GAN}(G_{S \to T}, D_T, I_S, I_T) + \mathcal{L}_{GAN}(G_{T \to S}, D_S, I_T, I_S)) \\
&+ (\mathcal{L}_{cycle}(G_{S \to T}, G_{T \to S}, S) + \mathcal{L}_{cycle}(G_{S \to T}, G_{T \to S}, S)) \\
&+ \lambda_2 * (\mathcal{L}_{structure}(G_{T \to S}, S) + \mathcal{L}_{structure}(G_{S \to T}, S)) \\
&+ \lambda_3 * \mathcal{L}_{GAN}(G_{T \to S}, D_{OUT}, S, I_S, I_T)
\end{aligned} \tag{8}$$

where λ_i are weight used to balance each item.

4   Experiment

4.1   Data

The dataset used in this paper is the Retinal Fundus Glaucoma Challenge of MICCAI 2018 [1], which contains 400 training samples (2124×2056 pixels) with manual annotation and 400 validation samples (1634×1634 pixels) without annotation. The training data is captured by Zeiss Visucam 500, and the validation data is taken by Canon CR-2. We set the training data as the source domain and the validation data as the target domain. In detail, 50 target domain samples are selected in the experiments. The raw images are first resized to 256 * 256 pixels for saving memory and the RGB pixel values are all normalized to a range of 0 to 1.

4.2   Network Architecture

The detail network architecture of our generators are shown in Table 1. We use instance normalization (IN) after each convolutional layer except for the last one. In addition, several notations are introduced: CONV[c, k, s, p]: convolutional layer with specific parameters, IN: instance normalization, LReLU: leaky rectified linear unit, GAvgPooling: global average pooling, c: output channel, k: kernel size, s: stride size, p: padding size.

4.3   Train Strategy

The network in this paper is implemented by PyTorch framework and trained with GTX 2080TI GPU. In the training stage, the generators are trained with Adam optimizer. The initial learning rate of the generator is 0.0002 and decrease 0.1 after every 20 epoch. The item weights in the loss is set to $\lambda_1 = 0.1, \lambda_2 = 0.5, \lambda_3 = 0.1$. The training parameters of discriminator are the same as generators, but the frequency of it is 5 times than the generators. All Images are performed data augmentation with random rotation and random flip. In the test stage, images in the target domain are transformed by specific generator and the output is fed into the segmentation network for prediction. In addition, we use dice score as our evaluation metric.

Table 1. Architecture of generator

Module	Input → output shape	Layer details
Down-sampling	$(w, h, 1) \rightarrow (w, h, 64)$	CONV[64, 7, 1, 3], IN, LReLU
	$(w, h, 64) \rightarrow (\frac{w}{2}, \frac{h}{2}, 128)$	CONV[128, 3, 2, 1], IN, LReLU
	$(\frac{w}{2}, \frac{h}{2}, 128) \rightarrow (\frac{w}{4}, \frac{h}{4}, 256)$	CONV[256, 3, 2, 1], IN, LReLU
Bottleneck	$(\frac{w}{4}, \frac{h}{4}, 256) \rightarrow (\frac{w}{4}, \frac{h}{4}, 256)$	Res-Block: CONV[256, 3, 1, 1], IN, LReLU
	$(\frac{w}{4}, \frac{h}{4}, 256) \rightarrow (\frac{w}{4}, \frac{h}{4}, 256)$	Res-Block: CONV[256, 3, 1, 1], IN, LReLU
	$(\frac{w}{4}, \frac{h}{4}, 256) \rightarrow (\frac{w}{4}, \frac{h}{4}, 256)$	Res-Block: CONV[256, 3, 1, 1], IN, LReLU
	$(\frac{w}{4}, \frac{h}{4}, 256) \rightarrow (\frac{w}{4}, \frac{h}{4}, 256)$	Res-Block: CONV[256, 3, 1, 1], IN, LReLU
	$(\frac{w}{4}, \frac{h}{4}, 256) \rightarrow (\frac{w}{4}, \frac{h}{4}, 256)$	Res-Block: CONV[256, 3, 1, 1], IN, LReLU
	$(\frac{w}{4}, \frac{h}{4}, 256) \rightarrow (\frac{w}{4}, \frac{h}{4}, 256)$	Res-Block: CONV[256, 3, 1, 1], IN, LReLU
	$(\frac{w}{4}, \frac{h}{4}, 256) \rightarrow (\frac{w}{4}, \frac{h}{4}, 256)$	Res-Block: CONV[256, 3, 1, 1], IN, LReLU
	$(\frac{w}{4}, \frac{h}{4}, 256) \rightarrow (\frac{w}{4}, \frac{h}{4}, 256)$	Res-Block: CONV[256, 3, 1, 1], IN, LReLU
	$(\frac{w}{4}, \frac{h}{4}, 256) \rightarrow (\frac{w}{4}, \frac{h}{4}, 256)$	Res-Block: CONV[256, 3, 1, 1], IN, LReLU
Up-sampling	$(\frac{w}{4}, \frac{h}{4}, 256) \rightarrow (\frac{w}{2}, \frac{h}{2}, 128)$	DeCONV[128, 3, 2, 1], IN, LReLU
	$(\frac{w}{2}, \frac{h}{2}, 128) \rightarrow (w, h, 64)$	DeCONV[64, 3, 2, 1], IN, LReLU
	$(w, h, 64) \rightarrow (w, h, 1)$	CONV[1, 3, 2, 1], Tanh

5   Result

In this section, we make qualitative and quantitative analysis for the proposed method. Three state-of-the-art methods are taken in comparison: Unet [15] is a popular architecture for medical image segmentation, MNet [5] is trained with supervised learning and achieve high score in optic disc segmentation task and SynSeg [8] is unsupervised method for cross domain image segmentation. Also, We did some ablation studies in our experiment to validate the practicability of proposed improvement means.

Table 2. Quantitative analysis for the proposed method

Methods	Dice score
Unet w/o adaption [15]	0.510
SynSeg [8]	0.885
MNet [5]	0.909
Ours w/o structure constrain	0.801
Ours w/o output space adaption	0.902
Ours	0.926

Figure 5 shows the results of some hard samples predicted by these methods. We find that the generated images have high similarity with the source image in appearance and keep the structure unchanged. Table 2 describes the detail evaluation of the results. Experiments show that the proposed method can effectively reduce the gap between source and target domains.

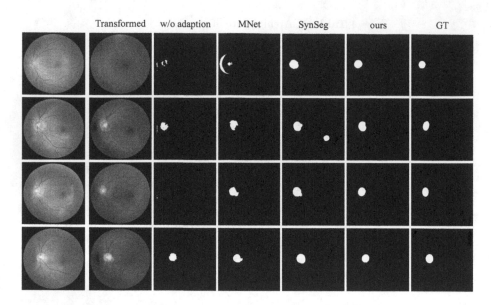

Fig. 5. Qualitative analysis for the proposed method

6 Conclusion

In this paper, we propose a new method that reuses the state-of-the-art segmentation model and ensures its high generalization ability when fed new samples with domain shift. The proposed method performs domain adaptation from the image space and the output space, eliminates the domain gap, and maximize the advantage of the pre-training model. In addition, to ensure the structural consistency between the segmentation result and the target domain image, we propose a gradient-based structure consistency constraint to suppress the geometric distortion of the generator effectively. Experiments show that the proposed method achieves the considerable performance of supervised segmentation model in the unsupervised segmentation of cross-source fundus images.

Acknowledgement. This work is supported by the National Natural Science Foundation of China (No. U1605254).

References

1. https://refuge.grand-challenge.org/Home/
2. Almotiri, J., Elleithy, K., Elleithy, A.: An automated region-of-interest segmentation for optic disc extraction. In: 2018 IEEE Long Island Systems, Applications and Technology Conference, LISAT 2018, pp. 1–6 (2018). https://doi.org/10.1109/LISAT.2018.8378019

3. Chen, L.C., Papandreou, G., Kokkinos, I., Murphy, K., Yuille, A.L.: Deeplab: semantic image segmentation with deep convolutional nets, atrous convolution, and fully connected CRFs. IEEE Trans. Pattern Anal. Mach. Intell. **40**(4), 834–848 (2017)
4. Dou, Q., Ouyang, C., Chen, C., Chen, H., Heng, P.A.: Unsupervised cross-modality domain adaptation of convnets for biomedical image segmentations with adversarial loss. arXiv preprint arXiv:1804.10916 (2018)
5. Fu, H., Cheng, J., Xu, Y., Wong, D.W.K., Liu, J., Cao, X.: Joint optic disc and cup segmentation based on multi-label deep network and polar transformation. IEEE Trans. Med. Imaging **37**(7), 1597–1605 (2018). https://doi.org/10.1109/TMI.2018.2791488
6. Goodfellow, I., et al.: Generative adversarial nets. In: Advances in Neural Information Processing Systems, pp. 2672–2680 (2014)
7. He, K., Zhang, X., Ren, S., Sun, J.: Deep residual learning for image recognition. In: Proceedings of the IEEE Conference on Computer Vision and Pattern Recognition, pp. 770–778 (2016)
8. Huo, Y., et al.: SynSeg-Net: synthetic segmentation without target modality ground truth. IEEE Trans. Med. Imaging **38**, 1016–1025 (2018)
9. Krizhevsky, A., Sutskever, I., Hinton, G.E.: Imagenet classification with deep convolutional neural networks. In: Advances in Neural Information Processing Systems, pp. 1097–1105 (2012)
10. LeCun, Y., Bottou, L., Bengio, Y., Haffner, P., et al.: Gradient-based learning applied to document recognition. Proc. IEEE **86**(11), 2278–2324 (1998)
11. Liu, W., Shen, X., Wang, C., Zhang, Z., Wen, C., Li, J.: H-net: neural network for cross-domain image patch matching. In: IJCAI, pp. 856–863 (2018)
12. Pan, S.J., Yang, Q.: A survey on transfer learning. IEEE Trans. Knowl. Data Eng. **22**(10), 1345–1359 (2009)
13. Redmon, J., Divvala, S., Girshick, R., Farhadi, A.: You only look once: unified, real-time object detection. In: Proceedings of the IEEE Conference on Computer Vision and Pattern Recognition, pp. 779–788 (2016)
14. Ren, S., He, K., Girshick, R., Sun, J.: Faster R-CNN: towards real-time object detection with region proposal networks. In: Advances in Neural Information Processing Systems, pp. 91–99 (2015)
15. Ronneberger, O., Fischer, P., Brox, T.: U-Net: convolutional networks for biomedical image segmentation. In: Navab, N., Hornegger, J., Wells, W.M., Frangi, A.F. (eds.) MICCAI 2015. LNCS, vol. 9351, pp. 234–241. Springer, Cham (2015). https://doi.org/10.1007/978-3-319-24574-4_28
16. Sedai, S., Roy, P.K., Mahapatra, D., Garnavi, R.: Segmentation of optic disc and optic cup in retinal fundus images using shape regression. In: Proceedings of the Annual International Conference of the IEEE Engineering in Medicine and Biology Society, EMBS 2016, pp. 3260–3264 (2016). https://doi.org/10.1109/EMBC.2016.7591424
17. Simonyan, K., Zisserman, A.: Very deep convolutional networks for large-scale image recognition. arXiv preprint arXiv:1409.1556 (2014)
18. Szegedy, C., et al.: Going deeper with convolutions. In: Proceedings of the IEEE Conference on Computer Vision and Pattern Recognition, pp. 1–9 (2015)
19. Tsai, Y.H., Hung, W.C., Schulter, S., Sohn, K., Yang, M.H., Chandraker, M.: Learning to adapt structured output space for semantic segmentation. In: Proceedings of the IEEE Conference on Computer Vision and Pattern Recognition, pp. 7472–7481 (2018)

20. Zhu, J.Y., Park, T., Isola, P., Efros, A.A.: Unpaired image-to-image translation using cycle-consistent adversarial networks. In: Proceedings of the IEEE International Conference on Computer Vision, pp. 2223–2232 (2017)
21. Zilly, J., Buhmann, J.M., Mahapatra, D.: Glaucoma detection using entropy sampling and ensemble learning for automatic optic cup and disc segmentation. Comput. Med. Imaging Graph. **55**, 28–41 (2017). https://doi.org/10.1016/j.compmedimag.2016.07.012

Low Resolution Person Re-identification by an Adaptive Dual-Branch Network

Zhanxiang Feng[1], Wenxiao Zhang[1], Jianhuang Lai[1,2]($\boxtimes$), and Xiaohua Xie[1,2]

[1] School of Data and Computer Science, Sun Yat-sen University, Guangzhou, China
{fengzhx7,stsljh,xiexiaoh6}@mail.sysu.edu.cn, zhwenx3@mail2.sysu.edu.cn
[2] Guangdong Key Laboratory of Machine Intelligence and Advanced Computing,
Ministry of Education, Guangzhou, China

Abstract. Low-resolution person re-identification (LR-REID) refers to matching cross-view pedestrians from varying resolutions, which is common and challenging for realistic surveillance systems. The LR-REID is largely under-study and the performance of the existing methods drops dramatically in LR domain. In this paper, we propose a novel adaptive dual-branch network (ADBNet) to recover HR pedestrians and learn discriminative resolution-robust representations. The ADB-Net employs a two-branch structure to obtain HR images, of which the SR-branch adopts the reconstruction loss to get the global image and the GAN-branch employs the generative loss to generate sharp details. These branches are combined into a universe model and the combination weights are adaptively learned. Furthermore, we propose a relative loss to guarantee shaper edges of the output results. Experimental results on two large-scale benchmarks prove the validity of the proposed approach.

Keywords: Person re-identification · Low resolution · Generative adversarial network · Super resolution

1 Introduction

Person re-identification (re-id) refers to matching pedestrians from non-overlapping cameras, which is very challenging because of the visual variations across changing views. With the ubiquity of surveillance systems, re-id has attracted increasing research attentions, and recent studies have achieved remarkable progresses. The existing approaches focus on addressing the issues of illumination variations, viewpoint changes and occlusions. However, the low-resolution (LR) problem is generally ignored by the current literature, which is common in the surveillance systems because of poor imaging quality and long-distance monitoring, arising the LR-REID problem. The performance of traditional re-id methods encounters a serious degradation when recognizing LR images because these methods fail to capture discriminative information from the LR pedestrians. Therefore, learning a resolution-free model is essential for overcoming the drawbacks of the existing methods and enhancing the realistic surveillance systems.

© Springer Nature Switzerland AG 2019
Y. Zhao et al. (Eds.): ICIG 2019, LNCS 11901, pp. 735–746, 2019.
https://doi.org/10.1007/978-3-030-34120-6_60

Some studies have made pioneering efforts to tackle with LR-REID. These methods first reduce the cross-resolution discrepancies by reconstructing HR pedestrians using super-resolution techniques or generative adversarial networks (GAN). Then learn a classification model is learned to distinguish the identity of the reconstructed outputs. Although the latest researches have considerably improved the robustness of re-id models against different resolutions, the performance of the LR-REID models is far beneath the traditional re-id techniques. An important reason is that the reconstruction methods fail to capture accurate and discriminative details. On the one hand, super-resolution methods tend to generate blurred images which are not superior for recognition. On the other hand, although GAN-based networks can obtain sharp details, the generated pedestrians contain annoying artifacts, which may confuse the discriminator and result in wrong decisions. To address this problem, we intend to desire a model which recovers articulate and accurate details. Furthermore, the adversarial loss, which is designed to determine whether the image is real or fake, is not always appropriate for reconstructing HR images because both LR and HR images are real images. A suitable loss function is to determine whether the reconstructed output contains sharper details than the input image.

Based on the above discussions, we propose an adaptive dual-branch network (ADBNet) structure to reconstruct HR images. The ADBNet consists of two branches, namely the SR-branch and the GAN-branch. The SR-Branch adopts the reconstruction loss to recover accurate outputs. The GAN-Branch imposes the adversarial loss to obtain sharp details. The ADBNet structure adaptively learns the weights of each branch and fuses the SR-Branch and GAN-Branch in a weighted summation manner to obtain the final HR output. Besides, we propose a resolution-aware adversarial loss named relative loss to determine whether the generated images have sharper edges than the LR images. Eventually, a classification module is attached behind the dual-branch structure to learn distinguishing features in an end-to-end manner. Experimental results prove the effectiveness of the proposed approach.

2   Related Works

2.1   Tradition Re-id Methods

Re-id has attracted widespread research attention in the last decade. The existing works mainly focus on feature extraction and metric learning. Deep learning [1,3,4,7,15,19,23] is playing an increasingly significant role in the re-id task because of the strengths in discriminative feature learning. With the development of deep learning techniques, deep networks have achieved remarkable progress in re-id, and deep models significantly outperform hand-crafted features. Furthermore, researchers implement metric learning [2,5,10,13] on the deep features to learn robust features. Although recent studies have reported high recognition accuracy for re-id, the existing methods are focused on handling the challenges of cross-view variations including illumination, pose, and occlusion, where the re-id models are trained and tested with HR images. The performance of traditional re-id methods will be dramatically affected by LR images, rising a LR-REID problem, which is largely under-study.

2.2 Low-Resolution Re-id Methods

Recently, some researches have paid attention to the LR problem and proposed pioneering researches [11,14,22] to handle the LR-REID task. The basic idea is first synthesising the HR images and then extracting blur-robust representations. Jiao et al. [11] proposed the SING model which first implemented super-resolution using SRCNN and learned a classification network to conduct person re-identification. Wang et al. [22] proposed a cascaded SR-GAN model to conduct scale-adaptive image super-resolution and used a re-id network to learn discriminative features. Although the latest researches have achieved remarkable progresses considering the resolution variations, the performances of the LR-REID models are still far from satisfactory. In this paper, we follow the framework to first synthesize HR images and then extract robust feature representations using the reconstructed images. Notably, we propose a novel adaptive dual-branch network to improve the image enhancement process.

3 Proposed Methodology

3.1 Overview

Figure 1 illustrates the structure of the proposed approach. The proposed method jointly optimizes two sub-networks, namely the adaptive dual-branch sub-network and re-id sub-network, to conduct image enhancement and recognition. The ADBNet consists of two branches, including the super-resolving branch (SR-branch) and the cycle-generative branch (GAN-branch). The weights W_{SR} and W_{GAN} are adaptively learned to combine the SR-branch and GAN-branch to obtain the desired HR reconstruction results. Finally, the re-id sub-network is integrated into the ADBNet to learn discriminative resolution-robust representations.

The notations of the proposed approach are as follows. Given the training data as $(\boldsymbol{x}_i^l, \boldsymbol{x}_i^h, y_i), i = 1, 2, ..., N$, where $\boldsymbol{x}_i^l$ and $\boldsymbol{x}_i^l$ refer to the LR and HR images, y_i is the corresponding identity label, and N denotes the number of training samples, we use the ADBNet to learn an enhancing function $F_{ADB}(.)$ which compensates high-frequency components for the LR input. Then a feature extraction function $F_{FE}(.)$ is adopted to learn discriminant features for LR-REID.

3.2 Adaptive Dual-Branch Network

As described above, the ADBNet combines the SR-branch together with the GAN-branch to generate HR pedestrians. The SR-branch adopts the reconstruction loss to produce accurate details whereas the GAN-branch imposes the generative adversarial loss to generate sharp edges. Finally, the ADBNet learns the

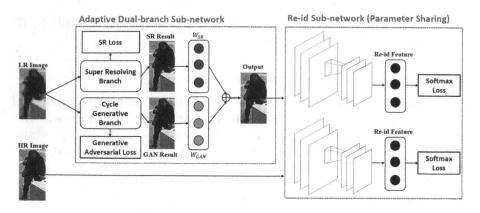

Fig. 1. The proposed LR-REID framework. A novel adaptive dual-branch network is proposed to recover HR images. Then a re-id network is attached by the reconstructed images to learn discriminative feature representations.

adaptive weights to combine the SR-branch and GAN-branch together to merit from the complementary information in the outputs of these branches.

Super-Resolving Branch. The SR-branch recovers the high-frequency details using a traditional image super-resolution manner. We adopt the resnet-based structure used by Johnson et al. [12] as the generator network of the SR-branch. Particularly, we exploit the structure which contains 6 resnet blocks to extract deep features from the LR images. Then a reconstruction loss is imposed to guarantee that the output is as close to the ground-truth HR image as possible. Denote $F_{SR}(.)$ as the function of the SR-branch, the reconstruction loss can be formulated as:

$$\mathcal{L}_{SR} = \frac{1}{N} \sum_{i=1}^{N} \parallel F_{SR}(\boldsymbol{x}_i^l) - \boldsymbol{x}_i^h \parallel_1 . \tag{1}$$

The researches of super-resolution techniques have verified that the reconstruction loss tends to produce blurry outputs, which is detrimental for recognition. Therefore, we employ the generative network to obtain clear outputs.

Cycle-Generative Branch. We adopt the generative networks to extract sharp details, which is supplementary to the SR-branch. Inspired by [25], we use a cycle structure to generate HR images. The proposed cycle-generative branch is shown in Fig. 2. Our model consists of two generators and three discriminators, including Generator L2H, Generator H2L, Discriminator H, Discriminator L and Discriminator R. For LR images, we generate HR outputs using generator L2H. Then the HR outputs are mapped back to LR domain using generator H2L. Similar implementation is conducted for HR images. The cycle-generative branch includes three losses: adversarial loss, cycle loss, and relative loss.

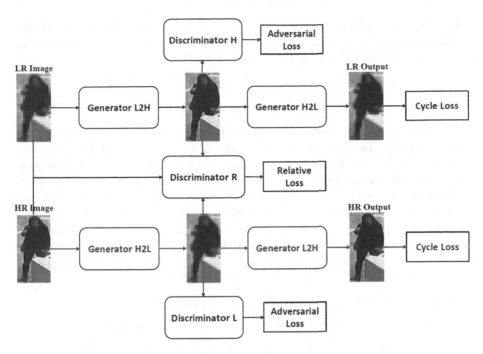

Fig. 2. The structure of the cycle-generative branch.

We apply the adversarial loss [8] to both Generator L2H and Generator H2L to improve the quality of the generative outputs. For Generator L2H and Discriminator H, the adversarial loss can be expressed as:

$$\mathcal{L}_{ADV}^{L2H} = E_{\boldsymbol{x}_i^h \sim p_{data}(\boldsymbol{x}_i^h)}[log(D_H(\boldsymbol{x}_i^h))] \\ + E_{\boldsymbol{x}_i^l \sim p_{data}(\boldsymbol{x}_i^l)}[log(1 - D_H(G_{L2H}(\boldsymbol{x}_i^l)))], \tag{2}$$

where $G_{L2H}(.)$ refers to the function of Generator L2H and $D_H(.)$ refers to the function of Discriminator H. In this manner, Generator L2H tries to generate more realistic images to fool the Discriminator H while Discriminator H aims to distinguish the real and fake samples. We also introduce a similar adversarial loss for Generator H2L and Discriminator L, which can be formulated as:

$$\mathcal{L}_{ADV}^{H2L} = E_{\boldsymbol{x}_i^l \sim p_{data}(\boldsymbol{x}_i^l)}[log(D_L(\boldsymbol{x}_i^l))] \\ + E_{\boldsymbol{x}_i^h \sim p_{data}(\boldsymbol{x}_i^h)}[log(1 - D_L(G_{H2L}(\boldsymbol{x}_i^h)))]. \tag{3}$$

The adversarial training has many potential solutions. The generative network can produce any HR image for the input LR image, which is harmful to re-id because the identity information may be changed by the generator. Therefore, we impose the cycle loss on the cycle-generative branch to ensure that the mapping functions G_{L2H} and G_{H2L} are cycle-consistent. The cycle loss can reduce the space of possible mapping functions and force the generated output to maintain the intrinsic information of the input image. The cycle loss for Generator L2H and Generator H2L can be formulated as:

$$
\begin{aligned}
\mathcal{L}_{CYC} = & E_{\boldsymbol{x}_i^l \sim p_{data}(\boldsymbol{x}_i^l)}[\|\, G_{H2L}(G_{L2H}(\boldsymbol{x}_i^l)) - \boldsymbol{x}_i^l \,\|_1] \\
& + E_{\boldsymbol{x}_i^h \sim p_{data}(\boldsymbol{x}_i^h)}[\|\, G_{L2H}(G_{H2L}(\boldsymbol{x}_i^h)) - \boldsymbol{x}_i^h \,\|_1].
\end{aligned}
\tag{4}
$$

The adversarial loss focuses on determining whether the input images are realistic samples or generated ones, which may be insufficient for LR-REID because both LR and HR images are realistic images. For the sake of improving the recognition performance of LR-REID, the generator not only needs to produce realistic outputs, but also needs to generate images whose details are sharper than the input images. Based on the above discussions, we propose a relative loss to guarantee that the generator produces images with higher resolution. The relative loss is designed to determine the relative relations of two images in terms of sharpness, or whether the output image is more clear than the input image. The formula of the relative loss is as follows:

$$
\begin{aligned}
\mathcal{L}_{REL} = & E[log(D_R(\boldsymbol{x}_i^l, \boldsymbol{x}_i^h))] + E[log(1 - D_R(\boldsymbol{x}_i^h, \boldsymbol{x}_i^l))] \\
& + E[log(D_R(\boldsymbol{x}_i^l, G_{L2H}(\boldsymbol{x}_i^l)))] + E[log(1 - D_R(G_{L2H}(\boldsymbol{x}_i^l)), \boldsymbol{x}_i^l)] \\
& + E[log(D_R(G_{H2L}(\boldsymbol{x}_i^h)), \boldsymbol{x}_i^h)] + E[log(1 - D_R(\boldsymbol{x}_i^h, G_{H2L}(\boldsymbol{x}_i^h)))],
\end{aligned}
\tag{5}
$$

where D_R represents the function of the Discriminator R. The Discriminator R first learns the relative relationship considering sharpness between the LR and HR images. Then the discriminator is utilized to determine the relative relationship between the input images and the outputs of the generators.

Notably, we follow [12] and adopt the structure of 6 resnet blocks to extract features for the Generator L2H and Generator H2L. For the discriminators, we follow [25] and use Patch-GANs.

Adaptive Weights Learning. The outputs of the SR-branch and the GAN-branch are complementary since the SR-branch produces accurate but blurry images whereas the GAN-branch generates sharp images with artifacts. In this paper, we integrate these branches in an united framework and learn an adaptive weight for each branch to combine the reconstruction results. Particularly, we attach a convolution layer for each branch, denoted as W_{SR} and W_{GAN} respectively, to learn the suitable weights. Then the final output of the ADBNet can be formulated as:

$$
\boldsymbol{x}_{RE} = W_{SR} F_{SR}(\boldsymbol{x}_i^l) + W_{GAN} G_{L2H}(\boldsymbol{x}_i^l).
\tag{6}
$$

Finally, we employ a reconstruction loss to guarantee that the output x_{RE} is the same as the ground-truth HR images, which can be formulated as:

$$
\mathcal{L}_{RE} = \frac{1}{N} \sum_{i=1}^{N} \|\, \boldsymbol{x}_{RE} - \boldsymbol{x}_i^h \,\|_1 .
\tag{7}
$$

3.3   Feature Extraction Network

In this paper, we propose an end-to-end structure to conduct image enhancement and classification. A classification network is attached behind the ADBNet to learn distinguishing features for both the HR images and reconstructed outputs. Denote $L_S(.)$ as the classification function, the objective function of the feature extraction network can be formulated as:

$$\mathcal{L}_{Re-id} = \frac{1}{N} \sum_{i=1}^{N} (L_S(F_{FE}(\boldsymbol{x}_i^h), y_i) + L_S(F_{FE}(F_{ADB}(\boldsymbol{x}_i^l)), y_i)). \qquad (8)$$

Here we adopt ResNet50 [9] as the feature extraction network and impose Softmax loss [7] to train the classifier.

3.4   Overall Objective Function

Consequently, the objective function of the proposed approach can be formulated as:

$$\mathcal{L} = \mathcal{L}_{Re-id} + \lambda_1\mathcal{L}_{SR} + \lambda_2\mathcal{L}_{CYC} + \lambda_3\mathcal{L}_{RE} + \lambda_4(\mathcal{L}_{ADV}^{L2H} + \mathcal{L}_{ADV}^{H2L} + \mathcal{L}_{REL}). \qquad (9)$$

4   Experiment

4.1   Experimental Settings

Datasets. We conduct experiments on two simulated large-scale LR datasets, namely LR-Market-1501 and LR-CUHK03, to evaluate the effectiveness of the ADBNet. We obtain the LR images by down-scaling the HR images with a ratio of 0.25 for both datasets. *LR-Market-1501* is built from the Market-1501 dataset [24]. The market-1501 dataset contains 32,668 annotated bounding boxes of 1,501 identities from six cameras in an open system. The images are automatically detected by a deformable parts model detector. *LR-CUHK03* is based on the CUHK03 dataset [15]. The CUHK03 dataset contains more than 14,000 images of 1,467 people captured from six cameras. The images of each person are from two disjoint cameras.

Evaluation Protocols. We follow the standard evaluation protocol to ensure fair comparisons between the ADBNet and the other approaches. For **LR-Market-1501** benchmark, we train the models with the standard training set (750 identities) and match 3,368 query images with the standard testing set (751 identities). Notably, we adopt the single-query mode for testing. We conduct matching using the LR query images and the HR testing gallery images. Rank-1, Rank-5, Rank-10 accuracies and mean average precision (mAP) are computed to evaluate the performance of all the methods. For **LR-CUHK03** benchmark, we follow the protocol used by [11]: randomly split the samples into 100 people for testing and the remainder for training. We randomly select one image from the gallery for each identity to form the gallery set and construct the probe set with all LR images. The cumulative matching characteristic (CMC) is used to evaluate the performance of the compared methods.

Table 1. Comparisons on LR-Market-1501

Rank (%)	1	5	10	mAP
PCB$_{HR}$ [18]	18.1	31.7	38.2	11.7
MGN$_{HR}$ [20]	30.3	51.7	61.3	33.5
ResNet50$_{HR}$ [9]	15.5	28.2	35.2	11.1
SRCNN [6]+ResNet50$_{HR}$	51.2	72.7	79.8	33.7
ADBNet+ResNet50$_{HR}$	60.5	80.1	85.9	40.2
ResNet50	60.0	80.3	86.1	38.5
ADBNet	72.1	86.6	91.0	48.6

Implementation Details. We adopt ResNet50 as the baseline feature extraction model, use 6 resnet blocks [12] to form the generators, and employ the Patch-GANs [25] as the discriminators. The proposed method is implemented using the PyTorch [17] framework. We set the learning rate as 10^{-2} for the re-id sub-network and 2×10^{-4} for the ADBNet. The proposed approach is tuned to the best efficiency, and we set $\lambda_1 = 10$, $\lambda_2 = 10$, $\lambda_3 = 10$, and $\lambda_4 = 1$ for all the datasets. The training ends when the number of epcho reaches 100.

4.2   Comparisons with Other Methods

In this section, we compare the ADBNet with the state-of-the-art methods to validate its effectiveness. The compared methods include traditional re-id methods, such as DGD [23], MGN [20], PCB [18], and ResNet50, and LR-REID methods, such as JUDEA [16], SDF [21], and SING [11].

LR-Market-1501. The researchers have not yet studied LR-REID on the Market-1501 dataset, we compare the ADBNet with some popular re-id methods. Furthermore, we also evaluate the performance of reconstruction-based methods which reconstruct HR images and conduct recognition on the generated outputs. Table 1 shows the comparison results, where the subscript HR means training merely on HR images. Obviously, the performance of the popular re-id methods (PCB and MGN) degrades significantly when dealing with LR images. Table 1 also verifies that pre-processing is valuable for addressing the LR-REID task. The SRCNN approach significantly improves the Rank-1/mAP of ResNet50 from 15.1%/11.1% to 51.2%/33.7%. Notably, recovering HR images by ADBNet dramatically outperforms the SRCNN methods, by a margin of 9.3%/6.5% in terms of Rank-1/mAP. Finally, the ADBNet remarkably outperforms the compared methods. The ADBNet achieves the highest Rank-1/mAP statistics of 72.1%/48.6, which is very closed to the traditional re-id performance. Compared with the baseline ResNet50, the ADBNet achieves an improvement of 12.1% in Rank-1 accuracy and 10.1% in mAP.

Table 2. Comparisons on LR-CUHK03

Rank (%)	1	5	10
JUDEA [16]	26.2	58.0	73.4
SDF [21]	22.2	48.0	64.0
DGD [23]	58.5	86.0	92.2
SRCNN [6]+XQDA	49.0	74.8	85.0
SING [11]	67.7	90.7	94.7
ADBNet	68.9	88.4	92.5

LR-CUHK03. Table 2 illustrates the comparison results between the ADB-Net and the state-of-the-art LR-REID methods. The ADBNet is proven to be competitive against other methods and achieve the highest Rank-1 accuracy in LR-CUHK03. Note that the Rank-5 and Rank-10 accuracies of ADBNet is lower than SING. However, SING is tested with images down-scaled by magnification factors including 2, 3, and 4. Therefore, the ADBNet deals with images of much lower resolutions in average, and the performance of ADBNet can be further improved when dealing with smaller magnification factors.

Table 3. Ablation experiments on LR-Market-1501

Method	Rank-1	Rank-5	Rank-10	mAP
ADBNet w/o SR-branch	60.7	79.2	85.4	37.1
ADBNet w/o GAN-branch	66.0	83.5	88.7	44.1
ADBNet w/o RL	69.2	85.1	90.0	45.3
ADBNet	72.1	86.6	91.0	48.6

4.3 In-Depth Analysis

Ablation Analysis. In this section, we conduct ablation experiments on LR-Market-1501 to evaluate the effects of each component in ADBNet. Table 3 shows the results of ablation experiments. Without the SR-branch, although we can obtain sharp details by merely using the GAN-branch, the Rank-1 accuracy/mAP is only 60.7%/37.1% because of the annoying artifacts. On the other hand, we can only achieve a Rank-1 accuracy/mAP of 66.0%/44.1% with blurry HR images generated by the GAN-branch. Combining the SR-branch and the GAN-branch results in a better Rank-1 accuracy/mAP of 69.2%/45.3%. Finally, using the proposed relative loss can further improve the performance of ADBNet and reports the best Rank-1 accuracy/mAP of 72.1%/48.6%.

LR   SR   GAN   ADBNet   HR   LR   SR   GAN   ADBNet   HR

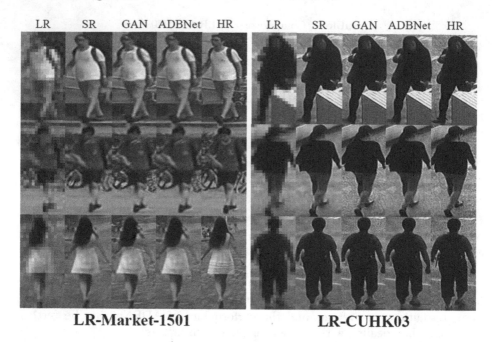

LR-Market-1501   **LR-CUHK03**

Fig. 3. The reconstructed HR images on LR-Market-1501 and LR-CUHK03.

Reconstruction Results. Figure 3 illustrates the reconstruction results of each component in ADBNet. Obviously, the reconstructed images of the SR-branch are much blurry than the HR outputs of the GAN-branch, particularly in the contours of pedestrians. Although GAN-branch can produce high-quality results in LR-CUHK03 dataset, the outputs of GAN-branch in LR-Market-1501 contain many artifacts, which is detrimental for recognition. Notably, the ADBNet can achieve a better balance between improving the image quality and reducing inappropriate artifacts, and generate accurate sharp details for the pedestrians.

5   Conclusion

In this paper, we propose a novel ADBNet to deal with the LR-REID issue, which is largely under-study in the current literature. The proposed ADBNet consists of a SR-branch and a GAN-branch, and adaptively learns the combination weights to obtain sharp and accurate reconstruction outputs. Furthermore, we design a relative loss to generate HR outputs with sharper details. Finally, a re-id sub-network is integrated into the proposed approach to learn discriminative features. The experimental results on two large-scale simulated LR benchmarks validate the effectiveness of the proposed approach.

Acknowledgment. This project was supported by the NSFC 61573387.

References

1. Ahmed, E., Jones, M., Marks, T.K.: An improved deep learning architecture for person re-identification. In: CVPR, pp. 3908–3916 (2015)
2. Chen, D., Yuan, Z., Chen, B., Zheng, N.: Similarity learning with spatial constraints for person re-identification. In: CVPR, pp. 1268–1277 (2016)
3. Chen, S., Guo, C., Lai, J.: Deep ranking for person re-identification via joint representation learning. TIP **25**(5), 2353–2367 (2016)
4. Chen, Y.C., Zhu, X., Zheng, W.S., Lai, J.H.: Person re-identification by camera correlation aware feature augmentation. TPAMI **40**(2), 392–408 (2018)
5. Chen, Y., Zheng, W., Lai, J., Yuen, P.: An asymmetric distance model for cross-view feature mapping in person re-identification. TCSVT **27**(8), 1661–1675 (2016)
6. Dong, C., Loy, C.C., He, K., Tang, X.: Learning a deep convolutional network for image super-resolution. In: Fleet, D., Pajdla, T., Schiele, B., Tuytelaars, T. (eds.) ECCV 2014. LNCS, vol. 8692, pp. 184–199. Springer, Cham (2014). https://doi.org/10.1007/978-3-319-10593-2_13
7. Feng, Z., Lai, J., Xie, X.: Learning view-specific deep networks for person re-identification. TIP **27**(7), 3472–3483 (2018)
8. Goodfellow, I., et al.: Generative adversarial nets. In: NIPS, pp. 2672–2680 (2014)
9. He, K., Zhang, X., Ren, S., Sun, J.: Deep residual learning for image recognition. In: CVPR, pp. 770–778 (2016)
10. He, W., Chen, Y., Lai, J.: Cross-view transformation based sparse reconstruction for person re-identification. In: ICPR, pp. 3410–3415 (2016)
11. Jiao, J., Zheng, W.S., Wu, A., Zhu, X., Gong, S.: Deep low-resolution person re-identification. In: AAAI (2018)
12. Johnson, J., Alahi, A., Fei-Fei, L.: Perceptual losses for real-time style transfer and super-resolution. In: Leibe, B., Matas, J., Sebe, N., Welling, M. (eds.) ECCV 2016. LNCS, vol. 9906, pp. 694–711. Springer, Cham (2016). https://doi.org/10.1007/978-3-319-46475-6_43
13. Koestinger, M., Hirzer, M., Wohlhart, P., Roth, P.M., Bischof, H.: Large scale metric learning from equivalence constraints. In: CVPR, pp. 2288–2295 (2012)
14. Li, K., Ding, Z., Li, S., Fu, Y.: Discriminative semi-coupled projective dictionary learning for low-resolution person re-identification. In: AAAI (2018)
15. Li, W., Zhao, R., Xiao, T., Wang, X.: Deepreid: deep filter pairing neural network for person re-identification. In: CVPR, pp. 152–159 (2014)
16. Li, X., Zheng, W.S., Wang, X., Xiang, T., Gong, S.: Multi-scale learning for low-resolution person re-identification. In: ICCV, pp. 3765–3773 (2015)
17. Paszke, A., et al.: Automatic differentiation in pytorch (2017)
18. Suh, Y., Wang, J., Tang, S., Mei, T., Lee, K.M.: Part-aligned bilinear representations for person re-identification. In: Ferrari, V., Hebert, M., Sminchisescu, C., Weiss, Y. (eds.) Computer Vision – ECCV 2018. LNCS, vol. 11218, pp. 418–437. Springer, Cham (2018). https://doi.org/10.1007/978-3-030-01264-9_25
19. Wang, G., Lai, J., Xie, X.: P2snet: can an image match a video for person re-identification in an end-to-end way? TCSVT **28**(10), 2777–2787 (2018)
20. Wang, G., Yuan, Y., Chen, X., Li, J., Zhou, X.: Learning discriminative features with multiple granularities for person re-identification, pp. 274–282. ACM (2018)
21. Wang, Z., Hu, R., Yu, Y., Jiang, J., Liang, C., Wang, J.: Scale-adaptive low-resolution person re-identification via learning a discriminating surface. In: IJCAI, pp. 2669–2675 (2016)

22. Wang, Z., Ye, M., Yang, F., Bai, X., Satoh, S.: Cascaded SR-GAN for scale-adaptive low resolution person re-identification. In: IJCAI, pp. 3891–3897 (2018)
23. Xiao, T., Li, H., Ouyang, W., Wang, X.: Learning deep feature representations with domain guided dropout for person re-identification. In: CVPR, pp. 1249–1258 (2016)
24. Zheng, L., Shen, L., Tian, L., Wang, S., Wang, J., Tian, Q.: Scalable person re-identification: a benchmark. In: ICCV, pp. 1116–1124 (2015)
25. Zhu, J.Y., Park, T., Isola, P., Efros, A.A.: Unpaired image-to-image translation using cycle-consistent adversarial networks. In: ICCV, pp. 2223–2232 (2017)

Siamese Network for Pedestrian Group Retrieval: A Benchmark

Ling Mei[1], Jianhuang Lai[2,3]($\boxtimes$), Xiaohua Xie[2,3], and Zeyu Chen[2]

[1] School of Electronics and Information Technology, Sun Yat-sen University,
Guangzhou, China
mei13@mail2.sysu.edu.cn
[2] School of Data and Computer Science, Sun Yat-sen University, Guangzhou, China
stsljh@mail.sysu.edu.cn, xiexiaoh6@mail.sysu.edu.cn,
chenzy5@mail2.sysu.edu.cn
[3] Guangdong Key Laboratory of Machine Intelligence and Advanced Computing,
Ministry of Education, Beijing, China

Abstract. In many artificial intelligence applications such as security field, it is important to identify if a specific group of pedestrians has been observed over a network of other surveillance cameras, which ascribes to the pedestrian group retrieval problem. To address this issue, this paper contributes a novel dataset for the pedestrian group retrieval named "SYSU-Group". We collect diverse images with various pose from every camera for each group, which brings kinds of realistic challenges to the dataset, such as viewpoint variations, illumination changes and internal exchanges of group members. Moreover, we propose the Siamese Verification-Identification based Group Retrieval (SVIGR) method, which combines verification and identification modules in a Siamese network to extract person feature and follows the principle of minimum distance matching to measure the distance among pedestrian groups, and eventually gets the ranking of each query pedestrian group image. Experimental results demonstrate the superiority of SVIGR on the proposed group retrieval dataset.

Keywords: Pedestrian group retrieval · Siamese network · Minimum distance matching · Group distance vector

1 Introduction

Group is several stable pedestrians gathering with high motion collectiveness for a sustained period of time, which constitutes the primary unit of crowd. In this paper, we focus on a novel task, Pedestrian Group Retrieval, which aims to re-identify a specific group of persons when they appear in another regions. Namely, given an image of group in some scene, the task is to search for the images of the same group under different scenes from the gallery. Group retrieval is of great help to many public security applications including anomaly detection, suspicious activity surveillance, tracking, and crowd behavior analysis in real life.

© Springer Nature Switzerland AG 2019
Y. Zhao et al. (Eds.): ICIG 2019, LNCS 11901, pp. 747–759, 2019.
https://doi.org/10.1007/978-3-030-34120-6_61

In recent years, the technology of surveillance has been developed dramatically rapidly in many public place such as airports, train stations, and supermarkets *etc.*, which provides convenience to achieve the goal of group retrieval [25,35].

Group retrieval is a tremendously difficult problem: Cameras are placed quite far from each other without any overlap among their views, which results in notoriously difficult problem such as different illumination conditions [20,21] and large changes of viewpoint. Moreover, group retrieval requires to identify multiple persons who have variation in pose and interactive movement of internal members. As the group is the basic unit of crowd, a crowd with changing members can be constituted by some unrepeated groups with unchanging members, we assume the members of a group do not change and unrepeated under different cameras during the group retrieval period. Since it is at the very beginning in this field, we do not consider the changing case in this paper.

Group retrieval and person re-id differ fundamentally. Person re-id [6] re-identifies the individual person while group retrieval re-identifies specific several persons as a whole. In addition, persons tend to move together as a collective group in real life, members in a group identity have much more internal interactions than person re-id such as exchanging position, chatting or other interactive actions. Since person re-id fails to deal with the internal relationships of groups, the group retrieval task can't be solved with traditional person re-id methods directly. To address these issues, group retrieval focuses more on the relationships among persons than person re-id.

Considering the issues above, this paper makes three contributions: (1) We propose the new pedestrian group retrieval problem and its definition. Then we create a new group retrieval dataset named "SYSU-Group", which contains 524 persons that constitute 208 unrepeated groups under 8 high definition (HD) cameras. In order to provide a reliable benchmark, we propose mean Average Precision (mAP) and Cumulated Matching Characteristics (CMC) [13] for evaluation. (2) We proposed a Siamese Verification-Identification based Group Reidentification (SVIGR) method to solve the group retrieval. Two convolutional neural networks of verification module and identification module are combined to get all the pedestrian features, then we use the cosine distance to measure the pedestrian distance. Moreover, we adopt the minimum distance matching principle to generate the distance vector of each group. In this way, we can get a ranking list according to the ascending order of the distance. (3) Extensive experiments show the superiority of the proposed SCIGR method compared with state-of-the-art methods for achieving group retrieval, and validate the generalization of proposed SVIGR in terms of extracting original features of person re-id. We report competitive results of SVIGR on the proposed dataset for group retrieval task.

The article is organized as follows. In Sect. 1, we first introduce the definition, challenges of the group retrieval task. In Sect. 2, we review some related works about the proposed group retrieval method. In Sect. 3, we describe the SYSU-Group dataset and its evaluation protocol. Section 4 presents how to retrieval the group under multiple cameras by the proposed SVIGR framework. In Sect. 5, experimental results of the proposed method and other comparable methods are presented. In Sect. 6, we give the conclusion.

2   Related Works

In this section, we summarize the work on different aspects of group retrieval.

Pedestrian Detection. The group retrieval algorithms need pedestrian detection for preprocessing. Traditional pedestrian detection methods locate a person by means of appearance cues. [3] ultilizes HOG feature with SVM for prediction. [4] uses cascaded Adaboost [5] on integral channel features of Haar to improve detection precision and efficiency. [12] uses cues from body parts to address occlusion in crowded scene. Recently, deep learning methods like Faster RCNN [7], SSD [16] and YOLO [23] improve the performance of pedestrian detection by a large margin. Faster RCNN detect piles of sliding windows within two stages, while YOLO and SSD focus more on the global view in an end-to-end way. YOLO achieves an excellent efficiency among these methods by splitting an image into grids to obtain possible bounding box (bbox) with a fully CNN. Recently, YOLOv3 [24] replaces softmax function with logistic regression threshold and is characterized by a higher accuracy than previous versions while retaining an outstanding speed.

Feature Extraction. Most image retrieval tasks are quite associated with the method of feature extraction. WHOS [15] is a handcrafted and localized feature representation method to mitigate problems caused by background clutter and noise, WHOS features are weighted based on their distance to the center. Another similar method is GOG [19], it weights patches according to the distance to the central line, but it is not reasonable for the issues like occlusion. On the other hand, LDFV [17] uses local descriptors that include pixel spatial location, gradient, and intensity information to encode the Fisher vector [26] representation. Moreover, gBiCov [18] is a multi-scale biologically inspired feature that uses covariance descriptors to encode, it computes the distance between two persons by the Euclidean distance of their signatures. In texture based models, LOMO [14] uses HSV color histograms and extracts a scale-invariant LBP [22] texture operator, then maximizes the occurrence to make a robust representation against viewpoint changes. ELF [8] is also a texture based method, it combines color histograms in the RGB, HSV and YCbCr color spaces to obtain the texture histograms. Furthermore, HistLBP [32] also encodes the RGB, HSV, YCbCr color spaces to color histograms, and combines them with the texture histograms of LBP features. All above feature extraction methods have already applied on the person re-id task [10], we use them to compare with our proposed method to evaluate the performance of the group retrieval task in this paper.

Deeply Learned Modules. Except the handcrafted color and texture based modules, the deep learning based discrimination methods play more important roles in pedestrian retrieval like [29] which learns robust features by a refined part pooling (RPP). There are two types of mainstream deep learning modules:

Table 1. The composition of the SYSU-Group group retrieval dataset.

Size	Training set			Testing set			Total dataset		
	Groups	Persons	Samples	Groups	Persons	Samples	Groups	Persons	Samples
2	73	146	2546	65	130	2238	138	276	4784
3	23	69	767	20	60	675	43	129	1442
4	5	20	160	14	56	469	19	76	629
5	2	10	55	3	15	86	5	25	141
6	1	6	30	2	12	45	3	18	75
Total	104	251	3558	104	273	3513	208	524	7071

the verification module and the identification module. The verification module proposed in [2] is a deep metric learning that used in signature verification, which takes a pair of images as input and output the similarity according to the cosine distance. Recently, researchers have applied the verification modules in the image retrieval task such as person re-id. Yi *et al.* [33] used the verification module to divide the person image into three horizontal parts and trained three part-CNNs to extract features. Wu *et al.* [31] improved the verification module by using smaller filters and deeper network than the similar work of Ahmed *et al.* [1]. Verification modules are limited as the query image need to pair every gallery image, which makes the computation not efficient. On the other hand, some researchers have tried to combine the verification module with the identification module, the DeepID networks [28] optimize the network by both verification and identification modules, the work has performed well in terms of face recognition. Additionally, Zheng *et al.* [36] combined the two modules and applied them to person re-id task efficiently. Motivated by these works above, we propose our SVIGR method to solve the group retrieval problem.

3   SYSU-Group Dataset

In this section, a new group retrieval dataset, the "SYSU-Group" dataset is introduced. We begin with the description of the proposed dataset, then we introduce the criterion of annotated group identity, and we give the criterion about how to generate ground truth of group retrieval results, finally the evaluation protocol is introduced.

3.1   Description of the Dataset

As shown in Fig. 1, all the group images are collected from a total of 8 cameras, including three 1920 × 1080 full High Definition (HD) cameras that placed in diverse indoor campus scenes (cam1 - cam3), and five 1280 × 1080 HD cameras that located in diverse outdoor campus scenes (cam4 - cam8). All cameras do not overlap with each other, the group images are obtained by screenshots of the whole pedestrian group area from original video frames, each group image is resized to 256 × 256 pixels. The dataset contains 7071 fully annotated bboxes of 208 different group identities, the number of persons in each group is from 2 to 6, images of each group identity are captured by at most eight cameras. More detailed statistic about our dataset is demonstrated in Table 1.

cam1   cam2   cam3   cam4   cam5   cam6   cam7   cam8

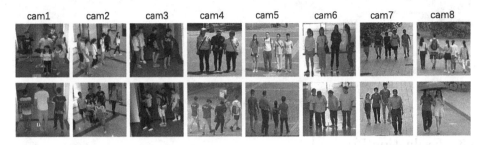

Fig. 1. Sample queries in SYSU-Group dataset. All the samples are hand-cropped bboxes. Each group has at most 8 queries, one for each camera.

The target of group retrieval is different from person re-id: Group retrieval re-identifies several persons who have interaction with others such as exchanging internal position and chatting, while person re-id has only one single person in an image patch almost without interaction with others. Therefore, group retrieval identifies persons in a group not only by their appearance but also via their correlations.

To avoid ambiguity, according to the assumption of unchanging group members in Sect. 1, the same group has fixed members under all the cameras. However, the sizes of different groups are diverse which is similar to the reality. Since it is at the very beginning in this field, in order to pay more attention to group retrieval rather than detection, all the images of the proposed group retrieval dataset are hand-cropped and annotated from the original video, these hand-cropped boxes contain each whole group without noise data such as irrelevant passerby. The SYSU-Group dataset is challenging for its diversity in pose, age, illumination, occlusion and actions, which is similar to scenarios in real life. For better presentation, we choose some query images in the dataset as exemplar to show its variety in Fig. 1.

3.2 Ground Truth

Second, in addition to the standard cropped group boxes, we also provide the ground truth of the group retrieval result. We notice that the Market1501 dataset [34] also provides the good and junk indexes of all cropped boxes to evaluate. Considering this, for a hand-cropped annotated group image id_cam (id and cam represent the group identity and the collected camera of the image, respectively). We first partition it as the corresponding collected camera and give it a sequence number. Then mark the images with same id and different cam as "good" images; and other images are marked as "junk" images, meaning these images are of zero influence to the group retrieval accuracy.

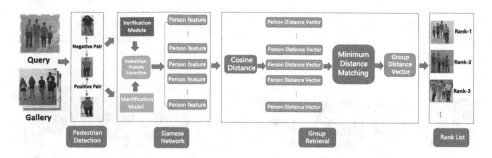

Fig. 2. Overview of the proposed SVIGR method for pedestrian group retrieval.

3.3 Evaluation Protocol

In this paper, we use mean Average Precision (mAP) and Cumulated Matching Characteristics (CMC) to evaluate the performance of group retrieval, which are common evaluation metrics in person re-id task. CMC gives cumulating accuracies that a query group identity appears in gallery candidate rank lists.

The group retrieval dataset is divided into training and testing sets by splitting the *id* numbers of all groups randomly, there are 104 training groups contain 251 persons and 104 testing groups contain 273 persons, and the number of training and testing images are 3558 and 3513, respectively. In the testing set, we select one query for each group from its appeared cameras to make up the query set. In this way, a group has at most 8 queries. The rest of the testing set is sorted out to form the gallery set, there are 704 query images and 2809 gallery images in total. Our dataset is an ideal benchmark for group retrieval methods, which can evaluate their generalization capacities for practical usages.

4 Our Method

4.1 Overall Framework

Our method has four steps: First, we use an advancing pedestrian detection to extract all the individual persons from groups. Second, we adopt a convolutional Siamese network to extract the features of all individual persons, then generate all the mutual distances of individuals by calculating the cosine distance. Third, for each group, we merge all the mutual distances that belong to the group to generate the GDV. Finally, the group distance vectors are concatenated into a group distance matrix, then we get the similarity score of groups according to the distance matrix and find the group retrieval rank list. The overall framework is shown in Fig. 2.

4.2 Pedestrian Detection

We use an advancing pedestrian detection method named YOLOv3 [24] to detect the members in each group image. YOLOv3 is a fast and accurate detector which adopt an efficient Darknet-53 network [24] that can better utilizes the GPU. We employ YOLOv3 to detect all the bboxes of pedestrians in each group and divide them into training set and testing set according to their *id* mentioned in Sect. 3.3. In addition, all the pedestrian bboxes can be used as the person re-id data to compare the robustness of feature extraction methods, which will be described in Sect. 5.2.

4.3   Feature Extraction

After obtaining the bbox images, we train a Siamese network proposed in [36] to extract the person re-id features of individuals. The Siamese network combines a verification module with an identification module simultaneously. Particularly, in training stage, we use the identity of group (Group-ID) as the class label for training classifier, which reflects the specific relationships among groups.

The identification module is used to treat the group retrieval as a kind of classification task with M classes of different groups, but its identification loss L_I does not influence the inter-class and the intra-class discrepancy of groups. To address the problem, the verification module is introduced to treat the group retrieval as a kind of binary classification task to indicate if they belong to a same group or not, the verification loss L_V takes a pair of images as input and outputs a label according to their similarity loss, which enlarges the inter-class discrepancy of different groups and narrows the intra-class discrepancy of same groups. We use the verification module only in the training stage to learn the proper network parameters for the testing stage to extract pedestrian features. Moreover, the class of verification is group rather than individual person, its computing cost is low. The total training loss L in the Siamese network are computed as

$$L = L_I(f, t, \theta_I) + L_V(f_{pos}, f_{neg}, t, \theta_V)$$
$$= \sum_{t=1}^{M} -u_t \log(\hat{u}) + \sum_{c=0}^{1} -v_c \log(\hat{v}) \tag{1}$$

Here, f, f_{pos}, f_{neg} define the output of the identification module, the verification module with positive pair and negative pair in Fig. 2, respectively. In Eq. 1, we use cross entropy loss both in L_I for identity prediction and L_V for pair verification. t is the target class, θ_I and θ_V are the parameters of the convolutional layer of the identification module and the verification module, respectively. $\hat{u}$ and u_c are predicted and probability for class respectively, where $u_t = 1$ for the target class and $u_t = 0$ otherwise. Similarly, so do the definitions of $\hat{v}$ and v_c for pair verification. For the positive pair, $v_0 = 1, v_1 = 0$. Otherwise, $v_0 = 0, v_1 = 1$.

For the aforementioned reasons, the Siamese network combines both the verification loss and the identification loss to make up for the shortage of each other, which can predict the identity of group and give the similarity of a pair of person bboxes. After the loss function optimization by the Siamese network in training stage, we can use the optimal parameters θ_I and θ_V to extract the pedestrian features in the testing stage.

Query	Rank-1	Rank-2	Rank-3	Rank-4	Rank-5

Fig. 3. The visual examples of group retrieval results that using the proposed method with ResNet50 network and MMD strategy on the SYSU-Group dataset.

4.4   Group Distance Vector (GDV)

In group retrieval, we adopt the principle of the minimum distance matching to generate the GDV. For a query group q with m members and a gallery group g with n members, we denote $d_{q,g}^{i,j}$ as the pedestrian distance between member i in q and member j in g by computing the cosine distance between the pedestrian features. For member i in group p, it is quite intuitional for us to match the member in group g with the minimum distance between them, therefore the optimal member-group distance between member i in group q and group g is denoted as

$$d_{q,g}^i = \min\{d_{q,g}^{i,j}\}_{j=1}^n, n = 2, 3, \dots, 6 \tag{2}$$

In order to obtain a robust expression of the group distance to handle the intra-class discrepancy caused by reasons like internal position exchanges within group. We adopt the minimum distance matching principle in Eq. 2 to find the most similar member pairs. Here we give two strategies to measure the group distance. One is the Minimum Member Distance (MMD). Since there are not same member in different groups which is defined in Sect. 1, we attempt to denote the group distance as

$$d_{q,g} = \min\{d_{q,g}^i\}_{i=1}^m, m = 2, 3, \dots, 6 \tag{3}$$

Here, the most minimum distance among all similar member pairs can verify if there is same member in the group i and j. An alternative strategy is to use the Average Distance (AD) which defined as

$$d_{q,g} = \frac{1}{mn} \sum_{i=1}^m \sum_{j=1}^n \{d_{q,g}^{i,j}\} \tag{4}$$

In this way, we can metric distance between groups no matter their sizes are equal or not, then sort these distances by an ascending order to get GDV as

$$\mathbf{d}_q = (d_{q,1}, \ldots, d_{q,g}, \ldots, d_{q,G}) \tag{5}$$

where G is the number of gallery images, then we can obtain the group distance matrix D by merging GDV as

$$\mathbf{D} = [\mathbf{d}_1, \ldots, \mathbf{d}_q, \ldots, \mathbf{d}_Q]^T \tag{6}$$

where Q is the number of query images. In this way, the group distance matrix is concatenated by vectors of all query images. When two group images are needed to match, the similarity $d_{q,g}$ of them can be found by the row index q and the column index g of $\mathbf{D}$. Therefore, the rank list of the query group image can be generated. No matter how the group members exchange their positions, as the members of the group remain the same, both the group distance $d_{q,g}$ defined in Eqs. 3 and 4 are invariant. Therefore, the disturbance of the exchanges of internal members in a group can be avoided.

Table 2. Comparison with the state-of-the-art image retrieval methods on the SYSU-Group group retrieval dataset. Rn means the top Rank-n accuracy. Rank-1 and mAP (%) are more representative to the performance in all metrics.

Method	R1	R5	R10	mAP	Method	R1	R5	R10	mAP
CaffeNet (AD)	50.7	70.9	77.4	32.5	HistLBP [32]	9.7	19.5	26.1	5.5
VGG16 (AD)	65.3	84.2	88.1	46.7	ELF [8]	18.6	36.7	45.9	7.1
GoogleNet (AD)	76.9	91.1	92.8	63.4	LDFV [17]	24.2	41.9	51.3	10.9
ResNet50 (AD)	72.9	87.9	93.2	58.0	gBiCov [18]	27.3	54.0	65.1	10.3
CaffeNet (MMD)	78.4	90.6	92.8	46.2	LOMO [14]	37.1	59.4	69.0	15.0
VGG16 (MMD)	88.6	96.6	98.4	62.3	GOG [19]	63.5	82.0	87.2	30.8
GoogleNet (MMD)	92.2	**98.4**	**99.4**	71.9	WHOS [15]	68.2	84.5	90.3	37.4
ResNet50 (MMD)	**94.5**	97.9	98.6	**76.7**	RPP [29]	74.4	90.3	94.0	46.2

5 Experiments

5.1 Performance of SVIGR Framework

For purposes of evaluation, We implement our Siamese network with four backbone CNN models: CaffeNet [11], VGG16 [27], GoogleNet [30], and ResNet50 [9]. In our experiment, all person re-id methods are used as basic feature extractors which use Group-IDs as the class labels for training classifier, there are both 104 different Group-IDs in training and testing stage, as shown in Table 1. After that we generate GDV from person re-id features following the minimum distance matching strategy, which is commonly used to metric the similarity between query and gallery. We compare MMD strategy (Eq. 3) with AD strategy (Eq. 4) in our SVIGR method. As Table 2 shows, MMD surpasses AD with four kinds of CNN backbone networks and all handcrafted methods, especially for the rank-1 precision and mAP. Therefore, we use MMD to obtain GDV in the following experiments.

Moreover, our method surpass all 8 state-of-the-art re-id methods including handcrafted and deep learning ones in Table 2. Although RPP [29] method is a strong deep learning baseline on person re-id, our method outperforms it in all metrics, which reveals the superiority of our proposal. The best rank-1 precision and mAP of all methods are our proposed method with ResNet50 that reaches 94.5% and 76.7% which exceeds RPP by 20.1%, 30.5%, respectively. Furthermore, Fig. 3 illustrates the group retrieval results under different complex scenarios including occlusion and illumination changes, which is detrimental to retrieval. Notably, the results show that our method not only re-identify the correct group in other cameras, but also can overcome the challenge of huge changes of viewpoint that bring out difficulty of illumination variation.

Table 3. Results of ablation experiment on SYSU-Group dataset by using identification module (I) and verification module (V) individually and jointly (%).

Method	Rank-1	mAP	Method	Rank-1	mAP
CaffeNet (V)	9.7	3.3	VGG16 (V)	9.5	2.4
CaffeNet (I)	76.6	45.3	VGG16 (I)	80.0	50.2
CaffeNet (V+I)	**78.4**	**46.2**	**VGG16 (V+I)**	**88.6**	**62.3**
GoogleNet (V)	72.2	39.1	ResNet50 (V)	87.1	62.0
GoogleNet (I)	87.4	65.3	ResNet50 (I)	91.2	71.2
GoogleNet (V+I)	**92.2**	**71.9**	**ResNet50 (V+I)**	**94.5**	**76.7**

In addition, to figure out the contribution of each module, we perform ablation experiment with our proposed method on the dataset. In addition to jointly trainning networks with both module, we also train the verification module and identification module individually. Table 3 gives the quantitative results of rank-1 precision and mAP, results of all networks show that the identification is more important to the accuracy as the module do the task of classification, and the combination of the two modules can improve the performance furtherly. The results validate the effectiveness of combing two modules in our SVIGR method.

5.2   Further Experiment on Person Re-id

Besides, we evaluate the generalized ability of the proposed SVIGR on traditional person re-id task. We use the original person re-id features extracted by SVIGR with the four backbone networks that are shown as bold font in Table 4. We use the person re-id data extracted from the group images by YOLOv3 detector [24] in SYSU-Group dataset. As is divided in Table 1, persons in the training and the testing set are 251 and 273, respectively.

Table 4. Results of person re-id task on SYSU-Group dataset (%).

Method	mAP	Rank-1	Rank-5	Rank-10
LOMO [14]	13.9	34.3	54.1	64.2
RPP [29]	27.9	55.5	73.5	81.0
CaffeNet	38.6	68.3	82.9	88.1
VGG16	47.4	74.4	88.8	92.8
GoogleNet	55.8	80.7	91.4	94.2
ResNet50	**61.9**	**84.9**	**93.0**	**95.4**

In this experiment, we compare the performance of SVIGR with two kinds of representative person re-id methods: One is the handcraft method such as LOMO [14], the other is the state-of-the-art method such as RPP [29].

Particularly, in training stage, we follow our group retrieval task to use Group-IDs as labels for all methods to extract original person re-id features. In testing stage, the evaluation metrics are mAP and Rank-1, which follows traditional person re-id task. Experimental results in Table 4 show that SVIGR raises by a large margin compared with traditional handcraft method, and exceeds the state-of-the-art deep learning method as well, the results verify the robustness of SVIGR in terms of feature extraction.

6   Conclusion

This paper first defines the problem of the group retrieval, then introduces a high quality group retrieval dataset and its evaluation protocol. Furthermore, an efficient SVIGR method is proposed to solve the group retrieval. In the future, more robust feature learning methods will be considered to make group retrieval benefit person re-id further, then research more innovative solution to solve this problem.

Acknowledgements. This work was supported by National Key Research and Development Program of China (2016YFB1001003), the NSFC (61573387).

References

1. Ahmed, E., Jones, M., Marks, T.K.: An improved deep learning architecture for person re-identification. In: CVPR, pp. 3908–3916 (2015)
2. Bromley, J., Guyon, I., LeCun, Y., Säckinger, E., Shah, R.: Signature verification using a "siamese" time delay neural network. In: NIPS, pp. 737–744 (1994)

3. Dalal, N., Triggs, B.: Histograms of oriented gradients for human detection. In: CVPR, vol. 1, pp. 886–893. IEEE Computer Society (2005)
4. Dollár, P., Tu, Z., Perona, P., Belongie, S.: Integral channel features (2009)
5. Friedman, J., Hastie, T., Tibshirani, R., et al.: Additive logistic regression: a statistical view of boosting (with discussion and a rejoinder by the authors). Ann. Stat. **28**(2), 337–407 (2000)
6. Gheissari, N., Sebastian, T., Hartley, R.: Person reidentification using spatiotemporal appearance. In: CVPR, vol. 2, pp. 1528–1535. IEEE (2006)
7. Girshick, R.: Fast R-CNN. In: ICCV, pp. 1440–1448 (2015)
8. Gray, D., Tao, H.: Viewpoint invariant pedestrian recognition with an ensemble of localized features. In: Forsyth, D., Torr, P., Zisserman, A. (eds.) ECCV 2008. LNCS, vol. 5302, pp. 262–275. Springer, Heidelberg (2008). https://doi.org/10.1007/978-3-540-88682-2_21
9. He, K., Zhang, X., Ren, S., Sun, J.: Deep residual learning for image recognition. In: CVPR, pp. 770–778 (2016)
10. Karanam, S., Gou, M., Wu, Z., Rates-Borras, A., Camps, O., Radke, R.J.: A systematic evaluation and benchmark for person re-identification: features, metrics, and datasets. arXiv preprint arXiv:1605.09653 (2016)
11. Krizhevsky, A., Sutskever, I., Hinton, G.E.: ImageNet classification with deep convolutional neural networks. In: NIPS, pp. 1097–1105 (2012)
12. Leibe, B., Seemann, E., Schiele, B.: Pedestrian detection in crowded scenes. In: CVPR, vol. 1, pp. 878–885. IEEE (2005)
13. Li, W., Zhao, R., Xiao, T., Wang, X.: DeepReID: deep filter pairing neural network for person re-identification. In: CVPR, pp. 152–159 (2014)
14. Liao, S., Hu, Y., Zhu, X., Li, S.Z.: Person re-identification by local maximal occurrence representation and metric learning. In: CVPR, pp. 2197–2206 (2015)
15. Lisanti, G., Masi, I., Bagdanov, A.D., Del Bimbo, A.: Person re-identification by iterative re-weighted sparse ranking. TPAMI **37**(8), 1629–1642 (2015)
16. Liu, W., et al.: SSD: single shot multibox detector. In: Leibe, B., Matas, J., Sebe, N., Welling, M. (eds.) ECCV 2016. LNCS, vol. 9905, pp. 21–37. Springer, Cham (2016). https://doi.org/10.1007/978-3-319-46448-0_2
17. Ma, B., Su, Y., Jurie, F.: Local descriptors encoded by fisher vectors for person re-identification. In: Fusiello, A., Murino, V., Cucchiara, R. (eds.) ECCV 2012. LNCS, vol. 7583, pp. 413–422. Springer, Heidelberg (2012). https://doi.org/10.1007/978-3-642-33863-2_41
18. Ma, B., Su, Y., Jurie, F.: Covariance descriptor based on bio-inspired features for person re-identification and face verification. IVC **32**(6–7), 379–390 (2014)
19. Matsukawa, T., Okabe, T., Suzuki, E., Sato, Y.: Hierarchical Gaussian descriptor for person re-identification. In: CVPR, pp. 1363–1372 (2016)
20. Mei, L., Lai, J., Xie, X., Zhu, J., Chen, J.: Illumination-invariance optical flow estimation using weighted regularization transform. TCSVT 1 (2019)
21. Mei, L., Chen, Z., Lai, J.: Geodesic-based probability propagation forefficient optical flow. Electron. Lett. **54**, 758–760 (2018)
22. Ojala, T., Pietikäinen, M., Harwood, D.: A comparative study of texture measures with classification based on featured distributions. PR **29**(1), 51–59 (1996)
23. Redmon, J., Divvala, S., Girshick, R., Farhadi, A.: You only look once: unified, real-time object detection. In: CVPR, pp. 779–788 (2016)
24. Redmon, J., Farhadi, A.: Yolov3: an incremental improvement. arXiv preprint arXiv:1804.02767 (2018)
25. Ristani, E., Tomasi, C.: Features for multi-target multi-camera tracking and re-identification. In: CVPR, pp. 6036–6046 (2018)

26. Sánchez, J., Perronnin, F., Mensink, T., Verbeek, J.: Image classification with the fisher vector: theory and practice. IJCV **105**(3), 222–245 (2013)
27. Simonyan, K., Zisserman, A.: Very deep convolutional networks for large-scale image recognition. arXiv preprint arXiv:1409.1556 (2014)
28. Sun, Y., Wang, X., Tang, X.: Deeply learned face representations are sparse, selective, and robust. In: CVPR, pp. 2892–2900 (2015)
29. Sun, Y., Zheng, L., Yang, Y., Tian, Q., Wang, S.: Beyond part models: person retrieval with refined part pooling (and a strong convolutional baseline). In: ECCV, pp. 480–496 (2018)
30. Szegedy, C., et al.: Going deeper with convolutions. In: CVPR, pp. 1–9 (2015)
31. Wu, L., Shen, C., van den Hengel, A.: PersonNet: person re-identification with deep convolutional neural networks. arXiv preprint arXiv:1601.07255 (2016)
32. Xiong, F., Gou, M., Camps, O., Sznaier, M.: Person re-identification using kernel-based metric learning methods. In: Fleet, D., Pajdla, T., Schiele, B., Tuytelaars, T. (eds.) ECCV 2014. LNCS, vol. 8695, pp. 1–16. Springer, Cham (2014). https://doi.org/10.1007/978-3-319-10584-0_1
33. Yi, D., Lei, Z., Liao, S., Li, S.Z.: Deep metric learning for person re-identification. In: ICPR, pp. 34–39. IEEE (2014)
34. Zheng, L., Shen, L., Tian, L., Wang, S., Wang, J., Tian, Q.: Scalable person re-identification: a benchmark. In: CVPR, pp. 1116–1124 (2015)
35. Zheng, L., Yang, Y., Hauptmann, A.G.: Person re-identification: past, present and future. arXiv preprint arXiv:1610.02984 (2016)
36. Zheng, Z., Zheng, L., Yang, Y.: A discriminatively learned CNN embedding for person reidentification. TOMM **14**(1), 13 (2018)

Compression of Deep Convolutional Neural Networks Using Effective Channel Pruning

Qingbei Guo[1,2], Xiao-Jun Wu[1(✉)], and Xiuyang Zhao[2]

[1] Jiangsu Provincial Engineering Laboratory of Pattern Recognition
and Computational Intelligence, Jiangnan University, Wuxi 214122, China
wu_xiaojun@jiangnan.edu.cn
[2] Shandong Provincial Key Laboratory of Network Based Intelligent Computing,
University of Jinan, Jinan 250022, China

Abstract. Pruning is a promising technology for convolutional neural networks (CNNs) to address the problems of high computational complexity and high memory requirement. However, there is a principal challenge in channel pruning. Although the least important feature-map is removed each time based on one pruning criterion, these pruning may produce considerable fluctuation in classification performance, which easily results in failing to restore its capacity. We propose an effective channel pruning criterion to reduce redundant parameters, while significantly reducing such fluctuations. This criterion adopts the loss-approximating Taylor expansion based on not the pruned parameters but the parameters in the subsequent convolutional layer, which differentiates our method from existing methods, to evaluate the importance of each channel. To improve the learning effectivity and efficiency, the importance of these channels is ranked using a small proportion of training dataset. Furthermore, after each least important channel is pruned, a small fraction of training dataset is used to fine-tune the pruned network to partially recover its accuracy. Periodically, more proportion of training dataset is used for the intensive recovery in accuracy. The proposed criterion significantly addresses the aforementioned problems and shows outstanding performance compared to other criteria, such as Random, APoZ and Taylor pruning criteria. The experimental results demonstrate the excellent compactness performances of our approach, using several public image classification datasets, on some popular deep network architectures. Our code is available at: https://github.com/QingbeiGuo/Based-Taylor-Pruning.git.

Keywords: Deep neural network · Classification · Pruning

1 Introduction

Convolutional neural networks or their variants have obtained state-of-the-art accuracy in computer vision applications, e.g. image classification [9,18,20,23],

© Springer Nature Switzerland AG 2019
Y. Zhao et al. (Eds.): ICIG 2019, LNCS 11901, pp. 760–772, 2019.
https://doi.org/10.1007/978-3-030-34120-6_62

objection detection [1,2,17] and vision tracking [21,22], depending on their deep architectures and a large number of parameters, which leads to considerable consumptions in computation and storage resources. However, the huge computation and storage requirements hinder CNNs from being deployed on more extensive applications, for example, smart mobile devices due to their limited computation and memory resources. Fortunately, deep models have been found to be considerably redundant in weights, filters, channels, and even layers [4,6,13,19]. Network pruning provides a practical solution to remove these redundant parameters for the decreased requirement in computation and storage resources, while remaining a comparable level of recognition capacity of original deep network models.

Many previous researchers have explored the pruning techniques based on various granularities [4,6,13,14,19]. Although non-structured pruning *e.g.* intra-kernel pruning can achieve the comparable compression ratio, structured pruning, such as filter and channel pruning, brings more regular sparsity structure, which is desirable for network compression and inference speedup [14]. Various magnitude-based pruning criteria have been developed for more effective network compression, including minimum magnitude and sensitivity analysis, *etc.* [5,7,12,16]. Han *et al.* described a simple non-structured pruning criterion which removes all connections with weights below a given threshold [5]. Hu *et al.* evaluated the importance of neurons on a large validation dataset using the average percentage of zero (APoZ) activations, and pruned redundant neurons according to APoZ [7]. Sensitivity analysis was pioneered by LeCun *et al.*, and they simplified the approximate saliency of parameters by the second-order Taylor expansion to remove its redundant parameters, called Optimal Brain Damage (OBD) [12]. Molchanov *et al.* further applied Taylor expansion to deep models for network compression [16]. Instead of the second-order terms, only the first-order terms were remained to simplify the approximate saliency of each parameter due to the usage of ReLU [3] activation function.

Although these methods have experimentally achieved good compression ratio with little loss in accuracy, even without loss of classification accuracy, they easily suffer from dramatic fluctuations in accuracy during the compressing process, especially on a small training dataset. Figure 1 explains our observations. Such fluctuations disclose two main problems. (1) The importance of many parameters pruned based on these criteria are not reasonably estimated, thus leading to an obviously inconsistent effect rather than minimizing the effect on the network performance. (2) As the pruning result each time provides the initialization parameters for the subsequent retraining, the drastic fluctuations easily impede the performance recovery of neural networks. Through a large amount of experiments, Mishkin & Matas indicate the importance of initializations for the classification performance in very deep network [15]. Consequently, the fluctuations, especially the drastic drops, provide a very bad initialization for the subsequent retraining, which could result in failing to restore original classification accuracy. We suggest that the pruning methods producing fewer fluctuations is desirable to more easily preserve the capacity of original deep network model, even achieve better performance, which means more accurate estimation for the parameter importance.

Inspired by the Taylor expansion [12,16], we present a novel pruning criterion, namely based-Taylor pruning, based on Taylor expansion to prune deep network models. To the best of our knowledge, different from the dependance on the pruned parameters [12,16], our proposed Taylor expansion depends on not the pruned parameters but the parameters in their subsequent convolutional layer, considering the effect on both the layer output and the network output. In particular, for the last convolutional layer, the following fully-connected layer is selected as its subsequent layer due to no following convolutional layer. The change caused by pruning a parameter propagates from its current layer to its subsequent layer, finally causing the objective function to change. For each parameter in the subsequent layer, we use the Taylor expansion to directly approximate the change of the network output. In this paper, the importance of each parameter is evaluated by the absolute sum of all the approximate changes from the changes in its subsequent layer. Our pruning criterion attaches importance to minimizing the accumulation of change amplitude coming from each change of feature-maps in the subsequent layer, which is desired to address the problem of great fluctuations, especially drastic drops, during the pruning process.

In order to reduce the computational cost, we rank the importance of parameters on a small proportion of training dataset. Furthermore, retaining the learning capacity of the compressed network is performed mostly on a very small training dataset, which is periodically replaced on the whole training dataset. Using these two different sizes of retraining datasets has individual purposes. Retraining on a small proportion of training dataset restricts a drastic drop in classification performance in order to prevent the compressed network from failing to recover original learning capacity. On the other hand, retraining on a higher proportion of training dataset is to intensively recover the learning capacity of the compressed network on the basis of the result of previous retraining on the small training dataset.

In summary, the main contributions of this paper are presented as follows:

- We present a channel pruning criterion, which is used for more accurate importance estimation of feature-maps, thus leading to a more stable pruning process;
- Several experiments on multiple popular vision benchmarks demonstrate that our method achieves the higher recognition performance than other methods, such as Random, APoZ, and Taylor, under the same compression rate.

The rest of this paper is organized as follows: Sect. 2 formulates our proposed based-Taylor pruning method, and further analyzes the essential difference from other based-Taylor pruning criteria. In Sect. 3, several deep models on multiple popular vision classification datasets are performed to extensively validate the compression effectiveness of our method. Finally, Sect. 4 presents the conclusion.

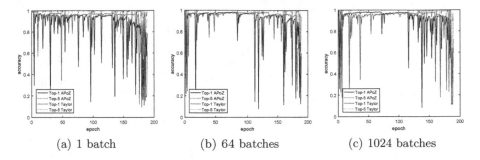

(a) 1 batch   (b) 64 batches   (c) 1024 batches

Fig. 1. Observations of different pruning methods on MNIST with different batches for the importance estimation of feature-maps. After each pruning, 32 batches are used for retraining.

2   Approach

In this section, we will comprehensively introduce our pruning approach. First, the deep CNN model is formulated. Next, our pruning criterion is presented in detail. Finally, compared with other based-Taylor criteria, our pruning approach is further analyzed to reveal our essence of excellent performances.

2.1   Formulation

Consider a well-trained deep convolutional neuron network with L layers, as shown in Fig. 2, each layer l consists of N_l channels, where $1 \leq l \leq L$, and implements a specific operation, such as linear, non-linear, normalization and pooling operations, with parameters $\theta_l = \{\theta_l^1, \theta_l^2, ..., \theta_l^{N_l}\}$, where θ_l^i denotes the parameters of the ith filter in the lth layer. The parameters of each filter consist of the weight matrix and the bias. All these parameters collectively form the entire network parameter set $\Theta = \{\theta_1, \theta_2, ..., \theta_L\}$. Let E be the objective function of this deep model. Given an input image samples $x = (x_1, x_2, ..., x_n)$, the output of the network can be written as $y = f(\Theta, x)$, where f represents an operation series including linear, non-linear, normalization and pooling operations. Since this study focuses on channel pruning, let a_l be the set of feature-maps of the lth layer such that $a_l = \{a_l^1, a_l^2, ..., a_l^{N_l}\}$, where a_l^i denotes the ith feature-map in the lth layer. Each feature-map is the input of the subsequent layer as well as the output of the current layer, thus $a_{l+1} = f_l(\theta_l, a_l)$, where f_l denotes the operation function of the lth layer.

Our goal is to prune the least significant parameters for the compression of deep CNN networks with minimizing the effect on its original performance in recognition accuracy. To achieve this goal, this compression problem becomes the following optimization problem:

$$\min_{\Theta'} |E(\Theta', x) - E(\Theta, x)| \quad s.t. \quad |\Theta'| \leq |\Theta| \cdot p_c, \tag{1}$$

where $|\Theta|$ and $|\Theta'|$ are the number of the parameters in original network and the pruned network, respectively, and p_c is a pre-defined compression rate.

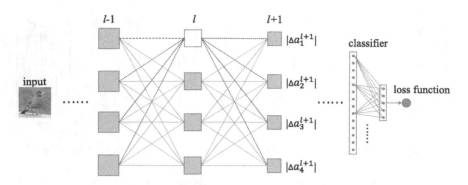

Fig. 2. A simple CNN architecture for pruning networks. □ denotes the pruned feature maps, and dotted line the pruned connections.

2.2 Pruning Criterion

We will introduce our based-Taylor channel pruning criterion for compressing the deep CNN networks. The importance of a parameter is evaluated by using Taylor expansion to approximate the change in the objective function caused by removing that parameter [12,16]. By contrast with [12,16], our criterion also approximates the importance of a parameter using the change in the objective function by Taylor expansion based on not the parameter but the parameters of its subsequent layer. The difference of criterion among these three methods will be further analyzed in the following subsection.

Specially, for any feature-map a_l^i, its pruning will lead to the change of the feature-maps in its subsequent layer. Hence, the following equation will hold.

$$\delta a_{l+1} = f_{l+1}(\theta_{l+1}, a_l) - f_{l+1}(\theta'_{l+1}, a'_l), \tag{2}$$

where $\theta'_{l+1} = \theta_{l+1} - \{\theta^i_{l+1}\}$ and $a'_l = a_l - \{a_l^i\}$.

The multivariate Taylor expansion of the objective function with respect to a_{l+1} and δa_{l+1} is:

$$E(a_{l+1}^1 + \delta a_{l+1}^1, ..., a_{l+1}^{N_{l+1}} + \delta a_{l+1}^{N_{l+1}}) = E(a_{l+1}^1, ..., a_{l+1}^{N_{l+1}})$$
$$+\delta a_{l+1}^1 \frac{\partial E}{\partial a_{l+1}^1} + ... + \delta a_{l+1}^{N_{l+1}} \frac{\partial E}{\partial a_{l+1}^{N_{l+1}}} + R_1(\delta a_{l+1}^1, ..., \delta a_{l+1}^{N_{l+1}}). \tag{3}$$

For a pruning deep network without convergence, we only preserve the first-order term and neglect the first-order Lagrange remainder to simplify the approximation because of the usage of ReLU activation functions in deep network model. Thus, the change of the objective function caused by the pruned parameter a_l^i will be approximated to as follow:

$$\delta E = E(a_{l+1}^1 + \delta a_{l+1}^1, ..., a_{l+1}^{N_{l+1}} + \delta a_{l+1}^{N_{l+1}}) - E(a_{l+1}^1, ..., a_{l+1}^{N_{l+1}}) =$$
$$\delta a_{l+1}^1 \frac{\partial E}{\partial a_{l+1}^1} + ... + \delta a_{l+1}^{N_{l+1}} \frac{\partial E}{\partial a_{l+1}^{N_{l+1}}} = \sum_{k=1}^{N_{l+1}} \delta a_{l+1}^k \frac{\partial E}{\partial a_{l+1}^k}. \tag{4}$$

According to the axiom of triangle inequality, the absolute value of δE can be amplified as

$$|\sum_{k=1}^{N_{l+1}} \delta a_{l+1}^k \frac{\partial E}{\partial a_{l+1}^k}| \leq \sum_{i=k}^{N_{l+1}} |\delta a_{l+1}^k \frac{\partial E}{\partial a_{l+1}^k}|. \tag{5}$$

Finally, considering the dimension difference of feature map tensors between convolutional layers, our pruning criterion is addressed based on Eq. (5) as follows:

$$C(a_l^i) = \frac{1}{||a_{l+1}||} \sum_{k=1}^{N_{l+1}} |\delta a_{l+1}^k \frac{\partial E}{\partial a_{l+1}^k}|, \tag{6}$$

where $C(a_l^i)$ represents the importance-evaluation value for any feature map a_l^i, and $||a_{l+1}||$ denotes the dimension of tensor a_{l+1}.

Obviously, our proposed criterion focuses on minimizing each change amplitude in the objective function which is resulted from each change of feature-maps in the subsequent layer caused by pruning the estimated feature-map, considering the effect on the output of layers and the output of networks, which benefits the more accurate evaluation of parameter importance.

2.3 Criterion Analysis

In this subsection, we will further analyze our proposed criterion to identify the essential difference from Taylor expansion [16]. Although both criteria based on the Taylor expansion approximately evaluate the importance of a parameter, there is a fundamental difference among them.

We will convert the estimated parameter in Taylor expansion criteria to the parameters of its subsequent layer, just like our pruning criterion, in order to more clearly show their difference.

The criterion of Taylor expansion is written as

$$C(a_l^i) = \frac{1}{||a_l^i||} |\delta a_l^i \frac{\partial E}{\partial a_l^i}|. \tag{7}$$

We have

$$\delta a_l^i \frac{\partial E}{\partial a_l^i} = \sum_{k=1}^{N_{l+1}} \delta a_l^i \frac{\partial E}{\partial a_{l+1}^k} \frac{\partial a_{l+1}^k}{\partial a_l^i} = \sum_{k=1}^{N_{l+1}} \delta a_{l+1}^k \frac{\partial E}{\partial a_{l+1}^k}. \tag{8}$$

Thus, we can rewrite the criterion of Taylor expansion as follows:

$$C(a_l^i) = \frac{1}{||a_l^i||} |\sum_{k=1}^{N_{l+1}} \delta a_{l+1}^k \frac{\partial E}{\partial a_{l+1}^k}|. \tag{9}$$

This pruning criterion is very similar in form but different in essence to our proposed criterion. Our proposed pruning criterion is the sum of the absolute of each change, so empirically focuses on minimizing not only the total change amplitude but also every change amplitude, which implies the essential difference from Taylor expansion criteria. Furthermore, it is easily found out that our method considers the effect on the layer output as well as the network output.

2.4 Strategy Optimization

Our pruning approach mainly contains three steps: (1) obtain a well-trained baseline network; (2) rank parameters using our pruning criterion, and remove the least important parameter one by one. (3) retrain the pruned network for the recovery of recognition capacity.

In this study, many strategies are adopted to greatly improve the learning efficiency with little loss, even without loss of the network capacity as follows:

(1) To obtain the well-trained deep network, we fine-tune the deep network based on the pre-trained model using less training time for optimal performance.

(2) The saliency of a parameter depends on not only the structure and parameters of the network but also the training dataset. A moderate fraction of training dataset can reasonably and efficiently rank the importance of parameters.

(3) After pruned, the deep network is subsequently retrained to compensate for the accuracy loss on account of previous pruning. The training dataset with two different sizes are conjunctively exploited to try to restore the pruned network to the original capacity. Each time after the least significant parameter is pruned, the pruned network is initialized using the weights before just pruning instead of from scratch, and then is retrained on a partial training dataset, which is randomly chosen in the full training dataset, for the recovery of the network performance. Not entirely recovering the network capacity, such recovery aims to avert a drastic accuracy drop which easily drives the deep model into a bad local minimum, thus preserves sufficient capacity from failing to recovering to its original performance. After such several iterations of pruning and retraining, the pruned network is retrained several epochs on the more proportion of or the full training dataset to intensively recover its recognition accuracy. Therefore, the combination of two retraining strategies guarantees the recovery of the network capacity and the learning efficiency simultaneously.

3 Experiments

For showing the effectiveness of our proposed method, we experiment on two popular network architectures which include LeNet [11] and AlexNet [9], and analyze the performance of our method on three different dataset *e.g.* MNIST [10] and CIFAR10/CIFAR100 [8]. All the experiments are performed in pyTorch (0.1.12), and the experiment results are compared with existing pruning methods.

3.1 LeNet on MNIST Dataset

We examine our proposed pruning criterion on the MNIST dataset without data augmentation, which is one of the most popular datasets for handwritten digit recognition. The MNIST dataset consists of 50,000 training images and 10,000 testing images, uniformly classified into 10 digital labels, and every sample is 28 × 28 grayscale image.

A modified version of LeNet is used as our baseline model. It has two convolutional layers with 32 and 64 feature-maps, respectively, and each convolutional layer is followed by one max-pooling layer. The baseline model is attained by training 60 epochs from scratch on MNIST, finally achieving a top-1 accuracy of 99.28% and a top-5 accuracy of 100%.

At each pruning iteration, we first rank the importance of all the parameters through one epoch on 32 batches with batch-size 32, accounting for about $p_r = 2.0\%$ of the full training dataset, and then prune the least important feature-map from the baseline model. For retraining, we deploy two kinds of strategies with different proportion of the full training dataset and constant hyper-parameters of batch-size 32, learning rate 0.0001, momentum 0.9 and weight decay 0.001. One retraining strategy is to use one epoch on 32 batches to partly recover the network capacity after one feature-map is pruned. At the end of each pruning interval $n_i = 8$, the other intensively retraining strategy is implemented to strengthen the capacity recovery of the pruned network by retraining $n_t = 5$ epoches on the partial training dataset with 32 batches. The process of pruning and retraining alternately continues until the desirable pruning network is obtained with an expected compression rate and without loss of accuracy.

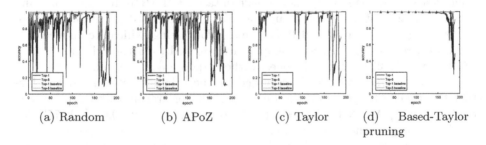

| (a) Random | (b) APoZ | (c) Taylor | (d) Based-Taylor pruning |

Fig. 3. Comparison with Random, APoZ and Taylor in LeNet on MNIST. (Color figure online)

In Fig. 3, we shows the performance changes in accuracy during the complete pruning process. The red horizontal line is a baseline whose value is the recognition accuracy of the baseline network, and the periodic triangles are the point where the pruned network is intensively retrained. The continual and drastic drops are explicitly observed on the methods of Random, APoZ and Taylor Expansion in Figs. 3(a), (b) and (c). Although each pruned feature-map is estimated as the most unimportant one in the deep network, its pruning produces a large negative effect on accuracy. Especially, at the beginning of the pruning process, since only a few redundant feature-maps are pruned, the learning capacity of the pruned network is not impaired. Thus, the reason for these continual drops in accuracy is that the estimated importance based on these three criteria is inconsistent with its actual importance. In the following process, several drops drastically decrease the recognition accuracy to about 10% and 20% for APoZ and Taylor Expansion, respectively, which is so far below the accuracy baseline.

Compared with these two criteria, no one large drop occurs on our proposed pruning method except for the end process, as is shown in Fig. 3(d). Each pruning has little effect on accuracy, so the result of each pruning provides the good initialization for the subsequent retraining, which guarantees the pruned model to maintain the stable performance in accuracy, reaching and even exceeding the accuracy baseline. As for the large drops at the end process, the learning capacity of the pruned network has been impaired due to pruning a large amount of feature-maps, which leads the pruned network to be vulnerable to the reduction of any feature-map. In this case, regardless of the method used, pruning any feature-map will result in a large drop in accuracy. Moreover, the learning capacity of the pruned network is unlikely to reach to its accuracy baseline. Therefore, our proposed pruning criterion is verified to be more stable and effective than the other three methods. Furthermore, it reveals that our proposed pruning criterion more accurately represent the importance of each feature-map.

3.2   AlexNet on CIFAR10 Datasets

Another deep CNN model AlexNet well-pretrained on the large scale dataset ImageNet is transferred to the small scale datasets CIFAR10, further examining the performance of our pruning method. For the CIFAR10 datasets, there are 50,000 images for training and 10,000 images for testing, which are divided into 10 classes, respectively. All the samples are rescaled from the original 32×32 RGB images to 224×224 RGB ones, without data augmentation.

The pre-trained AlexNet is regarded as the original network, and is further fine-tuned 20 epochs on the target domain for transferring learning. The final fine-tuned AlexNet achieves a top-1/top-5 accuracy of 90.08%/99.75% for CIFAR10, used as the baseline model.

We compress the baseline model with the same process of pruning and retraining as LeNet but with different hyper-parameters. The importance of feature-maps is estimated through one epoch on 256 batches with batch-size 32, about 16.38% of the full target training dataset in size. The hyper-parameters of retraining on CIFAR10 are set with batch-size 32, learning rate 0.0001, momentum 0.9 and weight decay 0.001. The first kind of retraining strategy is deployed using one epoch on 256 batches, and after the pruning interval $n_i = 32$, the other kind of intensively retraining strategy is deployed using $n_t = 10$ epoches on the full training dataset.

Figure 4 shows the changes of recognition accuracy in the entire pruning process of AlexNet on CIFAR10. As is shown in Figs. 4(a), (b) and (c), a large number of drastic drops continually occur during the entire pruning process, especially at the beginning and end, even some of them drop to just close to 10% in Top-1 accuracy. At the beginning, although a few parameters are pruned, the pruned network still has sufficiently redundant parameters which keep it enough

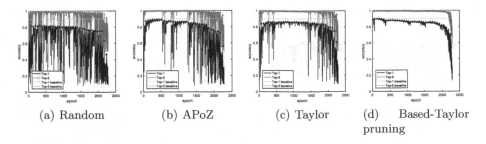

(a) Random   (b) APoZ   (c) Taylor   (d)   Based-Taylor
                                                                 pruning

Fig. 4. Comparison with Random, APoZ and Taylor in AlexNet on CIFAR10.

learning capacity. Hence, such drops reveal that these three methods of Random, APoZ and Taylor Expansion do not accurately reflect the importance of each pruned parameter. Although each retraining tries to recover the performance of networks, the subsequent drops result in the performance in Top-1 accuracy falling to a lower level. At the end, pruning a large number of parameters makes the pruned network to be very vulnerable for any new pruning. Further, inaccurately pruning aggravates the performance degradation in accuracy.

By contrast, in Fig. 4(d), only a few little drops occur in our pruning method. The performance of the deep network in accuracy first increases to slightly above the baseline level. After that, the pruned network gradually falls below the baseline level without large fluctuations. At the end, the performance quickly drops like as Random, APoZ and Taylor Expansion, but with little fluctuations. Therefore, our proposed method more accurately reflect the inherent importance of each parameter than Random, APoZ and Taylor Expansion.

3.3   AlexNet on CIFAR100 Datasets

CIFAR100 is another popular vision dataset, just like CIFAR10, except for representing 100 classes instead of 10 classes. We obtain the well-pretrained AlexNet on CIFAR100 as the baseline model using the same training process as CIFAR10, finally achieving 69.26% and 90.9% in Top-1 and Top-5 accuracy, respectively. In the pruning and retraining process, there are some different hyper-parameters from CIFAR10. The hyper-parameter of 512 batches is set for the dataset of ranking and retaining, and each intensively retraining is implemented 16 epoches.

Similar to the pruning result of AlexNet on CIFAR10, Fig. 5 shows the pruning process of Random, APoZ, Taylor Expansion and our based-ref Taylor Expansion. In Figs. 5(a), (b) and (c), there are a large number of drastic drops in the pruning process. In these three methods, no one contributes to recovering to the original level of accuracy by retraining from beginning to end, and several continual and dramatic drops evidently lead the performance in recognition accuracy to a lower level. Figure 5(d) shows the pruning result of our Base-ref Taylor method. At the beginning, our method can recover to the original level of accuracy. There are some little drops, but the following retraining operations manage to avoid the larger drops.

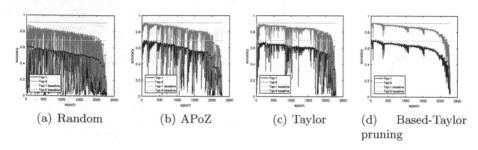

(a) Random   (b) APoZ   (c) Taylor   (d)   Based-Taylor
                                                               pruning

Fig. 5. Comparison with Random, APoZ and Taylor in AlexNet on CIFAR100.

4   Conclusion

One of the most key factors of pruning criteria for compression is to represent the accurate importance of each parameter. An inaccurate pruning can impair the capacity of networks, thus causing the drastic fluctuations of its accuracy performance. We introduced a pruning criterion, called as based-Taylor pruning, which more accurately evaluates the importance of feature-maps using the approximation of Taylor expansion based on not the pruned parameters but the parameters of the subsequent layer. Our proposed criterion focuses on the effect on not only the layer output but also the network output. Both retraining strategies complementarily manage to recover the performance of networks in accuracy. We experimented our pruning method using LeNet and AlexNet on several popular vision datasets. The experimental results demonstrated that our pruning criterion more accurately ranks the importance of feature-maps than Random, APoZ and Taylor expansion, and thus making the full pruning process more stable. Finally, we achieve the reduction of AlexNet in size by 10× and 4× on CIFAR10 and CIFAR100 with 4% and 6% loss in accuracy, respectively. Especially, in LeNet on MNIST, we obtain the reduction in size by more 5× almost without loss in accuracy.

Acknowledgment. This work is supported by the National Key R&D Program of China (Grant No. 2018YFB1004901), by the National Natural Science Foundation of China (Grant No.61672265, U1836218), by the 111 Project of Ministry of Education of China (Grant No. B12018), by UK EPSRC GRANT EP/N007743/1, MURI/EPSRC/DSTL GRANT EP/R018456/1.

References

1. Girshick, R., Donahue, J., Darrell, T., Malik, J.: Rich feature hierarchies for accurate object detection and semantic segmentation. In: Proceedings of the IEEE Conference on Computer Vision and Pattern Recognition (CVPR), Columbus, OH, USA, pp. 580–587 (2014)
2. Girshick, R., Donahue, J., Darrell, T., Malik, J.: Region-based convolutional networks for accurate object detection and segmentation. IEEE Trans. Pattern Anal. Mach. Intell. **38**(1), 142–158 (2016)

3. Glorot, X., Bordes, A., Bengio, Y.: Deep sparse rectifier neural networks. In: 14th International Conference on Artificial Intelligence and Statistics, Fort Lauderdale, United States, vol. 15, pp. 315–323 (2011)
4. Han, S., Mao, H., Dally, W.J.: Deep compression: compressing deep neural network with pruning, trained quantization and Huffman coding. CoRR abs/1510.00149 (2015). http://arxiv.org/abs/1510.00149
5. Han, S., Pool, J., Tran, J., Dally, W.J.: Learning both weights and connections for efficient neural networks. CoRR abs/1506.02626 (2015). http://arxiv.org/abs/1506.02626
6. He, Y., Zhang, X., Sun, J.: Channel pruning for accelerating very deep neural networks. CoRR abs/1707.06168 (2017). http://arxiv.org/abs/1707.06168
7. Hu, H., Peng, R., Tai, Y., Tang, C.: Network trimming: a data-driven neuron pruning approach towards efficient deep architectures. CoRR abs/1607.03250 (2016). http://arxiv.org/abs/1607.03250
8. Krizhevsky, A., Hinton, G.: Learning multiple layers of features from tiny images. Technical report, Science Department, University of Toronto, Tech (2009)
9. Krizhevsky, A., Sutskever, I., Hinton, G.E.: ImageNet classification with deep convolutional neural networks. In: Proceedings of the 25th International Conference on Neural Information Processing Systems, NIPS 2012, USA, vol. 1, pp. 1097–1105 (2012)
10. LeCun, Y., Bottou, L., Bengio, Y.: Gradient-based learning applied to document recognition. Proc. IEEE **86**(11), 2278–2324 (1998)
11. LeCun, Y., et al.: Handwritten digit recognition with a back-propagation network. In: Advances in Neural Information Processing Systems, vol. 2, pp. 396–404 (1990)
12. LeCun, Y., Denker, J.S., Solla, S.A.: Optimal brain damage. In: Proceedings of the 2nd International Conference on Neural Information Processing Systems, NIPS 1989, Cambridge, MA, USA, pp. 598–605 (1989)
13. Li, H., Kadav, A., Durdanovic, I., Samet, H., Graf, H.P.: Pruning filters for efficient convnets. CoRR abs/1608.08710 (2016). http://arxiv.org/abs/1608.08710
14. Mao, H., et al.: Exploring the regularity of sparse structure in convolutional neural networks. CoRR abs/1705.08922 (2017). http://arxiv.org/abs/1705.08922
15. Mishkin, D., Matas, J.: All you need is a good init. CoRR abs/1511.06422 (2015). http://arxiv.org/abs/1511.06422
16. Molchanov, P., Tyree, S., Karras, T., Aila, T., Kautz, J.: Pruning convolutional neural networks for resource efficient transfer learning. CoRR abs/1611.06440 (2016). http://arxiv.org/abs/1611.06440
17. Ren, S., He, K., Girshick, R., Sun, J.: Faster R-CNN: towards real-time object detection with region proposal networks. In: Proceedings of the 28th International Conference on Neural Information Processing Systems, NIPS 2015, Cambridge, MA, USA, vol. 1, pp. 91–99 (2015)
18. Simonyan, K., Zisserman, A.: Very deep convolutional networks for large-scale image recognition. CoRR abs/1409.1556 (2014). http://arxiv.org/abs/1409.1556
19. Wen, W., Wu, C., Wang, Y., Chen, Y., Li, H.: Learning structured sparsity in deep neural networks. CoRR abs/1608.03665 (2016). http://arxiv.org/abs/1608.03665
20. Wu, C., Wen, W., Afzal, T., Zhang, Y., Chen, Y., Li, H.: A compact DNN: approaching GoogLeNet-level accuracy of classification and domain adaptation. CoRR abs/1703.04071 (2017). http://arxiv.org/abs/1703.04071
21. Xu, T., Feng, Z.H., Wu, X.J., Kittler, J.: Learning adaptive discriminative correlation filters via temporal consistency preserving spatial feature selection for robust visual tracking. arXiv preprint arXiv:1807.11348 (2018)

22. Xu, T., Wu, X.J., Kittler, J.: Non-negative subspace representation learning scheme for correlation filter based tracking. In: 2018 24th International Conference on Pattern Recognition (ICPR), pp. 1888–1893. IEEE (2018)
23. Zhang, X., Zou, J., He, K., Sun, J.: Accelerating very deep convolutional networks for classification and detection. IEEE Trans. Pattern Anal. Mach. Intell. **38**(10), 1943–1955 (2016)

Hybird Single-Multiple Frame Super-Resolution Reconstruction of Video Face Image

Jianbin Gao[1,2(✉)], Huan Tang[1,2], and James C. Gee[2]

[1] School of Resources and Environment,
University of Electronic Science and Technology of China, Chengdu, China
gaojianbin0769@gmail.com, huantang@gmail.com
[2] Center for Digital Health, University of Electronic Science and Technology
of China, Chengdu, China
gee@uestc.edu.cn

Abstract. The goal of super-resolution (SR) face image reconstruction is to generate a high quality and high resolution (HR) image based on low resolution (LR) input face images, to support video face recognition. This paper proposes a novel hybrid SR reconstruction framework of combining a multiframe SR interpolation method and a learning based single frame SR method. The proposed framework first formulates a simple but effective frame selection strategy as a preprocessing step. A weight map based on canonical correlation analysis (CCA) theory is then estimated to blend the results from single frame SR method and multiframe SR approach. Experimental results demonstrate that the proposed method achieves competitive performance against the state-of-the-art both simulated and real LR sequences.

Keywords: Face super resolution · Frame selection · CCA

1 Introduction

Among many biological identification features, face is the most commonly used one as evidenced by recently increasing interest of face recognition in computer vision community. Faces often appear in monitoring system of banks, shopping malls, and other public areas, as important clues for public security systems. However, under uncontrolled natural conditions, nonfrontal, noisy low resolution (LR) images tend to be produced. These LR images can neither be reliably evaluated by human readers, nor effectively used as direct input of face recognition systems. Super resolution (SR) reconstruction technology represents great

This work was supported in part by the programs of International Science and Technology Cooperation and Exchange of Sichuan Province under Grant 2017HH0028, Grant 2018HH0102 and Grant 2019YFH0014.

Y. Zhao et al. (Eds.): ICIG 2019, LNCS 11901, pp. 773–784, 2019.
https://doi.org/10.1007/978-3-030-34120-6_63

potential to solve this problem. Face super resolution, which is also called face hallucination, is firstly proposed by Baker [1].

Generally, from the number of input images, existing face image SR methods can distinguish into two types, namely, multiframe SR (MFSR) and single frame SR (SFSR) techniques. The MFSR approaches include two kinds of typical algorithms: (i) interpolation based and (ii) reconstruction based methods. Non-uniform interpolation approaches [2,3] intuitively generate improved resolution images by fusing the information of a sequence of consecutive LR frames with the same scenario. The reconstruction based methods [4,5] in addition to the use of multiframe information, but also join various priori information to regularize HR solution space based on the Bayesian framework. These methods ensure restored HR image is consistent with the original LR face. However, they are usually subtle to inappropriate blurring operations and rely on severely the accurate image registration.

In the other hand, learning based approaches are the most representative face SR algorithms in SFSR methods for latest decade. Learning based technique intends to estimate an HR image according to the relationship learned from the LR and HR training image pairs, which can distinguish into global face based and local patch based methods, from the manner of processing face image. The global face based approaches usually regard the holistic face as a whole to hallucinate the LR face images, which have a certain degree of robustness. Some commonly used machine learning models include the Principal Component Analysis (PCA) [6], the locality preserving projections [7], and the canonical correlation analysis (CCA) [8]. Though these methods can preserve the structure of global face variations, they ignore some individual facial details beyond principal components. Compared with the global methods, local patch based approaches achieve better visual quality since they divide the whole face into several overlapped patches to derive HR blocks. Ma *et al.* [9] present a position patch based face SR technique, where all the training patches located at the same positions are combined to produce the query patch, and the reconstruction weights are obtained by solving the least square regression (LSR) problem. Recently, the idea of locality constrained representation (LCR) has been proposed by Jiang *et al.* [10] to constrain the least square reconstruction and give more robust results to input noise. These methods are easy to get the embedding dictionary and yield content results. Two limitations of these methods are that they rely on aligning same size inputs as the training set LR image, which results in performance degradation without accurate alignment, otherwise, they fail to preserve the fine details of a face when the high quality training samples are not sufficient, video sequences dataset, for example.

Recently, some methods based on deep learning has been developed [13–15]. These methods have achieved good results in an poor environment. However, the neural network based methods have to input large-scale data during training stage, which will inevitably result in higher time and space complexity.

To tackle this problem, this paper presents a novel hybrid SR reconstruction framework to super-resolve LR faces in video sequences. Firstly, we propose a simple frame selection mechanism to discard nonfrontal fuzzy frames and reduce registration calculating pressure. Then, we utilize a weighted map based on CCA theory to combine single frame local learning based method and multiple frames non linear interpolation two different approaches, which are conductive to preserve local facial details and the global face consistency with original LR input simultaneously.

The rest of the paper is organized as follows. Section 2 briefly describes related methods. In Sect. 3, the face frame selection criterion and hybrid mechanism are both explained. We present experimental results and implementation details in Sect. 4 and draw conclusions and future work in Sect. 5.

2   Related Work

Towards our proposed face reconstruction method, we select representative global method and local method: traditional multiframe interpolation method and popular learning based single frame algorithm to simultaneously enhance local and global faces features. Among the many possible SFSR and MFSR methods, we choose a specific multiframe ANC (adaptive normalized convolution) interpolation method [3] and a single frame learning based approach [11] as a concrete implementation of our proposed approach. Both the two methods have the prominent advantage of performance stability and low computational cost. The chosen SR methods are briefly described in the following sections.

2.1   Multiframe Global Non-uniform Interpolation SR

In order to make a trade-off between computational complexity and reconstruction performance of the overall algorithm, the selected global method is a kind of nonlinear interpolation method based on normalized convolution (NC) framework by TQ Pham et al. [3], which suggests a solution for fusion of irregularly sampled images using adaptive normalized convolution. According to their algorithm, the window function of adaptive NC is adapted to local linear structures, and each sample would carry its own robust certainty which can minimize smoothing of sharp corners and tiny details by ignoring samples from other intensity distributions in local analysis. Besides, the registration method applied before reconstruction multiple images is Keren et al. [12].

2.2   Single Frame Local Learning Based SR

For single frame SR approach, the fast direct learning based SFSR method proposed in [11] are selected to preserve local feature of images by creating effective mapping functions. They first split the original input image feature space into numerous subspaces and then learn mapping priors between HR and LR images for each subspace by collecting sufficient training exemplars. All learned regression functions are simple and fast to handle subspace image reconstruction problem. The novelty of this approach lies in subspace mapping functions which facilitate feasibility of fast resolving HR patches and generates high quality SR images with sharp edges and rich textures.

3   Proposed Method

3.1   Frame Selection

In a continuous collection of facial image sequences, due to the movement of the human body, a variety of head poses, facial expressions may be present. While side face image, fuzzy, occlusive and low light face image cannot provide useful information for multiframe face super-resolution reconstruction. In [16], the authors proposed a new method to measure the canonical view face images via rank and symmetry of matrix. The frontal state and sharpness of face image can be measured easily and efficiently based on this metric, but the change of the picture brightness is not taken into account. We adopt this method in this paper, and propose a new selection mechanism by adding the brightness factor to remove relatively dark frames. The new mechanism is formulated as shown below. Let a matrix $I_j^i \in \mathbb{R}^{M \times N}$ denote the j-th face image of person i. S_i is the set of all face images of person i, $I_j^i \in S_i$. The canonical view, sharpness and brightness measurement index can be written as follows.

$$SF(I_j^i) = \|I_j^i A - I_j^i B\|_F^2 - (\alpha \|I_j^i\|_* + \beta I_j^i\_lu) \tag{1}$$

where α, β are both contant coefficient, respectively control weight of sharpness and brightness factor. $\| \cdot \|_F$ is the Frobenius norm, and $\| \cdot \|_*$ denotes the nuclear norm, which is the sum of the singular values of a matrix. $I_j^i\_lu$ denotes brightness factor of face image I_j^i which is calculated as a mean of illumination component of the image in YCbCr color space. $A, B \in \mathbb{R}^{M \times N}$ are two diagonal constant matrixes with elements $A = diag([1_{N/2}, 0_{N/2}])$ and $B = diag([0_{N/2}, 1_{N/2}])$. The first term in Eq. (1) measures the facial symmetry, which is the difference between the left half and the right half of the face. We can evaluate the facial pose based on face's symmetry. The second term measures the rank of the face, which reflects the sharpness degree of face because the rank of an image can characterize the structure information of the image and larger rank indicates sharper images. The third term measures brightness of the face image, notice that larger mean of illumination component of the image in YCbCr color space corresponds to brighter faces in our formulation. Smaller value of $SF(I_j^i)$ indicates that the face is more likely to be useful for MFSR. The selected several frontal face images when input into multiple images registration is hypothesized to result in more reliable estimates of motion parameters. The face with smallest value of Eq. (1) is chosen as reference image of registration process. In this paper, experimentally, we select five face images to participate in the MFSR reconstruction.

3.2   Combination Weight Map Estimation

In this part, we discuss the estimation of weight map for effectively combining multiframe interpolation and single frame learning-based method results. For face images, we often make resolved HR faces as input of a series of important identification, such as face recognition. So we need to guarantee the reconstructed

HR face image is similar to the ground truth face as far as possible. As mentioned above, the HR image information recovered by the MFSR method is mainly derived from fusing LR images of the same scene, while the SFSR method is mostly from learned mapping relation between HR and LR example images from training datasets. In other words, the MFSR method ensures the global consistency of the super-resolved face but with confining performance, while the local SFSR method can effectively protect facial details but with low reliability of the global face. Hence, the idea of making use of resolved HR face by MFSR to constrain the HR face from SFSR method is proposed, compensating for the effects of insufficient high quality training samples in video sequences at the same time.

We propose a fusion weight map based on CCA theory, because when CCA theory is used for feature extraction, it can effectively ensure the maximum consistency between high and low resolution facial features. CCA theory is first proposed by Hotelling [17] in 1936, which is a classical multivariate data processing method of interdependent vector. The same pair of high and low resolution images is different in resolution and dimension while they have a common internal structure. In a correlated subspace, the correlation of same pair of high and low resolution image can be maximized. CCA for feature extraction can effectively ensure the highest pertinence between high and low resolution face features and now is widely used in the field of face SR [8,18]. The following experiment demonstrates that it can enhance the consistency between the HR face image and the original LR face, simultaneously improve SR reconstruction performance.

Suppose there is a pair of high and low resolution image, after removing the mean, they respectively are $X_{LR} \in \Re$, $Y_{HR} \in \Re^{P \times Q}$. The aim of CCA is to find the projection direction $\mu \in \Re^{M \times d}$ and $\nu \in \Re^{P \times Q}$ to form the CCA subspace, d is the CCA subspace dimension, so that the LR image in CCA subspace $X_{LR\_CCA} = \mu^T X_{LR}$ and the HR image in CCA subspace $Y_{HR\_CCA} = \nu^T Y_{HR}$ have the largest correlation. Therefore, the projection direction μ, ν can be obtained by maximizing the correlation coefficient:

$$\max_{\mu,\nu} \rho = \frac{E[X_{LR-CCA} * Y_{HR-CCA}{}^T]}{\sqrt{E[(X_{LR-CCA})^2]E[(Y_{HR-CCA})^2]}}$$

$$= \frac{E[\mu^T X_{LR} Y_{HR}^T \nu^T]}{\sqrt{E[\mu^T X_{LR} X_{LR}\mu]E[Y_{HR}Y_{HR}\nu]}} \qquad (2)$$

$$= \frac{\mu^T S_{XY} \nu}{\sqrt{\mu^T S_{XX}\mu \cdot \nu^T S_{YY}\nu}}$$

where, $E[\cdot]$ denotes mathematical expectation, S_{XX}, S_{YY}, S_{XY} represent the autocovariance matrix and the mutual crosscovariance matrix between X_{LR} and Y_{HR}, respectively. More details about the solution of Eq. (2) can be found in [17].

In this paper, an LR face image sequence $I_{LR,k} \in \Re^{M \times N}$ is used as input, k denotes the number of LR face images. To super-resolve these face frames $I_{LR,k}$

by our proposed means, five face frames with acceptable canonical view, sharpness and brightness are obtained firstly by proposed frame selection mechanism, then both HR_M reconstructed by MFSR method and HR_S resolved by SFSR method are calculated. In the end, enhanced HR face image HR_F is computed by our algorithm described in the following.

For achieving greatest relevance between the final reconstructed HR face image and the original GT (Ground Truth) face, we first designate a fixed size sliding window, HR_M reconstructed by MFSR is typically subdivided into overlapping patches with this window. Then, a CCA subspace is computed to calculate the maximum correlation coefficients between the segmented image blocks and the original LR reference face image in this subspace. The CCA subspace and correlation coefficient can be calculated according to (2). The highest correlation coefficients between the obtained image blocks and the original LR face image are arranged from high to low, and a certain number of HR_M image blocks are selected and added to the HR_S face image as constraint information of keeping consistent face features with the LR face images. Figure 1 gives a block diagram illustration. The final HR face image HR_F can be calculated by the following formula.

$$HR_F = \sum_{1 \leq j \leq P} \sum_{1 \leq j \leq Q} [\lambda(i,j) * HR_M(i,j) + (1 - \lambda(i,j)) * HR_S(i,j)] \quad (3)$$

$$\lambda(i,j) = \begin{cases} \omega, (i,j) \in windows_{max} - correlation \\ 0, others \end{cases} \quad (4)$$

where λ represents the weight map to control the area that needs to be fused and the fusion weight, $windows_{max-correlation}$ denotes checked HR_M mage blocks corresponding window positions, ω is constant coefficient, controlling the weight of HR_M face image blocks blended into corresponding area of HR_S. The proposed fusion weight map strategy can effectively improve the consistency between the reconstructed HR face and the original GT face. Moreover, the quality of the final HR_F image is improved too because of the addition of HR_M image information.

4   Experiments

In this section, we show experimental results of proposed hybrid method and compare its objective quality and visual impression to other state of the art methods, using both the PSNR and SSIM metrics. Experiments were carried out to test SR performance of the proposed method in both simulated and real LR image sequences. Simulated LR image sequences super-resolution experiments are conducted on three datasets namely, Prima head pose image dataset [19], YUV Video Sequences [20], and Head Tracking Image Sequences [21]. The proposed algorithm is also evaluated on the Choke Point database [22] to further demonstrate its efficacy in real world surveillance scenarios.

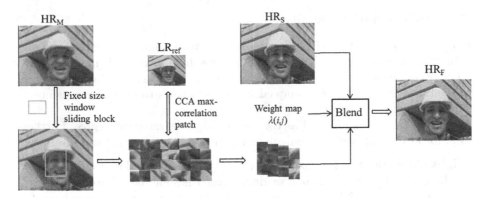

Fig. 1. Detailed block diagram of proposed combining method based on CCA theory.

Table 1. Comparison based on correlation coefficients, PSNR (dB) and SSIM.

Input sequence	Correlation coefficients			PSNR (dB) and SSIM				
	$\rho(MF)$	$\rho(SF)$	$\rho(Ours)$	Bicubic	MFSR	SFSR	LCR	Ours
Person01,	0.0694	0.0030	0.0463	28.29	27.84	31.21	29.15	**31.36**
#45–49(×4)				0.937	0.930	0.957	0.942	**0.958**
Person03,	0.0225	0.0121	0.0451	26.40	26.07	27.50	28.09	**28.17**
#45–49(×4)				0.910	0.906	0.932	0.921	**0.934**
Person08,	0.0307	0.0293	0.1179	27.15	26.74	27.17	27.49	**27.68**
#41–45(×4)				0.924	0.919	0.940	0.920	**0.936**
foreman,	0.0975	0.0161	0.0796	28.93	25.62	28.97	29.11	**29.13**
#17–21(×2)				0.951	0.882	0.952	0.955	**0.959**
suzie,	0.1634	0.0470	0.0954	35.64	27.34	36.56	36.59	**36.71**
#40–44(×2)				0.973	0.865	0.978	0.971	**0.978**
villains1,	0.2333	0.0636	0.4072	31.78	23.22	32.37	32.25	**32.48**
#182–186(×2)				0.950	0.840	0.951	0.949	**0.952**

4.1   Implementation Details

In this section, we show experimental results of proposed hybrid method and compare its objective quality and visual impression to other state of the art methods, using both the PSNR and SSIM metrics. Experiments were carried out to test SR performance of the proposed method in both simulated and real LR image sequences. Simulated LR image sequences super-resolution experiments are conducted on three datasets namely, Prima head pose image dataset [19], YUV Video Sequences [20], and Head Tracking Image Sequences [21]. The proposed algorithm is also evaluated on the Choke Point database [22] to further demonstrate its efficacy in real world surveillance scenarios.

4.2   Implementation Details

For color images, we apply the proposed algorithm on brightness channel (Y) and magnify color channels (UV) by Bicubic because human vision is more sensitive to brightness change. HR face sequences are first blurred by Gaussian kernel with variance σ and then subsampled by scale factor s to obtain a sequence of simulated multiple LR faces.

We take $\sigma = 2.0$ and $s = 4$ as the case of high degradation degree. Furthermore, for evaluating a resolution enhancement by $s = 2$, $\sigma = 0.4$ was employed.

Weight Coefficient ω. Extensive experiments are performed to decide the value of ω, based on these experiments, we draw some rules for the value of ω. Figure 2 shows the average PSNR gained over SFSR [11] method for various values of ω from 0 to 1. It should be noted that other parameters were fixed during the course of changing ω. As can be seen from the diagram, the test images in YUV dataset achieve the best performance at ω value of 0.1 or 0.3, while the best results of the Prima dataset images are obtained at a value of ω of 0.75. It is noteworthy that the size of YUV dataset picture is $176*144$ does not exceed 200, while the picture size of Prima dataset is $384*288$, which exceeds 200. Thus, we can conclude some ω value rules which are suitable for most of cases. When the ground truth image size does not exceed 200, ω value of 0.1–0.3 can get better results; on the contrary ω value of 0.7–0.9 is more appropriate. The special case is that when the performance of MFSR or SFSR methods is particularly poor, the value of w tends to be marginalized around 0 or 1, which is not applicable to the above rule.

Sliding Window Size and Overlapping Rate. The number of HR_M image blocks hinges on the sliding window size and overlapping rate. We keep sliding window width and height a same value to make the algorithm runs faster. During the experiment, it is found that setting window size around 0.6 or 0.84 of the original HR imge small edge size can achieve better results than HR_S. According to experience when the sliding window size exceeds half of small edge size of HR_M image, the overlap rate is better to be 0.25, otherwise, set it to 0.6. Specially, for very small window size 1–15, setting overlap rate as 1 is advisable. Theses parameter values yield not only appropriate number of blocks but excellent performance.

The CCA subspace dimension d, is empirically set to 0.3 of input LR face images small size edge to gererate pleasing HR results. The number of selected image patches superimposed on HR_S image will be given by evaluation criterion of making the final reconstructed image optimal quality. For frame selection part, we empirically set $\alpha = 0.03$ and $\beta = 1.5$ in most cases.

4.3   Simulated LR Sequences

Our method is compared with the MFSR, SFSR method reported in [3, 11] and state of the art patch-based approach, Jiang's locality-constrained representation (LCR) [10]. All implementation parameters setting and code are adopted

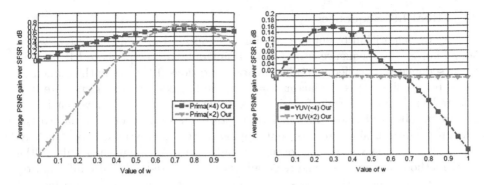

Fig. 2. Average PSNR gain over SFSR method for a varying value of ω. Results are given both for upscaling factor of 2 (red) and 4 (blue). From left to right: results of sequences in Prima dataset [19] and YUV dataset [20] (Color figure online)

according to their papers to achieve their best possible results. Visual results comparison is depicted in Fig. 3. In order to substantiate the objective quality results, we also give Table 1, which is used quantitatively to measure the results of different SR methods. Each cell 2 results in Table 1 shows: Top - image PSNR (dB), Bottom-SSIM index, which shows the improvement in PSNR and SSIM index by applying our method versus the other methods. The best result for each sequence is highlighted. It can be seen that the proposed method outperforms theses method in [3, 10, 11]. Moreover, as shown in Table 1, correlation coefficient between HR_M and input LR face in CCA subspace is indeed larger than HR_S do. Furthermore, correlation coefficients of our hybrid method are higher than that of SFSR method, which show that the results of the multiframe method do bring consistency constraint information to the blended HR image.

In Fig. 3, it can be seen that our images are visually better such as the left face of the girl has more details in our proposed case compared to Bicubic, SFSR and even LCR method. Besides, in the results produced by the LCR method, we can observe some fake edges on the edge of the girl's hair and the boy's shoulder, which are clearly visible in our results. Similar observations can also be made for the other sequence. Our results gain in PSNR of up to 0.67 dB and 0.51 dB over advanced SFSR method, which can be found in Table 1 for 'Person03' and 'Person08' sequence respectively. Our results as well as improve over SFSR algorithm in SSIM, which demonstrate that the proposed method indeed increased the correspondence between resolved and original HR face. In Table 1, notice that the result of MFSR method is very bad for 'suzie' and 'villains1', because 'villains1' present substantial movement and 'suzie' contain interference object (telephone handset) nearby the faces, which may impact the reconstruction of face parts. In this case, our results also exceed 0.15 dB over advanced SFSR method for 'suzie'. Similarly, we obtain the max value in SSIM.

4.4   Real LR Sequences

We use our algorithm to enlarge face images captured in challenging real world surveillance video sequence from Choke Point database [22], where LR faces in those sequences are collected under natural conditions. LR faces in the sequences are super-resolved by scale factor $s = 2$, where five consecutive frames are first selected via frame selection function. Examples of results are shown in Fig. 4. It can be observed that while in hair regions our results look similar to the conventionally obtained results, but appear better in textured areas such as facial contours, glasses and the corner of mouth, eyes. Although the results of the LCR method appear to have more detail, they are infused with false information such as the outline of the glasses around the eyes of the girl and the artifacts of the boy's face. This shows that our framework is promising in the field of face reconstruction problem in video surveillance scenario.

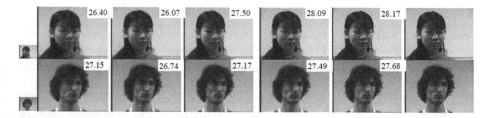

Fig. 3. The results on face images super-resolution the Prima dataset (scaling factor: 4). From left to right: the reference LR frame, reconstructed images by Bicubic, MFSR [3], SFSR [11], LCR [10], our method and ground truth image. The upper right corner values are PSNR in dB.

Fig. 4. The results on face images super-resolution the Choke Point database (scaling factor:2). From left to right: all input LR face images, reconstructed images by Bicubic, MFSR [3], SFSR [11], LCR [10], and our method.

5   Conclusions and Future Work

This paper presented a new hybrid SR reconstruction framework for super-resolution multiple LR face image inputs. Contrary to the conventional multi-frame reconstruction strategy of numerous complex iterations and heavily dependent on the registration link, our algorithm, first introduces a simple frame selection mechanism as a preprocessing step to simplify the registration problem, then utilizes proposed CCA-based fusion strategy to combine SFSR and MFSR estimates to obtain our reconstructed HR face. The proposed method improves face SR performance by adding trained prior from external datasets and enhanced consistency restriction information via CCA fusion theory. We have demonstrated that our algorithm outperforms several state of the art SFSR algorithms.

References

1. Baker, S., Kanade, T.: Hallucinating faces. In: The Fourth International Conference on Automatic Face and Gesture Recognition, pp. 83–88. IEEE, Grenoble (2000)
2. Park, S.C., Park, M.K., Kang, M.G.: Super-resolution image reconstruction: a technical overview. IEEE Signal Process. Mag. **20**(3), 21–36 (2003)
3. Pham, T.Q., et al.: Robust fusion of irregularly sampled data using adaptive normalized convolution. EURASIP J. Adv. Signal Process. **2006**(1), 236–236 (2006)
4. Dai, S., Han, M., Xu, W., et al.: SofCuts: a soft edge smoothness prior for color image super-resolution. IEEE Trans. Image Process **18**(5), 969–981 (2009)
5. Sun, J., Sun, J., Xu, Z., et al.: Gradient profile prior and its applications in image super-resolution and enhancement. IEEE Trans. Image Process. **20**(6), 1529–1542 (2011)
6. Park, J.S., Lee, S.W.: An example-based face hallucination method for single-frame, low-resolution facial images. IEEE Trans. Image Process **17**(10), 1806–1816 (2008)
7. Zhang, X., Peng, S., Jiang, J.: An adaptive learning method for face hallucination using locality preserving projections. In: The 8th IEEE International Conference on Automatic Face and Gesture Recognition, pp. 1–8. IEEE, Amsterdam (2008)
8. Huang, H., He, H., Fan, X., Zhang, J.: Super-resolution of human face image using canonical correlation analysis. Pattern Recogn. **43**(7), 2532–2543 (2010)
9. Ma, X., Zhang, J., Qi, C.: Hallucinating face by position-patch. Pattern Recogn. **43**(6), 2224–2236 (2010)
10. Jiang, J., Hu, R., Wang, Z., Han, Z.: Noise robust face hallucination via locality-constrained representation. IEEE Trans. Multimed. **16**(5), 1268–1281 (2014)
11. Yang, C.Y., Yang, M.H.: Fast direct super-resolution by simple functions. In: ICCV, pp. 561–568. IEEE (2013)
12. Keren, D., Peleg, S., Brada, R.: Image sequence enhancement using sub-pixel displacements. In: CVPR, pp. 742–746. IEEE (1988)
13. Bulat, A., Tzimiropoulos, G.: Super-FAN: integrated facial landmark localization and super-resolution of real-world low resolution faces in arbitrary poses with GANs. In: CVPR, pp. 109–117. IEEE (2018)
14. Hui, Z., Wang, X., Gao, X.: Fast and accurate single image super-resolution via information distillation network. In: CVPR, pp. 723–731. IEEE (2018)
15. Yu, X., Fernando, B., Hartley, R., Porikli, F.: Super-resolving very low-resolution face images with supplementary attributes. In: CVPR, pp. 908–917. IEEE (2018)
16. Zhu, Z., Luo, P., Wang, X., Tang, X.: Recover canonical-view faces in the wild with deep neural networks. arXiv preprint (2014)
17. Hotelling, H.: Relations between two sets of variates. In: Kotz, S., Johnson, N.L. (eds.) Breakthroughs in Statistics. SSS, pp. 162–190. Springer, New York (1992). https://doi.org/10.1007/978-1-4612-4380-9_14

18. An, L., Bhanu, B.: Face image super-resolution using 2D CCA. Signal Process **103**, 184–194 (2014)
19. Gourier, N., Hall, D., Crowley, J.L.: Estimating face orientation from robust detection of salient facial structures. In: FG Net Workshop on Visual Observation of Deictic Gestures, Cambridge (2004)
20. YUV video sequences (QCIF). http://trace.eas.asu.edu/yuv/index.html
21. BMP Image Sequences for Elliptical Head Tracking. http://cecas.clemson.edu/~stb/research/headtracker/seq/
22. Wong, Y., Chen, S., Mau, S., Sanderson, C., Lovell, B.C.: Patch-based probabilistic image quality assessment for face selection and improved video-based face recognition. In: CVPR, pp. 74–81. IEEE (2011)

A Fast Adaptive Subpixel Extraction Method for Light Stripe Center

Wei Zou and Zhenzhong Wei[✉]

Key Laboratory of Precision Opto-mechatronics Technology, Ministry of Education,
Beihang University, No. 37 Xueyuan Road, Haidian District, Beijing 100191, China
zhenzhongwei@buaa.edu.cn

Abstract. Aiming at the problem of light stripe distribution uneven and large curvature variation, which results in wrong stripe center extraction, a fast light stripe center extraction method based on the adaptive template is proposed. Firstly, the adaptive threshold method is used to reduce the image convolution area, and the multi-thread parallel operation is used to improve the speed of extracting the light stripe center. Secondly, the multi-direction template method is used to estimate the width of the light stripe along the normal direction, so that the size of the Gaussian template can be automatically obtained. Finally, the Hessian matrix eigenvalues are normalized to eliminate the multiple light stripe centers at both ends of the light stripe, and avoid extracting the wrong light stripe centers at the intersection position or the large curvature change, thus ensuring the continuity of the light stripe. This method has fast processing speed, good robustness, and high precision. It is very suitable for vision measurement image, medical image, and remote sensing image.

Keywords: Light stripe center · Image extraction · Hessian matrix · Normalization · Gussian filter

1 Introduction

In the field of vision measurement [1–6], such as geometric dimension measurement, 3D profilometry measurement, 3D object classification recognition, and so on, it usually uses the camera to obtain the modulated light stripe to solve and reconstruct the contour data of the measured object. Structured light vision sensor is widely used because of its simple structure, low cost, non-contact, high efficiency. Thus, extracting the center line of the light stripe is a key step for 3D reconstruction. Further, in the field of medical image analysis [7–11], such as blood vessel detection in retina or fundus images, pulmonary nodule detection in CT images, information enhancement of ultrasound images, and so on. All of these need to detect the edge contour or the center for medical diagnosis. Moreover, in the field of remote sensing, the center of curvilinear structures are extracted from aerial and satellite images to determine key information such as

© Springer Nature Switzerland AG 2019
Y. Zhao et al. (Eds.): ICIG 2019, LNCS 11901, pp. 785–800, 2019.
https://doi.org/10.1007/978-3-030-34120-6_64

rivers and roads [12–14]. Therefore, fast and accurate center extraction of lines are extremely important for visual measurement, and center detection of lines algorithm can also be applied to medical images and remote sensing images.

At present, the methods of extracting the center of light stripe can be divided into two categories according to the extracting accuracy. One is pixel-level accuracy, such as extreme value method [15–18], skeleton thinning method [19]; the other is sub-pixel accuracy, including direction template method [20,21], gray centroid method [22,23], curve fitting method [24,25] and Hessian matrix method (Steger method) [26–28].

Extremum method and threshold method can detect and extract the light stripe center quickly, but the extraction accuracy is not high and is vulnerable to noise interference. The skeleton refinement method has fast extraction speed, but it has poor noise resistance and is prone to burrs. Directional template method has fast extraction speed, but its robustness and accuracy are not as good as curve fitting method and Hessian matrix method. Although the curve fitting method also has high accuracy, high stability and good robustness, the detection and extraction speed is slow, and it is not suitable for the detection of light stripe in complex background images. Gray centroid method is generally used in simple light stripe images because of its high accuracy, fast speed and good stability. Steger method is widely used to extract of the center of light stripe image in visual measurement, the center line of blood vessel in medical detection image and the center line of road in remote sensing image because of its high accuracy and good anti-noise ability.

Steger's method has high accuracy, good stability and strong anti-noise ability, but the computational complexity caused by multiple convolutions is huge, resulting in a slower speed. In addition, if the whole image only uses the same Gauss convolution kernel size, it is difficult to accurately extract the center line and even lead to the discontinuity of the light stripe when the light stripe width distribution is uneven and the curvature varies greatly. At both ends of the light stripe, it is easy to generate multiple centers point of the light stripe because of discontinuity of light stripe and abrupt change of gray level. Similarly, if the light stripe is located at the edge of the image, there will also be multiple centers. If there are intersections in the light stripe, the center line of the light stripe near the intersection will often have a larger error.

For different problems, many papers also present various improvement methods. For example, aiming at the problem that multi-convolution of Gauss template increases the time of extracting the center of the light stripes, the processing range of stripe image is reduced by morphology and image recognition [29,30], and the convolution times of image by Gauss template are reduced, so as to improve the speed of image extraction, or to improve the speed of image extraction by virtue of the performance of hardware [31]. In order to solve the problem of uneven distribution of light stripe width, adaptive template is often used to solve the problem of strip discontinuity caused by the change of light stripe width. However, adaptive template increases the number of convolutions, thus increasing the extraction time.

To solve these problems, we also propose an improved method based on the Hessian matrix. First, the original image is divided into several sub-blocks according to the number of CPU threads on the computer, and the extraction speed of the light stripe center is improved by multi-threaded parallel computing. Secondly, in sub-block images, morphological methods (corrosion, expansion, regional connectivity) and adaptive threshold methods are used to segment the light stripe image. Thirdly, the width of the light stripe is estimated by binary image using the multi-directional template. Thus, the size of the convolution template for Gaussian filter is determined and image convolution is carried out by adaptive template method, and candidate points of light stripe center are determined by Steger method using Hessian matrix. Finally, the eigenvalues of the Hessian matrix are normalized by the data normalization method. According to the discriminant conditions, the candidate points of the light stripe center are judged to be the center points of the light stripe. Finally, the light stripe center of the whole image is extracted.

2   Algorithms and Principles

2.1   Image Sub-block and Parallel Computing

When structured light is used for dimension measurement or three-dimensional topographic scanning, the modulated light stripe projected on the measured object will change with the change of its surface shape. When the surface curvature of the object is large or at the corner of the object, it is easy to produce uneven width distribution of the cross section of the light stripe. The method based on Hessian matrix needs to estimate the light stripe width along the normal direction as the convolution template size of the Gauss filtering. Therefore, the best method is to use many different sizes of Gauss convolution templates, which usually leads to the reduction of image processing speed. At present, with the improvement of hardware performance, it is easy to improve the speed of the algorithm using the performance of hardware itself. So, we use a multi-threaded parallel computing method, as shown in Fig. 1.

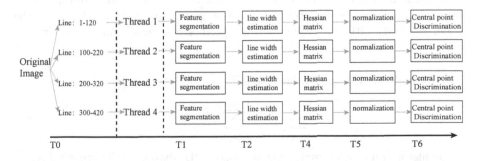

Fig. 1. The flow chart of multi-threaded parallel computing presented in this paper.

We use the method shown in Fig. 2 to segment the image. Different sub-images are obtained according to the number of threads in the CPU. There are some overlapping images between adjacent sub-images, so it is convenient to use the Gauss convolution template for convolution calculation. Of course, the width of overlapping area is larger than that of convolution template. Image segmentation is not only helpful to improve the speed of the algorithm, but also helpful to get more accurate thresholds for binary image processing.

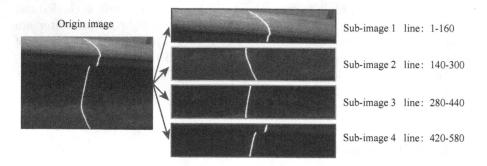

Fig. 2. An image segmentation method proposed in this paper.

According to the characteristics of the image, the upper edge noise of the image is more serious, while the other areas are more uniform. In addition, the light stripe in the image are generally vertical distribution. We divide the image into four horizontal sub-images.

2.2 Feature Segmentation of Light Stripe

Before image convolution, the region segmentation theory is used to segment the image automatically. Because of the great difference between the gray value of light stripe and background, the threshold of image segmentation is automatically determined by the method of maximum class square (OSTU method). In order to improve the running speed, we simplify it as follows.

For an image, if the maximum gray value is $L_{\max}$ and the minimum gray value is $L_{\min}$, then the distribution range of gray value is $[L_{\min}, L_{\max}]$. If the number of pixels with gray value of L is n_L, then the total number of pixels is
$$N = \sum_{i=L_{\min}}^{i=L_{\max}} n_i.$$

By normalizing the gray value, the following results can be obtained:

$$\sum_{i=L_{\min}}^{i=L_{\max}} p_i = \sum_{i=L_{\min}}^{i=L_{\max}} \frac{n_i}{N} = 1 \tag{1}$$

The histogram defined by Eq. (1) is called a 2D histogram. There are $L_{\max} - L_{\min}$ straight lines that are perpendicular to the main diagonal line in a 2D histogram, as shown in Fig. 3.

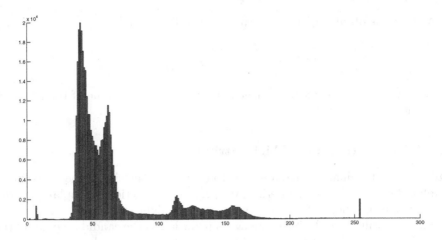

Fig. 3. Straight-line intercept histogram.

Assuming a gray value threshold is τ. The gray value threshold divides the pixels in the stripe image into two categories, namely $[L_{\min}, \tau]$, and $[\tau, L_{\max}]$. Then the probability of two kinds of occurrence is

$$
\begin{cases}
\rho_1 = \rho(\tau) = \sum\limits_{i=L_{\min}}^{\tau} p_i \\
\rho_2 = \rho(\tau) = \sum\limits_{i=L_{\min}}^{\tau} p_i
\end{cases}
\tag{2}
$$

Then, the mean value is

$$
\begin{cases}
\mu_1(\tau) = \sum\limits_{i=0}^{\tau} i p_i / \omega_1(\tau) = \mu(\tau)/\omega_1(\tau) \\
\mu_2(\tau) = \sum\limits_{\tau+1}^{L_{\max}} i p_i / \omega_2(\tau) = \frac{\mu_\tau - \mu(\tau)}{1 - \omega_1(\tau)}
\end{cases}
\tag{3}
$$

Where, $\mu(\tau) = \sum\limits_{i=L_{\min}}^{\tau} i p_i, \mu_\tau = \sum\limits_{i=L_{\min}}^{L_{\max}} i p_i.$

Then, the variance of the two groups of data is:

$$
\begin{cases}
\sigma_1 = \sum\limits_{i=L_{\min}}^{\tau} (i - \mu_1)^2 p_i / \rho_1 \\
\sigma_2 = \sum\limits_{i=\tau+1}^{L_{\max}} (i - \mu_2)^2 p_i / \rho_2
\end{cases}
\tag{4}
$$

Between-class variance σ_B^2 can be computed using:

$$
\sigma_B^2(\tau) = \rho_1[\mu_1(\tau) - \mu_\tau]^2 + \rho_2[\mu_2(\tau) - \mu_\tau]^2 = \rho_1\rho_2[\mu_1(\tau) - \mu_2(\tau)]^2
\tag{5}
$$

The optimal threshold τ^* can be selected from

$$
\sigma_B^2(\tau^*) = \max_{L_{\min} \leq \tau \leq L_{amx}} \sigma_B^2(\tau)
\tag{6}
$$

After $\tau'*$ is obtained, all pixels can be classified using

$$f(x,y) = \begin{cases} 0 & if\ I(x,y) \le \tau^* \\ 1 & if\ I(x,y) > \tau^* \end{cases} \tag{7}$$

where $f(x,y)$ is the segmented image, $I(x,y)$ is the gray value of the image at the point (i,j).

2.3 Width Estimation of Light Stripe

After we get the binary image, we need to determine the size of the convolution template for the next Gauss filtering. We propose a directional template method to determine the width of the cross section of the light stripe.

The gray value 1 in the binary image is regarded as the candidate point of the image light stripe. Four directional templates are used to convolute the candidate point in the light stripe image, and the minimum value of the four convolutions is taken as the light stripe width through the candidate point (Fig. 4).

Fig. 4. Light stripe width estimation based on multi-directional template

$$I_w(x,y) = s \cdot \min(I(x,y) * T_i) \quad i = 1,2,3,4 \tag{8}$$

Where, s is the compensation coefficient, and the ratio of the light stripe width in gray image to that in binary image is regraded as the value s. In this paper, s is $\sqrt{2}$.

2.4 Calculation of the Hessian Matrix

In general, the intensity profile $f(x)$ of a light stripe can be described by a Gaussian function as shown in Fig. 5. The Gauss function is as follow.

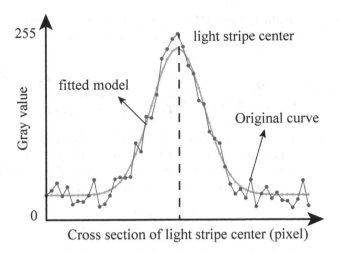

Fig. 5. Light stripe line-width estimation and Gaussian curve fitting

$$f(x) = \frac{A}{\sqrt{2\pi\sigma}} \exp\left[-\frac{(x-\mu)^2}{2\sigma^2}\right] \tag{9}$$

where A represents the intensity of the light stripe, and σ represents the standard deviation width of light stripe.

For filtering the noise in the image, the image $I(x, y)$ is convolved with Gaussian kernel $g_\sigma(x, y)$ and differential operators to calculate the derivatives of $I(x, y)$. So, the center point of the light stripe is given by the first-order zero-crossing-point of the convolution result, which also reaches local extreme points in the second-order derivatives.

Therefore, we use the Gauss filter to eliminate the noise in the image. The Gauss filter is as follow.

$$g_\sigma(x, y) = \frac{1}{2\pi\sigma^2} \exp\left(-\frac{x^2 + y^2}{2\sigma^2}\right) \tag{10}$$

The first-order partial derivatives of the Gaussian function are expressed by

$$\begin{cases} g_x(x, y) = \left(-\frac{x}{2\pi\sigma^4}\right) \exp\left(-\frac{x^2+y^2}{2\sigma^2}\right) \\ g_y(x, y) = \left(-\frac{y}{2\pi\sigma^4}\right) \exp\left(-\frac{x^2+y^2}{2\sigma^2}\right) \end{cases} \tag{11}$$

The second-order partial derivatives of the Gaussian function are denoted by

$$\begin{cases} g_{xx}(x, y) = \left(-\frac{1}{2\pi\sigma^4}\right)\left(1 - \frac{x^2}{\sigma^2}\right) \exp\left(-\frac{x^2+y^2}{2\sigma^2}\right) \\ g_{yy}(x, y) = \left(-\frac{1}{2\pi\sigma^4}\right)\left(1 - \frac{y^2}{\sigma^2}\right) \exp\left(-\frac{x^2+y^2}{2\sigma^2}\right) \\ g_{xy}(x, y) = \left(\frac{xy}{2\pi\sigma^6}\right) \exp\left(-\frac{x^2+y^2}{2\sigma^2}\right) \end{cases} \tag{12}$$

According to the above Eqs. (11) and (12), the first-order derivative and the second-order derivative of the image can be convoluted by image and Gaussian kernel function.

$$\begin{cases} I_x\left(x,y\right) = I\left(x,y\right)*g_x\left(x,y\right) \\ I_y\left(x,y\right) = I\left(x,y\right)*g_y\left(x,y\right) \end{cases}, \begin{cases} I_{xx}\left(x,y\right) = I\left(x,y\right)*g_{xx}\left(x,y\right) \\ I_{yy}\left(x,y\right) = I\left(x,y\right)*g_{yy}\left(x,y\right) \\ I_{xy}\left(x,y\right) = I\left(x,y\right)*g_{xy}\left(x,y\right) \end{cases} \quad (13)$$

where $I_x\left(x,y\right)$, $I_y\left(x,y\right)$ are the first-order partial derivatives with the image $I\left(x,y\right)$ along the x and y directions, and $I_{xx}\left(x,y\right)$, $I_{yy}\left(x,y\right)$, and $I_{xy}\left(x,y\right)$ represent the second-order partial derivatives with the image $I\left(x,y\right)$, respectively.

So, the Hessian matrix of any point in an image is given by

$$H\left(x,y\right) = \begin{bmatrix} I_{xx}\left(x,y\right) I_{xy}\left(x,y\right) \\ I_{xy}\left(x,y\right) I_{yy}\left(x,y\right) \end{bmatrix} \quad (14)$$

The eigenvalues of the Hessian matrix are the maximum and minimum of the second-order directional derivatives of the image at this point. The corresponding eigenvectors are the directional vectors of the two extremes, and the two vectors are orthogonal. For light stripe images, the normal direction is the eigenvector corresponding to the larger absolute eigenvalue of the Hessian matrix, and the eigenvector corresponding to the smaller absolute eigenvalue of the Hessian matrix is perpendicular to the normal direction.

Suppose λ_1 and λ_2 are the two eigenvalues of the Hessian matrix, and n_x and n_y are their corresponding eigenvectors. The eigenvalue λ_1 approaches zero, and the other eigenvalue λ_2 is far less than zero because the gray value of the edge of the light stripe is much larger than that of the background. Namely, $\|\lambda_1\| \approx 0, \lambda_2 \ll 0$.

Obviously, whether the above conditions are satisfied is a necessary condition for judging that the image point is the center of the light stripe. Therefore, we should choose two thresholds to judge the two eigenvalues. If the threshold is larger, the effective center of the light stripe will be lost, and the discontinuity of the light stripe will occur. If the threshold is smaller, multiple light stripe centers will be generated on the same cross-section of the light stripe.

In order to solve the above problems, we use the data normalization method commonly used in the neural network to normalize the above two eigenvalues. Therefore, we propose a normalized model based on Gauss function and an arctangent function.

2.5 Normalization of Hessian Matrix Eigenvalues and Discrimination of Light Stripe Centers

For smaller eigenvalues λ_1 tending to zero, we use standard Gauss function to normalize it. So, when the eigenvalue λ_1 tends to zero, the normalized value tends to 1. For the eigenvalue λ_2 with larger absolute value, we use the arc-tangent function to normalize it. When the absolute value of the eigenvalue λ_1 increases, the normalized value approaches to 1. These two normalization functions are as follows.

$$h_1 (\lambda_1) = \exp\left(-\frac{\lambda_1^2}{a}\right) \tag{15}$$

where λ_1 is the eigenvalue of the Hessian matrix corresponding to the axis direction of the light stripe. The gray value of the light stripe center varies tinily in its axis direction. a is a constant that we need to set.

$$h_2 (\lambda_2) = \begin{cases} 0 & \lambda_2 \geq 0 \\ \frac{2}{\pi}\text{atan}\left(\frac{\lambda_2}{b}\right) & \lambda_2 < 0 \end{cases} \tag{16}$$

where λ_2 is the eigenvalue of the Hessian matrix corresponding to the normal direction of the light stripe. The gray value of the light stripe center varies greatly in its normal direction. b is also a constant that we need to set.

We use the product of these two normalized functions as the final normalized function. The proposed function is illustrated in Fig. 6.

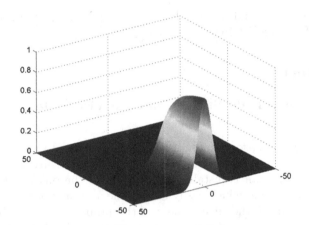

Fig. 6. The normalization function $h (\lambda_1, \lambda_2)$ for the eigenvalues λ_1, λ_2 of the Hessian matrix

$$\begin{cases} h (\lambda_1, \lambda_2) = 0 & \lambda_2 \geq 0 \\ h (\lambda_1, \lambda_2) = \frac{2}{\pi} \exp\left(-\frac{\lambda_1^2}{a}\right) a\tan\left(\frac{\lambda_2}{b}\right) & \lambda_2 < 0 \end{cases} \tag{17}$$

It should be noted that the values of a and b here are related to the stripe width of the actual image. We usually use the following formula to get the value.

$$\begin{cases} a = 2 * \sum_{i=1}^{N} I_w (x, y)/N \\ b = - \sum_{i=1}^{N} I_w (x, y)/N \end{cases} \tag{18}$$

Where, N is the sum of the gray value 1 of the segmented light stripe image. For this paper, a is set to 20 and b is set to -0.1.

To obtain the sub-pixel center point of light stripe, let $(x + tn_x, y + tn_y)$ be the sub-pixel coordinates of the center coordinate (x, y) along the normal direction of the unit vector (n_x, n_y) deduced by the Hessian matrix. The Taylor expansion of $I(x + tn_x, y + tn_y)$ at $I(x, y)$ is

$$I(x + tn_x, y + tn_y) = I(x, y) + tn_x I_x(x, y) + tn_y I_y(x, y) + \\ +1/2\left(t^2 n_x^2 I_{xx}(x, y) + 2t^2 n_x n_y I_{xy}(x, y) + t^2 n_y^2 I_{yy}(x, y)\right) \tag{19}$$

where $I_x(x, y)$, $I_y(x, y)$ are the convolution results of the first-order partial derivatives in x and y directions, and t is an unknown value. Because of the sub-pixel center exists on the normal vector of light stripe, the derivative of Eq. (19) is equal to 0, then

$$t = -\frac{n_x I_x(x, y) + n_y I_y(x, y)}{n_x^2 I_{xx}(x, y) + 2n_x n_y I_{xy}(x, y) + n_y^2 I_{yy}(x, y)} \tag{20}$$

If $tn_x \in [-1/2, 1/2]$ and $tn_y \in [-1/2, 1/2]$, $(x + tn_x, y + tn_y)$ is the sub-pixel center point of the light stripe.

3   Experiment

3.1   Extraction of Light Stripe Center in Visual Measurement of Rails

Line-structure vision sensor is often used in rail dimension vision measurement. Fast measurement of rail dimension is very important to ensure the safety of train running. Usually the rail is installed outdoors, and the outdoor light environment is always changing, which easily affects the accuracy of light stripe center extraction. We validate the effectiveness of the proposed method by extracting light strips from the image of rail size measurement. The light stripe image of size measurement of rails is shown in Fig. 7.

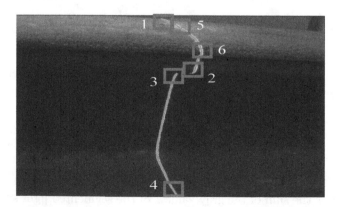

Fig. 7. Light stripe image of real object for size measurement of rails

From Fig. 7, we can see that the background noise of the stripe image is more complex, and the light stripe width distribution is not uniform. The width of the light stripe at both ends is obviously thinner. In the area labeled 5 in Fig. 7, the curvature of the light stripe changes greatly. It is impossible to extract the center line of light stripe directly by using traditional Hessain matrix.

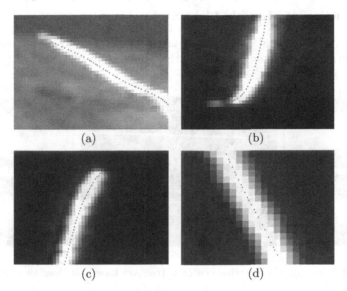

Fig. 8. (a) (b) (c) (d) Extraction of the center line of the light stripe in the area image labeled 1, 2, 3 and 4 in Fig. 7, respectively

From Fig. 8(a), we can see that the background noise of the stripe area labeled 1 in Fig. 8 is more serious. The method proposed in this paper can still ensure the correct extraction of the center line of the light stripe.

From Fig. 8(b), we can see that in the area labeled 2 in Fig. 8, the light stripe fines obviously at the end. The method proposed in this paper can still ensure the correct extraction of the center line of the light stripe.

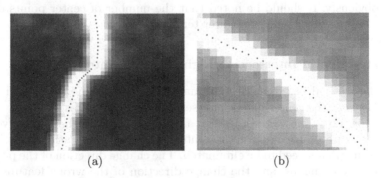

Fig. 9. (a) Center line extraction with great curvature change of light stripe. (b) Center line extraction with uneven change of light stripe width.

From Fig. 9(a), we can see that when the curvature of the stripe curve changes greatly, the center line of the light stripe can still be correctly extracted.

From Fig. 9(b), we can see that when the width of the stripe curve is unevenly distributed, the center line of the light stripe can still be correctly extracted.

3.2   Extraction of Multiple Light Stripe Centers

We use the method based on Hessian matrix and the method proposed in this paper to extract the center lines of multiple light stripes in the image to verify the extraction accuracy.

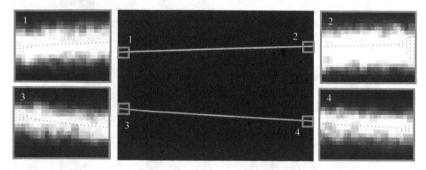

Fig. 10. Multiple light stripe center extraction based on Hessian matrix

From Fig. 10, we can see that it is obviously incorrect to extract the central points of two light stripes in the image by using the Hessian matrix method. There are obviously many center points of the light stripes at both ends of the light stripes. We need to delete the wrong center points of the light stripes. We can directly remove the center points of the stripes at both ends of the image, this method is simple, but it is easy to retain or delete the wrong center points of the stripes too much.

From Fig. 11, we can see that after normalizing the eigenvalues of Hessian matrix proposed in this paper, excessive light stripe centers at both ends can be deleted obviously. It should be noted that the number of center points deleted depends on the size of the Gauss filter template.

3.3   Extraction of Intersecting Light Stripe Centers

We use the method based on Hessian matrix and the method proposed in this paper to extract the center line of intersecting light strips, respectively.

From Fig. 12(a), we can see that at the intersection of the two curves, the extracted feature points are not the center of the two quadratic curves, so the wrong feature points need to be eliminated. The change direction of the points on each conic is continuous, and the change direction of the wrong feature points extracted at the intersection will change abruptly, and the angle between the mutation position and the change vector of the adjacent points will increase.

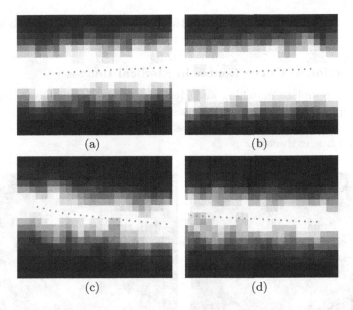

Fig. 11. (a) Extraction of the center line of the light stripe in the area image labeled 1 in Fig. 10. (b) Extraction of the center line of the light stripe in the area image labeled 2 in Fig. 10. (c) Extraction of the center line of the light stripe in the area image labeled 3 in Fig. 10. (d) Extraction of the center line of the light stripe in the area image labeled 4 in Fig. 10.

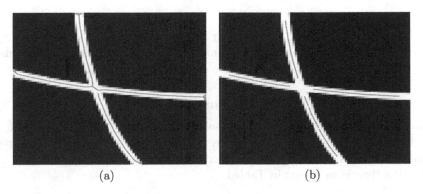

Fig. 12. (a) Extraction of the center line of intersecting light stripes based on Hessian matrix. (b) Extraction of the center line of intersecting light stripes based on the method proposed in this paper.

As can be seen from Fig. 12(b), the proposed method can not only filter out the multiple centers of the four endpoints of two curved strips, but also filter out the wrong centers at the intersection. If we need to get the center of the light strip, we can use the right center of the light strip which has been extracted to fit the curve. After fitting, we can get the intersection point, and then we can get the correct center line of the intersection.

3.4   Extraction of Curve Center in Medical Images

The adaptive template matching and Hessian matrix eigenvalue normalization method proposed in this paper are used to extract the central line of color retinal vessels to verify the effectiveness of the proposed method in complex multi-central line extraction.

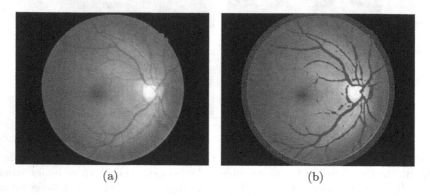

(a)   (b)

Fig. 13. (a) Original retinal vascular image. (b) Extracted vascular images. (Color figure online)

As shown in Fig. 13(a), it is a color image of retinal vessels. Figure 13(b) shows the extracted image. From the image, it can be seen that for the image without lesions, the blood vessels can be segmented completely.

3.5   Comparisons of Extraction Time

We use the configuration of the notebook computer: i5-3210M dual-core four-threaded CPU, 8G memory, compiled by Visual studio 2015 software, the resolution of 768*576 light stripe image to extract, that is, the image in Fig. 7. The calculating time is as shown in Table 1.

Table 1. Comparisons of extraction time between different method.

Methods	Steger method	The proposed method
Time	54 ms	66 ms

Although the time of the proposed method is comparable to that of Steger's method (the speed of the proposed method is a little slower), the calculation time can be reduced by increasing the number of threads in the CPU or using GPU acceleration.

4   Conclusion

A fast sub-pixel center extraction method based on adaptive template is proposed. The adaptive threshold method is used to reduce the image convolution area, and the multi-thread parallel operation is used to improve the speed of stripe center extraction. The multi-directional template is used to estimate the width of the light stripe along the normal direction. The size of the Gaussian convolution template is determined according to the actual width. The Hessian matrix is calculated and its eigenvalues are normalized to establish the criterion for the center of the light stripe. This method has fast processing speed, good robustness and high precision. It is suitable for vision measurement image, medical image, remote sensing image and other fields.

Acknowledgement. National Science Fund for Distinguished Young Scholars of China (51625501).

References

1. Liu, Z., Li, X.J., Yin, Y.: On-site calibration of line-structured light vision sensor in complex light environments. Opt. Express **23**(23), 248310–248326 (2015)
2. O'Toole, M., Mather, J., Kutulakos, K.N.: 3D shape and indirect appearance by structured light transport. IEEE Trans. Pattern Anal. Mach. Intell. **38**(7) (2016)
3. Lilienblum, E., Al-Hamadi, A.: A structured light approach for 3-D surface reconstruction with a stereo line-scan system. IEEE Trans. Instrum. Meas. **64**(5) (2015)
4. Zhang, S., Yau, S.: Absolute phase-assisted three-dimensional data registration for a dual-camera structured light system. Appl. Opt. **47**(17), 3134–3142 (2008)
5. Liu, Z., Li, F., Huang, B., Zhang, G.: Real-time and accurate rail wear measurement method and experimental analysis. J. Opt. Soc. Am. A **31**(8), 1721–1729 (2014)
6. Gong, Z., Sun, J., Zhang, G.: Dynamic structured-light measurement for wheel diameter based on the cycloid constraint. Appl. Opt. **55**(1), 198–207 (2016)
7. Xiong, G., Zhou, X., Degterev, A., Ji, L.: Automated neurite labeling and analysis in fluorescence microscopy images. Cytometry Part A **69**, 494–505 (2006)
8. Faizal, K.Z., Kavitha, V.: An effective segmentation approach for lung CT images using histogram thresholding with EMD refinement. In: Sathiakumar, S., Awasthi, L., Masillamani, M., Sridhar, S. (eds.) Internet Computing and Information Communications. AISC, vol. 216, pp. 483–489. Springer, New Delhi (2014). https://doi.org/10.1007/978-81-322-1299-7_45
9. Bae, K.T., Kim, J.S., Na, Y.H., et al.: Pulmonary nodules: automated detection on CT images with morphologic matching algorithm-preliminary results. Radiology **236**(1), 286–293 (2005)
10. Jonathan, L., Evan, S., Trevor, D.: Fully convolutional networks for semantic segmentation. IEEE Trans. Pattern Anal. Mach. Intell. **39**(4), 640–651 (2017)
11. Idris, S.A., Jafar, F.A.: Image enhancement filter evaluation on corrosion visual inspection. In: Sulaiman, H.A., Othman, M.A., Othman, M.F.I., Rahim, Y.A., Pee, N.C. (eds.) Advanced Computer and Communication Engineering Technology. LNEE, vol. 315, pp. 651–660. Springer, Cham (2015). https://doi.org/10.1007/978-3-319-07674-4_61

12. Cai, W., Dong, A., Zhang, X.: Crack width detection of the concrete surfaced based on images. In: Wu, Y. (ed.) Advances in Computer, Communication, Control and Automation. LNEE, vol. 121, pp. 625–632. Springer, Heidelberg (2011). https://doi.org/10.1007/978-3-642-25541-0_79

13. Hinz, S., Baumgartner, A.: Automatic extraction of urban road networks from multi-view aerial imagery ISPRS. J. Photogramm. Remote Sens. **23**, 83–98 (2003)

14. Hinz, S., Wiedemann, C.: Increasing efficiency of road extraction by self-diagnosis photogramm. Eng. Remote Sens. **70**, 1457–1466 (2004)

15. Canny, J.: A computational approach to edge detection. IEEE. Trans. Pattern Anal. **6**, 679–698 (1986)

16. Perona, P., Malik, J.: Scale-space and edge detection using anisotropic diffusion. IEEE. Trans. Pattern Anal. **12**, 629–639 (1990)

17. Weijer, J.V.D., Gevers, T., Geusebroek, J.M.: Edge and corner detection by photometric quasi-invariants. IEEE. Trans. Pattern Anal. **27**, 625–630 (2005)

18. Tsai, L.W., Hsieh, J.W., Fan, K.C.: Vehicle detection using normalized color and edge map. IEEE Trans. Image Process. **16**, 850–864 (2007)

19. Suárez, J.P., Carey, G.F., Plaza, A.: Graph-based data structures for skeleton-based refinement algorithms. Commun. Numer. Methods Eng. **17**(12), 903–910 (2001)

20. Ziou, D.: Line detection using an optimal IIR filter. Pattern Recogn. **24**, 465–478 (1991)

21. Laligant, O., Truchetet, F.: A nonlinear derivative scheme applied to edge detection. IEEE. Trans. Pattern Anal. **32**, 242–257 (2010)

22. Shortis, M.R., Clarke, T.A., Short, T.: A comparison of some techniques for the subpixel location of discrete target images. In: Proceedings of the SPIE, vol. 2350, pp. 239–250 (1994)

23. Luengo-Oroz, M.A., Faure, E., Angulo, J.: Robust iris segmentation on uncalibrated noisy images using mathematical morphology. Image Vis. Comput. **28**, 278–284 (2010)

24. Xu, G.S.: Sub-pixel edge detection based on curve fitting. In: Proceedings of the Second International Conference on Information and Computing Science (Ulsan, Korea, September 2009), pp. 373–375 (2009)

25. Goshtasby, A., Shyu, H.L.: Edge detection by curve fitting. Image Vis. Comput. **13**, 169–177 (1995)

26. Steger, C.: An unbiased detector of curvilinear structures. IEEE Trans. Pattern Anal. **20**, 113–125 (1998)

27. Qi, L., Zhang, Y., Zhang, X., Wang, S., Xie, F.: Statistical behavior analysis and precision optimization for the laser stripe center detector based on Steger's algorithm. Opt. Express **21**, 13442–13449 (2013)

28. Lemaitre, C., Perdoch, M., Rahmoune, A., Matas, J., Miteran, J.: Detection and matching of curvilinear structures. Pattern Recogn. **44**, 1514–1527 (2011)

29. Hu, K., Zhou, F.Q., Zhang, G.J.: Fast extraction method for sub-pixel center of strcutured light stripe. Chin. J. Sci. Instrum. **27**(10), 1326–1329 (2006)

30. Wang, S., Ye, A., Hao, G.: Autonomous pallet localization and picking for industrial forklifts based on the line structured light. In: IEEE International Conference on Mechatronics and Automation (2016)

31. Xie, F., Zhang, Y., Wang, S., et al.: Robust extrication method for line structured-light stripe. Optik **124**(23), 6400–6403 (2013)

32. Guan, X., Sun, L., Li, X., Su, J., Hao, Z., Lu, X.: Global calibration and equation reconstruction methods of a three dimensional curve generated from a laser plane in vision measurement. Optics Express **22**(18), 22043 (2014)

Help LabelMe: A Fast Auxiliary Method for Labeling Image and Using It in ChangE's CCD Data

Yunfan Lu, Yifan Hu, and Jun Xiao[✉]

University of Chinese Academy of Sciences, Beijing 100049, China
xiaojun@ucas.ac.cn

Abstract. Automatic detection of regions of interest in lunar image data is a basic subject for understanding the moon. In recent years, deep learning has greatly promoted the development of detection algorithms, but it requires a lot of markup data. In addition, the current common full-mark method has a large amount of work, a high degree of repetition and complex operation steps. In view of this situation, we combine the target detection technology with the whole marking method to design a new auxiliary marking method. We performed a labeling comparison experiment on the universal data set Pascal voc2012/2007. The results show that our method improves 2.5 times in efficiency with consistent accuracy. Based on the Common datasets, we also conducted experiments on the Chang'E series lunar CCD datasets, which also showed good performance and improved 2.8 times compared with the full mark indicates that our method can be applied to both general data sets and private data sets and also provides a research direction for the current auxiliary marking methods.

Keywords: Chang'E lunar CCD data · Image marking · Object detection

1 Introduction

Vision is the most important way for humans to perceive the outside world. Common graph data sets generated in daily life, have laid an important data foundation for visual understanding methods of deep learning, such as Microsoft COCO [16], Pascal VOC [5], etc. The earth remote sensing data set represented by DOTA [24] and the autonomous driving data set represented by KITTI [7] have greatly promoted the development of perception algorithms in related fields.

The full markup tools represented by LabelMe [22] have been widely used in object detection, semantic segmentation, instance segmentation and other visual understanding tasks of data marking. The marked data makes it possible to establish a differential relationship between data and tags, and also provides a basis for the evaluation of algorithms on large-scale data.

© Springer Nature Switzerland AG 2019
Y. Zhao et al. (Eds.): ICIG 2019, LNCS 11901, pp. 801–810, 2019.
https://doi.org/10.1007/978-3-030-34120-6_65

The full mark method is a method of marking all object areas in an image from scratch. Namely, simply enter a picture and mark it with tools like LabelMe. However, the full markup method is very redundant for large-scale data.

Take VOC as an example. For the detection task, there are 11540 pictures and 27,450 objects belonging to 20 categories. There are less than 1024 × 1024 pixels for each image and less than 12100567040 pixels for the entire dataset. However, for the DOM-7M dataset of Chang'E-2 [1], there are 844 sub-frames, each sub-frame has 29055 × 29005 pixels, with a total of 712498913100 pixels, which is greater than 59 times of VOC. It means that the labeling of lunar data is very costly (Fig. 1).

(a) Initia   (b) Ground Truth   (c) Auxiliary

(d) Initia   (e) Ground Truth   (f) Auxiliary

Fig. 1. (a) and (d) are original image of VOC, (c) and (f) proposal from the auxiliary method, (b) and (e) are ground truth.

On the other hand, the semantic information on the surface of the moon is less rich in everyday life. According to the manifold law of data distribution, it can be seen that the same type of high-dimensional data in nature is concentrated in the vicinity of a low-dimensional manifold [13,14]. Intuitively speaking, in pedestrian detection, although the objective detection frame of each person is different because of the human body type, the position and shape of the detection frame are consistent with a certain rule. For example, the length is mostly greater than the width, and the proportion is within a small interval [15]. However, the distribution of such data itself is difficult to express through formal methods, and its features are often high-dimensional.

Therefore, based on such objective facts, the work of the auxiliary mark for the detection task is to obtain a function $f(X) = Y$, where $X \in R^{W \cdot H \cdot 3}$ is an

RGB picture, $y_i \in Y$ is a sequence. For $y_i \in R^4$, $y_i = (x, y, w, h)$ represents a rectangle box, which is the smallest rectangle box containing all pixels of the object.

A common method for fitting the function f is to use detection networks such as the series of R-CNN [2,8,9]. This method has achieved great success on the COCO dataset but requires strong computing resources, and the marking task is often performed on the desktop computer. Our method uses a lightweight network to get real-time feedback in the case of only using CPU.

In general, our method adds the auxiliary mark module on the basis of the existing full mark method, and forms a human-friendly design for the label organization. In the experiment, it is found that with our method, there is a 2.5 multiplier on the general dataset and a 2.8 multiplier on the lunar dataset.

2 Related Work

The auxiliary marking task can be started from three aspects: data source, auxiliary method and marking tool. Our method was tested simultaneously on the general data set and Chang'E series dataset. The auxiliary method adopted the method of object detection, and the marking tool adopted the full marking method represented by LabelMe. These three parts will be introduced below.

For the marking of object detection task, researchers have proposed a large number of detection algorithms. Due to the limited length of the paper, we only introduce the detection algorithm most closely related to the work of our paper. Interested readers can refer to the more detailed overview [17].

2.1 Object Detection Algorithm

Object detection is a fundamental topic in computer vision research. Its basic definition can be simply considered as using a minimum horizontal rectangle to include all the pixels where the object is located. For the improvement schemes such as polygon detection, this article will not go into details. In recent years, the Deep Convolutional Neural Networks (DCNN) have achieved great success in object detection tasks. The object detection models need to consider the diversity of the appearance and scale of the objects. The classic methods are the Two-stage method represented by Faster-RCNN [20] and the One-stage method by YOLO [19] SSD [6,18].

Two-Stage Method. Faster-RCNN achieves high-precision object detection through feature sharing and region proposal network (RPN). In recent years, the methods such as R-FCN [4] Cascade RCNN [2] MS RCNN [11] have reached new heights in general data sets. But these high-precision detection algorithms are not applicable to the scene of auxiliary marking because of the huge. Computations introduced by he multi-step dispersion method and cascade form. Therefore, this method adopts a more lively one-stage method. With the development of current network quantization methods such as ONNX [21]TVM [3], it will be of practical value to use two-stage method as an auxiliary network in the future.

One-Stage Method. Compared to Two-stage, One-stage method lacks the Proposal process and directly returns to the Bounding Box on the Feature Map. It has a great advantage in speed and is therefore often used in devices with less computational resources, such as various kinds of embedded machine. The SSD method is representative of this series, which directly predicts the detection box and its categories, and does not generate a proposal process. It uses the anchor mechanism to solve the problem about different size object detection blocks. In recent years, many research scholars have proposed R-SSD [12] on the basis of SSD, and its use of multi-feature fusion has a higher improvement on small objects. At present, this field has a rapid development due to the research of many scholars. But the basic idea of the SSD series is not fundamentally changed (Fig. 2).

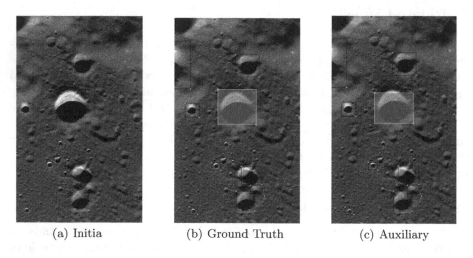

(a) Initia   (b) Ground Truth   (c) Auxiliary

Fig. 2. The three stages of moon picture.

Lightweight Neural Network. Mini networks are usually divided into two kinds: compression model and direct training small model. The compression model refers to transforming a large network into a network with fewer parameters and less computation by methods such as quantization, compression, hashing, pruning, and vector coding. The second method is to train small models directly. The MobileNets [10] used in our method focuses on optimizing the delay while considering small networks.

MobileNet is designed for lightweight and efficient network for the mobile device. Its basic architecture is manifold architecture (Streamlined), and it uses the Separable convolution (Depthwise Separable Convolutions) to build a lightweight neural network. What's more, it also has two orders of magnitude fewer parameters than the original VGG [23]. In recent years, lightweight neural network is also an important research direction and the proposal of ShuffleNet [25] etc. also provides an alternative to MobileNet.

2.2   Full Mark Tool

LabelMe is a full mark tool which is currently widely used. It has pixel-level operations and uses JSON as a form of exchange data. The binary stream of the edited image is characterized in the JSON data. The organization of the label is a sequence of points. The area enclosed by the closed polygon is the area where the label is obtained. This approach groups different data types together and provides good robustness across different data types.

The full mark operation is all operated by the marker, which inevitably depends on the marker. Marker's attention and efficiency in a large number of repetitive markings will decrease.

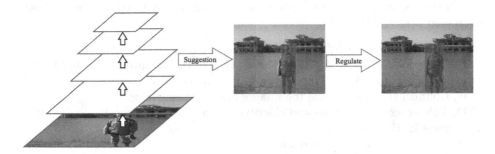

Fig. 3. Architecture of our model

3   Proposed Methods

The task of the auxiliary mark can be simplified to the progress of the following Fig. 3. First using a suggested network to get a suggestion box for a set of interesting objects. Next fine tune the results of the proposed network.

3.1   The Algorithm Processes

Let $I \in R^{W \times H \times 3}$ is a RGB image. Manual marking time as following.

$$T_{completely} = Time(I, H_{completely}) \tag{1}$$

$H_{completely}$ mean the process of completely manual marking. While for the auxiliary process, The time can be express as

$$T_{proposal} = Time(P(I), H_{proposal}) \tag{2}$$

Function P is the proposal network. $H_{proposal}$ is the process of adjusting for proposal object box. In the case of consistent accuracy, the sum of the time of $P(i)$ and adjusting should less than the time of completely marking.

3.2 Quantification

For such a pipeline process, we use a quantitative method to measure time. Completely mark is Fig. 4. Auxiliary method is Fig. 5.

Fig. 4. Full mark method pipeline

For the completely mark method it can be quantized into the above pipeline. Each step is as follows.

– A, Marker software loads images.
– B, Marker makes a preliminary observation of the picture, according to the experimental principle of human-computer interaction.
– C, Confirm the goals inside the images and start marking.
– D, This process is marking and classifying. Specifically, it can be divided into serveral E, F processes.
– E, Confirm the target boundary.
– F, Put a label on the confirmed boundary.
– G, Save and mark the next one.

It can be seen form the actual operation the there is no method for parallelizing pipeline stages in the traditional completely method. And the pipeline for the process with auxiliary marks is show in the Fig. 5. Each step is as follows.

Fig. 5. Auxiliary mark method pipeline

– 1, Time of the proposal network generate recommendations.
– 2, Fine-tune the suggested label. Such as target error.
– 3, Confirm the goals inside the images and start marking.

3.3 Propose Network Design

Propose Network uses the SSD framework, while we replace the backbone of VGG to MobileNets-v1, which has fewer parameters and can run in PC, only has a CPU, in real time. In the experiment, our training process were run on the GPU server and the test process were run on the CPU personal notebook.

The SSD algorithm is to put the feature block in the feature map of each layer to a classifier and a regressor. Its main computational complexity is in the feature extraction part. Because in this task, the mAP has a non-linear relationship with the mark time of auxiliary process. In response to this problem, this paper demonstrates in the experimental part.

3.4 Segmentation and Combine the Chang'E Image

The original image has a length and width of 2e5 pixels. It is too large too load. We layer it with 4-Node tree and each node is an image, which is 512×512 pixels, like Fig. 6(b).

The visual field of each node is 4 times that of the next layer. In this way we can get different object with large and small size as different levels. In Fig. 6(a), the bound box has robust size for crater in lunar.

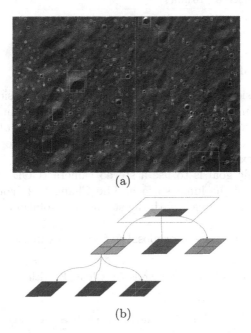

Fig. 6. A, Combine the same image to a complete image. B, Using N-node tree to divide a large picture into small picture.

4   Experiments

4.1 Pascal VOC Dataset

We comprehensively evaluate our method on the PASCAL VOC 2007. This dataset consists of about 5k train-val images and 5k test images over 20 object categories. Table 1 show the mark time with SSD-MobildNet proposal network.

In this experiment, three marker personnel mark three set of photos per person and we average their time. The three set of photo is completely set, auxiliary set and observe set. Completely set means marker should label an image without any supplementary information. Auxiliary set means the input for marker is the result of proposal net, which has some weak mark. Observe set means marker only observer the image but do not mark it.

Table 1. The time for different methods in VOC with SSD-MobildNet which mAP@0.5 is 75.4

Marking type	Times	Cumber of pictures	Average time
Completely	39 min	50	47 s
Auxiliary	20 min	50	23 s
Only observe	16 min	50	19 s

4.2   CCD Data of Chang'E Series

The Chang'E-1 mission is China's first lunar exploration mission. The Chang'E-2 mission is based on the backup satellite of the lunar exploration project. After technical improvement, Chang'E-2 serves as the pilot satellite of the second phase of the lunar exploration project and tests some key technologies of the second phase of the lunar exploration project.

Its main scientific goal is to use a stereo camera (CCD) to capture three-dimensional images of the lunar surface. The Chang 'e-2 orbit is 100 km high, and the distance between pixels with the strongest resolution of the CCD stereo camera is 7 m.

Table 2 show the improve of proposal net for mark in Chang'E.

Table 2. The time for different methods in Chang'E with SSD-MobileNet which mAP@0.5 is 71.0

Marking type	Times	Cumber of pictures	Average time
Completely	33 min	50	39 s
Auxiliary	17 min	50	20 s
Only observe	13 min	50	15 s

4.3   The Relationship Between mAP and Auxiliary Cost

In this experiment we explore the relationship between the auxiliary net with different mAP and the cost time. In Fig. 7, the x axis is the marking time and the y axis is the mAP of different proposal network.

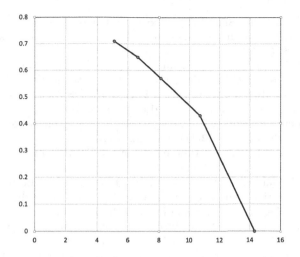

Fig. 7. Time with mAP.

5 Conclusion

In this paper, we proposed the useful suggested network in image marking process and practiced in common dataset and Chang'E dataset. In order to solve the problem of too large to load Chang'E CCD data, we propose a method of layering by quad tree. There has been an increase in the time stamping of generic data sets and private data. It also explores the relationship between time and mAP. Improvements to the network and interactive methods in feature are good research directions.

Acknowledgments. This work is supported by the Strategic Priority Research Program of the Chinese Academy of Sciences (No. XDA23090304), National Natural Science Foundation of China (No. 61471338), Youth Innovation Promotion Association CAS (2015361), Key Research Program of Frontier Sciences CAS (QYZDY-SSW-SYS004), Beijing Nova program (Z171100001117048), Beijing science and technology project (Z181100003818019), and the Open Research Fund of Key Laboratory of Space Utilization, Chinese Academy of Sciences.

References

1. Data release and information service system of China's lunar exploration program, national astronomical observatories of China. http://moon.bao.ac.cn. Accessed 19 May 2019
2. Cai, Z., Vasconcelos, N.: Cascade R-CNN: delving into high quality object detection. In: Proceedings of the IEEE Conference on Computer Vision and Pattern Recognition, pp. 6154–6162 (2018)
3. Chen, T., et al.: TVM: an automated end-to-end optimizing compiler for deep learning. In: 13th USENIX Symposium on Operating Systems Design and Implementation (OSDI 2018), pp. 578–594 (2018)

4. Dai, J., Li, Y., He, K., Sun, J.: R-FCN: Object detection via region-based fully convolutional networks. In: Advances in Neural Information Processing Systems, pp. 379–387 (2016)
5. Everingham, M., Van Gool, L., Williams, C.K., Winn, J., Zisserman, A.: The pascal visual object classes (VOC) challenge. Int. J. Comput. Vis. **88**(2), 303–338 (2010)
6. Fu, C.Y., Liu, W., Ranga, A., Tyagi, A., Berg, A.C.: DSSD: deconvolutional single shot detector. arXiv preprint arXiv:1701.06659 (2017)
7. Geiger, A., Lenz, P., Stiller, C., Urtasun, R.: Vision meets robotics: the KITTI dataset. Int. J. Robot. Res. **32**(11), 1231–1237 (2013)
8. Girshick, R.: Fast R-CNN. In: Proceedings of the IEEE International Conference on Computer Vision, pp. 1440–1448 (2015)
9. Girshick, R., Donahue, J., Darrell, T., Malik, J.: Rich feature hierarchies for accurate object detection and semantic segmentation. In: Proceedings of the IEEE Conference on Computer Vision and Pattern Recognition, pp. 580–587 (2014)
10. Howard, A.G., et al.: MobileNets: efficient convolutional neural networks for mobile vision applications. arXiv preprint arXiv:1704.04861 (2017)
11. Huang, Z., Huang, L., Gong, Y., et al.: Mask scoring R-CNN (2019)
12. Jeong, J., Park, H., Kwak, N.: Enhancement of SSD by concatenating feature maps for object detection. arXiv preprint arXiv:1705.09587 (2017)
13. Lei, N., Luo, Z., Yau, S.T., Gu, D.X.: Geometric understanding of deep learning. arXiv preprint arXiv:1805.10451 (2018)
14. Lei, N., Su, K., Cui, L., Yau, S.T., Gu, X.D.: A geometric view of optimal transportation and generative model. Comput. Aided Geom. Des. **68**, 1–21 (2019)
15. Li, J., Liang, X., Shen, S., Xu, T., Feng, J., Yan, S.: Scale-aware fast R-CNN for pedestrian detection. IEEE Trans. Multimed. **20**(4), 985–996 (2017)
16. Lin, T.-Y., et al.: Microsoft COCO: common objects in context. In: Fleet, D., Pajdla, T., Schiele, B., Tuytelaars, T. (eds.) ECCV 2014. LNCS, vol. 8693, pp. 740–755. Springer, Cham (2014). https://doi.org/10.1007/978-3-319-10602-1_48
17. Liu, L., et al.: Deep learning for generic object detection: a survey. arXiv preprint arXiv:1809.02165 (2018)
18. Liu, W., et al.: SSD: single shot multibox detector. In: Leibe, B., Matas, J., Sebe, N., Welling, M. (eds.) ECCV 2016. LNCS, vol. 9905, pp. 21–37. Springer, Cham (2016). https://doi.org/10.1007/978-3-319-46448-0_2
19. Redmon, J., Farhadi, A.: Yolo: real-time object detection. Pjreddie.com (2016)
20. Ren, S., He, K., Girshick, R., Sun, J.: Faster R-CNN: towards real-time object detection with region proposal networks. In: Advances in Neural Information Processing Systems, pp. 91–99 (2015)
21. Rotem, N., et al.: Glow: graph lowering compiler techniques for neural networks. arXiv preprint arXiv:1805.00907 (2018)
22. Russell, B.C., Torralba, A., Murphy, K.P., Freeman, W.T.: LabelMe: a database and web-based tool for image annotation. Int. J. Comput. Vis. **77**(1–3), 157–173 (2008)
23. Simonyan, K., Zisserman, A.: Very deep convolutional networks for large-scale image recognition. arXiv preprint arXiv:1409.1556 (2014)
24. Xia, G.S., et al.: DOTA: a large-scale dataset for object detection in aerial images. In: Proceedings of the IEEE Conference on Computer Vision and Pattern Recognition, pp. 3974–3983 (2018)
25. Zhang, X., Zhou, X., Lin, M., Sun, J.: ShuffleNet: an extremely efficient convolutional neural network for mobile devices. In: Proceedings of the IEEE Conference on Computer Vision and Pattern Recognition, pp. 6848–6856 (2018)

Author Index

Printed in the United States
By Bookmasters